ALGEBRAIC *K*-THEORY

The Homotopy Approach of Quillen and an Approach from Commutative Algebra

ALGEBRAIC *K*-THEORY

The Homotopy Approach of Quillen and an Approach from Commutative Algebra

SATYA MANDAL

University of Kansas, USA

World Scientific

NEW JERSEY • LONDON • SINGAPORE • BEIJING • SHANGHAI • HONG KONG • TAIPEI • CHENNAI • TOKYO

Published by

World Scientific Publishing Co. Pte. Ltd.
5 Toh Tuck Link, Singapore 596224
USA office: 27 Warren Street, Suite 401-402, Hackensack, NJ 07601
UK office: 57 Shelton Street, Covent Garden, London WC2H 9HE

Library of Congress Control Number: 2023015302

British Library Cataloguing-in-Publication Data
A catalogue record for this book is available from the British Library.

Cover photo: Daniel Quillen
Photo credit: Cynthia Cohen

ALGEBRAIC *K*-THEORY
The Homotopy Approach of Quillen and an Approach from Commutative Algebra

ISBN 978-981-126-938-7 (hardcover)
ISBN 978-981-126-939-4 (ebook for institutions)
ISBN 978-981-126-940-0 (ebook for individuals)

For any available supplementary material, please visit
https://www.worldscientific.com/worldscibooks/10.1142/13236#t=suppl

Desk Editors: Balasubramanian Shanmugam/Rok TingTan

Typeset by Stallion Press
Email: enquiries@stallionpress.com

Dedicated to my mother!

Preface

In this book, I take a pedagogic approach to Algebraic K-theory. I tried to find the shortest route possible, with complete details, to arrive at the homotopy approach of Quillen [Q] to Algebraic K-theory, with a simple goal to produce a self-contained and comprehensive pedagogic document in Higher Algebraic K-theory, which is accessible to upper-level graduate students as well as a broader mathematics community. With the intent to launch this project, I gave a course on Algebraic K-theory in Fall 2017. The lecture notes for the course constituted the first draft of this manuscript, and I could cover only about half of what is in this current version of the manuscript. While preparing this manuscript I always kept these students in my mind. In all sincerity, throughout I remained mindful of what it may take to have a meaningful communication with these students. I made a rule for myself that I would provide the full background that these students would require. Most of the students were in their second year. Two of them were bold enough to concede that they did not take a course in general topology. So, the background on Homotopy theory (Appendix B) begins with a section (Section B.1) on the basic jargon of Topology. I tried to provide complete proofs, and left little that a student may get stuck on and abandon. However, I was able to trim some of the materials that appear to be almost integral parts of the required background. In particular, I avoided singular homology and the plus construction. Homology theory is not needed for such a course in Quillen K-theory.

The homotopy approach to Algebraic K-theory, in its early part, sprouted out of the study of projective modules, which has been my specialization. I was always fascinated by the homotopy approach of Quillen [Q] to Algebraic K-theory, and I faced a literature gap, in the pedagogic sense. I am sure I was not alone in that. I overcame this gap, and eventually published in K-theory [M]. I concluded that a sizable readership would benefit if I wrote down the roadmap that I took. My prior experience of publishing a book [M2] gave me the confidence that this would be a doable endeavor. In this book I wrote down whatever literature gap I faced, to build this bridge. In fact, I was able to crystallize the required information, out of mountains of literature [GJ, Ha, H, Hu, S, St, W], to name some.

A small portion of this book includes my own research contributions, on Nori Homotopy and Chow–Witt groups (Sections 9.5.1 and 10.5). However, strictly speaking, this manuscript is to be treated as a purely pedagogic document. The targeted readership of this book is the upper level graduate students and those academics who simply like to read good mathematics and sometimes decide to teach the same.

The book concludes (Section 10.5) by relating the Nori homotopy obstruction set $\pi_0(\mathcal{LO}(P))$ [MM], for projective A-modules P to split off a free direct summand, with the Chow–Witt groups $\widetilde{CH}^n(A, \det P)$ of Barge–Morel [BM], where A is a regular ring, containing a field κ with $1/2 \in \kappa$, and $rank(P) = n$. The Chow-Witt group $\widetilde{CH}^n(A, \det P)$ is the cohomology, at degree n, of the Gersten GW-complex [M]. This reconnects the homotopy approach of Quillen to K-theory to projective modules. I submit that my work [M, M5] in Section 10.5, provides an antithesis to any skepticism that Quillen K-theory cannot generate any new research. From the point of view of commutative algebra, the research in Quillen K-theory was never explored. This book fills the literature gap that the commutative algebra community faced, which was solely responsible for this non occurrence. I expect that this book will generate more research in Quillen K-theory from the viewpoint of commutative algebra.

Among the available literature I would like to mention two books [Wc2, Sv] and a survey article [Sm1]. I benefited enormously from all of them. These are more suited for specialists. My approach in this book has been fully pedagogic. Barring some exceptions, none of the topics reported in Chapters 8–11 appeared in any book. Ample literature is available on classical K-theory [B, Mi, GM, R, Sv, Wc2], so it is not discussed in this book, except in Chapter 4 on Agreement, because that is not the point of this project. This takes us to the heart of the matter quicker. For motivational purposes, readers are referred to the books of Milnor [Mi] and Rosenberg [R]. The book of Swan [Sw3] is an important landmark, as well. The book of Bass [B] provided the early foundation of classical K-theory.

As the subtitle of the book indicates, my targeted readership includes the Algebra and the Algebraic geometry community. The nature of the topics and methods span out so broadly (e.g. Chapters 5–8), it brings commutative algebra, algebraic geometry, and K-theory under the same umbrella. The book is written in a commutative algebra style. I am hopeful that a faculty member in any mid-level university will be able to pick up this book and teach a graduate course in Algebraic K-Theory. There is a harmony and a melody in the way the results and the methods build up in this theory, from chapter to chapter. There is a complete story to tell in Algebraic K-theory. I am hopeful that this will be evident in this exposition.

I divide the materials in this book into three parts. The original goal of this project was to provide an exposition, fully faithful to the pedagogic approach, of the paper of Quillen [Q], along with its agreement with classical K-theory. At some point, the inclusion of the work of Swan on K-theory of quadrics [Sw1] became an integral part of the project, in my mind. I consider these (Chapters 2–8) as the main body of this book. Chapters 2, 3, 5, 6, and 7 provide expositions of the paper [Q], and Chapter 4 is on agreement of classical K-theory and Quillen K-theory, also due to Quillen [Gd1]. Chapter 8 is an exposition of [Sw1]. I deviate a little from [Sw1] by assuming 2 is invertible in the base ring and by avoiding Azumaya algebras.

In Chapters 9–11 I provide some Epilogues on the progress made since the paper [Q] was published. My original thinking was that there should be two more books on these, with the same pedagogic approach, one on K-theory and one on Hermitian K-theory. However, it became clear that an abrupt termination at the end of Chapter 8 would seem unnatural. In fact, it would definitely be beneficial to point out to readers how much of the subsequent developments is based on the foundation already built in [Q], and background and techniques already established in the remaining 11 chapters. The approach in these Chapters (9–11) differs from the rest of the chapters. These are more of an overview, and I do not attempt to give complete proofs, as I do in the other chapters. In Chapter 9, I give an overview of Waldhausen K-Theory [Wh, TT] and the homotopy approach to the Negative K-Theory, which is mostly an exposition of some of the papers of Schlichting. In Chapter 10, I give an exposition of homotopy approach to Hermitian K-theory, which is also an exposition of some other papers of Schlichting. In Chapter 11, I provide an exposition of the Witt groups of triangulated categories along with basic definitions regarding triangulated categories. The Quillen K-theory still remains the foundation of Algebraic K-theory. The Waldhausen K-theory is a natural progression of the Quillen K-theory. An exact sequence is determined by its inflation $M \hookrightarrow N$, barring a choice of the cokernel. So, the extra information in an exact sequence, beyond its inflation, is superfluous. In Waldhausen K-theory the "cofibrations" take the place of inflations. The remainder in Waldhausen K-theory is an improvisation of Quillen K-theory. Similar comments can be made regarding developments in the Hermitian K-theory. In other words, subsequent developments (Chapters 9–11) are a natural progression of the work of Quillen [Q]. This justifies my decision to explain Quillen K-theory thoroughly, and to give only an overview on the Epilogue chapters. Readers get a chance to work out the details in these chapters, improvising the methods learned in the preceding chapters.

The background, preliminaries, and tools needed to deal with the main body and the Epilogues (Chapters 2–11) are developed in Chapters A, B, C, and 1. A part of these materials is classical. In order to get to the heart of the matter as quickly as possible, I had to make a decision to put the Chapters A, B, and C in the Appendix. However, I want to caution readers that these contain the basic jargon and tools that are absolutely necessary for the rest of the book. The other rationale for this decision is that, some of it being classical, some readers will know some of these chapters, and some parts are used much later in the book. I take some credit for being able to crystallize everything that is needed in these four chapters. It is particularly so because not everything in these four chapters is classical. Some comes directly from [Q], as it falls within the context of the respective chapters. This includes Section A.4 on Exact categories, Section A.5 on Localization and Quotient Categories, Section B.5 on Quasifibrations, Section 1.3 on Bisimplicial sets, and others.

I provide a set of exercises and problems. To deal with some of the problems, readers may need to do some extra work, while exercises should be within the reach of the intended readership.

I comment on prerequisites and related issues of instructions. The author being a commutative algebraist, the book was written from an algebraic viewpoint. Accordingly, it is assumed that readers will have a reasonable familiarity with commutative algebra (and scheme theory). Many in algebra being less prepared with the jargon of Topology, familiarity with basic Topology will not be a required prerequisite. A complete and required background information on Topology is developed in Appendix B. We avoid Homology theory because it is not needed for our purpose. Appendix B is devoted mainly to Homotopy theory, which is absolutely essential for our purposes. It is less likely that intended readership in Algebra or K-theory will be familiar with the Dold–Thom theory on Excision and Quasifibrations (Sections B.2.2 and B.5). It would be a stretch to assume that the average readership would be familiar with Appendix C, on CW complexes. The final result in Appendix C is the theorem of Whitehead (C.3.8), which is one of the most frequently used tools in this book. I cover category theory in Appendix A. For completeness, and convenience of readers, I include basic jargon and facts on categories, abelian categories, pullback and pushforward, and other topics. The most basic object that we work with in this theory is exact categories, defined by Quillen [Q]. These are covered in Sections A.4 and A.5.

The book is fully self-contained, for the intended readership. A prepared audience may skip some of what is in these appendix chapters. However, an unprepared audience must go through all of it, while some may choose to skip some of the detailed proofs. Instructors may like to apply proper judgment and cover the materials in these appendix chapters, as and when needed. The idea in this book is to provide all the necessary tools for the algebra community, so that the homotopy approach to K-theory may look like another tangible theory. For a prepared reader in Algebra, the gradient is properly set, for a climb from the most basic concepts, like homotopy, to the Epilogue chapters. This should be viewed more like an upper-level graduate course in Algebra. Depending on the preparedness of the audience, in a one semester course, a feasible goal may be to finish up to Chapter 8.

About the Author

Satya Mandal joined University of Kansas in 1988 and currently holds the position of a professor. He completed his graduate studies in Indian Statistical Institute, Kolkata and Tata Institute of Fundamental Research, Mumbai. He completed his PhD under the supervision of Amit Roy in Tata Institute and got his degree in 1985. He served as a Wissenschaftliche Hilfskraft (Research Assistant) in Universität Regensburg, Germany, during the year 1983–1984, under the supervision of Earnst Kunz. He further served as a Postdoctoral Fellow, during the year 1987–1988, in Mathematical Sciences Research Institute (MSRI), Berkeley.

The author is a self-made academic. He devoted the early part of his research career in specializing in Classical K-Theory and Projective Modules. In the later years, his interest shifted to the algebra of Quillen K-theory.

The author has published approximately 50 papers including his other book published in 1997. He gave numerous lectures in national and international conferences. He served his department in the capacity of the Chairman of the department. He has been active with experimentations with online pedagogy, since the inception of Internet around 2000.

Acknowledgment

I communicated and consulted with many, in person or by email, on this project. Among them are Sankar P. Dutta, Daniel Grayson, Peter May, M. P. Murthy, Madhav V. Nori, Marco Schlichting, Richard Swan, Charles Weibel, and others. I am truly thankful to all of them for their patience, sincerity, and inputs. Let me also acknowledge some of the teachers who trained me. Among them are Professors Ashok Maitra (truly a legend), Amiya Mukherjee, B. V. Rao at Indian Statistical Institute, and Professors Amit Roy (my advisor), R. C. Cowsik, N. Mohan Kumar, M. S. Raghunathan, K. Santaram, R. R. Simha, R. Sridharan, and Balwant Singh at Tata Institute. The whole manuscript was professionally copyedited by Dr. Tim Lantz. I am immensely indebted to him for his patience with the manuscript and for his impressive contributions. Funding for this came from some internal grants from KU and from an NSA grant.

I would like to take this opportunity to express my gratitude to the State of Kansas and students at my university for supporting my position for more than three decades by tax dollars and tuition. Numerous internal resources from the University of Kansas, postdoctoral support, and travel and visitor support, were the keys for this project to be conceived and completed.

I express my gratitude to my extended family for the mutual loving support we share. I am thankful to my recently departed father-in-law Injachan and brother-in-law Samir for all that they did for the family. My loving thanks to my wife Elsit, my daughter Nila and also to Arunima, Sulangna, Debmalya, and Nairit for their inspirational presence and support.

Contents

Chapter 1

Simplicial Sets

Triangulation of topological spaces are among the foundational pillars of classical mathematics. The standard n-simplex $\Sigma_n \subseteq [0,1]^{n+1}$ is defined to be the set of all points $(t_0, t_1, \ldots, t_n) \in [0,1]^{n+1}$ such that $\sum t_i = 1$. So, they are, essentially, the points, closed intervals, triangles, tetrahedrons, and so on. A triangulation of a topological space X is a process of dissecting X into simplexes, where all the faces of an $n+1$-dimensional simplex σ fit along the existing n-simplexes. Triangulable spaces X are convenient to study, because it is easier to compute the (co)Homology and homotopy groups. In a sense, the theory of Simplicial Sets does the reverse engineering to this process of tringulation. The n-simplexes of a simplicial set K and the gluing rules are given, and its geometric realization $|K|$ is to be constructed. Needless to say that it comes with a lot of modern twists to the classical theory, and is a very rich theory by its own rights.

In a sense, Quillen's homotopy approach to K-theory is built on the foundation of the geometry of simplicial sets. We begin this chapter by introducing and developing the basic concepts of simplicial sets and the geometric realizations. The other main goal of this chapter is to prove Quillen's theorem on geometric realization of Bisimplicial sets, in Section 1.3.

1.1. Simplicial Sets

Definition 1.1.1. For integers $n \geq 0$, denote the set $[n] := \{0, 1, 2, \ldots, n\}$. Let Δ be the category whose objects are $\{[n] : n = 0, 1, 2, \ldots\}$ and the morphisms are the order-preserving maps. So,

$$Mor_\Delta([m],[n]) := \Delta([m],[n]) :=$$
$$\{f : [m] \longrightarrow [n] : f(i) \leq f(j)\ \forall\ 0 \leq i \leq j \leq m\}.$$

This category is referred to as the **Simplicial category**, or sometimes the **Δ-category**.

1. Δ has two distinguished subcategories Δ_+, Δ_- whose objects are the same as that of Δ.
   The morphisms of Δ_+ are the order-preserving injective maps.
   The morphisms of Δ_- are the order-preserving surjective maps.
2. For integers $n \geq 1$, and $0 \leq i \leq n$, define the morphisms in Δ_+

   $d^i : [n-1] \longrightarrow [n]$ by

   $$d^i(k) = \begin{cases} k & \text{if } k < i \\ k+1 & \text{if } i \leq k \leq n-1 \end{cases} \qquad \text{by skipping } i.$$

   Also, for integers $n \geq 1$, and $0 \leq i \leq n-1$, define the morphisms in Δ_-

   $s^i : [n] \longrightarrow [n-1]$ by

   $$s^i(k) = \begin{cases} k & \text{if } k \leq i \\ k-1 & \text{if } i < k \leq n \end{cases} \qquad \text{by identifying } i, i+1.$$

3. The following are the basic cosimplicial identities:

$$\begin{aligned} d^j d^i &= d^i d^{j-1} && \forall i < j \\ s^j d^i &= d^i s^{j-1} && \forall i < j \\ &= id && i = j, j+1 \\ &= d^{i-1} s^j && \forall\ i > j+1 \\ s^j s^i &= s^{i-1} s^j && \forall i > j \end{aligned} \tag{1.1}$$

The following is a useful lemma.

Lemma 1.1.2. *Let $f \in Mor_\Delta([n],[m])$. Then f can be written (uniquely) as*

$$f = d^{i_k} d^{i_{k-1}} \cdots d^{i_1} s^{j_1} s^{j_2} \cdots s^{j_l} \quad \textit{with} \quad \begin{cases} 0 \leq j_1 < j_2 < \cdots < j_l < n \\ 0 \leq i_1 < i_2 < \cdots < i_k \leq m \end{cases}$$

Proof. Left to the readers! [Mc] ∎

Definition 1.1.3. Given a category $\mathscr{C}$, a **Simplicial Object in $\mathscr{C}$** is a contravariant functor $\Delta \longrightarrow \mathscr{C}$. Likewise, a covariant functor $\Delta \longrightarrow \mathscr{C}$ is called a **Cosimplicial Object in $\mathscr{C}$**. So, a **simplicial set** is a contravariant functor $K : \Delta \longrightarrow$ **Sets**.

Such a simplicial set K is given by the following data: (1) A sequence of sets $K_n := K[n]$. An **n-simplex** is an element $x \in K_n$, and we write $\dim x = n$. An element $x_0 \in K_0$ is called a **vertex**. (2) The **face maps** $d_i : K_n \longrightarrow K_{n-1}$, which are dual to the maps d^i above. (3) The **degeneracy maps** $s_i : K_{n-1} \longrightarrow K_n$, which are dual to the maps s^i above. (4) The face and degeneracy maps are subject to the **simplicial identities**

$$\begin{aligned} d_i d_j &= d_{j-1} d_i && \forall i < j \\ d_i s_j &= s_{j-1} d_i && \forall i < j \\ &= id && i = j, j+1 \\ &= s_j d_{i-1} && \forall\, i > j+1 \\ s_i s_j &= s_j s_{i-1} && \forall i > j \end{aligned} \tag{1.2}$$

These identities are obtained by reversing the arrows in (1.1). Conversely, because of (1.1.2), any association $n \mapsto K_n$, satisfying the simplicial identities (1.2), determines a simplicial set.

A **map** $\tau : K \longrightarrow L$ of simplicial sets is a natural transformation. Such a map is given by a sequence of set theoretic maps

$\tau_n : K_n \longrightarrow L_n$ commuting with the face and degeneracy maps:

$$\begin{array}{ccc} K_n & \xleftarrow{s_i} & K_{n-1} \\ \tau \downarrow & & \downarrow \tau \\ L_n & \xleftarrow[s_i]{} & L_{n-1} \end{array} \qquad \begin{array}{ccc} K_n & \xrightarrow{d_i} & K_{n-1} \\ \tau \downarrow & & \downarrow \tau \\ L_n & \xrightarrow[d_i]{} & L_{n-1} \end{array}$$

The simplicial sets, together with the maps of simplicial sets, form a category, to be denoted by **SSet**. For $x \in K_n$, images of x under arbitrary iterations of face maps d_is are called **faces** of x, and images of x under iterated degeneracies are called **degeneracies** of x. Since 0-iteration is included, x is both a face and degeneracy of itself.

A simplex $x \in K_n$ is called **nondegenerate** if it is a degeneracy of itself only. Given any simplex $x \in K_m$, it is the degeneracy of a unique nondegenerate simplex $y \in K_n$ for some $n \leq m$. So, any $x \in K_0$ is nondegenerate, but its images $s(x) \in K_1$ are not.

The following are some of the usual definitions in the category **SSet**.

Definition 1.1.4. Let $K \in$ **SSet**. A **simplicial subset** of K is a map $L \longrightarrow K$ in **SSet** such that $L_n \longrightarrow K_n$ is one to one for all $n \geq 0$. In other words, $L_n \subseteq K_n$, and the inclusion map $L \longrightarrow K$ is a map of simplicial sets. Given simplicial subset $L \subseteq K$, the **quotient** $\frac{K}{L}$ is defined as

$$\left(\frac{K}{L}\right)_n = \frac{K_n}{L_n}, \quad \text{i.e., by identifying } L_n \text{ to one point.}$$

For $K \in Obj(\mathbf{SSet})$, and a subset $A \subseteq \cup K_n$, the simplicial subset $K(A) \subseteq K$, **generated by** A, is defined to be the smallest simplicial subset of K, containing A. So, $K(A)$ consists of all the iterated degeneracies and faces of simplexes in A.

For $K, L \in$ **SSet**, the **product** $K \times L \in$ **SSet** is defined by

$$\begin{cases} \forall\, n \in \mathbb{N} & (K \times L)_n := K_n \times L_n \\ f \in Mor_\Delta([m],[n]) & (K \times L)(f) := K(f) \times L(f) : \\ & K_n \times L_n \longrightarrow K_m \times L_m \end{cases}$$

It is worthwhile to point out that the product is the diagonal of the bifunctor $K \times L : \Delta \times \Delta \longrightarrow$ **Sets**, sending $([m],[n]) \mapsto K_m \times L_n$. Such bifunctors are called bisimplicial sets, to be discussed in Section 1.3.

The following are examples of some objects in **SSet** that may be considered among the basic building blocks of the category **SSet**.

Example 1.1.5. Fix an integer $n \geq 0$.

1. The representable functor

$$\Delta(-,[n]) : \Delta \longrightarrow \mathbf{Sets} \qquad \text{given by} \quad \Delta[n]([k]) = \Delta([k],[n])$$

is a simplicial set. This is denoted by $\Delta[n] := \Delta(-,[n]) \in \mathbf{SSet}$. For integers $k \geq 0$, it can be described by

$$\Delta[n]([k]) = \{(0 \leq n_1 \leq n_2 \leq \cdots \leq n_k \leq n)\}$$

A nondegenerate k simplex is a strictly increasing sequence $0 \leq n_1 < n_2 < \cdots < n_k \leq n$. For $k \geq n+1$, all k-simplexes of $\Delta[n]$ are degeneracies.
2. The **boundary** $\partial\Delta[n]$ of $\Delta[n]$ is defined to be the smallest simplicial subset of $\Delta[n]$, whose nondegenerate k simplexes are those of $\Delta[n]$, except the identity $1_n : [n] \xrightarrow{\sim} [n]$.
3. For integers $0 \leq r \leq n$, the **r-horn** $\Lambda^r[n] \in \mathbf{SSet}$ is defined to be the smallest simplicial subset of $\Delta[n]$, whose nondegenerate k-simplexes

$$= \{\iota \in \Delta([k],[n]) : \iota \text{ is injective}, \quad \iota \neq 1_n, \iota \neq d_r\}$$

So, $\Lambda^r[n] \subseteq \partial\Delta[n]$ has one fewer nondegenerate simplex, namely the face d_r opposite of the vertex r.
4. Define the n-sphere, $S^n := \frac{\Delta[n]}{\partial\Delta[n]} \in \mathbf{SSet}$. It has two nondegenerate simplexes, $\star \in S^n_0$ and $\sigma_n \in S^n_n$.

Cosimplicial sets (objects) are defined by reversing the arrows.

Definition 1.1.6. A **cosimplicial set** is a covariant functor $K : \Delta \longrightarrow \mathbf{Set}$. This is given by a sequence of sets K^n, and for each $f \in Mor_\Delta([m],[n])$ in Δ, there is a map $K(f) : K^m \longrightarrow K^n$ satisfying obvious laws of compositions. With $d^i := K(d^i)$ and $s^i := K(s^i)$, a cosimplicial set is determined by cosimplicial identities (1.1).

Likewise, a covariant functor $K : \Delta \to \mathbf{Top}$ is called a **cosimplicial space**. For any category $\mathscr{C}$, **cosimplicial objects** are defined likewise.

The most natural example of a cosimplicial set (in fact, a space) would be the following.

Example 1.1.7. As usual $\mathbb{N} = \{0, 1, 2, \ldots\}$. Let $e_0, e_1, e_2, \cdots \in \mathbb{R}^{\mathbb{N}}$ be the standard basis of $\mathbb{R}^{\mathbb{N}}$. Define,

$$\begin{cases} \Sigma^n = \{\sum_{i=0}^n \lambda_i e_i : \forall\ i\ 0 \leq \lambda_i \leq 1 \text{ and } \sum_{i=0}^n \lambda_i = 1\} \\ \forall\ f \in Mor_\Delta([m],[n]), \qquad \Sigma(f) : \Sigma^m \longrightarrow \Sigma^n \text{ is defined by} \\ \Sigma(f)\left(\sum_{i=0}^m \lambda_i e_i\right) = \sum_{k=0}^n \left(\sum_{f(i)=k} \lambda_i\right) e_k \end{cases}$$

Then $\Sigma^n \subseteq [0,1]^{n+1}$ inherits the usual topology, and $\Sigma(f)$ is a map in **Top**. Therefore, $\Sigma : \Delta \longrightarrow$ **Top** is a covariant functor, or a cosimplicial space.

The space Σ^n is called the **Standard n-simplex**.

1.2. Geometric Realization

Given a simplicial set K, we would define its geometric realization $|K|$.

Definition 1.2.1. Let $K \in$ **SSet**. Let Σ be the cosimplicial space, as in (1.1.7). With discrete topology on K_n, consider the disjoint (disconnected) union of spaces:

$$\mathcal{X}(K) := \coprod_{n=0}^{\infty} K_n \times \Sigma^n$$

We define a relation $\sim$ on $\mathcal{X}(K)$. For $(\tau, y) \in K_m \times \Sigma^m$ and $(\sigma, x) \in K_n \times \Sigma^n$, define

$$\begin{aligned} (\sigma, x) \sim (\tau, y) &\ \text{ if } \exists\ f \in Mor_\Delta([m],[n]) \\ \ni \quad K(f)(\sigma) = \tau, &\quad \Sigma(f)(y) = x. \end{aligned}$$

The relationship $\sim$ is reflexive and transitive, but not necessarily symmetric. Let $\simeq$ denote the equivalence relation on $\mathcal{X}(K)$, generated by $\sim$. Note, this equivalence relation can be described using the face and degeneracy maps only.

The geometric realization of K is defined to be [Mi2]

$$|K| := \frac{\mathcal{X}(K)}{\simeq} = \frac{\coprod_{n=0}^{\infty} K_n \times \Sigma^n}{\simeq} \qquad \text{with quotient topology (B.6.8).} \tag{1.3}$$

So, a subset $U \subseteq |K|$ is open, if and only if its inverse image in $K_n \times \Sigma^n$ is open. Since K_n has discrete topology, U is open if and only if its inverse image in $\sigma \times \Sigma^n$ is open, $\forall\ \sigma \in K_n$, $\forall\ n \geq 0$. It follows immediately,

$$\forall\ n \geq 0 \qquad |\Delta[n]| = \Sigma^n \tag{1.4}$$

To describe the topology, for $\sigma \in K_n$, consider the natural maps $\Phi_\sigma : \Sigma^n \longrightarrow |K|$, given by $\Phi_\sigma(t) = [(\sigma, t)]$. Then a set theoretic map $f : |K| \longrightarrow Z$ is continuous if and only if $f\Phi_\sigma$ is continuous, for all $\sigma \in K$.

The following is on the behavior of geometric realizations as a functor.

Lemma 1.2.2. *This association $K \mapsto |K|$ (1.2.1), defines a functor*

$$|\ -\ | : \mathbf{SSet} \longrightarrow \mathbf{Top}$$

Further, this functor has a right adjoint, defined as follows:

$$S : \mathbf{Top} \longrightarrow \mathbf{SSet} \qquad \text{by } S(Y)_n = Mor_{\mathbf{Top}}(\Sigma^n, Y)$$

(This functor S is called the singular functor.)

Proof. It is easy to see that both $|\ -\ |$ and S are functors. What remains is to define the natural bijections:

$$\Phi : Mor_{\mathbf{Top}}(|K|, Y) \xrightarrow{\sim} Mor_{\mathbf{SSet}}(K, S(Y))$$
$$\forall\ K \in Obj(\mathbf{SSet}),\ Y \in Obj(\mathbf{Top})$$

For $f \in Mor_{\mathbf{Top}}(|K|, Y)$ and $\forall\ n \geq 0$, consider the commutative triangle of maps in **Top**, as follows:

$$\begin{array}{ccc} K_n \times \Sigma^n & \longrightarrow & |K| \\ & \searrow & \downarrow f \\ & & Y \end{array}$$

This gives a map $\Phi(f)_n$: $K_n \longrightarrow Mor_{\mathbf{Top}}(\Sigma^n, Y) = S(Y)_n$

Define $\Phi : Mor_{\mathbf{Top}}(|K|, Y) \longrightarrow Mor_{\mathbf{SSet}}(K, S(Y))$ by $\Phi(f) = \{\Phi(f)_n : n \geq 0\}$.

It is obvious that Φ is natural with respect to both the coordinates. Conversely, given $F \in Mor_{\mathbf{SSet}}(K, S(Y))$, we have a sequence of maps

$$F_n : K_n \longrightarrow S(Y)_n = Mor_{\mathbf{Top}}(\Sigma^n, Y)$$

This defines a map

$$\widetilde{F}_n : K_n \times \Sigma^n \longrightarrow Y \quad \text{in } \mathbf{Top}$$

One checks that $\widetilde{F}_n$ factors through a continuous map $\Psi(F) : |K| \longrightarrow Y$. Now, Ψ defines a map

$$\Psi : Mor_{\mathbf{SSet}}(K, S(Y)) \longrightarrow Mor_{\mathbf{Top}}(|K|, Y)$$

It is clear that Φ and Ψ are inverse of each other. The proof is complete. ■

The following is on geometric realizations and pushforward diagrams in **SSet**.

Lemma 1.2.3. *Let*

$$\begin{array}{ccc} A & \xrightarrow{f_K} & K \\ {\scriptstyle f_L}\downarrow & & \downarrow{\scriptstyle \iota_K} \\ L & \xrightarrow[\iota_L]{} & K\coprod_A L \end{array}$$

be a pushforward diagram in **SSet** (or in the category of simplicial spaces). *Then the diagram of the geometric realizations:*

$$\begin{array}{ccc} |A| & \xrightarrow{|f_K|} & |K| \\ {\scriptstyle |f_L|}\downarrow & & \downarrow{\scriptstyle |\iota_K|} \\ |L| & \xrightarrow[|\iota_L|]{} & |K\coprod_A L| \end{array}$$

is a pushforward diagram in **Top**.

Proof. Clearly, the diagram of geometric realization commutes. Consider the commutative diagram of continuous maps.

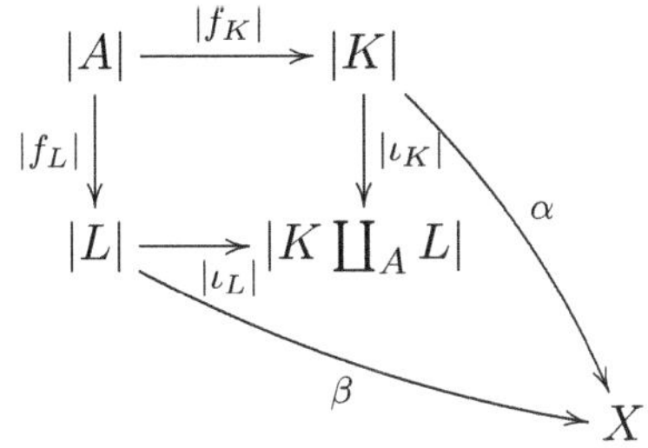

We have $|K \coprod_A L| = \dfrac{(K_n \coprod_{A_n} L_n) \times \Sigma^n}{\sim}$

Define,

$$\varphi : \left|K \coprod_A L\right| \longrightarrow X \text{ by } \varphi([(\sigma,t)]) = \begin{cases} \alpha([(\sigma,t)]) & \text{if } (\sigma,t) \in K_n \times \Sigma^n \\ \beta([(\sigma,t)]) & \text{if } (\sigma,t) \in L_n \times \Sigma^n \end{cases}$$

One checks that φ is well defined and continuous. It is clear that $\varphi\,|\iota_K| = \alpha$ and $\varphi\,|\iota_L| = \beta$. Uniqueness of such a map φ is obvious. ∎

1.2.1. *The CW Structure on* $|K|$

Given a simplicial set $K \in$ **SSet**, the geometric realization $|K| \in$ **Top**. In fact, $|K|$ has a CW structure, which we describe in this section. First, we define the skeletons of $|K|$.

Notation 1.2.4. Suppose $K \in$ **SSet**. For integers $n \geq 0$, let $\mathcal{K}^n(K) \subseteq K$ denote the simplicial subset generated by the set $\cup_{i=0}^n K_i$. We would often write $\mathcal{K}^n := \mathcal{K}^n(K)$ when K is understood. We would establish that $|K|$ has a CW structure, $|\mathcal{K}^n(K)| \subseteq |K|$ being the n-skeleton.

For completeness, we prove the following, which are some set theoretic issues.

Lemma 1.2.5. *Let* $K \in \mathbf{SSet}$ *and* $n \geq 1$ *be an integer. Let* $(\sigma, x), (\sigma', x') \in K_{n-1} \times \Sigma^{n-1}$ *be distinct,* $(\tau, y) \in K_n \times \Sigma^n$*, such that*

$$(\sigma, x) \sim (\tau, y) \sim (\sigma', x') \quad in$$

$$|\mathcal{K}^n| \quad \ni \text{ “}\sim\text{” } \textit{are induced by face or degeneracy maps.}$$

Then either $\tau \in \mathcal{K}^{n-1}$ *or there is a chain of equivalences, of length two,*

$$(\sigma, x) \sim (\tau', y') \sim (\sigma', x') \quad where\ (\tau', y') \in K_{n-2} \times \Sigma^{n-2}$$

and the equivalences are induced by face or degeneracy maps.

Proof. Suppose one of the two equivalences is induced by a degeneracy. Assume $(\sigma, x) \sim (\tau, y)$ is given by a degeneracy $s^i : [n-1] \longrightarrow [n]$. In this case, $\tau = K(s^i)(\sigma) \in \mathcal{K}^{n-1}$. So, we assume both relations are induced by face maps. So,

$$\exists\ d^i, d^j \in Mor_\Delta([n-1],[n]) \ni \begin{cases} K(d^i)\tau = \sigma & \Sigma(d^i)(x) = y \\ K(d^j)\tau = \sigma' & \Sigma(d^j)(x') = y \end{cases}$$

Denote the faces in $Mor_\Delta([n-2],[n-1])$ by δ^i. If $i = j$, then $\sigma = \sigma'$ and $x = x'$. So, we assume $i < j$. We have

$$\begin{cases} K(d^j\delta^i) = K(\delta^i)K(d^j) = & K(\delta^{j-1})K(d^i) = K(d^i\delta^{j-1}) \\ K(\delta^{j-1})K(d^i)\tau = K(\delta^{j-1})(\sigma) & \Sigma(d^i)(x) = y \\ K(\delta^i)K(d^j)\tau = K(\delta^i)(\sigma') & \Sigma(d^j)(x') = y \end{cases}$$

$$\begin{cases} \text{With } y = (y_0, y_1, \dots, y_n) \quad \text{we have} \quad y_i = y_j = 0. \qquad \text{Let} \\ y' = (y_0, \dots, y_{i-1}, y_{i+1}, \dots, y_{j-1}, y_{j+1}, \dots, y_n) \in \Sigma^{n-2} \\ \text{and} \qquad \tau' = K(\delta^{j-1})(\sigma) = K(\delta_i)(\sigma') \in K_{n-2} \end{cases}$$

We have

$$\begin{cases} \Sigma(d^i)(\Sigma(\delta^{j-1})(y')) = \Sigma(d^i)(x) = y & \Longrightarrow \quad \Sigma(\delta^{j-1})(y') = x \\ \Sigma(d^j)(\Sigma(\delta^i)(y')) = |d^j|(x') = y & \Longrightarrow \quad \Sigma|\delta^i|(y') = x' \end{cases}$$

It follows

$$\begin{cases} (\tau', y') = (K(\delta^{j-1})(\sigma), y') \simeq (\sigma, x) \\ \qquad\qquad \text{given by} \quad \delta^{j-1} : [n-2] \longrightarrow [n-1] \\ (\tau', y') = (K(\delta^{i})(\sigma'), y') \simeq (\sigma', x')] \\ \qquad\qquad \text{given by} \quad \delta^{i} : [n-2] \longrightarrow [n-1] \end{cases}$$

So, we have $(\sigma, x) \sim (\tau', y') \sim (\sigma', x')$, with $(\tau', y') \in K_{n-2} \times \Sigma^{n-2}$. This completes the proof. ■

Proposition 1.2.6. *Suppose $K \in$ **SSet**. For integers $m \leq n$, consider the following commutative diagram of geometric realizations:*

$$\begin{array}{ccc} |\mathcal{K}^m| & \overset{\iota_{mn}}{\hookrightarrow} & |\mathcal{K}^n| \\ & \underset{\iota_m}{\searrow} & \downarrow \iota_n \\ & & |K| \end{array}$$

For all integers $m \leq n$, the maps $\iota_{mn}, \iota_n, \iota_n$ are injective.

Proof. It is enough to prove, for integers $n \geq 1$, the map $\iota := \iota_{(n-1)n} : |\mathcal{K}^{n-1}| \longrightarrow |\mathcal{K}^n|$ is injective. Let $\omega, \omega' \in |\mathcal{K}^{n-1}|$, with $\iota(\omega) = \iota(\omega')$. We can write $\omega = [(\sigma, x)]$, $\omega' = [(\sigma', x')]$ for some $(\sigma, x), (\sigma', x') \in K_{n-1} \times \Sigma^{n-1}$. There is a chain of equivalences

$$(\sigma, x) = (\tau_0, y_0) \sim (\tau_1, y_1) \sim \cdots \sim (\tau_r, y_r) = (\tau, y)$$

where $(\tau_i, y_i) \in K_{n_i} \times \Sigma^{n_i}$ and the relation $(\tau_{i-1}, y_{i-1}) \sim (\tau_i, y_i)$ is given by a degeneracy map or a face map. We use induction on r, to prove $\omega = \omega'$ in $|\mathcal{K}^{n-1}|$. For $r = 2$, it follows from (1.2.5). If $\tau_i \in \mathcal{K}^{n-1}$, for some $i = 1, \ldots, r-1$, then by induction, it follows $\omega = \omega'$ in $|\mathcal{K}^{n-1}|$. So, we assume $\tau_i \notin \mathcal{K}^{n-1}$, and hence $n_i \geq n$ for all $i = 1, \ldots, r-1$. Let $N = \max\{n_i : i = 1, \ldots, r-1\}$. Now, again, we use induction on N. If $N = n$, it follows $(\tau_1, y_1) = \cdots = (\tau_{r-1}, y_{r-1})$, and hence it follows from case $r = 2$ that $\omega = \omega'$ in $|\mathcal{K}^{n-1}|$. Now, assume $N \geq n + 1$. By the same argument it follows $\omega = \omega'$ in $|\mathcal{K}^{N-1}|$. So, the proof is complete by induction on N. ■

Notation 1.2.7. Let $K \in \mathbf{SSet}$ and $\sigma \in K_n$. Let q_n and Φ_σ denote the maps, as in the commutative diagram

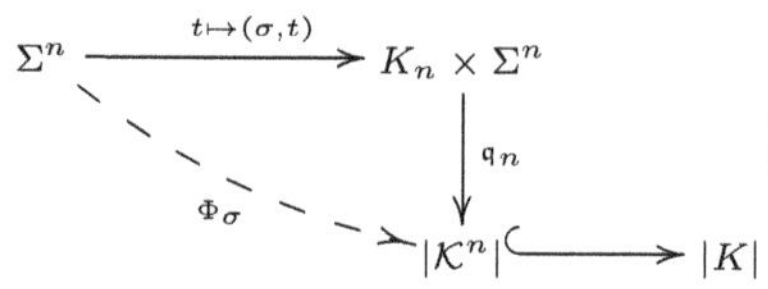

Further,
$$\begin{cases} \mathcal{X}^n(K) := \coprod_{k \leq n} K_k \times \Sigma^k \subseteq \mathcal{X}(K) \\ \partial\Sigma^n \text{ denotes the boundary of } \Sigma^n \\ Int(\Sigma^n) \text{ denotes the interior of } \Sigma^n \\ \text{We Prove (1.2.9) } \Phi_\sigma(\partial\Sigma^n) \subseteq |\mathcal{K}^{n-1}| \\ \text{and denote } \varphi_\sigma := (\Phi_\sigma)_{|\partial\Sigma^n} : \\ \partial\Sigma^n \longrightarrow |\mathcal{K}^{n-1}| \end{cases}$$

Also, $\rho_\sigma : \Sigma^m \longrightarrow \frac{\mathcal{X}^n(K)}{\simeq}$ would denote the natural map.

The following lemma is intuitive.

Lemma 1.2.8. *Let $K \in \mathbf{SSet}$, and $n \geq 0$. Then the map $\mathfrak{p} : \frac{\mathcal{X}^n(K)}{\simeq} \longrightarrow |\mathcal{K}^n|$ is a homeomorphism.*

Proof. Clearly $\mathfrak{p}$ is a well-defined surjective and continuous map. It also follows from (1.2.5) that $\mathfrak{p}$ is bijective. We prove that $\mathfrak{p}$ is an open map. Let $U \subseteq \frac{\mathcal{X}^n(K)}{\simeq}$ be open. For $\tau \in (\mathcal{K}^n(K))_k \subseteq K_k$, we have $m \leq n$, $\sigma \in K_m$ (nondegenerate and unique), and a degeneracy $s \in Mor_\Delta([k],[m])$, such that $\tau = K(s)\sigma$. We have a commutative diagram

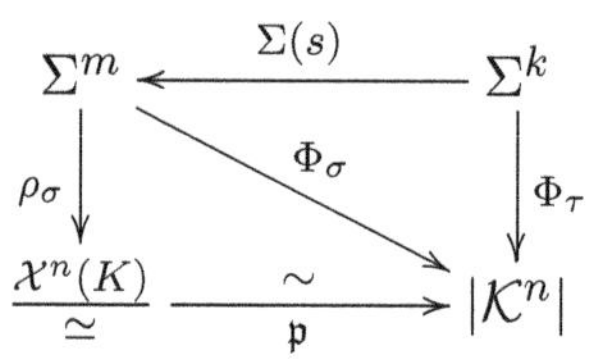

and $\Phi_\tau^{-1}\mathfrak{p}(U) = \Sigma(s)^{-1}\Phi_\sigma^{-1}\mathfrak{p}U$
$= \Sigma(s)^{-1}\rho^{-1}U$ is open.

The proof is complete. ■

Lemma 1.2.9. *Let $K \in \mathbf{SSet}$, and $\sigma \in K_n$ be a nondegenerate n-simplex, and Φ_σ be as in (1.2.7). Write $\mathfrak{e}_\sigma = \Phi_\sigma(Int(\Sigma^n))$. Then the restriction $(\Phi_\sigma)_{|Int(\Sigma^n)} : Int(\Sigma^n) \longrightarrow \mathfrak{e}_\sigma$ is a homeomorphism. Further, $\Phi_\sigma(\partial(\Sigma^n)) \subseteq |\mathcal{K}^{n-1}|$. Consequently,*

$$|\mathcal{K}^n| = |\mathcal{K}^{n-1}| \bigcup \left(\bigcup_{\sigma \in K_n,\ nondegenerate} \mathfrak{e}_\sigma \right) \quad \textit{is a disjoint union.}$$

Proof. Note $Int(\Sigma^n) = \{(t_0, t_1, \ldots, t_n) : 0 < t_i < 1 \; \forall \; i, \; \sum t_i = 1\}$. It follows that the restriction $(\rho_\sigma)_{|Int(\Sigma^n)} : Int(\Sigma^n) \longrightarrow \rho_\sigma(Int(\Sigma^n))$ is bijective and open, hence is a homeomorphism. By (1.2.8), $(\Phi_\sigma)_{|Int(\Sigma^n)} : Int(\Sigma^n) \longrightarrow \mathfrak{e}_\sigma$ is a homeomorphism.

Now, suppose $t = (t_0, t_1, \ldots, t_n) \in \partial(\Sigma^n)$. So, $t_i = 0$ for some i. Assume $t_0 = 0$. Then $t = d^0(t_1, \ldots, t_n)$. So,

$$\Phi_\sigma(t) = [(\sigma, t)] = [(K(d^0)\sigma, (t_1, \ldots, t_n))] \in \left|\mathcal{K}^{n-1}\right|$$

The proof is complete. ■

Proposition 1.2.10. *Let $K \in$ **SSet**, and $K'_n \subseteq K_n$ be the set of all nondegenerate n-simplexes. Then for all integers $n \geq 1$, the following*

$$\begin{array}{ccc} \coprod_{\sigma \in K'_n} \partial\Sigma^n & \xrightarrow{\coprod \varphi_\sigma} & \left|\mathcal{K}^{n-1}\right| \\ {\scriptstyle i}\downarrow & & \downarrow{\scriptstyle \iota_{(n-1)n}} \\ \coprod_{\sigma \in K'_n} \Sigma^n & \xrightarrow[\coprod \Phi_\sigma]{} & \left|\mathcal{K}^n\right| \end{array}$$

is a pushforward diagram in **Top** *where* $\varphi_\sigma := (\Phi_\sigma)_{|\partial\Sigma^n}$.

Proof. We check that the diagram satisfies the universal property of pushforward in **Top**. Extend the above diagram as follows:

$$\begin{array}{ccccc} \coprod_{\sigma \in K'_n} \partial\Sigma^n & \xrightarrow{\coprod \varphi_\sigma} & \left|\mathcal{K}^{n-1}\right| & & \\ {\scriptstyle i}\downarrow & & \downarrow{\scriptstyle \iota_{(n-1)n}} & \searrow{\scriptstyle f} & \\ \coprod_{\sigma \in K'_n} \Sigma^n & \xrightarrow[\coprod \Phi_\sigma]{} & \left|\mathcal{K}^n\right| & \overset{h}{\dashrightarrow} & Z \\ & \searrow_{\coprod g_\sigma} & & & \end{array}$$

In this diagram, $f, \coprod g_\sigma$ are given, such that $f(\coprod \varphi_\sigma) = (\coprod g_\sigma)i$. We need to prove that there is a unique map h, such that the outer triangles commute. When it exists, the uniqueness of h follows, because $|\mathcal{K}^n|$ is a union of the images of $\coprod \Phi_\sigma$ and $\iota_{(n-1)n}$. Use the partition

(1.2.9) of $|\mathcal{K}^n|$ to define h, as follows:

$$h(\omega) = \begin{cases} f(\omega) & \text{if } \omega \in \left|\mathcal{K}^{n-1}\right| \\ g_\sigma(t) & \text{if } \omega = \Phi_\sigma(t) \in \mathfrak{e}_\sigma, \ \sigma \in K'_n \end{cases}$$

Since (1.2.9) provides a partition, h is well defined. The upper right triangle commutes by definition. The lower left triangle commutes because g_σ and f agree on $\partial\Sigma^n$.

To check continuity, recall by (1.2.8) $\mathfrak{p} : \mathcal{X}^n(K) \xrightarrow{\sim} |\mathcal{K}^n|$ is a homeomorphism. We need to check, $\forall\ k \leq n$, $\sigma \in K_k$, the composition $h\Phi_\sigma$ is continuous. If $k \leq n-1$, then $h\Phi_\sigma = f\Phi_\sigma$, which is continuous. If $k = n$ and $\sigma \in K'_n$, then also $h\Phi_\sigma = g_\sigma$ is continuous. Suppose $\sigma \in K_n - K'_n$. Then $\sigma = K(s)\tau$ for some $s \in Mor_\Delta([n],[m])$, with $m \leq n-1$. Then the diagram

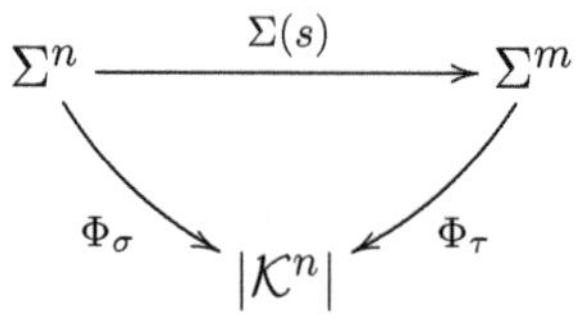

commutes. So, $h\Phi_\sigma = h\Phi_\tau\Sigma(s) = f\Phi_\tau\Sigma(s)$ is continuous.

The proof is complete. ■

The following is the main result [Mi2, Thm. 1] on CW structures on the geometric realizations.

Theorem 1.2.11. *Let* $K \in \mathbf{SSet}$. *Then the geometric realization* $|K|$ *has a CW complex structure. With notation as in* (1.2.7), *in this CW structure,* $|\mathcal{K}^n|$ *would be the* n*-skeleton, and the patching maps are* $\{\varphi_\sigma : \sigma \in K_n, \text{ nondegenerate}\}$. *Further, the cell decomposition is given in* (1.2.9).

Proof. It follows from (1.2.10) $|\mathcal{K}^n|$ is obtained by pushforward as required. Now, let $C \subseteq |K|$ be such that $C \cap |\mathcal{K}^n|$ is closed in $|\mathcal{K}^n|$ for all n. Again, it follows that the preimage of C in $\sigma \times \Sigma^n$ is closed, for all n and $\sigma \in K_n$. So, C is closed in $|K|$. The converse is also obvious. The proof is complete. ■

The following is a basic result on the geometric realizations.

Corollary 1.2.12. *Suppose* K *is a simplicial set and* $L \subseteq K$ *is a simplicial subset. Then the map* $|L| \longrightarrow |K|$ *is injective. Further,* $|L|$ *is a closed subset of* $|K|$.

Proof. This follows immediately from cell decomposition (1.2.9). ■

1.2.2. *Simplicial Spaces*

We recall the definition of simplicial spaces.

Definition 1.2.13. The category of Simplicial Topological spaces would be denoted by **SimTop**. So, an object of **SimTop** is a contravariant functor

$$\Delta \longrightarrow \mathbf{Top}, \qquad \text{equivalently, a covariant functor} \quad \Delta^{op} \longrightarrow \mathbf{Top}$$

$$\begin{aligned} &\forall\ K, L \in \mathbf{SimTop}, \quad Mor_{\mathbf{SimTop}}(K, L) \\ &= \text{Set of all natural transformations } \tau : K \longrightarrow L \end{aligned}$$

For such a $\tau : K \longrightarrow L$, the maps $\tau_n : K_n \longrightarrow L_n$ are required to be continuous.

Note that there is a forgetful functor **SimTop** $\longrightarrow$ **SSet**. On the other hand, any set X with its discrete topology can be considered a topological space. In this way, there is a natural functor **SSet** $\longrightarrow$ **SimTop**.

For $K \in$ **SimTop**, its geometric realization $|K|$ is defined exactly as in (1.2.1), while the skeletal structure $|K|^n$ is defined with some extra care, as follows [Mp, p. 100].

Definition 1.2.14. Let $K \in$ **SimTop**. As in (1.2.1),

$$\mathcal{X}(K) = \coprod_{r=0}^{\infty} K_r \times \Sigma^r, \qquad \text{and} \qquad |K| = \frac{\mathcal{X}(K)}{\simeq}$$

where $\simeq$ is as in (1.2.1). The **geometric realization** of K is defined to be the set $|K|$ with the quotient topology. Now, let

$$\mathcal{X}^n(K) = \coprod_{r=0}^{n} K_r \times \Sigma^r, \qquad \text{and let}$$

$$|K|^n = \frac{\mathcal{X}^n(K)}{\simeq} \quad \text{with quotient topology.}$$

$$\text{By (1.2.6, 1.2.8)}, \quad |K| = \bigcup_{n=0}^{\infty} |K|^n \quad \text{as a set.}$$

It follows that a subset $A \subseteq |K|$ is closed if and only if $|K| \cap A$ is closed in $|K|^n$ for all $n \geq 0$. However, without further hypothesis, $|K|^n$ may not be closed in $|K|$. Set theoretically, nothing has changed from (1.2.1). So, all the set theoretic aspects discussed above (Section 1.2) remain valid.

We remark that [Mp] works exclusively in the subcategory $\mathscr{U} \subseteq$ **Top**, of all compactly generated T_2-spaces. So, $K = \{K_n\} \in$ **SimTop**, considered in [Mp], are actually simplicial objects in $\mathscr{U}$, meaning $K_n \in \mathscr{U}$.

For subsequent references, we introduce the following notation.

Notation 1.2.15. Let $K \in$ **SSet**. The quotient map would be denoted by

$$\Phi_n : K_n \times \Sigma^n \longrightarrow |K|,$$

and may be referred to as a characteristic map.

For $n \geq 1$, let $\mathcal{L}_n := \mathcal{L}_n(K) := \cup_{i=0}^{n-1} K(s^i)K_{n-1} \subseteq K_n$, and let $\mathcal{L}_0$ be the empty set. So, $\mathcal{L}_n$ is the set of degeneracies, which is also called the **Latching set**. Subsequently, it should be taken as a standing hypothesis that $\mathcal{L}_n$ is closed in K_n for all n.

Lemma 1.2.16. *Let $K \in$ **SimTop**. Then*

1. *The image $\Phi_n(K_n \times \Sigma^n)$ is closed in $|K|^n$.*
2. *If $\mathcal{L}_n$ is closed in K_n, then $|K|^{n-1}$ is closed in $|K|^n$. Consequently, if $\mathcal{L}_n$ is closed in K_n for all n, then $|K|^n$ is closed in $|K|$, for all n.*

Proof. For $k \leq n$, it is easy to see that $\Phi_k^{-1}\Phi_n(K_n \times \Sigma^n) = K_k \times \Sigma^k$. This establishes (1). Also, $\Phi_n^{-1}(|K|^n - |K|^{n-1}) = (K_n - \mathcal{L}_n) \times (\Sigma^n - \partial\Sigma^n)$, which is open. This establishes (2). The proof is complete. ■

We quote the following from [P], which we used in (1.2.16).

Lemma 1.2.17. *Let $K \in$ **SimTop**, and assume K_n is a T_2-space $\forall\, n \geq 0$. Then the geometric realization $|K|$ is a T_2 space.*

Proof. It is similar to the proof of (C.1.10), while much more tedious. ■

Lemma 1.2.18. *Let $K \in$ **SimTop**, and $\mathcal{L}_n$ be closed in K_n. Then $|K|^n = |K|^{n-1} \cup \Phi_n(K_n \times \Sigma^n)$ is the union of two closed sets.*

Further, with $B_n = (\mathcal{L}_n \times \Sigma^n) \cup (K_n \times \partial\Sigma^n)$, *the diagram*

$$\begin{array}{ccc} B_n & \longrightarrow & |K|^{n-1} \\ \big\downarrow & & \big\downarrow \\ K_n \times \Sigma^n & \xrightarrow[\Phi_n]{} & |K|^n \end{array}$$

is a pushforward diagram in **Top**.

In particular, if $|K|^{n-1}$ *is compactly generated, and* K_n *is locally compact, then* $|K|^n$ *is compactly generated.*

Proof. It is immediate, because $|K|^{n-1}$ and $\Phi_n(K_n \times \Sigma^n)$ are closed in $|K|^n$, by (1.2.16). We use (1.2.17) to prove $|K|^n$ is a weak T_2 space. ■

Remark 1.2.19. For $K \in$ **SimStop**, the geometric realizations $|K|$ enjoy most of the usual topological properties, as in the case of simplicial sets. We left some of that as exercise (1.5.5). While $|K|^n$ behave like skeletons, as such there is no CW structure on $|K|^n$. We state the following [Mp] on CW structure.

Proposition 1.2.20. *Let* $K \in$ **SimTop**. *Assume, for all* n, K_n *is a CW complex and the face and degeneracy maps* $K(d^i)$, $K(s^i)$ *are cellular maps. Then the geometric realization* $|K|$ *has a CW structure, with one* $n+m$*-cell for each* m*-cell in* $K_n - \mathcal{L}_n(K)$.

Proof. In the general case, the proof would be an extension of the proof (1.2.11), which is the case when $K \in$ **SSet**. Our interest in this is limited to the case when $K \in$ **SSet**. We skip the proof for the general case [Mp, p. 103]. ■

For completeness, the following lemma would be helpful.

Lemma 1.2.21. *Let* $K \in$ **SimTop**, *and* X *be a compactly generated* T_2 *space. Assume* $\mathcal{L}_n(K)$ *is closed in* K_n, *and* K_n *is locally compact* (T_2) *space* $\forall\ n$. *Let* $X \times K$ *denote the simpicial space* $n \mapsto X \times K_n$. *Then* $k(X \times |K|) \cong |X \times K|$ *is homeomorphic.*

Proof. It follows from definition (1.2.14) that $|X \times K|^n \cong X \times |K|^n$. Write $Y^n = X \times |K|^n$, $Y = X \times |K|$. Note $X \times \mathcal{L}_n(K)$ is closed in $X \times K_n$. So, Y^n is closed in Y. Also, since $X \times K_n$ is compactly generated (B.6.7), by induction, Y^n is compactly generated (1.2.18). Consider the commutative diagram

$$\begin{array}{ccc} & & k(X \times |K|) \\ & \nearrow & \downarrow \\ |X \times K| & \longrightarrow & X \times |K| \end{array}$$

where the horizontal and vertical arrows are known to be continuous, we want to prove that the broken arrow is a homeomorphism.

Suppose $A \subseteq Y = X \times |K|$ is compactly closed. Let $C \subseteq Y^n$ be compact in Y^n. Then $A \cap C$ is closed in Y. So, $A \cap Y^n \cap C$ is closed in Y. So, $A \cap Y^n$ is compactly closed in Y^n. Therefore, $A \cap Y^n$ is closed in Y^n. Hence, A is closed in $|X \times K|$.

What remains to be shown is, if $A \subseteq |X \times K|$ is closed, then it is compactly closed in $X \times |K|$. Let $C \subseteq X \times |K|$ be compact. Then $C \subseteq C_1 \times C_2$, with C_1, C_2 compact. It follows $C_2 \subseteq |K|^n$ for some $n \geq 0$ (1.5.5). We have $A \cap (X \times |K|^n)$ is closed in $X \times |K|^n$. So, $A \cap C$ is closed in $X \times |K|^n$, which is closed in $X \times |K|$. Therefore, $A \cap C$ is closed in $X \times |K|$. So, A is compactly closed. The proof is complete. ■

1.3. Bisimplicial Sets

In this section, we discuss a lemma on bisimplicial sets.

Definition 1.3.1. A **bisimplicial set** $\mathbb{T}$ is a functor

$$\mathbb{T} : \Delta^o \times \Delta^o \longrightarrow \mathbf{Set}. \quad \text{For } ([m],[n]) \in Obj(\Delta^o \times \Delta^o)$$
$$\text{we write} \quad \mathbb{T}_{mn} = \mathbb{T}([m],[n]).$$

Similarly, a functor $\mathbb{T} : \Delta^o \times \Delta^o \longrightarrow \mathbf{Top}$.
is called a **bisimplicial space**

Immediate examples of bisimplicial sets are obtained by combining two simplicial sets $K, L \in \mathbf{SSet}$. We would mainly be interested in bisimplicial sets only. However, for the rest of this section, it would be convenient to treat Bisimplicial Sets $\mathbb{T}$ as bisimplicial topological spaces, with discrete topology on $\mathbb{T}_{mn}$. Given a bisimplicial space $\mathbb{T}$, we define several simplicial spaces as follows:

1. First, the diagonal $\mathbf{D}\mathbb{T} : \Delta^o \longrightarrow \mathbf{Top}$, is defined by $(\mathbf{D}\mathbb{T})_n := (\mathbf{D}\mathbb{T})([n]) := \mathbb{T}_{nn}$. Then $\mathbf{D}\mathbb{T} \in \mathbf{SimTop}$. We call it the diagonal of $\mathbb{T}$.
   For a fixed integer $m \geq 0$, define $\mathbf{R}\mathbb{T}_{m\bullet} : \Delta^o \longrightarrow \mathbf{Top}$, by $\mathbf{R}\mathbb{T}_{m\bullet}([n]) = \mathbb{T}_{mn}$. So, $\mathbf{R}\mathbb{T}_{m\bullet} \in \mathbf{SimTop}$. Likewise, for a fixed integer $n \geq 0$, define $\mathbf{L}\mathbb{T}_{\bullet n} : \Delta^o \longrightarrow \mathbf{Top}$, by $\mathbf{L}\mathbb{T}_{\bullet n}([m]) = \mathbb{T}_{mn}$. So, $\mathbf{L}\mathbb{T}_{\bullet n} \in \mathbf{SimTop}$.
   If $\mathbb{T}$ is a bisimplicial set, then $\mathbf{D}\mathbb{T}, \mathbf{R}\mathbb{T}_{m\bullet}, \mathbf{L}\mathbb{T}_{\bullet n} \in \mathbf{SSet}$.
2. Now, define two functors

$$\mathcal{L}\mathbf{R}\mathbb{T}, \mathcal{R}\mathbf{L}\mathbb{T} : \Delta^o \longrightarrow \mathbf{Top} \quad \text{by} \begin{cases} (\mathcal{L}\mathbf{R}\mathbb{T})_m := (\mathcal{L}\mathbf{R}\mathbb{T})[m] := |\mathbf{R}\mathbb{T}_{m\bullet}| \\ (\mathcal{R}\mathbf{L}\mathbb{T})_n := (\mathcal{R}\mathbf{L}\mathbb{T})[n] := |\mathbf{L}\mathbb{T}_{\bullet n}| \end{cases}$$

   Therefore, $\mathcal{L}\mathbf{R}\mathbb{T}, \mathcal{R}\mathbf{L}\mathbb{T} \in \mathbf{SimTop}$ are simplicial topological spaces. We will compare the geometric realizations of $\mathbf{D}\mathbb{T}, \mathcal{L}\mathbf{R}\mathbb{T}, \mathcal{R}\mathbf{L}\mathbb{T} \in \mathbf{SimTop}$, in Lemma (1.3.2).

The following lemma is from the paper of Quillen [Q, p. 94].

Lemma 1.3.2. *Let* $\mathbb{T} : \Delta^o \times \Delta^o \longrightarrow \mathbf{Top}$ *be a bisimplicial space and* $\mathbf{D}\mathbb{T}$, $\mathcal{L}\mathbb{T}$, $\mathcal{R}\mathbb{T}$ *be as above* (1.3.1). *Then there are natural homeomorphisms:*

$$|\mathcal{L}\mathbf{R}\mathbb{T}| \xleftarrow[\sim]{} |\mathbf{D}\mathbb{T}| \xrightarrow[\sim]{} |\mathcal{R}\mathbf{L}\mathbb{T}|$$

Proof. We establish only the first homeomorphism. We have

$$\begin{cases} |\mathbf{D}\mathbb{T}| = \dfrac{\coprod_{m\geq 0} \mathbb{T}_{mm} \times \Sigma^m}{\sim} \\ |\mathcal{L}\mathbf{R}\mathbb{T}| = \dfrac{\coprod_{m\geq 0} |\mathbf{R}\mathbb{T}_{m\bullet}| \times \Sigma^m}{\sim} = \dfrac{\coprod_{m\geq 0} \left(\frac{\coprod_{\geq} \mathbb{T}_{mn} \times \Sigma^n}{\sim} \right) \times \Sigma^m}{\sim} \end{cases}$$

There is a natural map

$$\varphi : |\mathbf{D}\mathbb{T}| \longrightarrow |\mathcal{L}\mathbf{R}\mathbb{T}| \quad \text{where } \varphi([(\tau, t)]) = [((\tau, t)], t)] \\ \forall\ (\tau, t) \in \mathbb{T}_{mm} \times |\Delta[m]|$$

Now, one proves that φ is a homeomorphism. This is done in two stages. For integers $r, s \geq 0$, let $\Omega[r, s] := \Delta[r] \times \Delta[s]$ be the bisimplicial set. So,

$$\Omega[r, s]([p], [q]) = \Delta([p], [r]) \times \Delta([q], [s]) \quad \forall\ p, q \geq 0$$

Fix $S \in \mathbf{Top}$ and integers $r, s \geq 0$. Let $\mathbb{T}^{r,s,S} := \Omega[r,s] \times S$ be defined by

$$\mathbb{T}^{r,s,S}_{mn} = \Omega[r,s]([m],[n]) \times S = \Delta([m],[r]) \times \Delta([n],[s]) \times S$$

The diagram
$$\begin{array}{ccc} \Delta^o \times \Delta^o & \xrightarrow{\Omega[r,s]} & \mathbf{Set} \\ & \searrow^{\mathbb{T}^{r,s,S}} & \downarrow{\times S} \\ & & \mathbf{Top} \end{array}$$
commutes.

We establish the lemma for such a bisimplicial set $\mathbb{T}^{r,s,S}$. We have

$$\left|\mathbf{D}\mathbb{T}^{r,s,S}\right| = \frac{\coprod_{m\geq 0} \Delta([m],[r]) \times \Delta([m],[s]) \times S \times \Sigma^m}{\sim} \xrightarrow{\sim} S \times \Sigma^r \times \Sigma^s$$

Similarly, for a fixed integer $m \geq 0$, we have

$$\left|\mathbf{R}\mathbb{T}^{r,s,S}_{m\bullet}\right| = \frac{\coprod_{n\geq 0} \Delta([m],[r]) \times \Delta([n],[s]) \times S \times \Sigma^n}{\sim}$$
$$\xrightarrow{\sim} \Delta([m],[r]) \times S \times \Sigma^s$$

Therefore,

$$\begin{aligned} \left|\mathcal{L}\mathbf{R}\mathbb{T}^{r,s,S}\right| &= \frac{\coprod_{m\geq 0} |\mathbf{R}\mathbb{T}_{m\bullet}| \times \Sigma^m}{\sim} \\ &= \frac{\coprod_{m\geq 0} \Delta([m],[r]) \times S \times \Sigma^s \times \Sigma^m}{\sim} \xrightarrow{\sim} S \times \Sigma^r \times \Sigma^s \end{aligned}$$

This establishes that φ is a homeomorphism, in this case. Now, we establish the same for general bisimplicial spaces $\mathbb{T}$.

1. Fix integers $r, s \geq 0$.

$$\begin{aligned} &\forall\ (f, g, \tau) \in \Delta([m],[r]) \times \Delta([n],[s]) \times \mathbb{T}_{rs} \\ &\varepsilon_{rs}(f, g, \tau) := \mathbb{T}(f, g)(\tau) \in \mathbb{T}_{mn} \end{aligned}$$

   This defines maps (natural transformation) of bisimplicial sets,

$$\varepsilon_{rs} : \Omega([r,s]) \times \mathbb{T}_{rs} = \Delta[r] \times \Delta[s] \times \mathbb{T}_{rs} \longrightarrow \mathbb{T}$$

We note the following:

(a) ε_{rs} is a map of bisimplicial spaces.
(b) ε_{rs} is determined by the equation $\varepsilon_{rs}(1_{[r]}, 1_{[s]}, \tau) = \tau$, while $1_{[r]}$ is the nondegenerate simplex of $\Delta[r]$.
(c) Therefore, the images under ε_{rs} are the faces and degeneracies of elements in $\mathbb{T}_{rs}$, of the third coordinate.

2. Combining, we get a surjective map

$$\varepsilon := \coprod_{r\geq 0, s\geq 0} \varepsilon_{rs} : \coprod_{r\geq 0, s\geq 0} \Omega[r,s] \times \mathbb{T}_{rs} \longrightarrow \mathbb{T}$$

3. For $(f, g) \in \Delta([r],[r']) \times \Delta([s],[s'])$, we have two maps of bisimplicial spaces:

$$\begin{array}{ccc} \Delta[r] \times \Delta[s] \times \mathbb{T}_{r's'} & \xrightarrow{(f,g,1)} & \Delta[r'] \times \Delta[s'] \times \mathbb{T}_{r's'} \\ & \searrow^{(1,1,\mathbb{T}(f,g))} & \\ & & \Delta[r] \times \Delta[s] \times \mathbb{T}_{rs} \end{array}$$

In fact, the diagram

$$\begin{array}{ccc} \Omega[r,s] \times \mathbb{T}_{r's'} & \xrightarrow{(f,g,1)} & \Omega[r',s'] \times \mathbb{T}_{r's'} \\ {\scriptstyle (1,1,\mathbb{T}(f,g))}\downarrow & & \downarrow{\scriptstyle \varepsilon_{r's'}} \\ \Omega[r,s] \times \mathbb{T}_{rs} & \xrightarrow[\varepsilon_{rs}]{} & \mathbb{T} \end{array}$$

commutes.

4. Taking the coproduct, we have a commutative diagram

$$\begin{array}{ccc} \coprod_{(r,s)} \Omega[r,s] \times \mathbb{T}_{r's'} & \xrightarrow{\coprod(f,g,1)} & \coprod_{(r',s')} \Omega[r',s'] \times \mathbb{T}_{r's'} \\ {\scriptstyle \coprod(1,1,\mathbb{T}(f,g))}\downarrow & & \downarrow{\scriptstyle \varepsilon} \\ \coprod_{(r,s)} \Omega[r,s] \times \mathbb{T}_{rs} & \xrightarrow[\varepsilon]{} & \mathbb{T} \end{array} \tag{1.5}$$

The coproduct is taken over the set

$$\mathscr{I} = \{(f, g) \in \Delta([r],[r']) \times \Delta([s],[s']) : r, r', s, s' \geq 0, \}$$

One checks, that this is a pushforward diagram of bisimplicial spaces. To check this, one needs to check that for all integers,

$p, q \geq 0$, the (p,q)-component of the above diagram is a pushforward in **Top**:

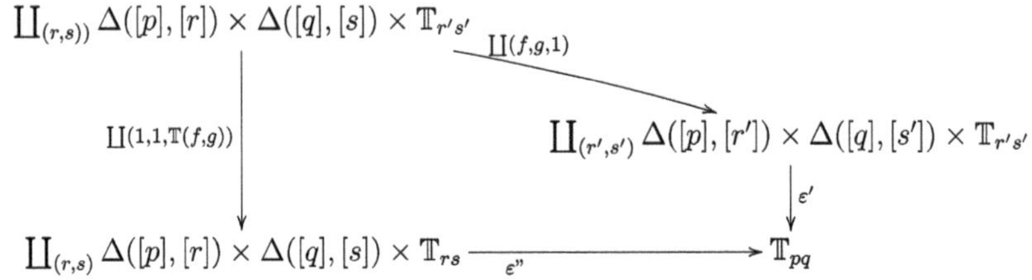

where $\varepsilon', \varepsilon$" denote the restrictions of ε. In fact, both $\varepsilon', \varepsilon$" are surjective maps. It follows from construction of pushforward (in **Set**) that the above rectangle is a pushforward diagram.

5. Therefore, it is established that the diagram (1.5) is a pushforward diagram of bisimplicial sets. Hence, it follows that the diagonals **D** of the diagram (1.5) are also pushforward diagrams. Further, it follows from (1.2.3) that the $\mathcal{L}\mathbf{R}$ of the above diagram (1.5) are also pushforward diagrams.

Now, consider the commutative diagram

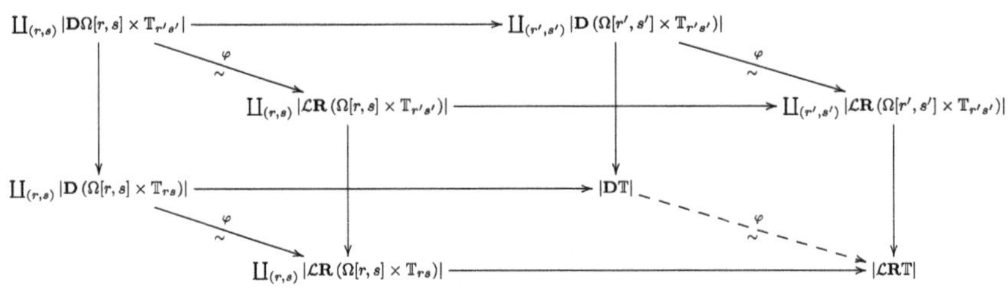

We need to prove that the lower right hand map φ is a bijection. Since the other three are bijections, by the properties of pushforward, so is the fourth one. The proof is complete. ■

The following corollary is of some importance to us.

Corollary 1.3.3. *Let* $K, L \in$ **SSet**. *Then* $|K \times L| \cong k(|K| \times |L|)$ *are homeomorphic.*

Proof. Let $\mathbb{T}([m],[n]) = K_m \times L_n$. Then $\mathbb{T}$ is a bisimplicial set. Then $R\mathbb{T}_{m\bullet}([n]) = K_m \times L_n$. So, the geometric realization $|R\mathbb{T}_{m\bullet}| = K_m \times |L|$. Note, $|L|$ is a compactly generated T_2-space, by (1.2.11,

C.1.10, C.1.11). By (1.2.21), $|\mathcal{L}R\mathbb{T}| = k(|K| \times |L|)$. By (1.3.2),

$$|K \times L| = |\mathbb{DT}| \cong |\mathcal{L}R\mathbb{T}| = k(|K| \times |L|)$$

The proof is complete. ■

In fact, Corollary 1.3.3 works in a much further generality, which we state in two steps.

Corollary 1.3.4. *Suppose $K, L \in$ **SimTop**, such that $\mathcal{L}_n(K)$, $\mathcal{L}_n(L)$ are closed in K_n, L_n, for all $n \geq 0$. Further, assume that K_n, L_n are compactly generated, locally compact T_2 spaces, for all $n \geq 0$. Then $|K \times L| \cong k(|K| \times |L|)$ are homeomorphic.*

Proof. Same as the proof of (1.3.3). Use Exercises 1.5.5 and B.6.7. ■

Remark 1.3.5. Corollary 1.3.4 is valid in further generality [Mp, p. 103]. In fact, a more elementary set theoretic proof is possible, while tedious.

1.4. The Homotopy Groups of Simplicial Sets

For $K \in$ **SSet**, the easiest way to define **homotopy groups** of K is

$$\pi_n(K, v_0) := \pi_n(|K|, v_0) \quad \forall\, n \geq 0, \quad \forall\, v \in K_0 \tag{1.6}$$

There is a desire to define $\pi_n(K, v_0)$ fully combinatorially, without any reference to Topological Homotopy Theory. This works, to an extent [H, Chapter 3], mainly when K is fibrant (1.4.1). In the context of Quillen K-Theory, this may not be of much use, because the simplicial sets we consider (*nerve of a category*) may not be fibrant. For our purpose, the above definition (1.6) would suffice. While we avoid going into deeper details of the homotopy theory of Simplicial Sets, we provide a flavor.

We define the Kan extension property, which is a key concept in this theory.

Definition 1.4.1. A simplicial set K, is said to satisfy the **Kan extension condition** if $\forall\, n \in \mathbb{N}, 0 \leq k \leq n$, and maps $f : \Lambda^k[n] \longrightarrow K$ in **SSet**; f extend to a map $\varphi \in Mor_{\mathbf{SSet}}(\Delta[n], K)$ as in the commutative diagram

$$\begin{array}{ccc} \Lambda^k[n] & \xrightarrow{f} & K \\ \big\downarrow & \nearrow_{\varphi} & \\ \Delta[n] & & \end{array}$$

Such a simplicial set K is also called **Fibrant**. More generally, a map $g : K \longrightarrow L$ is called a **fibration**, if g has the right lifting property, as in the diagram

$$\begin{array}{ccc} \Lambda^k[n] & \xrightarrow{f} & K \\ \big\downarrow & \nearrow_{\varphi} & \big\downarrow g \\ \Delta[n] & \xrightarrow[F]{} & L \end{array}$$

Meaning, if the outer diagram commutes in **SSet**, then there is a map φ such that the inner triangles commute in **SSet**. In particular, a simplicial set K is fibrant $\iff$ the terminal map $K \longrightarrow \Delta[0]$ is a fibration.

So, a fibration in **SSet** is analogous to Weak fibrations (B.3.1) in **Top**. Recall that every object $X \in$ **Top** is fibrant. The following provides a contrast.

Example 1.4.2. The simplicial set $\Delta[1]$ does not satisfy Kan extension condition. Before we proceed to demonstrate this, we give a complete list of nondegenerate simplexes (which we denote with $'$), in the respective simplicial sets:

$$\begin{cases} (\Delta[2])'_0 = \{(0), (1), (2)\} \\ (\Delta[2])'_1 = \{(0,1), (1,2), (0,2)\} \\ (\Delta[2])'_2 = \{(0,1,2)\} \end{cases} \qquad \begin{cases} (\Lambda^0[2])'_0 = \{(0), (1), (2)\} \\ (\Lambda^0[2])'_1 = \{(0,1), (0,2)\} \end{cases}$$

Define $f \in Mor_{\mathbf{SSet}}(\Lambda^0[2], \Delta[1])$ as follows:

$$\begin{cases} f(0,1) = (0,1) \\ f(0,2) = (0,0) \end{cases} \qquad \text{and consider} \qquad \begin{array}{ccc} \Lambda^0[2] & \xrightarrow{f} & \Delta[1] \\ \big\downarrow & & \\ \Delta[2] & & \end{array}$$

So, f is defined on a set of generators of $\Lambda^0[2]$, and extends to $\Lambda^0[2]$. We show f does not extend to $\Delta[2]$. In deed,

$$f(0) = f(d_1(0,1)) = d_1(f(0,1)) = d_1(0,1) = 0,$$
$$\text{Likewise, } f(1) = 1, \ f(2) = 0.$$

Now, suppose there is an extension $F : \Delta[2] \longrightarrow \Delta[1]$, of f. Write $F(0,1,2) = (i,j,k)$ with $0 \le i \le j \le k \le 1$. Then

$$\begin{cases} F(1,2) = F(d_0(0,1,2)) = d_0(i,j,k) = (j,k) \\ 1 = F(1) = F(d_1(1,2)) = d_1(F(1,2)) = d_1(j,k) = j \\ 0 = F(2) = F(d_0(1,2)) = d_0(F(1,2) = d_0(j,k) = k \end{cases}$$

This contradicts that $j \le k$.

Example 1.4.3. Let $Y \in Obj(\mathbf{Top})$, and $SY \in \mathbf{SSet}$ be the singular functor (1.2.2). Then $S(Y)$ has the Kan extension property. Consequently, for $K \in \mathbf{SSet}$, by adjunction (1.2.2), we obtain an universal arrow

$$K \longrightarrow S\,|K| \qquad \text{in} \quad \mathbf{SSet}, \text{ while } S\,|K| \text{ is fibrant.}$$

(*One may say* $S\,|K|$ *is a fibrant replacement of* K.)
Proof. Let $f \in Mor_{\mathbf{SSet}}(\Lambda^k[n], SY)$ be any arrow. By adjunction (1.2.2)

$$Mor_{\mathbf{Top}}(|\Lambda^k[n]|, Y) \cong Mor_{\mathbf{SSet}}(\Lambda^k[n], SY)$$

Corresponding to f, there is a continuous map $\tilde{f} : |\Lambda^k[n]| \longrightarrow Y$. Recall $|\Delta[n]| = \Sigma^n$. Further, $\left|\Lambda^k[n]\right| \hookrightarrow |\Delta[n]|$ has a retraction $r : |\Delta[n]| \hookrightarrow \left|\Lambda^k[n]\right|$ (see 1.5.3). Hence, $\tilde{f}r : |\Delta[n]| \longrightarrow Y$. By adjunction, there is an arrow $F \in Mor_{\mathbf{SSet}}(\Delta[n], SY)$ corresponding to $\tilde{f}r$, completing the commutative diagram

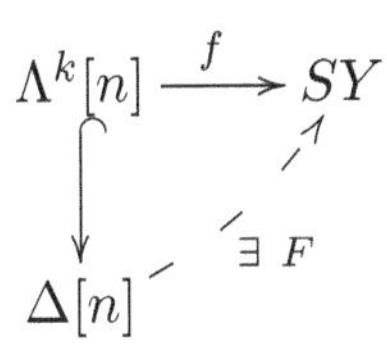

The proof is complete. ∎

Definition 1.4.4. Let $K \in \mathbf{SSet}$ be fibrant. For $x, y \in K_0$, we say that x, y are **homotopic**, and write $x \sim y$, if there is $z \in K_1$ such that $d_1 z = x$, $d_0 z = y$.

Further, let $\simeq$ be the equivalence relations on K_0, generated by $\sim$. Define,

$$\pi_0(K) = \frac{K_0}{\simeq} \tag{1.7}$$

Lemma 1.4.5. *Let* $K \in \mathbf{SSet}$. *Then there is a natural bijection* $\pi_0 K \cong \pi_0 |K|$.

Proof. A vertex $v \in K_0$ determines a point in $|K|$, which we denote by the same notation $v \in |K|$. For a point $x \in |K|$, let $[x] \in \pi_0 |K|$ denote the path component of x. We have

$$|K| = \frac{\coprod_n K_n \times \Sigma^n}{\simeq}.$$ Consider the diagram

$$\begin{array}{ccc} K_0 & \xrightarrow{\ \mathfrak{p}_0\ } & \\ \downarrow & \searrow & \\ \pi_0 K & \dashrightarrow_{\mathfrak{p}} & \pi_0 |K| \end{array} \qquad \text{where} \quad \mathfrak{p}_0(v) = [v]$$

It follows that $\mathfrak{p}_0$ factors through a surjective map $\mathfrak{p}$, as in the diagram. To see that $\mathfrak{p}$ is injective, let $[v] \in \pi_0 K$, with $v \in K_0$. Let $K([v])$ be the subset of K, consisting of $\sigma \in K$, such that σ has a vertex in $[v]$ (*hence all the vertices in* $[v]$). Clearly, $K([v])$ is a simplicial subset of K, and K is a disjoint union of $K([v])$. So, $|K| = \coprod_{[v] \in \pi_0 K} |K([v])|$ is a disjoint (disconnected) union. This established that $\mathfrak{p}$ is injective. The proof is complete. ■

Irrespective of whether $\sim$ is an equivalence relation or not, $\pi_0 K$ and $\pi_0 |K|$ coincide (1.4.5). However, for higher homotopy groups, one may require further caution and need to work with fibrant objects.

Lemma 1.4.6. *Let* $K \in \mathbf{SSet}$ *be fibrant. Then homotopy of vertices is an equivalence relation on* K_0. *We denote the set of equivalence classes by* $\pi_0 K$.

Proof. We prove reflexivity, symmetry and transitivity of $\sim$, as follows.

1. (**Reflexivity**): Let $x \in K_0$. We have the commutative diagrams,

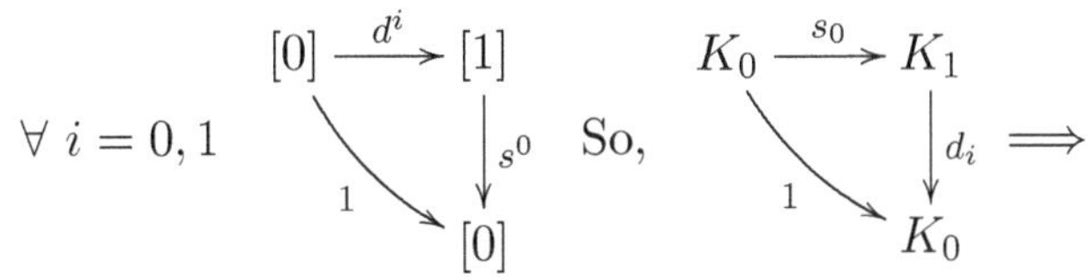

$$d_0(s_0(x)) = x = d_1(s_0(x))$$

2. (**Symmetry**): Suppose $x \sim y \in K_0$. So, $\exists\ \mathfrak{e} \in K_1 \ni d_1\mathfrak{e} = x, d_0\mathfrak{e} = y$. We give a complete list of nondegenerate simplexes, in the respective simplicial sets:

$$\Delta[1]' = \begin{cases} 0, 1 \\ (0,1) \end{cases} \quad \Delta[2]' = \begin{cases} 0, 1, 2 \\ d_2 := (0,1), d_1 := (0,2), \\ d_0 := (1,2), \quad (0,1,2) \end{cases} \quad \Lambda^0[2]' = \begin{cases} 0, 1, 2 \\ d_1, d_2 \end{cases}$$

Note $s_0x, \mathfrak{e} \in K_1$. Nondegenerate edges $d^1, d^2 \in \Lambda^0[2]$ generate $\Lambda^0[2]$. The association

$$\begin{cases} d_1 \mapsto s_0x \\ d_2 \mapsto \mathfrak{e} \end{cases} \quad \text{defines an arrow} \quad f : \Lambda^0[2] \longrightarrow K \quad \text{in } \mathbf{SSet}.$$

For clarity, we represent $\Lambda^0[2] \subseteq \Delta[2]$ as follows:

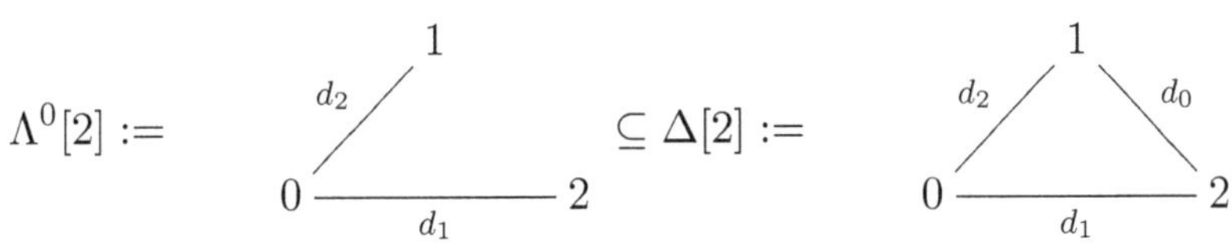

The map $f : \Lambda^0[2] \longrightarrow K$ is given by

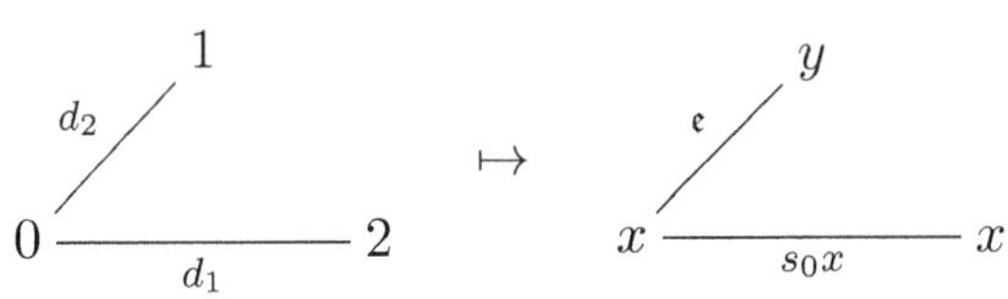

Since K is fibrant, f extends to $F : \Delta[2] \longrightarrow K$. Write $\sigma := F(1_{[2]}) \in K_2$ and $\tilde{\mathfrak{e}} = d_0\sigma = F(d_0)$. So, the diagrams above extend to

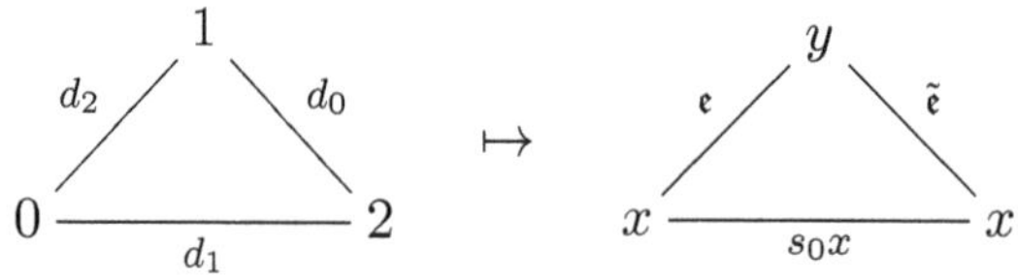

$\Longrightarrow d_0\tilde{\mathfrak{e}} = y, d_1\tilde{\mathfrak{e}} = x$

3. (**Transitivity**): Suppose $x \sim y \sim z$. By symmetry, we assume $z \sim y$.

So, $\exists\ \mathfrak{e}_2, \mathfrak{e}_0 \in K_1,\ \ni\ d_1\mathfrak{e}_2 = x, d_0\mathfrak{e}_2 = y = d_1\mathfrak{e}_0, d_0\mathfrak{e}_0 = z$.
The nondegenerate edges d_0, d_2 generate $\Lambda^1[2]$. The association

$$\begin{cases} d_0 \mapsto \mathfrak{e}_0 \\ d_2 \mapsto \mathfrak{e}_2 \end{cases} \text{defines an arrow} \quad f : \Lambda^1[2] \longrightarrow K \quad \text{in } \mathbf{SSet}$$

For clarity, the association is displayed as follows:

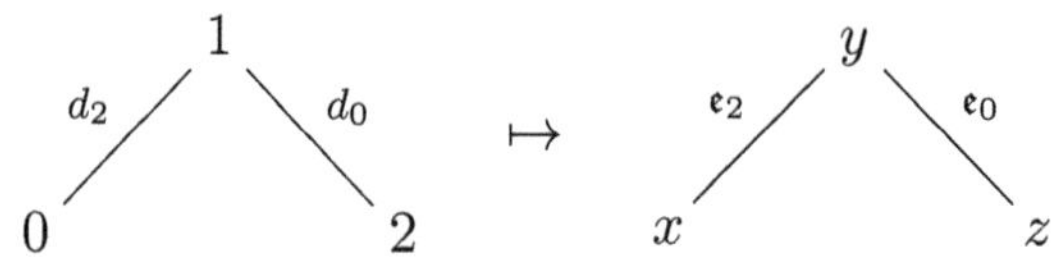

Since K is fibrant, f extends to $F : \Delta[2] \longrightarrow K$. Let $\sigma = F(1_{[2]})$ and $\mathfrak{e}_1 = F(d_1) = d_1(\sigma)$. We have

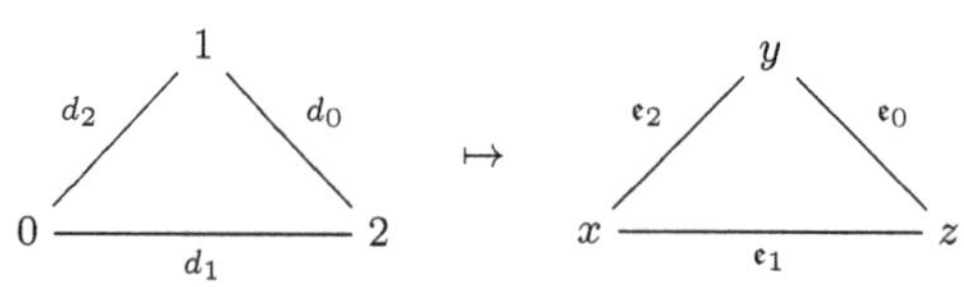

$\Longrightarrow d_1(\mathfrak{e}_1) = x, d_0\mathfrak{e}_1 = z$

The proof is complete. ∎

1.4.1. *Higher Homotopy Groups*

We define higher homotopy groups for fibrant objects K in **SSet**. First, we define homotopy.

Two vertices of $0, 1$, lead to two maps $\mathbf{0}, \mathbf{1} : \Delta[0] \hookrightarrow \Delta[1]$ in **SSet**.

Definition 1.4.7. Suppose $K, L \in$ **SSet** and $f, g : K \longrightarrow L$ are two maps in **SSet**. We say that f is **homotopic** to g, if

$$\begin{gathered} \exists\ H : K \times \Delta[1] \longrightarrow L \ \text{ in } \mathbf{SSet}\ \ni \\ H(x, 0) = f(x),\ H(x, 1) = g(x)\ \ \forall\ x \in K_n,\ n \in \mathbb{N} \end{gathered}$$

where H is a map of simplicial sets, and $H(-, 0)$ and $H(-, 1)$ are defined as follows:

$$\begin{array}{ccccc} K & \xrightarrow{\sim} & K \times \Delta[0] & \xrightarrow{1_K \times \mathbf{0}} & K \times \Delta[0] \\ & & & & \downarrow{\scriptstyle H} \\ & & & \overset{H(-,0)}{\dashrightarrow} & L \end{array} \quad \text{and} \quad \begin{array}{ccccc} K & \xrightarrow{\sim} & K \times \Delta[0] & \xrightarrow{1_K \times \mathbf{1}} & K \times \Delta[0] \\ & & & & \downarrow{\scriptstyle H} \\ & & & \overset{H(-,1)}{\dashrightarrow} & L \end{array}$$

Further, for simplicial subsets $A \subseteq K$, two maps $f, g : K \longrightarrow L$ in **SSet** are said to be homotopic, relative to A, if there is a homotopy

$$H : K \times \Delta[1] \longrightarrow L\ \ni \begin{cases} H(\sigma, 0) = f(\sigma) & \forall\ \sigma \in K \\ H(-, 1) = g(\sigma) & \forall\ \sigma \in K \\ H(\sigma, \eta) = f(\sigma) = g(\sigma) & \forall\ (\sigma, \eta) \in A \times \Delta[1] \end{cases}$$

To define higher homotopy groups of simplicial sets, we introduce the following lemma, together with notation and definitions.

Lemma 1.4.8. *Let $K \in$ **SSet**, $v \in K_0$ and $n \geq 1$ be an integer. Let*

$$\mathfrak{F}^n(K, v) = \{f \in Mor_{\mathbf{SSet}}(\Delta[n], K) : f(\partial\Delta[n]) = v\}$$

(Recall, $Mor_{\mathbf{SSet}}(\Delta[n], K) \xrightarrow{\sim} K_n$ given by $f \mapsto f(1_{[n]})$. So, an n-simplex $\sigma \in K_n$ is in $\mathfrak{F}^n(K, v)$, if all its faces collapse to v.) *For $f, g \in \mathfrak{F}^n(K, v)$, define $f \sim g$, if f is homotopic to g, relative to $\partial\Delta[n]$. So, $f \sim g$, if*

$$H : \Delta[n] \times \Delta[1] \longrightarrow K\ \ni \begin{cases} H(\sigma, 0) = f(\sigma) & \forall\ \sigma \in \Delta[n] \\ H(\sigma, 1) = g(\sigma) & \forall\ \sigma \in \Delta[n] \\ H(\sigma, \eta) = v & \forall\ (\sigma, \eta) \in \partial\Delta[n] \times \Delta[1] \end{cases}$$

*If $K \in$ **SSet** is fibrant, then $\sim$ is an equivalence relation on $\mathfrak{F}^n(K, v)$.*

Proof. We skip the proof. See [H, Thm. 3.3.1] and others. ■

Definition 1.4.9. Let $K \in$ **SSet** be fibrant, $v \in K_0$ and $n \geq 1$ be an integer. With notation, as in (1.4.8), define

$$\pi_n(K, v) = \frac{\mathfrak{F}^n(K, v)}{\sim} \qquad \text{the set of all equivalence classes.}$$

We also quote the following [H, p. 85] without proof.

Lemma 1.4.10. *Let $K \in$ **SSet** be fibrant, $v \in K_0$ be a vertex, $n \geq 0$. For $\sigma \in \mathfrak{F}^n(K, v)$, $[\sigma] = [v] \in \pi_n(K, v)$ if and only if there is a simplex $\tau \in K_{n+1}$ such that $d_{n+1}\tau = \sigma$ and $d_i\tau = v$ for all $i \leq n$.*

The following is an important correspondence.

Lemma 1.4.11. *Let $K \in$ **SSet** be fibrant and $v \in K_0$ be a vertex. Then there is a natural bijection*

$$\varpi : \pi_n(K, v) \xrightarrow{\sim} \pi_n(|K|, v) \qquad \forall \quad n \geq 0$$

Therefore, $\pi_n(K, v)$ is a group for $n \geq 1$ and is abelian for $n \geq 2$.

Outline of the Proof. The case $n = 0$ was proved above (1.4.5). For $n \geq 1$, we only define the natural map. Let $f \in \mathfrak{F}^n(K, v)$. That means $f : \Delta[n] \longrightarrow K$ such that $f|\partial\Delta[n]) = v$. So, the geometric realization $|f|$ is a map $|f| : (|\Delta[n]|, |\partial\Delta[n]|) \longrightarrow (|K|, v)$. Define

$$\varpi_0 : \mathfrak{F}^n(K, v) \longrightarrow \pi_n(|K|, v) \quad \text{by} \quad \varpi(f) = [|f|]$$

One notes that ϖ_0 respects homotopy and the ϖ_0 factors through a map

$$\varpi : \pi_n(K, v) \longrightarrow \pi_n(|K|, v)$$

Now, one checks that ϖ is bijective. A complete proof is available in [H, Prop. 3.6.3]. ■

Remark 1.4.12. Suppose $p : K \longrightarrow L$ is a map of fibrant $K, L \in$ **SSet** and $v \in K_0$ is a vertex. Then p induces a map of the homotopy groups (or sets) $\pi_n(K, v) \longrightarrow \pi_n(L, p(v))$, making these associations functorial:

$$\begin{cases} \pi_0 : \mathbf{SSet} \longrightarrow \mathbf{Sets} \\ \pi_1 : \mathbf{SSet} \longrightarrow \mathbf{Gr} \\ \pi_n : \mathbf{SSet} \longrightarrow \mathbf{Ab} \quad n \geq 2 \end{cases}$$

We proceed to develop a long exact sequence of homotopy groups.

Definition 1.4.13. Let $L, K \in \mathbf{SSet}$ be fibrant, $p : L \longrightarrow K$ be a fibration in $\mathbf{SSet}$, and $v \in L_0$, a vertex. Let $[g] \in \pi_n(K, p(v))$, with $g \in \mathfrak{F}^n(K, p(v))$. Consider the diagram

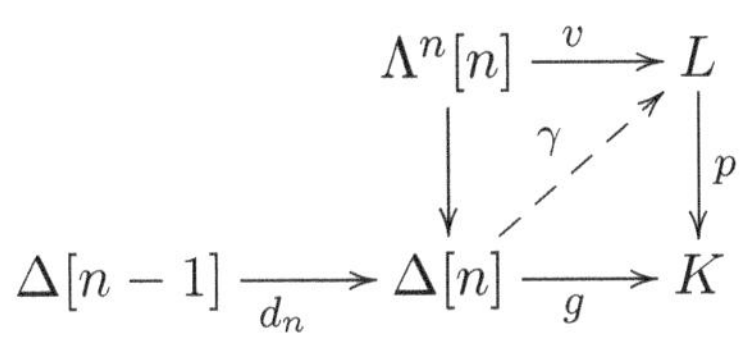

In this diagram, the rectangle commutes. Since p is a fibration, g lifts to a map γ in $\mathbf{SSet}$. By definition $pd_n\gamma = d_np\gamma = d_ng = p(v)$. Let $F = p^{-1}(pv)$ be the fiber of p over $p(v)$. Consider the diagram

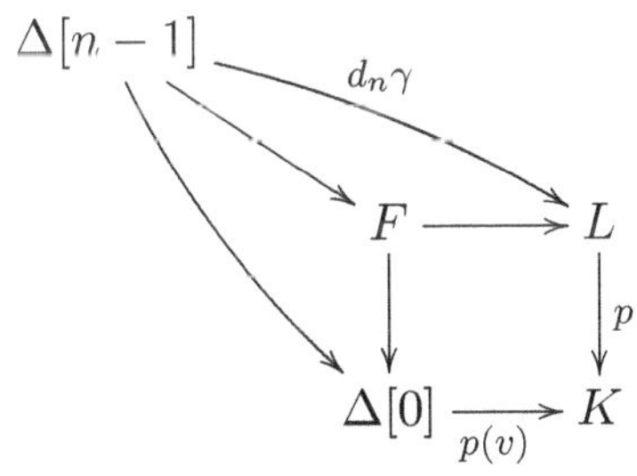

So, $d_n\gamma$ lies in F. Being pullback of p, F is fibrant. Further, $d_n\gamma \in \mathfrak{F}^{n-1}(F, v)$. So, $[d_n\gamma] \in \pi_{n-1}(F, v)$. Define

$$\partial : \pi_n(K, p(v)) \longrightarrow \pi_{n-1}(F, v) \qquad \text{by} \qquad \partial([g]) = [d_n\gamma]$$

One checks that $[d_n\gamma]$ is independent of the choice of the representative g and the lift γ. So, ∂ is a well-defined map.

Lemma 1.4.14. *Let $p : L \longrightarrow K$ be a fibration of fibrant simplicial sets L, K and v be a vertex of L. Let $i : F \longrightarrow L$ be the fiber of p over $p(v)$. Then the sequence*

$$\begin{array}{ccccccc}
\cdots \xrightarrow{\partial} & \pi_n(F,v) & \longrightarrow & \pi_n(L,v) & \longrightarrow & \pi_n(K,p(v)) \\
\xrightarrow{\partial} & \pi_{n-1}(F,v) & \longrightarrow & \pi_{n-1}(L,v) & \longrightarrow & \pi_{n-1}(K,p(v)) \\
 & \cdots & & \cdots & & \cdots \\
\xrightarrow{\partial} & \pi_1(F,v) & \longrightarrow & \pi_1(L,v) & \longrightarrow & \pi_1(K,p(v)) \\
\xrightarrow{\partial} & \pi_0(F,v) & \longrightarrow & \pi_0(L,v) & \longrightarrow & \pi_0(K,p(v))
\end{array} \tag{1.8}$$

is exact.

Proof. By (1.4.11), $\pi_n(F,v) = \pi_n(|F|,v)$, $\pi_n(L,v) = \pi_n(|L|,v)$, $\pi_n(K,pv) = \pi_n(|K|,pv)$. It is known (see [H, Cor. 3.6.2], [GJ]) that the geometric realization $|p| : |L| \longrightarrow |K|$ is a weak fibration. Now, the exactness of (1.8) follows from the exactness of (B.20). The proof is complete. ■

1.5. Exercises

Exercise 1.5.1. Let $K \in \mathbf{SSet}$. For integers $n \geq 0$, prove $Mor_{\mathbf{SSet}}(\Delta[n], K) \cong K_n$.

Exercise 1.5.2. Let $K, L \in \mathbf{SSet}$. Let $B \subseteq K$ be the set of generators of K, consisting of the nondegenerate simplexes. Let $f : B \longrightarrow L$ be a set theoretic map such that (1) $\forall\ \sigma \in B$, $\dim \sigma = \dim f(\sigma)$, (2) for $\sigma, \tau \in B$, if $K(\delta_1)\sigma = K(\delta_2)\tau$ for some face maps δ_1, δ_2 in Δ_-, then $L(\delta_1)f(\sigma) = L(\delta_2)f(\tau)$. Then f extends to a unique map $F \in Mor_{\mathbf{SSet}}(K, L)$.

Exercise 1.5.3. Interpret the geometric realizations of the examples in (1.1.5), and prove. In particular, (1) $|\Delta[n]|$ is homeomorphic to the standard n-simplex $\Sigma^n \cong I^n$ (see 1.1.7), (2) $|\partial\Delta[n]|$ is homeomorphic to the boundary $\partial\Sigma^n \cong \partial I^n \cong \mathbb{S}^{n-1}$, (2) $|\Lambda^0[n]|$ is homeomorphic to $J^{n-1} := (I^{n-1} \times 1) \cup (\partial I^{n-1} \times I)$.

Exercise 1.5.4. Let $X \in \mathbf{SSet}$ and $\ell \in X_1$ be an 1-simplex. Then there is a path $\gamma(\ell) : I \longrightarrow |X|$, such that $\gamma(0) = d_0(\ell)$ and $\gamma(1) = d_1(\ell)$.

Exercise 1.5.5. Let $K \in \mathbf{SimTop}$. Assume K_n is Hausdorff and $\mathcal{L}_n$ is closed in K_n, for all $n \geq 0$. Then

1. Prove if $Y \subseteq |K|$ is compact, then $Y \subseteq |K|^n$ for some n. (**Hint:** The proof of (C.1.8) works.)
2. If K_n is compactly generated Hausdorff, for all $n \geq 0$, then $|K|$ is compactly generated. (**Hint:** We need to prove that $|K|^n$ is compactly generated. Note, $K_n \times \Sigma^n$ is compactly generated. Use the pushforward diagram in (1.2.18).)
3. Prove $|K|$ is Hausdorff. (**Hint:** [P], Proof would be similar to (C.1.10).)

The following is from [GG], which we formulate as exercises. Before that, we give the following definition.

Definition 1.5.6. There is a convenient way to look at the Δ category (see [GG]). Its objects are defined to be nonempty finite totally ordered sets. For $A, B \in Obj(\Delta)$, define $AB := A \coprod B \in Obj(\Delta)$ to be the disjoint union, and $a < b$, $\forall a \in A$, $b \in B$. It is clear, this category is the same as Δ (i.e., equivalent). For integers, $m, n \geq 0$, we have $[m][n] = [m+n+1]$. Given $A, B \in Obj(\Delta)$, there are natural arrows $AB \longrightarrow A$, $AB \longrightarrow B$, $A \longrightarrow AB$, $B \longrightarrow AB$.

Let $F : X \longrightarrow Y$ be a map of simplicial sets, and $\rho \in Y_m$. Define

$$(\rho \mid F)([n]) = \left\{(\sigma, \tau) \in X([n]) \times Y([m][n]) : \tau_{|[m]} = \rho, F(\sigma) = \tau_{|[n]}\right\} \tag{1.9}$$

Exercise 1.5.7. Let $F : X \longrightarrow Y$ be a map of simplicial sets, and $\rho \in Y([m])$. Prove that $(\rho \mid F) \in \mathbf{SSet}$. **Hint:** See [GG].

Exercise 1.5.8. Let $X \in \mathbf{SSet}$ and $\rho \in X_n$. Prove that the geometric realization $|(\rho \mid 1_X)|$ is contractible. **Hint:** See [GG].

Chapter 2

Classifying Spaces of Categories

To each small category $\mathscr{C}$, we associate a topological space $\mathbb{B}\mathscr{C} \in$ **Top**, to be called the classifying space of $\mathscr{C}$ in this chapter. Then we adapt all the topological and homotopy properties of $\mathbb{B}\mathscr{C}$ to the categories $\mathscr{C}$. First, we introduce the following notation.

Notation 2.0.1. Recall a category $\mathscr{C}$ is called a small category if $Obj(\mathscr{C})$ is a set. Most of what follows would make sense only for small categories or those categories that are equivalent to a small category. For this reason, unless stated otherwise, subsequently all the categories are assumed to be equivalent to a small category.

1. The category of small categories and functors will be denoted as **Cat**. We will often work with categories $\mathscr{C}$ such that there is an equivalence $F : \mathscr{C}_s \xrightarrow{\sim} \mathscr{C}$ where $\mathscr{C}_s$ is in **Cat**. Such a category, $\mathscr{C}$, is said to have a set of isomorphism classes of objects.
2. The category of CW complexes would be denoted as **CW**. The category of pointed CW complexes would be denoted as $\mathbf{CW}_\bullet$. Also, the category of pairs of CW complexes would be denoted as **CWPair**.

2.1. The Classifying Spaces of Categories

We define the nerve of a category.

Definition 2.1.1. Suppose $\mathscr{C}$ is a **small** category. To any such category, we associate a simplicial set $\mathscr{N}\mathscr{C} \in$ **SSet**, as follows:

1. Let the set of vertices $\mathscr{N}\mathscr{C}_0$ be the set of all objects in $\mathscr{C}$. For integers $n \geq 1$, an n-simplex is a sequence of composable arrows

$$\sigma: \quad X_0 \xrightarrow{f_1} X_1 \xrightarrow{f_2} X_2 \xrightarrow{f_3} \cdots \longrightarrow X_{n-1} \xrightarrow{f_n} X_n \tag{2.1}$$

and $\mathscr{N}\mathscr{C}_n$ denotes the set of all n-simplexes.
2. The face maps $d_i : \mathscr{N}\mathscr{C}_n \longrightarrow \mathscr{N}\mathscr{C}_{n-1}$ for $i \neq 0, n$ are given by sending an n-simplex σ, as in (2.1), by dropping X_i and

$$\text{replacing} \quad X_{i-1} \xrightarrow{f_{i-1}} X_i \xrightarrow{f_i} X_{i+1} \quad \text{with} \quad X_{i-1} \xrightarrow{f_i f_{i-1}} X_{i+1}$$

Further, d_0 and d_n are obtained by dropping X_0, X_n, respectively.
3. The degeneracy maps $s_i : \mathscr{N}\mathscr{C}_n \longrightarrow \mathscr{N}\mathscr{C}_{n+1}$ are given by sending an n-simplex σ, as in (2.1), by

$$\text{replacing} \quad X_{i-1} \xrightarrow{f_{i-1}} X_i \xrightarrow{f_i} X_{i+1}$$

$$\text{with} \quad X_{i-1} \xrightarrow{f_{i-1}} X_i \overset{1_{X_i}}{\dashrightarrow} X_i \xrightarrow{f_i} X_{i+1}$$

The simplicial set $\mathscr{N}\mathscr{C} \in \mathbf{SSet}$ is called the **nerve** of $\mathscr{C}$.

Definition 2.1.2. Let $\mathscr{C}$ be a small category. The **Classifying space** $\mathbb{B}\mathscr{C}$ of $\mathscr{C}$ is defined to be the geometric realization $|\mathscr{N}\mathscr{C}| \in$ **Top**. Therefore, $\mathbb{B}\mathscr{C} := |\mathscr{N}\mathscr{C}|$. So, objects $X \in Obj(\mathscr{C})$ correspond to a vertex in $\mathbb{B}\mathscr{C}$. For $X \in Obj(\mathscr{C})$ and integers $n \geq 0$, define

$$\pi_n(\mathscr{C}, X) := \pi_n(\mathbb{B}\mathscr{C}, X)$$

to be called the **Homotopy groups** (or set, when $n = 0$) of $\mathscr{C}$.

Remark: Without further conditions, there is no reason that the nerve $\mathscr{N}\mathscr{C}$ would be a Fibrant as a simplicial set.

The following obvious lemma would be very useful.

Lemma 2.1.3. *Suppose $\mathscr{C}$ is a category and $f : X \longrightarrow Y$ is a morphism. Then there is a path $\gamma := \gamma(f) : [0,1] \longrightarrow \mathbb{B}\mathscr{C}$ such that $\gamma(f)(0) = X$ and $\gamma(f)(1) = Y$.*

Proof. Follows from Lemma 1.5.4. ∎

Following are two useful examples of classifying spaces.

Example 2.1.4. Let $\mathbb{I} := \{0,1\}$ be the partial ordered set. Then $\mathbb{B}\mathbb{I} = [0,1]$. Note, the nerve $\mathscr{N}(\mathbb{I}) = \Delta[1]$.

Example 2.1.5. Suppose G is a group. Let $\underline{G}$ denote the category with one object $\star$ and $Mor(\star, \star) = G$. Then

$$\pi_i\left(\underline{G}, \star\right) = \begin{cases} G & \text{if } i = 1 \\ 0 & \text{if } i \neq 1 \end{cases}$$

Here the isomorphism $G \xrightarrow{\sim} \pi_i\left(\underline{G}, \star\right)$ sends $g \in G$ to the loop corresponding to $g : \star \longrightarrow \star$ (see Lemma 1.5.4).

Proof. This is very classical and we give a proof. Write $K = \mathscr{N}\underline{G}$. So, $K_n = G^n$, and an n-simplex looks like

$$\star \xrightarrow{g_1} \star \xrightarrow{g_2} \star \cdots \qquad \star \xrightarrow{g_n} \star \quad \text{with} \quad g_i \in G \tag{2.2}$$

We show $K \in \mathbf{SSet}$ is fibrant. It is clear any map $\Lambda^0[1] \longrightarrow K$, $\Lambda^1[1] \longrightarrow K$, extends to a map $\Delta[1] \longrightarrow K$. Now let $g : \Lambda^2[2] \longrightarrow K$ be any map. The nondegenerate simplexes of $\Lambda^2[2]$ are $(0, 2)$, $(1, 2)$. The map g is represented as follows:

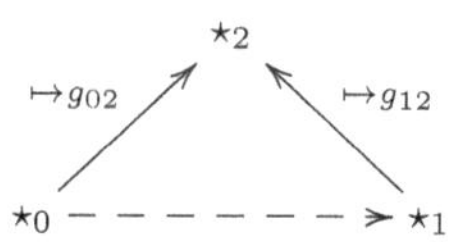

$\Lambda^2[2]$ misses the dotted arrow. Let $g_{01} = g_{12}^{-1} g_{02}$

The map $g : \Lambda^2[2] \longrightarrow K$ is determined by $(1, 2) \mapsto g_{12}$, $(0, 2) \mapsto g_{02}$. This extends to a map $\Delta[2] \longrightarrow K$ by sending $1_{[2]} \mapsto (g_{01}, g_{02})$. We extend maps $g : \Lambda^\epsilon[2] \longrightarrow K$ similarly. Now, for $n \geq 3$, let $g : \Lambda^n[n] \longrightarrow K$ be a map. We have $g((0, 1)) = g_1 \in K_1$, and $g(d_0) \in K_{n-1}$ is given by

$$\star_1 \xrightarrow{g_2} \star_2 \xrightarrow{g_3} \star \cdots \longrightarrow \star_{n-1} \xrightarrow{g_n} \star_n$$

Now, g extends to a map $\Delta[n] \longrightarrow K$ by sending $1_{[n]} \mapsto (g_1, g_2, \ldots, g_n)$. More generally, for $\epsilon = 0, 1, \ldots, n$, and for a map $g : \Lambda^\epsilon[n] \longrightarrow K$, all the edges $g_{i,i+1} \in G$ are available. So, any such map extends to a map $\Delta[n] \longrightarrow K_\bullet$. This completes the proof that K is fibrant.

By (1.4.11), $\pi_n(K, \star) \xrightarrow{\sim} \pi_n(|K|, \star)$. Use the notation, as in Section 3.3. Fix $n \neq 1$, $\mathfrak{F}^n(K, \star)$ is the set of all $\sigma \in X_n$ such that $d_i\sigma = \star$ for all $i = 0, \ldots, n$. It checks easily that (2.2), $\mathfrak{F}^n(K, \star)$ is a singleton. Hence, $\pi_n(|K|, \star) \cong \pi_n(K, \star) = 0$, for $n \neq 1$.

In the case $n = 1$, $\mathfrak{F}^1(K, \star) \cong G$, we have the commutative diagram

$$\begin{array}{ccc} G & \xrightarrow{\iota} & \pi_1(K, \star) \\ & \searrow & \wr\downarrow \\ & & \pi_n(|K|, \star) \end{array} \qquad \text{write} \quad \iota(g) = [g]$$

Suppose $\iota(g) = 0$, meaning $[g] = [\star]$. By unwinding the definition of equivalence (1.4.8), it follows $g = e$ is the identity. Alternately, by (1.4.10), there is a 2-simplex

$$\sigma := \star_0 \xrightarrow{g_1} \star_1 \xrightarrow{g_2} \star_2 \qquad \ni \qquad d_2\sigma = g, d_0\sigma = d_1\sigma = \star$$

This means

$$\sigma := \star_0 \xrightarrow{g_1 = g} \star_1 \xrightarrow{g_2 = e} \star_2 \quad \text{and} \quad d_1\sigma = eg = e \Longrightarrow g = e$$

So, ι is injective. The proof is complete. ■

2.1.1. *Properties of the classifying spaces*

In this section, we summarize some of the basic properties of the classifying spaces.

Proposition 2.1.6. *Let $\mathscr{C}, \mathscr{D}$ be two small categories and $F : \mathscr{C} \longrightarrow \mathscr{D}$ be a functor. Then F induces a continuous map $\mathbb{B}(F) : \mathbb{B}\mathbb{C} \longrightarrow \mathbb{B}\mathscr{D}$. This association*

$$\begin{cases} \mathscr{C} \mapsto \mathbb{B}\mathscr{C} \\ F \mapsto \mathbb{B}(F) \end{cases} \quad \textit{defines a functor } \mathbb{B} : \mathbf{Cat} \longrightarrow \mathbf{Top}$$

Proof. Follows from construction. ■

Proposition 2.1.7. *Let $\mathscr{C}$ be a small category. Then there is a (cellular) homeomorphism $\mathbb{B}\mathscr{C} \xrightarrow{\sim} \mathbb{B}\mathscr{C}^o$, where $\mathscr{C}^o$ is the dual (opposite) category. This homeomorphism is not induced by any natural functor.*

Proof. Follows from construction. ■

Proposition 2.1.8. *Let $\mathscr{C}, \mathscr{D}$ be two small categories. It follows easily that $\mathscr{N}(\mathscr{C} \times \mathscr{D}) \xrightarrow{\sim} \mathscr{N}(\mathscr{C}) \times \mathscr{N}(\mathscr{D})$ is a bijection.*

So, there is a continuous bijection $\mathbb{B}(\mathscr{C} \times \mathscr{D}) \longrightarrow \mathbb{B}\mathscr{C} \times \mathbb{B}\mathscr{D}$. It follows from (1.3.3) that the map $\mathbb{B}(\mathscr{C} \times \mathscr{D}) \xrightarrow{\sim} k(\mathbb{B}\mathscr{C} \times \mathbb{B}\mathscr{D})$ is a homeomorphism, where k indicates compactly generated topology. Consequently, if $\mathbb{B}\mathscr{C}$ or $\mathbb{B}\mathscr{D}$ is locally compact, then $\mathbb{B}(\mathscr{C} \times \mathscr{D}) \longrightarrow \mathbb{B}\mathscr{C} \times \mathbb{B}\mathscr{D}$ is a homeomorphism (B.6.7). This would be the case if one of them has only finitely many distinct skeletons (1.2.18).

In particular, $\mathbb{B}(\mathscr{C} \times \mathbb{I}) \cong \mathbb{B}\mathscr{C} \times [0,1]$ are homeomorphic, with notation as in (2.1.4).

Proposition 2.1.9. *Let $\mathscr{C}, \mathscr{D}$ be two small categories and $F, G : \mathscr{C} \longrightarrow \mathscr{D}$ be two functors. Suppose $\theta : F \longrightarrow G$ is a natural transformation. Then θ induces a homotopy*

$$H : \mathbb{B}\mathscr{C} \times I \longrightarrow \mathbb{B}\mathscr{D} \quad \ni \quad H(0) = \mathbb{B}(F),\ H(1) = \mathbb{B}(G)$$

Proof. For a morphism $f \in Mor_{\mathscr{C}}(X, Y)$, we have the commutative diagram

$$\begin{array}{ccc} F(X) & \xrightarrow{F(f)} & F(Y) \\ {\scriptstyle \theta_X}\downarrow & & \downarrow{\scriptstyle \theta_Y} \\ G(X) & \xrightarrow[G(f)]{} & G(Y) \end{array}$$

With $\mathbb{I} = \{0, 1\}$, define a functor

$\mathcal{H} : \mathscr{C} \times \mathbb{I} \longrightarrow \mathscr{D} \quad \ni$

$$\begin{cases} \mathcal{H}(X,0) = F(X),\ \mathcal{H}(X,1) = G(X) & \forall\ X \in Obj(\mathscr{C}) \\ \mathcal{H}(f,\iota) = F(f) : F(X) \longrightarrow F(Y) & \forall\ f \in Mor_{\mathscr{C}}(X,Y), \iota \in Mor_{\mathbb{I}}(0,0) \\ \mathcal{H}(f,\iota) = G(f) : G(X) \longrightarrow G(Y) & \forall\ f \in Mor_{\mathscr{C}}(X,Y), \iota \in Mor_{\mathbb{I}}(1,1) \\ \mathcal{H}(f,\iota) = \theta_Y F(f) = G(f)\theta_X & \forall\ f \in Mor_{\mathscr{C}}(X,Y), \iota \in Mor_{\mathbb{I}}(0,1) \end{cases}$$

By (2.1.6, 2.1.8), there is a homotopy

$$H := \mathcal{H} : \mathbb{B}\mathscr{C} \times [0,1] \equiv \mathbb{B}(\mathscr{C} \times \mathbb{I}) \longrightarrow \mathbb{B}\mathscr{D}$$

It follows that $H(-,0) = \mathbb{B}F$ and $H(-,1) = \mathbb{B}G$. The proof is complete. ■

Corollary 2.1.10. *Let $F : \mathscr{C} \longrightarrow \mathscr{D}$, $G : \mathscr{D} \longrightarrow \mathscr{C}$ be two functors of small categories, and F be left adjoint to G. Then $\mathbb{B}(F) : \mathbb{B}\mathscr{C} \longrightarrow \mathbb{B}\mathscr{D}$ is a homotopy equivalence.*

Proof. There is a natural transformation $1_{\mathscr{C}} \longrightarrow GF$ as follows, with $A \in Obj(\mathscr{C})$, $X \in Obj(\mathscr{D})$:

$$Mor_{\mathscr{D}}(FA, X) \xrightarrow[\sim]{\eta_{AX}} Mor_{\mathscr{C}}(A, GX)$$

$$\text{So,} \quad \begin{cases} Mor_{\mathscr{D}}(FA, FA) \xrightarrow{\sim} Mor_{\mathscr{C}}(A, GFA) \\ Mor_{\mathscr{D}}(FGX, X) \xrightarrow{\sim} Mor_{\mathscr{C}}(GX, GX) \end{cases}$$

Let $\theta(A) = \eta(1_{FA}) : A \longrightarrow GFA$. Then $\theta : 1_{\mathscr{C}} \longrightarrow GF$ is a natural transformation. By (2.1.9), $1_{\mathbb{B}\mathscr{C}} \simeq \mathbb{B}(G)\mathbb{B}(F)$ are homotopic. Likewise, $\mathbb{B}(F)\mathbb{B}(G) \simeq 1_{\mathbb{B}\mathscr{C}}$ are homotopic. So, $\mathbb{B}(F)$ is a homotopy equivalence. ■

Corollary 2.1.11. *Suppose $\mathscr{C}$ is a category with an initial object or a final object. Then $\mathbb{B}\mathscr{C}$ is contractible.*

Proof. Assume $\mathscr{C}$ has an initial object 0. Let $\mathbf{0}$ denote the full subcategory of $\mathscr{C}$ consisting of the zero object. Let $F : \mathbf{0} \longrightarrow \mathscr{C}$ be the inclusion functor, and $G : \mathscr{C} \longrightarrow \mathbf{0}$ be the constant functor. It follows that F is a left adjoint to G. Hence, $\mathbb{B}\mathbf{0} \longrightarrow \mathbb{B}\mathscr{C}$ is a homotopy equivalence, by (2.1.10). ■

Definition 2.1.12. From now on, terminology and concepts from Topology are borrowed and adapted freely to the Category Theory. A category $\mathscr{C}$ is said to have a "topological property" if the classifying space $\mathbb{B}\mathscr{C}$ has the same property. For example,

1. Two functors $F, G : \mathscr{C} \longrightarrow \mathscr{D}$ are called **homotopic** if $\mathbb{B}(F)$ and $\mathbb{B}(G)$ are homotopic.
2. A functor $F : \mathscr{C} \longrightarrow \mathscr{D}$ is called a **homotopy equivalence** if $\mathbb{B}\mathscr{C} \longrightarrow \mathbb{B}\mathscr{D}$ is a homotopy equivalence.
3. A category $\mathscr{C}$ is called **contractible** if $\mathbb{B}\mathscr{C}$ is contractible.

4. A commutative diagram of categories

$$\begin{array}{ccc} \mathcal{C} & \longrightarrow & \mathcal{D} \\ \downarrow & & \downarrow l \\ \mathcal{E}' & \longrightarrow & \mathcal{D}' \end{array} \quad \text{is called } \textbf{Homotopy Cartesian} \text{ if}$$

$$\begin{array}{ccc} \mathbb{B}\mathcal{C} & \longrightarrow & \mathbb{B}\mathcal{D} \\ \downarrow & & \downarrow \\ \mathbb{B}\mathcal{E}' & \longrightarrow & \mathbb{B}\mathcal{D}' \end{array} \quad \text{is so, as in (B.5.1).}$$

In this case, if $\mathcal{E}'$ is contractible, we say $\mathcal{C} \longrightarrow \mathcal{D} \longrightarrow \mathcal{D}'$ is a homotopy fibration. Consequently, a long exact sequence of homotopy groups/sets follows (B.5.5).

2.1.2. *Directed and filtering limit*

In this section, we establish some definitions in category theory before adapting them to classifying spaces.

Definition 2.1.13. A (nonempty) partially ordered set $(\mathcal{P}, \leq)$ is called a **directed set**, if $\forall\ i, j \in \mathcal{P}$ there is a $k \in \mathcal{P}$ such that $i \leq k, j \leq k$.

A small nonempty category I is defined to be a **filtering category**, if

1. $\forall\ i, j \in Obj(I)$ there are two arrows:

$$\begin{array}{ccc} i & & \\ & \overset{f}{\searrow} & \\ & & k \\ & \underset{g}{\nearrow} & \\ j & & \end{array} \tag{2.3}$$

2. Given two arrows $\lambda, \beta : i \longrightarrow k$, there is an arrow $\gamma : k \longrightarrow \kappa$ such that $\gamma\lambda = \gamma\beta$. Diagramatically,

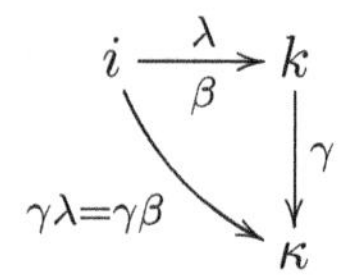

It follows that a directed set is a filtering category.

Definition 2.1.14. Suppose I is a filtering small category. A functor $A : I \longrightarrow$ **Set**, also written as $\{A_i : i \in I\}$, is called a **filtering family** of sets. Also, a functor $\Theta : I \longrightarrow$ **Cat** is called a **filtering family** of small categories. With $\mathscr{C}_i = \Theta(i)$, such a family may be written as $\{\mathscr{C}_i : i \in I\}$.

Proposition 2.1.15. *Suppose I is a filtering category and $\Theta : I \longrightarrow$* **Cat** *is a filtering family of small categories, indexed by I. We will write $\mathscr{C}_i = \Theta(i)$. A (small) category $\mathscr{C}$, together with a family $\{\theta_i : \mathscr{C}_i \longrightarrow \mathscr{C} \ \forall \ i \in I\}$ of functors, is defined to be* **the limit of** Θ *if this family $\{\theta_i\}$ satisfies the* **universal property** *in the sense of the following diagram:*

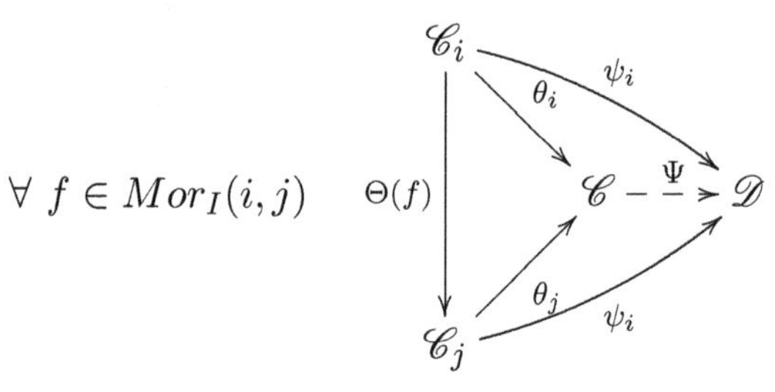

(2.4)

In this diagram, the left hand-triangle commutes. Further, given a family of the commutative diagram of the outer triangles, there is a unique functor, Ψ, such that all the triangles commute. We write $\mathscr{C} = \lim_I \mathscr{C}_i$ or $\mathscr{C} = \lim_\Theta \mathscr{C}_i$.

Since we are working with small categories, this is, indeed, a limit in the category of sets, as described subsequently. In particular, such

limits exist and are defined up to equivalences of categories. The limit $\mathscr{C}$ *is constructed as follows:*

$$\begin{cases} Obj(\mathscr{C}) = \lim_I Obj(\mathscr{C}_i) = \frac{\bigsqcup_{i\in I} Obj(\mathscr{C}_i)}{\sim} \\ \text{Denote} \qquad \theta_i : Obj(\mathscr{C}_i) \longrightarrow Obj(\mathscr{C}) \\ Mor_{\mathscr{C}}(X,Y) := \lim_{i\in I, \theta_i(X_i)=X, \theta_i(Y_i)=Y} Mor_{\mathscr{C}_i}(X_i, Y_i) \end{cases} \tag{2.5}$$

1. *It follows that the nerve* $\mathscr{N}(\mathscr{C}) = \lim_I \mathscr{N}(\mathscr{C}_i)$.
2. *Let* $X_i \in Obj(\mathscr{C}_i)$ *be a family such that* $X = \theta_i(X_i)$. *Then* θ_i *induces the maps*

$$\pi_n(\mathscr{C}_i, X_i) \longrightarrow \pi_n(\mathscr{C}, X) \qquad \forall\, n \geq 0$$

Proposition 2.1.16. *With notation as in (2.1.15), we have*

$$\lim_I \pi_n(\mathscr{C}_i, X_i) = \pi_n(\mathscr{C}, X) \qquad \forall\, n \geq 0$$

Proof. We have $\mathscr{N}\mathscr{C} = \lim_I \mathscr{N}\mathscr{C}_i$. The rest of the proof follows from the corresponding result on simplicial sets (2.1.17), as follows. ∎

Proposition 2.1.17. *Let* I *be a filtering category and* $\{K(i) : i \in I\}$ *be a filtering family of simplicial sets. Let* $K = \lim_I K(i)$ *and* $v = \lim_I v_i \in K$, *where* $v_i \in K(i)_0$ *are vertices* $\forall\, i \in I$. *Then*

$$\lim_{i\in I} \pi_n(|K(i)|, v_i) = \pi_n(|K|, v) \quad \forall\, n \geq 0$$

Proof. Let $f(i) : K(i) \longrightarrow K$ denote the canonical maps and $|f(i)| : |K(i)| \longrightarrow |K|$ be the map of geometric realizations. Fix $n \geq 0$. The map $\lim_{i\in I} \pi_n(|K(i)|, v_i) \longrightarrow \pi_n(|K|, v)$ is induced by the maps $|f(i)|$. Let $\gamma : (\mathbb{S}^n, \star) \longrightarrow (|K|, v)$ be a map representing an element $[\gamma] \in \pi_n(|K|, v)$. Since $\gamma(\mathbb{S}^n) \subseteq |K|$ is compact, by (C.1.8) and (1.2.11) there is a finite subcomplex $\mathfrak{F} \subseteq K$ such that $\mathfrak{F}$ is generated by finitely many nondegenerate simplexes and $\gamma(\mathbb{S}^n) \subseteq |\mathfrak{F}|$. There is an $i_0 \in I$, large enough, and a finite subcomplex $\mathfrak{F}(i_0) \subseteq K(i_0)$ such that $\mathfrak{F}(i_0) \longrightarrow \mathfrak{F}$ is a bijection. So, $|\mathfrak{F}(i_0)| \longrightarrow |\mathfrak{F}|$ is a homeomorphism. So, γ lifts to a map $\tilde{\gamma} : (\mathbb{S}^n, \star) \longrightarrow (|\mathfrak{F}(i_0)|, v_{i_0}) \subseteq (|K(i_0)|, v_{i_0}))$. Therefore, $[\tilde{\gamma}] \in \pi_n(|K(i_0)|, v_{i_0})$ maps to $[\gamma] \in \pi_n(|K|, v)$. This establishes that the map $\lim_{i \in I} \pi_n(|K(i)|, v_i) \longrightarrow \pi_n(|K|, v)$ is surjective.

Now suppose $[\gamma] \in \lim_I \pi_n(|K(i)|, v_i)$ maps to zero in $\pi_n(|K|, v)$, where γ is represented by a map $\gamma : (\mathbb{S}^n, \star) \longrightarrow (|K(i_0)|, v_{i_0})$ for some i_0. So, $[|f(i)|\gamma] = 0$ and there is a homotopy $H : \mathbb{S}^n \times I \longrightarrow |K|$ such that $H(-, 0) = c_v$, the constant map and $H(-, 1) = |f(i)|\gamma$. Since $H(\mathbb{S}^n \times I)$ is compact, by the same argument above, for some $i_1 \in I$ large enough, H lifts to a homotopy $\tilde{H} : \mathbb{S}^n \times I \longrightarrow |K(i_1)|$ from the constant map to γ_{i_1}, where $\gamma_{i_1}$is the image of γ. So, $[\gamma_{i_1}] = 0 \in \pi_n(|K(i_1)|, v_{i_1})$. So, $[\gamma] = [\gamma_{i_1}] = 0 \in \lim_I \pi_n(|K(i)|, v_i)$. So, the map $\lim_{i \in I} \pi_n(|K(i)|, v_i) \longrightarrow \pi_n(|K|, v)$ is injective. The proof is complete. ■

Corollary 2.1.18. *Use the notation as in (2.1.15). Assume* $\forall f : i \longrightarrow j$ *maps, the map* $\mathbb{B}\mathscr{C}_i \longrightarrow \mathbb{B}\mathscr{C}_j$ *is a homotopy equivalence. Then* $\mathbb{B}\mathscr{C}_i \longrightarrow \mathbb{B}\mathscr{C}$ *is a homotopy equivalence* $\forall i \in I$.

Proof. (*It would clearly be obvious if the homotopies* $\mathbb{B}\mathscr{C}_i \longrightarrow \mathbb{B}\mathscr{C}_j$ *were assumed to be compatible.*) However, for a fixed $i_0 \in I$, by (2.1.16), $\pi_n(\mathbb{B}\mathscr{C}_{i_0}, v_{i_0}) \xrightarrow{\sim} \pi_n(\mathbb{B}\mathscr{C}_i, v)$ are isomorphisms, for all v_{i_0}, and $n \geq 0$. Therefore, by Whitehead's theorem (C.3.8), $\mathbb{B}\mathscr{C}_{i_0} \longrightarrow \mathbb{B}\mathscr{C}$ is a homotopy equivalence. The proof is complete. ■

Corollary 2.1.19. *Suppose I is a filtering category. Then I is contractible.*

Proof. We need to prove that $\mathbb{B}I$ is contractible. In fact, $I = \lim_{i \in I}(I/i)$. Since I/i has a final object, $\mathbb{B}(I/i)$ is contractible. By (2.1.18), $\mathbb{B}I$ is contractible. The proof is complete. ■

2.1.3. *A key lemma on quasifibrations*

The following is a key needed to prove the Theorem A, B (2.2.2, 2.2.3) in [Q].

Lemma 2.1.20. *Let I be a small category and* $\iota : I \longrightarrow$ **Top** *be a functor. For objects i in I, we write* $X_i := \iota(i)$. *Let* $\mathscr{N}I$ *be the nerve* *of I. Subsequently, a p-simplex $\sigma \in \mathscr{N}I_p$ is denoted by*

$$\sigma = i_0 \xrightarrow{f_0} i_1 \xrightarrow{f_1} \cdots \longrightarrow i_p, \qquad \text{and} \quad \sigma(j) := i_j \tag{2.6}$$

For integers $p \geq 0$, write $\mathcal{X}_p := \coprod_{\sigma \in \mathscr{N}I_p} X_{\sigma(0)}$. Then $\mathcal{X} := \{\mathcal{X}_p : p \geq 0\}$ is an object in **SimTop**. *So, a p-simplex in $\mathcal{X}$ is a pair (σ, z), with $\sigma \in \mathscr{N}I_p$, and $z \in X_{i_0}$. We have the 0th face $d_0(\sigma, z) = (d_0\sigma, \iota(f_0)(z))$, and for $i \geq 1$, the ith face $d_i(\sigma, z) = (d_i\sigma, z)$.*

There is a natural map $\gamma : \{\mathcal{X}_p\} \longrightarrow \mathscr{N}I$, in **SimTop**, *defined by $\gamma(\sigma, z) = \sigma$. Assume $\forall\ f : i \longrightarrow j \in Mor_I$, the induced arrow $\iota(f) : X_i \longrightarrow X_j$, is a homotopy equivalence. Then the geometric realization $|\gamma| : |\mathcal{X}| \longrightarrow \mathbb{B}I$ is a quasifibration.*

Proof. We denote the geometric realization $X_I := |\mathcal{X}|$, and $g := |\gamma|$. Let F_p be the p-skeletons of $\mathbb{B}I$.

Let $\mathscr{N}I'_p$ denote the set of all nondegenerate simplexes in $\mathscr{N}I_p$. We have a commutative diagram of pushforward (co-cartesian) diagrams as follows:

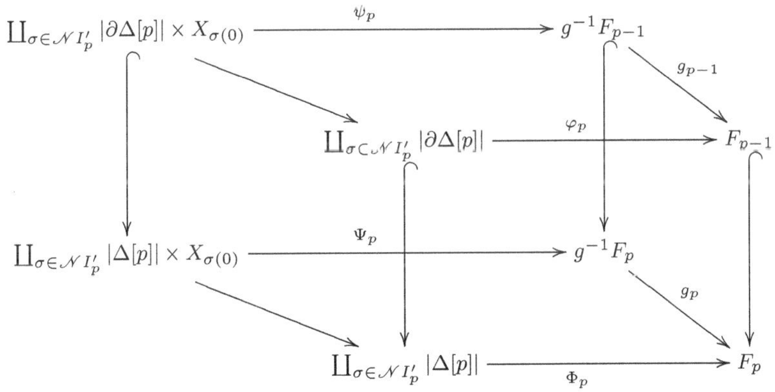

(2.7)

The front of the cube is the pushforward diagram defining the p-skeleton F_p (see 1.2.1, 1.2.11). For the back side, refer to (1.2.14). It is clear that $g^{-1}F_p = X_I^p$. At least set theoretically, the back side is a push forward diagram. One checks that, in this case, the back side is also a pushforward in **Top**, while it does not match exactly with the diagram in (1.2.18). The Proposition B.5.12 (Dold–Thom) applies to the map g. So, it would suffice to prove that the restriction map $g_p : g^{-1}F_p \longrightarrow F_p$, of g, is a quasifibration, for all $p \geq 0$. We do this by induction on p.

The mid-point $O_p = (1/(p+1), 1/(p+1), \ldots, 1/(p+1)) \in |\Delta[p]|$ is called its barycenter. Let $V_1 \subseteq F_p$ be the open subset obtained by removing all the barycenters in the p-cells, and $U_1 = g^{-1}V_1$. Let $V_2 = F_p \setminus F_{p-1}$ and $U_2 = g^{-1}V_2$. So, V_1, V_2 is an open cover of F_p, while $V_1 \cap V_2$ is the image of the lower horizontal map, barycenters removed. To prove that the restriction map $g_p : g^{-1}F_p \longrightarrow F_p$ is a quasifibration, we use Proposition B.5.11 (Dold–Thom), with respect to this open cover V_1, V_2. We need to prove that the restriction maps (*labels added*) $f_1 : U_1 \longrightarrow V_1$, $f_2 : U_2 \longrightarrow V_2$, and $f_{12} : U_1 \cap U_2 \longrightarrow V_1 \cap V_2$ are quasifibrtions. Let $(|\Delta[p]|)^{int} := |\Delta[p]| \setminus \partial |\Delta[p]|$ denote the interior of $|\Delta[p]|$. Note, f_2, f_{12} are, respectively, equivalent to the maps

$$\begin{cases} \coprod_{\sigma \in \mathcal{N} I'_p} (|\Delta[p]|)^{int} \times X_{\sigma(0)} \longrightarrow \coprod_{\sigma \in \mathcal{N} I'_p} (|\Delta[p]|)^{int} \\ \coprod_{\sigma \in \mathcal{N} I'_p} ((|\Delta[p]|)^{int} \setminus \{O_p\}) \times X_{\sigma(0)} \longrightarrow \coprod_{\sigma \in \mathcal{N} I'_p} (|\Delta[p]|)^{int} \setminus \{O_p\} \end{cases}$$

Since, projection maps are quasifibration, f_2, f_{12} are quasifibrations.

What remains to be established is that f_1 is a quasifibration. There is a strong deformation retraction $\mu : |\Delta[p]| \setminus \{O_p\} \longrightarrow \partial |\Delta[p]|$, corresponding to the retraction $r : \mathbb{D}^p \setminus \{0\} \longrightarrow \mathbb{S}^{p-1}$ sending $x \mapsto \frac{x}{\|x\|}$. Use the diagram (2.7). Then μ induces two deformation retracts ρ, η, as in the following commutative diagram:

$$\begin{array}{ccccc} g^{-1}F_{p-1} & \hookrightarrow & U_1 & \xrightarrow{\eta} & g^{-1}F_{p-1} \\ \downarrow g_{p-1} & & \downarrow f_1 & & \downarrow g_{p-1} \\ F_{p-1} & \hookrightarrow & V_1 & \xrightarrow[\rho]{} & F_{p-1} \end{array} \tag{2.8}$$

Fix $n \geq 0$. Let $b \in F_p$, $F := f_1^{-1}b$, and $x \in F$. First assume $b \in F_{p-1}$. We have $\pi_n(U_1, F, x) \cong \pi_n(F_{p-1}, F, x)$, $\pi_n(V_1, b) \cong \pi_n(F_{p-1}, b)$. Further, $\pi_n(g^{-1}F_{p-1}, F, x) \cong \pi_n(F_{p-1}, b)$, by induction. Therefore, $\pi_n(U_1, F, x_0) \cong \pi_n(U_1, b)$, as required.

Now, let $b \in V_1 \setminus F_{p-1}$. So, $b = \Phi_p(\sigma, y)$ for some $\sigma \in \mathcal{N} I'_p$, $y \in |\Delta[p]| \setminus \partial |\Delta[p]|$, and $y \neq O_p$. Also $x = \Phi_p(\sigma, y, \xi)$ for some $\xi \in X_{i_0}$. Use the description of σ, as in (2.6). It follows, $F \cong (\sigma, z) \times X_{i_0}$. We write $\mu(y) = z = (z_0, \ldots, z_p)$. Let $\{j : z_j \neq 0\} = \{j_0, j_1, \ldots, j_q\}$, with $j_0 < j_1 < \cdots < j_q$. Write $z' = (z_{j_0}, \ldots, z_{j_q}) \in |\Delta[q]|$. Let $\tau = i_{j_0} \longrightarrow i_{j_1} \longrightarrow \cdots \longrightarrow i_{j_q}$. Then $\tau \in \mathbb{N}I'_q$, $z' \in |\Delta[q]| \setminus \partial |\Delta[q]|$.

We have $c := \Phi_q(\tau, z') \in F_q \subseteq F_{p-1}$. We have $G := f_1^{-1}c \cong (\tau, z') \times X_{i_{j_0}}$, $c = \rho(b)$, $\eta(x) = \varphi_p(\tau, z', f_{i_0 i_{j_0}}(\xi))$, where $f_{i_0 i_{j_0}} : X_{i_0} \longrightarrow X_{j_0}$ is the composition map. Consider the commutative diagram

$$\begin{array}{ccc} (\sigma, y) \times X_{i_0} & \xrightarrow[\sim]{} & F \\ {\scriptstyle f_{i_0 i_{j_0}}} \big\downarrow & & \big\downarrow {\scriptstyle \eta_{|F}} \\ (\tau, z') \times X_{i_{j_0}} & \xrightarrow{\sim} & G \end{array}$$

By hypothesis, $f_{i_0 i_{j_0}}$ is a homotopy equivalence, and hence so is $\eta_{|F} : F \longrightarrow G$. Therefore, $\eta : (U_1, F, x) \longrightarrow (g^{-1}F_{p-1}, G, \eta(x))$ is a homotopy equivalence. For integers $n \geq 0$, consider the commutative diagram

$$\begin{array}{ccc} \pi_n(U_1, F, x) & \xrightarrow[\sim]{\eta_*} & \pi_n\left(g^{-1}F_{p-1}, G, \eta(x)\right) \\ \big\downarrow & & \big\downarrow \wr \\ \pi_n(U_1, b) & \xrightarrow[\rho_*]{\sim} & \pi_n(F_{p-1}, c) \end{array}$$

The second vertical map is an isomorphism, by induction. Also, ρ_*, η_* are isomorphisms. So, the first vertical map is an isomorphism. The proof is complete. ■

2.2. Exact Sequences of Homotopy Groups

Perhaps [Q, Theorems A, B] provide the foundation of Quillen K-Theory. Theorem A (2.2.2) provides a sufficient condition for a functor to be a homotopy equivalence, and Theorem B (2.2.3) provides a Long Exact sequence of homotopy groups. In analogy to the homotopy fiber $\mathscr{F}(f, b)$, for a continuous map $f : X \longrightarrow B$, and $b \in B$ (B.4.5), we introduce the following definitions.

Definition 2.2.1. Let $F : \mathscr{C} \longrightarrow \mathscr{D}$ be a functor. For $Y \in Obj(\mathscr{D})$, we define two categories.

1. Define the category Y/F as follows:

$$\begin{cases} Obj\ Y/F = \{(X,u) : X \in Obj\ \mathscr{C},\ u : Y \longrightarrow F(X) \in Mor_{\mathscr{D}}(Y, F(X))\} \\ Mor_{Y/F}\left((X_1,u_1),(X_2,u_2)\right) = \{\varphi \in Mor_{\mathscr{C}}\left(X_1, X_2\right) : u_2 = F(\varphi)u_1\} \end{cases}$$

Meaning the diagram
$$\begin{array}{ccc} Y & \xrightarrow{u_1} & FX_1 \\ & \searrow^{u_2} & \downarrow F(\varphi) \\ & & FX_2 \end{array}$$
commutes.

Further, for arrows $v : Y \longrightarrow Z$, in $\mathscr{D}$, there are functors

$$v^* : Z/F \longrightarrow Y/F \quad \text{sending} \quad \begin{cases} (X,u) \mapsto (X, uv) \\ \varphi \in Mor\left((X_1,u_1),(X_2,u_2)\right) \mapsto \varphi \end{cases} \tag{2.9}$$

2. Define the category F/Y as follows:

$$\begin{cases} Obj\ F/Y = \{(X,u) : X \in Obj\ \mathscr{C},\ u : F(X) \longrightarrow Y \in Mor_{\mathscr{D}}(F(X), Y)\} \\ Mor_{F/Y}\left((X_1,u_1),(X_2,u_2)\right) = \{\varphi \in Mor_{\mathscr{C}}\left(X_1, X_2\right) : u_2 F(\varphi) = u_1\}. \end{cases}$$

Meaning the diagram
$$\begin{array}{ccc} FX_1 & & \\ F(\varphi)\downarrow & \searrow^{u_1} & \\ FX_2 & \xrightarrow[u_2]{} & Y \end{array}$$
commutes.

Also, for arrows $v : Y \longrightarrow Z$, in $\mathscr{D}$, there are functors

$$v_* : F/Y \longrightarrow F/Z \quad \text{sending} \quad \begin{cases} (X,u) \mapsto (X, vu) \\ \varphi \in Mor\left((X_1,u_1),(X_2,u_2)\right) \mapsto \varphi \end{cases} \tag{2.10}$$

3. With $F = 1_{\mathscr{C}}$, we denote $Y/\mathscr{C} := Y/1_{\mathscr{C}}$ and $\mathscr{C}/Y := 1_{\mathscr{C}}/Y$.
4. **Remark:** These categories Y/F, F/Y are fairly analogous to the homotopy fibers (B.4.5) $\mathscr{F}(f,y)$ of continuous maps $f : X \longrightarrow Y$ and $y \in Y$. The functors (2.9, 2.10) are analogous to similar maps between homotopy fibers of continuous maps $f : X \longrightarrow Y$, obtained by adjoining paths when there is one.

2.2.1. *The Theorem A*

The following is the statement of celebrated [Q, Theorem A].

Theorem 2.2.2 (Theorem A). *Let $F : \mathscr{C} \longrightarrow \mathscr{D}$ be a functor such that Y/F is contractible, $\forall\ Y \in Obj(\mathscr{D})$. Then F is a homotopy equivalence. Dually (see Proposition 2.1.7), if F/Y is contractible, $\forall\ Y \in Obj(\mathscr{D})$, then F is a homotopy equivalence.*

Proof. (*Intuitively, the proof would be an application of the theorem of Whitehead (C.3.8), if it was applicable.*) Let $S(F)$ be the category, described as follows:

1. The objects are

$$Obj(S(F)) = \{(X, Y, v) : X \in Obj(\mathscr{C}), Y \in Obj(\mathscr{D}), v \in Mor_{\mathscr{D}}(Y, FX)\}$$

2. For $(X, Y, v), (X', Y', v') \in Obj(S(F))$, define $Mor_{S(F)}((X, Y, v), (X', Y', v'))$

$$= \left\{ (u, w) \in Mor_{\mathscr{C}}(X, X') \times Mor_{\mathscr{D}}(Y', Y) : \begin{array}{ccc} Y' & \xrightarrow{\ w\ } & Y \\ {\scriptstyle v'}\downarrow & & \downarrow{\scriptstyle v} \\ FX' & \xleftarrow[Fu]{} & FX \end{array} \quad \text{commutes} \right\}$$

There are functors:

$$\mathscr{D}^o \xleftarrow{\ p_2\ } S(F) \xrightarrow{\ p_1\ } \mathscr{C} \qquad \ni \qquad \begin{cases} p_1(X, Y, v) = X & p_1(u, w) = u \\ p_2(X, Y, v) = Y & p_2(u, w) = w \end{cases}$$

Let $T(F)$ be the bisimplicial set, namely the functor

$$T(F) : \Delta^o \times \Delta^o \longrightarrow \textbf{Sets} \qquad \text{defined by} \qquad T(F)_{pq} :=$$

$$T(F)\left(\underline{p}, \underline{q}\right) = \left\{ \begin{array}{ll} Y_p \xrightarrow{\ f_p\ } Y_{p-1} \longrightarrow \cdots \longrightarrow Y_0 \xrightarrow{\ v_0\ } FX_0 & \text{in } \mathscr{D} \\ & \\ X_0 \xrightarrow[u_1]{} X_1 \xrightarrow[u_2]{} \cdots \longrightarrow X_{q-1} \xrightarrow[u_q]{} X_q & \text{in } \mathscr{C} \end{array} \right\}$$

So, $T(F)_{pq} \subseteq (\mathscr{N}\mathscr{D})_{p+1} \times (\mathscr{N}\mathscr{C})_q$. For morphisms (f, g) in $\Delta^o \times \Delta^o$, define $T(F)(f, g)$ accordingly. Treat the simplicial set $\mathscr{N}(\mathscr{C})$ (*the nerve*) as a bisimplicial set, by $(\underline{p}, \underline{q}) \mapsto \mathscr{N}(\mathscr{C})_q$. Now consider the map of bisimplicial sets:

$$\tau_1 : T(F) \longrightarrow \mathscr{N}(\mathscr{C}) \quad \text{is the projection} \quad T(F)_{pq} \longrightarrow \mathscr{N}(\mathscr{C})_{pq} = \mathscr{N}(\mathscr{C})_q \tag{2.11}$$

Given an ordered pair in $T(F)_{pp}$, as described above, for $0 \leq k \leq p$, let

$$v_k = F(u_k) \cdots F(u_1) v_0 f_1 \cdots f_k : Y_k \longrightarrow FX_k. \quad \text{So,} \quad (X_k, Y_k, v_k) \in S(F), \quad \text{and}$$

$$(X_0, Y_0, v_0) \xrightarrow[(u_1, f_1)]{} (X_1, Y_1, v_1) \longrightarrow \cdots \longrightarrow (X_p, Y_p, v_p) \in \mathscr{N}(S(F))_p.$$

This shows that the nerve $\mathscr{N}(S(F))$ is the diagonal $\underline{p} \mapsto T(F)_{pp}$ (i.e. *in bijection*). The first part of the following commutative diagram leads to its geometric realization, as follows:

$$\begin{array}{ccc} \mathscr{N}S(F) \xrightarrow{\sim} \{T(F)_{pp}\} & \qquad & \mathbb{B}S(F) \xrightarrow{\sim} |\{T(F)_{pp}\}| \\ {\scriptstyle \mathscr{N}(p_1)} \searrow \quad \downarrow \tau_1 & & {\scriptstyle \mathbb{B}(p_1)} \searrow \quad \downarrow |\tau_1| \\ \mathscr{N}\mathscr{C} & & \mathbb{B}\mathscr{C} \end{array}$$

Now, for a fixed q

$$\begin{cases} T(F)_{pq} = \coprod_{X_\bullet \in \mathscr{N}(\mathscr{C})_q} \mathscr{N}(\mathscr{D}/FX_0)^o_p & \text{So,} \\ |p \mapsto T(F)_{pq}| = \coprod_{X_\bullet \in \mathscr{N}(\mathscr{C})_q} \mathbb{B}(\mathscr{D}/FX_0)^o \end{cases}$$

Since $\mathscr{D}/FX_0$ has a final object, namely, $1 : FX_0 \longrightarrow FX_0$, $\mathbb{B}(\mathscr{D}/FX_0) \simeq \star$ is contractible. So, for all $q \geq 0$, there is a homotopy equivalence

$$|p \mapsto T(F)_{pq}| = \coprod_{X_\bullet \in \mathscr{N}(\mathscr{C})_q} \mathbb{B}(\mathscr{D}/FX_0)^o \longrightarrow \coprod_{X_\bullet \in \mathscr{N}(\mathscr{C})_q} \star = \mathscr{N}\mathscr{C}_q$$

with discrete topology on $\mathscr{N}(\mathscr{C})_q$. Write $\mathcal{T}_q = |p \mapsto T(F)_{pq}|$. So, $\mathcal{T} = \{\mathcal{T}_q\}$ is a simplicial topological space. It follows from above, $\forall\, q \geq 0$, the map

$$\mathcal{T}_q \longrightarrow \mathscr{N}(\mathscr{C})_q \quad \text{is a homopoty equivalence, with discrete topology on } \mathscr{N}(\mathscr{C})_q.$$

Claim: $|\mathcal{T}| = |(q \mapsto |p \mapsto T(F)_{pq}|)| \simeq \mathbb{B}\mathscr{C}$ is a homotopy equivalence.

The claim follows, at least intuitively, because $\mathcal{T}_q \longrightarrow \mathscr{N}(\mathscr{C})_q$ is a homotopy equivalence. However, we give a proof using Lemma 2.1.20, with $I = \mathscr{C}$, and the functor $\mathscr{C} : X \mapsto \mathbb{B}(\mathscr{D}/FX)^o$. Note $\mathbb{B}(\mathscr{D}/FX)^o$ is contractible, hence they are homotopically equivalent. Therefore, Lemma 2.1.20 is applicable, and $|\mathcal{T}| \longrightarrow \mathbb{B}\mathscr{C}$ is a quasifibraton, with the fibers being contractible. It follows immediately that the map $|\mathcal{T}| \longrightarrow \mathbb{B}\mathscr{C}$ is a weak equivalence. Since both $|\mathcal{T}|$, $\mathbb{B}\mathscr{C}$ are CW complexes, it follows from Whitehead's Theorem (C.3.8) this map is a homotopy equivalence. So, the claim is established. We have the commutative diagram

$$\begin{array}{ccccc} \mathbb{B}S(F) & \xrightarrow{\sim} & |\{T(F)_{pp}\}| & \xrightarrow{\sim} & |\mathcal{T}| \\ & \searrow^{\mathbb{B}(p_1)} & \downarrow{\scriptstyle |\tau_1|} & \swarrow & \\ & & \mathbb{B}\mathscr{C} & & \end{array}$$

In this diagram, the second horizontal homeomorphism is obtained by Lemma 1.3.2. The right diagonal map is a homotopy equivalence, by the claim established above. Now, it follows that $\mathbb{B}(p_1)$ is a homotopy equivalence.

Now consider the functor $p_2 : S(F) \longrightarrow \mathscr{D}^o$. Using similar arguments, we establish that $\mathbb{B}(p_2)$ is a homotopy equivalence. As before, treat $\mathscr{N}(\mathscr{D}^o)$ as a bisimplicial set by $(\underline{p}, \underline{q})) \mapsto \mathscr{N}(\mathscr{D}^o)_p$. As in (2.11), consider the projection map to the Y-coordinate:

$$\tau_2 : T(F) \longrightarrow \mathscr{N}(\mathscr{D}^o) \quad \text{sending} \quad T(F)_{pq} \mapsto \mathscr{N}(\mathscr{D}^o)_p \tag{2.12}$$

As before, the first of the following commutative diagram leads to the second, as follows:

$$\begin{array}{ccc} \mathscr{N}S(F) & \xrightarrow{\sim} & \{T(F)_{qq}\} \\ & \searrow^{\mathscr{N}(p_2)} & \downarrow{\scriptstyle \tau_2} \\ & & \mathscr{N}\mathscr{D}^o \end{array} \qquad \begin{array}{ccc} \mathbb{B}S(F) & \xrightarrow{\sim} & |\{T(F)_{qq}\}| \\ & \searrow^{\mathbb{B}(p_2)} & \downarrow{\scriptstyle |\tau_2|} \\ & & \mathbb{B}\mathscr{D}^o \end{array}$$

In fact, for any fixed $Y_\bullet \in \mathscr{N}(\mathscr{D}^o)$, we have

$$\{(x_\bullet, y_\bullet) \in T(F)_{pq} : y_\bullet = Y_\bullet\} \simeq \mathscr{N}(Y_0/F)$$

$$\text{For a fixed } p, \quad \begin{cases} T(F)_{pq} = \coprod_{Y_\bullet \in \mathscr{N}(\mathscr{D})^o_p} \mathscr{N}(Y_0/F)_q \\ |q \mapsto T(F)_{pq}| = \coprod_{Y_\bullet \in \mathscr{N}(\mathscr{D})^o_p} \mathbb{B}(Y_0/F) \end{cases} \quad \text{So,}$$

By hypothesis, $\mathbb{B}(Y_0/F)$ is contractible. Therefore, for fixed p, the map

$$|q \mapsto T(F)_{pq}| = \coprod_{Y_\bullet \in \mathscr{N}(\mathscr{D}^0)_p} \mathbb{B}(Y_0/F) \longrightarrow \mathscr{N}(\mathscr{D}^0)_p$$

is a homotopy equivalence, $\forall\ p$, with discrete topology on $\mathscr{N}(\mathscr{D}^0)_p$. Write $\mathcal{F}_p = |q \mapsto T(F)_{pq}|$, and $\mathcal{F} = \{\mathcal{F}_p\}$. So, $\mathcal{F}_p \longrightarrow \mathscr{N}(\mathscr{D}^0)_p$ is a homotopy equivalence, $\forall\ p$. Arguing as before, using Lemma 2.1.20, it follows

$$|\mathcal{F}| = |(p \mapsto |q \mapsto T(F)_{pq}|)| \simeq \mathbb{B}\mathscr{D}^o \quad \text{is a homotopy equivalence.}$$

Now, we have the commutative diagram

$$\begin{array}{ccccc} \mathbb{B}S(F) & \xrightarrow{\sim} & |\{T(F)_{qq}\}| & \xrightarrow{\sim} & |\mathcal{F}| \\ & {\scriptstyle \mathbb{B}(p_2)} \searrow & \downarrow {\scriptstyle |\tau_2|} & \swarrow & \\ & & \mathbb{B}\mathscr{D}^o & & \end{array}$$

As before, the second horizontal homeomorphism is obtained by Lemma 1.3.2. Thus, we established that the right diagonal map is a homotopy equivalence. So, $\mathbb{B}(p_2)$ is a homotopy equivalence.

The above inferences on F apply to the identity functors as well. So, we have the following commutative diagram:

$$\begin{array}{ccccc} \mathscr{D}^o & \xleftarrow{p_2} & S(F) & \xrightarrow{p_1} & \mathscr{C} \\ \| & & \downarrow {\scriptstyle f} & & \downarrow {\scriptstyle F} \\ \mathscr{D}^o & \xleftarrow[\pi_2]{} & S(id) & \xrightarrow[\pi_1]{} & \mathscr{D} \end{array} \tag{2.13}$$

This induces the following corresponding commutative diagram of the classifying spaces:

$$\begin{array}{ccccc} \mathbb{B}(\mathscr{D}^o) & \xleftarrow{\mathbb{B}(p_2)} & \mathbb{B}(S(F)) & \xrightarrow{\mathbb{B}(p_1)} & \mathbb{B}(\mathscr{C}) \\ \| & & \downarrow {\scriptstyle \mathbb{B}(f)} & & \downarrow {\scriptstyle \mathbb{B}(F)} \\ \mathbb{B}(\mathscr{D}^o) & \xleftarrow[\mathbb{B}(\pi_2)]{} & \mathbb{B}(S(id)) & \xrightarrow[\mathbb{B}(\pi_1)]{} & \mathbb{B}(\mathscr{D}) \end{array} \tag{2.14}$$

All the horizontal maps are homotopy equivalences. Hence, $\mathbb{B}(f)$ is a homotopy equivalence. Consequently, $\mathbb{B}(F)$ homotopy equivalence. The proof is complete. ∎

2.2.2. *The Theorem B*

The [Q, Theorem B] provides the tool needed to use the long exact sequences in Homotopy theory (B.21) to develop the same for categories and functors.

Let $F : \mathscr{C} \longrightarrow \mathscr{D}$ be a functor. Fix $Y \in Obj(\mathscr{D})$. Consider the commutative diagram,

$$\begin{array}{ccc} Y/F & \xrightarrow{j} & \mathscr{C} \\ & \searrow^{Fj} & \downarrow F \\ & & \mathscr{D} \end{array} \qquad \text{where} \begin{cases} j(X,v) = X \ \forall\ (X,v) \in Obj(Y/F) \\ j(u) = u \qquad \forall\ u \in Mor_{Y/F}((X_1,v_1),(X_2,v_2)) \end{cases}$$

So, $\mathbb{B}(Fj) = \mathbb{B}(F)\mathbb{B}(j)$. Let $C_Y : Y/F \longrightarrow \mathscr{D}$ be the constant functor Y. Therefore, $c_Y := \mathbb{B}(C_Y) : \mathbb{B}(Y/F) \longrightarrow \mathbb{B}(\mathscr{D})$ is the constant map. Define a natural transformation

$$C_Y \longrightarrow Fj \quad \text{by } v : C_Y(X,v) = Y \longrightarrow Fj(X,v) = FX$$

By (2.1.9), this leads to a homotopy

$$H : \mathbb{B}(Y/F) \times [0,1] \longrightarrow \mathbb{B}(\mathscr{D}) \qquad c_Y \mapsto \mathbb{B}(Fj)$$

For $z \in \mathbb{B}(Y/F)$, $t \mapsto H(z,t)$ is a path starting from Y to $\mathbb{B}(Fj)(z)$. Using the opposite path $\overline{H(z,-)}$, we obtain a canonical map

$$\mathbb{B}(Y/F) \longrightarrow \mathscr{F}(\mathbb{B}(Fj), Y) \quad \text{sending } z \mapsto \overline{H(z,-)} \tag{2.15}$$

Now we state the celebrated Theorem B.

Theorem 2.2.3 (Theorem B). *Let $F : \mathscr{C} \longrightarrow \mathscr{D}$ be a functor. Assume $\forall\ v \in Mor_{\mathscr{D}}(Y,Z)$, the functors $v^* : Z/F \longrightarrow Y/F$, as defined in (2.9), are homotopy equivalences. Then $\forall\ Y \in Obj(\mathscr{D})$, the commutative diagram*

$$\begin{array}{ccc} Y/F & \xrightarrow{j} & \mathscr{C} \\ {\scriptstyle F'}\downarrow & & \downarrow{\scriptstyle F} \\ Y/\mathscr{D} & \xrightarrow[j']{} & \mathscr{D} \end{array} \quad \textit{is a homotopy cartesian, where} \begin{cases} j(X,v) = X \\ F'(X,v) = (FX,v) \\ j'(Y',v) = Y' \end{cases} \tag{2.16}$$

We further have the following:

1. *Note,* $\forall\ Y \in Obj(\mathscr{D})$, *the category* $Y/\mathscr{D}$ *has an initial object* $(Y, 1_Y)$. *So,* $Y/\mathscr{D}$ *is contractible. Therefore, the upper right part of the cartesian square* (2.16)

$$Y/F \xrightarrow{j} \mathscr{C} \xrightarrow{F} \mathscr{D} \qquad \text{is a homotopy fibration.}$$

2. *For all* $X \in Obj(\mathscr{C})$ *and* $FX = Y$, *there is a long exact sequence:*

$$\cdots \longrightarrow \pi_{n+1}(\mathscr{D}, Y) \longrightarrow \pi_n\left(Y/F, \tilde{X}\right)$$

$$\longrightarrow \pi_n(\mathscr{C}, X) \longrightarrow \pi_n(\mathscr{D}, Y) \longrightarrow \cdots$$

ending at $\pi_0(\mathscr{D}, Y)$, *where* $\tilde{X} := (X, 1_Y) \in Obj(Y/F)$).

3. *This theorem admits a dual formulation, replacing the categories* Y/F *by categories* F/Y, *etc., in the diagram* (2.16).

Proof. Let $S(F)$ denote the category, as defined in the proof of (2.2.2). Consider the projection functors

$$\mathscr{D}^o \xleftarrow{p_2} S(F) \xrightarrow{p_1} \mathscr{C} \quad \text{where} \quad \begin{cases} p_1(X,Y,v) = X & p_1(u,w) = u \\ p_2(X,Y,v) = Y & p_2(u,w) = w \end{cases}$$

As in the proof of (2.2.2), p_1 is a homotopy equivalence. Also, from the proof of (2.2.2), the map $\mathbb{B}(p_2) : \mathbb{B}S(F) \longrightarrow \mathbb{B}(\mathscr{D}^o)$ is the realization of the map

$$\coprod_{Y_\bullet \in \mathscr{N}\mathscr{D}^o_p} \mathbb{B}(Y_0/F) \longrightarrow \mathscr{N}(\mathscr{D}^0)_p$$

By hypothesis, $\forall\ v \in Mor_{\mathscr{D}^o}(Y, Y')$, the maps $\mathbb{B}(Y/F) \longrightarrow \mathbb{B}(Y'/F)$ are homotopy equivalences. So, Lemma 2.1.20 is applicable, with $I := \mathscr{D}^o$, to the functor $Y \mapsto \mathbb{B}(Y/F)$ from $\mathscr{D}^o \longrightarrow$ **Top**. By Lemma 2.1.20, we see that $\mathbb{B}p_2$ is a quasifibration. So, for all $Y \in Obj(\mathscr{D}^o)$, the maps $\mathbb{B}(Y/F) \longrightarrow \mathscr{F}(\mathbb{B}(p_2), Y)$ are weak equivalences. By Whitehead's Theorem C.3.8, the map is a homotopy equivalence. Therefore,

$$\forall\, Y \in Obj(\mathscr{D}^o) \quad \text{the diagram} \quad \begin{array}{ccc} Y/F & \longrightarrow & S(F) \\ \downarrow & & \downarrow{\scriptstyle p_2} \\ pt & \xrightarrow[Y]{} & \mathscr{D}^o \end{array} \quad \text{is a homotopy cartesian.}$$

Now, we have the following commutative diagram (*compare with diagrams* (2.13, 2.14)):

$$\begin{array}{ccccc} Y/F & \longrightarrow & S(F) & \xrightarrow[\sim]{p_1} & \mathscr{C} \\ \downarrow & & \downarrow{\scriptstyle f} & & \downarrow{\scriptstyle F} \\ Y/\mathscr{D} & \longrightarrow & S(1_{\mathscr{D}}) & \xrightarrow[\pi_1]{\sim} & \mathscr{D} \\ {\scriptstyle\wr}\downarrow & & {\scriptstyle\wr}\downarrow{\scriptstyle \pi_2} & & \\ pt & \xrightarrow[Y]{} & \mathscr{D}^o & & \end{array}$$

The combination of two rectangles on the left is a homotopy cartesian. Since two vertical maps in the lower left rectangle are homotopy equivalences, the upper left rectangle is a homotopy cartesian. Also, two horizontal maps on the right-hand rectangle are homotopy equivalences. Therefore, the combination of two upper rectangles is a homotopy cartesian. ■

2.2.3. *Fiberd and cofiberd category version of Theorems A and B*

Given a functor $F : \mathscr{C} \longrightarrow \mathscr{D}$, and an object $Y \in Obj(\mathscr{D})$, one can define the fiber category in a natural way (2.2.4). The goal of this section is to interpret Theorems A and B in terms of the fiber categories.

Definition 2.2.4. Let $F : \mathscr{C} \longrightarrow \mathscr{D}$ be a functor. For $Y \in Obj(\mathscr{D})$, refer to the definitions (2.2.1) of the categories Y/F and F/Y. Further, define the **fiber category** $F^{-1}Y$ as follows:

$$\begin{cases} Obj\ F^{-1}Y = \{X \in Obj\ \mathscr{C} : F(X) = Y\} \\ Mor_{F^{-1}Y}(X_1, X_2) = \{\varphi \in Mor_{\mathscr{C}}(X_1, X_2) : F(\varphi) = 1_Y\} \end{cases}$$

There are two natural (inclusion) functors $\Phi : F^{-1}Y \longrightarrow Y/F$ and $\Phi : F^{-1}Y \longrightarrow F/Y$, as follows. The functor $\Phi := \Phi_{Y2F} : F^{-1}Y \longrightarrow Y/F$ is given by

$$\begin{cases} Obj\ \Phi : F^{-1}Y \longrightarrow Obj(Y/F) \ X \mapsto (X, 1_Y) \\ \varphi \in Mor_{F^{-1}Y}(X_1, X_2) \qquad \varphi \mapsto \varphi \end{cases}$$

Likewise, define the inclusion functor $\Phi := \Phi_{F2Y} : F^{-1}Y \longrightarrow F/Y$. We take the liberty to use the same notation Φ when there is no confusion.

1. The functor F is called **prefiberd** if $\forall\ Y \in Obj\ \mathscr{D}$ the functor $\Phi_{Y2F} : F^{-1}Y \longrightarrow Y/F$ has a right adjoint $R_{Y2F} : Y/F \longrightarrow F^{-1}Y$. So, we have natural bijection

$$\eta : Mor_{F^{-1}Y}\left(X_1, R_{Y2F}(X_2, u)\right) \xrightarrow{\sim} Mor_{Y/F}\left((X_1, 1_Y), (X_2, u)\right)$$

Likewise, the functor F is called **precofiberd** if the functor $\Phi_{F2Y} : F^{-1}Y \longrightarrow F/Y$ has a left adjoint $L_{F2Y} : F/Y \longrightarrow F^{-1}Y$. So, we have natural bijections

$$\eta : Mor_{F^{-1}Y}\left(L_{F2Y}(X_1, u), X_2\right) \xrightarrow{\sim} Mor_{F/Y}\left((X_1, u), (X_2, 1_Y)\right)$$

2. Now suppose F is precofibered. So, $\Phi_{F2Y} : F^{-1}Y \longrightarrow (F/Y)$ has a left adjoint $L_{F2Y} : (F/Y) \longrightarrow F^{-1}Y$ for all $Y \in Obj\ \mathscr{C}$. So, $\forall v \in Mor_{\mathscr{D}}(Y, Y')$, we have a commutative diagram

$$\begin{array}{ccc} F^{-1}Y & \xrightarrow{\Phi_{F2Y}} & (F/Y) \\ {\scriptstyle v_\star}\downarrow & & \downarrow{\scriptstyle v_*} \\ F^{-1}Y' & \underset{L_{F2Y}}{\overset{\Phi_{F2Y'}}{\rightleftarrows}} & (F/Y') \end{array} \quad \text{where} \begin{cases} \text{2nd varetical } v_* \text{ is as in (2.10)} \\ \text{and, the 1st } v_\star := L_{F2Y'} v_* \Phi_{F2Y} \end{cases}$$

These functors $v_\star$ are called the cobase change maps.
If F is prefiberd, define $v^\star := R_{Y'2F} v^* \Phi_{Y'2F} : F^{-1}Y' \longrightarrow F^{-1}Y$, where v^* is as in (2.9). These functors $v^\star$ are called the base change maps. Subsequently, we may use the same notation v^* for both v^*, and same notation v_* for both v_*.

3. If F is precofibered, then F is defined to be **cofibered** if $(vw)_* = v_* w_*$ whenever v, w are composable.
If F is prefibered, F is defined to be **fibered**, likewise.

The following would be the prefibered or precofibered version of Theorem A (2.2.2).

Corollary 2.2.5. *Suppose $F : \mathscr{C} \longrightarrow \mathscr{D}$ is either prefibered or precofibered such that $F^{-1}(Y)$ is contractible, for all objects Y in $\mathscr{D}$. Then F is a homotopy equivalence.*

Proof. We prove the first case. By Corollary 2.1.10, $F^{-1}(Y) \longrightarrow Y/F$ is a homotopy equivalence. Since $F^{-1}(Y)$ is contractible, so is Y/F. Now, the corollary follows from Theorem A (2.2.2). ■

The following would be the prefibered or precofibered version of Theorem B (2.2.3).

Corollary 2.2.6. *Suppose $F : \mathscr{C} \longrightarrow \mathscr{D}$ is a prefibered* (resp. precofibered) *functor. Assume, for every arrow $u : Y \longrightarrow Y'$ in $\mathscr{D}$ that the base change functor $u^* : F^{-1}(Y') \longrightarrow F^{-1}(Y)$* (resp. cobase change functor $u_* : F^{-1}(Y) \longrightarrow F^{-1}(Y')$) *is a homotopy equivalence. Then $\forall\ Y \in Obj(\mathscr{D})$ the diagram*

$$\begin{array}{ccc} F^{-1}(Y) & \longrightarrow & \mathscr{C} \\ \downarrow & & \downarrow \\ pt & \longrightarrow & \mathscr{D} \end{array}$$

is a homotopy cartesian. (So, the upper right sequence is a homotopy fibration.) *Consequently, for any $X \in Obj(F^{-1}(Y))$, there is a long exact sequence as follows:*

$$\cdots \longrightarrow \pi_{i+1}(\mathscr{D}, Y) \longrightarrow \pi_i\left(F^{-1}(Y), X\right)$$

$$\longrightarrow \pi_i(\mathscr{C}, X) \longrightarrow \pi_i(\mathscr{D}, Y) \longrightarrow \cdots$$

Proof. Again, it follows from (2.1.10) that $Y'/F \longrightarrow Y/F$ is a homotopy equivalence. Now, the corollary follows by an application of Theorem B (2.2.3). ■

2.3. Exercises

Exercise 2.3.1. Let $\mathbf{n} = \{0, 1, \ldots, n\}$ be the partially ordered set. Compute the classifying space $\mathbb{B}\mathbf{n}$.

Exercise 2.3.2. Let G be a group. Let $\underline{G}$ denote the one object category $\star$, and $Mor(\star, \star) = G$. Describe the classifying space $\mathbb{B}\underline{G}$ and establish that

$$\pi_n(\mathbb{B}\underline{G}, \star) = \begin{cases} G & \text{if } n = 1 \\ 0 & \text{if } n \neq 1 \end{cases}$$

Exercise 2.3.3. Give an explicit proof of (2.1.11).

Exercise 2.3.4. Let $\mathscr{A}$ be a small category, and $\mathscr{B}, \mathscr{C} \subseteq \mathscr{A}$ be two full subcategories.

1. We say $\mathscr{B}$ and $\mathscr{C}$ are disjoint if $Mor_{\mathscr{A}}(B, C) = \phi$, for all $B \in Obj(\mathscr{B})$, $C \in Obj(\mathscr{C})$.
2. Given two disjoint subcategories $\mathscr{B}, \mathscr{C} \subseteq \mathscr{A}$, we say $\mathscr{A}$ is disjoint union of $\mathscr{B}, \mathscr{C}$ if, further, $Obj(\mathscr{A}) = Obj(\mathscr{B}) \cup Obj(\mathscr{C})$. We write $\mathscr{A} = \mathscr{B} \coprod \mathscr{C}$ to mean this is a disjoint union. In this case, prove $\mathbb{B}\mathscr{A} = \mathbb{B}\mathscr{B} \cup \mathbb{B}\mathscr{C}$ is a disconnected union, meaning $\mathbb{B}\mathscr{B}, \mathbb{B}\mathscr{C}$ are open and $\mathbb{B}\mathscr{B} \cap \mathbb{B}\mathscr{C} = \phi$.
3. Two objects $A_0, A \in \mathscr{A}$ are defined to be connected if there is a sequence of objects $A_0, A_1, \ldots, A_{n-1}, A_n = A$ such that $\forall i = 0, 1, \ldots, n-1$ either $Mor(A_i, A_{i+1}) \neq \phi$ or $Mor(A_{i+1}, A_i) \neq \phi$. This is an equivalence relation $\sim_{\mathscr{A}}$ on $Obj(\mathscr{A})$.
   If all pairs of objects in $\mathscr{A}$ are connected, we say that $\mathscr{A}$ is a **connected category**. In this case, prove that $\mathbb{B}\mathscr{A}$ is path connected.
4. Prove that $\mathscr{A}$ is a disjoint union

   $$\mathscr{A} = \coprod_{i \in I} \mathscr{B}_i \qquad \text{where} \quad \forall i \ \mathscr{B}_i \text{ is connected, } \forall i \neq j \quad \mathscr{B}_i,\ \mathscr{B}_j \text{ are disjoint.}$$

   Further, prove

   $$\mathbb{B}\mathscr{A} = \coprod_{i \in I} \mathbb{B}\mathscr{B}_i$$

   is a decomposition into path-connected components.

Exercise 2.3.5. Formulate the dual version of Theorem A (2.2.2) by replacing Y/F by F/Y; and derive it from Theorem A (2.2.2).

Exercise 2.3.6. Formulate the dual version of Theorem B (2.2.3) by replacing Y/F by F/Y; and derive it from Theorem B (2.2.3).

Exercise 2.3.7. Under the notation and hypotheses of Theorem A (2.2.2), prove that the homotopy fiber $\mathscr{F}(\mathbb{B}F, Y)$ is null homotopic.

Exercise 2.3.8. Under the notation and hypotheses of Theorem B (2.2.3), prove that the natural inclusion map $|Y/F| \hookrightarrow \mathscr{F}(\mathbb{B}F, Y)$ is a homotopy equivalence. (**Hint:** Use (C.3.8).)

The Theorems A and B (2.2.2, 2.2.3) constitute a central piece of argument in the subsequent chapters and, in general, in this theory. We provide two other subsequent versions of Theorems A, B from [Wh], [GG]. First, we give the versions from [Wh, Lemma 1.4.A, 1.4.B].

Let $f : L \longrightarrow K$ be a map in **SSet**. For integers $n \geq 0$ and $y_n \in K_n$, define the simplicial set $F(f, y_n)$, by the pullback diagram

$$\begin{array}{ccc} F(f,y_n) & \longrightarrow & L \\ \downarrow & & \downarrow f \\ \Delta^n & \xrightarrow[y]{} & K \end{array} \tag{2.17}$$

The following is a version of Theorem A (2.2.2).

Exercise 2.3.9 (Theorem AW). With notation as above, suppose $|F(f, y_n)|$ is contractible for all $n \geq 0$ and $y_n \in K_n$. Then $|f| : |K| \longrightarrow |L|$ is a homotopy equivalence. (**Hint:** See [Wh, Lemma 1.4.A] and apply Theorem A (2.2.2).)

The following is a version of Theorem B (2.2.3).

Exercise 2.3.10 (Theorem BW). Suppose $f : K \longrightarrow L$ is a map of simplicial sets. Assume that for all arrows $\iota : [m] \longrightarrow [n]$ in Δ, and $y_n \in L_n$, the induced maps $|F(f, \iota^* y_n)| \longrightarrow |F(f, y_n)|$ are homotopy equivalences. Then the diagram

$$\begin{array}{ccc} |F(f,y_n)| & \longrightarrow & |K| \\ \downarrow & & \downarrow |f| \\ |\Delta^n| & \xrightarrow[y]{} & |L| \end{array} \quad \text{is a homotopy cartesian, } \forall \mathrm{n},\ \mathrm{y_n} \in \mathrm{K_n}.$$

(**Hint:** See [Wh, Lemma 1.4.B] and apply (2.2.3).)

The following are from [GG].

Exercise 2.3.11 (Theorem AGG). Let $F : X \longrightarrow Y$ be a map of simplicial sets. Recall the notation $\rho \mid F$ from (1.9). Assume, for all $\rho \in Y([m])$, the geometric realization $|(\rho \mid F)|$ is contractible. Prove that the geometric realization $|F| : |X| \longrightarrow |Y|$ is a homotopy equivalence. (**Hint:** See [GG].)

Exercise 2.3.12 (Theorem BGG). Let $F : X \longrightarrow Y$ be a map of simplicial sets. Recall the notation $\rho \mid F$ from (1.9). Assume, for all $\rho \in Y([m])$, and maps $f : [n] \longrightarrow [m]$, the map $(\rho \mid F) \longrightarrow (f^*\rho \mid F)$ is a homotopy equivalence. Then for all integers $n \geq 0$ and $\rho \in Y[n]$ the square

$$\begin{array}{ccc} |(\rho \mid F)| & \longrightarrow & |X| \\ \downarrow & & \downarrow{\scriptstyle |F|} \\ |(\rho \mid 1_Y)| & \longrightarrow & |Y| \end{array} \quad \text{is a homotopy cartesian.}$$

Consequently, there is a long exact sequence of homotopy groups, with all appropriate choices of base points. (**Hint:** See [GG].)

Chapter 3

Quillen K-Theory

Given a small Exact Category $\mathscr{E}$, Quillen [Q] associates a category $\mathbb{Q}\mathscr{E}$ to be called Quillen's $\mathbb{Q}$-category. The homotopy groups of $\mathbb{BQ}\mathscr{E}$ are defined to be the K-groups of $\mathscr{E}$.

3.1. Quillen's $\mathbb{Q}$-Construction

Given an exact category $\mathscr{E}$, a new category $\mathbb{Q}\mathscr{E}$ is associated in [Q], as follows.

Definition 3.1.1. Suppose $\mathscr{E}$ is a small exact category. Define the category $\mathbb{Q}\mathscr{E}$ as follows.

1. The objects of $\mathbb{Q}\mathscr{E}$ are the same as that of $\mathscr{E}$. Given two objects $X, Y \in \mathbb{Q}\mathscr{E}$, we define morphisms $X \longrightarrow Y$ in $\mathbb{Q}\mathscr{E}$, as follows. Consider pairs (p, i) of arrows in $\mathscr{E}$, as in the diagram

$$X \overset{p}{\twoheadleftarrow} Z \overset{i}{\hookrightarrow} Y \quad \ni \exists \quad \text{conflations} \quad \begin{cases} K' \hookrightarrow Z \overset{p}{\twoheadrightarrow} X \\ Z \overset{i}{\hookrightarrow} Y \twoheadrightarrow C \end{cases} \quad \text{in } \mathscr{E}. \tag{3.1}$$

In other words, p is a deflation, and i is an inflation. Two such pairs

$$\begin{cases} X \overset{p}{\twoheadleftarrow} Z \overset{i}{\hookrightarrow} Y \\ X \underset{p'}{\twoheadleftarrow} Z' \underset{i'}{\hookrightarrow} Y \end{cases} \quad \text{are defined to be isomorphic if}$$

$$\exists \text{ an isomorphism } \tau : Z \xrightarrow{\sim} Z' \quad \ni \quad \begin{array}{ccccc} & & Z & & \\ & \swarrow^{p} & & \searrow^{i} & \\ X & & \wr\downarrow\tau & & Y \\ & \nwarrow_{p'} & & \nearrow_{i'} & \\ & & Z' & & \end{array} \quad \text{commutes.}$$

Note that such an isomorphism τ is unique because i' is monic. A **morphism** $X \longrightarrow Y$ in $\mathbb{Q}\mathscr{E}$ is an isomorphism class $[(p,i)]$ of such pairs, as in (3.1). A diagram, as in (3.1), will be denoted by (Z,p,i).

2. (**Compositions**): Let $X \longrightarrow Y$ and $Y \longrightarrow Z$ be two morphisms in $\mathbb{Q}\mathscr{E}$, represented by $X \overset{p}{\twoheadleftarrow} W \overset{i}{\hookrightarrow} Y$, $Y \overset{q}{\twoheadleftarrow} V \overset{j}{\hookrightarrow} Z$. The composition is represented by (U, pq', ji'), as in the diagram

$$\begin{array}{ccccccc} & & U & \overset{i'}{\hookrightarrow} & V & \overset{j}{\hookrightarrow} & Z \\ & & {\scriptstyle q'}\downarrow & & \downarrow{\scriptstyle q} & & \\ X & \underset{p}{\twoheadleftarrow} & W & \underset{i}{\hookrightarrow} & Y & & \end{array} \qquad \text{where } U = V \times_Y W \text{ is the pullback.} \qquad (3.2)$$

We remark, the composition is well defined and associative. It is assumed that the classes of such diagrams (3.2) form a set.

We introduce the following special morphisms and notation.

Given an inflation $\iota : X \hookrightarrow Y$ in $\mathscr{E}$, there is an arrow $\iota_! : X \longrightarrow Y$ in $\mathbb{Q}\mathscr{E}$ represented by $X \overset{1_X}{\twoheadleftarrow} X \overset{\iota}{\hookrightarrow} Y$. Such an arrow in $\mathbb{Q}\mathscr{E}$ is referred to as "*injective*" [Q]. Likewise, given a deflation $p : Y \twoheadrightarrow X$ in $\mathscr{E}$, there is an arrow $p^! : X \longrightarrow Y$ in $\mathbb{Q}\mathscr{E}$ represented by $X \underset{p}{\twoheadleftarrow} Y \underset{1_Y}{\hookrightarrow} Y$. Such an arrow in $\mathbb{Q}\mathscr{E}$ is referred to as "*surjective*" [Q]. (Such references as "*injective*" and "*surjective*" to these arrows are not meant to be confused with the injective and surjective maps, in the usual sense (A.1.2)).

We insert a few basic lemmas.

Lemma 3.1.2. *Let $\mathscr{E}$ be an exact category. Consider the commutative diagram*

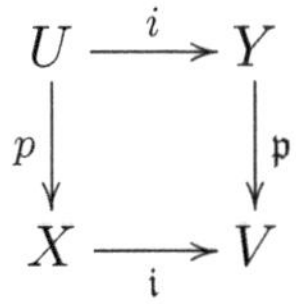

Suppose i is an inflation and p is a deflation. Then if V is a pushforward of i, p, then U is a pullback of $\mathfrak{i}, \mathfrak{p}$. Conversely, suppose $\mathfrak{i}$ is an inflation and $\mathfrak{p}$ is a deflation. Then if U is a pullback of $\mathfrak{i}, \mathfrak{p}$ then V is a pushforward of i, p.

Proof. We prove the first assertion. Consider the commutative diagram

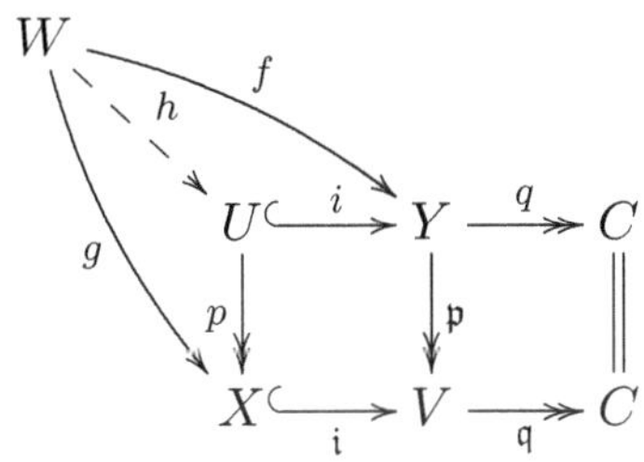

In this diagram, the first rectangle is a pushforward. Note $C = co\ker(i) = co\ker(i)$ coincide. Let f, g be given arrows such that $\mathfrak{p}f = \mathfrak{i}g$. We show that there is a unique arrow h, so that the diagram commutes. Uniqueness of h, if it exists, follows because i is monic. Now, note $qf = \mathfrak{q}\mathfrak{p}f = \mathfrak{q}\mathfrak{i}g = 0$. Since $U = \ker(q)$, $f = ih$ for some h. Again, $\mathfrak{i}ph = \mathfrak{p}ih = \mathfrak{p}f = \mathfrak{i}g$. Since $\mathscr{E}$ is exact $\mathfrak{i}$ is monic. So, $g = ph$. So, U is the pullback. The latter assertion follows similarly, or by reversing the arrows. The proof is complete. ■

Lemma 3.1.3. *Let $\mathscr{E}$ be an exact category and $u : X \longrightarrow Y$ be an arrow in $\mathbb{Q}\mathscr{E}$. Then there is a unique decomposition $u = i_! p^!$, where $i_! : U \longrightarrow Y$ is an injective map corresponding to an inflation $i : U \hookrightarrow Y$, and $p^! : X \longrightarrow U$ is a surjective map corresponding to a deflation $p : U \twoheadrightarrow X$.*

Further, there is another unique decomposition $u = \mathfrak{p}^! \mathfrak{i}_!$, where $\mathfrak{i}_! : X \longrightarrow V$ is an injective map, the map corresponding to an inflation $\mathfrak{i} : X \hookrightarrow V$, and $\mathfrak{p}^! : V \longrightarrow Y$ is a surjective map, the map corresponding to a deflation $\mathfrak{p} : Y \twoheadrightarrow V$.

Proof. Let u be represented by (U, i, p). It follows immediately, from the definition of composition, that $u = i_! p^!$, and the decomposition is unique. For the latter decomposition, consider the pushforward diagram

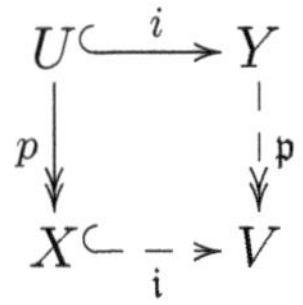

It follows from the definition of the composition that $u = \mathfrak{p}^! \mathfrak{i}_!$. Uniqueness of such a decomposition follows from (3.1.2). The proof is complete. ■

Lemma 3.1.4. *Let $\mathscr{E}$ be an exact category. Let $\mathfrak{p}$, ρ be deflations and $\mathfrak{i}$, ι be inflations in the following diagram:*

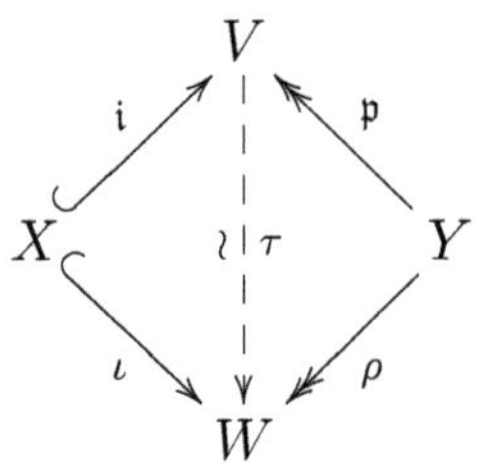

Then $\mathfrak{p}^! \mathfrak{i}_! = \rho^! \iota_! \iff \exists$ *isomorphism* $\tau \ni$ *the diagram commutes.*

Proof. Follows from (3.1.2)

We list the following observations:

1. Given a bicartesian square (i.e., pullback and pushout)

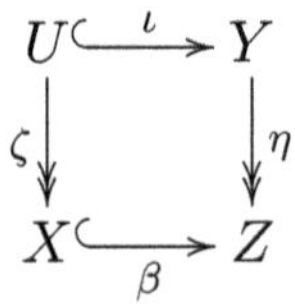

$$\Longrightarrow \quad \iota_! \zeta^! = \eta^! \beta_! \qquad (3.3)$$

2. **Lemma.** *If a map is both injective and surjective, then it is an isomorphism.*

Proof. Suppose $u = \Theta_! = \Psi^!$. Then we have the commutative diagram

$$\begin{array}{ccccc} X & \overset{1_X}{\twoheadleftarrow} & X & \overset{\Theta}{\hookrightarrow} & Y \\ \| & & \wr\big\downarrow\Phi & & \| \\ X & \underset{\Psi}{\twoheadleftarrow} & Y & \underset{1_Y}{\hookrightarrow} & Y \end{array}$$

This shows that $\Theta = \Psi^{-1}$. Now, suppose Θ is an isomorphism in $\mathscr{E}$. Then we have a commutative diagram

$$\begin{array}{ccccc} X & \overset{1_X}{\twoheadleftarrow} & X & \overset{\Theta}{\hookrightarrow} & Y \\ \| & & \big\downarrow\Theta & & \| \\ X & \underset{\Theta^{-1}}{\twoheadleftarrow} & Y & \underset{1_Y}{\hookrightarrow} & Y \end{array}$$

It follows from the diagram that $\Theta_!$ is an isomorphism in $\mathbb{Q}\mathscr{E}$ and its inverse is $(\Theta^{-1})^!$. The proof is complete. ■

3. **Lemma.** *Every morphism in $\mathbb{Q}\mathscr{E}$ is a monomorphism.*

Proof. Since every map in $\mathbb{Q}\mathscr{E}$ is a composition of an *injective* followed by a *surjective* map, we prove that such maps are monomorphisms in $\mathbb{Q}\mathscr{E}$. Suppose $\eta : Y \hookrightarrow W$ is an inflation, and $\beta : Y \twoheadrightarrow W$ is a deflation in $\mathscr{E}$. Consider two arrows in $\mathbb{Q}\mathscr{E}$:

$$u = \left(X \overset{p}{\twoheadleftarrow} Z \overset{\iota}{\hookrightarrow} Y \right), \qquad v = \left(X \overset{q}{\twoheadleftarrow} Z' \overset{\zeta}{\hookrightarrow} Y \right)$$

Now suppose $\eta_! u = \eta_! v$. Then we have the commutative diagram

$$\begin{array}{lccccccc} \eta_! u : & X & \overset{p}{\twoheadleftarrow} & Z & \overset{\iota}{\hookrightarrow} & Y & \overset{\eta}{\hookrightarrow} & W \\ & \| & & \wr\big\downarrow\varphi & & & & \| \\ \eta_! v : & X & \overset{q}{\twoheadleftarrow} & Z' & \overset{\zeta}{\hookrightarrow} & Y & \overset{\eta}{\hookrightarrow} & W \end{array}$$

where φ is an isomorphism in $\mathscr{E}$. Since η is a monomorphism in $\mathscr{E}$, $\zeta\varphi = \iota$ and hence $u = v$. Likewise, we check that $\beta^! u = \beta^! v$ implies $u = v$. The proof is complete. ■

4. **The Universal Property of $\mathbb{Q}\mathscr{E}$:** First, we have the following observations:

   (a) There is no natural functor from $\mathscr{E}$ to $\mathbb{Q}\mathscr{E}$. This is because, for $f \in Mor_{\mathscr{E}}(X, Y)$, there is no natural way to associate an arrow in $Mor_{\mathbb{Q}\mathscr{E}}(X, Y)$.

   (b) However, given an inflation $\iota : X \hookrightarrow Y$, in $\mathscr{E}$, we can associate $\iota_!$ in $\mathbb{Q}\mathscr{E}$. Given two composable inflations ι, ζ in $\mathscr{E}$, we have $(\zeta\iota)_! = \zeta_!\iota_!$ in $\mathbb{Q}\mathscr{E}$. Likewise, given a deflation $p : X \twoheadrightarrow Y$ in $\mathscr{E}$, we can associate $p^!$ in $\mathbb{Q}\mathscr{E}$. Given two composable deflations p, q in $\mathscr{E}$, we have $(pq)^! = q^!p^!$ in $\mathbb{Q}\mathscr{E}$. Further, $(Id_X)_! = (Id_X)^! = Id_X$.

   (c) Given the bicartesian square (3.3), where all the maps are admissible (i.e. *inflation or deflation*), then $\iota_!\zeta^! = \eta^!\beta_!$.

   Let $\mathscr{C}$ be a category. Let $h : \mathscr{E} \longrightarrow \mathscr{C}$ provide an association of objects to objects, inflations and deflations to maps (which is not necessarily a functor), as follows:

$$\begin{array}{ccc} \mathscr{E} & \longrightarrow & \mathbb{Q}\mathscr{E} \\ & \searrow^{h} & \\ & & \mathscr{C} \end{array} \qquad \ni \qquad \begin{cases} X \mapsto hX \in Obj(\mathscr{C}) & \forall\, X \in Obj(\mathscr{E}) \\ \iota \mapsto h_!(\iota) : hX \longrightarrow hY & \forall \text{ inflations } \iota : X \hookrightarrow Y \\ p \mapsto h^!(p) : hY \longrightarrow hX & \forall \text{ deflations } p : X \twoheadrightarrow Y \end{cases}$$

   If h respects composition of inflations and deflations (3.1), then

   (a) $h : \mathscr{E}_- \longrightarrow \mathscr{C}$ is a functor, where $\mathscr{E}_-$ is the category whose objects are the same as those of $\mathscr{E}$, and morphisms are the inflations of $\mathscr{E}$,

   (b) $h : \mathscr{E}_+ \longrightarrow \mathscr{C}$ is a functor, where $\mathscr{E}_+$ is the category whose objects are the same as those of $\mathscr{E}$, and morphisms are the deflations of $\mathscr{E}$.

   Finally, if $h_!$ and $h^!$ respect both the conditions 4b and 4c, then there is a unique functor $\Theta : \mathbb{Q}\mathscr{E} \longrightarrow \mathscr{C}$ such that the diagram

$$\begin{array}{ccc} \mathscr{E} & \longrightarrow & \mathbb{Q}\mathscr{E} \\ & \searrow^{h} & \downarrow{\scriptstyle\Theta} \\ & & \mathscr{C} \end{array} \qquad \text{commutes and is compatible with} \quad \begin{cases} \iota \mapsto \iota_! \\ p \mapsto p^! \end{cases}$$

The following is an obvious lemma.

Lemma 3.1.5. *For any exact category $\mathscr{E}$, we have an isomorphism of categories*

$$\mathbb{Q}\mathscr{E} = \mathbb{Q}(\mathscr{E}^o)$$

induced by the contra variant functor $\mathscr{E} \longrightarrow \mathscr{E}^o$.

3.1.1. *Admissible layers*

Let $\mathscr{E}$ be an exact category. There is an alternate and useful way to describe the morphisms in $\mathbb{Q}\mathscr{E}$, in terms of admissible layers. Since it is very useful, we consolidate it in this separate subsection.

Definition 3.1.6. Let $\mathscr{E}$ be an exact category and $M \in Obj(\mathscr{E})$. As usual, an inflation $K \hookrightarrow M$ in $\mathscr{E}$ is also called an admissible monomorphism, and similarly, deflations are called admissible epimorphism. Fix an object M in $\mathscr{E}$. Two inflations $\iota : K \hookrightarrow M$, $\iota' : K' \hookrightarrow M$, respectively, and two deflations $p : M \twoheadrightarrow C$, $p' : M \twoheadrightarrow C'$ are defined to be equivalent or isomorphic if there is a commutative diagram

$$\begin{array}{ccc} K & \overset{\iota}{\hookrightarrow} & M \\ \wr\downarrow & \nearrow_{\iota'} & \\ K' & & \end{array}, \quad \text{respectively,} \quad \begin{array}{ccc} M & \overset{p}{\twoheadrightarrow} & C \\ & \searrow^{p'} & \downarrow\wr \\ & & C' \end{array},$$

where the vertical arrows are isomorphisms. It follows that these vertical isomorphisms are unique. An **admissible sub-object** K of M is defined to be an isomorphism class of admissible monomorphisms $\iota : K \hookrightarrow M$. Likewise, an **admissible quotient** C of M, is defined to be an isomorphism class of admissible epimorphisms $p : M \twoheadrightarrow C$.

There is a bijection between the admissible sub-objects of M and admissible quotient objects of M. Note, for an inflation $\iota : K \hookrightarrow M$ and for an isomorphism $\tau : M \xrightarrow{\sim} M$, with $\tau \neq 1_M$, ι and $\tau\iota$ do not represent the same admissible sub-object of M. When M is understood, an admissible sub-object $\iota : K \hookrightarrow M$ would be denoted by $[(K, \iota)]$ or (K, ι), or K itself (*by abuse of notation*).

1. Given $M \in Obj(\mathscr{E})$, let $\mathcal{S}(M)$ denote the set of all admissible sub-objects of M. Given $(K_1, \iota_1), (K_2, \iota_2) \in \mathcal{S}(M)$, define $(K_1, \iota_1) \leq (K_2, \iota_2)$, if there is a (*unique*) inflation $\iota : K_1 \hookrightarrow K_2$, such that

$\iota_1 = \iota_2\iota$. We would abbreviate $K_1 \le K_2$. So, $\mathcal{S}(M)$ is a **partially ordered set** and hence a category.

Likewise, let $\mathcal{I}(M)$ denote the set of all admissible quotients. Then $\mathcal{I}(M)$ is also a partially ordered set.

2. An ordered pair $((K_1, \iota_1), (K_2, \iota_2))$ of admissible sub-objects in $\mathcal{S}(M)$ is defined to be an **admissible layer** if $(K_1, \iota_1) \le (K_2, \iota_2)$. We abbreviate and say (K_1, K_2) is an admissible layer.
3. The set of all admissible layers in M is denoted by $\mathcal{L}(M)$. As usual, $\mathcal{L}(M)$ is also a partially ordered set, where

$$\forall\ (K_1, K_2), (L_1, L_2) \in \mathcal{L}(M) \quad \text{define}$$
$$(K_1, K_2) \le (L_1, L_2) \Longleftrightarrow L_1 \le K_1 \le K_2 \le L_2$$

Lemma 3.1.7. *Let $\mathscr{E}$ be an exact category.*

1. *Let $f \in Mor_{\mathbb{Q}\mathscr{E}}(C, M)$ be represented by $f = \left(C \overset{p}{\twoheadleftarrow} K \overset{\iota}{\hookrightarrow} M\right)$. Then $\ker(p) \in \mathcal{S}(M)$ is well defined, and $\overline{p} : \frac{K}{\ker(p)} \xrightarrow{\sim} C$ is an isomorphism. Further, $(\ker(p), K)$ is a well-defined admissible layer in $\mathcal{S}(M)$, hence represents an arrow $\ker(p) \longrightarrow K$ in the partially ordered set $\mathcal{S}(M)$.*
2. *Conversely, let $(K_0, \iota_0), (K_1, \iota_1))$ be an admissible layer in $\mathcal{S}(M)$. So, $\iota_0 = \iota_1\iota$, for (unique) inflations $\iota : K_0 \hookrightarrow K_1$. Let $p(\iota) : K_1 \twoheadrightarrow C(\iota)$ be a choice of cokernel of ι. Then $f = \left(C(\iota) \overset{p(\iota)}{\twoheadleftarrow} K_1 \overset{\iota_1}{\hookrightarrow} M\right)$ represents a morphism $f \in Mor_{\mathbb{Q}\mathscr{E}}(C(\iota), M)$.*
3. *With notation as in (3.1.7), the association $f \mapsto (K, \ker(p)), \overline{p})$ is a bijection*

$$Mor_{\mathbb{Q}\mathscr{E}}(C, M) \xrightarrow{\sim} \left\{((K_1, K_2), \varphi) : (K_1, K_2) \in \mathcal{L}(M),\ \varphi \in Iso_{\mathscr{E}}\left(\frac{K_1}{K_0}, C\right)\right\}$$

where $Iso_{\mathscr{E}}\left(\frac{K_1}{K_0}, C\right)$ denotes the set of all the isomorphisms $\frac{K_1}{K_0} \xrightarrow{\sim} C$.

For another morphism, $g = \left(M \overset{q}{\twoheadleftarrow} L \overset{i}{\hookrightarrow} N\right)$, consider the pullback diagram to define gf:

$$\begin{array}{ccccccc}
K_0 \times_M L & \hookrightarrow & K \times_M L & \hookrightarrow & L & \xrightarrow{i} & N \\
\downarrow & & \downarrow{\scriptstyle q'} & & \downarrow{\scriptstyle q} & & \\
K_0 & \hookrightarrow & K & \hookrightarrow & M & & \\
 & & \downarrow{\scriptstyle p} & & & & \\
 & & C & & & &
\end{array} \qquad \text{with} \quad K_0 = \ker(p)$$

Then the bijection sends $gf \mapsto ((\ker(pq'), K \times_M L), \overline{pq'}) = (K_0 \times_M L, K \times_M L, \overline{pq'})$.

Proof. Obvious! ■

Given a category $\mathscr{C}$, there is a simple way to construct a category of arrows in $\mathscr{C}$ (see (3.1.9)). We describe the same for partially ordered sets, as follows.

Definition 3.1.8. Let $(S, \leq)$ be a partially ordered set. We define the category $Arr(S)$ of arrows of S. Let $Arr(S) = \{(x, y) \in S \times S : x \leq y\}$. For $(x_1, y_1), (x_2, y_2) \in Arr(S)$, define $(x_1, y_1) \leq (x_2, y_2)$ if $x_2 \leq x_1 \leq y_1 \leq y_2$. This defines a partial order relation on $Arr(S)$. So, $(Arr(S), \leq)$ is a category.

Lemma 3.1.9. *Let $\mathscr{E}$ be an exact category and $M \in Obj(\mathscr{E})$. Let $\mathcal{L}(M) := Arr(\mathcal{S}(M))$ be the category of admissible layers in M. For arrow $f = \left(C \overset{p}{\twoheadleftarrow} K \overset{\iota}{\hookrightarrow} M \right)$, in $\mathbb{Q}\mathscr{E}$, define $F(C, f) = (\ker(p), K) \in \mathcal{L}(M)$. Then $F : \mathbb{Q}\mathscr{E}/M \longrightarrow \mathcal{L}(M)$ is a an equivalence of categories.*

Proof. As described in (3.1.7), this association $(C, f) \mapsto F(C, f)$ is well defined on $Obj(\mathbb{Q}\mathscr{E}/M)$. For an arrow $u \in Mor_{\mathbb{Q}\mathscr{E}/M}((C_1, f_1), (C_2, f_2))$, by chasing the definition of compositions in $\mathbb{Q}\mathscr{E}$, one can see $F(C_1, f_1) \leq F(C_2, f_2)$. So, F defines a functor. It follows from the comments in (3.1.7) that F is fully faithful and essentially surjective. Therefore, F is an equivalence of categories. The proof is complete. ■

3.2. $K_0(\mathscr{E})$ and $\pi_1(\mathbb{BQ}\mathscr{E}, 0)$ of Exact Categories $\mathscr{E}$

For exact categories $\mathscr{E}$, we compare the classical Grothendieck group $K_0(\mathscr{E})$ and $\pi_1(\mathbb{BQ}\mathscr{E}, 0)$. First, we define the Grothendieck group of an exact category.

Definition 3.2.1. Suppose $\mathscr{E}$ is a small exact category. Let $\mathbb{Z}(\mathscr{E})$ be the free abelian group generated by the objects of $\mathscr{E}$. Let $\mathscr{R}(\mathscr{E})$ be the subgroup of $\mathbb{Z}(\mathscr{E})$ generated by the set

$$\{Y - X - Z \in \mathbb{Z}(\mathscr{E}) : X \hookrightarrow Y \twoheadrightarrow Z \quad \text{is a conflation.}\}$$

$$\text{Define the Grothendieck group} \quad K_0(\mathscr{E}) := \frac{\mathbb{Z}(\mathscr{E})}{\mathscr{R}(\mathscr{E})}$$

For $X \in Obj(\mathscr{E})$, its image in $K_0(\mathscr{E})$ is denoted by $[X]$.

We remark:

1. The conflation $0 \hookrightarrow 0 \twoheadrightarrow 0$ implies $0 = [0] \in K_0(\mathscr{E})$.
2. Suppose $\pi : X \xrightarrow{\sim} Y$ is an isomorphism. Then $0 \hookrightarrow X \overset{\pi}{\twoheadrightarrow} Y$ is a conflation. Hence, $[X] = [Y] \in K_0(\mathscr{E})$.
3. The map $\iota : Obj(\mathscr{E}) \longrightarrow K_0(\mathscr{E})$ has the following universal property: Suppose G is an abelian group and $\beta : Obj(\mathscr{E}) \longrightarrow G$ is a set theoretic map such that

$$\beta(Y) = \beta(X)\beta(Z) \quad \forall \text{ conflations} \quad X \hookrightarrow Y \twoheadrightarrow Z,$$

then there is a unique group homomorphism

$$\psi : K_0(\mathscr{E}) \longrightarrow G \qquad \ni \qquad \begin{array}{ccc} Obj(\mathscr{E}) & \xrightarrow{\iota} & K_0(\mathscr{E}) \\ & \searrow^{\beta} & \downarrow \psi \\ & & G \end{array} \quad \text{commutes.}$$

We say that $K_0(\mathscr{E})$ has the above universal property in the category **Ab** of abelian groups. We will see next that $K_0(\mathscr{E})$ has the same universal property in the category **Gr** of groups.
4. The above definition of $K_0(\mathscr{E})$ makes sense whenever $\mathscr{E}$ has a set of isomorphism classes of objects, by replacing $\mathscr{E}$ by an equivalent small exact subcategory $\mathscr{E}' \subseteq \mathscr{E}$. In this case, let E be the set

of all isomorphism classes of objects in $\mathscr{E}$, and $\mathbb{Z}(E)$ be the free abelian group generated by E. Let $\mathscr{R}(E) \subset \mathbb{Z}(E)$ be defined as above. Then

$$K_0(\mathscr{E}) = \frac{\mathbb{Z}(E)}{\mathscr{R}(E)}.$$

Lemma 3.2.2. *Let $\mathscr{E}$ be a small exact category. Then the map $\iota : Obj(\mathscr{E}) \longrightarrow K_0(\mathscr{E})$ has the following universal property: Given a group G (possibly non-commutative) and a set theoretic map $\zeta : Obj(\mathscr{E}) \longrightarrow G$ such that*

$$\zeta(Y) = \zeta(X)\zeta(Z) \quad \forall \text{ conflations} \quad X \hookrightarrow Y \twoheadrightarrow Z,$$

then there is a unique group homomorphism

$$\psi : K_0(\mathscr{E}) \longrightarrow G \qquad \ni \qquad \begin{array}{ccc} Obj(\mathscr{E}) & \xrightarrow{\iota} & K_0(\mathscr{E}) \\ & \searrow^{\zeta} & \downarrow{\scriptstyle\psi} \\ & & G \end{array} \quad \text{commutes.}$$

$$\text{In particular,} \qquad K_0(\mathscr{E}) := \frac{\mathbb{Z}(\mathscr{E})}{\mathscr{R}(\mathscr{E})} \cong \frac{\mathscr{F}(\mathscr{E})}{\mathscr{N}(\mathscr{E})}$$

where $\mathscr{F}(\mathscr{E})$ is the free (non-abelian) group generated by $Obj(\mathscr{E})$, and $\mathscr{N}(\mathscr{E})$ is the normal subgroup generated by the set

$$W = \left\{Y(XZ)^{-1} : X \hookrightarrow Y \twoheadrightarrow X \quad \text{is a conflation}\right\}.$$

Proof. Write $\mathscr{G}(\mathscr{E}) := \frac{\mathscr{F}(\mathscr{E})}{\mathscr{N}(\mathscr{E})}$. (*Recall elements of $\mathscr{F}(\mathscr{E})$ are "words".*) Note that the map $\eta : Obj(\mathscr{E}) \longrightarrow \mathscr{G}(\mathscr{E})$ has the universal property in the category **Gr** of groups. For $X, Y \in Obj(\mathscr{E})$, using the split conflation, it follows $\eta(X)\eta(Y) = \eta(X \oplus Y) = \eta(Y \oplus X) = \eta(Y)\eta(X)$. Form this, it follows, $\mathscr{G}(\mathscr{E})$ is abelian. Therefore, both $\mathscr{G}(\mathscr{E})$ and $K_0(\mathscr{E})$ have the same universal property in the category **Ab** of abelian groups. Hence, $\mathscr{G}(\mathscr{E}) \cong K_0(\mathscr{E})$. ■

The following proposition establishes the key connection between classical K-theory to homotopy groups of $\mathbb{BQ}\mathscr{E}$.

Proposition 3.2.3. *Let $\mathscr{E}$ be a small exact category. The classifying space $\mathbb{BQ}\mathscr{E}$ is considered as a pointed topological space, the base point being a fixed zero object 0 of $\mathscr{E}$. Then*

1. *The classifying space $\mathbb{BQ}\mathscr{E}$ is (path) connected. So, $\pi_0(\mathbb{BQ}\mathscr{E}) = 0$.*
2. *There is an isomorphism $K_0(\mathscr{E}) \xrightarrow{\sim} \pi_1(\mathbb{BQ}\mathscr{E}, 0)$.*

Proof. The following proof is drawn from [Sm1], which is more elaborate than the one in [Q]. Suppose X is an object of $\mathscr{E}$ or $\mathbb{Q}\mathscr{E}$. The diagram

$$0 \twoheadleftarrow 0 \underset{0_X}{\hookrightarrow} X \quad \text{represents a map} \quad 0 \longrightarrow X \quad \text{in} \quad \mathbb{Q}\mathscr{E}.$$

This gives a path from 0 to X. Since all the vertices of $\mathbb{BQ}\mathscr{E}$ are path connected to 0, $\mathbb{Q}\mathscr{E}$ is path connected. Hence, $\pi_0(\mathbb{BQ}\mathscr{E}) = 0$. (*We comment that $Mor_{\mathbb{Q}(\mathscr{E})}(0, X)$ is not necessarily a singleton.*). With notation, as Lemma 3.2.2,

$$K_0(\mathscr{E}) = \frac{\mathscr{F}(\mathscr{E})}{\mathscr{N}(\mathscr{E})}.$$

Define a group homomorphism $\varphi_0 : \mathscr{F}(\mathscr{E}) \longrightarrow \pi_1(\mathbb{BQ}\mathscr{E})$, as follows. For $f \in Mor_{\mathbb{Q}\mathscr{E}}(X, Y)$, the corresponding path from X to Y, in the classifying space $\mathbb{B}(\mathbb{Q}\mathscr{E})$, is denoted by $\gamma(f)$. For $X \in Obj(\mathscr{E})$, there are two natural arrows in $Mor_{\mathbb{Q}\mathscr{E}}(0, X)$, as follows:

$$(0, 0, 0_X) = \left(0 \twoheadleftarrow 0 \xhookrightarrow{0_X} X\right), \quad (X, 0, 1_X) = \left(0 \twoheadleftarrow X \xhookrightarrow{1_X} X\right).$$

By (2.1.3), these two arrows define two paths $\gamma_0^X := \gamma((0, 0, 0_X)$, $\gamma_1^X := \gamma((X, 0, 1_X)$ in $\mathbb{BQ}\mathscr{E}$, from 0 to X. So, $\ell_X := \overline{\gamma_0^X}\gamma_1^X$ is a loop at 0, where $\overline{\gamma_0^X}$ denotes the path opposite to γ_0^X. Define

$$\varphi_0 : \mathscr{F}(\mathscr{E}) \longrightarrow \pi_1(\mathbb{BQ}\mathscr{E}) \quad \text{by } \varphi(X) = [\ell_X]$$

Now, consider a conflation

$$X \xhookrightarrow{i} Y \xrightarrow{p} Z \quad \text{which corresponds to} \quad Y(XZ)^{-1} \in \mathscr{N}(\mathscr{E}).$$

$$\text{Now, } \ell_X = \overline{\gamma_0^X}\gamma_1^X : 0 \underset{\gamma_1^X}{\overset{\gamma_0^X}{\rightrightarrows}} X \text{ is homotopic to}$$

$$\overline{(\gamma(X, 1_X, i)\gamma_0^X)}\,(\gamma(X, 1_X, i)\gamma_1^X) : 0 \underset{\gamma_1^X}{\overset{\gamma_0^X}{\rightrightarrows}} X \xrightarrow{\gamma(X, 1_X, i)} Y.$$

It follows, $(X, 1_X, i)(0, 0, 0_X) = (0, 0, 0_Y) \in Mor_{\mathbb{Q}\mathscr{E}}(0, Y)$ and $(X, 1_X, i)(0, 0, 1_X) = (X, 0, i) \in Mor_{\mathbb{Q}\mathscr{E}}(0, Y)$. So, $[\ell_X] =$

$[\overline{\gamma}_0^Y\gamma(X,0,i)]$. Likewise, $[\ell_Z] = [\overline{\gamma(X,0,i)}\gamma_1^Y]$. So, $[\ell_X][\ell_Z] = [\ell_Y]$. Therefore, φ_0 factors through a well-defined homomorphism $\varphi : K_0(\mathscr{E}) \longrightarrow \pi_1(\mathbb{B}\mathbb{Q}\mathscr{E}, 0)$.

Now, we prove that φ is surjective. Let $\ell \in \pi_1(\mathbb{B}\mathbb{Q}\mathscr{E}, 0)$. By (C.3.9), the map $\pi_1(\mathbb{B}\mathbb{Q}\mathscr{E}^1, 0) \twoheadrightarrow \pi_1(\mathbb{B}\mathbb{Q}\mathscr{E}, 0)$ is surjective, where $\mathbb{B}\mathbb{Q}\mathscr{E}^1$ denotes the 1-skeleton. Therefore, ℓ is represented by a product of paths $\gamma_n\gamma_{n-1}\cdots\gamma_1$ such that

1. γ_i is a path from X_{i-1} to X_i, with $X_i \in Obj(\mathscr{E})$, and $X_0 = X_n = 0$.
2. Further,
   (a) either $\gamma_i = \gamma(f_i)$, in which case $f_i \in Mor_{\mathbb{Q}\mathscr{E}}(X_{i-1}, X_i)$,
   (b) Or $\gamma_i = \overline{\gamma}(f_i)$, in which case $f_i \in Mor_{\mathbb{Q}\mathscr{E}}(X_i, X_{i-1})$.
3. In fact, for any nonzero object X in $\mathscr{E}$, $Mor_{\mathbb{Q}\mathscr{E}}(X, 0) = \phi$. So, we can assume $f_1 : 0 \longrightarrow X_1$, and $f_n : 0 \longrightarrow X_{n-1}$.

Let $g_i = \gamma(X_i, 0, 1_{X_i})$. Up to homotopy, we have

$$\gamma_n\gamma_{n-1}\cdots\gamma_i\cdots\gamma_1 = (\gamma_n g_{n-1})\cdots(\overline{g}_i\gamma_i g_{i-1})\cdots(\overline{g}_2\gamma_2 g_1)(\overline{g}_1\gamma_1).$$

We will prove that, each $[(\overline{g}_i\gamma_i g_{i-1})]$ is in the image of φ. There are two cases,

1. $\gamma_i = \gamma(f_i)$, where $f_i : X_{i-1} \longrightarrow X_i$. So, f_i is represented by

$$X_{i-1} \xleftarrow[p]{} U_i \xhookrightarrow[\iota]{} X_i.$$

We have $g_{i-1} = \gamma(X_{i-1}, 0, 1_{X_{i-1}})$, given by the arrow

$$0 \twoheadleftarrow X_{i-1} \xrightarrow{1} X_{i-1}.$$

So, $\gamma_i g_{i-1}$ is given by the arrow

$$0 \twoheadleftarrow U_i \xhookrightarrow[\iota]{} X_i.$$

Therefore,

$$\begin{aligned}
\overline{g}_i\gamma_i g_{i-1} &= \overline{\gamma}(X_i, 0, 1_{X_i})\gamma(U_i, 0, \iota) \\
&\sim \overline{\gamma}(X_i, 0, 1_{X_i})\gamma\left((U_i, 1_{U_i}, \iota)o(U_i, 0, 1_{U_i})\right) \\
&\sim \overline{\gamma}(X_i, 0, 1_{X_i})\gamma(U_i, 1_{U_i}, \iota)\gamma(U_i, 0, 1_{U_i}) \\
&\sim \overline{\gamma}(X_i, 0, 1_{X_i})\gamma(U_i, 1_{U_i}, \iota)\gamma(0, 0, 0_{U_i})\overline{\gamma}(0, 0, 0_{U_i})\gamma(U_i, 0, 1_{U_i}) \\
&\sim \overline{\gamma}(X_i, 0, 1_{X_i})\gamma\left((U_i, 1_{U_i}, \iota)o(0, 0, 0_{U_i})\right)\gamma(\ell_{U_i}) \\
&\sim \overline{\gamma}(X_i, 0, 1_{X_i})\gamma(X_i, 0, 0_{X_i})\gamma(\ell_{U_i}) \sim \overline{\gamma(\ell_{X_i})}\gamma(\ell_{U_i})
\end{aligned}$$

So,

$$[(\overline{g}_i \gamma_i g_{i-1})] = \varphi(X_i)^{-1}\varphi(U_i)$$

2. $\gamma_i = \gamma(f_i)^{-1}$, where $f_i : X_i \longrightarrow X_{i-1}$. This case is dealt similarly, as above.

This establishes that φ is surjective. To prove that φ is injective, we define a map $\pi_1(\mathbb{B}\mathbb{Q}\mathscr{E}, 0) \longrightarrow K_0(\mathscr{E})$. As in (2.1.5), let $\underline{K_0(\mathscr{E})}$ denote the category with one object $\star$ and $Mor(\star,\star) = \underline{K_0(\mathscr{E})}$. Define a functor

$$\Phi : \mathbb{Q}\mathscr{E} \longrightarrow \underline{K_0(\mathscr{E})} \qquad \forall\, X \in Obj(\mathbb{Q}\mathscr{E}) \quad \Phi(X) = \star \quad \text{and,}$$

$$\text{for } f := \left(X \overset{p}{\twoheadleftarrow} W \overset{\iota}{\hookrightarrow} Y \right) \in Mor_{\mathbb{Q}\mathscr{E}}(X,Y), \quad \Phi(f) = [\ker(p)].$$

We check that Φ is a functor. For another morphism

$$g := \left(Y \overset{q}{\twoheadleftarrow} V \overset{\iota}{\hookrightarrow} Z \right) \in Mor_{\mathbb{Q}\mathscr{E}}(Y,Z)$$

we need to prove $\Phi(gf) = [\ker(p)] + [\ker(q)]$.

This follows, because $\ker(q) \hookrightarrow \ker(qp') \twoheadrightarrow \ker(p)$ is a conflation. Therefore, Φ is a functor, which induces a map of the classifying spaces $\mathbb{B}\mathbb{Q}\mathscr{E} \longrightarrow \mathbb{B}\underline{K_0(\mathscr{E})}$. Hence, it induces a map

$$\eta := \pi_1(\mathbb{B}\Phi) : \pi_1(\mathbb{B}\mathbb{Q}\mathscr{E}, 0) \longrightarrow \pi_1\left(\underline{K_0(\mathscr{E})}, \star\right) = K_0(\mathscr{E}).$$

Now, for $[X] \in K_0(\mathscr{E})$,

$$\eta(\varphi([X])) = \eta(\gamma(0,0,X)^{-1}\gamma(X,0,1_X)) = -[0] + [X] = [X].$$

Therefore, φ is a bijection. The proof is complete. ■

3.3. Higher K-Groups of Exact categories

Now we define higher K-groups.

Definition 3.3.1. Suppose $\mathscr{E}$ is a small exact category and 0 denotes the zero object.

$$\forall\ n \subset \mathbb{N} \quad \text{define } K_n(\mathscr{E}) = \pi_{n+1}(\mathbb{BQ}\mathscr{E}, 0)$$

The definition is independent of the choice of the zero object, because $\mathbb{BQ}\mathscr{E}$ is path connected. Further, the definition extends to any exact category $\mathscr{E}$ that has a set of isomorphism classes of objects.

Lemma 3.3.2. *Let $F : \mathscr{E} \longrightarrow \mathscr{D}$ be an exact functor of exact categories. Then F induces a functor $\mathbb{Q}F : \mathbb{Q}\mathscr{E} \longrightarrow \mathbb{Q}\mathscr{D}$. The map $\mathbb{Q}F$ sends an object $X \in \mathscr{E}$ to FX and a morphism*

$$f := \begin{array}{ccc} W & \hookrightarrow & Y \\ \downarrow & & \\ X & & \end{array} \in Mor_{\mathbb{Q}\mathscr{E}}(X, Y) \quad \mapsto \quad \begin{array}{ccc} FW & \hookrightarrow & FY \\ \downarrow & & \\ FX & & \end{array} \in Mor_{\mathbb{Q}\mathscr{D}}(FX, FY)$$

Consequently, a continuous map $\mathbb{BQ}F : \mathbb{BQ}\mathscr{E} \longrightarrow \mathbb{BQ}\mathscr{D}$ of pointed spaces is obtained. This induces homomorphisms of K groups

$$F_n := K_n(F) : K_n(\mathscr{E}) \longrightarrow K_n(\mathscr{D}) \qquad \forall \quad n \geq 0.$$

The following is a list of immediate consequences of (3.3.2).

1. *For each (fixed) $n \geq 0$, the association*

$$n \mapsto K_n(\mathscr{E})$$

is a functor from the category of small exact categories and exact functors to the category **Ab** *of abelian groups.*

2. *If $F : \mathscr{E} \xrightarrow{\sim} \mathscr{D}$ is an equivalence of exact categories, then F induces a natural equivalence $\mathbb{Q}\mathscr{E} \longrightarrow \mathbb{Q}\mathscr{D}$. By* (2.1.9)*, the induced maps $F_n : K_n(\mathscr{E}) \xrightarrow{\sim} K_n(\mathscr{D})$ are isomorphisms, $\forall\ n \geq 0$.*

   Further, suppose $F, G : \mathscr{E} \longrightarrow \mathscr{D}$ are two exact functors of exact categories. Let $\theta : F \longrightarrow G$ be a natural equivalence. Then θ induces a natural transformation (equivalence) $\mathbb{Q}F \longrightarrow \mathbb{Q}G$. Consequently, by (2.1.9)*, θ induces a homotopy between the maps $\mathbb{BQ}F$ and $\mathbb{BQ}G$. Therefore, $K_n(F) = K_n(G) : K_n(\mathscr{E}) \longrightarrow K_n(\mathscr{D})$ for all $n \geq 0$.* (For two functors $F, G : \mathscr{E} \longrightarrow \mathscr{D}$ of exact categories, a natural transformation $\theta : F \longrightarrow G$ need not induce a natural transformation from $\mathbb{Q}F$ to $\mathbb{Q}G$. So, they need not induce homotopic maps.).

3. *For any exact category $\mathscr{E}$, by* (3.1.5), *we have*

$$\forall n \geq 0 \quad K_n(\mathscr{E}^o) = \pi_{n+1}(\mathbb{B}\mathbb{Q}(\mathscr{E}^o)) = \pi_{n+1}(\mathbb{B}(\mathbb{Q}\mathscr{E})) = K_n(\mathscr{E}).$$

4. *Suppose $\mathscr{E}, \mathscr{D}$ are two exact categories. Then $\mathscr{E} \times \mathscr{D}$ is also an exact category. It follows $\mathbb{Q}(\mathscr{E} \times \mathscr{D}) = \mathbb{Q}(\mathscr{E}) \times \mathbb{Q}(\mathscr{D})$. The natural bijective map $\iota : \mathbb{B}(\mathbb{Q}(\mathscr{E} \times \mathscr{D})) \longrightarrow \mathbb{B}(\mathbb{Q}(\mathscr{E})) \times \mathbb{B}(\mathbb{Q}(\mathscr{D}))$ is continuous and is not necessarily a homeomorphism (see* (2.1.8)). *However, when the right side is endowed with compactly generated topology (see* (2.1.8)), *then ι is indeed a homeomorphism. Fix $n \geq 0$ and consider the commutative diagram*

$$\begin{array}{ccc}
\pi_n\left(\mathbb{B}\mathbb{Q}(\mathscr{E} \times \mathscr{D}), 0\right) & \xrightarrow{\ \sim\ } & \pi_n\left((\mathbb{B}\mathbb{Q}(\mathscr{E}) \times (\mathbb{B}\mathbb{Q}\mathscr{D}), (0,0)\right) \\
& \searrow{\scriptstyle (\text{dashed})} & \downarrow \wr \\
& & \pi_n\left(\mathbb{B}\mathbb{Q}(\mathscr{E}, 0\right) \oplus \pi_n\left(\mathbb{B}\mathbb{Q}\mathscr{D}, 0\right)
\end{array}$$

Here the horizontal map is induced by ι. Since homotopy groups are determined by the compactly generated topology, the horizontal map is an isomorphism. Further, the vertical map is induced by the two projection maps, which is also an isomorphism, by usual topological arguments. So, it follows, $\forall\ n \geq 0$ the maps

$$K_n(\mathscr{E} \times \mathscr{D}) \xrightarrow{\ \sim\ } K_n(\mathscr{E}) \oplus K_n(\mathscr{D}) \quad \text{sending } x \mapsto (p_{\mathscr{E}})_*(x) + (p_{\mathscr{D}})_*(x)$$

are isomorphisms, where $p_{\mathscr{E}}$, $p_{\mathscr{D}}$ denote the projection functors.

5. *Suppose $\mathscr{E}$ is an exact category. Then the direct sum functor $\mathscr{E} \times \mathscr{E} \longrightarrow \mathscr{E}$, sending $(M, N) \mapsto M \oplus N$, is an exact functor. This induces a group homomorphism*

$$\oplus_* : K_n(\mathscr{E}) \oplus K_n(\mathscr{E}) \longrightarrow K_n(\mathscr{E}) \qquad \forall\ n \geq 0.$$

The map coincides with the direct sum map because of the commutative diagram

$$\begin{array}{ccc}
K_n(\mathscr{E}) & & \\
{\scriptstyle i_1} \downarrow & \searrow^{1} & \\
K_n(\mathscr{E}) \oplus K_i(\mathscr{E}) & \xrightarrow[\oplus_*]{} & K_n(\mathscr{E}) \\
{\scriptstyle i_2} \uparrow & \nearrow_{1} & \\
K_n(\mathscr{E}) & &
\end{array}$$

where i_1, i_2 are induced, respectively, by the functors $M \mapsto (M, 0)$ and $M \mapsto (0, M)$.

Lemma 3.3.3. *Suppose I is a small filtering category and $j \mapsto \mathscr{E}_j$ is a functor from I to the category of small exact categories. Let $\lim_{j\in I} \mathscr{E}_j$ denote the filtered limit, which is an exact category. Then $\mathbb{Q}(\lim_{j\in I} \mathscr{E}_j) = \lim_{j\in I} (\mathbb{Q}(\mathscr{E}_j))$. Hence, by (2.1.16),*

$$K_n\left(\lim_{j\in I} \mathscr{E}_j\right) = \lim_{j\in I} K_n(\mathscr{E}_j) \quad \forall\ n \geq 0.$$

Proof. It is obvious that $\mathbb{Q}(\lim_{j\in I} \mathscr{E}_j) = \lim_{j\in I} (\mathbb{Q}(\mathscr{E}_j))$ is an exact category. The lemma is an immediate consequence of (2.1.16). ■

Example 3.3.4. Let $\mathscr{P}(A)$ denote the category of all finitely generated projective (left) A-modules, where A is a ring with unity. Write

$$K_n(A) := K_n(\mathscr{P}(A)) \quad \forall\ n \geq 0$$

1. Any homomorphism $A \longrightarrow B$ of rings induces an exact functor $\mathscr{P}(A) \longrightarrow \mathscr{P}(B)$, sending $P \mapsto B \otimes P$. This induces a well-defined group homomorphism
$$K_n(A) \longrightarrow K_n(B) \quad \forall\ n \geq 0.$$
So, $\forall n \geq 0$, the association $A \mapsto K_n(A)$ is a covariant functor from the category of rings with unity to the category of abelian groups.
2. It follows,
$$\forall\ n \geq 0 \qquad K_n(A \times B) \cong K_n(A) \oplus K_n(B)$$
are naturally isomorphic.
3. Suppose I is a small filtering category and $j \mapsto A_j$ is a functor from I to the category of rings with unity. It follows, from (3.3.3):
$$\forall\ n \geq 0 \qquad K_n\left(\lim_{j\in I} A_j\right) = \lim_{j\in I} K_n(A_j).$$

Example 3.3.5. Let X be a (noetherian) scheme. (For example $X = \operatorname{Spec}(A)$.)

1. The category $QCoh(X)$ of the quasi-coherent modules over X is an exact category. So, this gives rise to a K-theory.
2. The category $Coh(X)$ of the coherent modules over X is also an exact category. So, this gives rise to a K-theory. This is often referred to as G-theory of X, and the K-groups are denoted by $G_n(X)$.
3. Likewise, the category $\mathscr{P}(X)$ of locally free modules over X is an exact category (as in 3.3.4) . So, this gives rise to a K-theory. This is usually referred to as K-theory of X, and the K-groups are denoted by $K_n(X)$.

A comparison of the K-theory of these categories constitutes a big chunk of the literature. The classical K-theory mostly refers to the K-theory of the category $\mathscr{P}(A)$ of finitely generated projective A-modules, where A is a (commutative) ring (3.3.4). Classically, three groups were defined, which we (temporarily) denote by $K_0^c(A)$, $K_1^c(A)$, $K_2^c(A)$. It seems odd to use superscript c to denote classical K-groups, because it predates everything we have discussed so far. However, they all coincide with Quillen K-groups. We have already established that $K_0^c(A)$ agrees with Quillen $K_0(A)$ (3.2.3). The coincidence of the other two is dealt with in Chapter 4. However, we state them here.

Lemma 3.3.6. *Let A be a (commutative) ring. Refer to Sections* 4.7.1 *and* 4.7.2 *for further details and notation. The classical K_1-group of A is defined as follows* (*see* 4.7)*:*

$$K_1^c(A) = \frac{GL(A)}{EL(A)}. \quad \textit{We have} \quad K_1^c(A) \simeq K_1(\mathscr{P}(A)).$$

Proof. see (4.7.6). ■

Lemma 3.3.7. *Let A be a (commutative) ring. The classical K_2-group of A, denoted temporarily by $K_2^c(A)$, is defined as a subgroup of the so-called "Steinberg Group" $\mathrm{St}(A)$, given in* [Mi, §5], *and discussed in Section* 4.7.3. *We have,*

$$K_2^c(A) \simeq K_2(\mathscr{P}(A))$$

Proof. See Section 4.7.3. ■

3.4. Exact Sequences and Filtrations

Let $\mathscr{E}$ be an exact category, and let $\varepsilon(\mathscr{E})$ be the category of the exact sequences in $\mathscr{E}$. More precisely, objects X, Y of $\varepsilon(\mathscr{E})$ are the exact sequences (i.e. *conflations*) in $\mathscr{E}$, and a morphism $f : X \longrightarrow Y$ is a commutative diagram in $\mathscr{E}$:

$$\begin{array}{ccccccccc} X := & 0 \longrightarrow & K & \hookrightarrow & M & \twoheadrightarrow & C & \longrightarrow 0 \\ f\downarrow & & \downarrow\iota & & \downarrow\beta & & \downarrow\varphi & \\ Y := & 0 \longrightarrow & K' & \hookrightarrow & M' & \twoheadrightarrow & C' & \longrightarrow 0 \end{array} \tag{3.4}$$

1. Given $X \in Obj(\varepsilon(\mathscr{E}))$, as in (3.4), write $s(X) := K$, $t(X) := M$, and $q(X) := C$. Then $s, t, q : \varepsilon(\mathscr{E}) \longrightarrow \mathscr{E}$ define three functors. The images $s(X)$, $t(X)$, $q(X)$ would, respectively, be called sub, total, and quotient objects of X.
2. In fact, $\varepsilon(\mathscr{E})$ has a structure of an exact category, where a sequence $X \longrightarrow Y \longrightarrow Z$ is declared exact (i.e. *conflation*) if the corresponding vertical sequences, as in diagram (3.4), are exact.
3. It follows that $s, t, q : \varepsilon(\mathscr{E}) \longrightarrow \mathscr{E}$ are exact functors.

Theorem 3.4.1. *Let $\mathscr{E}$ be an exact category. Then the functor $(s, q) : \mathbb{Q}(\varepsilon(\mathscr{E})) \longrightarrow \mathbb{Q}(\mathscr{E}) \times \mathbb{Q}(\mathscr{E})$ is a homotopy equivalence.*

Proof. By Theorem A (2.2.2), it would be enough to show that $(s, q)/(K, C)$ is contractible, for any $(K, C) \in Obj\,(\mathbb{Q}(\mathscr{E}) \times \mathbb{Q}(\mathscr{E}))$. Fix such a pair (K, C), and let $\mathscr{C}(K, C) := (s, q)/(K, C)$.

$$\text{So,} \quad Obj(\mathscr{C}(K, C)) = (s, q)/(K, C)$$
$$= \{(X, u, v) : u \in Mor_{\mathbb{Q}\mathscr{E}}(sX, K), v \in Mor_{\mathbb{Q}\mathscr{E}}(qX, C)\}$$

More explicitly,

$$\begin{array}{l} u = \left(sX \twoheadleftarrow U \hookrightarrow K \right), \\ v = \left(qX \twoheadleftarrow V \hookrightarrow C \right) \end{array} \tag{3.5}$$

1. Let $\mathscr{C}'(K, C)$ be the full subcategory of $\mathscr{C}(K, C)$, consisting of objects (X, u, v) such that u is *surjective* (in the sense of Section 3.1), meaning u looks like $u = \left(sX \twoheadleftarrow K \xrightarrow{1} K \right)$.

2. Let $\mathscr{C}"(K, C)$ be the full subcategory of $\mathscr{C}(K, C)$, consisting of objects (X, u, v) such that u is *surjective* and v is *injective* (in the sense of Section 3.1), meaning u, v look like

$$u = \left(sX \twoheadleftarrow K \xrightarrow{1} K \right) \quad v = \left(qX \xleftarrow{1} qX \hookrightarrow C \right).$$

Now the theorem follows from (3.4.2) as follows. The proof is complete. ■

Lemma 3.4.2. *Use the notation, as in the proof of* (3.4.1). *Then the inclusion functors* $\mathscr{C}''(K, C) \longrightarrow \mathscr{C}'(K, C)$ *and* $\mathscr{C}'(K, C) \longrightarrow \mathscr{C}(K, C)$ *have left adjoints. Consequently, both are homotopy equivalences (by* (2.1.10)).

Further, $\mathscr{C}''(K, C)$ *is contractible, and hence so is* $\mathscr{C}(K, C)$.

Proof. We will abbreviate notations $\mathscr{C} := \mathscr{C}(K, C)$, $\mathscr{C}' := \mathscr{C}'(K, C)$, $\mathscr{C}'' := \mathscr{C}''(K, C)$. Given $\mathcal{X} = (X, u, v)$ in $\mathscr{C}$, we will establish that there is a universal arrow $\chi : \mathcal{X} \longrightarrow L(\mathcal{X})$, with $L(\mathcal{X}) \in \mathscr{C}'(K, C)$. This means, given $\mathcal{X}_1 \in \mathscr{C}'$ and $f : \mathcal{X} \longrightarrow \mathcal{X}_1$ in $\mathscr{C}(K, C)$, there is a unique morphism Φ, as in the commutative diagram

$$\begin{array}{ccc} \mathcal{X} & \xrightarrow{\chi} & L(\mathcal{X}) \\ & \searrow^{f} & \Big\downarrow{\scriptstyle \exists\,\Phi\ unique} \\ & & \mathcal{X}_1 \end{array} \tag{3.6}$$

By (A.1.15) this completes the proof that $\mathscr{C}'(K, C) \longrightarrow \mathscr{C}(K, C)$ has a left adjoint. The objects $E \in Obj(\varepsilon(\mathscr{C})$ are denoted by $E = \left(sE \overset{\iota}{\hookrightarrow} tE \overset{\varsigma}{\twoheadrightarrow} qE \right)$.

With $\mathcal{X} = (X, u, v) \in Obj(\mathscr{C})$ and u as in (3.5). Write $X = \left(sX \overset{\iota}{\hookrightarrow} tX \overset{\varsigma}{\twoheadrightarrow} qX \right)$. Rewrite $u = \left(sX \overset{i}{\hookrightarrow} M \overset{\mathfrak{p}}{\twoheadleftarrow} K \right)$ (compare with (3.5)). Consider the pushforward diagram

$$\begin{array}{ccccccc} & & sX & \overset{\iota}{\hookrightarrow} & tX & \overset{\varsigma}{\twoheadrightarrow} & qX \\ & {\scriptstyle u}\swarrow & \Big\downarrow{\scriptstyle i} & & \Big\downarrow{\scriptstyle \beta} & & \Big\| \\ K & \overset{\mathfrak{p}}{\twoheadrightarrow} & M & \underset{\iota'}{\hookrightarrow} & T & \underset{\varsigma'}{\twoheadrightarrow} & qX \end{array} \tag{3.7}$$

where T is the pushforward of i and ι, as shown. Define $\overline{X} = \left(M \overset{\iota'}{\hookrightarrow} T \overset{\zeta'}{\twoheadrightarrow} qX \right)$. We also denote $i_*X := \overline{X}$. Note, $s(\overline{X}) = M$, $t(\overline{X}) = T$, $q(\overline{X}) = qX$. Further, $\mathfrak{p} : K \twoheadrightarrow s(\overline{X})$ induces the (*surjective*) map $\mathfrak{p}^! : M \longrightarrow K$ in $\mathbb{Q}\mathscr{E}$.

Now, define $L(\mathcal{X}) := (\overline{X}, \mathfrak{p}^!, v)$. So, $\overline{\mathcal{X}}$ is in $\mathscr{C}'$. Note $(i, \beta, 1_{qX})_! : X \longrightarrow \overline{X}$ defines an (*injective*) arrow in $\mathbb{Q}\varepsilon(\mathscr{E})$. It follows that $(i, \beta, 1_{qX})_! : \mathcal{X} \longrightarrow L(\mathcal{X})$ is an arrow in $\mathscr{C}(K, C)$. Define $\chi := (i, \beta, 1_{qX})_! \in Mor_{\mathscr{C}(K,C)}(\mathcal{X}, L(\mathcal{X}))$.

Now, let $f : \mathcal{X} = (X, u, v) \longrightarrow \mathcal{X}_1$ be an arrow in $\mathscr{C}(K, C)$, as in (3.6). In particular, $\mathcal{X}_1 = (X_1, u_1, v_1)$, with $u_1 = j^!$, for some deflation $j : K \twoheadrightarrow sX_1$. We split the map f in $\mathbb{Q}\varepsilon\mathscr{E}$ as $f = h^! g_! : X \longrightarrow X_1$, where $h : X \hookrightarrow X_0$ is an inflation and $g : X_1 \twoheadrightarrow X_0$ is a deflation in $\varepsilon\mathscr{E}$. The condition $u = (sf)u_1$ is given by the diagram

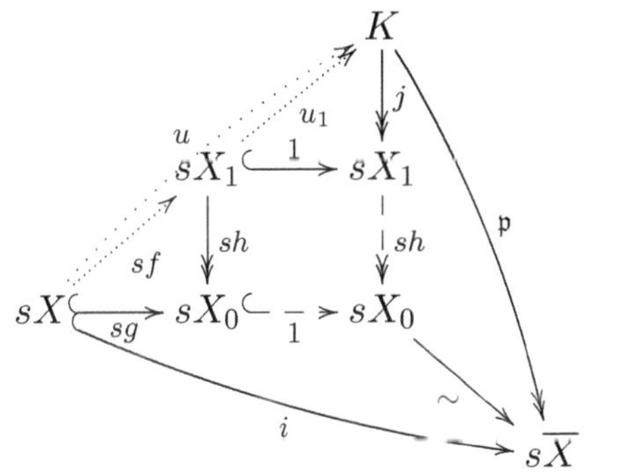

So, we can assume $\begin{cases} sX_0 = s\overline{X} = M \\ i = sg \end{cases}$

So, the splitting of f is displayed in the following commutative diagram:

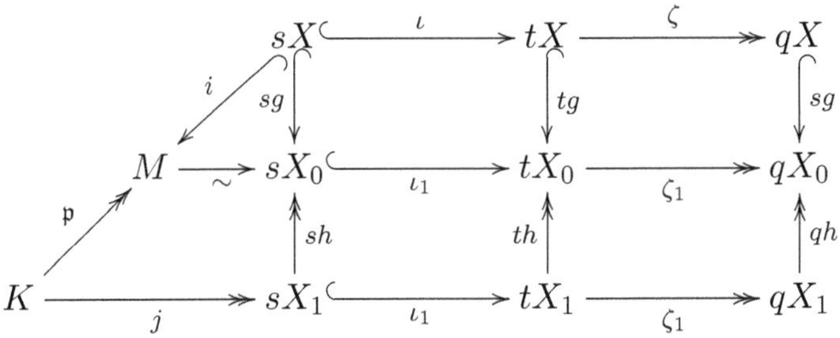

We construct a map $\overline{X} \longrightarrow X_1$, as follows:

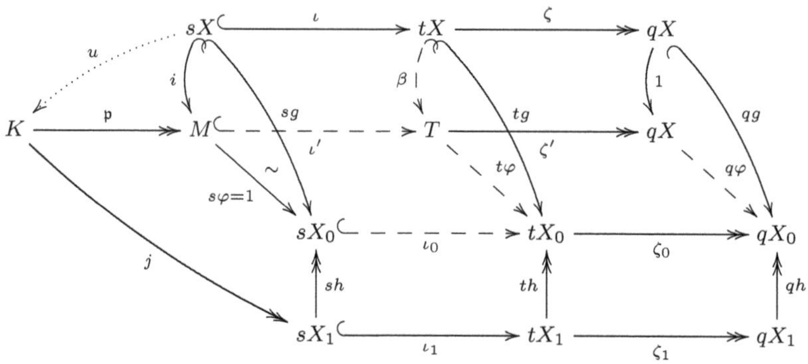

Here $t\varphi$ is defined by the pushforward property. The arrow $\varphi = (s\varphi, t\varphi, q\varphi) : \overline{X} \longrightarrow X_1$ in $\varepsilon\mathscr{E}$ is an inflation (because $s\varphi = 1$, $q\varphi = qg$ are monic). Define $\Phi : \overline{X} \longrightarrow \mathcal{X}_1$ in $\mathbb{Q}\varepsilon\mathscr{E}$ by combing φ and h. Note that $q\varphi = qg$ and hence $q\Phi = qf$. It follows that $\Phi : L(X) = (\overline{X}, \mathfrak{p}^!, v) \longrightarrow (X_1, j^!, v_1)$ and the diagram (3.6) commutes.

To see uniqueness, let $\Psi : L(X) \longrightarrow \mathcal{X}_1$ be another arrow in $\mathbb{Q}\varepsilon\mathscr{E}$ commuting the diagram (3.6). Write $\Psi = \gamma^! \alpha_! := \left(\overline{X} \overset{\alpha}{\hookrightarrow} Y \overset{\gamma}{\twoheadleftarrow} X_1\right)$. We have $f = \Phi\chi = \Psi\chi$, which is $h^! g_! \chi = \gamma^! \alpha_! \chi$. This equality is given by the isomorphism τ, as in the (partly) commutative diagram

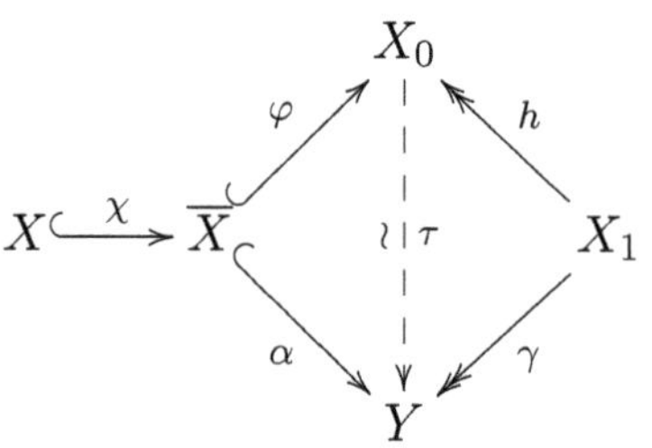

$$\ni \quad \gamma = \tau h, \quad \alpha\chi = \tau\varphi\chi$$

It remains to be shown that $\alpha = \tau\varphi$. Since $q\chi = 1$, it follows that $q\alpha = q(\tau\varphi)$. Since $\Psi : L(X) \longrightarrow \mathcal{X}_1$, we have $j^! s\Psi = \mathfrak{p}^!$. The following diagram shows $sY \cong M$.

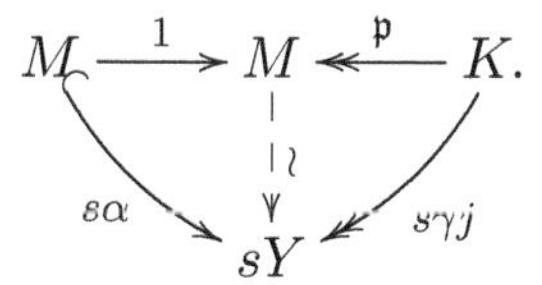

So, we can assume $sY = M = sX_0, s\alpha = 1, \mathfrak{p} = s\gamma j$.

We expand the above diagram as follows:

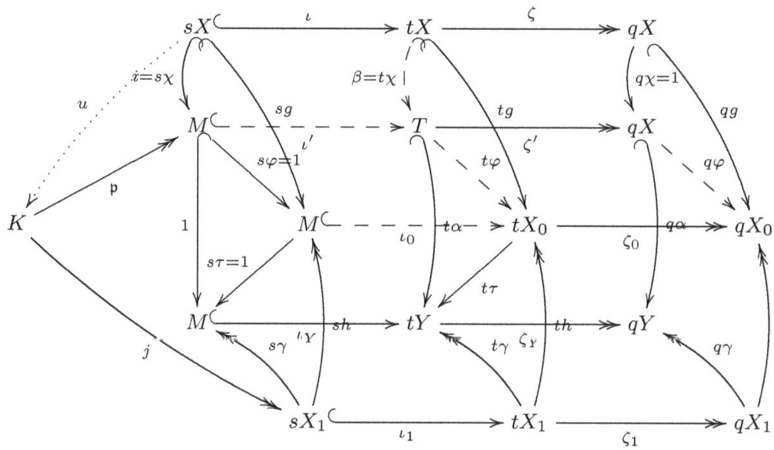

Note that $\mathfrak{p}$ is an epimorphism. We have

$$s(\tau h) = s\gamma \Longrightarrow s(\tau h)j = s\gamma j = \mathfrak{p} \Longrightarrow s\tau((sh)j) = s\gamma j = \mathfrak{p} \Longrightarrow s\tau\mathfrak{p} = \mathfrak{p} \Longrightarrow s\tau = 1.$$

Further,

$$(t\tau)(t\varphi)\iota' = (t\tau)\iota_0 = \iota_Y = (t\alpha)\iota', \quad \text{Also, } \tau\varphi\chi = \alpha\chi \Longrightarrow t(\tau\varphi\chi) = t(\alpha\chi).$$

Since T is a pushforward, we have $t(\tau\varphi) = t\alpha$. So, we have established that $\tau\varphi = \alpha$. This completes the uniqueness of Φ.

Now, we prove that $\mathscr{C}'' \longrightarrow \mathscr{C}'$ has a left adjoint. Let $\mathcal{X} = (X, u, v)$ be an object in $\mathscr{C}'$, where $u = p^! : sX \longrightarrow K$ is *surjective* and $v : qX \longrightarrow C$ is a morphism in $\mathbb{Q}\mathscr{E}$, which is given by $v := \left(qX \overset{j}{\twoheadleftarrow} M \overset{i}{\hookrightarrow} C \right)$. Consider the diagram

$$\begin{array}{cccccccccc} \overline{X} := & 0 & \longrightarrow & sX & \hookrightarrow & T & \longrightarrow & M & \longrightarrow & 0 \\ & & & \| & & \downarrow & & \downarrow j & & \\ X = & 0 & \longrightarrow & sX & \hookrightarrow & tX & \longrightarrow & qX & \longrightarrow & 0 \end{array}$$

where T is the pullback. We also write $j^*X := \overline{X}$. By a dual argument, similar to the above, it is established that there is a universal arrow $\mathcal{X} \longrightarrow L(\mathcal{X})$, with $L(\mathcal{X}) := (j^*X, u, i_!) \in Obj(\mathscr{C}'')$. This completes the proof that $\mathscr{C}'' \hookrightarrow \mathscr{C}'$ has a left adjoint and hence is a homotopy equivalence.

To prove $\mathscr{C}''$ is contractible, let $\mathbf{0} \in Obj\,(\varepsilon(\mathscr{E}))$ denote the zero conflation. Corresponding to the arrow $K \twoheadrightarrow 0$, let $\mathfrak{p}_K \in Mor_{\mathbb{Q}\mathscr{E}}\,(0, K)$ denote the *surjective* map, and corresponding to the $0 \hookrightarrow C$, let $\mathfrak{i}_K \in Mor_{\mathbb{Q}\mathscr{E}}\,(C, 0)$ denote the *injective* map. Consider the object $\mathcal{I} = (\mathbf{0}, \mathfrak{p}_K, \mathfrak{i}_C) \in \mathscr{C}''$. It is clear the $\mathcal{I}$ is an initial object in $\mathscr{C}''$. So, $\mathscr{C}''$ is contractible. The proof is complete. ■

3.4.1. *Additivity theorem*

In this section, we give several applications of Theorem 3.4.1. We define exact sequences of exact functors and admissible filtration of exact functors.

Definition 3.4.3. Let $\mathscr{E}, \mathscr{D}$ be two exact categories (small).

1. Let $\mathscr{C} = Fun(\mathscr{E}, \mathscr{D})$ be the category of all exact functors whose objects are exact functors and a morphism $\Phi : F \longrightarrow G$ is a natural transformation.
2. In fact, $\mathscr{C} = Fun(\mathscr{E}, \mathscr{D})$ is an exact category, where a sequence
$$0 \longrightarrow \Phi \longrightarrow \Psi \longrightarrow \Theta \longrightarrow 0$$
is declared exact (i.e. a conflation), if
$$\forall\ X \in Obj(\mathscr{E}) \quad 0 \longrightarrow \Phi(X) \longrightarrow \Psi(X) \longrightarrow \Theta(X) \longrightarrow 0 \quad \text{is exact.}$$
3. A sequence of natural transformations (morphisms), in $\mathscr{C}$,
$$0 = \Phi_0 \hookrightarrow \Phi_1 \hookrightarrow \Phi_2 \hookrightarrow \cdots \qquad \Phi_r = \Phi$$
is defined to be an **admissible filtration** of Φ if
$$\forall\ X \in Obj(\mathscr{E}),\ \forall\ i = 0, \ldots, r-1 \quad \Phi_i(X) \hookrightarrow \Phi_{i+1}(X) \quad \text{are inflations in } \mathscr{D}.$$

4. Given an admissible filtration, as above, the quotient morphisms $\frac{\Phi_{i+1}}{\Phi_i}$ are defined, and determined up to isomorphisms. There is no reason why such a quotient would be exact. However, if $\frac{\Phi_{i+1}}{\Phi_i}$ is exact $\forall\ i = 0, 1, \ldots, r-1$, then $\frac{\Phi_p}{\Phi_q}$ is exact, for all $0 \leq q \leq p \leq r$.

The following is known as the Additivity theorem.

Theorem 3.4.4. *Let $\mathscr{E}, \mathscr{D}$ be two exact categories. Let $G, F, H : \mathscr{E} \longrightarrow \mathscr{D}$ be three exact functors, such that*

$$0 \longrightarrow G \longrightarrow F \longrightarrow H \longrightarrow 0 \quad \textit{is also exact.} \tag{3.8}$$

$$\textit{Then} \qquad \forall\ n \geq 0 \quad F_* = G_* + H_* : K_n(\mathscr{E}) \longrightarrow K_n(\mathscr{D}).$$

Proof. As before, let $\varepsilon(\mathscr{D})$ denote the exact category of short exact sequences in $\mathscr{D}$. Let $s, t, q : \varepsilon(\mathscr{D}) \longrightarrow \mathscr{D}$ be the sub, total, and quotient functors. Then

$$0 \longrightarrow s \longrightarrow t \longrightarrow q \longrightarrow 0 \quad \text{is exact.}$$

First, we prove that $t_* = s_* + q_* : K_i(\varepsilon(\mathscr{D})) \longrightarrow K_i(\mathscr{D})$. Let $f : \mathscr{D} \times \mathscr{D} \longrightarrow \varepsilon(\mathscr{D})$ be the functor sending (K, C) to the split exact sequence

$$0 \longrightarrow K \longrightarrow K \oplus C \longrightarrow C \longrightarrow 0.$$

Note $(s, q)f = Id$. By Theorem 3.4.1, the functor $(s, q) : \mathbb{Q}(\varepsilon(\mathscr{D})) \longrightarrow \mathbb{Q}(\mathscr{D}) \times \mathbb{Q}(\mathscr{D})$ is a homotopy equivalence. Hence, $f : \mathbb{Q}(\mathscr{D}) \times \mathbb{Q}(\mathscr{D}) \longrightarrow \mathbb{Q}(\varepsilon(\mathscr{D}))$ is also a homotopy equivalence. Consider the commutative diagram of functors

$$\begin{array}{ccc} \mathbb{Q}\mathscr{D} \times \mathbb{Q}\mathscr{D} & \xrightarrow{f} & \mathbb{Q}\varepsilon(\mathscr{D}) \\ {\scriptstyle f}\downarrow & & \\ \mathbb{Q}\varepsilon(\mathscr{D}) & & \downarrow{\scriptstyle t} \\ {\scriptstyle (s,q)}\downarrow & & \\ \mathbb{Q}\mathscr{D} \times \mathbb{Q}\mathscr{D} & \xrightarrow[\oplus]{} & \mathbb{Q}\mathscr{D} \end{array} \qquad \left\{ \begin{array}{l} \text{which is} \quad tf = \oplus(s, q)f \\ \text{Therefore,} \quad (\mathbb{B}\mathbb{Q}t)(\mathbb{B}\mathbb{Q}f) = (\mathbb{B}\mathbb{Q}\oplus)(\mathbb{B}\mathbb{Q}(s, q))(\mathbb{B}\mathbb{Q}f) \end{array} \right.$$

Since, $\mathbb{B}\mathbb{Q}f$ is a homotopy equivalence, $\mathbb{B}\mathbb{Q}t$ and $(\mathbb{B}\mathbb{Q}\oplus)(\mathbb{B}\mathbb{Q}(s, q))$ are homotopic and hence induce the same morphisms on the homotopy groups. Therefore,

$$\forall\, n \geq 0 \quad K_n(t) = K_n(\oplus)K_n((s,q)) = K_n(s) + K_n(q) : K_n(\mathscr{D}) \longrightarrow K_n(\mathscr{D}).$$

With the notation of the theorem, this means $\forall\, n \geq 0, \quad t_* = s_* + q_* : K_n(\mathscr{D}) \longrightarrow K_n(\mathscr{D})$.

To prove the theorem, consider the exact functor

$$\Phi : \mathscr{E} \longrightarrow \varepsilon(\mathscr{D}) \text{ sending } X \mapsto \left(0 \longrightarrow G(X) \longrightarrow F(X) \longrightarrow H(X) \longrightarrow 0 \right)$$

Note $t\Phi = F$, $s\Phi = G$, and $q\Phi = H$. Therefore, $\forall\, n \geq 0$, we have the commutative diagram

$$\begin{array}{ccc} K_n(\mathscr{E}) & \xrightarrow{\Phi_*} & K_n(\varepsilon(\mathscr{D})) \\ & \searrow^{F_* = G_* + H_*} & \downarrow{\scriptstyle t_* = s_* + q_*} \\ & & K_n(\mathscr{D}) \end{array}$$

The proof is complete. ■

Corollary 3.4.5. *Let $\Phi_i : \mathscr{C} \longrightarrow \mathscr{D}$ be exact functors, for $i = 0, 1, \ldots, n$. Suppose*

$$0 \longrightarrow \Phi_0 \longrightarrow \Phi_1 \longrightarrow \Phi_2 \longrightarrow \cdots \longrightarrow \Phi_n \longrightarrow 0$$

is an exact sequence. Then

$$\sum_{p=0}^{n} (-1)^p \left(\Phi_p\right)_* = 0 : K_i(\mathscr{E}) \longrightarrow K_i(\mathscr{D}).$$

Proof. Follows from (3.4.4), by induction. ■

Corollary 3.4.6. *Let $\Phi : \mathscr{C} \longrightarrow \mathscr{D}$ be an exact functor. Suppose Φ admits an admissible filtration*

$$0 = \Phi_0 \hookrightarrow \Phi_1 \hookrightarrow \Phi_2 \hookrightarrow \cdots \hookrightarrow \Phi_r = \Phi$$

such that $\frac{\Phi_{p+1}}{\Phi_p}$ is exact $\forall\, p = 0, 1, \ldots, r-1$. Then

$$\Phi_* = \sum_{p=1}^{r} \left(\frac{\Phi_p}{\Phi_{p-1}} \right)_* : K_n(\mathscr{E}) \longrightarrow K_n(\mathscr{D}) \qquad \forall\, n \geq 0.$$

Proof. Follows from (3.4.4). ■

3.4.2. *Some examples*

Example 3.4.7. Let A be a commutative Noetherian ring and $\mathscr{P}(A)$ be the category finitely generated projective A-module. Then we write $K_i(A) := K_i(\mathscr{P}(A))$. This example was discussed in (3.3.4). Given a finitely generated projective A-module P, the association $Q \mapsto P \otimes Q$ defines an exact functor $P \otimes \ : \mathscr{P}(A) \longrightarrow \mathscr{P}(A)$. This will induce homomorphisms as follows:

$$[P \otimes -] : K_n(A) \longrightarrow K_n(A) \qquad \forall n \geq 0$$

In fact, short exact sequences

$$0 \longrightarrow P_{-1} \longrightarrow P_0 \longrightarrow P_1 \longrightarrow 0 \qquad \text{in } \mathscr{P}(A)$$

induce exact sequences of exact functors:

$$0 \longrightarrow P_{-1} \otimes - \longrightarrow P_0 \otimes - \longrightarrow P_1 \otimes - \longrightarrow 0.$$

So, it follows, from (3.4.4),

$$[P_0 \otimes -]_* = [P_{-1} \otimes -]_* + [P_1 \otimes -] : K_n(A) \longrightarrow K_n(A). \qquad \forall n \geq 0$$

The same works for any Noetherian scheme X (*more generally ringed spaces*). The above example can be formulated for the category $\mathscr{P}(X)$ of locally free sheaves, exactly in the same manner.

Before the next example, we give the following graded versions of Nakayama's Lemma.

Lemma 3.4.8. *Let $S = S_0 \oplus S_1 \oplus S_2 \oplus \cdots$ be a graded commutative ring and $S_+ = \bigoplus_{i \geq 1} S_i$. Let $M = \bigoplus_{n \in \mathbb{Z}} M_n$ be a finitely generated graded S-module. Assume $M = S_+ M$. Then $M = 0$.*

Proof. Let $m_1, m_2, \ldots, m_k$ be a set of homogeneous generators of M. We have

$$\begin{pmatrix} m_1 \\ m_2 \\ \cdots \\ m_k \end{pmatrix} = \begin{pmatrix} a_{11} & a_{12} & \cdots & a_{1k} \\ a_{21} & a_{22} & \cdots & a_{2k} \\ \cdots & \cdots & \cdots & \cdots \\ a_{k1} & a_{k2} & \cdots & a_{kk} \end{pmatrix} \begin{pmatrix} m_1 \\ m_2 \\ \cdots \\ m_k \end{pmatrix}$$

where $a_{ij} \in S$ are homogeneous elements of positive degree. Let $A = (a_{ij})$ be the $k \times k$ matrix, as above. So,

$$(I - A) \begin{pmatrix} m_1 \\ m_2 \\ \cdots \\ m_k \end{pmatrix} = \begin{pmatrix} 0 \\ 0 \\ \cdots \\ 0 \end{pmatrix}.$$

Multiplying by the $Adj(I - A)$, we have $\det(I - A)m_i = 0 \ \forall \ i = 1, \ldots, k$. It follows that $\det(A) = 1 + f_1 + \cdots + f_n$, where $f_j \in A_j$, for $j = 1, \ldots, n$. Therefore,

$$\forall \ i = 1, \ldots k \qquad (1 + f_1 + \cdots + f_n)m_i = 0; \quad \text{hence} \quad m_i = 0.$$

So, $M = 0$. The proof is complete. ■

The following is a graded analogue of the corresponding ungraded result.

Lemma 3.4.9. *Let S and S_+ be as in (3.4.8). Let $P = \bigoplus_{n \in \mathbb{Z}} P_n$, $Q = \bigoplus_{n \in \mathbb{Z}} Q_n$ be finitely generated graded projective S-modules. Let $S_+ = \bigoplus_{i \geq 1} S_i$. Let $\varphi_0 : \frac{P}{S_+P} \xrightarrow{\sim} \frac{Q}{S_+Q}$ be an isomorphism of S_0-modules. Then φ_0 lifts to an isomorphism $\varphi : P \xrightarrow{\sim} Q$ of graded S-modules.*

Proof. Similar to the proof of the corresponding ungraded result, using (3.4.8). ■

Definition 3.4.10. Let S be as in (3.4.8). We recall a standard notation. For a graded S-module $M = \oplus_{k \in \mathbb{Z}} M_k$, and integer d, define $M(d)$, by $M(d)_n = M(n + d)$, to be called the twisted module. Let $\mathscr{P}gr(S)$ denote the category of finitely generated graded projective S-modules, $P = \bigoplus_{n \in \mathbb{Z}} P_n$, and graded homomorphisms (*of degree zero*). (*One checks that P is graded projective if and only if $P \oplus Q \cong \oplus_{i=1}^n S(d_i)$ for some $Q \in Obj(\mathscr{P}gr(S))$*).

For $n \geq 0$, let $t : K_n(\mathscr{P}gr(S)) \longrightarrow K_n(\mathscr{P}gr(S))$ denote the map induced by the functor $P \mapsto P(-1)$. Then t has an inverse, given by $P \mapsto P(1)$. This defines a $\mathbb{Z}[t, t^{-1}]$-module structure on the groups $K_n(\mathscr{P}gr(S))$.

Proposition 3.4.11. *Let S and S_+ be as in (3.4.8). For integers $n \geq 0$, the map $K_n(S_0) \longrightarrow K_n(\mathscr{P}gr(S))$ induces an isomorphism*

$$\Lambda : \mathbb{Z}[t, t^{-1}] \otimes_{\mathbb{Z}} K_n(S_0) \xrightarrow{\sim} K_n(\mathscr{P}gr(S)) \qquad \text{of } \mathbb{Z}[t, t^{-1}]-\text{modules.}$$

Proof. As before, $\mathscr{P}(S_0)$ denotes the category of the finitely generated projective S_0-modules. Fix integer $n \geq 0$. We have natural functors $\iota : \mathscr{P}(S_0) \longrightarrow \mathscr{P}gr(S)$ sending $Q \longrightarrow S \otimes_{S_0} Q$. Composing, the following commutative diagrams of functors and the induced homomorphisms on K-groups are obtained:

$$\begin{array}{ccc} \mathscr{P}(S_0) & \xrightarrow{\iota} & \mathscr{P}gr(S) \\ & \searrow^{\iota_l} & \downarrow{\scriptstyle S(-l)\otimes -} \\ & & \mathscr{P}gr(S) \end{array}, \qquad \begin{array}{ccc} K_n(S_0) \times t^l & \xrightarrow{K_n(\iota)} & K_n(\mathscr{P}gr(S)) \\ & \searrow^{K_n(\iota_l)} & \downarrow{\scriptstyle t^l} \\ & & K_n(\mathscr{P}gr(S)) \end{array} \tag{3.9}$$

Thus, we obtain the $\mathbb{Z}[t, t^{-1}]$-homomorphism

$$\Lambda = (\oplus_{l \in \mathbb{Z}} K_n(\iota_l)) : \mathbb{Z}[t, t^{-1}] \otimes_{\mathbb{Z}} K_n(S_0) \longrightarrow K_n(\mathscr{P}gr(S)). \tag{3.10}$$

Consider S_0 as a graded ring concentrated at degree zero, and $Q \in Obj(\mathscr{P}(S_0))$ as an S_0-graded module concentrated at degree zero. Then $\forall l \in \mathbb{Z}$, $S(-l) \otimes_{S_0} Q \in \mathscr{P}gr(S)$. Observe $(S(-l) \otimes_{S_0} Q)_n = S_{n-l} \otimes_{S_0} Q$. In other words, $S(-l) \otimes_{S_0} Q = Q \oplus (S_1 \otimes Q) \oplus (S_2 \otimes Q) \oplus \cdots$, with the first term Q at degree l. For $P = \oplus P_l \in \mathscr{P}gr(S)$, let

$$\overline{P} := \frac{P}{S_+ P}. \quad \text{Then} \quad \overline{P} \in \mathscr{P}gr(S_0). \quad \text{So,} \quad \overline{P}_l = \frac{P_l}{(S_+ P)_l}.$$

We define functors

$$\begin{cases} T : \mathscr{P}gr(S) \longrightarrow \mathscr{P}gr(S_0) & \text{by } T(P) = \overline{P} \\ \forall\, l \in \mathbb{Z}, \quad T_l : \mathscr{P}gr(S) \longrightarrow \mathscr{P}(S_0) & \text{by } T_l(P) = \overline{P}_l \end{cases}$$

Since exact sequences in $\mathscr{P}gr(S)$ split, T, T_l are exact functors. Define the homomorphism Θ as follows:

$$\Theta : K_n\left(\mathscr{P}gr(S)\right) \longrightarrow \mathbb{Z}[t, t^{-1}] \otimes_{\mathbb{Z}} K_n\left(S_0\right) \quad \text{by} \quad \Theta(x) = \sum_{l \in \mathbb{Z}} t^l (T_l)_*(x).$$

We need to justify that Θ is well defined. Note, $T_l(P) \neq 0$ only for finitely many $l \in \mathbb{Z}$. Let $\mathscr{G}_k \subseteq \mathscr{P}gr(S)$ be the full subcategory, so that $T_l(P) = 0$ unless $-k \leq l \leq k$. Further, Θ is well defined on $K_n(\mathscr{G}_k)$. Since $\mathscr{P}gr(S) = \cup \mathscr{G}_k$, we conclude that Θ is well defined on $K_n\left(\mathscr{P}gr(S)\right)$. We will prove that Θ is an inverse to Λ.

By construction (3.9), the restriction of Λ on $t^r K_n(S_0)$ is induced by the functor

$$\iota_r(Q) = Q \otimes_{S_0} S(-r). \qquad \text{We have} \quad T_l \iota_r(Q) = \begin{cases} 0 & \text{if } l \neq r \\ Q & \text{if } l = r \end{cases}.$$

Therefore,

$$\Theta\Lambda(t^r x) = \Theta(\iota_{r*}(x)) = \sum_{l \in \mathbb{Z}} t^l (T_l)_*(\iota_{r*}(x)) = t^r (T_r \iota_r)_*(x) = t^r x.$$

This completes the proof that $\Theta\Lambda = 1$. It remains to be shown that $\Lambda\Theta = 1$.

Let $\mathscr{P}_\infty(S) \subseteq \mathscr{P}gr(S)$ denote the full subcategory of objects of the form $\oplus_{l \in \mathbb{Z}} S(-l) \otimes_{S_0} Q^l$, with $Q_l \in \mathscr{P}(S_0)$, and only finitely many $Q^l \neq 0$. Note that $\mathscr{P}_\infty(S)$ is closed under direct sum. We have a direct sum decomposition

$$S \otimes_{S_0} \overline{P} = \bigoplus_{l \in \mathbb{Z}} S(-l) \otimes_{S_0} \overline{P}_l \in Obj(\mathscr{P}_\infty(S)). \tag{3.11}$$

Note $\overline{P} = \overline{S \otimes_{S_0} \overline{P}}$. Therefore, by (3.4.9), P is isomorphic (*non-canonically*) to $S \otimes_{S_0} T(P)$. It follows that the inclusion functor $\mathscr{P}_\infty(S) \hookrightarrow \mathscr{P}gr(S)$ is essentially surjective. Consequently, $\mathbb{Q}\mathscr{P}_\infty(S) \hookrightarrow \mathbb{Q}\mathscr{P}gr(S)$ is essentially surjective, hence a homotopy equivalence. So, we would replace $\mathscr{P}gr(S)$ by $\mathscr{P}_\infty(S)$ to prove $\Lambda\Theta = 1$.

For integers $r \geq 0$, let $\mathscr{P}_r(S) \subseteq \mathscr{P}_\infty(S)$ be the full subcategory of objects of the form $\oplus_{l=-r}^{r} S(-l) \otimes_{S_0} Q^l$. Define the functors

$$\Phi_k : \mathscr{P}gr(S) \longrightarrow \mathscr{P}_\infty(S) \quad \text{by} \qquad \Phi_k\,(P) = \left(\oplus_{l \leq k} S(-l) \otimes_{S_0} \overline{P}_l\right).$$

In fact, Φ_k is obtained by the following composition:

$$\begin{array}{ccc} \mathscr{P}gr(S) & \xrightarrow{\ T\ } & \mathscr{P}gr(S_0) \\ & \searrow^{\Phi_k} & \downarrow{\scriptstyle \varphi_k} \\ & & \mathscr{P}_\infty(S) \end{array} \qquad \text{where} \quad \begin{cases} T(P) = \overline{P} \\ \varphi_k(\oplus_{l \in \mathbb{Z}} Q^l) = \oplus_{l \leq k} S(-l) \otimes_{S_0} Q^l) \end{cases} \tag{3.12}$$

Therefore, Φ_k is exact, because so is T and φ_k. The restriction of Φ_r to $\mathscr{P}_k(S)$ defines a functor on $\Phi_r : \mathscr{P}_k(S) \longrightarrow \mathscr{P}_k(S)$ for all $r \leq k$ (*for which we use the same notation* Φ_r).

$$\begin{cases} \frac{\Phi_k}{\Phi_{k-1}}(P) = S(-k) \otimes_{S_0} \overline{P}_k & \forall\ P \in \mathscr{P}gr(S) \\ \frac{\Phi_k}{\Phi_{k-1}}(P) = S(-k) \otimes_{S_0} Q^k & \forall\ P = \oplus_{l \in \mathbb{Z}} S(-l) \otimes Q^l \in \mathscr{P}_\infty(S) \end{cases}$$

Further, the restriction $(\Phi_k)_{|\mathscr{P}_k} = 1_{\mathscr{P}_k}$. Apply (3.4.6) to the restriction of Φ_k to $\mathscr{P}_k(S)$.

$$\forall\ n \geq 0 \quad \sum_{r=1}^{k} \left(\frac{\Phi_r}{\Phi_{r-1}}\right)_* = Id : K_n\left(\mathscr{P}_k(S)\right) \longrightarrow K_n\left(\mathscr{P}_k(S)\right).$$

Since $\mathscr{P}_\infty(S) = \cup \mathscr{P}_k(S)$, it follows that

$$\forall\ n \geq 0 \quad \sum_{r \in \mathbb{Z}} \left(\frac{\Phi_r}{\Phi_{r-1}}\right)_* = Id : K_n\left(\mathscr{P}_\infty(S)\right) \longrightarrow K_n\left(\mathscr{P}_\infty(S)\right). \tag{3.13}$$

Here the functor

$$\left(\frac{\Phi_r}{\Phi_{r-1}}\right)_* : \mathscr{P}_\infty(S) \longrightarrow \mathscr{P}_\infty(S) \quad \text{sends} \quad P = \oplus_{l \in \mathbb{Z}} S(-l) \otimes Q^l \mapsto S(-r) \otimes_{S_0} Q^r.$$

For $P = \oplus_{l \in \mathbb{Z}} S(-l) \otimes Q^l \in \mathscr{P}_\infty(S)$, we have $T_l(P) = Q^l$. Now $\left(\frac{\Phi_r}{\Phi_{r-1}}\right)$ is given by the composition

$$\begin{array}{ccc} \mathscr{P}_\infty(S) & \xrightarrow{\ T_r\ } & \mathscr{P}(S_0) \\ & \searrow^{\left(\frac{\Phi_r}{\Phi_{r-1}}\right)} & \downarrow{\scriptstyle S(-r)\otimes -} \\ & & \mathscr{P}_\infty(S) \end{array} \Longrightarrow \begin{array}{ccc} K_n\left(\mathscr{P}_\infty(S)\right) & \xrightarrow{\ (T_r)_*\ } & K_n\left(S_0\right) \\ & \searrow^{\left(\frac{\Phi_r}{\Phi_{r-1}}\right)_*} & \downarrow{\scriptstyle x \mapsto t^r K_n(\iota)(x)} \\ & & K_n\left(\mathscr{P}_\infty(S)\right) \end{array}$$

So,

$$\begin{array}{ccc} \mathscr{P}_\infty(S) & \xrightarrow{\iota T_r} & \mathscr{P}(S_0) \\ & \searrow^{\left(\frac{\Phi_r}{\Phi_{r-1}}\right)} & \downarrow{\scriptstyle S(-r)\otimes -} \\ & & \mathscr{P}_\infty(S) \end{array} \qquad \left(\frac{\Phi_r}{\Phi_{r-1}}\right)_*(x) = t^r(\iota T_r)_*(x)$$

Adding all these diagrams, over $r \in \mathbb{Z}$, by (3.13), we have

$$\begin{array}{ccc} K_n\left(\mathscr{P}_\infty(S)\right) & \xrightarrow{\Theta} & K_n\left(S_0\right) \otimes \mathbb{Z}[t, t^{-1}] \\ & \searrow^{1} & \downarrow{\scriptstyle \Lambda} \\ & & K_n\left(\mathscr{P}_\infty(S)\right) \end{array}$$

So, $\Lambda\Theta = 1$. The proof is complete. ■

3.5. The Resolution Theorem

The following from the Classical K-Theory motivates this section.

Proposition 3.5.1. *Let A be a Noetherian commutative ring. Let $\mathscr{P}(A)$ denote the category of finitely generated projective A-modules and let $\mathbb{H}(A)$ denote the category of finitely generated A-modules with finite projective dimension. Then $K_0(\mathscr{P}(A)) \cong K_0(\mathbb{H}(A))$. (Use Definition* 3.2.1 *only.)*

Proof. For definitions of the classical K_0, refer to Definition 3.2.1(4). We prove that the homomorphism $K_0(\iota) : K_0(\mathscr{P}(A)) \longrightarrow K_0(\mathbb{H}(A))$ induced by the inclusion functor $\iota : \mathscr{P}(A) \longrightarrow \mathbb{H}(A)$ is an isomorphism. We define the inverse of this map $K_0(\iota)$.

Let $\mathcal{H}$ denote the isomorphism classes of objects in $\mathbb{H}(A)$. For an object M in $\mathbb{H}(A)$, let $\overline{M}$ denote its isomorphism class in $\mathcal{H}$ and $[M]$ denote its class in $K_0(\mathbb{H}(A))$ (and likewise for $\mathscr{P}(A)$). For such an M, let

$$0 \longrightarrow P_n \longrightarrow P_{n-1} \longrightarrow \cdots \longrightarrow P_1 \longrightarrow P_0 \longrightarrow M \longrightarrow 0$$

be a $\mathscr{P}$-resolution of M. We say that $P_\bullet$ is a $\mathscr{P}(A)$-resolution of M. Now, define

$$\varphi_0 : \mathcal{H} \longrightarrow K_0(\mathscr{P}(A)) \quad \text{by} \quad \varphi(\overline{M}) = \sum_{i=0}^{n} (-1)^i [P_i].$$

We need to show that φ_0 is well defined. Let $Q_\bullet$ be another $\mathscr{P}(A)$-resolution of M (*by augmenting zeros at the tails, we can assume that both* $P_\bullet$ *and* $Q_\bullet$ *have the same length* n). It would be sufficient to prove that there is an exact sequence

$$\cdots \longrightarrow P_1 \oplus Q_2 \longrightarrow P_0 \oplus Q_1 \longrightarrow Q_0 \longrightarrow 0.$$

To see this, note that by properties of projective modules, there is a map of chain complexes $f_\bullet : P_\bullet \longrightarrow Q_\bullet$, as in the following commutative diagram:

$$\begin{array}{ccccccccccccc}
0 & \longrightarrow & P_n & \longrightarrow & P_{n-1} & \longrightarrow \cdots \longrightarrow & P_1 & \longrightarrow & P_0 & \longrightarrow & M & \longrightarrow & 0 \\
 & & \downarrow{\scriptstyle f_n} & & \downarrow{\scriptstyle f_{n-1}} & & \downarrow{\scriptstyle f_1} & & \downarrow{\scriptstyle f_0} & & \| & & \\
0 & \longrightarrow & Q_n & \longrightarrow & Q_{n-1} & \longrightarrow \cdots \longrightarrow & Q_1 & \longrightarrow & Q_0 & \longrightarrow & M & \longrightarrow & 0
\end{array}$$

Let $C(f_\bullet)$ denote the cone of $f_\bullet$ (see [W, pp. 18–19]). So, $C(f_\bullet)_n = Q_n \oplus P_{n-1}$ and the sequence

$$0 \longrightarrow Q_\bullet \longrightarrow C(f_\bullet) \longrightarrow P_\bullet[-1] \longrightarrow 0.$$

is exact. This gives rise to a long exact sequence of homologies

$$\cdots \longrightarrow H_1(P_\bullet) \longrightarrow H_1(Q_\bullet) \longrightarrow H_1(C(f_\bullet)) \longrightarrow$$
$$H_0(P_\bullet) \longrightarrow H_0(Q_\bullet) \longrightarrow H_0(C(f_\bullet)) \longrightarrow 0.$$

Since $H_0(P_\bullet) = H_0(Q_\bullet) = M$, and $H_i(P_\bullet) = H_i(Q_\bullet) = 0,\ \forall\, i \geq 1$, it follows $H_i(C(f_\bullet)) = 0\ \forall\, i$. This establishes the above, and hence φ_0 is a well-defined set theocratic map. So, φ_0 extends to a group homomorphism, as follows:

$$\begin{array}{ccc} \mathcal{H} & \xrightarrow{\varphi_0} & K_0(\mathscr{P}(A)) \\ \downarrow & \nearrow_{\overline{\varphi}_0} & \\ \mathbb{Z}(\mathcal{H}) & & \end{array}$$

Now, $K_0(\mathbb{H}(A)) = \frac{\mathbb{Z}(\mathcal{H})}{\mathscr{R}(\mathcal{H})}$. We show that $\overline{\varphi}_0$ respects the relations $\mathscr{R}(\mathcal{H})$. To see this, suppose

$$0 \longrightarrow K \longrightarrow M \longrightarrow C \longrightarrow 0 \quad \text{is exact in } \mathbb{H}(A).$$

Given finite $\mathscr{P}(A)$-resolutions $P_\bullet$ of K and $Q_\bullet$ of C (say of same length n), inductively, one can build a resolution

$$0 \longrightarrow Q_n \oplus P_n \longrightarrow Q_{n-1} \oplus P_{n-1} \longrightarrow \cdots \longrightarrow Q_1 \oplus P_1 \longrightarrow$$
$$Q_0 \oplus P_0 \longrightarrow M \longrightarrow 0.$$

Therefore,

$$\overline{\varphi}_0(\overline{M}) = \overline{\varphi}_0(\overline{K}) + \overline{\varphi}_0(\overline{C}).$$

Hence, $\overline{\varphi}_0$ factors through a homomorphism:

$$\begin{array}{ccc} \mathcal{H} & \xrightarrow{\varphi_0} & K_0(\mathscr{P}(A)) \\ \downarrow & \nearrow_{\overline{\varphi}_0} & \uparrow \varphi \\ \mathbb{Z}(\mathcal{H}) & \longrightarrow & K_0(\mathbb{H}(A)) \end{array}$$

It is easy to see that $K_0(\iota)$ and φ are inverse of each other. The proof is complete. ■

3.5.1. *Extension closed subcategories*

Suppose $\mathscr{E}$ is an exact category. For subsequent references, let

$$0 \longrightarrow K \longrightarrow M \longrightarrow C \longrightarrow 0 \quad \text{be an exact sequence in } \mathscr{E}. \tag{3.14}$$

Recall that short exact sequences (3.14) are also known as "**extensions**" (see [HS, Chapter III]). (*As always, assume that $\mathscr{E}$ has a set of isomorphism classes.*)

Refer to the definition (A.4.4) of an **extension closed subcategory** $\mathscr{P} \subseteq \mathscr{E}$ of an exact category $\mathscr{E}$. We have the following remarks:

1. If $\mathscr{P}$ is an extension closed subcategory of $\mathscr{E}$, then $\mathscr{P}$ is also an exact category, where a sequence in $\mathscr{P}$ is declared exact if and only if it is exact in $\mathscr{E}$. Consequently, $\mathscr{P} \hookrightarrow \mathscr{E}$ is an exact functor.
2. As discussed before, for an example let A be a Noetherian commutative ring. Let $Coh(A)$ be the category of finitely generated A-modules, and $\mathscr{P}(A) \subseteq Coh(A)$ be the full subcategory of finitely generated projective A-modules. Then $\mathscr{P}(A) \subseteq Coh(A)$ is an extension closed subcategory.
3. Further, if $\mathscr{P}$ is an extension closed subcategory of $\mathscr{E}$, then $\mathbb{Q}(\mathscr{P})$ is a subcategory of $\mathbb{Q}(\mathscr{E})$ while not necessarily a full subcategory. This is because an inflation $\iota : Q \hookrightarrow P$ in $\mathscr{E}$ with $Q, P \in Obj(\mathscr{P})$ need not be an inflation in $\mathscr{P}$. In the above example, $\mathscr{P}(A) \subseteq Coh(A)$, for $P, Q \in Obj(\mathscr{P}(A))$, an inclusion $Q \hookrightarrow P$ would not be an inflation in $\mathscr{P}(A)$, unless it splits.
4. For the convenience of subsequent discussions, for $M \in \mathscr{E}$, an exact sequence

$$0 \longrightarrow P_k \longrightarrow \cdots \xrightarrow{d_2} P_1 \xrightarrow{d_1} P_0 \xrightarrow{d_0} M \longrightarrow 0 \quad \text{with } P_i \in Obj(\mathscr{P}) \tag{3.15}$$

is called a $\mathscr{P}$-resolution of M of length k. By an exact sequence, we mean $d_r : P_r \twoheadrightarrow \ker(d_{r-1})$ is a deflation in $\mathscr{E}$ for $r = 0, 1, \ldots, k$.

The goal of this section is to extend (3.5.1) and prove $K_n(\mathscr{P}(A)) \xrightarrow{\sim} K_n(\mathbb{H}(A))$ for all $n \geq 0$. The proof depends on the length of the resolutions (3.15). The following is the first step.

Theorem 3.5.2. *Suppose $\mathscr{P}$ is a full subcategory of an exact category $\mathscr{E}$, which is extension closed. Assume the following:*

1. *Given a conflation* (3.14) *in $\mathscr{E}$, if $M \in Obj(\mathscr{P})$, then $K \in Obj(\mathscr{P})$.*
2. *For any $C \in Obj(\mathscr{E})$, there is a conflation* (3.14), *with $M \in Obj(\mathscr{P})$.*

Then the inclusion $\mathbb{Q}\mathscr{P} \hookrightarrow \mathbb{Q}\mathscr{E}$ is a homotopy equivalence. Consequently, $K_i(\mathscr{P}) \xrightarrow{\sim} K_i(\mathscr{E})$ $\forall\, i \geq 0$. (The conditions (1) and (2) are verbalized, respectively, as that $\mathscr{P}$ is closed under subobjects and any $C \in Obj(\mathscr{E})$ is the subquotient of an object in $\mathscr{P}$.)

Proof. Let $\mathscr{C}$ be the full subcategory of $\mathbb{Q}(\mathscr{E})$, whose objects are those of $\mathscr{P}$. Recall that $\mathbb{Q}(\mathscr{P})$ is not necessarily a full subcategory of $\mathbb{Q}(\mathscr{E})$. However, we have the factorization

$$\begin{array}{ccc} \mathbb{Q}(\mathscr{P}) & \xrightarrow{g} & \mathscr{C} \\ & \searrow & \downarrow f \\ & & \mathbb{Q}(\mathscr{E}) \end{array}$$

We prove that both f, g are homotopy equivalences. First, we show g is a homotopy equivalence. As usual, we use Theorem A (2.2.2) and show that $\forall\ P \in Obj(\mathscr{P})$, the category g/P is contractible. Refer to (3.1.9) for the definition of the category $\mathcal{L}(P)$ of admissible layers $((K, \iota_K), (L, \iota_L))$ in P, in the category $\mathscr{E}$. We abbreviate $(K, L) := ((K, \iota_K), (L, \iota_L))$. In fact, $\mathcal{L}(P)$ is a partially ordered set. By hypothesis (1), it follows that $K, L \in Obj(\mathscr{P})$. Let $\mathcal{L}(P, \mathscr{P}) = \{(K, L) \in \mathcal{L}(P) : L/K \in Obj(\mathscr{P})\}$. For $(K_0, L_0), (K_1, L_1) \in \mathcal{L}(P, \mathscr{P})$, we redefine ordering $(K_0, L_0) \leq_{\mathscr{P}} (K_1, L_1)$ if and only if $K_1 \leq K_0 \leq L_0 \leq L_1$ and $K_0/K_1, L_1/L_0 \in Obj(\mathscr{P})$. By reworking the arguments in (3.1.9), it follows that there is an equivalence of categories , $g/P \xrightarrow{\sim} \mathcal{L}(P, \mathscr{P})$. We prove that $\mathcal{L}(P, \mathscr{P})$ is contractible. For $(K, L) \in \mathcal{L}(P, \mathscr{P})$. We have $(K, L) \leq_{\mathscr{P}} (0, L) \geq_{\mathscr{P}} (0, 0)$. Define two functors

$$\begin{cases} F : \mathcal{L}(P, \mathscr{P}) \longrightarrow \mathcal{L}(P, \mathscr{P}) & \text{by } F(K, L) = (0, L) \\ C : \mathcal{L}(P, \mathscr{P}) \longrightarrow \mathcal{L}(P, \mathscr{P}) & \text{by } C(K, L) = (0, 0) \end{cases}$$

We define two natural transformations:

$$\begin{cases} \tau_1 : Id_{\mathcal{L}(P,\mathscr{P})} \longrightarrow F & \text{by the unique arrow } (K, L) \leq_{\mathscr{P}} (0, L) \\ \tau_2 : C \longrightarrow F & \text{by the unique arrow } \ (0, 0) \leq_{\mathscr{P}} (0, L) \end{cases}$$

By Proposition 2.1.9, there are homotopies: $Id_{\mathbb{B}\mathbb{Q}\mathcal{L}(P,\mathscr{P})} \mapsto \mathbb{B}F$ and $Constant \mapsto \mathbb{B}F$. So, $Id_{\mathbb{B}\mathbb{Q}\mathcal{L}(P),\mathscr{P}}$ is homotopic to the constant map. Hence, $\mathcal{L}(P, \mathscr{P})$ is contractible. Therefore, g/P is contactable. By Theorem 2.2.2, g is a homotopy equivalence.

It remains to be proven that $f : \mathscr{C} \longrightarrow \mathbb{Q}\mathscr{E}$ is a homotopy equivalence. To do this, we prove that M/f is contractible, $\forall\ M \in Obj(\mathbb{Q}\mathscr{E})$. Fix M. The objects in M/f are pairs $Y := (P, u)$, with $P \in Obj(\mathscr{P})$ and $u \in Mor_{\mathbb{Q}\mathscr{E}}(M, P)$. Let such a morphism u be represented as $u = \left(M \underset{j}{\twoheadleftarrow} Q \underset{i}{\hookrightarrow} P \right)$, where j is a deflation and i is an inflation in $\mathscr{E}$.

Write $\mathscr{F}(M) = M/f$. Let $\mathscr{F}_s(M) \subseteq \mathscr{F}(M)$ be the full subcategory of objects (P, v) in $\mathscr{F}(M)$, such that $v = p^!$ is *surjective*. We prove that the inclusion $\iota(M) : \mathscr{F}_s(M) \hookrightarrow \mathscr{F}(M)$ has a right adjoint. By (A.1.15), it suffices to show that $\forall\ Y \in Obj(\mathscr{F}(M))$ there is a universal arrow $\chi_Y : R(Y) \longrightarrow Y$, with $R(Y) \in Obj(\mathscr{F}_s(M))$. Consider $Y := (P, u) \in Obj(\mathscr{F}(M))$, with the notation as above. Let $R(Y) := (Q, j^!)$. Then $R(Y) \in Obj(\mathscr{F}_s(M))$. Define $\chi_Y := i_! \in Mor_{\mathscr{F}(M)}(R(Y), Y)$. Then $\chi_Y : R(Y) \longrightarrow Y$ has the universal property, as in (A.1.15). While it requires routine checking, we outline it. Let $X = (L, p^!) \in \mathscr{F}_s(M)$ and $\varphi \in Mor_{\mathscr{F}(M)}(X, Y)$. It follows $\varphi = \mathfrak{q}^! : L \longrightarrow Q$ is a *surjective* arrow, given by deflation $\mathfrak{q} : Q \twoheadrightarrow L$ with $j = p\mathfrak{q}$. Define $\Phi(\varphi) = \mathfrak{q}^! \in Mor_{\mathscr{F}_s(M)}(X, R(Y))$. One checks $\chi_Y \Phi(\varphi) = \varphi$. Note that any arrow in $\mathscr{F}_s(M)$ is *surjective*. So, the uniqueness of the morphism $\Phi(\varphi)$ follows from the fact that $\chi_Y \Phi(\varphi) = \varphi$ is a decomposition of φ as a composition of an *injective* and a *surjective* arrow. This completes the proof that χ_Y is universal. Hence, $\iota(M) : \mathscr{F}_s(M) \hookrightarrow \mathscr{F}(M)$ has a right adjoint. By (2.1.10), $\iota(M)$ is a homotopy equivalence.

So, it would be enough to prove that $\mathcal{F}_s(M)$ is contractible. We will prove that $\mathcal{F}_s(M)^{op}$ is contractible. Objects in $\mathcal{F}_s(M)^{op}$ are pairs (P, p), where $P \in Obj(\mathscr{P})$ and $p : P \twoheadrightarrow M$ are deflations. A morphism $j : (P, p) \longrightarrow (Q, q)$ is a deflation $j : P \twoheadrightarrow Q$ such that $p = qj$.

By hypothesis (2), there is a deflation $p_0 : P_0 \twoheadrightarrow M$, with $P_0 \in Obj(\mathscr{P})$. Fix such a deflation p_0. By hypothesis (1), $\ker(p_0) \in Obj(\mathscr{P})$. Let (P, p) be an object in $\mathcal{F}_s(M)^{op}$. Consider the pullback diagram

$$\begin{array}{ccc} P \times_M P_0 & \xrightarrow{p_1} & P_0 \\ {\scriptstyle p_2}\downarrow & & \downarrow{\scriptstyle p_0} \\ P & \xrightarrow[p]{} & M \end{array}$$

Since $\ker(p_2) = \ker(p_0)$, $\ker(p_2) \in Obj(\mathscr{P})$. By the extension closed property, it follows that $P \times_M P_0 \in Obj(\mathscr{P})$. We define the following functors:

$$\begin{cases} Id : \mathcal{F}_s(M)^{op} \longrightarrow \mathcal{F}_s(M)^{op}\ Id(P, p) = (P, p) \\ F : \mathcal{F}_s(M)^{op} \longrightarrow \mathcal{F}_s(M)^{op}\ \ F(P, p) = (P \times_M P_0, pp_2) \\ C : \mathcal{F}_s(M)^{op} \longrightarrow \mathcal{F}_s(M)^{op}\ \ C(P, p) = (P_0, p_0) \end{cases}$$

So, $p_2 : F \longrightarrow Id$ and $p_0 p_1 : F \longrightarrow C$ define two natural transformations. By (2.1.9), there are homotopies: $\mathbb{B}F \mapsto Id_{\mathbb{B}\mathcal{F}_s(M)^{op}}$. Therefore, $\mathscr{F}'$ is contractible. The proof is complete. ∎

Theorem 3.5.2 was formulated as an inductive step to the following (3.5.3).

Corollary 3.5.3. *Suppose $\mathscr{E}$ is an exact category and $\mathscr{P}$ is an extension closed full subcategory. For subsequent reference, let*

$$0 \longrightarrow K \longrightarrow M \longrightarrow C \longrightarrow 0 \quad \textit{be a conflation in } \mathscr{E}. \tag{3.16}$$

Assume the following:

1. *For any conflation* (3.16) *in $\mathscr{E}$, if $C, M \in Obj(\mathscr{P})$, then $K \in Obj(\mathscr{P})$.*
2. *Given a deflation $j : M \twoheadrightarrow P$ in $\mathscr{E}$, with $P \in Obj(\mathscr{P})$,*

$$\exists \begin{cases} g : P' \twoheadrightarrow P, & \textit{a deflation} \\ f : P' \longrightarrow M, & \textit{a morphism} \end{cases} \ni\ P' \in Obj(\mathscr{P}) \textit{ and } \begin{array}{ccc} & & P' \\ & \overset{f}{\swarrow} & \downarrow g \\ M & \underset{j}{\twoheadrightarrow} & P \end{array} \textit{ commute.}$$

(The condition is valid if $\forall\ M \in Obj(\mathscr{E})$, there is a deflation $P' \twoheadrightarrow M$, with $P' \in Obj(\mathscr{P})$.)

For integers $n \geq 0$, let $\mathscr{P}_n$ denote the full subcategory of objects M in $\mathscr{E}$ that have a $\mathscr{P}$-resolution of length $\leq n$ (see 3.15). *So, $\mathscr{P}_0 = \mathscr{P}$. Then*

1. *$\mathscr{P}_n$ is an extension closed full subcategory of $\mathscr{E}$. Therefore, $\mathscr{P}_n$ is an exact category.*
2. *Write $\mathscr{P}_\infty = \bigcup \mathscr{P}_n$. Then $\mathscr{P}_\infty$ is also an exact subcategory of $\mathscr{E}$.*
3. *The following*

$$\mathbb{Q}\mathscr{P} = \mathbb{Q}\mathscr{P}_0 \xrightarrow{\sim} \mathbb{Q}\mathscr{P}_1 \xrightarrow{\sim} \cdots \xrightarrow{\sim} \mathbb{Q}\mathscr{P}_n \xrightarrow{\sim} \cdots \xrightarrow{\sim} \mathbb{Q}\mathscr{P}_\infty$$

are homotopy equivalences. In particular,

$$K_r(\mathscr{P}) \xrightarrow{\sim} K_r(\mathscr{P}_n) \xrightarrow{\sim} K_r(\mathscr{P}_\infty) \quad \forall\ r \geq 0,\ n \geq 0.$$

Proof. The proof follows by an application of (3.5.2), to the inclusion $\mathscr{P}_n \subseteq \mathscr{P}_{n+1}$, with additional help from the following Lemma 3.5.4. Note, since $\mathscr{P} \subseteq \mathscr{P}_n$, any $C \in Obj(\mathscr{P}_{n+1})$ is a subquotient of an object in $\mathscr{P}_n$. We use (3.5.4) to show that $\mathscr{P}_n$ is closed under subobjects, in $\mathscr{P}_{n+1}$. Suppose $i : K \hookrightarrow M$ is an inflation in $\mathscr{E}$, and $M \in Obj(\mathscr{P}_{n+1})$. By hypothesis, there is a deflation $p_0 : P_0 \twoheadrightarrow M$ with $P \in Obj(\mathscr{P})$. Let $K_0 = \ker(p_0)$. By (3.5.4(1)), $K_0 \in Obj(\mathscr{P}_n)$. So, M has a $\mathscr{P}$-resolution $P_\bullet$ of length $n+1$. The pullback of $P_\bullet$ gives a $\mathscr{P}$-resolution of K of length $n+1$. So, $\mathscr{P}_n$ is closed under subobjects, in $\mathscr{P}_{n+1}$. Thus, (3.5.2) applies to the inclusion $\mathscr{P}_n \subseteq \mathscr{P}_{n+1}$, and $\mathbb{Q}\mathscr{P}_n \longrightarrow \mathbb{Q}\mathscr{P}_{n+1}$ is a homotopy equivalence. To see $\mathbb{Q}\mathscr{P} \longrightarrow \mathbb{Q}\mathscr{P}_\infty$ is a homotopy equivalence, use Whitehead Theorem (C.3.8). The proof is complete. ■

Lemma 3.5.4. *Suppose* $\mathscr{P}$, $\mathscr{E}$, *and* $\mathscr{P}_n$ *are as in* (3.5.3). *Suppose*

$$0 \longrightarrow K \xrightarrow{\ i\ } M \xrightarrow{\ j\ } C \longrightarrow 0 \quad \text{is exact in } \mathscr{E}. \tag{3.17}$$

Under the hypothesis of (3.5.3), *and for integers* $n \geq 0$, *we have the following:*

1. *If* $M \in Obj(\mathscr{P}_n)$ *and* $C \in Obj(\mathscr{P}_{n+1})$, *then* $K \in Obj(\mathscr{P}_n)$.
2. *If* $K, C \in Obj(\mathscr{P}_{n+1})$, *then* $M \in Obj(\mathscr{P}_{n+1})$. (So, $\mathscr{P}_n$ is extension closed in $\mathscr{E}$.)
3. *If* $M, C \in Obj(\mathscr{P}_{n+1})$, *then* $K \in Obj(\mathscr{P}_{n+1})$.

Proof. (*The proof is imitation of the standard arguments on projective dimension of modules* M, *over commutative Noetherian rings* A (see [Mh].). We prove the case $n = 0$. To prove (1), for $n = 0$, assume $M \in Obj(\mathscr{P}_0)$, $C \in Obj(\mathscr{P}_1)$. Consider the diagram

$$\begin{array}{ccccccccc}
 & & & & P_1 & = & P_1 & & \\
 & & & & \downarrow & & \downarrow{\scriptstyle g} & & \\
0 & \longrightarrow & K & \longrightarrow & P_1 \times_C M & \longrightarrow & P_0 & \twoheadrightarrow & 0 \\
 & & \| & & \downarrow & & \downarrow{\scriptstyle f} & & \\
0 & \longrightarrow & K & \xrightarrow[i]{} & M & \xrightarrow[j]{} & C & \longrightarrow & 0
\end{array}$$

where the last vertical line is a resolution of C, with $P_0, P_1 \in Obj(\mathscr{P})$, and $P_1 \times_C M$ is the pullback of (j, f). The middle vertical line and the middle horizontal line are conflations in $\mathscr{E}$. Since the middle vertical line is a conflation, $P_1 \times_C M \in Obj(\mathscr{P})$, by the extension closed property of $\mathscr{P}$. Now, by hypothesis (1) of (3.5.3), $K \in Obj(\mathscr{P})$. This establishes (1) for $n = 0$.

Now we prove (2), in the case $n = 0$. So, we prove that $\mathscr{P}_1$ is extension closed in $\mathscr{E}$. Assume $K, C \in \mathscr{P}_1$. Since $C \in \mathscr{P}_1$, there is deflation $g : P \twoheadrightarrow C$ in $\mathscr{E}$, with $P \in Obj(\mathscr{P})$. Consider the pullback

$$\begin{array}{ccc} P \times_C M & \overset{p}{\twoheadrightarrow} & P \\ {\scriptstyle q}\big\downarrow & & \big\downarrow{\scriptstyle g} \\ M & \underset{j}{\twoheadrightarrow} & C \end{array}$$

By hypothesis (2), we have the commuting diagram

$$\begin{array}{ccc} & & P_0 \\ & \overset{f}{\swarrow} & \big\downarrow{\scriptstyle g_0} \\ P \times_C M & \underset{p}{\twoheadrightarrow} & P \end{array}$$

where $P_0 \in Obj(\mathscr{P})$, $g_0 : P_0 \twoheadrightarrow P$ is a deflation, and $f \in Mor(P_0, P \times_C M)$. Then $d_0 := gg_0 : P_0 \twoheadrightarrow C$ is a deflation. Let $P_1 = \ker(d_0)$. By (1), $P_1 \in Obj(\mathscr{P})$. Also, let $\eta = qf : P_0 \longrightarrow M$. So, $j\eta = d_0$. With this, we construct the following commutative diagram:

$$\begin{array}{ccccccccc} & & 0 & & & & 0 & & \\ & & \big\downarrow & & & & \big\downarrow & & \\ 0 & \longrightarrow & Q_1 & \longrightarrow & N & \longrightarrow & P_1 & \longrightarrow & 0 \\ & & \big\downarrow & & \big\downarrow & & \big\downarrow & & \\ 0 & \longrightarrow & Q_0 & \longrightarrow & Q_0 \oplus P_0 & \longrightarrow & P_0 & \longrightarrow & 0 \\ & & {\scriptstyle \partial_0}\big\downarrow & & \big\downarrow{\scriptstyle \delta = (i\partial_0 + \beta)} & & \big\downarrow{\scriptstyle d_0} & & \\ 0 & \longrightarrow & K & \underset{i}{\longrightarrow} & M & \underset{j}{\longrightarrow} & C & \longrightarrow & 0 \\ & & \big\downarrow & & & & \big\downarrow & & \\ & & 0 & & & & 0 & & \end{array}$$

Here the first vertical line is a conflation, with $Q_0, Q_1 \in \mathscr{P}$, obtained by the hypothesis. The map $\beta : P_0 \longrightarrow M$ is obtained by hypothesis (2) of (3.5.3), with $j\beta = d$. Recall, by (A.4.5), $\mathscr{E}$ can be embedded as a fully exact extension closed subcategory of an abelian category

$\mathscr{C}$. Let $N = \ker(\delta)$ in $\mathscr{C}$. By Snake Lemma A.3.7,

$$Q_1 \longrightarrow N \longrightarrow P_1 \longrightarrow coKer(\partial) \longrightarrow coKer(\delta) \longrightarrow coKer(d)$$

is exact. Since $coKer(\partial) = coKer(d) = 0$, it follows $coKer(\delta) = 0$. Further,

$$Q_1 \longrightarrow N \longrightarrow P_1 \longrightarrow 0 \quad \text{is exact in } \mathscr{C}.$$

$$\text{Also, } \; 0 \longrightarrow Q_1 \longrightarrow N \quad \text{is exact in } \mathscr{C}.$$

Therefore,

$$0 \longrightarrow Q_1 \longrightarrow N \longrightarrow P_1 \longrightarrow 0 \quad \text{is exact in } \mathscr{C}.$$

Since $\mathscr{P} \hookrightarrow \mathscr{E} \hookrightarrow \mathscr{C}$ are extension closed, $N \in Obj(\mathscr{P})$. Now, the middle vertical line is a conflation in $\mathscr{C}$. Since $\mathscr{E} \hookrightarrow \mathscr{C}$ is fully exact, the middle vertical line is conflation in $\mathscr{E}$. Therefore, $M \in Obj(\mathscr{P}_1)$. This completes the proof that $\mathscr{P}_1$ is an extension closed subcategory of $\mathscr{E}$.

Now, we prove (3) for $n = 0$. So, let $M, C \in Obj(\mathscr{P}_1)$. Consider the pullback diagram

$$\begin{array}{ccccc}
P_1 & = & P_1 & \twoheadrightarrow & 0 \\
\downarrow & & \downarrow & & \downarrow \\
P_0 \times_M K & \hookrightarrow & P_0 & \overset{jd}{\twoheadrightarrow} & C \\
\downarrow & & \downarrow{\scriptstyle d} & & \| \\
K & \underset{i}{\hookrightarrow} & M & \underset{j}{\twoheadrightarrow} & C
\end{array}$$

Here the middle vertical line is a conflation, with $P_0, P_1 \in Obj(\mathscr{P})$. By (1), $P_0 \times_M K \in \mathscr{P}$. Since the first vertical map is also a conflation, $K \in \mathscr{P}_1$. So, (3) is established in this case.

So, the lemma is established in the case $n = 0$. Now, the proof is completed by induction. ■

Resolving Subcategories

In the commutative algebra literature, some authors [AB] have introduced the following concept of resolving categories, in this context of resolutions theorem.

Definition 3.5.5. Suppose $\mathscr{E}$ is an exact category (*with a set of isomorphism classes*). For subsequent reference, let

$$0 \longrightarrow K \longrightarrow M \longrightarrow C \longrightarrow 0 \quad \text{be an exact sequence in } \mathscr{E}. \tag{3.18}$$

Now suppose $\mathscr{P}$ is an extension closed subcategory of $\mathscr{E}$, (*hence an exact subcategory*). Then $\mathscr{P}$ is said to be a **resolving subcategory** of $\mathscr{E}$ if the following two conditions are satisfied:

1. Given an exact sequence (3.18) in $\mathscr{E}$, if M, C are in $\mathscr{P}$, then so is K.
2. For all objects M in $\mathscr{E}$, there is an exact sequence:

$$0 \longrightarrow P_n \longrightarrow P_{n-1} \longrightarrow \cdots \longrightarrow P_1 \longrightarrow P_0 \longrightarrow M \longrightarrow 0$$

with P_i in $\mathscr{P}$. In this case, we say $P_\bullet$ is a $\mathscr{P}$-resolution of M.

Remark 3.5.6. Before we start commenting on the K-theory of rings, we make a decision that our rings A are not necessarily commutative unless stated so. The methods do not really simplify in the case of commutative rings. However, there are only a limited number of applications in [Q] that concern noncommutative situations. For this reason, some readers may choose to assume all rings are commutative.

Example 3.5.7. Suppose A is a ring. By an A-module, we usually mean a left A-module. In analogy to the commutative situation, $Coh(A)$ will denote the category of finitely generated left A-modules. The category of finitely generated left projective A-modules is denoted by $\mathscr{P}(A)$.

1. Note, $\mathscr{P}(A)$ is an exact subcategory of $Coh(A)$. It is not a resolving subcategory of $Coh(A)$. However, if A is a regular ring (see 5.2.3), then $\mathscr{P}(A)$ is a resolving subcategory of $Coh(A)$.
2. Let $\mathbb{H}(A)$ be the full subcategory of $Coh(A)$ whose objects are the A-modules M in $Coh(A)$, with finite projective dimension. Then $\mathbb{H}(A)$ is an exact subcategory of $Coh(A)$. Further, $\mathscr{P}(A)$ is a resolving subcategory of $\mathbb{H}(A)$.

Proposition 3.5.8. *Let $\mathscr{E}$ be an exact category and $\mathscr{P} \subseteq \mathscr{E}$ be a resolving subcategory. Then*

$$\mathbb{Q}\mathscr{P} \longrightarrow \mathbb{Q}\mathscr{E} \quad \text{is a homotopy equivalence.}$$

In particular, $\forall\ n \geq 0$, $K_n(\mathscr{P}) \xrightarrow{\sim} K_n(\mathscr{E})$ is an isomorphism.

Proof. We apply (3.5.3). The first condition of (3.5.3) follows from the definition of resolving subcategory. Now suppose $j : M \twoheadrightarrow P$ in $\mathscr{E}$ is a deflation in $\mathscr{E}$, with $P \in Obj(\mathscr{P})$. Again by the definition of resolving category, there is a deflation $\iota : P_0 \twoheadrightarrow M$, with $P_0 \in Obj(\mathscr{P})$. Therefore, the diagram of deflations

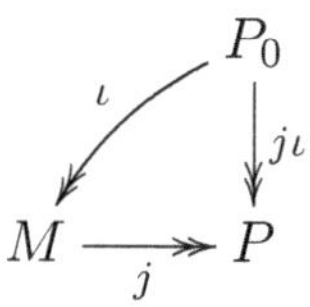

commutes.

So, the second condition of (3.5.3) is satisfied. By the second condition in the definition of resolving category, $\mathscr{P}_\infty = \mathscr{E}$. So, the proof is complete by (3.5.3). ■

The following is an extension of the classical Example 3.5.1.

Corollary 3.5.9. *Let A be a commutative Noetherian ring and $\mathscr{P}(A) \subseteq \mathbb{H}(A)$ be as in (3.5.7). Then $\mathbb{Q}\mathscr{P} \longrightarrow \mathbb{Q}\mathbb{H}(A)$ is a homotopy equivalence. Therefore, $\forall\ n \geq 0$, $K_n(\mathscr{P}(A)) \xrightarrow{\sim} K_n(\mathbb{H}(A))$ is an isomorphism. In particular, if A is a regular ring, then $K_n(\mathscr{P}(A)) \xrightarrow{\sim} K_n(Coh(A))$.*

More generally, suppose X is a Noetherian scheme. Let $Coh(X)$ denote the category of coherent sheaves on X, and $\mathscr{P}(X)$ denote the category of locally free sheaves on X. Let $\mathbb{H}(X)$ denote the full subcategory of $Coh(X)$, whose objects have finite $\mathscr{P}(X)$-resolution. Then $\forall\ n \geq 0$, $K_n(\mathscr{P}(X)) \xrightarrow{\sim} K_n(\mathbb{H}(X))$. In particular, if X is a regular scheme that is Noetherian, integral, and separated, then

$$K_n(\mathscr{P}(X)) \xrightarrow{\sim} K_n\,(Coh(X)) \qquad \forall n \geq 0.$$

Proof. Follows from (3.5.3 or 3.5.8). Recall that if $\dim A = d$, then $\mathbb{H}(A) = (\mathscr{P}(A))_d$ (see [Mh]). When X is Noetherian, integral, separated, and regular, then $Coh(X) = \mathscr{P}_\infty$ (see [Hr, Ex. III.6.9]). ■

A Transfer Map: Let $f : A \longrightarrow B$ be a homomorphism of commutative rings. Use the notation $\mathbb{H}(A) \hookrightarrow Coh(A)$ as above (3.5.7). Assume, as an A-module, $B \in \mathbb{H}(A)$. In this case, $M \in \mathbb{H}(B) \Longrightarrow M \in \mathbb{H}(A)$. (*This can be seen by double complexes* [Wc1, p. 99].) Therefore, f defines an exact functor $\mathbb{H}(B) \longrightarrow \mathbb{H}(A)$. By (3.5.9), this functor induces the homomorphisms f_*, as follows:

$$\begin{array}{ccc} K_n(B) & \overset{f_*}{\dashrightarrow} & K_n(A) \\ \wr\downarrow & & \downarrow\wr \\ K_n(\mathbb{H}(B)) & \longrightarrow & K_n(\mathbb{H}(A)) \end{array} \qquad \forall\, n \geq 0, \quad \text{to be called the } \textbf{Transfer map}.$$

Given another ring homomorphism $g : B \longrightarrow C$, with $C \in \mathbb{H}(B)$, we have

$$(gf)_* = f_* g_* : K_n(C) \longrightarrow K_n(A) \qquad \forall\, n \geq 0.$$

We have the natural functor

$$\mathscr{P}(A) \times \mathbb{H}(A) \longrightarrow \mathbb{H}(A) \quad \text{sending} \quad (P, M) \mapsto P \otimes_A M.$$

This is exact in both coordinates. By Additivity Theorem (3.4.4), we obtain the maps

$$\forall\, n \geq 0 \qquad \begin{array}{ccc} K_0(A) \otimes K_n(A) & \dashrightarrow & K_n(A) \\ \wr\downarrow & & \downarrow\wr \\ K_0(A) \otimes K_n(\mathbb{H}(A)) & \longrightarrow & K_n(\mathbb{H}(A)) \end{array}$$

where both horizontal maps are obtained, by sending $[P] \otimes z \mapsto (P \otimes_A -)_* z$. So, we have two products:

$$\begin{cases} K_0(A) \otimes K_n(A) \longrightarrow K_n(A) \\ K_0(B) \otimes K_n(B) \longrightarrow K_n(B) \end{cases} \quad \text{to be denoted by} \quad x \otimes y \mapsto x \cdot y$$

Further, we have the pullback maps

$$f^* : K_n(A) \longrightarrow K_n(B) \qquad \forall\, n \geq 0.$$

Now, we have the projection formula agreeable.

Theorem 3.5.10. *Under the above setup, we have*

$$f_*(f^*x \cdot y) = x \cdot f_*y \qquad x \in K_0A,\ y \in K_nB.$$

In other words, the diagram

$$\begin{array}{ccccc} K_0(A) \otimes K_n(B) & \xrightarrow{f^*\otimes 1} & K_0(B) \otimes K_n(B) & \longrightarrow & K_n(B) \\ \downarrow{\scriptstyle 1\otimes f_*} & & & & \downarrow{\scriptstyle f_*} \\ K_0(A) \otimes K_n(A) & & \longrightarrow & & K_n(A) \end{array} \quad \textit{commutes.}$$

Proof. For $P \in \mathscr{P}(A)$ and $M \in \mathbb{H}(B)$, we have the identity

$$M \mapsto P \otimes_A B \otimes M = P \otimes_A M.$$

Now, the theorem follows from additivity theorem (3.4.4). ■

Exact Connected Sequences

Corollary 3.5.11. *Suppose $\mathscr{E}$ is an exact category. Let $T = \{T_i : i \geq 1\}$ be an exact connected sequence of functors from an exact category $\mathscr{E}$ to an abelian category $\mathscr{A}$. Meaning*

1. $T_i : \mathscr{E} \longrightarrow \mathscr{A}$ *are functors* (not necessarily exact),
2. *For each conflation*

$$K \hookrightarrow M \twoheadrightarrow C \quad \text{in } \mathscr{E} \tag{3.19}$$

and $i \geq 2$, there is a morphism $\delta_i : T_i(C) \longrightarrow T_{i-1}(K)$ such that the following:

$$\cdots \longrightarrow T_2(K) \longrightarrow T_2(M) \longrightarrow T_2(C) \xrightarrow{\delta_2} T_1(K) \longrightarrow T_1(M) \longrightarrow T_1(C)$$

is exact.

Let $\mathscr{P}$ be the full subcategory of T-acyclic objects in $\mathscr{E}$. Meaning

$$Obj(\mathscr{P}) = \{M \in Obj(\mathscr{E}) : T_i(M) = 0\ \forall\ i \geq 1\} \quad \textit{is the full subcategroy } \mathscr{E}.$$

Then $\mathscr{P}$ is an extension closed subcategory of $\mathscr{E}$.

Further assume that

1. *For all $M \in Obj(\mathscr{E})$, there is a deflation $P \twoheadrightarrow M$ for some $P \in Obj(\mathscr{P})$.*
2. *For all $M \in Obj(\mathscr{E})$, $T_n(M) = 0 \quad \forall\ n \gg 0$.*

Then $K_i(\mathscr{P}) \xrightarrow{\sim} K_i(\mathscr{E})\ \forall\ i \geq 0.$

Proof. It follows immediately that $\mathscr{P}$ is extension closed and hence $K_i(\mathscr{P})$ is defined. For integers $n \geq 0$, define the full subcategories

$$\mathscr{P}_n = \{M \in Obj(\mathscr{E}) : T_i(M) = 0\ \forall\ i \geq n+1\}.$$

It follows that $\mathscr{P}_n \subseteq \mathscr{P}_{n+1}$ is an extension closed subcategory, for all $n \geq 0$. It follows that, given a conflation (3.19) in $\mathscr{P}_{n+1}$, if $M, C \in Obj(\mathscr{P}_n)$, then $K \in Obj(\mathscr{P}_n)$. Both the conditions of (3.5.3) are satisfied. Hence, for all integers $i \geq 0$, $K_i(\mathscr{P}_n) \xrightarrow{\sim} K_i(\mathscr{P}_{n+1})$ are isomorphisms. By hypothesis, $\mathscr{E} = \mathscr{P}_\infty = \cup_{n \geq 0} \mathscr{P}_n$. Therefore, $K_i(\mathscr{E}) = K_i(\mathscr{P}_0)$ for all $i \geq 0$. The proof is complete. ■

Remark 3.5.12. The **connected sequence of functors** defined above essentially comprises the **δ-functors** mentioned in [Hr, Section III.1, p. 205]. The δ-functors have much broader applications.

Example 3.5.13. For a Noetherian ring A (*not necessarily commutative*), the category of finitely generated left A-modules is denoted by $Coh(A)$. The K-groups would be denoted by $G_n(A) = K_n(Coh(A))$.

Let $f : A \longrightarrow B$ be a homomorphism of Noetherian rings. Define functors $T_n : Coh(A) \longrightarrow \mathbf{Ab}$ by $T_n(M) = Tor_n^A(B, M)$. Then $\{T_n : n \geq 0\}$ defines a connected sequence of functors. Assume $\forall\ M \in Coh(A),\quad T_n(M) = 0,\ \forall\ n \gg 0$. Let $\mathscr{P} \subseteq Coh(A)$ be the full subcategory objects M, with $Tor_n^A(M, B) = 0\ \forall\ n \neq 0$. Then $K_n(\mathscr{P}) \cong G_n(A)$ for all $n \geq 0$, by (3.5.11).

Note, the functor $\mathscr{P} \longrightarrow Coh(B)$, sending $M \mapsto M \otimes_A B$, is exact. So, there is a well-defined homomorphism

$$G_n(A) \cong K_n(\mathscr{P}) \longrightarrow G_n(B) \quad \forall\ n \geq 0.$$

The conditions above are satisfied if B has finite Tor/Flat dimension (5.4.1) over A, as a right A-module. When $f : A \longrightarrow B$ is flat, we have the same, without having to use (3.5.11).

3.6. Dévissage and Localization in Abelian Categories

To provide context, let A be a commutative Noetherian ring and $\overline{A} = \frac{A}{\sqrt{0}}$ be the reduced ring, where $\sqrt{0}$ denotes the nil radical of A.

Among the goals of the Dévissage theorems is to prove $G_n(A) \cong G_n(\overline{A})$ for all $n \geq 0$ (3.6.3).

We will be working on the following setup.

Definition 3.6.1. Let $\mathscr{A}$ be an abelian category, with a set of isomorphism classes of objects. (*Recall abelian categories are exact categories.*) Let $\mathscr{B} \subseteq \mathscr{A}$ be a subcategory such that (1) $\mathscr{B} \subseteq \mathscr{A}$ is a full subcategory, (2) $\mathscr{B}$ is closed under subobjects, quotient objects, and finite products in $\mathscr{A}$. In this case,

1. $\mathscr{B}$ is an abelian category, and the inclusion $\mathscr{B} \hookrightarrow \mathscr{A}$ is an exact functor.
2. Further, $\mathbb{Q}\mathscr{B} \hookrightarrow \mathbb{Q}\mathscr{A}$ is a full subcategory.

   Proof. Let $M, N \in Obj(\mathscr{B})$. The map $Mor_{\mathbb{Q}\mathscr{B}}(M, N) \longrightarrow Mor_{\mathbb{Q}\mathscr{A}}(M, N)$ is injective from the definition of morphisms. Now a map $f \in Mor_{\mathbb{Q}\mathscr{A}}(M, N)$ can be written as $f = p^! i_!$ where $p : U \twoheadrightarrow M$ is a deflation and $i : U \hookrightarrow N$ is an inflation. Since $\mathscr{B}$ is closed under subobjects, $U \in Obj(\mathscr{B})$. Since, $\mathscr{B}$ is also closed under quotient, $U \hookrightarrow N$ is an inflation in $\mathscr{B}$. Similarly, p is a deflation in $\mathscr{B}$. So, the map is a bijection. ■

The following is known as Dévissage Theorem.

Theorem 3.6.2 (Dévissage). *Let $\mathscr{B} \hookrightarrow \mathscr{A}$ be as in the setup above (3.6.1). Assume that every object $M \in Obj(\mathscr{A})$ has a filtration:*

$$0 = M_0 \hookrightarrow M_1 \hookrightarrow \cdots \hookrightarrow M_r =: M \quad \ni \ \forall\, j = 1, \ldots, r-1 \quad \frac{M_j}{M_{j-1}} \in Obj(\mathscr{B}).$$

Then $\mathbb{Q}\mathscr{B} \hookrightarrow \mathbb{Q}\mathscr{A}$ is a homotopy equivalence, and hence $K_n(\mathscr{B}) \xrightarrow{\sim} K_n(\mathscr{A})$ is an isomorphism for all integers $n \geq 0$.

Proof. Let $f : \mathbb{Q}\mathscr{B} \hookrightarrow \mathbb{Q}\mathscr{A}$ be the inclusion functor. Fix $M \in Obj(\mathscr{A})$. By Theorem A (2.2.2), it suffices to show that f/M is contractible. An object of f/M is given by a pair (N, u), where $N \in Obj(\mathscr{B})$ and $u \in Mor_{\mathbb{Q}\mathscr{A}}(N, M)$. Such a morphism u is given by an $\mathscr{A}$-admissible layer (M_0, M_1) in M, as in the diagram

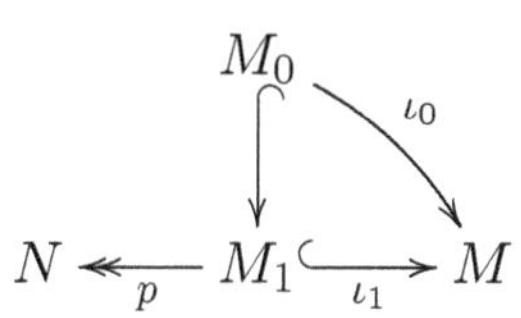

with $M_0 = \ker(p)$.

Let $\mathcal{L}(M)$ denote the category (partially ordered set) of all such $\mathscr{A}$-admissible layers $((M_0, \iota_0), (M_1, \iota_1))$ in M. We abbreviate $(M_0, M_1) := ((M_0, \iota_0), (M_1, \iota_1))$. Let $\mathcal{L}(M, \mathscr{B}) = \{(M_0, M_1) \in \mathcal{L}(M) : M_1/M_0 \in Obj(\mathscr{B})\}$ be the full subcategory of $\mathcal{L}(M)$. Further,

$$(M_0, M_1) \leq (M_0', M_1') \quad \text{if } M_0' \leq M_0 \leq M_1 \leq M_1' \qquad \text{(see 3.1.9).}$$

The bijection $Mor_{Q\mathscr{E}/M}(N, N') \xrightarrow{\sim} Mor_{\mathcal{L}(M)}((M_0, M_1), (M_0', M_1'))$ is clarified by the diagram

$$\begin{array}{ccccc}
M_0 & \hookleftarrow & M_0' & & \\
\downarrow & & \downarrow & & \\
M_1 & \hookrightarrow & M_1' & \hookrightarrow & M \\
{\scriptstyle q}\downarrow & & \downarrow{\scriptstyle p} & & \\
N & \twoheadleftarrow \frac{M_1}{M_0'} \hookrightarrow & N' & &
\end{array}
\qquad \text{with} \begin{cases} M_0' = \ker(p), \\ M_0 = \ker(q) \end{cases}$$

Since $\mathscr{B}$ is assumed to be closed under subobjects, $\frac{M_1}{M_0'} \in Obj(\mathscr{B})$. Since $\mathcal{L}(0, \mathscr{B})$ is trivial and M has a filtration, with quotients in $\mathscr{B}$, it is enough to prove that the inclusion $\iota : \mathcal{L}(M', \mathscr{B}) \hookrightarrow \mathcal{L}(M, \mathscr{B})$ is a homotopy equivalence, with $M' \subseteq M$ and $\frac{M}{M'} \in Obj(\mathscr{B})$. We define two functors:

$$\begin{cases} r : \mathcal{L}(M, \mathscr{B}) \longrightarrow \mathcal{L}(M', \mathscr{B}) & (M_0, M_1) \mapsto (M_0 \cap M', M_1 \cap M') \\ s : \mathcal{L}(M, \mathscr{B}) \longrightarrow \mathcal{L}(M, \mathscr{B}) & (M_0, M_1) \mapsto (M_0 \cap M', M_1) \end{cases}$$

$$r, s \text{ are well defined because } \frac{M_1 \cap M'}{M_0 \cap M'} \subseteq \frac{M_1}{M_0 \cap M'} \subseteq \frac{M_1}{M_0} \times \frac{M_1}{M'} \in Obj(\mathscr{B}).$$

Consider the following:

1. $r\iota = Id_{\mathcal{L}(M', \mathscr{B})}$.
2. There are natural transformations:

$$\begin{cases} \iota r \longrightarrow s & (M_0 \cap M', M_1 \cap M') \leq (M_0 \cap M', M_1) \\ Id_{\mathcal{J}(M)} \longrightarrow s & (M_0, M_1) \leq (M_0 \cap M', M_1) \end{cases}$$

Now, by (2.1.9), r is a homotopy inverse of ι. The proof is complete. ∎

The following is an important corollary of our primary interest.

Lemma 3.6.3. *Let A be a Noetherian ring (commutative) and $\sqrt{0}$ be the nil radical. Let $\overline{A} = \frac{A}{\sqrt{0}}$. As before, we denote $G_n(A) = K_n(Coh(A))$, for all $n \geq 0$. Then for all integers $n \geq 0$, we have $G_n(A) = G_n(\overline{A})$.*

Proof. Consider the inclusion functor $Coh\left(\overline{A}\right) \hookrightarrow Coh(A)$. Write $I = \sqrt{0}$. Then $I^r = 0$ for some integer $r \geq 0$. So, for any object M in $Coh(A)$, there is a filtration:

$$0 \subseteq I^{r-1}M \subseteq \cdots \subseteq IM \subseteq M. \qquad \text{with } \frac{I^k M}{I^{k+1}} \in Coh\left(\overline{A}\right).$$

The lemma follows from (3.6.2). The proof is complete. ■

The following provides a contrast.

Remark 3.6.4. The following are results of Richard Swan. Let A and $\overline{A}$ be as in (3.6.3). Let $U(A)$ denote the groups of units of A, and as before, $K_n(A) = K_n(\mathscr{P}(A))$, for all $n \geq 0$. In general, $K_1(A) \neq K_1(\overline{A})$. To see this, assume A is local. Matrices $\alpha \in GL(A)$ are diagonalizable, up to multiplication by some elementary matrices. So, $K_1(A) = U(A)$ and $K_1(\overline{A}) = U(\overline{A})$. So, $K_1(A) \neq K_1(\overline{A})$ unless $\sqrt{0} = 0$.

However, even when A is nonlocal, $K_0(A) = K_0(\overline{A})$. Define the natural functor $\mathfrak{q} : \mathscr{P}(A) \longrightarrow \mathscr{P}(\overline{A})$ by $\mathfrak{q}(P) := \overline{P} = \frac{P}{\sqrt{0}P}$. Since $\mathfrak{q}$ is an exact functor, it induces a homomorphism $\mathfrak{q}_0 : K_0(A) \xrightarrow{\sim} K_0(\overline{A})$. By Nakayama's lemma, $\overline{P} \cong \overline{Q} \Longrightarrow P \cong Q$. It follows from this that $\mathfrak{q}_0$ is an injective map.

To prove $\mathfrak{q}_0$ is surjective, let $\mathcal{P} \in Obj(\mathscr{P}(\overline{A}))$. We will construct $P \in Obj(\mathscr{P}(A))$, such that $\overline{P} \cong \mathcal{P}$. To see this, note that there is an idempotent matrix $\varphi \in GL_n(\overline{A})$ for some $n \geq 0$ such that $\mathcal{P} = \ker(\varphi)$. Write $F = A^n$ and $\overline{F} = \overline{A}^n$. Let $\Phi \in GL_n(A)$ be any lift of φ. Write $P = \ker(\Phi) \subseteq F$, $Q = \Phi(F) \subseteq F$, and $\mathcal{Q} = \varphi(\overline{F}) \subseteq \overline{F}$. We have the following commutative diagram of exact sequences:

$$\begin{array}{ccccccccc} 0 & \longrightarrow & P & \xrightarrow{\iota} & F & \xrightarrow{\Phi} & Q & \longrightarrow & 0 \\ & & \downarrow & & \downarrow & & \downarrow & & \\ 0 & \longrightarrow & \mathcal{P} & \xrightarrow[i]{} & \overline{F} & \xrightarrow[\varphi]{} & \mathcal{Q} & \longrightarrow & 0 \end{array}$$

where the vertical maps are surjective.

In fact, the second line splits. With $t = 1 - \varphi : \overline{F} \longrightarrow \mathcal{P}$, we have $ti = 1_{\mathcal{P}}$. Let $\eta : F \longrightarrow P$ be any lift of t. We have $(\eta\iota) \otimes \overline{A} = ti = 1$. By Nakayama's Lemma, $co\ker(\eta\iota) = 0$. So, $\eta\iota : P \twoheadrightarrow P$ is surjective. So, $\eta\iota$ is an isomorphism, by the Noetherian property of P. Therefore, ι splits and $P \in Obj(\mathscr{P}(A))$, with $\overline{P} = \mathcal{P}$. So, $\mathfrak{q}_0$ is surjective. ∎

3.6.1. *Semisimple Objects in abelian categories*

Definition 3.6.5. Let $\mathscr{A}$ be an abelian category.

1. An object M in $\mathscr{A}$ is called a **simple object** if only subobjects of M are $0, M$.
2. An object M in $\mathscr{A}$ is called a **semisimple object** if M is an image of a direct sum of simple objects. In other words, if there is a surjective map $\bigoplus_{i\in I} S_i \twoheadrightarrow M$, where $\{S_i :\in I\}$ is a (possibly infinite) set of simple objects.
3. We say that an object M has finite length if M has a finite filtration

$$0 = M_0 \hookrightarrow M_1 \hookrightarrow M_2 \hookrightarrow \cdots \hookrightarrow M_n =: M \quad \ni \ \forall\, j = 1, \ldots, n \quad \frac{M_j}{M_{j-1}} \text{ is simple.}$$

(*Define length* $\ell(M) := n$. *It is left as an exercise (3.8.4) that* $\ell(M)$ *is well defined.*)

Lemma 3.6.6. *Let* $\mathscr{A}$ *be an abelian category and* M *be an object. Then the following are equivalent:*

1. M *is semisimple.*
2. $M \cong \oplus_{i\in I} S_i$, *where* $\{S_i :\in I\}$ *is a set of simple objects. (So,* $\ell(M) = Card(I)$.*)*

Proof. (2) $\Longrightarrow$ (1) is obvious. To prove the converse, let $f : \oplus_{i\in I} S_i \twoheadrightarrow M$ be a surjective map, where $\{S_i : i \in I\}$ is a set of simple objects. Let $\mathscr{I} := \{I_0 \subseteq I : f_{|\cup_{i\in I_0} S_i}$ is injective$\}$. By Zorn's lemma, $\mathscr{I}$ has a maximal element $J \subseteq I$. Write $N = \oplus_{i\in J} S_i \subseteq \oplus_{i\in I} S_i$. Claim $f_0 := f_{|N} : N \xrightarrow{\sim} M$ is an isomorphism. If not, there is $i_1 \in I \setminus J$ such that $f(S_{i_1}) \not\subseteq f_0(N)$. Let $f_1 := f_{|S_{i_1}} : S_{i_1} \longrightarrow M$. We have $f_0 : N \longrightarrow M$ and $f_1 : S_1 \longrightarrow M$ are injective. Write $g = f_{|(N\oplus S_{i_1})} : N \oplus S_{i_1} \longrightarrow M$. Then $g = f_0 \oplus f_1$. We prove g is injective. Let $\alpha : W \longrightarrow N \oplus S_{i_1}$ be a map such that $g\alpha = 0$. Since $N \oplus S_{i_1} \cong N \times S_{i_1}$, it follows that $f_0 p_0 \alpha = 0$ and $f_1 p_1 \alpha = 0$, where $p_0 : N \oplus S_{i_1} \longrightarrow$

N and $p_1 : N \oplus S_{i_1} \longrightarrow S_{i_1}$ are the projections. Since, f_0, f_1 are injective, we have $p_1\alpha = 0$, $p_0\alpha = 0$. Hence, $\alpha = 0$. So, g is injective. This is a contradiction to the maximality of J. Therefore, f_0 is an isomorphism. The proof is complete. ■

Corollary 3.6.7. *Let $\mathscr{A}$ be an abelian category* (with a set of isomorphism classes of objects) *such that every object has finite length. Then*

$$\forall\ n \geq 0 \qquad K_n(\mathscr{A}) = \bigoplus_{j \in J} K_n(D_j)$$

where $\{S_j : j \in J\}$ is a set of representatives, of isomorphism classes of simple objects of $\mathscr{A}$, and $D_j = End(S_j)^o$. (Note that D_j is a division ring.)

Proof. Let $\mathscr{B}$ be the full subcategory of semisimple objects in $\mathscr{A}$. Then by Dévissage (3.6.2), $K_n(\mathscr{B}) \xrightarrow{\sim} K_n(\mathscr{A})$ is an isomorphism, $\forall\ n \geq 0$.

For $j \in J$, let $\mathscr{B}_j$ be the full subcategory of $\mathscr{B}$ consisting of all objects which are isomorphic to a direct sum of S_j (3.6.6). So, $\mathscr{B} = \prod_{j \in J} \mathscr{B}_j = \lim\left(\prod_{k=1}^m \mathscr{B}_{j_k}\right)$ is the limit of the finite direct product. By Lemma 3.3.2(4), 3.3.3, it follows that

$$\forall\ n \geq 0 \quad K_n(\mathscr{B}) = \begin{cases} K_n\left(\lim\left(\prod_{k=1}^m \mathscr{B}_{j_k}\right)\right) \xrightarrow{\sim} \lim\left(K_n\left(\prod_{k=1}^m \mathscr{B}_{j_k}\right)\right) \\ \xrightarrow{\sim} \lim\left(\bigoplus_{k=1}^m K_n\left(\mathscr{B}_{j_k}\right)\right) \xrightarrow{\sim} \bigoplus_{j \in J} K_n(\mathscr{B}_j) \end{cases}$$

are isomorphisms. Now, the proof follows from the following Lemma (3.6.8). ■

Lemma 3.6.8. *Suppose $\mathscr{B}$ is an abelian category such that (1) each object is semisimple and has finite length, and* (2) *$\mathscr{B}$ has only a single simple object X, up to isomorphism. Let $D := Mor(X, X)^o$ and $\mathscr{P}(D)$ denote the category of finite dimensional left D-vector spaces. Then the functor*

$$F : \mathscr{B} \longrightarrow \mathscr{P}(D) \quad \ni \quad F(M) = Mor_{\mathscr{B}}(X, M)$$

is an equivalence of categories.

Proof. It follows from Lemma 3.6.6 that, each object $M \cong X^n$ for some integer $n \geq 0$. ■

3.6.2. *Quillen's localization theorem*

The following is the celebrated localization theorem of Quillen.

Theorem 3.6.9 (Localization Theorem). *Let $\mathscr{A}$ be an abelian category and $\mathscr{B} \subseteq \mathscr{A}$ be a Serre subcategory* (A.5.6), *$\iota : \mathscr{B} \hookrightarrow \mathscr{A}$ being the inclusion functor. Let $\frac{\mathscr{A}}{\mathscr{B}}$ be the quotient abelian category and $\mathfrak{q} : \mathscr{A} \longrightarrow \frac{\mathscr{A}}{\mathscr{B}}$ be the canonical functor. Then the sequence*

$$\mathscr{B} \xrightarrow{\iota} \mathscr{A} \xrightarrow{\mathfrak{q}} \frac{\mathscr{A}}{\mathscr{B}} \qquad \textit{is a homotopy fibration.}$$

Consequently, there is an exact sequence

$$\cdots \xrightarrow{\mathfrak{q}_*} K_1\left(\tfrac{\mathscr{A}}{\mathscr{B}}\right) \longrightarrow K_0(\mathscr{B}) \xrightarrow{\iota_*} K_0(\mathscr{A}) \xrightarrow{\mathfrak{q}_*} K_0\left(\tfrac{\mathscr{A}}{\mathscr{B}}\right) \longrightarrow 0.$$

Proof. Fix a zero object 0 in $\mathscr{A}$ and let 0 denote its image in $\frac{\mathscr{A}}{\mathscr{B}}$, which is also a zero object in $\frac{\mathscr{A}}{\mathscr{B}}$. From the definition of Serre subcategories (A.5.6) and the construction (A.5.7) of $\frac{\mathscr{A}}{\mathscr{B}}$, it follows that $\mathscr{B}$ is the full subcategory of objects M in $\mathscr{A}$ such that $\mathfrak{q}M \simeq 0$. The diagram

$$\begin{array}{ccc} \mathbb{Q}\mathscr{B} & \xrightarrow{\mathbb{Q}\iota} & \mathbb{Q}\mathscr{A} \\ & \searrow^{0} & \downarrow \mathbb{Q}\mathfrak{q} \\ & & \mathbb{Q}\frac{\mathscr{A}}{\mathscr{B}} \end{array} \qquad \text{commutes up to natural equivalence.}$$

(*Sometimes, we may ignore distinguishing between* $\mathfrak{q}$ *and* $\mathbb{Q}\mathfrak{q}$. *It will be clear from the context whether we mean* $\mathfrak{q}$ *or* $\mathbb{Q}\mathfrak{q}$.) There is an obvious natural functor $F : \mathbb{Q}\mathscr{B} \longrightarrow 0/\mathbb{Q}\mathfrak{q}$, and a functor G, as follows:

$$\begin{cases} F : \mathbb{Q}\mathscr{B} \longrightarrow 0/\mathbb{Q}\mathfrak{q} & F(M) = (M, 0 \xrightarrow{\sim} \mathfrak{q}M) \\ G : 0/\mathbb{Q}\mathfrak{q} \longrightarrow \mathbb{Q}\mathscr{A} & G(N, u) = N \end{cases}$$

$$\ni \quad \text{the diagram} \qquad \begin{array}{ccc} \mathbb{Q}\mathscr{B} & \xrightarrow{F} & 0/\mathbb{Q}\mathfrak{q} \\ & \searrow^{\mathbb{Q}\iota} & \downarrow G \\ & & \mathbb{Q}\mathscr{A} \end{array}$$

commutes. To establish the theorem, we check the hypotheses of Theorem B (2.2.3) to $\mathbb{Q}\mathfrak{q}$, and establish that F is a homotopy equivalence. The following would settle the proof of the theorem:

1. To apply Theorem B (2.2.3) to $\mathbb{Q}\mathfrak{q}$, we prove that for all $u : V' \longrightarrow V$ in $\mathbb{Q}\frac{\mathscr{A}}{\mathscr{B}}$, the base change functor $u^* : V/\mathbb{Q}\mathfrak{q} \longrightarrow V'/\mathbb{Q}\mathfrak{q}$ is a homotopy equivalence. Recall the functor u^* sends $(W, v) \mapsto (W, vu)$. Such morphisms u in $\mathbb{Q}\frac{\mathscr{A}}{\mathscr{B}}$ factors $u = i_! p^!$, as composition of *surjective* and *injective* maps. Therefore, it suffices to prove the same when $u = i_!$ is *injective* and when $u = p^!$ is *surjective*, in $\mathbb{Q}\frac{\mathscr{A}}{\mathscr{B}}$. We will prove this for *injective* maps $u = i_!$ only. The case when $u = p^!$ is *surjective* follows by similar arguments or by reversing the arrows, which changes *surjective* maps to *injective* maps.
   Further, let $i_V : 0 \hookrightarrow V$ denote the unique arrow in $\frac{\mathscr{A}}{\mathscr{B}}$. For an inflation $i : V' \hookrightarrow V$ in $\frac{\mathscr{A}}{\mathscr{B}}$, we have $i_V = i i_{V'}$. So, we only prove that for $u = (i_V)_! : 0 \longrightarrow V$ in $\mathbb{Q}\frac{\mathscr{A}}{\mathscr{B}}$, the base chance functor $u^* : V/\mathbb{Q}\mathfrak{q} \longrightarrow 0/\mathbb{Q}\mathfrak{q}$ is a homotopy equivalence. This will be completed in the following (3.6.19).
2. For objects V in $\mathbb{Q}\frac{\mathscr{A}}{\mathscr{B}}$, let $\mathscr{F}_V$ be the full subcategory of $V/\mathbb{Q}\mathfrak{q}$, consisting of objects (M, u) such that $u : V \xrightarrow{\sim} \mathfrak{q}M$ is an isomorphism in $\mathbb{Q}\frac{\mathscr{A}}{\mathscr{B}}$. It is clear $\mathbb{Q}\mathscr{B} \xrightarrow{\sim} \mathscr{F}_0$ is an equivalence of categories. We prove in the following (3.6.10) that the inclusion $\mathscr{F}_V \hookrightarrow V/\mathbb{Q}\mathfrak{q}$ is a homotopy equivalence. In particular, $F : \mathbb{Q}\mathscr{B} \longrightarrow 0/\mathbb{Q}\mathfrak{q}$ is a homotopy equivalence.

So, it remains to establish (3.6.10) and prove $(i_V)_!^*$ is a homotopy equivalence (3.6.19). Thus, the proof of the theorem will be complete. ■

Lemma 3.6.10. *Use the notation form* (3.6.9). *For* V *in* $\mathbb{Q}\frac{\mathscr{A}}{\mathscr{B}}$, *the inclusion functor* $\iota_V : \mathscr{F}_V \hookrightarrow V/\mathbb{Q}\mathfrak{q}$ *is a homotopy equivalence.*

Proof. Fix $(N, v) \in Obj(V/\mathbb{Q}\mathfrak{q})$. We prove that $\iota_V/(N, v)$ is contractible. By Theorem A (2.2.2), this will complete the proof. The map $v : V \longrightarrow \mathfrak{q}N$ in $\mathbb{Q}\frac{\mathscr{A}}{\mathscr{B}}$ is represented by an admissible layer (N_0, N_1) in $\mathfrak{q}N$, and an isomorphism $\theta(v) : \frac{N_1}{N_0} \xrightarrow{\sim} V$, in $\frac{\mathscr{A}}{\mathscr{B}}$ (*that is*, $N_0 \hookrightarrow N_1 \hookrightarrow \mathfrak{q}N$).

Write $\mathscr{C} = \iota_V/(N, v)$. An object in $\mathscr{C} = \iota_V/(N, v)$ is given by $((M, \mu), w)$ with $(M, \mu) \in Obj(\mathscr{F}_V)$ and $w \in Mor_{\mathscr{C}}((M, \mu), (N, v))$. Indeed, $w \in Mor_{\mathbb{Q}\mathscr{A}}(M, N)$ with $(\mathfrak{q}w)\mu = v$ in $\mathbb{Q}\frac{\mathscr{A}}{\mathscr{B}}$. Further,

$\mu : V \xrightarrow{\sim} \mathfrak{q}M$ is an isomorphism. As an admissible layer, μ is given by $(0, \mathfrak{q}M)$ and isomorphism $\theta(\mu) : \mathfrak{q}M \xrightarrow{\sim} V$ in $\frac{\mathscr{A}}{\mathscr{B}}$. As an admissible layer, w is represented by $(M_0(w), M_1(w))$ in N and an isomorphism $\theta(w) : \frac{M_1(w)}{M_0(w)} \xrightarrow{\sim} M$ in $\mathscr{A}$ (i.e. $M_0(w) \hookrightarrow M_1(w) \hookrightarrow N$ in $\mathscr{A}$). To translate the identity $(\mathfrak{q}w)\mu = v$ as admissible layers in N, consider the diagram

$$\begin{array}{ccc} \frac{\mathfrak{q}M_1(w)}{\mathfrak{q}M_0(w)} & \xrightarrow[\sim]{\mathfrak{q}\theta(w)} & \mathfrak{q}M \\ & & \wr\big\downarrow \theta(\mu) \\ \frac{N_1}{N_0} & \xrightarrow[\theta(v)]{\sim} & V \end{array} \qquad \text{in } \frac{\mathscr{A}}{\mathscr{B}}.$$

Here the composition $(\mathfrak{q}w)\mu$ represents the composition $\theta(\mu)(\mathfrak{q}\theta(w))$, and $\theta(v)$ represents v. Since $(\mathfrak{q}w)\mu = v$, we have $(N_0, N_1) = (\mathfrak{q}M_0(w), \mathfrak{q}M_1(w))$.

As before, let $\mathcal{L}(N)$ denote the set of admissible layers $((M_0, \iota_0), (M_1, \iota_1))$ in N in the category $\mathscr{A}$. We will abbreviate $(M_0, M_1) := ((M_0, \iota_0), (M_1, \iota_1))$. Write

$$\mathcal{L}(N, v) = \{(M_0, M_1) \in \mathcal{L}(N) : (\mathfrak{q}M_0, \mathfrak{q}M_1) = (N_0, N_1)\}.$$

The partial order relation in $\mathcal{L}(N)$ induces a partial order relation in $\mathcal{L}(N, v)$. It follows from the above discussions that the following functors:

$$\begin{cases} \iota_V/(N, v) \longrightarrow \mathcal{L}(N, v) \quad ((M, \mu), w) \mapsto (M_0(w), M_1(w)) \\ \mathcal{L}(N, v) \longrightarrow \iota_V/(N, v) \quad (M_0, M_1) \mapsto \left(\left(\frac{M_1}{M_0}, \mu\right), w\right) \\ \text{where } \mu = v : \mathfrak{q}\frac{M_1}{M_0} = \frac{N_1}{N_0} \xrightarrow{\sim} V, \ \text{ and } \ w \text{ is defined by } (M_0, M_1). \end{cases}$$

are inverses of each other. So, $\iota_V/(N, v)$ is equivalent to the partially ordered set of layers $\mathcal{L}(N, v)$. Now, for $(M_0, M_1), (M_0', M_1') \in \mathcal{L}(N, v)$, we have

$$(M_0, M_1) \leq (M_0 \cap M_0', M_1 + M_1'), \geq (M_0', M_1').$$

Recall that $M_0 \cap M_0'$, $M_1 + M_1'$ are defined as follows:

$$M_0 \cap M_0' = \ker\left(N \longrightarrow \frac{N}{M_0} \oplus \frac{N}{M_0'} \right),$$

$$M_1 + M_1' = \text{Image}\left(M_1 \oplus M_1' \longrightarrow N \right).$$

Since $\mathfrak{q}$ is an exact functor, it follows that $\mathfrak{q}(M_0 \cap M_0') = N_0$ and $\mathfrak{q}(M_1 + M_1') = N_1$. So, $(M_0 \cap M_0', M_1 + M_1') \in \mathcal{L}(N, v)$. Therefore, $\mathcal{L}(N, v)$ is a directed set. So, by (2.1.19), $\mathcal{L}(N, v)$ is contractible. The proof is complete. ∎

Next, we introduce the following categories and notation.

Definition 3.6.11. Use all the notation in (3.6.9). Let $N \in Obj(\mathscr{A})$. Define the category $\mathbf{E}_N$ as follows: (1) Objects of $\mathbf{E}_N$ are the pairs (M, h), where $h \in Mor_{\mathscr{A}}(M, N)$ such that $\ker(h), co\ker(h) \in \mathscr{B}$, or equivalently $\mathfrak{q}h$ is an isomorphism in $\frac{\mathscr{A}}{\mathscr{B}}$ *(such an arrow h is also called mod-$\mathscr{B}$ isomorphism).* (2) For objects (M_1, h_1), (M_2, h_2), let

$$
\begin{aligned}
&Mor_{\mathbf{E}_N}\left((M_1, h_1), (M_2, h_2)\right)\\
&\quad = \left\{ u \in Mor_{\mathbb{Q}\mathscr{A}}(M_1, M_2) : u = i_! j^!, \text{ and } \quad h_2 i = h_1 j \text{ in } \mathscr{A} \right\}.
\end{aligned}
$$

So, a morphism $u \in Mor_{\mathbb{Q}\mathscr{A}}(M_1, M_2)$, represented by $u = \left(M_1 \overset{j}{\twoheadleftarrow} U \overset{i}{\hookrightarrow} M_2 \right)$, is a morphism $u \in Mor_{\mathbf{E}_N}\big((M_1, h_1), (M_2, h_2)\big)$, if the diagram

$$
\begin{array}{ccc}
U & \overset{i}{\hookrightarrow} & M_2 \\
{\scriptstyle j}\downarrow & & \downarrow{\scriptstyle h_2} \\
M_1 & \underset{h_1}{\longrightarrow} & N
\end{array}
\qquad \text{commutes.} \tag{3.20}
$$

One checks that commutativity of (3.20) is independent of the factorization $u = i_! j^!$. Further, the diagram (3.20) induces the following morphism in $\mathscr{A}$:

$$
\kappa(u) := \left(\ker(h_1) \twoheadleftarrow \ker(h_1 j) \hookrightarrow \ker(h_2) \right) \in Mor_{\mathbb{Q}\mathscr{A}}(\ker(h_1), \ker(h_2))
$$

Since $\mathscr{B}$ is closed under subobjects in $\mathscr{A}$, $\ker(h_1 j) \in Obj(\mathscr{B})$. So, $\kappa(u) \in Mor_{\mathbb{Q}\mathscr{B}}(\ker(h_1), \ker(h_2))$. Define the functor

$$
\kappa_N : \mathbf{E}_N \longrightarrow \mathbb{Q}\mathscr{B} \quad \text{by} \quad \begin{cases} \kappa((M, h)) = \ker(h) \ \forall \ (M, h) \in Obj\left(\mathbf{E}_N\right) \\ \kappa(u) = \qquad\qquad \text{as above} \end{cases}
$$

So, $\kappa((M, h)) = \ker(h)$ is determined up to canonical isomorphisms, by the exact sequence

$$
0 \longrightarrow \kappa\left(M, h\right) \longrightarrow M \longrightarrow N.
$$

Before we proceed, we provide a roadmap toward the completion of the proof of the Localization theorem (3.6.9). For any choice (*and we would have some*) of a pair (N, φ) with $N \in Obj(\mathscr{A})$ and an isomorphism $\varphi : \mathfrak{q}N \xrightarrow{\sim} V$ in $\frac{\mathscr{A}}{\mathscr{B}}$, we develop the following diagram of functors:

$$\begin{array}{ccccc} \mathbf{E}_N & \xrightarrow{\psi_{(N,\varphi)}} & \mathscr{F}_V & \overset{\sim}{\hookrightarrow} & V/\mathbb{Q}\mathfrak{q} \\ \downarrow{\scriptstyle \kappa_N} & & & & \downarrow{\scriptstyle ((i_V)_!)^*} \\ \mathbb{Q}\mathscr{B} & \xrightarrow{\sim} & \mathscr{F}_0 & \overset{\sim}{\hookrightarrow} & 0/\mathbb{Q}\mathfrak{q} \end{array} \qquad (3.21)$$

The functor $\psi_{(N,\varphi)}$ will be defined, and we will prove that it is a homotopy equivalence (3.6.18). All the other functors have been defined. We observed that $\mathbb{Q}\mathscr{B} \xrightarrow{\sim} \mathscr{F}_0$ is a homotopy equivalence. The last two inclusions are homotopy equivalences, by (3.6.10). We also prove that κ_N is a homotopy equivalence (3.6.13). The diagram is commutative up to a natural transformation. So, the diagram is homotopy commutative. It follows from all the above that $(i_V)_!)^*$ is a homotopy equivalence. This will complete the proof of (3.6.9). Now, we proceed, according to this plan.

Lemma 3.6.12. *Use notation as in* (3.6.9, 3.6.11). *Let* $\mathbf{E}_N^d$ *denote to full subcategory of* $\mathbf{E}_N$ *consisting of objects* (M, h) *such that* h *is a deflation (i.e. epimorphism). Then the restriction map* $\kappa_N^d : \mathbf{E}_N^d \longrightarrow \mathbb{Q}\mathscr{B}$ *of* κ_N *is a homotopy equivalence.*

Proof. As usual, by (2.2.2), it is enough to prove that κ_N^d/T is contractible $\forall\, T \in Obj(\mathbb{Q}\mathscr{B})$. So, N and T remain fixed for the rest of the proof, and we write $\mathscr{C} = \kappa_N^d/T$. Also, let $\mathscr{C}' \subseteq \mathscr{C}$ be the full subcategory of objects $((M, h), u) \in \mathscr{C}$, with $u \in Mor_{\mathbb{Q}\mathscr{B}}(\ker(h), T)$ such that u is *surjective.* We will prove that the inclusion $\mathscr{C}' \subseteq \mathscr{C}$ has a left adjoint.

Let $X = ((M, h), u)$ be an object in $\mathscr{C}$. So, $(M, h) \in Obj(\mathscr{C})$, $u \in Mor_{\mathbb{Q}\mathscr{B}}(\ker(h), T)$. Factorize $u = j^! i_!$, as in $u = j^! i_! = \left(\ker(h) \overset{i}{\hookrightarrow} T_0 \overset{j}{\twoheadleftarrow} T \right)$, where i is an inflation and j is a deflation in $\mathscr{B}$. Further, $h : M \twoheadrightarrow N$ being a deflation in $\mathscr{A}$, we have the

pushforward diagram

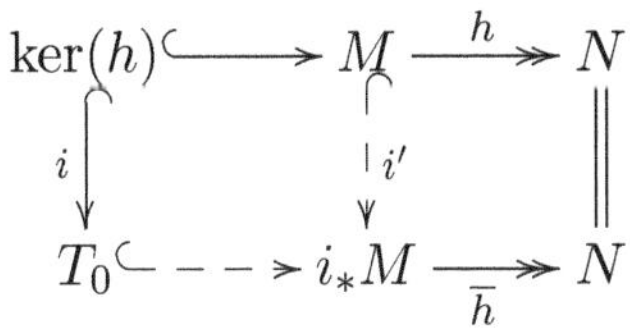

where i_*M is the pushforward.

(3.22)

Write $\overline{X} = \big((i_*M, \overline{h}), j^!\big)$. Note $\overline{X} = \big((i_*M, \overline{h}), j^!\big)$ is an object in $\mathscr{C}'$. From diagram (3.20), it is clear that $i'_! : (M, h) \longrightarrow (i_*M, \overline{h})$ is an arrow in $\mathbf{E}_N$. It follows that $i'_! : X \longrightarrow \overline{X}$ is an arrow in $\mathscr{C}$. As in (3.5.2), one checks that the arrow $X \longrightarrow \overline{X}$ is universal. Therefore, the inclusion $\mathscr{C}' \hookrightarrow \mathscr{C}$ has left adjoint (A.1.15). Hence, this inclusion is a homotopy equivalence (2.1.10). However, $\mathscr{C}'$ has an initial element, namely $((N, 1_N), j_T^j)$. So, $\mathscr{C}'$ is contractible, and so is $\mathscr{C}$. The proof is complete. ■

Lemma 3.6.13. *Use notation as in* (3.6.9, 3.6.11, 3.6.12). *The functor $\kappa_N : \mathbf{E}_N \longrightarrow \mathbb{Q}\mathscr{B}$ is a homotopy equivalence.*

Proof. Due to (3.6.12), it suffices to prove that the inclusion $\mathbf{E}_N^d \hookrightarrow \mathbf{E}_N$ is a homotopy equivalence. Recall that $\mathcal{S}(N)$ denotes the partially ordered set of subobjects $\iota : I \hookrightarrow N$ (we abbreviate $I \subseteq N$). Let $\mathcal{S}(N, \mathscr{B}) = \{I \in \mathcal{S}(N) : N/I \in \mathscr{B}\}$. Let $F : \mathbf{E}_N \longrightarrow \mathcal{S}(N, \mathscr{B})$ be the functor $F(M, h) = Im(h)$. For $I \in \mathcal{S}(N, \mathscr{B})$, it is clear that $F^{-1}I = \mathbf{E}_I^d$. Further, $I/F = \{(M, h) \in \mathbf{E}_N : I \leq h(M)\}$. There is a right adjoint functor $R : I/F \longrightarrow F\ ^{1}I$ defined by $R(M, h) = M \times_N I$ (*more simplistically*, $R(M, h) := \big(h^{-1}(I) \twoheadrightarrow I\big)$). So, F is prefibered. For $I \leq J$, the base change map $F^{-1}J \longrightarrow F^{-1}I$ sends $(M, h) \mapsto (M \times_I J \twoheadrightarrow J)$ (*more simplistically*, $(M, h) \mapsto \big(h^{-1}(I) \twoheadrightarrow I\big)$). So, it is established that F is fibered, with fiber $F^{-1}I = \mathbf{E}_I^d$. We have the commutative diagram

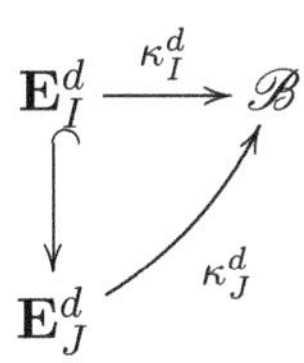

By (3.6.12), κ_J^d and κ_I^d are homotopy equivalences, and hence so is the vertical inclusion. So, the fibered version of Theorem B (2.2.6) applies to $F : \mathbf{E}_N \longrightarrow \mathcal{S}(N, \mathscr{B})$. However, $\mathcal{S}(N, \mathscr{B})$ is contractible because it has a terminal object N. Therefore, $\mathbf{E}_I^d \hookrightarrow \mathbf{E}_N$ is a homotopy equivalence for all $I \in \mathcal{S}(N, \mathscr{B})$. In particular, $\mathbf{E}_N^d \hookrightarrow \mathbf{E}_N$ is a homotopy equivalence. The proof is complete. ∎

Lemma 3.6.14. *Use notation as in* (3.6.9, 3.6.11, 3.6.12). *Let* $g : N \longrightarrow N_1$ *be an arrow in* $\mathscr{A}$*, which is a mod-*$\mathscr{B}$ *isomorphism (i.e.* $\ker(g), co\ker(g) \in Obj(\mathscr{B})$*). Then the functor* $g_* : \mathbf{E}_N \longrightarrow \mathbf{E}_{N_1}$ *sending* $(M, h) \mapsto (M, gh)$ *is a homotopy equivalence.*

Proof. For $(M, h) \in Obj(\mathbf{E}_N)$, there is a natural inflation $\iota : \kappa_N(M, h) \longrightarrow \kappa_{N'}(g_*(M, h))$, defined as follows:

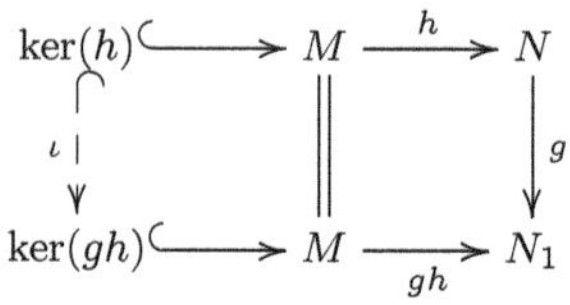

This induces a natural transformation $\iota : \kappa_N \longrightarrow \kappa_{N_1}$ ∋

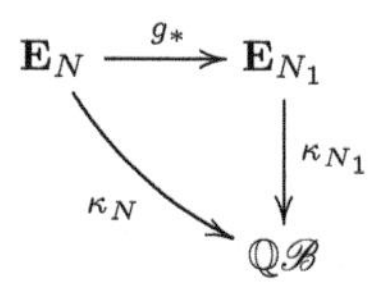

commutes via ι. Since κ_N, κ_{N_1} are homotopy equivalences (3.6.13), so is g_*. ∎

We will prove that $\mathscr{F}_V$ is homotopically equivalent to $\mathbf{E}_N$, when $\mathfrak{q}N \cong V$. For convenience of our discussions, recall (Ex. 3.8.1) that there is a bijection

$$Iso_{\frac{\mathscr{A}}{\mathscr{B}}}(M, N) \xrightarrow{\sim} Iso_{\mathbb{Q}\frac{\mathscr{A}}{\mathscr{B}}}(\mathfrak{q}M, \mathfrak{q}N) \qquad \text{sending } \varphi \mapsto \varphi_!.$$

For this reason, subsequently we may not distinguish these two sets of isomorphisms. Now, we further define a filtering category.

Definition 3.6.15. Use notation as in (3.6.9, 3.6.11, 3.6.12). Let $V \in Obj\left(\frac{\mathscr{A}}{\mathscr{B}}\right)$. Define the category $\mathscr{I}_V$ as follows: (1) Objects of $\mathscr{I}_V$ are pairs (N, φ) where $N \in Obj(\mathscr{A})$ and $\varphi : \mathfrak{q}N \xrightarrow{\sim} V$ is an isomorphism in $\frac{\mathscr{A}}{\mathscr{B}}$, (2) and $\forall\ (N_0, \varphi_0), (N_1, \varphi_1) \in \mathscr{I}_V$, define

$$Mor_{\mathscr{I}_V}((N_0, \varphi_0), (N_1, \varphi_1)) = \{g \in Mor_{\mathscr{A}}(N_0, N_1) : \varphi_0 = (\mathfrak{q}g)\varphi_1\}.$$

Lemma 3.6.16. *The category* $\mathscr{I}_V$*, as defined in* (3.6.15)*, is a filtering category.*

Proof. The category $\mathscr{I}_V$ is nonempty because $(V, \mathfrak{q}1_V)$ is in it (*considering* V *as an object in* $\mathscr{A}$). Now, let (N_0, φ_0), (N_1, φ_1) be objects in $\mathscr{I}_V$. From the construction of $\frac{\mathscr{A}}{\mathscr{B}}$, we can write $\varphi_0 = s_0^{-1} g_0 = \left(N_0 \xrightarrow{g_0} M_0 \xleftarrow{s_0} V \right)$, where s_0 is a *mod* $- \mathscr{B}$ isomorphism. Now, $g_0 : (N_0, \varphi_0) \longrightarrow (M_0, (\mathfrak{q}s_0)^{-1})$ is a morphism in $\mathscr{I}_V$. Replacing (N_0, φ_0) by $\left(M_0, (\mathfrak{q}s_0)^{-1}\right)$, we can assume $\varphi_0 = (\mathfrak{q}s_0)^{-1}$ for some mod-$\mathscr{B}$ isomorphism $s_0 : V \longrightarrow N_0$. Similarly, we can assume $\varphi_1 = (\mathfrak{q}s_1)^{-1}$ for some mod-$\mathscr{B}$ isomorphism $s_1 : V \longrightarrow N_1$. Mod-$\mathscr{B}$-isomorphisms satisfy both right and left Ore conditions (A.5.2). So, we have a commutative diagram

$$\begin{array}{ccc} V & \xrightarrow{s_0} & N_0 \\ {\scriptstyle s_1}\downarrow & & \downarrow{\scriptstyle t_0} \\ N_1 & \xrightarrow[t_1]{} & N \end{array} \qquad \text{where } t_0, t_1 \text{ are mod}-\mathscr{B} \text{ isomorphisms.}$$

Let $\varphi = \mathfrak{q}(t_0 s_0)^{-1} = \mathfrak{q}(t_1 s_1)^{-1} : N \xrightarrow{\sim} V$. Then (N, φ) is an object in $\mathscr{I}_V$. The arrows t_0, t_1 define the arrows $t_i : (N_0, \varphi_i) \longrightarrow (N, \varphi)$ in $\mathscr{I}_V$ for $i = 1, 2$.

Now suppose $g_1, g_2 : (N_1, \varphi_1) \longrightarrow (N_2, \varphi_2)$ are arrows in $\mathscr{I}_V$. So, $g_1, g_2 \in Mor_{\mathscr{A}}(N_1, N_2)$ such that $\varphi_2 \mathfrak{q} g_1 = \varphi_1 = \varphi_2 \mathfrak{q} g_2$. Therefore, $\mathfrak{q}(g_1 - g_2) = 0$. So, $I := Image(g_1 - g_2) \in \mathscr{B}$. Let $\overline{N} = \frac{N_2}{I}$ and $g : N_2 \longrightarrow \overline{N}$ be the quotient map, and $\overline{\varphi} := (\mathfrak{q}g)^{-1}\varphi_2$. Then g represents an arrow $g : (N_2, \varphi_2) \longrightarrow \left(\overline{N}, \overline{\varphi}\right)$ such that $gg_1 = gg_2$. So, g equalizes g_1 and g_2. The proof is complete. ■

Lemma 3.6.17. *Continue to use the notation in* (3.6.9, 3.6.11, 3.6.12, 3.6.15). *We define a functor* $\Theta : \mathscr{I}_V \longrightarrow \underline{Cat}$ by

$$\begin{cases} \Theta(N, \varphi) = \mathbf{E}_N & \forall\ (N, \varphi) \in Obj(\mathscr{I}_V) \\ \Theta(g) = g_* : \mathbf{E}_{N_0} \longrightarrow \mathbf{E}_{N_1} & \forall\ g \in Mor_{\mathscr{I}_V}\left((N_0, \varphi_0), (N_1, \varphi_1)\right) \end{cases}$$

So, Θ *is a filtering family of categories.*

Also, for (N, φ) *in* $\mathscr{I}_V$*, define the functor* $\psi_{(N,\varphi)} : \mathbf{E}_N \longrightarrow \mathscr{F}_V$*, as follows:*

$$\begin{cases} \psi_{(N,\varphi)}(M, h) = \left(M, (\varphi \mathfrak{q} h)^{-1}\right) & \forall\ (M, h) \in Obj(\mathbf{E}_N) \\ \psi_{(N,\varphi)}(u) = u & \forall\ u = i_! j^! \in \\ & Mor_{\mathbf{E}_N}\left((M_0, h_0), (M_1, h_1)\right). \end{cases} \tag{3.23}$$

Note that $\forall\ g \in Mor_{\mathscr{I}_V}((N_0,\varphi_0),(N_1,\varphi_1))$, *we have* $\psi_{(N_0,\varphi_0)} = \Phi_{(N_0,\varphi_0)}g_*$. *By the universal property (2.1.15), there is a functor* Ψ, *as follows:*

$$\begin{array}{ccc} \Theta(N,\varphi) & \xrightarrow{\quad\psi_{(N,\varphi)}\quad} & \\ \Big\downarrow{\scriptstyle \theta_{(N,\varphi)}} & \searrow & \\ \lim_{\mathscr{I}_V}\Theta(N,\varphi) & \overset{}{\underset{\Psi}{-\,-\,\to}} & \mathscr{F}_V \end{array}$$

Then Ψ *is a homotopy isomorphism.*

Proof. Given $(M,\varphi) \in Obj(\mathscr{F}_V)$, we have $\psi_{(M,\varphi^{-1})}(M,1_M) = (M,\varphi)$. So, Ψ is surjective on objects. By construction (2.1.15), objects in $\lim_{\mathscr{I}_V}\Theta(N,\varphi) = \lim_{\mathscr{I}_V}\mathbf{E}_N$ are represented by those in $\Theta(N,\varphi)$. We prove Ψ is injective. Suppose $\Psi(\theta_{(N,\varphi)}(M,h)) = \Psi(\theta_{(N_1,\varphi_1)}(M_1,h_1))$, for some $(M,h) \in Obj\,(\Theta(N,\varphi))$, $(M_1,h_1) \in Obj\,(\Theta(N_1,\varphi_1))$. Since $\mathscr{I}_V$ is directed, we can assume $(N_1,\varphi_1) = (N,\varphi)$. Therefore, $\psi_{(N,\varphi)}(M,h) = \psi_{(N,\varphi)}(M_1,h_1)$. So, $M = M_1$, and $\mathfrak{q}h = \mathfrak{q}h_1$ in $\mathbb{Q}\frac{\mathscr{A}}{\mathscr{B}}$. So, $\mathfrak{q}h = \mathfrak{q}h_1$ in $\frac{\mathscr{A}}{\mathscr{B}}$. Let $\overline{N} = \frac{N}{Im(h-h_1)}$ and $g : N \longrightarrow \overline{N}$ be the quotient map. Let $\overline{\varphi} = (\mathfrak{q}g)^{-1}\varphi$. Then $\mathfrak{q}g : (N,\theta) \longrightarrow (\overline{N},\overline{\varphi})$ is an arrow in $\mathscr{I}_V$, which induces $g_* : \mathbf{E}_N \longrightarrow \mathbf{E}_{\overline{N}}$. Now, $g_*(M,h) = (M,gh) = (M,gh_1) = g_*(M,h_1)$. So, it is established that Ψ is bijective on the objects.

The morphisms in $\lim_{\mathscr{I}_V}\Theta(N,\varphi)$ are represented by morphisms in $\Theta(N,\varphi) = \mathbf{E}_N$. Suppose $u_1,u_2 \in Mor_{\Theta(N,\varphi)}((M_1,h_1),(M_2,h_2))$ are such that $\psi_{(N,\varphi)}(u_1) = \psi_{(N,\varphi)}(u_2)$ in $Mor_{\mathscr{F}_V}(M_1, (\varphi\mathfrak{q}h_1)^{-1}), (M_2,(\varphi\mathfrak{q}h_2)^{-1}))$. By definition (3.23) $u_1 = \psi(u_1) = \psi(u_2) = u_2$. So, Ψ is injective on the morphisms.

We already proved that Ψ is bijective on the objects. So, to prove surjectivity, let $u \in Mor_{\mathscr{F}_V}\left(\psi_{(N,\varphi)}(M_1,h_1)), \psi_{(N,\varphi)}(M_2,h_2))\right)$, where $(M_1,h_1),(M_2,h_2)$ are objects in $\mathbf{E}_N$ for some $(N,\varphi) \in \mathscr{I}_V$. Then $u \in Mor_{\mathbb{Q}\mathscr{A}}(M_1,M_2)$ such that the diagram

$$\begin{array}{ccc} V & \xrightarrow{(\varphi\mathfrak{q}h_1)^{-1}} & \mathfrak{q}M_1 \\ & \searrow^{(\varphi\mathfrak{q}h_2)^{-1}} & \Big\downarrow{\scriptstyle \mathfrak{q}u} \\ & & \mathfrak{q}M_2 \end{array} \quad \text{commutes in } \frac{\mathscr{A}}{\mathscr{B}}, \quad \text{and} \quad \begin{cases} \mathfrak{q}u & \text{is isomorphism,} \\ \mathfrak{q}h_2\mathfrak{q}u = \mathfrak{q}h_1 \end{cases}$$

We write $u = i_!j^! = \left(M_1 \overset{j}{\twoheadleftarrow} M_0 \overset{i}{\hookrightarrow} M_2 \right)$. It follows that we have a commutative diagram

$$\begin{array}{ccc} \mathfrak{q}M_0 & \overset{\mathfrak{q}i}{\hookrightarrow} & \mathfrak{q}M_2 \\ {\scriptstyle \mathfrak{q}j}\downarrow & & \wr\downarrow{\scriptstyle \mathfrak{q}h_2} \\ \mathfrak{q}M_1 & \underset{\mathfrak{q}h_1}{\overset{\sim}{\longrightarrow}} & \mathfrak{q}N \end{array}$$

So, $\mathfrak{q}(h_2i - h_1j) = 0$, hence $Im(h_2i - h_1j) \in \mathscr{B}$,

Let $\overline{N} = \frac{N}{Im(h_2i-h_1j)}$ and $g : N \twoheadrightarrow \overline{N}$ be the quotient map. With $\overline{\varphi} = \varphi(\mathfrak{q}g)^{-1}$, we have an arrow $\mathfrak{q}g : (N,\varphi) \longrightarrow (\overline{N},\overline{\varphi})$ in $\mathscr{I}_V$. Consider the diagram

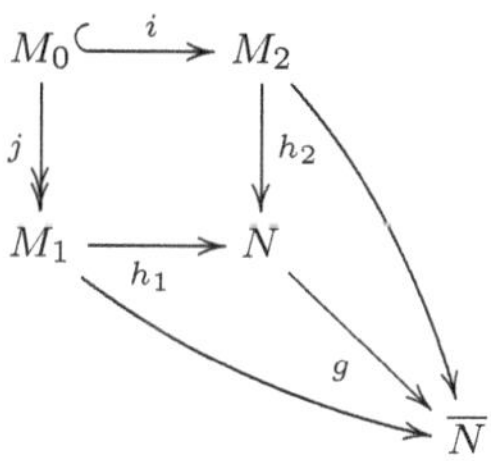

Outer square commutes, inner square may not.
$g_*(M_1, h_1) = (M_1, gh_1)$
$g_*(M_2, h_2) = (M_2, gh_2)$

(3.24)

Further, $u = i_!j^!$ represents an arrow $g_*(M_1,h_1) \longrightarrow g_*(M_2,h_2)$ in $\mathbf{E}_{\overline{N}}$. And, $\psi_{(\overline{N},\overline{\varphi})}(u) = u$, as required. The proof is complete. ■

Lemma 3.6.18. *For any isomorphism* $\varphi : \mathfrak{q}N \xrightarrow{\sim} V$ *is* $\frac{\mathscr{A}}{\mathscr{B}}$, *the functor* $\psi_{(N,\varphi)} : \mathbf{E}_N \longrightarrow \mathscr{F}_V$ *is a homotopy equivalence.*

Proof. By (3.6.17) $\mathscr{F}_V = \lim_{\mathscr{I}_V} \Theta(N,\varphi)$. Also, $\forall\ g \in Mor_{\mathscr{I}_V}((N_0,\varphi_0),(N_1,\varphi_1))$, the functors $\Theta(g) : \Theta(N_0,\varphi_0) \longrightarrow \Theta(N_1,\varphi_1)$ are homotopy equivalences, by (3.6.14). Now the lemma follows, by an application of (2.1.18). The proof is complete. ■

To finish the proof of the Localization Theorem (3.6.9), we prove the following Lemma.

Lemma 3.6.19. *For the inflation* $i_V : 0 \hookrightarrow V$ *in* $\frac{\mathscr{A}}{\mathscr{B}}$, *and* $(i_V)_! \in Mor_{\mathbb{Q}\frac{\mathscr{A}}{\mathscr{B}}}(0,V)$, *the functor* $((i_V)_!)^* : V/\mathbb{Q}\mathfrak{q} \longrightarrow 0/\mathbb{Q}\mathfrak{q}$ *is a homotopy equivalence.*

Proof. Let $(N, \varphi) \in \mathscr{I}_V$, be any object (e.g. $(N, \varphi) = (V, q1_V)$). So, $\varphi : \mathfrak{q}N \xrightarrow{\sim} V$ in $\frac{\mathscr{A}}{\mathscr{B}}$. Recall the diagram (3.21).

$$\begin{array}{ccccc} \mathbf{E}_N & \xrightarrow{\psi_{(N,\varphi)}} & \mathscr{F}_V & \overset{\sim}{\hookrightarrow} & V/\mathbb{Q}\mathfrak{q} \\ \downarrow{\scriptstyle \kappa_N} & & & & \downarrow{\scriptstyle ((i_V)_!)^*} \\ \mathbb{Q}\mathscr{B} & \xrightarrow{\sim} & \mathscr{F}_0 & \overset{\sim}{\hookrightarrow} & 0/\mathbb{Q}\mathfrak{q} \end{array}$$

As promised, the functors $\psi_{(N,\varphi)}$ were defined and proven to be homotopy equivalences (3.6.18). The functor κ_N is also a homotopy equivalence (3.6.13). The lower left and upper right compositions are, respectively, given by

$$\begin{cases} (M, h) \mapsto \left(\ker(h), 0 \xrightarrow{\sim} \mathfrak{q}(\ker(h))\right) \\ (M, h) \mapsto \left(M, (i_{\mathfrak{q}M})_! : 0 \xrightarrow{\sim} \mathfrak{q}M\right) \quad \text{because } \mathfrak{q}h\varphi \text{ is an isomorphism.} \end{cases}$$

The inclusion $\ker(h) \hookrightarrow M$ induces a natural transformation, from the first one to the second one. So, the diagram is homotopy commutative. Therefore, $((i_V)_!)^*$ is a homotopy equivalence. The proof is complete. ■

With this, the proof of the Localization Theorem (3.6.9) is complete.

Remark 3.6.20. We comment on the conclusion of the proof of the Localization theorem in Lemma 3.6.19. In the proof of (3.6.19), the only tangible choice of (N, φ) in $\mathscr{I}_V$ was $(N, \varphi) = (V, \mathfrak{q}1_V)$. Then we asserted that $\psi_{(V,\mathfrak{q}1_V)} : \mathbf{E}_V \longrightarrow \mathscr{F}_V$ was a homotopy equivalence, using (3.6.18). One could ask why was it not possible to give a direct proof of this fact, by defining a homotopy inverse $\mathscr{F}_V \longrightarrow \mathbf{E}_V$? We attempt to provide some clarity.

1. In the proof of (3.6.17), preimage of $(M, \varphi) \in \mathscr{F}_V$ was first defined as $(M, 1_M) \in \Theta(M, \varphi^{-1}) = \mathbf{E}_M$. However, $\varphi \in Iso_{\mathbb{Q}\frac{\mathscr{A}}{\mathscr{B}}}(V, M) \cong Iso_{\frac{\mathscr{A}}{\mathscr{B}}}(V, M)$. So, we can write $\varphi = (\mathfrak{q}t)^{-1}(\mathfrak{q}s)$, where $s : V \longrightarrow N$ and $t : M \longrightarrow N$ are mod-$\mathscr{B}$ isomorphisms. By (3.6.14), these

give zigzag homotopy equivalences

$$\mathbf{E}_M \xrightarrow{t^*} \mathbf{E}_N \xleftarrow{s_*} \mathbf{E}_V \tag{3.25}$$

This is still discomforting, in that this depends on (M, φ). In any case, given an object (N, φ) in $\mathscr{I}_V$, we have zigzag homotopy equivalences, as in (3.25).

2. Now suppose $u \in Mor_{\mathscr{F}_V}((M_1, \varphi_1), (M_2, \varphi_2))$ and $\varphi_i \in Mor_{\frac{\mathscr{A}}{\mathscr{B}}}(V, M_i)$ for $i = 1, 2$. Then $u \in Mor_{\mathbb{Q}\mathscr{A}}(M_1, M_2)$. For $i = 1, 2$, write $\varphi_i^{-1} = (\mathfrak{q}t_i)^{-1}(\mathfrak{q}h_i)$ for some mod-$\mathscr{B}$ isomorphisms $h_i : M \longrightarrow N_i$ and $t_i : V \longrightarrow N_i$. By the Ore condition, $s_1t_1 = s_2t_2 =: t$ for some mod-$\mathscr{B}$ isomorphisms $s_i : N_i \longrightarrow N$. So, $\varphi_i^{-1} = (\mathfrak{q}(t_is_i))^{-1}(\mathfrak{q}(s_ih_i))$. Adjusting $h_i := s_ih_i$ and $t_i := s_it_i = t$, we can assume $N_1 = N_2 = N$, and $t = t_1 = t_2$. We reproduce the diagram (3.24), with same notation, for this new situation:

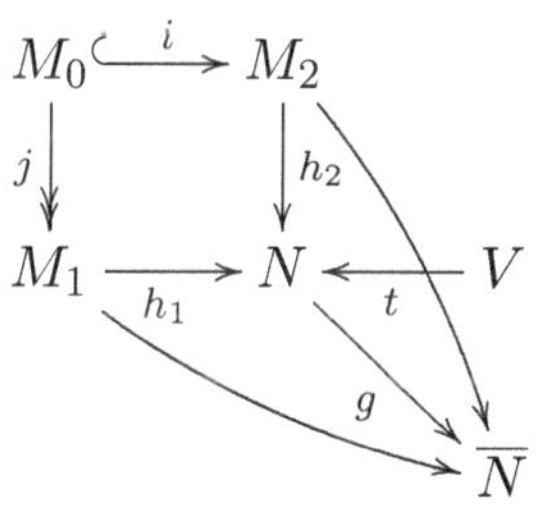

$$\begin{cases} g_*(M_1, h_1) = (M_1, gh_1) \\ g_*(M_2, h_2) = (M_2, gh_2) \\ \text{And,} \quad u \in Mor_{\mathbf{E}_{\overline{N}}}((M_1, h_1), (M_2, h_2)) \end{cases}$$

Now, $(gt)_* : \mathbf{E}_V \longrightarrow Mor_{\mathbf{E}_{\overline{N}}}(g_*(M_1, h_1), g_*(M_2, h_2))$ is a homotopy equivalence. The preimage of $u \in Mor_{\mathscr{F}_V}\big((M_1, \varphi_1), (M_2, \varphi_2)\big)$ in $\mathbf{E}_V$ is the preimage of u, via $(gt)_*$. For further insights for such a homotopy inverse of $(gt)_*$, see the proof of (3.6.9).

3.7. *K*-Theory Spaces and Reformulations

For an exact category $\mathscr{E}$, the K-groups are defined as $K_n(\mathscr{E}) := \pi_{n+1}(\mathbb{BQ}\mathscr{E}, 0)$. Conversely, for CW complexes Y, the homotopy groups $\pi_n(Y, y)$ determine Y, up to homotopy equivalences (C.3.8). From this perspective, it may make more sense to define K-Theory as a topological space, as follows.

Definition 3.7.1. Let $\mathscr{E}$ be an exact category. The ***K*-theory space** of $\mathscr{E}$ is defined to be the pointed topological space

$$\mathbf{K}(\mathscr{E}) := \Omega(\mathbb{BQ}\mathscr{E}) \quad \text{the loop space of } \mathbb{BQ}\mathscr{E} \quad \text{at } 0$$

with the constant loop (see B.4.7) as the base point. Therefore, we have

$$K_n(\mathscr{E}) = \pi_{n+1}(\mathbb{BQ}\mathscr{E}, 0) = \pi_n(\mathbf{K}(\mathscr{E}), 0) \qquad \forall\, n \geq 0.$$

Comparing the homotopy groups, by the theorem of Whitehead (C.3.8), most of the above can be reformulated in terms of the K-Theory spaces.

Lemma 3.7.2. *Let $\iota : \mathscr{P} \longrightarrow \mathscr{E}$ be an exact functor of exact categories. Then ι induces a map $\mathbf{K}(\mathscr{P}) \longrightarrow \mathbf{K}(\mathscr{E})$ of K-theory spaces.*

The Resolution Theorem is reformulated, as follows.

Theorem 3.7.3. *Let $\mathscr{E}$ be an exact category and $\mathscr{P} \subseteq \mathscr{E}$ be a resolving subcategory. Then the map*

$$\mathbf{K}(\mathscr{P}) \longrightarrow \mathbf{K}(\mathscr{E}) \quad \textit{is a homotopy equivalence.}$$

Proof. For integers $n \geq 0$, by (3.5.8), it follows that $\pi_n(\mathbf{K}(\mathscr{P}, 0) = \pi_{n+1}(\mathbf{K}(\mathbb{Q}\mathscr{P}, 0)) = \pi_{n+1}(\mathbf{K}(\mathbb{Q}\mathscr{E}, 0)) = \pi_n(\mathbf{K}(\mathscr{E}), 0))$. The proof is complete, by (C.3.8). ■

Theorem 3.7.4 (Dévissage). *Under the hypotheses of Theorem* 3.6.2, *the map*

$$\mathbf{K}(\mathscr{B}) \longrightarrow \mathbf{K}(\mathscr{A}) \quad \textit{is a homotopy equivalence.}$$

Proof. Same as above. ■

Theorem 3.7.5 (Localization). *Under the hypotheses of the Localization Theorem* (3.6.9)

$$\mathbf{K}(\mathscr{B}) \longrightarrow \mathbf{K}(\mathscr{A}) \longrightarrow \mathbf{K}\left(\tfrac{\mathscr{A}}{\mathscr{B}}\right)$$

is a homotopy fibration.

Proof. This follows from the Localization Theorem (3.6.9) in conjunction with (C.3.10). ■

3.8. Exercises

Exercise 3.8.1. Let $\mathscr{E}$ be an exact category. Then

1. Let $\theta : N \xrightarrow{\sim} M$ be an isomorphism. Then $\theta^! = (\theta^{-1})_! \in Mor_{\mathbb{Q}\mathscr{E}}(M, N)$ is an isomorphism.
2. There are bijections

$$\begin{cases} Iso_{\mathscr{E}}(N, M) \xrightarrow{\sim} Iso_{\mathbb{Q}\mathscr{E}}(M, N)\ \theta \mapsto \theta^! \\ Iso_{\mathscr{E}}(M, N) \xrightarrow{\sim} Iso_{\mathbb{Q}\mathscr{E}}(M, N)\ \theta \mapsto \theta_! \end{cases}$$

   where $Iso_{\mathscr{C}}(M, N)$ denotes the set of isomorphisms in the respective category $\mathscr{C}$.
3. Suppose $u = i_! p^! \in Mor_{\mathbb{Q}\mathscr{E}}(M, N)$. If u is an isomorphism, then so are i and j.

Exercise 3.8.2. Let $F : \mathscr{P} \longrightarrow \mathscr{E}$ be an exact functor, of exact categories. Then the following conditions are equivalent:

1. The map $\mathbb{B}\mathbb{Q}\mathscr{P} \longrightarrow \mathbb{B}\mathbb{Q}\mathscr{E}$ is a homotopy equivalence.
2. The map $\mathbf{K}\mathscr{P} \longrightarrow \mathbf{K}\mathscr{E}$ is a homotopy equivalence.
3. The maps $K_n(\mathscr{P}) \longrightarrow K_n(\mathscr{E})$ are isomorphisms $\forall n \geq 0$.

Exercise 3.8.3. Suppose $\mathscr{E}$ is an exact category and $\mathscr{P} \subseteq \mathscr{E}$ is a resolving subcategory (3.5.5). Give an elementary proof that $K_0(\mathscr{P}) \xrightarrow{\sim} K_0(\mathscr{E})$. (*Do not use the resolution theorem* (3.5.3).)

Exercise 3.8.4. Let $\mathscr{A}$ be an abelian category. Suppose $M \in Obj(\mathscr{A})$. Then M is said to have length n, and write $\ell(M) = n$ if there is a filtration

$$0 = M_0 \hookrightarrow M_1 \hookrightarrow M_2 \hookrightarrow \cdots \hookrightarrow M_n =: M$$
$$\ni \ \forall\, j = 1, \ldots, n \quad \frac{M_j}{M_{j-1}} \text{ is simple.}$$

Prove that the length $\ell(M)$ is well defined. (**Hint:** See [Wc2, p. 137].)

Exercise 3.8.5. [Q, p.105] Suppose D is a Dedekind domain, with quotient field F. Then there is an exact sequence

$$\cdots \longrightarrow \bigoplus_{\mathfrak{m}\in\max(D)} K_0\left(\frac{D}{\mathfrak{m}}\right) \longrightarrow K_1(D) \longrightarrow K_1(F) \longrightarrow$$

$$\bigoplus_{\mathfrak{m}\in\max(D)} K_0\left(\frac{D}{\mathfrak{m}}\right) \longrightarrow K_0(D) \longrightarrow K_0(F) \longrightarrow 0$$

(**Hint**: Use Ex. A.6.9.)

Exercise 3.8.6. Let $(A, \mathfrak{m})$ be a Noetherian local ring and $\kappa = \frac{A}{\mathfrak{m}}$. Let $\mathbb{M}_\ell(A)$ denote the category of A-modules of finite length and $\mathscr{V}(\kappa)$ denote category of the κ-vector spaces of finite dimension. Prove that map $\mathbf{K}(\mathbb{M}_\ell(A)) \longrightarrow \mathbf{K}(\mathscr{V}(\kappa))$ is a homotopy equivalence.

Problem 3.8.7. Let $\mathscr{E}$ be an exact category. Give an elementary description of $K_1(\mathscr{E})$. In particular, for a commutative Noetherian ring A, obtain an elementary description of $K_1(Coh(A))$. (**Hint:** Consult [GG, N].)

Remark 3.8.8. A major part of our remaining interest concerns K-theory schemes, some of what could be included as exercise in this section. However, we postpone this till Chapter 6.

Chapter 4

The Agreement with Classical K-Theory

Perhaps central to the study of algebraic K-theory has always been the K-theory of (commutative) rings A, that is the study of the K-theory of the exact category $\mathscr{P}(A)$ of finitely generated projective A-modules. As was mentioned at the end of Section 3.3, originally three groups, which we denoted (temporarily) by $K_0^c(A)$, $K_1^c(A)$, $K_2^c(A)$, were defined. Some exact sequences of these three groups already existed in the early part of development of the theory [Mi, Sw3]. The desire to define higher K-groups and to extend these exact sequences is explicitly mentioned in the book of Swan [Sw3, p. 213]. It would not be far from the truth to say that higher algebraic K-theory, discussed in the previous chapters, evolved out of this same desire.

The whole theory works for non-commutative rings, with unity. So, we will not assume that rings are necessarily commutative in this chapter. For such a ring A the category of finitely generated projective (right) A-modules will be denoted by $\mathscr{P}(A)$. We already saw that $K_0^c(A)$ coincides with $K_0(A)$ (3.3.6). The main goal of this chapter is, for $i = 1, 2$, to establish that $K_i^c(A)$ coincides with Quillen $K_i(A) := K_i(\mathscr{P}(A))$. However, the chapter is rich with various constructions, approaches to higher K-theory, and jargon. The original source for this chapter is the paper of Grayson [Gd1], and the results are mostly due to Quillen.

4.1. Symmetric Monoidal Categories

The direct sum operation on $\mathscr{P}(A)$ gives it the structure of a symmetric monoidal category in the following sense.

Definition 4.1.1. A **symmetric monoidal category** is a triple $(S, \odot, \mathbf{e})$, where S is a category, $\odot : S \times S \longrightarrow S$ is a functor (*to be referred to as "product", "addition", or "sum"*), and $\mathbf{e}$ is a distinguished object (*which would behave like an identity for the addition* $\odot$). These three components satisfy the following conditions.

1. (**Associativity**) The addition $\odot$ is associative, in the sense that there is a natural equivalence

$$\alpha : \odot(1 \times \odot) \xrightarrow{\sim} \odot(\odot \times 1)$$

More precisely, $\forall\ A, B, C \in Obj(S)$, there are natural isomorphisms

$$\alpha_{ABC} : A \odot (B \odot C) \xrightarrow{\sim} (A \odot B) \odot C$$

In other words, the diagram

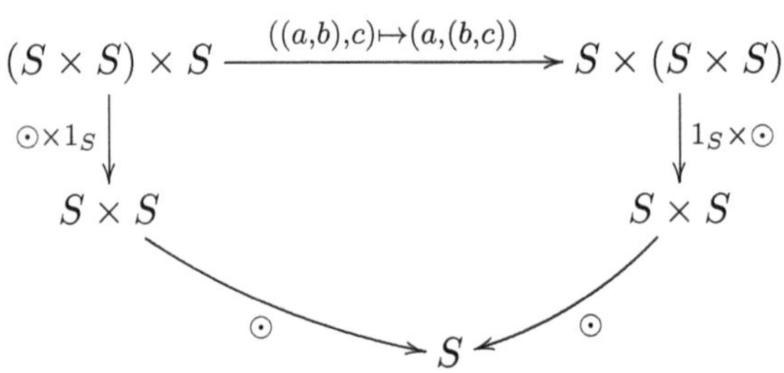

commutes, via α.

2. (**Identity**) There are natural equivalences

$$\begin{cases} \mathbf{e}\odot : \mathbf{e} \odot X \xrightarrow{\sim} X \\ \odot\mathbf{e} : X \odot \mathbf{e} \xrightarrow{\sim} X \end{cases} \ni$$

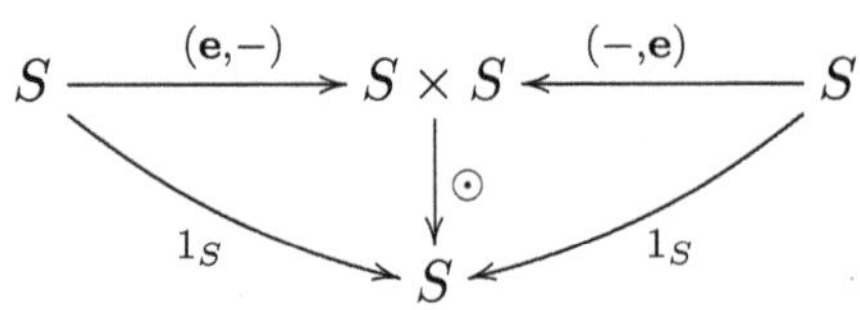

commute via $\mathbf{e}\odot, \odot\mathbf{e}$.

3. (**Symmetry**) There are natural equivalences

$$\chi : A \odot B \xrightarrow{\sim} B \odot A \quad \ni \quad \begin{array}{ccc} S \times S & \xrightarrow{(a,b)\mapsto(b,a)} & S \times S \\ & \searrow^{\odot} & \downarrow \odot \\ & & S \end{array} \quad \text{commute via } \chi.$$

4. The following two diagrams commute up to natural equivalences:

$$\begin{array}{ccccc} A \odot (B \odot (C \odot D)) & \xrightarrow[\sim]{\alpha} & (A \odot B) \odot (C \odot D) & \xrightarrow[\sim]{\alpha} & ((A \odot B) \odot C) \odot D \\ \wr\downarrow A\odot\alpha & & & & \alpha\odot D \downarrow\wr \\ A \odot ((B \odot C) \odot D) & & \xrightarrow[\alpha]{\sim} & & (A \odot (B \odot C)) \odot D \end{array}$$

$$\begin{array}{ccc} A \odot (\mathbf{e} \odot B) & \xrightarrow[\sim]{\alpha} & (A \odot \mathbf{e}) \odot B \\ \wr\downarrow & \swarrow_{\sim} & \\ A \odot B & & \end{array}$$

We only consider symmetric monoidal categories, as defined here, even when we drop the qualification "symmetric". However, if the triple $(S, \odot, \mathbf{e})$ satisfies only the conditions (1, 2, 4), it should be referred to as a **monoidal category**. (*While our definition already incorporates associativity* (1) *and symmetry* (3) *in it, the standard literature on monoidal categories may not always do so. So, all our (symmetric) monoidal categories are what the standard literature may call associative, symmetric monoidal categories.*)

Before we provide some examples, we make two observations.

Definition 4.1.2. Let $(\mathscr{M}, \odot)$ be a symmetric monoidal category such that the isomorphism classes of objects in $\mathscr{M}$ form a set $\mathscr{M}^{Iso}$. Then $\mathscr{M}^{Iso}$ is an abelian monoid. The Grothendieck group of this monoid is denoted by $K_0^{\odot}\mathscr{M}$. It is a group given by the following universal property:

$$\begin{array}{ccc} \mathscr{M}^{Iso} & \longrightarrow & K_0^{\odot}\mathscr{M} \\ & \searrow_{f} & \downarrow \varphi \\ & & G \end{array}$$

Meaning that given a group G and a monoid homomorphism f, there is a unique group homomorphism φ, so that the above diagram commutes. (*Since $\mathscr{M}$ is assumed to be symmetric, it does not make any difference whether the diagram has the universal property in the category of groups* **Gr**, *or of abelian groups* **Ab**.)

Corollary 4.1.3. *Suppose $(S, \odot, \mathbf{e})$ is a symmetric monoidal category. Then the addition $\odot : S \times S \longrightarrow S$ induces a product $\mu : k(\mathbb{B}S \times \mathbb{B}S) \longrightarrow \mathbb{B}S$ on the classifying space, where $k(\mathbb{B}S \times \mathbb{B}S)$ denotes the compactly generated topology on the product space* (2.1.8). *It also follows that, up to homotopy μ is commutative and associative and $\mathbf{e} \in \mathbb{B}S$ acts as an identity. This endows $\mathbb{B}S$ structure of a H_c-space (see Section* 4.8).

Given a symmetric monoidal category S, the following is a standard way to construct a new symmetric monoidal category, which is of our primary interest.

Definition 4.1.4. Let S be a symmetric monoidal category. Let $Iso(S)$ be the subcategory of S, whose objects are the same as that of S, and morphisms are the isomorphisms of S. The following are some useful observations:

1. For objects P in S, let $\overline{P}$ denote the isomorphism classes of objects in S.
$$\forall\ P, Q \in Iso(S),\ \overline{P} \neq \overline{Q} \Longrightarrow Mor_{Iso(S)}(P, Q) = \phi,$$
$$Mor_{Iso(S)}(P, P) = Aut_S(P).$$
2. Let $\{\overline{P}_i : i \in I\}$ be the set of all isomorphism classes of objects in S, where I is an indexing set, and P_i is a choice of representative of its isomorphism class $\overline{P}_i$. Then the inclusion $\coprod_{i\in I} Aut_S(P_i) \longrightarrow Iso(S)$ is an equivalence of categories, where $Aut_S(P_i)$ denotes the one-object category.
3. Consequently, we have the homotopy equivalence of the classifying spaces
$$\mathbb{B}Iso(S) \simeq \coprod_{i\in I} \mathbb{B}Aut_S(P_i) \quad \text{disjoint (disconnected) union in } \mathbf{Top}.$$

Example 4.1.5. The main example of our interest is the category $\mathscr{P}(R)$, of finitely generated (right) projective R-modules, where R is a ring, with $\odot := \oplus$, being the direct sum and $\mathbf{e} := R^0 = \{\mathbf{0}\}$ being the free R-module of rank zero. The following related examples are also among our primary interest.

1. Let $S := Iso\mathscr{P}(R) \subseteq \mathscr{P}(R)$ be the monoidal subcategory, as above. It follows

$$\mathbb{B}S \simeq \coprod_{i \in I} \mathbb{B}Aut(P_i) \quad \text{disjoint (disconnected) union in } \mathbf{Top}$$

where $\coprod$ runs through an indexing set I, with choices of representatives $P_i \in \overline{P}_i$, as in (4.1.4).
2. Let $\mathscr{F}(R)$ be the subcategory of $S = Iso\mathscr{P}(R)$, whose objects are based free R-modules $\{R^n : n \geq 0\}$, and

$$Mor_{\mathscr{F}(R)}(R^n, R^m) = \begin{cases} \phi & \text{if } m \neq n \\ GL_n(R) & \text{if } m = n \end{cases}$$

Then $\mathscr{F}(R)$ inherits the symmetric monoidal category structure from $S = Iso(\mathscr{P}(R))$. Here $GL_n(R)$ denotes the group of automorphisms of R^n. Since $R^n \in \mathscr{P}(R)$ is treated as right module (of columns), matrices $\alpha \in GL_n(R)$ act from the left side of R^n.
   We caution that in a non-commutative situation, rank of free modules may not make sense [We2, p. 2]. It is possible that $R^m \cong R^n$ but $n \neq m$. However, such a possibility is ruled out in the category $\mathscr{F}(R)$.

$$\text{It follows that} \quad \mathbb{B}\left(\mathscr{F}(R)\right) \simeq \coprod_{n \geq 0} \mathbb{B}\underline{GL_n(R)}$$

disjoint (disconnected) union in **Top**.

$$\text{We also write} \qquad \mathscr{F}(R) := \coprod \underline{GL_n(R)}.$$

Note that, ignoring the monoidal structure on $\mathscr{F}(R)$, the right-hand side can be viewed as a disjoint union of the one-element categories $\underline{GL_n(R)}$.
3. Note that, since $\mathscr{P}(R)$ has an initial object, the classifying space $\mathbb{B}\mathscr{P}(R)$ is contractible.

4.2. $S^{-1}S$-Construction

For a symmetric monoidal category S, we define categories $\mathscr{X}$ with S-action. Then we define localization $S^{-1}\mathscr{X}$ in this section.

Definition 4.2.1. Let $(S, \odot, \mathbf{e})$ be a symmetric monoidal category. An action of S on a category $\mathscr{X}$ *(from left)* is a functor $\odot : S \times \mathscr{X} \longrightarrow \mathscr{X}$ *(we use the same notation* $\odot$) such that

1. Action of $\mathbf{e}$ on $\mathscr{X}$ is a natural identity, meaning that $\forall\ E \in Obj(\mathscr{X})$, there are natural equivalences

$$\iota_E : \mathbf{e} \odot E \xrightarrow{\sim} E \ni \quad \begin{array}{ccc} \mathscr{X} & \xrightarrow{\mathbf{e}\odot -} & \mathscr{X} \\ & \searrow^{1_{\mathscr{X}}} & \downarrow \iota \\ & & \mathscr{X} \end{array} \quad \text{commute via } \iota.$$

2. The action $\odot$ is naturally associative. This means that $\forall\ P, Q \in Obj(S)$, $\forall\ E \in Obj(\mathscr{X})$ there are natural equivalences α_{PQE} : $(P \odot Q) \odot E \xrightarrow{\sim} P \odot (Q \odot E)$ such that

$$\begin{array}{ccc} (S \times S) \times \mathscr{X} & \xrightarrow{((P,Q),E)\mapsto((P,Q),E)} & S \times (S \times \mathscr{X}) \\ \downarrow{\odot\times 1} & & \downarrow{1\times\odot} \\ S \times \mathscr{X} & & S \times \mathscr{X} \\ \downarrow{\odot} & & \downarrow{\odot} \\ \mathscr{X} & \xrightarrow[\alpha_{PQE}]{\sim} & \mathscr{X} \end{array} \quad \text{commute via } \alpha.$$

3. Further, ι_E and α_{PQE} are required to be coherent, as well with four factors associativity $P_1 \odot P_2 \odot P_3 \odot E$, where $P_1, P_2, P_3 \in Obj(S)$, and $E \in Obj(\mathscr{X})$.

In this case, we say that $\mathscr{X}$ is an S-**category**, or $\mathscr{X}$ is a category with S-action.

We would be mainly concerned with examples $\mathscr{X}$ of S-categories, which would have the flavor of the following.

Example 4.2.2. Let R be a ring, and $\mathscr{P}(R)$ be the category of finitely generated (right) projective R-modules. Consider the symmetric monoidal category $S = Iso(\mathscr{P}(R))$. Let $\mathscr{X}$ be the category of exact sequences $E := \big(K \hookrightarrow M \twoheadrightarrow C \big)$ of finitely generated (right) R-modules. The arrows in $\mathscr{X}$ are maps of such sequences. Then S has an action on $\mathscr{X}$, defined by $P \odot E := \big(P \oplus K \hookrightarrow P \oplus M \twoheadrightarrow C. \big)$

Definition 4.2.3. Given (symmetric) monoidal categories $(S, \odot, \mathbf{e}_S)$, $(T, \odot, \mathbf{e}_T)$, a functor $f : S \longrightarrow T$ is defined to be a **monoidal functor** if

$$f(P \odot Q) \xrightarrow{\sim} f(P) \odot f(Q), \quad f(\mathbf{e}_S) \xrightarrow{\sim} \mathbf{e}_T \quad \text{are naturally isomorphic.}$$

In other words, the diagrams

$$\begin{array}{ccccc} S \times S & \xrightarrow{f \times f} & & & T \times T \\ \odot \big\downarrow & & & & \big\downarrow \odot \\ S & \xrightarrow[f]{} & T & \overset{\sim}{\dashrightarrow} & T \end{array} \qquad \begin{array}{ccccc} S & \xrightarrow{\mathbf{e}_S \odot} & & & S \\ f \big\downarrow & & & & \big\downarrow \mathbf{e}_T \odot f \\ T & \xrightarrow[\mathbf{e}_T \odot]{} & T & \overset{\sim}{\dashrightarrow} & T \end{array}$$

are naturally isomorphic.

Further, f must respect the associativity in S and T, meaning the functors

$$\left\{ \begin{array}{l} (P_1, P_2, P_3) \mapsto f(P_1) \odot (f(P_2) \odot f(P_3)) \\ (P_1, P_2, P_3) \mapsto (f(P_1) \odot f(P_2)) \odot f(P_3) \end{array} \right. \text{must be naturally equivalent.}$$

Definition 4.2.4. Let $(S, \odot, \mathbf{e})$ be a symmetric monoidal category, and $\mathscr{X}$, $\mathscr{Y}$ be categories with S-actions. A functor $f : \mathscr{X} \longrightarrow \mathscr{Y}$, of such S-categories, is said to preserve the S-action, if

$$\forall\ P \in ObjS,\ E \in Obj\mathscr{X} \quad f(P \odot E) \cong P \odot f(E)$$
$$\text{are naturally isomorphic.}$$

Further, f must also respect associativity, as in (4.2.3).

We proceed to do a few basic and important constructions of categories.

Definition 4.2.5. Let $(S, \odot, \mathbf{e})$ be a symmetric monoidal category, and $\mathscr{X}$ be an S-category. Define a new category $\langle S, \mathscr{X} \rangle$, as follows:

1. Objects of $\langle S, \mathscr{X} \rangle$ are the same as the objects of $\mathscr{X}$.
2. Given $E, F \in Obj\mathscr{X}$, an arrow $f : E \longrightarrow F$ in $\langle S, \mathscr{X} \rangle$ is an equivalence class of pairs (P, f), where $P \in ObjS$ and $f : P \odot E \longrightarrow F$ is an arrow in $\mathscr{X}$. Two such pairs (P, f), (Q, g) are

defined to be equivalent if there is an isomorphism $\tau : P \xrightarrow{\sim} Q$ in S such that the diagram

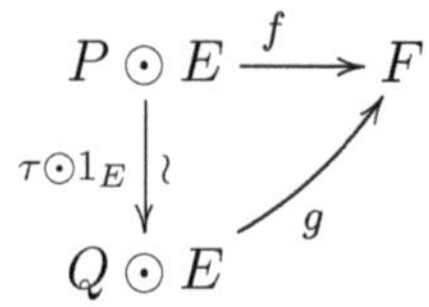

commutes.

In this case, we write $f := (P, f)$ or $f := [(P, f)]$, while P is understood to be the $\odot$-summand of the domain of f.

3. Given two arrows, $\varphi : E \longrightarrow F$, $\gamma : F \longrightarrow G$ in $\langle S, \mathscr{X} \rangle$, represented by (P, f) and (Q, g), the composition $\gamma\varphi$ is defined by the representation $(Q \odot P, h)$, where h is defined by the composition, as in the commutative diagram

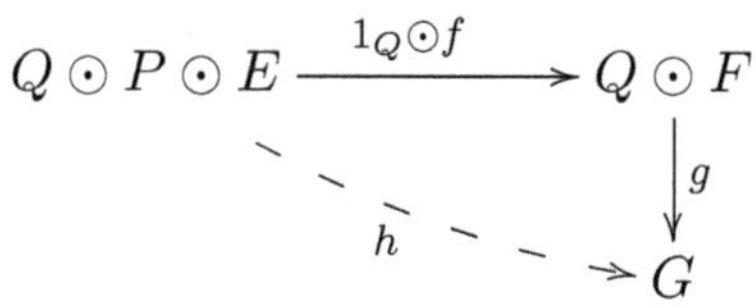

4. For $E \in Obj\langle S, \mathscr{X} \rangle = Obj\ \mathscr{X}$, the identity arrow 1_E in $\langle S, \mathscr{X} \rangle$ is represented by $(\mathbf{e}, 1_E)$.

Example 4.2.6. Suppose $(S, \odot, \mathbf{e})$ is a symmetric monoidal category and all morphisms of S are isomorphisms. Then $\forall\ P \in Obj(S)$, we have $Mor_{\langle S,S \rangle}(\mathbf{e}, P) = \{(P, 1_P)\}$. So, $\mathbf{e}$ is an initial object in $\langle S, S \rangle$. The morphism $(P, 1_P)$ is equivalent to all other of its representations (U, φ), as in the diagram

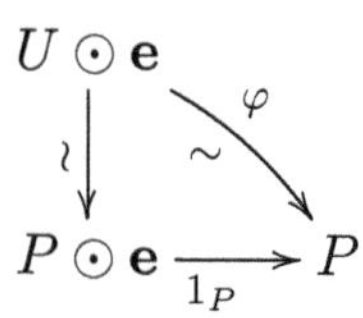

So, $\langle S, S \rangle$ is contractible.

The following construction will be a basic object of study in the subsequent sections.

Definition 4.2.7. Let $(S, \odot, \mathbf{e})$ be a symmetric monoidal category, and $\mathscr{X}$ be an S-category. Then S-acts on $S \times \mathscr{X}$ diagonally (from left). So, $S \times \mathscr{X}$ is also an S-category. Define

$$S^{-1}\mathscr{X} := \langle S, S \times \mathscr{X} \rangle$$

We may write that $F := (V, \varphi_1, \varphi_2) : (Q_1, E_1) \longrightarrow (Q_2, E_2)$ is an arrow in $S^{-1}\mathscr{X}$, to mean F is represented by $(V, \varphi_1, \varphi_2)$, with $V \in Obj(S)$, $\varphi_1 \in Mor_S(V \odot Q_1, Q_2)$, $\varphi_2 \in Mor_{\mathscr{X}}(V \odot E_1, E_2)$.

We remark that $S^{-1}\mathscr{X}$ has two S-actions, one from left $\odot_l$ and one from right $\odot_r$. For $P, Q \in Obj(S)$ and $E \in Obj(\mathscr{X})$, the actions are defined by

$$P \odot_l (Q, E) := (P \odot Q, E), \qquad \text{and} \qquad P \odot_r (Q, E) := (Q, P \odot E).$$

Given an arrow $f : P_1 \longrightarrow P_2$ in S, and an arrow $F := (V, \varphi_1, \varphi_2) : (Q_1, E_1) \longrightarrow (Q_2, E_2)$ in $S^{-1}\mathscr{X}$, define arrows

$$\begin{cases} f \odot_l F : (P_1 \odot Q_1, E_1) \longrightarrow (P_2 \odot Q_2, E_2) & \text{with } f \odot_l F := (V, f \odot \varphi_1, \varphi_2) \\ f \odot_r F : (Q_1, P_1 \odot E_1) \longrightarrow (Q_2, P_2 \odot E_2) & \text{with } f \odot_r F := (V, \varphi_1, f \odot \varphi_2) \end{cases}$$

See (4.3.8) for further information.

Remark 4.2.8. In particular, $S^{-1}S$ is defined in (4.2.7), for any symmetric monoidal category S. For a ring R, let $S = \mathscr{F}(R)$, $T = Iso(\mathscr{P}(R))$, as in (4.1.5). Our main interest in this chapter is the study of the classifying spaces $\mathbb{B}S^{-1}S$, $\mathbb{B}S^{-1}T$, and $\mathbb{B}T^{-1}T$, and their relationship with Quillen K-theory. All these will culminate in Sections 4.5 and 4.6.

4.3. The Projection Functors

For a symmetric monoidal category S and an S-category $\mathscr{X}$, in this section, we investigate the fibered and cofibered properties of the projection functors $S^{-1}\mathscr{X} \longrightarrow \langle S, S \rangle$ and $S^{-1}\mathscr{X} \longrightarrow \langle S, \mathscr{X} \rangle$.

Definition 4.3.1. Let $(S, \odot, \mathbf{e}_S)$ be a symmetric monoidal category, and $\mathcal{X}$ be an S-category. Then the projection $p_1 : S^{-1}\mathcal{X} \longrightarrow \langle S, S \rangle$, defined by

$$\begin{cases} p_1(P, X) = P & \forall\ (P, X) \in Obj\ S^{-1}\mathcal{X} \\ p_1(U, f, g) = (U, f) & \forall\ (U, f, g) \in Mor_{S^{-1}\mathcal{X}}((P, X), (Q, Y)) \end{cases} \tag{4.1}$$

defines a functor. For $P \in Obj(\langle S, S \rangle)$, the fiber category $p_1^{-1}P$ was defined in (2.2.4). For future reference, we introduce the notation p_{1P}, $\overline{p}_{1P}$ for two obvious functors. Define

$$p_{1P} : \mathscr{X} \longrightarrow p_1^{-1}P \quad \text{by} \begin{cases} p_{1P}(X) = (P, X) & X \in Obj\ \mathscr{X} \\ p_{1P}(g) = (\mathbf{e}_S, 1_P, g) & g \in Mor_{\mathscr{X}}(X, Y) \end{cases} \tag{4.2}$$

Further, define $\overline{p}_{1P}$ by the commutative diagram

$$\begin{array}{ccc} p_{1P} : \mathscr{X} & \longrightarrow & p_1^{-1}P \\ & \searrow^{\overline{p}_{1P}} & \downarrow \\ & & S^{-1}\mathscr{X} \end{array} \tag{4.3}$$

The vertical functor is a subcategory, but not necessarily a full subcategory.

Lemma 4.3.2. *Let* $(S, \odot, \mathbf{e})$ *be a symmetric monoidal category. Consider an arrow* $F = (U, f) = (V, g) : P \longrightarrow Q$ *in* $\langle S, S \rangle$, *represented by* $f : U \odot P \longrightarrow Q$, *as well as* $g : V \odot P \longrightarrow Q$. *Then*

$$\exists \text{ isomorphism } \tau : U \xrightarrow{\sim} V \text{ in } S \ni \begin{array}{ccc} U \odot P & \xrightarrow{f} & Q \\ {\scriptstyle \tau \odot 1_P} \downarrow \wr & \nearrow_{g} & \\ V \odot P & & \end{array} \quad \text{commutes.} \tag{4.4}$$

$$\text{Assume} \begin{cases} U \odot P \longrightarrow Q & \text{is monic, and the map} \\ Mor_S(U, U) \longrightarrow Mor_S(U \odot P, U \odot P) & \text{is injective } (\textit{\textbf{faithful}}) \end{cases} \tag{4.5}$$

The latter property is read as that every translation is **faithful**. (Subsequently, it will be assumed that all morphisms in S are monic.) *Under these hypotheses, it follows that there is only a* **unique choice** *of the isomorphism* τ *in* (4.4), *when* U, V *are given. In particular, if* $U = V$, *then* $\tau = 1_U$ *and* $f = g$.

Lemma 4.3.3. *Let $(S, \odot, \mathbf{e})$ be a symmetric monoidal category, and $\mathcal{X}$ be an S-category. Let $p_1 : S^{-1}\mathcal{X} \longrightarrow \langle S, S\rangle$ be the projection functor (4.1). Under the conditions of monicity and faithfulness (4.3.2), the functor $p_{1P} : \mathscr{X} \longrightarrow {p_1}^1 P$ is an equivalence of categories, for all $P \in Obj(S)$.*

Proof. We have $Obj(p_1^{-1}P) = \{(P, X) : X \in Obj\ \mathcal{X}\}$, and

$$\begin{aligned} &Mor_{p_1^{-1}P}((P, X), (P, Y)) \\ &= \left\{[(U, f, g)] : \begin{cases} f : U \odot P \longrightarrow P, \\ g : U \odot X \longrightarrow Y \end{cases} \ni f \sim 1_P\right\} \\ &= \{[(\mathbf{e}_S, 1_P, g)] : g : X \longrightarrow Y \in \mathscr{X}\} \\ &\cong Mor_{\mathscr{X}}(X, Y). \end{aligned} \tag{4.6}$$

The proof is complete. ∎

In preparation for further discussions on $S^{-1}\mathscr{X}$, we set up some notation and list some preliminary observations.

Notation 4.3.4. Let S be a symmetric monoidal category, and $\mathscr{X}$ be an S-category. As indicated before, we make a notational convention that we will write $v := (V, v) \in Mor_{\langle S, \mathscr{X}\rangle}(X, Y)$, or $v := (V, v) : X \longrightarrow Y$ in $\langle S, \mathscr{X}\rangle$, to mean that the arrow $v : X \longrightarrow Y$ in $\langle S, \mathscr{X}\rangle$ is represented by the arrow $v : V \odot X \longrightarrow Y$ in $\mathscr{X}$. By definition, V is unique up to isomorphism. Further, under the hypotheses of monicity and faithfulness (4.3.2), $v : V \odot X \longrightarrow Y$ in $\mathscr{X}$ is uniquely determined.

We have the following lemma.

Lemma 4.3.5. *Let S be a symmetric monoidal category, and $\mathscr{X}$ be an S-category. Assume that every morphism in S is monic and translations in S are faithful (4.3.2). Then the projection functor $p_1 : S^{-1}\mathscr{X} \longrightarrow \langle S, \mathscr{X}\rangle$ is cofibered.*

Proof. For $P \in Obj(\langle S, S\rangle)$, we proceed to describe p_1/P.

$$Obj\ (p_1/P) =$$
$$\left\{((Q, X), v) : (Q, X) \in Obj S^{-1}\mathcal{X}, v := (V, v) \in Mor_{\langle S,S\rangle}((Q, P)\right\}.$$

For $i = 1, 2$, and let $((Q_i, X_i), v_i) \in Obj(p_1/P)$, where $v_i := (V_i, v_i) \in Mor_{\langle S,S\rangle}(Q_i, P)$. A morphism, $\mathbf{u} := (W, f, g) \in$

$Mor_{p_1/P}(((Q_1, X_1), v_1), ((Q_2, X_2), v_2))$ is given by $f \in Mor_S(W \odot Q_1, Q_2)$, $g \in Mor_{\mathscr{X}}(W \odot X_1, X_2)$ such that the diagram

$$\begin{array}{ccc} Q_1 & \xrightarrow{v_1} & P \\ & \searrow^{\rho\Gamma = f} & \uparrow v_2 \\ & & Q_2 \end{array}$$

commutes in $\langle S, S\rangle$. Equivalently,

there is an isomorphism $\tau : V_1 \xrightarrow{\sim} V_2 \odot W$ in S such that the diagram

$$\begin{array}{ccccc} V_1 \odot Q_1 & & \xrightarrow{\quad v_1 \quad} & & P \\ \downarrow {\scriptstyle \tau \odot 1_{Q_1}} & & & & \| \\ V_2 \odot W \odot Q_1 & \xrightarrow{1_{V_2} \odot f} & V_2 \odot Q_2 & \xrightarrow{v_2} & P \end{array}$$

commutes in S. (4.7)

In fact, $\tau : V_1 \xrightarrow{\sim} V_2 \odot W$ is unique, under the monicity and faithfulness hypotheses (4.3.2).

1. Note that for $P \in Obj\ S = Obj(\langle S, S\rangle)$, the identity morphism $1_P \in Mor_{\langle S,S\rangle}(P, P)$ is represented by $1_P := (\mathbf{e}_S, 1_P)$.
2. For $X_1, X_2 \in Obj(\mathscr{X})$, we describe $Mor_{p_1/P}(((P, X_1), 1_P), ((P, X_2), 1_P))$. Let $\mathbf{u} := (W, f, g) \in Mor_{\rho/P}(((P, X_1), 1_P), ((P, X_2), 1_P))$. With $v_1 = v_2 = 1_P := (\mathbf{e}_S, 1_P)$, we have $V_1 = V_2 = \mathbf{e}_S$. It follows $\tau : \mathbf{e}_S \xrightarrow{\sim} W \cong \mathbf{e}_S \odot W \xrightarrow{\sim} W$. Reinterpreting (4.7), we have
$$\begin{aligned} &Mor_{p_1/P}(((P, X_1), 1_P), ((P, X_2), 1_P)) \\ &= \{(\mathbf{e}_S, 1_P, g) \in Iso(S) \times Mor_S(P, P) \times Mor_{\mathscr{X}}(X_1, X_2)\} \\ &\cong Mor_{\mathscr{X}}(X_1, X_2). \qquad (4.8) \end{aligned}$$
3. We describe $Mor_{p_1/P}(((Q_1, X_1), v_1), ((P, X_2), 1_P))$, for $((Q_1, X_1), v_1) \in Obj(p_3/P)$, and $X_2 \in Obj(\mathscr{X})$. So, $v_2 := (V_2, v_2) = 1_P := (\mathbf{e}_S, 1_P)$. Consequently, $V_2 = \mathbf{e}_S$ and $\tau : V_1 \xrightarrow{\sim} W$. So, $f = v_1$ in $\langle S, S\rangle$. Reinterpreting (4.7), we have
$$\begin{aligned} &Mor_{p_1/P}(((Q_1, X_1), v_1), ((P, X_2), 1_P)) \\ &= \{(V_1, v_1, g) \in Iso(S) \times Mor_S(V_1 \odot P, P) \times Mor_{\mathscr{X}}(V_1 \odot X_1, X_2)\} \\ &\cong Mor_{\mathscr{X}}(V_1 \odot X_1, X_2). \qquad (4.9) \end{aligned}$$

4. Consider the natural (inclusion) functor $\Phi : p_1^{-1}P \longrightarrow p_1/P$, defined by

$$\begin{cases} \Phi(P,X) = ((P,X),1_P) & \forall (P,X) \in Obj\,(p_1{}^1P) \\ \Phi(\mathbf{u}) = \mathbf{u} & \forall\ \mathbf{u} := (\mathbf{e}_S, 1_P, g) \in \\ & Mor_{p_1/P}\,((P,X_2),(P,X_2))\,. \end{cases}$$

Note that

$$\Phi : Mor_{p_1^{-1}P}\,((P,X_2),(P,X_2)) \xrightarrow{\sim} Mor_{p_1/P}(((P,X_1),1_P),((P,X_2),1_P))$$

is a bijection (4.6), (4.8).

5. In the opposite directions, define the functor $L : (p_1/P) \longrightarrow p_1^{-1}P$, with notation as in (4.3.4)

$$\begin{cases} L((Q_1,X_1),v_1) = (P, V_1 \odot X_1) \in Obj(\rho^{-1}P), \\ L(\mathbf{u}) = L(W,f,g) := (\mathbf{e}_S, 1_P, 1_{Q_2} \odot g) \in \\ Mor_{p_1^{-1}P}\,((P,Q_1\odot,X_1),(P,Q_2\odot,X_2))\,. \end{cases}$$

6. With notation as in (4.3.4), for $((Q_1,X_1),v_1) \in Obj(p_1/P)$, with $v_1 : V_1 \odot Q_1 \longrightarrow P$ and $(P,X_2) \in Obj(p_1^{-1}P)$, we have two columns of isomorphisms

$$\begin{array}{ccc} Mor_{p_1^{-1}P}\,(L\,((Q_1,X_1),v_1)\,,(P,X_2)) & & Mor_{p_1/P}\,(((Q_1,X_1),v_1),\Phi(P,X_2)) \\ \Big\| & & \Big\| \\ Mor_{p_1^{-1}P}\,((P,V_1\odot X_1),(P,X_2)) & \overset{\eta}{\dashrightarrow} & Mor_{p_1/P}\,(((Q_1,X_1),u_1),((P,X_2),1_P)) \\ \wr\downarrow & & \downarrow\wr \\ Mor_{\mathscr{X}}\,(V_1\odot X_1,X_2) & & Mor_{\mathscr{X}}\,(V_1\odot X_1,X_2) \end{array}$$

The first vertical bijection follows from (4.6) and the second vertical bijection follows from (4.9). This establishes that L is left adjoint to Φ, given by the following bijection:

$$\eta(\mathbf{e}_S, 1_P, G) = (U_1, u_1, G).$$

The bijection η is natural with respect to either coordinates. This establishes that p_1 is **precofibered**. We will prove that p_1 is **cofibered**.

7. Let $u = (U, u) : P \longrightarrow P'$ be an arrow in $\langle S, S \rangle$, with $u : V \odot P \longrightarrow P'$ in S. The functor

$$u_* : p_1/P \longrightarrow p_1/P' \quad \text{given by}$$

$$\begin{cases} u_*((Q_1, X_1), v_1) = ((Q_1, X_1), uv_1) \\ u_*(W, f, g) = (W, f, g) \end{cases}$$

for $((Q_1, X_1), v_1) \in Obj(p_1/P)$, with $v_1 := (V_1, v_1) : V_1 \odot Q_1 \longrightarrow P$ in S.

$$\begin{cases} \text{Let } (\mathbf{e}_S, 1_P, g) \in Mor_{p_1^{-1}P}((P, X_1), (P, X_2)) \\ \Phi(\mathbf{e}_S, 1_P, g) = (\mathbf{e}_S, 1_P, G) \in \\ \qquad Mor_{p_1/P}(((P, X_1), 1_P), ((P, X_2), 1_P)) \\ u_*\Phi(\mathbf{e}_S, 1_P, G) = (\mathbf{e}_S, 1_P, g) \in \\ \qquad Mor_{p_1/P'}(((P, X_1), u), ((P, X_2), u)) \\ Lu_*\Phi(\mathbf{e}_S, 1_P, G) = (\mathbf{e}_S, 1_{P'}, 1_P \odot g) \in \\ \qquad Mor_{p_1^{-1}P'}((P', P \odot X_1), (P', P \odot X_2)). \end{cases}$$

Using the same notation, for the functor $u_* : p_1^{-1}P \longrightarrow p_1^{-1}P$, we have

$$\begin{array}{ccc} p_1^{-1}P & \xrightarrow{\Phi} & (p_1/P) \\ {\scriptstyle u_*}\downarrow & & \downarrow{\scriptstyle u_*} \\ p_1^{-1}P' & \underset{L}{\overset{\Phi}{\rightleftarrows}} & (p_1/P') \end{array} \qquad \begin{cases} u_*(P, X_1) = (P', V \odot X_1) \\ u_*(\mathbf{e}_S, 1_P, g) = (\mathbf{e}_S, 1_{P'}, 1_P \odot g) \end{cases} \tag{4.10}$$

It follows that, for the composable arrows

$$\begin{array}{ccc} P_0 & \xrightarrow{u} & P_1 \\ & {\scriptstyle wu}\searrow & \downarrow{\scriptstyle w} \\ & & P_2 \end{array} \quad \text{in } \langle S, S \rangle, \qquad w_* u_* = (wu)_*$$

Therefore, p_1 is **cofibered**.

The proof is complete. ∎

We introduce the following definitions before we state some of the results.

Definition 4.3.6. Let S be a symmetric monoidal category, and $\mathscr{X}$ be an S-category. For $P \in Obj\ S$, define the **translation functor**

$$T_P : \mathscr{X} \longrightarrow \mathscr{X} \qquad \text{by} \qquad \begin{cases} T_P(X) = P \odot X & X \in Obj\ \mathcal{X} \\ T_P(f) = 1_P \odot f & f \in Mor_{\mathscr{X}}(X_1, X_2) \end{cases}$$

We say that S acts **invertibly**, on $\mathscr{X}$, if the translation maps T_P are homotopy equivalences, $\forall\ P \in Obj\ S$.

The following is a direct application of the cofibered versions of Theorem B (2.2.6).

Proposition 4.3.7. *Let S be a symmetric monoidal category, and $\mathscr{X}$ be an S-category. Assume that every morphism in S is monic and translations in S are faithful* (4.3.2). *Consider the projection functor $p_1 : S^{-1}\mathscr{X} \longrightarrow \langle S, S\rangle$. For $P \in Obj(S)$, recall the notation* (4.3.1), *as in the commutative diagram*

$$\begin{array}{ccc} \mathscr{X} & \xrightarrow[\sim]{p_{1P}} & p_1^{-1}P \\ & \overline{p}_{1P} \searrow & \downarrow \\ & & S^{-1}\mathscr{X} \end{array} \qquad \textit{see } (4.3.3).$$

1. *Suppose S acts invertibly on $\mathscr{X}$. Then the sequence*

$$\mathscr{X} \xrightarrow{\overline{p}_{1P}} S^{-1}\mathscr{X} \xrightarrow{p_1} \langle S, S\rangle \qquad \textit{is a homotopy fibration.}$$

$$\forall\ P \in Obj(\langle S, S\rangle)$$

   Consequently, if all maps in S are isomorphisms, then $\overline{p}_{1P} : \mathscr{X} \longrightarrow S^{-1}\mathscr{X}$ is a homotopy equivalence. (There is no obvious functor, representing the homotopy inverse. The projection is not a functor.)
2. *Conversely, if $\overline{p}_{1P}$ is homotopy equivalence $\forall\ P \in Obj(S)$, then S acts invertibly on $\mathscr{X}$.*

Proof. Assume S acts invertibly on $\mathscr{X}$. By (4.3.5), p_1 is precofibred (indeed, cofibered). For $v := (V, v) \in Mor_{\langle S,S\rangle}(P, P')$, we have the commutative diagram (4.10)

$$\begin{array}{ccc} \mathscr{X} & \xrightarrow{T_V} & \mathscr{X} \\ {\scriptstyle \overline{p}_{1P}}\downarrow & & \downarrow{\scriptstyle p_{1P'}} \\ p_1^{-1}P & \xrightarrow[v_*]{} & p_1^{-1}P' \end{array}$$

The vertical functors are homotopy equivalences, by Lemma 4.3.3, and so is T_V, by hypothesis. Therefore, v_* is a homotopy equivalence. So, the precofibered version of Theorem B (2.2.6) applies. Therefore, the sequence

$$p_1^{-1}P \longrightarrow S^{-1}\mathscr{X} \xrightarrow{p_1} \langle S, S\rangle \quad \text{is a homotopy fibration.}$$
$$\forall\ P \in Obj(\langle S, S\rangle)$$

Using the identification $p_{1P} : \mathscr{X} \xrightarrow{\sim} p_1^{-1}P$, the first part of (1) is established. If all maps in S are isomorphisms, then $\langle S, S\rangle$ has an initial object, and $\langle S, S\rangle$ is contractible. Therefore, $\overline{p}_{1P}$ is homotopy equivalence, for all $P \in Obj(\langle S, S\rangle)$.

To see the converse, assume that $\overline{p}_{1P} : \mathscr{X} \xrightarrow{\sim} S^{-1}\mathscr{X}$ is a homotopy equivalence, $\forall\ P \in Obj(\langle S, S\rangle)$. Now, for $P \in Obj(Obj(\langle S, S\rangle))$, consider the commutative diagram of functors

$$\begin{array}{ccc} \mathscr{X} & \xrightarrow{\overline{p}_{1e_S}} & S^{-1}\mathscr{X} \\ {\scriptstyle T_P}\downarrow & & \downarrow{\scriptstyle T_{rP}} \\ \mathscr{X} & \xrightarrow[\overline{p}_{1e_S}]{} & S^{-1}\mathscr{X} \end{array}$$

where T_{rP} is the translation, by the right action on $S^{-1}\mathscr{X}$, defined in (4.3.8). In fact, T_{rP} is a homotopy equivalence (4.3.8). Since the horizontal functors are homotopy equivalences, so is T_P. So, the converse (2) is established. ■

The following would be of our particular interest.

Lemma 4.3.8. *Let S be a symmetric monoidal category, and $\mathscr{X}$ be an S-category. There is a left S-action $\odot_l$ and a right S-action $\odot_r$ on $S^{-1}\mathscr{X}$, as defined in* (4.2.7). *Then both actions act invertibly on $S^{-1}\mathscr{X}$.*

Proof. For $P \in Obj\ S$, let $T_{lP}, T_{rP} : S^{-1}\mathscr{X} \longrightarrow S^{-1}\mathscr{X}$ be the translations, with respect to $\odot_l, \odot_r$, defined as, $T_{lP}(Q,X) = (P \odot Q, X)$, and $T_{rP} = (Q, P \odot X)$. Now,

$$T_{lP}T_{rP}(Q,X) = T_{rP}T_{lP}(Q,X) = (P \odot Q, P \odot X) = P \odot (Q,X).$$

There is a natural arrow $(Q,X) \longrightarrow T_{rP}T_{lP}(Q,X)$ in $S^{-1}\mathscr{X}$, represented by $(P, 1_{(P\odot Q, P\odot X)})$. This defines a natural transformation $1_{S^{-1}\mathscr{X}} \longrightarrow T_{rP}T_{lP} = T_{lP}T_{rP}$. Hence, by (2.1.9), both $T_l(P)$, $T_r(P)$ are homotopy equivalences. ■

Proposition 4.3.7 has a useful application (4.3.10), by iterating the process of $S^{-1}\mathscr{X}$-construction. This requires straightening some scope of confusions.

Remark 4.3.9. Let $(S, \odot)$ be a symmetric monoidal category, and $(\mathscr{X}, \odot)$ be an S-category. The availability of two S-actions $\odot_l$ and $\odot_r$, on $S^{-1}\mathscr{X}$, provides flexibility. It also creates confusion if we attempt to iterate the process. Namely, we obtain two localized categories

$$S_l^{-1}S^{-1}\mathscr{X} := (S, \odot_l)^{-1}(S^{-1}\mathscr{X}) \qquad \text{and}$$
$$S_r^{-1}S^{-1}\mathscr{X} := (S, \odot_r)^{-1}(S^{-1}\mathscr{X}).$$

Corollary 4.3.10. *Let S be a symmetric monoidal category and $\mathscr{X}$ be an S-category. Assume that every morphism in S is an isomorphism (i.e., $S = Iso(S)$) and translations in S are faithful* (4.3.2). *As before, $p_1 : S_l^{-1}S^{-1}\mathscr{X} \longrightarrow \langle S, S\rangle$ denotes projection to the 1st-coordinate. With other notation as in* (4.3.1), *the functors*

$$S^{-1}\mathscr{X} \xrightarrow{\ \overline{p}_{1P}\ } S_l^{-1}S^{-1}\mathscr{X} \qquad \textit{are homotopy equivalences}$$

$$\forall\ P \in Obj(S).$$

Further, the homotopy inverse $\Gamma_P : S_l^{-1}S^{-1}\mathscr{X} \longrightarrow S^{-1}\mathscr{X}$ of $\overline{p}_{1P}$ is given by

$$\Gamma_P(L,(Q,X)) = (P \odot Q, L \odot X), \quad \forall\ (L,(Q,X)) \in Obj(S_l^{-1}S^{-1}\mathscr{X})$$

which is a well-defined functor.

Proof. By (4.3.8), the left S-action $\odot_l$ acts invertibly on $S^{-1}\mathscr{X}$. By Proposition 4.3.7 it is established that $\overline{p}_{1P}$ is a homotopy equivalence. That Γ_P is a well-defined functor, and is the homotopy inverse, follows from (4.3.11), which follows. The proof is complete. ∎

Now we proceed to define the homotopy inverse functor Γ_P, mentioned above.

Definition 4.3.11. Let S be a symmetric monoidal category and $\mathscr{X}$ be an S-category. Let $(P_i, (Q_i, X_i)) \in Obj(S_l^{-1}S^{-1}\mathscr{X})$ for $i = 1, 2$. An arrow $\mathbf{u} \in Mor_{S_l^{-1}S^{-1}\mathscr{X}}((P_1, (Q_1, X_1)), (P_2, (Q_2, X_2)))$, is first represented by $\mathbf{u} = (U_0, u_0, f) : (P_1, (Q_1, X_1)) \longrightarrow (P_2, (Q_2, X_2))$, where $U_0 \in Obj(S)$, $u_0 : U_0 \odot P_1 \longrightarrow P_2$ is in S and $f : U_0 \odot_l (Q_1, X_1) = (U_0 \odot Q_1, X_1) \longrightarrow (Q_2, X_2)$ is in $S^{-1}\mathscr{X}$.
So, f is represented by $f = (U_1, u_1, u_2)$, with $U_1 \in Obj(S)$, $u_1 : U_1 \odot U_0 \odot Q_1 \longrightarrow Q_2$ in S and $u_2 : U_1 \odot X_1 \longrightarrow X_2$ in $\mathscr{X}$. So, we write $\mathbf{u} = (U_0, U_1, u_0, u_1, u_2)$. It follows,

$$
\begin{aligned}
&Mor_{S_l^{-1}S^{-1}\mathscr{X}}((P_1, (Q_1, X_1)), (P_2, (Q_2, X_2))) = \\
&\cup_{U_0, U_1 \in Obj S} Mor_S(U_0 \odot P_1, P_2) \times Mor_S(U_1 \odot U_0 \odot Q_1, Q_2) \\
&\quad \times Mor_{\mathscr{X}}(U_1 \odot X_1, X_2)
\end{aligned}
$$

where U_0, U_1 runs through all the isomorphism classes of objects in S. Define

$$
\begin{cases}
\Gamma(P, (Q, X)) = (Q, P \odot X) & \forall\ (P, (Q, X)) \in Obj(S_l^{-1}S^{-1}\mathscr{X}) \\
\Gamma(\mathbf{u}) = (U_0 \odot U_1, u_1, u_0 \odot u_2) & \forall\ \mathbf{u} = (U_0, U_1, u_0, u_1, u_2).
\end{cases}
$$

Note $\Gamma(\mathbf{u}) : (Q_1, P_1 \odot X_1) \longrightarrow (Q_2, P_2 \odot X_2)$ is an arrow in $S^{-1}\mathscr{X}$. One checks that the composition rule works. It follows that $\Gamma : S_l^{-1}S^{-1}\mathscr{X} \longrightarrow S^{-1}\mathscr{X}$ is a well-defined functor.

The following provides a homotopy inverse Γ_P to $\overline{p}_{1P} : S_l^{-1}S^{-1}\mathscr{X} \longrightarrow S^{-1}\mathscr{X}$ in (4.3.10).

Lemma 4.3.12. *Let S be a symmetric monoidal category and $\mathscr{X}$ be an S-category. As before, $p_1 : S_l^{-1}S^{-1}\mathscr{X} \longrightarrow \langle S, S\rangle$ denotes the first projection map and other notation, as in (4.3.1). Let $P \in Obj(S)$,* *and define the functor Γ_P by the commutative diagram*

$$\begin{array}{ccc} S_l^{-1}S^{-1}\mathscr{X} & \xrightarrow{\Gamma} & S^{-1}\mathscr{X} \\ & \searrow^{\Gamma_P} & \downarrow T_{lP} \\ & & S^{-1}\mathscr{X} \end{array} \qquad \textit{where } \Gamma \textit{ is as in } (4.3.11).$$

Then Γ_P is a homotopy left-inverse of $\overline{p}_{1P} : S^{-1}\mathscr{X} \longrightarrow S_l^{-1}S^{-1}\mathscr{X}$.

Proof. For $(Q, X) \in Obj(S^{-1}\mathscr{X})$, we have

$$\begin{aligned} \Gamma_P\overline{p}_{1P} &= T_{lP}\Gamma\overline{p}_{1P}(Q, X) = T_{lP}\Gamma(P, (Q, X) = T_{lP}(Q, P \odot X) \\ &= (P \odot Q, P \odot X) = P \odot (Q, X). \end{aligned}$$

The natural arrows $(P, 1_{P\odot(Q,X)}) : (Q, X) \longrightarrow T_{lP}\Gamma p_{1P}(Q, X)$ in $S^{-1}\mathscr{X}$ lead to a natural transformation $1_{S^{-1}\mathscr{X}} \longrightarrow \Gamma_P\overline{p}_{1P}$. Therefore, $\mathbb{B}1_{S^{-1}\mathscr{X}}$ and $\mathbb{B}\Gamma_P\overline{p}_{1P}$ are homotopic. So, Γ_P is a homotopy left-inverse of $\overline{p}_{1P}$. The proof is complete. ■

4.4. The $\mathscr{X}$-Coordinate Functor

Having dealt with the first coordinate map $p_1 : S^{-1}\mathscr{X} \longrightarrow \langle S, \mathscr{X}\rangle$ in the last section (Section 4.3), we consider the $\mathscr{X}$-coordinate functor in this section.

Lemma 4.4.1. *Let S be a symmetric monoidal category and $\mathscr{X}$ be an S-category. Suppose every map in $\mathscr{X}$ is monic, and $\forall\ P \in Obj(S)$, $\forall X \in Obj(\mathscr{E})$ the translation map $Aut_S(P) \longrightarrow Aut_{\mathscr{X}}(P \odot X)$ is an injection (faithfulness, analogous to (4.3.2)). Consider the 2nd-coordinate projection $p : S^{-1}\mathscr{X} \longrightarrow \langle S, \mathscr{X}\rangle$. Then p is cofibered. Also, $p^{-1}Y = S \times Y \cong S$, and for morphisms $u = (U, u) : Y_1 \longrightarrow Y_2$ in $\langle S, \mathscr{X}\rangle$, represented by $u : U \odot Y_1 \longrightarrow Y_2$ in $\mathscr{X}$, the induced functor $u_* : p^{-1}Y_1 \longrightarrow p^{-1}Y_2$ (i.e., the cobase change map) is given by*

$$u_*(P, Y_1) = (U \odot P, Y_2). \quad \textit{So, the diagram,}$$

$$\begin{array}{ccc} S & \xrightarrow{P \mapsto (P, Y_1)} & p^{-1}Y_1 \\ {\scriptstyle T_U}\downarrow & & \downarrow{\scriptstyle u_*} \\ S & \longrightarrow & p^{-1}Y_2 \end{array} \qquad \textit{commutes,}$$

where $T_U : S \longrightarrow S$ is the translation functor $P \mapsto U \odot P$.

Proof. Given $Y \in Obj\langle S, \mathscr{X}\rangle = Obj\ \mathscr{X}$, we describe the categories p/Y and $p^{-1}Y$. To simplify notation, we may write $(P, X; v) := ((P, X), v)$ for $((P, X), v) \in Obj(p/Y)$.

$$\begin{cases} Obj(p/Y) = \\ \qquad \{(P, X; v) : P \in Obj(S),\ X \in Obj(\mathscr{X}), v \in Mor_{\langle S,\mathscr{X}\rangle}(X, Y)\} \\ Mor_{p/Y}((P_1, X_1; v_1), (P_2, X_2; v_2)) = \\ \qquad \{(u_1, u_2) \in Mor_{S^{-1}\mathscr{X}}((P_1, X_1), (P_2, X_2)) \ni v_2u_2 = v_1 \in \langle S, \mathscr{X}\rangle\} \\ Obj(p^{-1}Y) = \{(P, Y) : P \in Obj(S)\} \\ Mor_{p^{-1}Y}((P_1, Y), (P_2, Y)) = \\ \qquad \{(u_1, u_2) \in Mor_{S^{-1}\mathscr{X}}((P_1, Y), (P_2, Y)) \ni u_2 = (\mathbf{e}_S, 1_Y) \in \langle S, \mathscr{X}\rangle\} \\ = Mor_S(P_1, P_2) \end{cases}$$

With $v_1 := (V_1, v_1) : X_1 \longrightarrow Y$, in $\langle S, \mathscr{X}\rangle$, represented by $v_1 : V_1 \odot X_1 \longrightarrow Y$ in S, we have

$$\begin{cases} Mor_{p/Y}((P_1, X_1; v_1), (P_2, Y; 1_Y)) \\ = \{(u_1, u_2) \in Mor_{S^{-1}\mathscr{X}}((P_1, X_1), (P_2, Y)) \ \ni u_2 = v_1 \text{ in } \langle S, \mathscr{X}\rangle\} \\ \cong Hom_S(V_1 \odot P_1, P_2). \end{cases}$$

An arrow $(u_1, u_2) \in Mor_{p/Y}((P_1, X_1; v_1), (P_2, X_2; v_2))$ is represented by $u_1 : U \odot P_1 \longrightarrow P_2$ in S and $u_2 : U \oplus X_1 \longrightarrow X_2$ in $\mathscr{X}$. Also, $v_1 = (V_1, v_1)$, $v_2 = (V_2, v_2)$ are represented by $v_1 : V_1 \odot X_1 \longrightarrow Y$ in S, $v_2 : V_2 \odot X_2 \longrightarrow Y$ in $\mathscr{X}$. And, $v_2u_2 = u_1$ means, there is an isomorphism $\tau : P_1 \xrightarrow{\sim} V_2 \odot U$ such that the following diagram commutes:

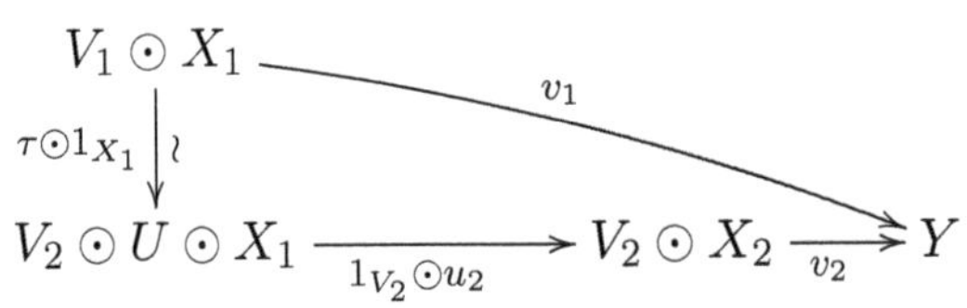

We compose

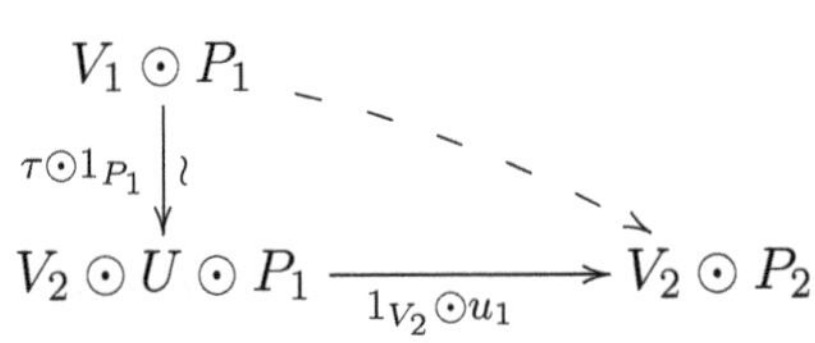

By the monicity and faithfulness hypotheses, τ is uniquely determined. Define $L(u_1, u_2) = ((1_{V_2} \odot u_1)(\tau \odot 1_{P_1}), 1_Y) \in Mor_{p^{-1}Y}$

$((V_1 \odot P_1, Y)), (V_2 \odot P_2, Y))$. With such notation, define, $L : p/Y \longrightarrow p^{-1}Y$ by

$$\begin{cases} L(P, X; v) = (V \odot P, Y) \qquad \forall\ (P, X; v) \in Obj(p/Y) \\ L(u_1, u_2) = ((1_{V_2} \odot u_1)(\tau \odot 1_{P_1}), 1_Y) \\ \qquad\qquad \forall\ (u_1, u_2) \in Mor_{p/Y}\left((P_1, X_1; v_1), (P_2, X_2; v_2)\right). \end{cases}$$

Then $L : p/Y \longrightarrow p^{-1}Y$ is a functor. We check that L is left adjoint to $\Phi : p^{-1}Y \longrightarrow p/Y$, as follows. Let $((P_1, X_1), v_1) \in Obj\ (p/Y)$, with $v_1 = (V_1, v_1) : X_1 \longrightarrow Y$ in $\langle S, \mathscr{X}\rangle$, represented by $v_1 : V_1 \odot X_1 \longrightarrow Y$ in $\mathscr{X}$ and $(P_2, Y) \in Obj\ p^{-1}Y$. Then

$$\begin{array}{ccc}
Mor_{p^{-1}Y}\left(L\left((P_1, X_1), v_1\right), (P_2, Y)\right) & & Mor_{p/Y}\left(((P_1, X_1), v_1), \Phi(P_1, Y)\right) \\
\Big\| & & \Big\| \\
Mor_{p^{-1}Y}\left((P_1 \odot V_1, Y), (P_2, Y)\right) & \overset{\eta}{\dashrightarrow} & Mor_{p/Y}\left(((P_1, X_1), v_1), ((P_2, Y), 1_Y)\right) \\
\wr\Big\downarrow & & \Big\downarrow\wr \\
Mor_S\left(P_1 \odot V_1, P_2\right) & & Mor_S\left(V_1 \odot P_1, P_2\right)
\end{array}$$

This establishes that L is left adjoint to Φ, by defining the following bijection:

$$\eta(\mathbf{e}_S, f, 1_Y) = (V_1, f, v_1)$$

Therefore, p is **precofibered**. Further routine checking completes the proof that p is **cofibered**, and of the rest of the assertions. The proof is complete. ■

Using essentially the same arguments in (4.4.1), we have the following.

Lemma 4.4.2. *Let S be a symmetric monoidal category, and $\mathscr{X}$ be an S-category. Assume that every map in $\mathscr{X}$ is monic, and $\forall\ P \in Obj\ S$, $\forall X \in Obj\ \mathscr{X}$ the translation maps $Aut_S(P) \longrightarrow Aut_{\mathscr{X}}(P \odot X)$ are injections (faithfulness). Treat $S^{-1}\mathscr{X}$ as an S-category, with its left action $\odot_l$. Then the $\mathscr{X}$-coordinate map*

$p_3 : S_l^{-1}S^{-1}\mathscr{X} \longrightarrow \langle S, \mathscr{X}\rangle$ is cofibered, and the sequence

$$S^{-1}S \xrightarrow{\quad \iota_Y \quad} S_l^{-1}S^{-1}\mathscr{X} \xrightarrow{p_3} \langle S, \mathscr{X}\rangle$$

is a homotopy fibration, $\forall Y \in Obj\ \mathscr{X}$,

where ι_Y is the functor, defined by

$$\begin{cases} \iota_Y(P,Q) = (P,(Q,Y)) & \forall\ (P,Q) \in Obj(S^{-1}S) \\ \iota_Y(U_0,u_0,u_1) = (U_0,\mathbf{e}_S,u_0,(u_1,1_Y)) & \\ \qquad \forall\ (U_0,u_0,u_1) \in Mor_{S^{-1}S}((P_1,Q_1),(P_2,Q_2)). & \end{cases} \tag{4.11}$$

Proof. To simplify notation, we write $(P,Q,X) := (P,(Q,X)) \in Obj\ S_l^{-1}S^{-1}\mathscr{X}$. An arrow $\mathbf{u} : (P_1,Q_1,X_1) \longrightarrow (P_2,Q_2,X_2)$ in $S_l^{-1}S^{-1}\mathscr{X}$ is given by $(U_0,(u_0,f))$, with $u_0 : U_0 \odot P_1 \longrightarrow P_2$ in S, and $f : U_0 \odot (Q_1,X_1) = (U_0 \odot Q_1, X_1) \longrightarrow (Q_2,X_2)$ in $S^{-1}\mathscr{X}$. So, $f = (U_1,u_1,u_2)$, with $u_1 : U_0 \odot U_1 \odot Q_1 \longrightarrow Q_2$ in S and $u_2 : U_1 \odot X_1 \longrightarrow X_2$ in $\mathscr{X}$. In summary, the arrow $\mathbf{u} : (P_1,Q_1,X_1) \longrightarrow (P_2,Q_2,X_2)$ in $S_l^{-1}S^{-1}\mathscr{X}$ is represented by

$$\mathbf{u} := (U_0,U_1,u_0,u_1,u_2) \text{ with } U_0,U_1 \in Obj\ S$$
$$\begin{cases} u_0 \in Mor_S(U_0 \odot P_1, P_2) \\ u_1 \in Mor_S(U_0 \odot U_1 \odot Q_1, Q_2) \\ u_2 \in Mor_{\mathscr{X}}(U_1 \odot X_1, X_2) \end{cases} \tag{4.12}$$

Let $IsoCl(S)$ denote the set of all isomorphism classes of objects in S. Then we have

$$\begin{aligned} &Mor_{S_l^{-1}S^{-1}\mathscr{X}}((P_1,Q_1,X_1),(P_2,Q_2,X_2)) = \\ &\bigcup_{[U_0],[U_1]\in IsoCl(S)} Mor_S(U_0 \odot P_1, P_2) \times Mor_S(U_0 \odot U_1 \odot Q_1, Q_2) \\ &\qquad\qquad \times Mor_{\mathscr{X}}(U_1 \odot X_1, X_2) \end{aligned}$$

So, the union runs over choices of one U_0, U_1 from each isomorphism class in $Obj(S)$. The third coordinate projection $p_3(\mathbf{u}) = u_2 \in Mor_{\mathscr{X}}(U_1 \odot X_1, X_2)$ is to be considered as a morphism $X_1 \longrightarrow X_2$ in

$\langle S, \mathscr{X} \rangle$. Let $Y \in Obj(\langle S, \mathscr{X} \rangle) = Obj(\mathscr{X})$. We describe the categories p_3/Y and $p_3^{-1}Y$, as follows:

$$\begin{cases} p_3/Y = \\ \{(P, Q, X; v) : P, Q \in Obj S,\ X \in Obj\ \mathscr{X}, v \in Mor_{\langle S, \mathscr{X} \rangle}(X, Y)\} \\ Mor_{p_3/Y}((P_1, Q_1, X_1; v_1), (P_2, Q_2, X_2; v_2)) = \\ \bigcup_{[U_0],[U_1] \in IsoCl(S)} \\ \Big\{(U_0, U_1, u_0, u_1, u_2) \in Mor_{S_l^{-1}S^{-1}\mathscr{X}}((P_1, (Q_1, X_1)), (P_2, (Q_2, X_2))) : \\ \qquad\qquad\qquad v_2 u_2 = v_1 \in \langle S, \mathscr{X} \rangle\} \end{cases}$$

Let $v_i := (V_i, v_i)$ be represented by the maps $v_i : V_i \odot X_i \longrightarrow Y$ in $\mathscr{X}$ for $i = 1, 2$. The equality, $v_2 u_2 = v_1$, above, is given by the commutative diagram

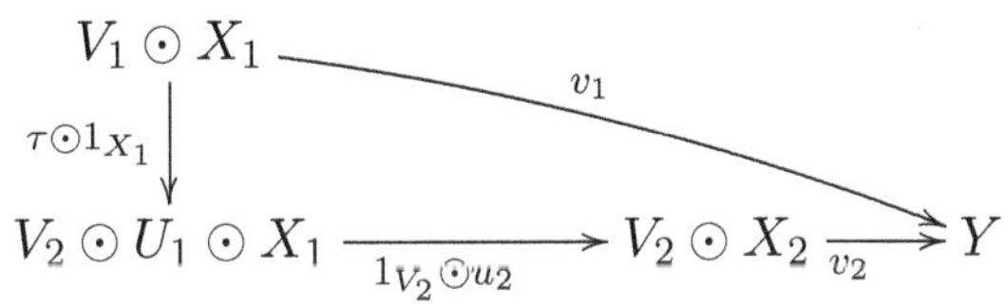

under the monic and faithfulness hypotheses, for a unique isomorphism $\tau : V_1 \xrightarrow{\sim} V_2 \odot U_1$ in S. (4.13)

So, we have

$$\begin{aligned} & Mor_{p/Y}((P_1, Q_1, X_1; v_1), (P_2, Q_2, X_2; v_2)) \\ & \quad = \bigcup_{[U_0],[U_1] \in IsoCl(S)} \{(u_0, u_1, u_2) \in \\ & Mor_S(U_0 \odot P_1, P_2) \times Mor_S(U_0 \odot U_1 \odot Q_1, Q_2) \\ & \quad \times Mor_{\mathscr{X}}(U_1 \odot X_1, X_2) : \\ & \exists \tau : V_1 \xrightarrow{\sim} V_2 \odot U_1, v_1 = v_2(1_{V_2} \odot u_2)(\tau \odot 1_{X_1})\Big\} \end{aligned}$$

Further, if $X_2 = Y$ and $v_2 = (V_2, v_2) = 1_Y := (\mathbf{e}_S, 1_Y)$, then the above diagram reduces to

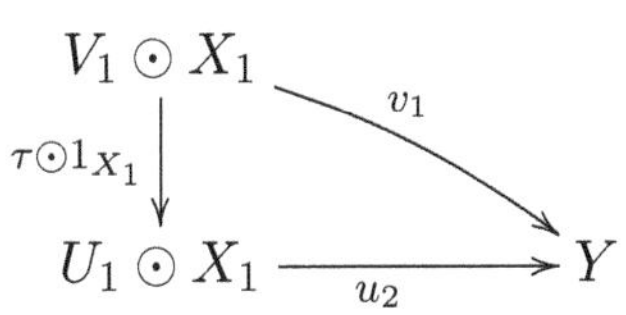

$V_2 = \mathbf{e}_S$, under the hypotheses, for a unique isomorphism $\tau : V_1 \xrightarrow{\sim} U_1$ in S.

(*Not surprisingly, it precisely means* $v_1 = u_2$ *in* $\langle S, \mathscr{X} \rangle$.) Therefore,

$$
\begin{aligned}
&Mor_{p/Y}((P_1, Q_1, X_1; v_1), (P_2, Q_2, Y; 1_Y)) \\
&\quad = \bigcup_{[U_0],[U_1] \in IsoCl(S)} \{(u_0, u_1, u_2) \in \\
&Mor_S(U_0 \odot P_1, P_2) \times Mor_S(U_0 \odot U_1 \odot Q_1, Q_2) \\
&\quad \times Mor_{\mathscr{X}}(U_1 \odot X_1, Y) : \\
&\exists \tau : V_1 \xrightarrow{\sim} U_1, v_1 = u_2(\tau \odot 1_{X_1}) \text{ i.e., } u_2 = v_1 \Big\} \\
&= Mor_S(U_0 \odot P_1, P_2) \times Mor_S(U_0 \odot V_1 \odot Q_1, Q_2) \times v_1 \\
&\quad \text{up to isomorphisms of } U_0 \\
&\cong Mor_{S^{-1}S}((P_1, Q_1 \odot V_1), (P_2, Q_2))
\end{aligned}
$$

If $X_1 = Y$ and $v_1 = (V_1, v_1) = (\mathbf{e}_S, 1_Y)$, then $V_1 = \mathbf{e}_S$. From this, the description of $p^{-1}Y$ follows:

$$
\begin{cases}
p^{-1}Y = \{(P, Q, Y) : P, Q \in ObjS\} \\
Mor_{p^{-1}Y}((P_1, Q_1, Y), (P_2, Q_2, Y)) = \\
Mor_S(U_0 \odot P_1, P_2) \times Mor_S(U_0 \odot Q_1, Q_2) \times 1_Y \\
\quad \text{up to isomorphisms of } U_0 \\
\cong Mor_{S^{-1}S}((P_1, Q_1), (P_2, Q_2))
\end{cases}
\tag{4.14}
$$

Let $\Phi : p_3^{-1}Y \longrightarrow p_3/Y$ be the natural functor. We proceed to define a left adjoint $L : p_3/Y \longrightarrow p_3^{-1}Y$, of Φ, as follows. With notation, as in (4.12), (4.13), let $\mathbf{u} = (u_0, u_1, u_2) : ((P_1, (Q_1, X_1)), v_1) \longrightarrow ((P_2, (Q_2, X_2)), v_2)$ be an arrow in p_3/Y. From the description above, we have arrows $u_0 : U_0 \odot P_1 \longrightarrow P_2$ and $u_1 : U_0 \odot U_1 \odot Q_1 \longrightarrow Q_2$, and $\tau : V_1 \xrightarrow{\sim} V_2 \odot U_1$, in S, commuting the respective diagram. So, we obtain a commutative diagram

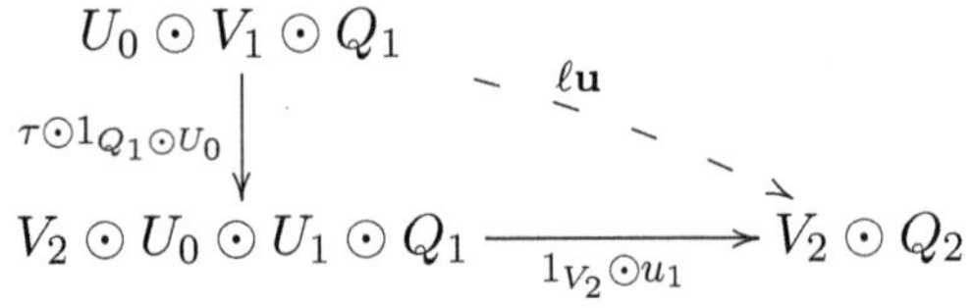

where $\ell\mathbf{u}$ is the composition in S.

Then $\mathbf{Lu} := (U_0, \mathbf{e}_S, u_0, \ell\mathbf{u}, 1_Y) : (P_1, V_1 \odot Q_1, Y) \longrightarrow (P_2, V_2 \odot Q_2, Y)$ is a map in $p_3^{-1}Y$. Accordingly, define

$$\begin{cases} L((P, Q, X), v) = (P, (V \odot Q, Y)), \quad v := (V, v) : V \odot X \longrightarrow Y \\ L(\mathbf{u}) = (U_0, \mathbf{e}_S, u_0, \ell\mathbf{u}, 1_Y) : (P_1, V_1 \odot Q_1, Y) \longrightarrow (P_2, V_2 \odot Q_2, Y) \end{cases}$$

Let $(P_1, (Q_1, X_1), v_1) \in Obj\ (p_3/Y)$, $(P_2, (Q_2, Y)) \in Obj\ p^{-1}Y$, with $v_1 : V_1 \odot X_1 \longrightarrow Y$ in S. Then we have two columns of bijections:

$$\begin{array}{ccc} Mor_{p^{-1}A}(L((P_1, (Q_1, X_1), v_1)), (P_2, (Q_2, Y))) & & Mor_{p/Y}((P_1, (Q_1, X_1), v_1), \Phi((P_2, (Q_2, Y))) \\ \| & & \| \\ Mor_{p^{-1}A}((P_1, (Q_1 \odot V_1, Y)), (P_2, (Q_2, Y))) & \overset{\eta}{\dashrightarrow} & Mor_{p/Y}((P_1, (Q_1 \odot V_1, Y)), ((P_2, (Q_2, Y)), 1_Y)) \\ \wr\downarrow & & \downarrow\wr \\ Mor_{S^{-1}S}((P_1, Q_1 \odot V_1), (P_2, Q_2)) & & Mor_{S^{-1}S}((P_1, Q_1 \odot V_1), (P_2, Q_2)) \end{array}$$

According to this diagram, a natural bijection η (the dotted arrow) is defined, as follows:

$$\eta(U_0, \mathbf{e}_S, u_0, u_1, 1_Y) = (U_0, V_1, u_0, u_1, v_1)$$

Therefore, p_3 is **precofibered**. Let $v := (V, v) : Y \longrightarrow Y'$, $w = (W, w) : Y' \longrightarrow Y$" be two composable arrows in $\langle S, \mathscr{X} \rangle$. The cobase change functor v_* is given by

$$\begin{aligned} v_*(P, (Q, Y)) &:= Lv_*\Phi(P, (Q, Y)) = Lv_*((P, (Q, Y)), 1_Y) \\ &= L(((P, (Q, Y)), v) = (P, (V \odot Q, Y'). \end{aligned}$$

It follows $(wv)_*(P, (Q, Y)) = w_*v_*(P, (Q, Y))$. An arrow $\mathbf{u} \in Mor_{p_3^{-1}Y}((P_1, (Q_1, Y)), (P_2, (Q_2, Y)))$, is given by $\mathbf{u} = (U_0, \mathbf{e}_S, u_0, u_1, 1_Y)$, with $u_0 : U_0 \odot P_1 \longrightarrow P_2$ in S, and $u_1 : U_0 \odot Q_1 \longrightarrow Q_2$ in $\mathscr{X}$.

$$\begin{cases} \Phi(\mathbf{u}) = v_*((U_0, \mathbf{e}_S, u_0, u_1, 1_Y)) : \\ \qquad\qquad ((P_1, (Q_1, Y)), 1_Y)) \longrightarrow ((P_2, (Q_2, Y)), 1_Y)) \\ v_*\Phi(\mathbf{u}) = (U_0, \mathbf{e}_S, u_0, u_1, 1_Y) : \\ \qquad\qquad ((P_1, (Q_1, Y)), v)) \longrightarrow ((P_2, (Q_2, Y)), v) \\ v_*(\mathbf{u}) := Lv_*\Phi(\mathbf{u}) = (U_0, \mathbf{e}_S, u_0, 1_V \odot u_1, 1_{Y'}) \\ \qquad\qquad \in Mor_{p_3^{-1}Y'}((P_1, (V \odot Q_1, Y')), (P_2, (V \odot Q_2, Y'))) \end{cases}$$

From this it follows that $(wv)_*(\mathbf{u}) = w_* v_*(\mathbf{u})$. So, p_3 is **cofibered**. We have the commutative diagram of functors

$$\begin{array}{ccc} S^{-1}S & \longrightarrow & p_3^{-1}Y \\ {\scriptstyle T_{rV}}\downarrow & & \downarrow{\scriptstyle v_*} \\ S^{-1}S & \longrightarrow & p_3^{-1}Y' \end{array} \qquad \text{where}$$

$$T_{rV}(P,Q) = V \odot_r (P,Q) = (P, V \odot Q).$$

By (4.14), the horizontal functors are homotopy equivalences, and so is the translation $T_r(V)$ (4.3.10). Therefore, the cobase change functors v_* are homotopy equivalences. Hence, the cofibered version of Theorem B (2.2.6) is applicable, and the lemma is established. The proof is complete. ■

The following is the final theorem in this section.

Theorem 4.4.3. *Let S be a symmetric monoidal category, and $\mathscr{X}$ be an S-category.*

1. *Assume that every morphism in S is an isomorphism (i.e., $S = Iso(S)$), and translations are faithful* (4.3.2).
2. *Also assume that all maps in $\mathscr{X}$ are monic, and the translation maps $Aut_S(P) \longrightarrow Aut_{\mathscr{X}}(P \odot X)$ are injective (i.e., faithful), $\forall\, P \in Obj\ S$, $\forall X \in Obj\ \mathscr{X}$.*

Let $p_2 : S^{-1}\mathscr{X} \longrightarrow \langle S, \mathscr{X}\rangle$ be the $\mathscr{X}$-coordinate projection. Then the sequence

$$S^{-1}S \xrightarrow{\ \nu_Y\ } S^{-1}\mathscr{X} \xrightarrow{\ p_2\ } \langle S, \mathscr{X}\rangle \quad \textit{is a homotopy fibration,}$$

$$\forall Y \in Obj\ \mathscr{X},$$

where $\nu_Y : S^{-1}S \longrightarrow S^{-1}\mathscr{X}$ denotes the functor $(P,Q) \mapsto (Q, P \odot Y)$.

Proof. Consider the diagram

$$\begin{array}{ccccc} & & S^{-1}\mathscr{X} & \xrightarrow{\ p_2\ } & \langle S, \mathscr{X}\rangle \\ & & \downarrow{\scriptstyle \overline{p}_{1\mathbf{e}_S}} & & \| \\ S^{-1}S & \xrightarrow[\ \iota_Y\ ]{} & S_l^{-1}S^{-1}\mathscr{X} & \xrightarrow[\ p_3\ ]{} & \langle S, \mathscr{X}\rangle \end{array}$$

The vertical arrow $\overline{p}_{1_{\mathbf{e}_S}}$ is the functor $\overline{p}_{1_{\mathbf{e}_S}}(Q, X) = (\mathbf{e}_S, (Q, X))$, which is a homotopy equivalence (4.3.10). The homotopy inverse of $\overline{p}_{1_{\mathbf{e}_S}}$ is given by $\Gamma_{\mathbf{e}_S}(P, (Q, X)) = (Q, P \odot Q)$. With ι_Y, as in (4.4.2), we have $\nu_Y = \Gamma_{\mathbf{e}_S}\iota_Y$. The sequence in the 2nd-line is a homotopy fibration, by Lemma 4.4.2. Therefore, the proof is complete by identifying $S^{-1}\mathscr{X}$ with $S_l^{-1}S^{-1}\mathscr{X}$, via $\overline{p}_{1_{\mathbf{e}_S}}$. ■

4.5. Split Exact Categories

Given an exact category $\mathscr{E}$, we define a new category $\mathbf{E}\mathscr{E}$, whose objects are conflations. This is not to be confused with the category of extensions $\varepsilon(\mathscr{E})$ defined before (Section 3.4).

Definition 4.5.1. Let $\mathscr{E}$ be an exact category. We define the category $\mathbf{E}\mathscr{E}$ as follows:

1. The objects of $\mathbf{E}\mathscr{E}$ are the admissible exact sequences (conflations), in $\mathscr{E}$.
2. Let $E_i := \left(K_i \overset{\iota_i}{\hookrightarrow} M_i \overset{\zeta_i}{\twoheadrightarrow} C_i \right) \in Obj\ \mathbf{E}\mathscr{E}$, for $i = 1, 2$. A morphism $f := (\alpha, \beta) : E_1 \longrightarrow E_2$ is given by an equivalence class of diagrams

$$\begin{array}{ccccccc} E_1 & & K_1 & \overset{\iota_1}{\hookrightarrow} & M_1 & \overset{\zeta_1}{\twoheadrightarrow} & C_1 \\ \Big\downarrow & & \alpha \uparrow & & \uparrow 1_{M_1} & & \uparrow \\ f & & K_2 & \underset{\iota_1\alpha}{\hookrightarrow} & M_1 & \twoheadrightarrow & C \\ \Big\downarrow & & 1_{K_2} \downarrow & & \downarrow \beta & & \downarrow \\ E_2 & & K_2 & \underset{\iota_2}{\hookrightarrow} & M_2 & \underset{\zeta_2}{\twoheadrightarrow} & C_2 \end{array} \tag{4.15}$$

Two such diagrams are defined to be equivalent if there is an isomorphism, with all the horizontal maps equal to identity, except C. Indeed, up to a choice of C, the diagram (4.15) is determined by the commutative diagram

$$\begin{array}{ccc} K_1 & \overset{\iota_1}{\hookrightarrow} & M_1 \\ \alpha \uparrow & & \downarrow \beta \\ K_2 & \underset{\iota_2}{\hookrightarrow} & M_2 \end{array}. \quad \text{We also write} \quad f = (\alpha, \beta). \tag{4.16}$$

Having defined the category $\mathbf{E}\mathscr{E}$, we have the following observations:

1. We have not said the $\mathbf{E}\mathscr{E}$ is an exact category.
2. With the above notation in (4.15), define

$$t(f) = (\ C_1 \twoheadleftarrow C \hookrightarrow C_2\ ) \in Mor_{\mathbb{Q}\mathscr{E}}(C_1, C_2).$$

   Define the target functor, $t : \mathbf{E}\mathscr{E} \longrightarrow \mathbb{Q}\mathscr{E}$ by

$$\begin{cases} t(E_1) = C_1 & E_1 \in Obj\ \mathbf{E}\mathscr{E} \text{ as in (4.15)} \\ t(f) = \text{as above} & f \in Mor_{\mathbf{E}\mathscr{E}}(E_1, E_2) \text{ as in (4.15)} \end{cases}$$

3. For $C \in Obj(\mathbb{Q}\mathscr{E})$, we sometimes denote the fiber subcategory $\mathbf{E}\mathscr{E}_C := t^{-1}C$. Recall that the identity map $1_C \in Mor_{\mathbb{Q}\mathscr{E}}(C, C)$ is given by $C \overset{1_C}{\twoheadleftarrow} C \overset{1_C}{\hookrightarrow} C$. It follows from diagram (4.15), the morphisms in $t^{-1}C = \mathbf{E}\mathscr{E}_C$ are given by commutative diagrams:

$$\begin{array}{ccc} K_1 & \overset{\iota_1}{\hookrightarrow} & M_1 \\ \alpha \uparrow \wr & & \wr \downarrow \beta \\ K_2 & \underset{\iota_2}{\hookrightarrow} & M_2 \end{array} \qquad \alpha, \beta \text{ being isomorphisms in } \mathscr{E}.$$

4. Further, $t^{-1}\mathbf{0} = \mathbf{E}\mathscr{E}_{\mathbf{0}} \xrightarrow{\sim} Iso(\mathscr{E})$, where $Iso(\mathscr{E})$ denotes the category whose objects are the same as the objects in $\mathscr{E}$, and morphisms are isomorphisms (4.1.4). More explicitly, the equivalence is defined as follows:
   (a) For $M \in Obj\ \mathscr{E}$, let $F(M)$ denote the conflation $(\ M \overset{1_M}{\hookrightarrow} M \twoheadrightarrow \mathbf{0}\ )$.
   (b) For an isomorphism $f : M_1 \xrightarrow{\sim} M_2$, define the morphism $F(M_1) \longrightarrow F(M_2)$, by the diagram

$$\begin{array}{ccc} M_1 & \overset{1_{M_1}}{\hookrightarrow} & M_1 \\ f^{-1} \uparrow \wr & & \wr \downarrow f \\ M_2 & \underset{1_{M_2}}{\hookrightarrow} & M_2 \end{array}$$

Lemma 4.5.2. *Let $\mathscr{E}$ be an exact category, $S = Iso(\mathscr{E})$ and $C \in Obj\ \mathscr{E}$. Note that S is a symmetric monoidal category under the direct sum $\oplus$. Then the fiber category $\mathbf{E}\mathscr{E}_C = t^{-1}C$ has a natural symmetric monoidal category structure $\odot$ (given in the proof). Further,*

1. *Every arrow in* $\mathbf{E}\mathscr{E}_C = t^{-1}C$ *is monic.*
2. *There is a natural monoidal functor* $\eta_C : S \longrightarrow \mathbf{E}\mathscr{E}_C$. *Further, for* $P \in Obj(S)$, $E \in Obj(\mathbf{E}\mathscr{E}_C)$ *the map* $Aut_S(P) \longrightarrow Aut_{\mathbf{E}\mathscr{E}_C}(P \odot E)$ *is injective (faithfulness).* (As is customary, for $P \in Obj(S)$, $E \in Obj(\mathbf{E}\mathscr{E}_C)$, we denote $P \odot E := \eta_C(P) \odot E$.)

Proof. An arrow $f := E_1 \longrightarrow E_2$ in $\mathbf{E}\mathscr{E}_C$ is given by $f = (\alpha, \beta)$, as in (4.16). As both α and β are monic (*in fact, isomorphisms*), so is f.

Now, we define the monoidal structure on $\mathbf{E}\mathscr{E}_C$. Let $E_i = \left(K_i \overset{\iota_i}{\hookrightarrow} M_i \overset{\zeta_i}{\twoheadrightarrow} C \right) \in Obj\ \mathbf{E}\mathscr{E}_C$, for $i = 1, 2$. Consider the pull-back diagram

$$\begin{array}{ccccc} & & K_2 & = & K_2 \\ & & \downarrow & & \downarrow \iota_2 \\ K_1 & \hookrightarrow & M & \twoheadrightarrow & M_2 \\ \| & & \downarrow & & \downarrow \zeta_2 \\ K_1 & \underset{\iota_1}{\hookrightarrow} & M_1 & \underset{\zeta_1}{\twoheadrightarrow} & C \end{array} \qquad \text{where} \quad M := M_1 \times_C M_2 \quad \text{is the pullback.}$$

$$\text{Define,} \quad E_1 \odot E_2 := \left(K_1 \oplus K_2 \underset{\iota}{\hookrightarrow} M \underset{\zeta}{\twoheadrightarrow} C \right) \in Obj\ \mathbf{E}\mathscr{E}_C.$$

This addition operation $\odot$ on $\mathbf{E}\mathscr{E}_C$ defines a symmetric monoidal structure on $\mathbf{E}\mathscr{E}_C$, where $\mathbf{e}_C := \left(\mathbf{0} \hookrightarrow C \overset{1_C}{\twoheadrightarrow} C \right)$ would be the zero-object.

We proceed to define the monoidal functor $\eta_C : S \longrightarrow \mathbf{E}\mathscr{E}_C$. For $P \in Obj(\mathscr{E})$, define $\eta_C(P) := \left(P \hookrightarrow P \oplus C \twoheadrightarrow C \right) \in Obj\ (\mathbf{E}\mathscr{E}_C)$. For an isomorphism $f : P \xrightarrow{\sim} Q$ in $\mathscr{E}$, define $\eta_C(f) = (f^{-1}, f \oplus 1_C) \in Mor_{\mathbf{E}\mathscr{E}}(\eta_C(P), \eta_C(Q))$. We display the same by the diagram,

$$\begin{array}{c} \eta_C(P) \\ \Big\downarrow \eta_C(f) \\ \eta_C(Q) \end{array} \qquad \begin{array}{ccccc} P & \hookrightarrow & P \oplus C & \twoheadrightarrow & C \\ f^{-1} \uparrow \wr & & \uparrow 1 & & \uparrow 1_C \\ Q & \hookrightarrow & P \oplus C & \twoheadrightarrow & C \\ 1_Q \downarrow & & \wr \downarrow f \oplus 1_C & & \downarrow 1_C \\ Q & \hookrightarrow & Q \oplus C & \twoheadrightarrow & C \end{array}$$

These associations $P \mapsto \eta_C(P)$, $f \mapsto \eta_C(f)$ define the functor $\eta_C : S \longrightarrow \mathbf{E}\mathscr{E}_C$. It is also clear that $\eta_C(P \oplus Q) = \eta_C(P) \odot \eta_C(Q)$. Hence, η_C is a monoidal functor.

In an additive category, the map $Aut(P) \longrightarrow Aut(P \oplus K)$ is injective. It follows from this that the map $Aut_S(P) \longrightarrow Aut_{\mathbf{E}\mathscr{E}_C}(P \odot E)$ is injective. The proof is complete. ■

In this chapter, our primary interest has been the category of projective modules, in which every exact sequence splits. We define split exact category in the obvious way.

Definition 4.5.3. An exact category $\mathscr{E}$ is said to be a **split exact category** if every exact sequence splits.

Remark 4.5.4. Let $\mathscr{E}$ be a split exact category and $C \in Obj\ \mathscr{E}$. Let $\mathbf{X} := \left(K \overset{\iota}{\hookrightarrow} P \overset{\zeta}{\twoheadrightarrow} C \right) \in Obj(\mathbf{E}\mathscr{E}_C)$. Then $\mathbf{X} \cong \eta_C(K)$ in $\mathbf{E}\mathscr{E}_C$. Further, $\langle S, \mathbf{E}\mathscr{E}_C \rangle$ is connected.

Proof. Consider the following commutative diagram:

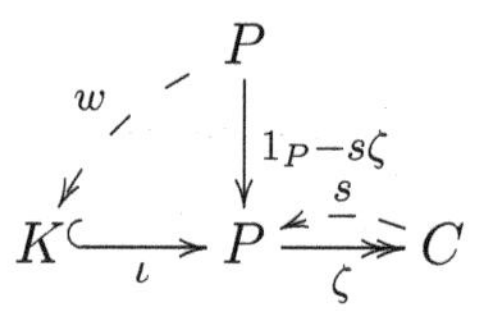

where $\zeta s = 1_C$.

Since $\iota = \ker(\zeta)$, there is an arrow w so that the triangle commutes. Since $\iota w \iota = \iota$, it follows that $w\iota = 1_K$. Define two arrows in $\mathbf{E}\mathscr{E}_C$, as follows:

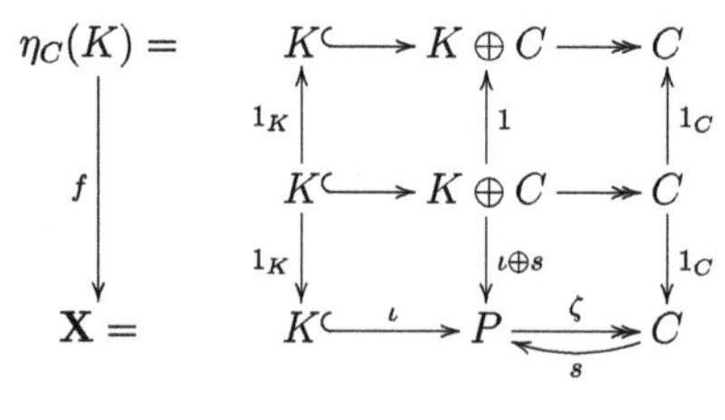

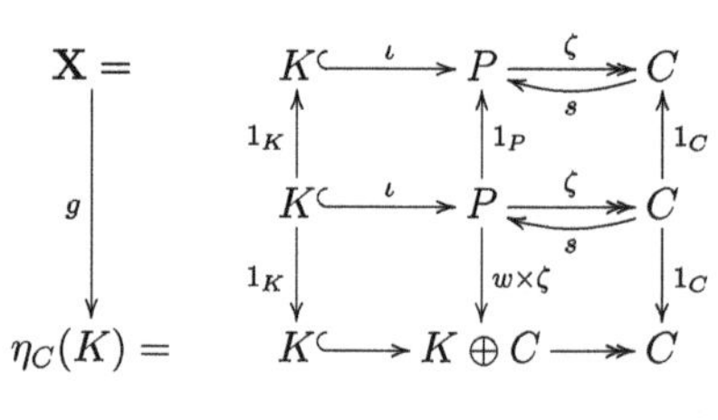

One checks that $gf = 1$ and $fg = 1$. So, $\eta_C(K) \cong \mathbf{X}$ in $\mathbf{E}\mathscr{E}_C$, hence also in $\langle S, \mathbf{E}\mathscr{E}_C \rangle$. Further, in $\langle S, \mathbf{E}\mathscr{E}_C \rangle$, we have an obvious arrow $\mathbf{e}_C \longrightarrow \eta_C(K)$ in $\langle S, \mathbf{E}\mathscr{E}_C \rangle$, represented by a map $1_{\eta_C(K)} : \mathbf{e}_C \odot \eta_C(K) \longrightarrow \eta_C(K)$ in $\mathbf{E}\mathscr{E}_C$. Therefore, $\langle S, \mathbf{E}\mathscr{E}_C \rangle$ is connected. The proof is complete. ■

The following construction will be of some use to us in the proof of the next theorem.

Definition 4.5.5. Suppose $\eta : (S, \odot_S, \mathbf{e}_S) \longrightarrow (E, \odot_E, \mathbf{e}_E)$ is a functor of symmetric monoidal categories. Then $\langle S, \mathbf{E} \rangle$ has a natural symmetric monoidal category structure, as follows.

Write $T = \langle S, \mathbf{E} \rangle$. For, $X_1, X_2 \in Obj(T) = Obj(\mathbf{E})$, define $X_1 \odot_T X_2 := X_1 \odot_E X_2$. For the morphisms $f_1 = (P_1, f_1), f_2 = (P_2, f_2) \in Mor_T(X_1, Y_1)$, with $P_1, P_2 \in Obj(S)$, define $f_1 \odot_T f_2 := (P_1 \odot_S P_2, f_1 \odot_E f_2)$. This defines the monoidal structure in T.

Theorem 4.5.6. *Let $\mathscr{E}$ be a split exact category, and $S = Iso(\mathscr{E})$. Fix $C \in Obj(\mathscr{E})$ and $E = \left(K \hookrightarrow M \xrightarrow{\zeta_E} C \right) \in Obj(\mathbf{E}\mathscr{E}_C)$. Then the functor*

$$\nu_E : S^{-1}S \longrightarrow S^{-1}\mathbf{E}\mathscr{E}_C \quad \textit{is a homotopy equivalence},$$

where $\nu_E(P, Q) = (Q, P \odot E)$ *with* $P \odot E := \left(P \oplus K \hookrightarrow P \oplus M \xrightarrow{0 \oplus \zeta_E} C\right)$ (*see* 4.4.3).

Proof. Because of (4.5.2), Theorem 4.4.3 is applicable. So, the sequence

$$S^{-1}S \xrightarrow{\nu_E} S^{-1}\mathbf{E}\mathscr{E}_C \longrightarrow \langle S, \mathbf{E}\mathscr{E}_C \rangle \quad \text{is a homotopy fibration.}$$

So, we need to prove $\langle S, \mathbf{E}\mathscr{E}_C \rangle$ is contractible. Write $T = \langle S, \mathbf{E}\mathscr{E}_C \rangle$. By (4.5.5), T has the monoidal product $\odot_T : T \times T \longrightarrow T$. This induces a map $\mu : \mathbb{B}(T \times T) \longrightarrow \mathbb{B}T$, of the classifying spaces. However, $\mathbb{B}(T \times T) \xrightarrow{\sim} k(\mathbb{B}T \times \mathbb{B}T)$ is a homeomorphism (2.1.8). This provides an H_c-space structure on $\mathbb{B}T$ (*refer to* Section 4.8 *for H_c-spaces*). Further, $\mathbb{B}T$ is connected (4.5.4). By Theorem 4.8.6, $\mathbb{B}T$ is group like. In particular, there is a map $j : 1_{\mathbb{B}T} \longrightarrow 1_{\mathbb{B}T}$ such that $j \cdot 1_{\mathbb{B}T}$ and $1_{\mathbb{B}T} \cdot j$ are null homotopic.

Suppose $E = \left(K \overset{\iota}{\hookrightarrow} M \overset{\zeta}{\twoheadrightarrow} C \right) \in Obj\,(\mathbf{E}\mathscr{E}_C)$. To establish notation, consider the

pullback diagram
$$\begin{array}{ccccc}
M & & K & = & K \\
 & \searrow^{\Delta} & \downarrow{\scriptstyle \iota_2} & & \downarrow{\scriptstyle \iota} \\
K & \underset{\iota_1}{\hookrightarrow} & M \times_C M & \underset{q_2}{\twoheadrightarrow} & M \\
\| & & {\scriptstyle q_1}\downarrow & & \downarrow{\scriptstyle \zeta} \\
K & \underset{\iota}{\hookrightarrow} & M & \underset{\zeta}{\twoheadrightarrow} & C
\end{array}$$

Δ being the diagonal arrow.

We have $$E \odot E = \left(K \oplus K \overset{(\iota_1,\iota_2)}{\hookrightarrow} M \times_C M \overset{\zeta q_1}{\twoheadrightarrow} C \right)$$

Define the functor $2_T : T \longrightarrow T$ by the composition, as follows:

$$\begin{array}{ccccc}
T & \overset{\Delta}{\longrightarrow} & T \times T & \overset{1_T \times 1_T}{\longrightarrow} & T \times T \\
 & \searrow_{2_T} & & & \downarrow{\scriptstyle \odot_T} \\
 & & & & T
\end{array}$$

So, $2_T(E) = E \odot_T E$.

Recall that $\mathbb{B}1_T \cdot \mathbb{B}1_T := \mathbb{B}2_T$ (consult Section 4.8). (*Ideally, in multiplicative notation, 2_T would be denoted by* $(1_T)^2$.) Consider the following arrow in $\mathbf{E}\mathscr{E}_C$:

$$\begin{array}{cccccccc}
K \odot E & & K \oplus K & \overset{(1_K,\iota)}{\hookrightarrow} & K \oplus M & \overset{(0,\zeta)}{\twoheadrightarrow} & C & \\
\Big| & & {\scriptstyle \beta}\uparrow & & {\scriptstyle 1}\uparrow & & {\scriptstyle 1_C}\uparrow & \\
{\scriptstyle \tau_E} & & K \oplus K & \overset{\alpha}{\hookrightarrow} & K \oplus M & \overset{(0,\zeta)}{\twoheadrightarrow} & C & \text{where} \\
\downarrow & & {\scriptstyle 1}\downarrow & & {\scriptstyle \wr}\downarrow{\scriptstyle (\iota_1,\Delta)} & & \downarrow{\scriptstyle 1_C} & \\
E \odot_T E & & K \oplus K & \underset{(\iota_1,\iota_2)}{\hookrightarrow} & M \times_C M & \underset{\zeta q_1 = \zeta q_2}{\twoheadrightarrow} & C &
\end{array}$$

$$\begin{cases} \Delta \text{ is the diagonal map,} \\ \alpha = \begin{pmatrix} 1_K & -1_K \\ 0 & \iota \end{pmatrix}, \\ \beta = \begin{pmatrix} 1_K & -1_K \\ 0 & 1_K \end{pmatrix} \end{cases}$$

Then (K, τ_E) represents a natural arrow $\tau_E : E \longrightarrow E \odot_T E =: 2_T(E)$ in T.

We established that there is a natural transformation $\tau : 1_T \longrightarrow 2_T$. So, the two maps $1_{\mathbb{B}T} \simeq \mathbb{B}2_T = \mathbb{B}1_T \cdot \mathbb{B}1_T$ are homotopic. So, $1_{\mathbb{B}T} = \mathbb{B}1_T \cdot \mathbb{B}1_T$ in $[\mathbb{B}T, \mathbb{B}T]$. Therefore, $j \cdot 1_{\mathbb{B}T} = j \cdot \mathbb{B}1_T \cdot \mathbb{B}1_T$ in $[\mathbb{B}T, \mathbb{B}T]$. Since $j \cdot 1_{\mathbb{B}T}$ is null homotopic (i.e., the identity element in $[\mathbb{B}T, \mathbb{B}T]$), it follows $\mathbb{B}1_T$ is null homotopic. Therefore, T is contractible. The proof is complete. ■

4.5.1. $S^{-1}S$ and $\mathbb{Q}\mathscr{E}$

For a split exact category $\mathscr{E}$, and $S = Iso(\mathscr{E})$, in this section, we prove that the sequence $S^{-1}S \longrightarrow S^{-1}\mathbf{E}\mathscr{E} \longrightarrow \mathbb{Q}\mathscr{E}$ is a homotopy fibration. To prepare for this, we prove the following.

Lemma 4.5.7. *As in* (4.4.3), *suppose S is a symmetric monoidal category, with $S = Iso(S)$. Assume that translations in S are faithful. Let $\mathscr{X}$ be a S-category. Let $F : \mathscr{X} \longrightarrow \mathscr{Y}$ be any functor, such that* $F(P \odot X) = F(X)$ (in other words, with trivial S-action on $\mathscr{Y}$, F is a functor of S-categories.). *Let $S^{-1}F : S^{-1}\mathscr{X} \longrightarrow \mathscr{Y}$ be the functor induced by F, sending $(P, X) \mapsto F(X)$. Then*

1. *For $Y \in Obj(\mathscr{Y})$, $F^{-1}Y$ is S-subcategory of $\mathscr{X}$. Further, $(S^{-1}F)^{-1}Y \cong S^{-1}(F^{-1}Y)$ are equivalent.*
2. *If F is fibered, then $S^{-1}F : S^{-1}\mathscr{X} \longrightarrow \mathscr{Y}$ is fibered.*

Proof. The proof of (1) is routine. So, we prove (2). Let $Y \in \mathscr{Y}$, $\Phi_Y : F^{-1}Y \longrightarrow Y/F$ be the natural functor and $R_Y : Y/F \longrightarrow F^{-1}Y$ be its right adjoint. So, for $X_1 \in Obj(\mathscr{X})$ and (X_2, v_2), we have a natural isomorphism

$$\eta_Y : Mor_{F^{-1}Y}(X_1, R(X_2, v_2)) \xrightarrow{\sim} Mor_{Y/F}((X_1, 1_{X_1}), (X_2, v_2)).$$

Write $\overline{F} = S^{-1}F$. Then $\overline{F}^{-1}Y = \{(P, X) : F(X) = Y\} = S \times F^{-1}Y$. Note $f \in Mor_{S^{-1}\mathscr{X}}((P_1, X_1), (P_2, X_2))$ is given by $f = (U, f_1, f_2)$ such that $f_1 \in Mor_S(U \odot P_1, P_2)$, $f_2 \in Mor_{\mathscr{X}}(U \odot X_1, X_2)$. So,

$$\begin{aligned} &Mor_{S^{-1}\mathscr{X}}((P_1, X_1), (P_2, X_2)) \\ &\quad = \bigcup_{U \in S} Mor_S(U \odot P_1, P_2) \times Mor_{\mathscr{X}}(U \odot X_1, X_2) \end{aligned}$$

the union runs through choices of representatives U from each isomorphism class. So,

$$\begin{cases} Mor_{\overline{F}^{-1}Y}((P_1,X_1),(P_2,X_2)) = \\ \bigcup_{U\in S}\{(f_1,f_2)\in Mor_S(U\odot P_1,P_2) \\ \quad \times Mor_{\mathscr{X}}(U\odot X_1,X_2): F(f_2)=1_Y\} \\ \text{So, the 2nd-coordinate} \quad f_2\in Mor_{F^{-1}Y}(U\odot X_1,X_2) \\ Mor_{Y/\overline{F}}(((P_1,X_1),v_1),((P_2,X_2),v_2)) = \\ \bigcup_{U\in S}\{(f_1,f_2)\in Mor_S(U\odot P_1,P_2) \\ \quad \times Mor_{\mathscr{X}}(U\odot X_1,X_2): F(f_2)v_1=v_2\}. \end{cases}$$

Define, the functor

$$\overline{R}_Y : Y/\overline{F} \longrightarrow \overline{F}^{-1}Y = S\times F^{-1}Y \quad \text{by}$$

$$\begin{cases} \overline{R}_Y((P,X),v) = (P,R_Y(X,v)) \\ \overline{R}_Y(U,f_1,f_2) = (U,f_1,R_Y(f_2)) \end{cases}$$

Define,

$$\overline{\eta}_Y : Mor_{\overline{F}^{-1}Y}((P_1,X_1),(P_2,R_Y(X_2,v_2))) \\ \longrightarrow Mor_{Y/\overline{F}}(((P_1,X_1),1_{X_1}),((P_2,X_2),v_2))$$

by $\overline{\eta}_Y(U,f_1,f_2) = (U,f_1,\eta(f_2))$. Since η_Y is a natural bijection, so is $\overline{\eta}_Y$. This establishes that $\overline{F}$ is prefibred. For $v : Y \longrightarrow Z$ in $\mathscr{Y}$, the functor $v^* : \overline{F}^{-1}Z \longrightarrow \overline{F}^{-1}Y$ is given by $v^*(P,X) = (P,R_Y(X,v))$. Since F is fibered, it follows that $\overline{F}$ is fibered. The proof is complete. ∎

Lemma 4.5.8. *Let $\mathscr{E}$ be an exact category and $S = Iso(\mathscr{E})$. Then natural functors*

$$\begin{cases} t : \mathbf{E}\mathscr{E} \longrightarrow \mathbb{Q}\mathscr{E} \quad \textit{and} \\ \mathfrak{q} : S^{-1}\mathbf{E}\mathscr{E} \longrightarrow \mathbb{Q}\mathscr{E} \end{cases} \qquad \textit{are fibered.}$$

Proof. By Lemma 4.5.7, it is enough to prove that $t : \mathbf{E}\mathscr{E} \longrightarrow \mathbb{Q}\mathscr{E}$ is fibered. For $C \in Obj(\mathbb{Q}\mathscr{E})$, we have

$$\begin{cases} Obj(C/t) = \{(E, v) : E \in Obj(\mathbf{E}\mathscr{E}),\ v \in Mor_{\mathbb{Q}\mathscr{E}}(C, t(E))\} \\ Mor_{C/t}((E_1, v_1), (E_2, v_2)) = \{f \in Mor_{\mathbb{Q}\mathscr{E}}(E_1, E_2) : t(f)v_1 = v_2\} \\ \quad \forall\ (E_1, v_1), (E_2, v_2) \in Obj(C/t) \\ \text{In particular,} \\ Mor_{C/t}((E_1, 1_C), (E_2, v_2)) = \{f \in Mor_{\mathbb{Q}\mathscr{E}}(E_1, E_2) : t(f) = v_2\} \end{cases}$$

Such an arrow $f \in Mor_{C/t}((E_1, v_1), (E_2, v_2))$ is presented by the following diagrams:

$$\forall\ i = 1, 2\ \ v_i := \left(C \overset{p_i}{\twoheadleftarrow} V_i \overset{q_i}{\hookrightarrow} C_i \right),$$

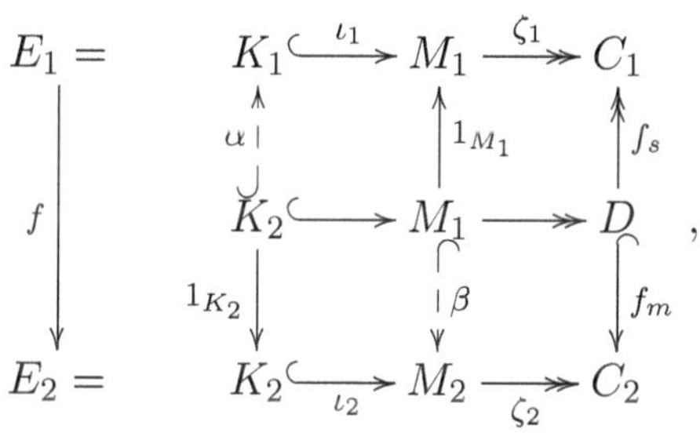

$$, \quad \ni \quad t(f)v_1 = v_2 \qquad (4.17)$$

The equality $t(f)v_1 = v_2$ is given by the commutative diagram

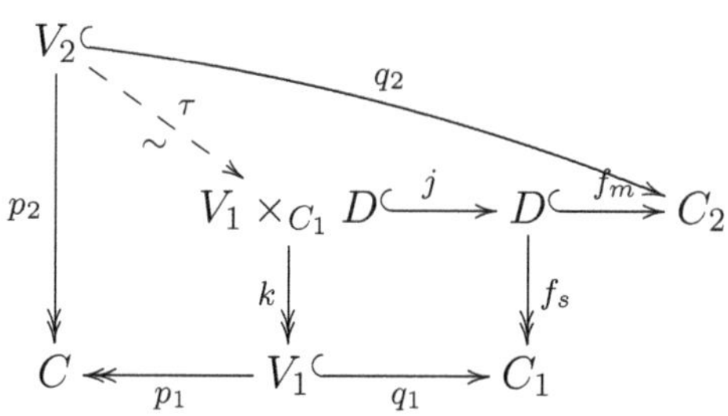

where τ is a unique isomorphism. We define a right adjoint $R : C/t \longrightarrow t^{-1}C$ of Φ. For $(E, v) \in Obj(C/t)$, consider the pullback

diagram

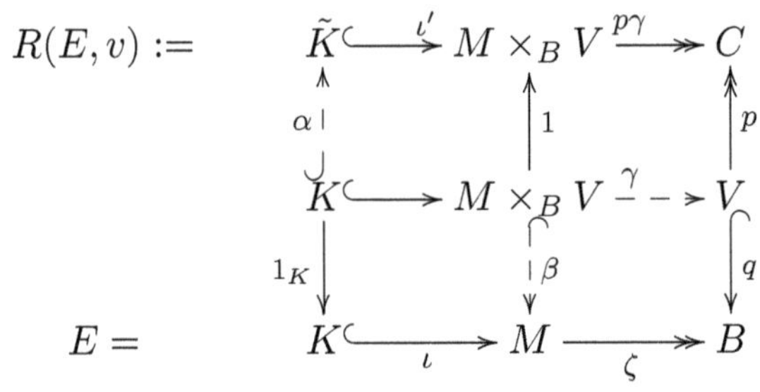

(4.18)

In this diagram, $(E, v) \in Obj(C/t)$, where E is represented in the third horizontal line and v by the last vertical line, and $M \times_B V$ is the pullback. Further, $\tilde{K} := \ker(p\gamma)$. Then the top line is defined to be $R(E, v) \in t^{-1}C$. Let $f \in Mor_{C/t}((E_1, v_1), (E_2, v_2))$, as displayed in what follows, fixing the notation:

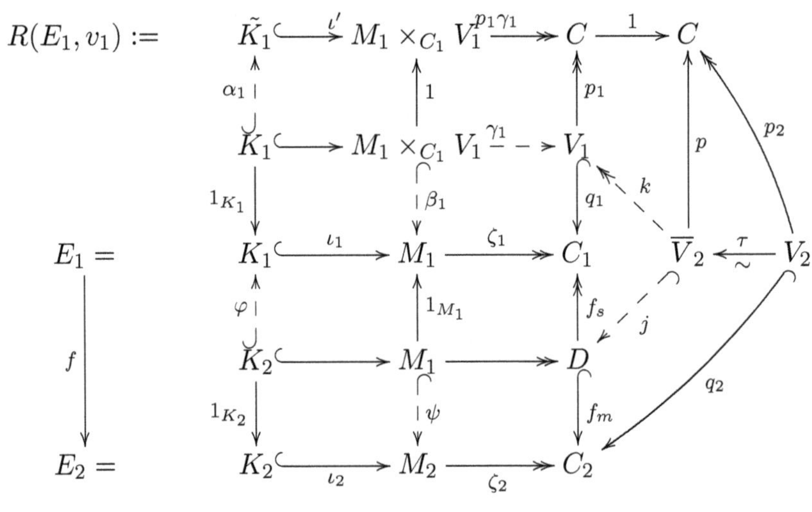

with $\overline{V}_2 := D \times_{C_1} V_1$ (4.19)

In this diagram, the right-hand side displays the equation $t(f)v_1 = v_2$. We develop the arrow $R(f)$ by a number of diagrams. We extract the following diagram from (4.19) as follows:

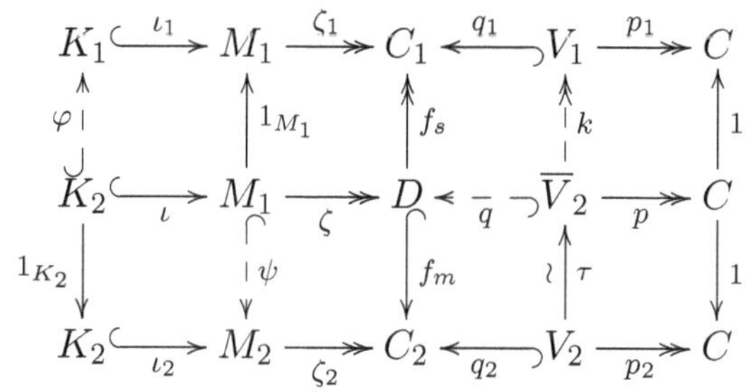

(4.20)

Fold the diagram along the middle line, with the left half horizontally, and the right half vertically. We raise three pullback diagrams (*three vertical walls*) as follows:

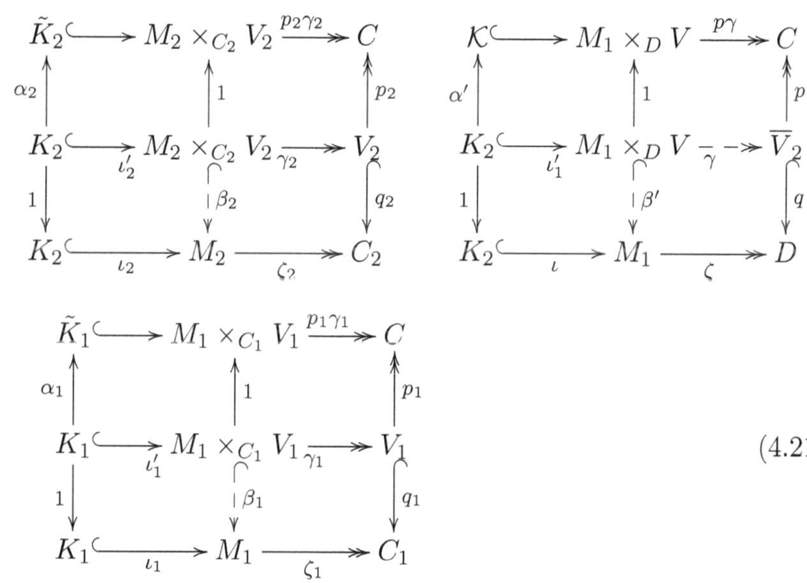

(4.21)

A cubical diagram is formed from (4.21), and (4.20) (*think of them as three vertical walls, one behind another*). For typographical reasons, display this cube in two cubical diagrams (4.22, 4.23) (*sharing the middle wall*). Note $M_1 \times_D V = M_1 \times_{C_1} V_1$. And, from the first diagram

(4.22), it would also be clear that $\tilde{K}_2 = \mathcal{K}$.

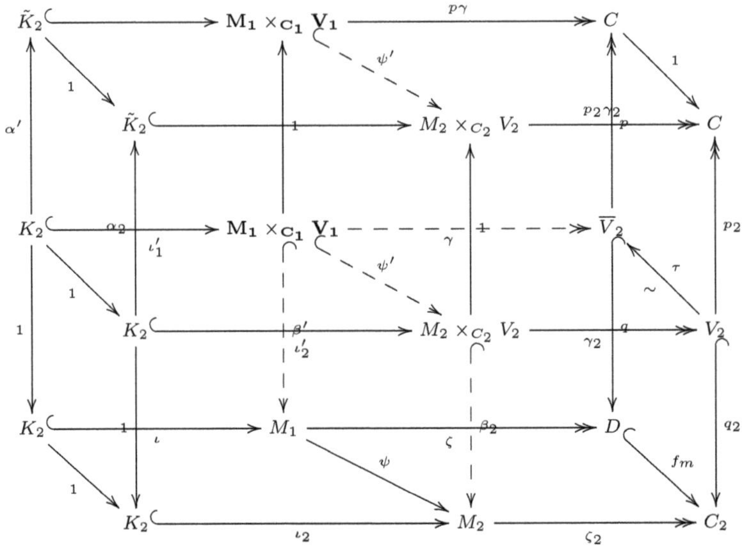

(4.22)

In this diagram, we used the identification $\mathcal{K} = \tilde{K}_2$. This follows, because ψ' is an isomorphism, and $p_2\gamma_2 = \tau p\gamma$. So, $\tilde{K}_2 = \ker(p_2\gamma_2) \cong \ker(p\gamma) = \mathcal{K}$.

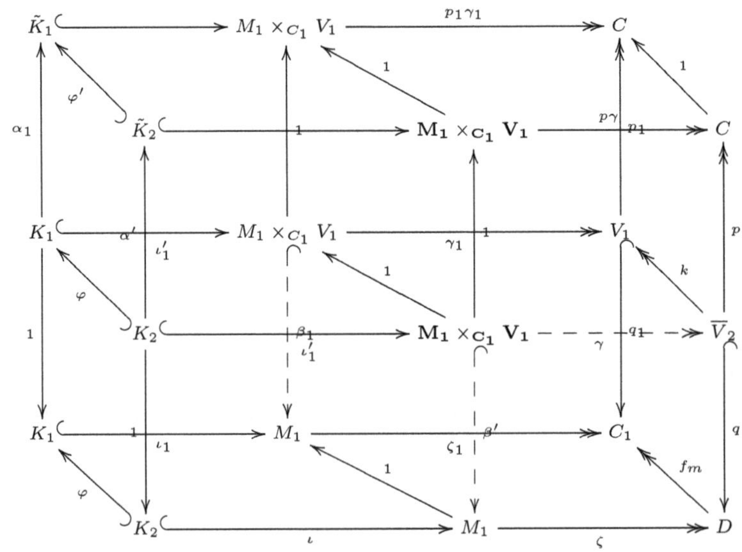

(4.23)

The ceiling of the cube (4.22, 4.23) is defined to be the morphism $R(f)$, as follows:

$$\begin{array}{ccccccc}
R(E_1, v_1) = & & \tilde{K}_1 & \hookrightarrow & M_1 \times_{C_1} V_1 & \xrightarrow{p_1\gamma_1} & C \\
\Big\downarrow & & \varphi' \Big\uparrow & & \Big\uparrow 1 & & \Big\uparrow 1 \\
R(f) & & \tilde{K}_2 & \hookrightarrow & \mathbf{M_1 \times_{C_1} V_1} & \xrightarrow{p\gamma} & C \\
\Big\downarrow & & 1 \Big\downarrow & & \Big\downarrow \psi' & & \Big\downarrow 1 \\
R(E_2, v_2) = & & \tilde{K}_2 & \hookrightarrow & M_2 \times_{C_2} V_2 & \xrightarrow[p_2\gamma_2]{} & C
\end{array} \tag{4.24}$$

This completes the definition of the functor $R : C/t \longrightarrow t^{-1}C$. We check that R is a right adjoint to the natural functor $\Phi : t^{-1}C \longrightarrow C/t$.

Let $E_1 = \left(K_1 \xhookrightarrow[\iota_1]{} M_1 \xrightarrow[\zeta_1]{} C \right) \in Obj\,(t\ ^{1}C)$, $(E_2, v_2) \in Obj\,(C/t)$ with $E_2 = \left(K_2 \xhookrightarrow[\iota_2]{} M_2 \xrightarrow[\zeta_2]{} C_2 \right)$ and $v_2 = \left(C \xleftarrow[p_2]{} V_2 \xhookrightarrow[q_2]{} C_2 \right)$. We have

$$\begin{cases}
Mor_{t^{-1}C}\,(E_1, R(E_2, v_2)) = \\
\qquad\qquad \{f \in Mor_{\mathbf{E}\mathscr{E}}\,(E_1, R(E_2, v_2)) : t(f) = 1_C\} \\
Mor_{C/t}\,(\Phi(E_1), (E_2, v_2)) = Mor_{C/t}\,((E_1, 1_C), (E_2, v_2)) \\
\qquad\qquad = \{f \in Mor_{\mathbf{E}\mathscr{E}}\,(E_1, E_2) : t(f) = v_2\}
\end{cases}$$

From construction, there is a natural arrow $\eta : R(E_2, v_2) \longrightarrow E_2$ in $\mathbf{E}\mathscr{E}$. By composition, there is a map

$$\overline{\eta} : Mor_{t^{-1}C}\,(E_1, R(E_2, v_2))$$

$$\longrightarrow Mor_{C/t}\,((E_1, 1_C), (E_2, v_2)) \qquad \begin{array}{ccc}
E_1 & \xrightarrow{f} & R(E_2, v_2) \\
& \overline{\eta}(f) \searrow & \Big\downarrow \eta \\
& & E_2
\end{array}$$

More explicitly, $\overline{\eta}(f)$ is the composition, as follows:

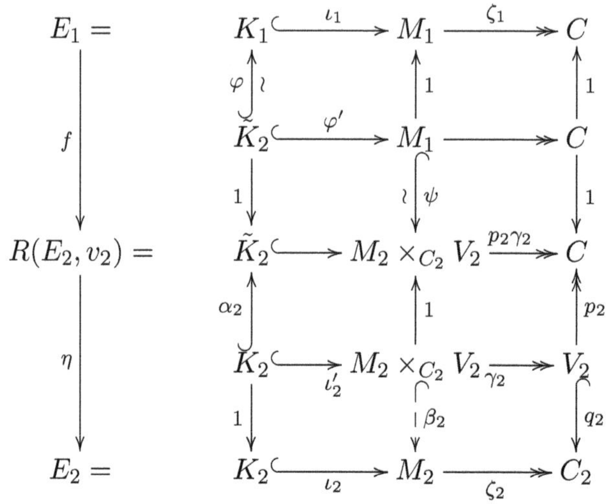

First, by diagram chasing, it follows that $\overline{\eta}$ is injective. Further, chasing the pullback properties, it follows that $\overline{\eta}$ is surjective. So, it is established that R is a right adjoint of Φ. So, t is prefibered.

For $v \in Mor_{\mathbb{Q}\mathscr{E}}(C, D)$, we consider the diagram

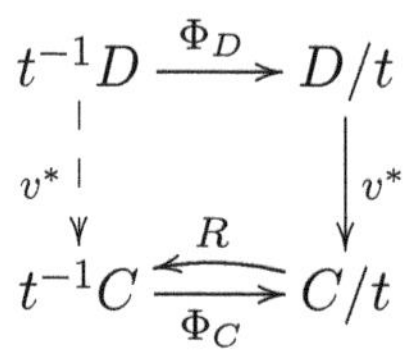

where the same notation v^* is used for two functors.

Let $v = \left(C \overset{s_1}{\twoheadleftarrow} V \overset{m_1}{\hookrightarrow} D \right)$ and $w = \left(B \overset{s_2}{\twoheadleftarrow} W \overset{m_2}{\hookrightarrow} C \right)$ be two morphisms in $\mathbb{Q}\mathscr{E}$. Let $E := \left(K \underset{\iota}{\hookrightarrow} M \underset{\zeta}{\twoheadrightarrow} D \right) \in Obj\left(t^{-1}D\right)$. To establish $w^*v^*(E) = (wv)^*(E)$, consider the

commutative diagram

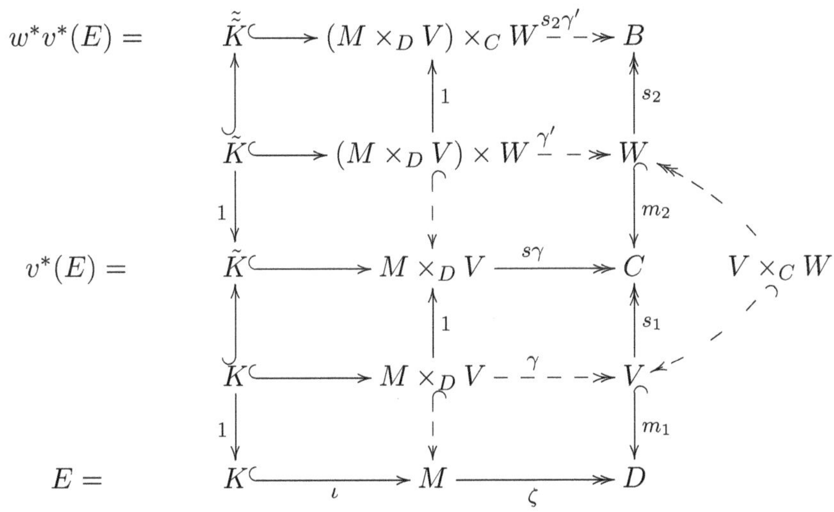

Since $M \times_D (V \times_C W) \cong (M \times_D V) \times_C W$, it follows that $w^* v^*(E) = (vw)^*(E)$. Therefore, t is fibered. ■

Corollary 4.5.9. *Let $\mathscr{E}$ be an exact category. Consider the functor $t : \mathbf{E}\mathscr{E} \longrightarrow \mathbb{Q}\mathscr{E}$. Let $v \in Mor_{\mathbb{Q}\mathscr{E}}(C, B)$ be represented by $v = \left(C \overset{}{\underset{p}{\twoheadleftarrow}} M \overset{}{\underset{q}{\hookrightarrow}} B \right)$. Then the base change functor $v^* : t^{-1}C \longrightarrow t^{-1}B$ is given by the diagram $v^*(E) = R(E, v)$, where the right adjoint R was defined in diagram (4.18). As is obvious in (4.18), there is a natural arrow $\eta_E \in Mor_{\mathbf{E}\mathscr{E}}(v^*(E), E)$.*

Theorem 4.5.10. *Let $\mathscr{E}$ be a split exact category, and $S = Iso(\mathscr{E})$. Then the sequence*

$$S^{-1}S \xrightarrow{\nu_E} S^{-1}\mathbf{E}\mathscr{E} \xrightarrow{\mathfrak{q}} \mathbb{Q}\mathscr{E} \quad \text{is a homotopy fibration} \quad \forall\, E \in Obj(\mathbf{E}\mathscr{E})$$

where $\nu_E(P, Q) = (Q, P \odot E)$.

Proof. By (4.5.8), $\mathfrak{q}$ is fibered. Let $C \in Obj(\mathbb{Q}\mathscr{E})$. By Lemma 4.5.7, $\mathfrak{q}^{-1}C = S^{-1}\mathbf{E}\mathscr{E}_C$. Let $v \in Mor_{\mathbb{Q}\mathscr{E}}(C, B)$ be represented by $v := \left(C \overset{p}{\twoheadleftarrow} M \overset{q}{\hookrightarrow} B \right)$. We will prove that $v^* : \mathfrak{q}^{-1}B \longrightarrow \mathfrak{q}^{-1}C$

is a homotopy equivalence. Consider the following diagram of compositions of arrows in $\mathbb{Q}\mathscr{E}$:

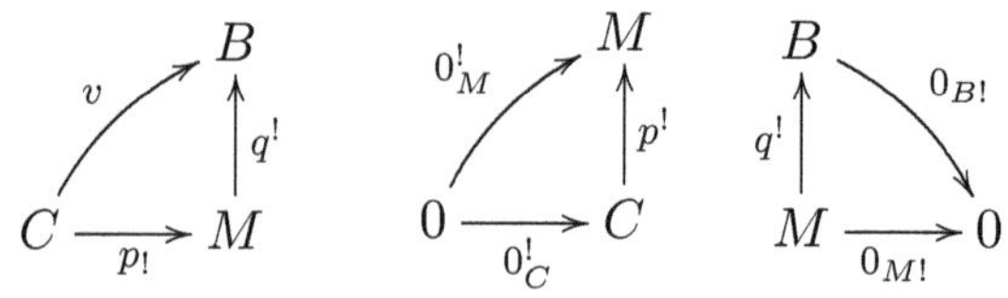

Therefore, to prove v^* is a homotopy equivalence, it is enough to prove the same for $v = 0^!_C$ and $v = 0_{C!}$, where $0^!_C = \left(0 \twoheadleftarrow C \hookrightarrow_1 C\right)$, and $0_{C!} = \left(0 \twoheadleftarrow 0 \hookrightarrow C\right)$.

1. Write $v = 0^!_C$. Now, $\mathfrak{q}^{-1}0 = S^{-1}t^{-1}0 = S^{-1}S$ and $\mathfrak{q}^{-1}0 = S^{-1}t^{-1}0 = S^{-1}\mathbf{E}\mathscr{E}_C$. The equivalence

$$S^{-1}S \longrightarrow \mathfrak{q}^{-1}0 \qquad \text{sends} \quad (P,Q) \mapsto \Big(P, \big(\, Q \longrightarrow Q \longrightarrow 0\,\big)\Big).$$

By definition,

$R(E,v) :=$

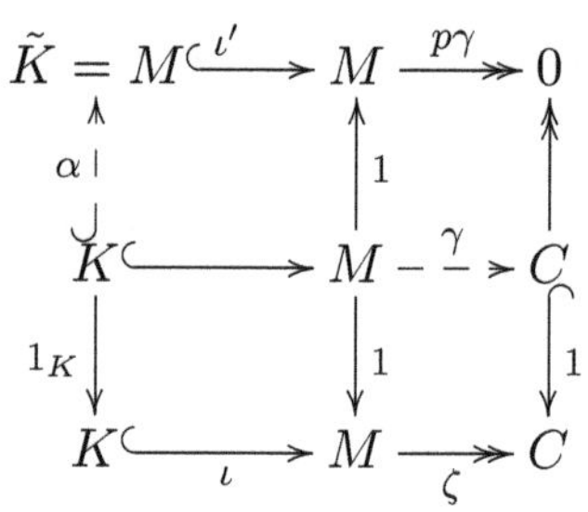

$E =$

So, $v^*(P,E) = (P,M)$

Now, with $\mathfrak{1} = \left(0 \longrightarrow C \xrightarrow[1]{} C\right) \in Obj(\mathbf{E}\mathscr{E}_C)$, consider the commutative diagram of functors

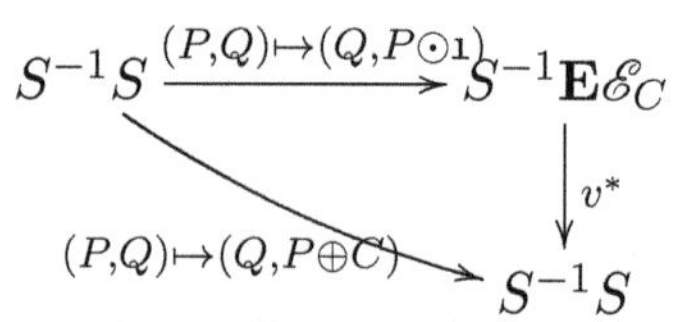

The horizontal functor is a homotopy equivalence by (4.5.6). Up to a switch, the diagonal functor is a translation, and hence a homotopy equivalence (4.3.8). So, v^* is a homotopy equivalence.

2. Now, let $v = 0_{C!} = \left(0 \longleftarrow 0 \longrightarrow C \right)$. By definition,

$$\begin{array}{lccccc}
R(F,v) := & \tilde{K} = K & \hookrightarrow & K & \twoheadrightarrow & 0 \\
& \uparrow{\scriptstyle \iota} & & \uparrow{\scriptstyle 1} & & \uparrow{\scriptstyle p} \\
& K & \hookrightarrow & K & \dashrightarrow & 0 \\
& \downarrow{\scriptstyle 1_K} & & \downarrow{\scriptstyle \iota} & & \downarrow{\scriptstyle q} \\
E = & K & \xrightarrow{\iota} & M & \xrightarrow{\zeta} & C
\end{array}
\qquad \text{So,} \quad v^*(P,E) = (P,K)$$

Again, consider the commutative diagram of functors

$$\begin{array}{ccc}
S^{-1}S & \xrightarrow{(P,Q)\mapsto(Q,P\odot 1)} & S^{-1}\mathbf{E}\mathscr{E}_C \\
& \searrow^{(P,Q)\mapsto(Q,P)} & \downarrow{\scriptstyle v^*} \\
& & S^{-1}S
\end{array}$$

Since the horizontal and the diagonal functors are homotopy equivalences, so is v^*.

Therefore, the proof is complete by an application of the fibered version of Theorem B (2.2.6). ■

The following exercise would be useful for proof of the next theorem.

Exercise 4.5.11. Let $\mathscr{X}$ be a category. Define a new category $Ar(\mathscr{X})$, whose objects are the arrows in $\mathscr{X}$. To define arrows in $Ar(\mathscr{X})$, consider the following commutative diagram of arrows in $\mathscr{X}$,

$$\begin{array}{ccc}
M_1 & \xrightarrow{f} & N_1 \\
\uparrow{\scriptstyle u} & & \downarrow{\scriptstyle v} \\
M_2 & \xrightarrow[g]{} & N_2
\end{array}
\qquad \text{which is} \quad g = ufv$$

Then $Mor_{Ar(\mathscr{X})}(f,g)$ is defined to be the set of all such pairs (u,v) as in the diagram. Then $\mathscr{X}$ and $Ar(\mathscr{X})$ are homotopically equivalent, by the functors defined in the proof.

Proof. Define $F : Ar(\mathscr{X}) \longrightarrow \mathscr{X}$, by $F(f : M \longrightarrow N) = N$. And, define $G : \mathscr{X} \longrightarrow Ar(\mathscr{X})$, by $G(N) = 1_N$. Clearly, $FG = 1_{\mathscr{X}}$. Also, $GF(f : M \longrightarrow N) = 1_N$. A natural transformation $1_{Ar(\mathscr{X})} \longrightarrow GF$ is given by the diagram

$$\begin{array}{ccc} N & \xrightarrow{1} & N \\ {\scriptstyle f}\uparrow & & \downarrow{\scriptstyle 1} \\ M & \xrightarrow[f]{} & N \end{array}$$

So, GF is homotopic to $1_{Ar(\mathscr{X})} \longrightarrow GF$, by (2.1.9). The proof is complete. ∎

Theorem 4.5.12. *Let $\mathscr{E}$ be a split exact category, and $S = Iso(\mathscr{E})$. Then $S^{-1}\mathbf{E}\mathscr{E}$ is contractible.*

Proof. Let $\mathscr{X}$ be the subcategory of $\mathbb{Q}\mathscr{E}$ whose objects are the same as that of $\mathbb{Q}\mathscr{E}$, and $Mor_{\mathscr{X}}(P, Q) := \{\iota_! \in Mor_{\mathbb{BQ}}(P, Q) : \iota : N \hookrightarrow M \text{ is an inflation in } \mathscr{E}\}$. Then $0 \in \mathscr{X}$ is an initial object, and hence $\mathscr{X}$ is contractible. Let $Ar(\mathscr{X})$ be the category arrows of $\mathscr{X}$, as in (4.5.11). There is a natural functor

$$F : Ar(\mathscr{X}) \longrightarrow \mathbf{E}\mathscr{E} \quad \text{sending } \iota_! \mapsto (\ K \overset{\iota}{\hookrightarrow} M \twoheadrightarrow co\ker(\iota)\ )$$
$$\forall \text{ inflations } (\iota : K \hookrightarrow M)$$

Suppose $(\iota_!^i : K_i \longrightarrow M_i) \in Obj\,(Ar(\mathscr{X}))$ for $i = 1, 2$. An arrow $f := (\alpha_!, \beta_!) : \iota_!^1 \longrightarrow \iota_!^2$ is given by the commutative diagram of inflations:

$$\begin{array}{ccc} K_1 & \overset{\iota^1}{\hookrightarrow} & M_1 \\ {\scriptstyle \alpha}\uparrow & & \downarrow{\scriptstyle \beta} \\ K_2 & \underset{\iota^2}{\hookrightarrow} & M_2 \end{array}$$

This corresponds to an arrow $F(\iota_!^1) \longrightarrow F(\iota^2)$ in $\mathbf{E}\mathscr{E}$ (see Definition 4.5.1). Define $F(f) : F(\iota_!^1) \longrightarrow F(\iota^2)$ to be this arrow. So, the functor $F : Ar(\mathscr{X}) \longrightarrow \mathbf{E}\mathscr{E}$ is defined. It is clear from (4.5.1) that F is an equivalence of categories. The inverse of F is given by the conflation E sent to its own inflation. Since $\mathscr{X}$ is contractible, so is $\mathbf{E}\mathscr{E}$. Since all maps in S are isomorphisms, by Proposition 4.3.7, $\mathbf{E}\mathscr{E} \xrightarrow{\sim} S^{-1}\mathbf{E}\mathscr{E}$ is a homotopy equivalence. Therefore, $S^{-1}\mathbf{E}\mathscr{E}$ is contractible. The proof is complete. ∎

The following theorem is one of our main goals in this chapter.

Theorem 4.5.13. *Let $\mathscr{E}$ be a split exact category and $S = Iso(\mathscr{E})$. Then there is a homotopy equivalence $\mathbb{B}(S^{-1}S) \xrightarrow{\sim} \Omega\mathbb{B}(\mathbb{Q}\mathscr{E})$, where $\Omega\mathbb{B}(\mathbb{Q}\mathscr{E})$ is the loop space. In particular, $K_n(\mathscr{E}) \cong \pi_n\left(\mathbb{B}S^{-1}S, \star\right)$, for all $n \geq 0$.*

Proof. Refer to Section B.5. By Theorem 4.5.10, we have homotopy fibration

$$\mathbb{B}S^{-1}S \longrightarrow \mathbb{B}(S^{-1}\mathbf{E}\mathscr{E}) \xrightarrow{|\mathfrak{q}|} \mathbb{B}\mathbb{Q}\mathscr{E} \quad \text{where} \quad |\mathfrak{q}| := \mathbb{B}\mathfrak{q}.$$

In other words, the map $\varphi : \mathbb{B}S^{-1}S \xrightarrow{\sim} \mathcal{F}(|\mathfrak{q}|, \mathbf{0})$ is a homotopy equivalence, where the latter term denotes the homotopy fiber. By (4.5.12), $\mathbb{B}(S^{-1}\mathbf{E}\mathscr{E})$ is contractible. So, $\Omega\mathbb{B}(\mathbb{Q}\mathscr{E}) \xrightarrow{\sim} \mathscr{F}(|\mathfrak{q}|, \mathbf{0})$ is a homotopy equivalence. The proof is complete. ∎

4.6. Cofinality

First we define Cofinality.

Definition 4.6.1. Suppose $f : (S, \odot_S) \longrightarrow (T, \odot_T)$ is a functor of symmetric monoidal categories. We say that f is **cofinal** if $\forall\ P \in Obj(T)$, there is an object $F \in Obj(S)$ such that $F \cong P \odot_T Q$ for some $Q \in Obj(T)$.

Theorem 4.6.2. *Suppose $f : (S, \odot_S) \longrightarrow (T, \odot_T)$ is a functor of symmetric monoidal categories, and f is cofinal. Assume $S = Iso(S)$ and $T = Iso(T)$, and every translation in S, T are faithful.*

1. *Suppose $\mathscr{X}$ is a T-category. Then $S^{-1}\mathscr{X} \simeq T^{-1}\mathscr{X}$ is a homotopy equivalence. In particular, $S^{-1}T \cong T^{-1}T$.*
2. *Assume $Aut_S(F) \cong Aut_T(F)$ for all $F \in Obj(S)$. Then the component of $(0,0)$ in $\mathbb{B}S^{-1}S$ is homotopically equivalent to the component of $(0,0)$ in $\mathbb{B}T^{-1}T \cong \mathbb{B}S^{-1}T$.*

Proof. It is easy to see that $S_l^{-1}T^{-1}\mathscr{X} \cong T_l^{-1}S^{-1}\mathscr{X}$. Now S acts invertibly on $T^{-1}\mathscr{X}$ (from left). So, $T^{-1}\mathscr{X} \longrightarrow S_l^{-1}T^{-1}\mathscr{X}$ is a homotopy equivalence (4.3.7, 4.3.10). In fact, T also acts invertibly on $S^{-1}\mathscr{X}$. To see this, let $F \cong P \odot Q$ for $F \in Obj(S)$ and $P, Q \in Obj(T)$. Now consider the diagram

$$\begin{array}{ccc} S^{-1}\mathscr{X} & \xrightarrow{T_P} & S^{-1}\mathscr{X} \\ & \searrow^{T_F} & \downarrow{T_Q} \\ & & S^{-1}\mathscr{X} \end{array} \qquad \text{So, } T_Q T_P \text{ is a homotopy equivalence.}$$

So, T_P has left homotopy inverse, and T_Q has right homotopy inverse. Hence, T_P is a homotopy equivalence. This establishes that T acts invertibly on $S^{-1}\mathscr{X}$. Therefore, $S^{-1}\mathscr{X} \longrightarrow T_l^{-1}S^{-1}\mathscr{X}$ is a homotopy equivalence (4.3.7, 4.3.10). In the following commutative diagram,

$$\begin{array}{ccc} S^{-1}\mathscr{X} & \xrightarrow{\sim} & S_l^{-1}T^{-1}\mathscr{X} \\ \downarrow & & \downarrow{\wr} \\ T^{-1}\mathscr{X} & \xrightarrow{\sim} & T_l^{-1}S^{-1}\mathscr{X} \end{array} \qquad \text{three functors are homotopy equivalences.}$$

Hence, the fourth one $S^{-1}\mathscr{X} \simeq T^{-1}\mathscr{X}$ is also a homotopy equivalence. This completes the proof of (1).

Now, we prove (2).

1. For $(F_1, P_1), (F_2, P_2) \in Obj(S^{-1}T)$, define $(F_1, P_1) \sim (F_2, P_2)$ to be equivalent if there are two arrows, as follows $(F_1, P_1) \longrightarrow (F, P) \longleftarrow (F_2, P_2)$. So, there are $U, V \in Obj(S)$, and the diagram of isomorphisms:

$$U \odot (F_1, P_1) \xrightarrow{\sim} (F, P) \xleftarrow{\sim} V \odot (F_2, P_2)$$

$$\Longleftrightarrow \begin{cases} U \odot F_1 \cong F \cong V \odot F_2 \\ U \odot P_1 \cong P \cong V \odot P_2 \end{cases} \tag{4.25}$$

   It follows that this is an equivalence relation on $S^{-1}T$. For $(F, P) \in S^{-1}T$, its equivalence class is denoted by $[(F, P)]_T$. This equivalence relation restricts to $S^{-1}S$. For $(F, G) \in S^{-1}S$, its equivalence class in $S^{-1}S$ is denoted by $[(F, P)]_S$. Let $\mathscr{I}_T$ be an indexing set, and $\{[(F_i, P_i)]_T : i \in \mathscr{I}_T\}$ be the set of all equivalence classes of the subcategories of $S^{-1}T$. Likewise, let $\mathscr{I}_S$ be an indexing set, and $\{[(K_i, G_i)]_S : i \in \mathscr{I}_S\}$ be the set of all equivalence classes of the subcategories of $S^{-1}S$.

2. It follows that $S^{-1}T = \coprod_{i\in\mathscr{I}_T}[(F_i,P_i)]_T$ is a disjoint union of subcategories. So, $\mathbb{B}S^{-1}T = \coprod_{i\in\mathscr{I}_T}\mathbb{B}[(F_i,P_i)]_T$. Since $\mathbb{B}[(F_i,P_i)]_T$ are (path) connected, these are the connected components of $\mathbb{B}S^{-1}T$. Likewise, $\mathbb{B}S^{-1}S = \coprod_{i\in\mathscr{I}_S}\mathbb{B}[(K_i,G_i)]_S$, and $\mathbb{B}[(K_i,G_i)]_S$ are its connected components of $\mathbb{B}S^{-1}S$.
3. The inclusion $[(0,0)]_S \longrightarrow [(0,0)]_T$, induces the map ι : $\mathbb{B}[(0,0)]_S \longrightarrow \mathbb{B}[(0,0)]_T$. It remains to prove that this map is a homotopy equivalence.

We write $\mathbf{0} = (0,0) \in [(0,0)]_S, [(0,0)]_T$. We use Whitehead's theorem (C.3.8) to prove that ι is a homotopy equivalence. By (4.25), $\pi_0(\mathbb{B}[(0,0)]_S = \pi_0(\mathbb{B}[(0,0)]_T = 0$ are trivial. So, we prove that the maps $\pi_n([(0,0)]_S,\mathbf{0}) \longrightarrow \pi_n([(0,0)]_T,\mathbf{0})$ are isomorphisms $\forall n\geq 1$.

To prepare for the proof, we make a few observations. By (4.25), for $(F,P)\in[(0,0)]_T$ there are $U,V\in Obj(S)$ such that $U\odot F\cong V$ and $U\odot P\cong V$. Observe that, given finitely many objects $(F_i,P_i)\in[(0,0)]_T$, for $i=0,1,\ldots,n$, there are $U_i,V_i\in Obj(S)$ such that $U_i\odot F_i\cong V_i\cong U_i\odot P_i$. Replacing U_i by $U := U_0\odot\cdots\odot U_n$, we have $U\odot F_i\cong V_i\cong U\odot P_i$, with $U,V_i\in Obj(S)$.

Now for $U\in Obj(S)$, consider the functor $\tau_U^T : [(0,0)]_T \longrightarrow [(0,0)]_T$ sending $(F,P) \mapsto (U\odot F, U\odot P)$. The arrows $(U,1_{U\odot F},1_{U\odot P}) : (F,P)\longrightarrow(U\odot F,U\odot P)$ in $S^{-1}T$ define a natural transformation $1_{[(0,0)]_T}\mapsto\tau_U^T$. So, τ_U^T is a homotopy equivalence. Likewise, its restriction $\tau_U^S : [(0,0)]_S\longrightarrow[(0,0)]_S$ is also a homotopy equivalence. The following diagram of functors:

$$\begin{array}{ccc} [(0,0]_S & \longrightarrow & [(0,0)]_T \\ {\scriptstyle\tau_U^S}\downarrow & & \downarrow{\scriptstyle\tau_U^T} \\ [(0,0]_S & \longrightarrow & [(0,0)]_T \end{array} \qquad \text{commutes.}$$

It induces the commutative diagram

$$\begin{array}{ccc} \pi_n([(0,0]_S,\mathbf{0}) & \xrightarrow{\pi_n(\iota)} & \pi_n[(0,0)]_T,\mathbf{0}) \\ {\scriptstyle\pi_n(|\tau_U^S|)}\downarrow\wr & & \wr\downarrow{\scriptstyle\pi_n(|\tau_U^T|)} \\ \pi_n[(0,0]_S,(U,U)) & \xrightarrow[\pi_n(\iota)]{} & \pi_n[(0,0)]_T,(U,U)) \end{array} \qquad \forall n\geq 1. \tag{4.26}$$

where the vertical maps are isomorphisms.

Now fix, $n \geq 1$. We prove $\pi_n(\iota) : \pi_n(\mathbb{B}[(0,0)]_S) \longrightarrow \pi_n(\mathbb{B}[(0,0)]_T)$ is an isomorphism. Let $\gamma : (\mathbb{S}^n, \star) \longrightarrow (\mathbb{B}[(0,0)]_T, \mathbf{v_0})$ be a map. Since $\gamma(\mathbb{S}^n)$ is compact, by (C.1.8) and (1.2.11), there is a finite simplicial subcomplex $K_\bullet \subseteq \mathscr{N}_\bullet([0,0]]_T)$ such that $K_\bullet$ is generated by finitely many nondegenerate simplexes and $\gamma(\mathbb{S}^n) \subseteq |K_\bullet|$. Consider such a nondegenerate set of generators $\{\sigma^j\}$ of $K_\bullet$. Such a generator can be written as

$$\sigma^j := \left((F_0^j, P_0^j) \xrightarrow{f_0^j} \cdots \longrightarrow (F_k^j, P_k^j) \xrightarrow{f_k^j} \cdots \longrightarrow (F_{p-1}^j, P_{p-1}^j) \xrightarrow{f_{p-1}^j} (F_p^j, P_p^j) \right).$$

By the above observation, there are objects U and V_i^j in S, such that $U \oplus F_i^j \cong V_i^j \cong U \oplus P_i^j$ for all i, j. Write

$$U \odot \sigma^j := \left((V_0^j, V_0^j) \xrightarrow{F_0^j} \cdots \longrightarrow (V_k^j, V_k^j) \xrightarrow{F_k^j} \cdots \longrightarrow (V_{p-1}^j, V_{p-1}^j) \xrightarrow{F_{p-1}^j} (V_p^j, V_p^j) \right)$$

where $F_i^j := U \odot f_i^j$. Now $\tau_U^T(K_\bullet)$ is generated by $\{U \odot \sigma^j\}$. Let $L_\bullet \subseteq \mathscr{N}_\bullet[(0,0)]_S$ be the simplicial set generated by $\{U \odot \sigma^j\}$. Then $L_\bullet \longrightarrow K_\bullet$ is an isomorphism of simplicial sets and the restriction $\iota_{||L_\bullet|} : |L_\bullet| \xrightarrow{\sim} |K_\bullet|$ is a homeomorphism. So, $\left|\tau_U^T\right| \gamma$ lifts to a map $\beta : (\mathbb{S}^n, \star) \longrightarrow (|L|, (U,U))$. So, it follows that $\pi_n([\beta]) = [\left|\tau_U^T\right| \gamma] \in \pi_n\left[(0,0)\right]_T, (U,U))$. Now it follows from diagram (4.26) that $[\gamma]$ has a preimage in $\pi_n\left[(0,0]_S, (U,U)\right)$. So, the map $\pi_n\left[(0,0]_S, (U,U)\right) \longrightarrow \pi_n\left[(0,0)\right]_T, (U,U))$ is surjective.

To prove injectivity, let $\gamma : (\mathbb{S}^n, \star) \longrightarrow [(0,0)]_S$ be such that $\pi_n(\iota)([\gamma]) = 0$. So, there is a homotopy $H : (\mathbb{S}^n, \star) \times I \longrightarrow [(0,0)]_T$ such that $H(-,0) = c_{(0,0)}$ the constant map and $H(-,1) = \iota\gamma$. Again, $H(\mathbb{S}^n \times I)$ is compact. So, using the same argument as above, we can find a simplicial subset $L_\bullet \subseteq \mathscr{N}_\bullet[(0,0)]_S$, $K_\bullet \subseteq \mathscr{N}_\bullet[(0,0)]_T$ such that $L_\bullet \xrightarrow{\sim} K_\bullet$ is an isomorphism. And hence, H lifts to a homotopy in $\tilde{H} : (\mathbb{S}^n, \star) \times I \longrightarrow ([(0,0)]_S, (U,U))$, from the constant path to a lift of $\left|\tau_U^T\right| \gamma$. Since the first vertical map in (4.26) is an isomorphism, it follows $[\gamma] = 0$. This establishes that the maps $\pi_n\left([(0,0)_S, \mathbf{0}\right) \longrightarrow \pi_n\left([(0,0)_T, \mathbf{0}\right)$ are isomorphisms, $\forall n \geq 0$.

Therefore, by Whitehead's theorem (C.3.8) the map ι is a homotopy equivalence. This completes the proof of (2). ■

Remark 4.6.3. The proof of part (2) of Theorem 4.6.2 is more transparent for the case of our main interest. That is, when A is a ring.

Fix such a ring A. $S = \{A^n : n \geq 0\}$, where A^n is treated as a based free (right) module, as defined in (4.1.5 (2)) and was denoted by $\mathscr{F}(A)$. And $T = Iso(\mathscr{P}(A))$, where $\mathscr{P}(A)$ is the category of finitely generated (right) projective A-modules. The main advantage is that it is easier to work with modules. While in the commutative case the $rank(P)$ provides a handle, it does not make much sense in the non-commutative situation [Wc2, p. 2]. However, from the structure of $S = \mathscr{F}(A)$, the ranks of the free modules A^n make sense, because $Mor_S(A^n, A^m) = \phi$, unless $n = m$.

1. For $P, Q \in Obj(T)$, define P and Q to be equivalent, and write $P \sim Q$, if $F_1 \oplus P \cong F_2 \oplus Q$, for some $F_1, F_2 \in Obj(S)$. This is an equivalence relation, and let $[P]$ denote the equivalence class of P. Indeed, $P \sim Q$ if and only if there is a sequence of arrows in $\langle S, T\rangle$ in either direction, starting from P, ending at Q. With $0 := A^0$, $[0]$ is the class of all stably free modules.
   We fix an indexing set I so that $\{[P_i] : i \in I\}$ is the set of all equivalence classes of projective modules in $\mathscr{P}(A)$, and $P_i \in [P_i]$.
2. As in (4.25), for $(F_1, P_1), (F_2, P_2) \in Obj(S^{-1}T)$ $(F_1, P_1) \sim (F_2, P_2)$ defined to be equivalent, if there are two arrows, as follows $(F_1, P_1) \longrightarrow (F, P) \longleftarrow (F_2, P_2)$. So, there are $U, V \in Obj(S)$, and the diagram of isomorphisms:
$$U \oplus (F_1, P_1) \xrightarrow{\sim} (F, P) \xleftarrow{\sim} V \oplus (F_2, P_2)$$
$$\Longleftrightarrow \quad \begin{cases} U \oplus F_1 \cong F \cong V \oplus F_2 \\ U \oplus P_1 \cong P \cong V \oplus P_2 \end{cases}$$
   As above, this is an equivalence relation. The second line means $P_1 \sim P_2$ in the sense of (1). Since the first isomorphism is in $S^{-1}S$, $U, V, F_1, F_2 \in S = \{A^n : n \geq 0\}$ and they have well-defined rank. In particular, $rank(U) + rank(F_1) = rank(V) + rank(F_2)$.
3. Now consider the category $S = \{A^n : n \geq 0\}$, as defined in (4.1.5 (2)), and the category $S^{-1}S$. Write $F_n = A^n$. In $S^{-1}S$ there is an arrow $(F_{m_1}, F_{n_1}) \longrightarrow (F_{m_2}, F_{n_2})$ if and only if $m_2 - m_1 = n_2 - n_1 \geq 0$. So,
$$S^{-1}S = \left(\coprod_{n\geq 0} [(F_n, 0)]_S\right) \coprod \left(\coprod_{n\geq 1} [(0, F_n)]_S\right) \tag{4.27}$$

is a disjoint union, and they correspond to the connected components of $\mathbb{B}S^{-1}S$. Therefore, $[(0,0)]_S = \{(F_n, F_n) : n \geq 0\}$. We denote $\mathbb{B}[(0,0)]_S = \mathbb{B}\{(F_n, F_n) : n \geq 0\}$.

4. For clarity, assume that $rank(P)$ is well defined (as in the case when A is commutative and connected) and is well behaved. With $r_1 = rank(P_1)$, $r_2 = rank(P_2)$, the first line means

$$\begin{cases} F_1 = F_2 \oplus A^{r_1 - r_2}\ if \quad r_1 \geq r_2 \text{ equivalently, if } rank(F_1) \geq rank(F_2) \\ F_2 = F_1 \oplus A^{r_2 - r_1}\ if \quad r_2 \geq r_1 \text{ equivalently, if } rank(F_2) \geq rank(F_1) \end{cases}$$

Let $\{[P_i] : i \in I\}$ be the set of all equivalence classes (1) of projective modules. Choose $P_i \in [P_i]$ such that $r_i := rank(P_i) = \min\{rank(P) : P \in [P_i]\}$. Such a choice is unique up to isomorphism. With notation as above,

$$\begin{cases} [(F, P_i)]_T = \{(F \oplus A^{rank(P) - r_i}, P) : P \in [P_i]\} \\ \text{In particular,} \quad [(0,0)]_T = \{(A^{rank(P)}, P) : P \in [0]\} \end{cases}$$

So, the correspondence $(n, i) \mapsto [(A^n, P_i)]_T$ is a bijective correspondence from $\mathbb{Z} \times I$ to the set of equivalence classes in $S^{-1}T$.

Theorem 4.6.4. *Let A be a ring, and $\mathscr{P}(A)$ be the category of finitely generated (right) projective A-modules. Let $S = \{A^n : n \geq 0\} = \mathscr{F}(A)$ be as defined in* (4.1.5(2)), *and $T = Iso(\mathscr{P}(A))$. We use some of the notation in the proof of* (4.6.2). *Then*

1. *All the connected components of $T^{-1}T$ are homotopically equivalent.*
2. *There are homotopy equivalences: $\mathbb{B}T^{-1}T \cong K_0(A) \times \mathbb{B}\left[(\mathbf{0},\mathbf{0})\right]_T \cong K_0(A) \times \mathbb{B}\left[(\mathbf{0},\mathbf{0})\right]_S$. In particular, $K_n(A) \cong \pi_n\left(\mathbb{B}\left[(0,0)\right]_T, 0\right) \cong \pi_n\left(\mathbb{B}\left[(0,0)\right]_S, 0\right)$ for all $n \geq 1$.*

Proof. By (4.6.2), $S^{-1}T \cong T^{-1}T$ are homotopically equivalent. So, we work with $S^{-1}T$. We use some of the notation in (4.6.3) and in the proof of (4.6.2). Let $P_i \in [0]_T$ be as in (4.6.3(1)) and $F \in Obj(S)$. The map $\tau_{(F,P_i)} : S^{-1}T \longrightarrow S^{-1}T$ sending $(P, Q) \mapsto (F \oplus P, P_i \oplus Q)$ is the composition of a left translation and a right translation. Since both are homotopy equivalences (4.3.8), so is $\tau_{(F,P_i)}$. Now, $\mathbb{B}\tau_{(F,P_i)}$ maps the connected component $\mathbb{B}\left[(0,0)\right]_T$ into $\mathbb{B}\left[(F, P_i)\right]_T$, and hence they are homotopically equivalent. So, (1) is established. Now, we prove (2). The second equivalence follows from (4.6.2). To prove the first equivalence, we only need to prove $K_0(A)$ is in bijection to $\pi_0\left(\mathbb{B}T^{-1}T\right)$. Define $\alpha : K_0(A) \longrightarrow \pi_0\left(\mathbb{B}T^{-1}T\right)$ by $\alpha([P]-[Q]) = [(P,Q)] \in \pi_0\left(\mathbb{B}T^{-1}T\right)$, which is well defined. The map $\beta : \pi_0\left(\mathbb{B}T^{-1}T\right) \longrightarrow K_0(A)$ given by $\beta([(P,Q)]) = [P]-[Q]$ is also well defined. So, α is an isomorphism. The descriptions of $K_n(A)$ follow from Theorem 4.5.13. The proof is complete. ∎

Remark 4.6.5. We avoided the plus construction of Quillen. Given a pointed (path) connected space $(X, 0)$ and a perfect normal subgroup $E \subseteq \pi_1(X, x)$, another pointed space $(X^+, 0^+)$ is constructed, satisfying certain universal properties. Given a commutative ring A, this construction is applied to the space $(\mathbb{B}GL(A), \star)$, and $(\mathbb{B}GL(A)^+, \star^+)$ is obtained. Plus construction predates the $S^{-1}S$ construction, and higher K-groups of A were defined as the homotopy groups $\pi_n(\mathbb{B}GL(A)^+, \star^+)$. Refer to, for example, [Ha] , for plus construction. It turns out $(\mathbb{B}GL(A)^+, \star^+)$ is homotopically equivalent to $\mathbb{B}\left[(\mathbf{0},\mathbf{0})\right]_S$, as above (see [Wc2, Thm. 4.9]).

4.7. Agreement of Modern and Classical *K*-Theory

The goal of this important section is to prove that the Classical and Modern (Quillen) K-Theory coincide. Let A be a ring. Other than $K_0(A)$, classically two other groups were defined, which we denote (temporarily) by $K_1^c(A)$ and $K_2^c(A)$. Among these two, more commonly discussed (and taught) was $K_1^c(A)$, which is also known as the Whitehead group. In this section, we prove that $K_i^c(A)$ coincides with $K_i(A)$, for $i = 1, 2$. In fact, we will prove $K_i^c(A) \cong \pi_i\left(\mathbb{B}S^{-1}S, \mathbf{0}\right)$, for $i = 1, 2$, where S is as above (4.6.4). First, we recall the classical definition.

4.7.1. *The Whitehead Group*

Let A be a ring. As usual, $GL_n(A)$ denotes the group of all invertible matrices of order n (to be viewed as automorphisms of A^n). The association $\alpha \mapsto \begin{pmatrix} \alpha & 0 \\ 0 & 1 \end{pmatrix}$ defines an inclusion map $GL_n(A) \hookrightarrow$

$GL_{n+1}(A)$. The (infinite) General Linear group is defined to be $GL(A) = \bigcup_{n\geq 0} GL_n(A)$.

Given $\lambda \in A$, and $1 \leq i, j \leq n$, with $i \neq j$, let $e_{ij}(\lambda)$ be the square matrix (of order n) whose only nonzero entry is λ, at (i,j) position. Let $EL_n(A) \subseteq GL_n(A)$ be defined as the subgroup generated by the set $\{1 + e_{ij}(\lambda) : \lambda \in A, 1 \leq i, j \leq n, i \neq j\}$. Define $EL(A) := \bigcup_{n\geq 0} EL_n(A) \subseteq GL(A)$ as the (infinite) elementary group.

Before we define Whitehead groups, we insert the following basic lemma.

Lemma 4.7.1. *Let A be a ring. Let $\alpha, \beta \in GL_n(A)$. Then*

$$\begin{pmatrix} \beta & 0 \\ 0 & \beta^{-1} \end{pmatrix} \in EL_{2n}(A) \qquad \text{and}$$

$$\begin{pmatrix} \alpha & 0 \\ 0 & \beta \end{pmatrix} \equiv \begin{pmatrix} \alpha\beta & 0 \\ 0 & 1_n \end{pmatrix} \equiv \begin{pmatrix} \beta\alpha & 0 \\ 0 & 1_n \end{pmatrix} \qquad \text{mod } EL_{2n}(A).$$

Proof. Write

$$\begin{aligned} \varepsilon &= \begin{pmatrix} 1_n & \beta - 1_n \\ 0 & 1_n \end{pmatrix} \begin{pmatrix} 1_n & 0 \\ 1_n & 1_n \end{pmatrix} \begin{pmatrix} 1_n & \beta^{-1} - 1_n \\ 0 & 1_n \end{pmatrix} \begin{pmatrix} 1_n & 0 \\ -\beta & 1_n \end{pmatrix} \\ &= \begin{pmatrix} \beta & 0 \\ 0 & \beta^{-1} \end{pmatrix}. \end{aligned}$$

All the matrices on the left-hand side are in $EL_{2n}(A)$, and hence so is the one on the right-hand side. This settles the first assertion. It follows that

$$\epsilon \begin{pmatrix} \alpha & 0 \\ 0 & \beta \end{pmatrix} = \begin{pmatrix} \beta\alpha & 0 \\ 0 & 1_n \end{pmatrix}, \quad \begin{pmatrix} \alpha & 0 \\ 0 & \beta \end{pmatrix} \epsilon = \begin{pmatrix} \alpha\beta & 0 \\ 0 & 1_n \end{pmatrix}.$$

The proof is complete. ∎

Recall, for a group G, and elements $f, g \in G$, we write $[f, g] := fgf^{-1}g^{-1}$, to be called the commutator. The commutator subgroup $[G, G]$ is defined as the subgroup generated by all the commutators $[f, g]$. The following would be used to define the Whitehead group (4.7.3).

Lemma 4.7.2. *Let A be a ring. Then*

$$[GL(A), GL(A)] = [EL(A), EL(A)] = EL(A).$$

Proof. For $\lambda \in A$ and $1 \leq i \neq j \leq n$. We use the notation $\epsilon_{ij}(\lambda) = 1 + e_{ij}(\lambda) \in EL_n(A)$. For distinct $1 \leq i, j, k \leq n$, it is easy to check $[\epsilon_{ij}(\lambda), \epsilon_{jk}(\mu)] = \epsilon_{ik}(\lambda\mu)$, which can be checked with $i = 1$, $j = 2$, $k = 3$. It follows from this that $EL_n(A) = [EL_n(A), EL_n(A)]$ for all $n \geq 3$, and hence $[EL(A), EL(A)] = E(A)$.

Given $\alpha, \beta \in GL_n(A)$, by (4.7.2),

$$\begin{pmatrix} \alpha\beta\alpha^{-1}\beta^{-1} & 0 \\ 0 & 1_n \end{pmatrix} = \begin{pmatrix} \alpha\beta & 0 \\ 0 & \beta^{-1}\alpha^{-1} \end{pmatrix} \begin{pmatrix} \alpha^{-1} & 0 \\ 0 & \alpha \end{pmatrix} \begin{pmatrix} \beta^{-1} & 0 \\ 0 & \beta \end{pmatrix}$$
$$\in EL_{2n}(A).$$

Therefore, $[GL(A), GL(A)] = E(A)$. The proof is complete. ∎

Definition 4.7.3. With notation as above, the **Whitehead group** is defined as follows:

$$K_1^c(A) = \frac{GL(A)}{EL(A)}.$$

We use superscript c (*temporarily*), to distinguish $K_1^c(A)$ from the Quillen group $K_1(A) := K_1(\mathscr{P}(A))$. However, most readers know that classically K_1-group of A was defined to be the Whitehead group. And, the notation $K_1(A)$ rightfully belongs to the Whitehead group.

4.7.2. *The Agreement of $K_1^c(A)$ and $K_1(A)$*

In this section, we prove that Quillen $K_1(A)$ agrees with the classical group $K_1^c(A)$.

Definition 4.7.4. Let A be a ring. Let $S = \{A^n : n \geq 0\} = \mathscr{F}(A) \subseteq Iso(\mathscr{P}(A))$ be the symmetric monoidal category, as defined in (4.1.5(2)). Denote $\mathbf{v}_n := [(A^n, A^n))] \in \mathbb{B}S^{-1}S$ and

$1_n := 1_{A^n}$. As before, $\mathbb{B}\left[(0,0)\right]_S$ denotes the connected component of $(0,0)$ in $\mathbb{B}S^{-1}S$. Given $\alpha \in GL_n(A) \subseteq GL(A)$, there is a loop $\ell_n(\alpha)$ in $\mathbb{B}\left[(\mathbf{0},\mathbf{0})\right]_S$, at the vertex $\mathbf{v}_n$, corresponding to the arrow $\left(A^0; \alpha, 1_n\right) : (A^n, A^n) \xrightarrow{\sim} (A^n, A^n)$ in $S^{-1}S$. The arrow $(A^n; 1_n, 1_n) : (\mathbf{0}, \mathbf{0}) \longrightarrow (A^n, A^n)$ in $S^{-1}S$ defines a path from $\mathbf{0} \longrightarrow \mathbf{v}_n$ in $\mathbb{B}S^{-1}S$. So, we have a commutative diagram of isomorphisms:

$$\begin{array}{ccc} \pi_1\left(\mathbb{B}\left[(0,0)\right]_S, (0,0)\right) & \xrightarrow[\sim]{\iota_n} & \pi_1\left(\mathbb{B}\left[(0,0))\right]_S, \mathbf{v}_n\right) \\ \| & & \downarrow \\ \pi_1\left(\mathbb{B}\left[(0,0)\right]_S, (0,0)\right) & \xrightarrow[\iota_{n+1}]{\sim} & \pi_1\left(\mathbb{B}\left[(0,0)\right]_S, \mathbf{v}_{n+1}\right). \end{array}$$

The second vertical arrow is obtained by using the path $\mathbf{v}_n \longrightarrow \mathbf{v}_{n+1}$ corresponding to the arrow $(A; 1_{n+1}, 1_{n+1}) : (A^n, A^n) \longrightarrow (A^{n+1}, A^{n+1})$ in $S^{-1}S$. Define, $\ell_0(\alpha) = \iota_n^{-1}(\ell_n(\alpha))$, which is a loop at $(0,0)$.

$$\text{Therefore,} \quad \iota_n([\ell_0(\alpha)]) = [\ell_n(\alpha)] = \left[\ell_{n+1}\begin{pmatrix} \alpha & 0 \\ 0 & 1 \end{pmatrix}\right] = \left[\ell_0\begin{pmatrix} \alpha & 0 \\ 0 & 1 \end{pmatrix}\right].$$

$$\text{Define,} \quad \eta : GL(A) \longrightarrow \pi_1\left(\mathbb{B}\left[(0,0)\right]_S, (0,0)\right) \quad \text{by}$$
$$\eta(\alpha) = \iota_0(\alpha) \quad \forall\, \alpha \in GL_n(A). \tag{4.28}$$

It follows from the above discussions that η is a well-defined group homomorphism.

Lemma 4.7.5. *With the notation as in (4.7.4), the map η is surjective.*

Proof. As before, let $\mathbb{B}[(0,0)]_S^k$ denote the k-skeleton of $\mathbb{B}[(0,0)]_S$. By (C.3.9), the map $\pi_1\left(\mathbb{B}[(0,0)]_S^1, (0,0)\right) \twoheadrightarrow \pi_1\left(\mathbb{B}[(0,0)]_S, (0,0)\right)$ is surjective. The space $\mathbb{B}[(0,0)]_S^1$ consists of lines $\ell(F) : (A^n, A^n) \longrightarrow (A^{n+r}, A^{n+r})$, one for each morphism $F : (A^r, f, g) : (A^n, A^n) \longrightarrow (A^{n+r}, A^{n+r})$. However, for two composable arrows F, G, the concatenation $\ell(G)\ell(F)$ cannot be treated as the same (homotopic) path $\ell(GF)$, because there is no nondegenerate 2-simplex in $\mathbb{B}[(0,0)]_S^1$. So, $\ell(F)$, $\ell(G)$, $\ell(GF)$ form a loop, which is not null homotopic in $\mathbb{B}[(0,0)]_S^1$. We argue to mitigate this situation.

A loop α at $(0,0)$ in $\mathbb{B}\left[(0,0)\right]_S$ is given by finitely many (use (C.1.8)) composable arrows (A^r, f, g), as in the following diagram, in either direction, and they can be joined to $(0,0)$, back and forth, as follows:

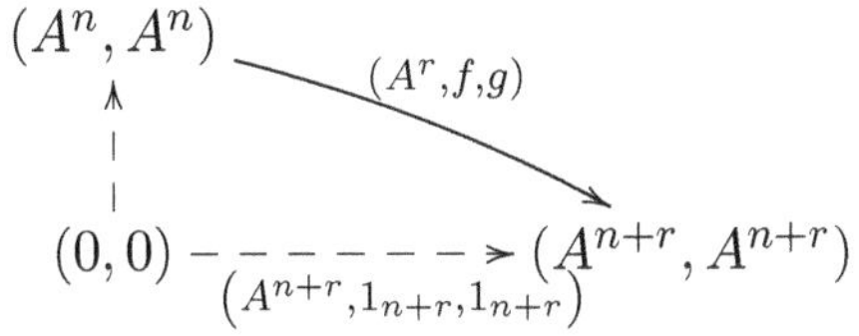

$f, g \in GL_{n+r}(A)$

This triangle corresponds to a loop at $(0,0)$ and the loop α is homotopic to concatenation of such loops, given by a single arrow (A^r, f, g). Therefore, $\pi_1\left(\mathbb{B}\left[(0,0))\right]_S, (0,0)\right)$ is generated by such loops given by such single arrows. Note that the arrow (A^r, f, g) in $S^{-1}S$ is a composition as in the following diagram, and which extends as follows:

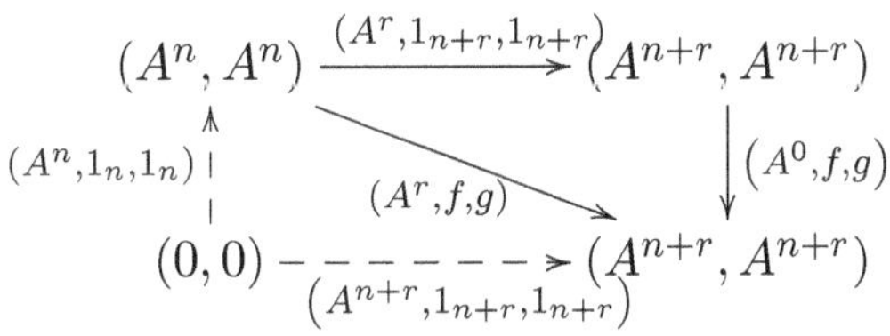

Since the upper triangle corresponds to a 2-simplex in $\mathbb{B}\left[(0,0)\right]_S$, the second vertical edge can be homotoped to coincide with the diagonal edge. Writing $N := n + r$, it follows from this that $\pi_1\left(\mathbb{B}\left[(0,0)\right]_S, (0,0)\right)$ is generated by loops, given by single arrows, as follows:

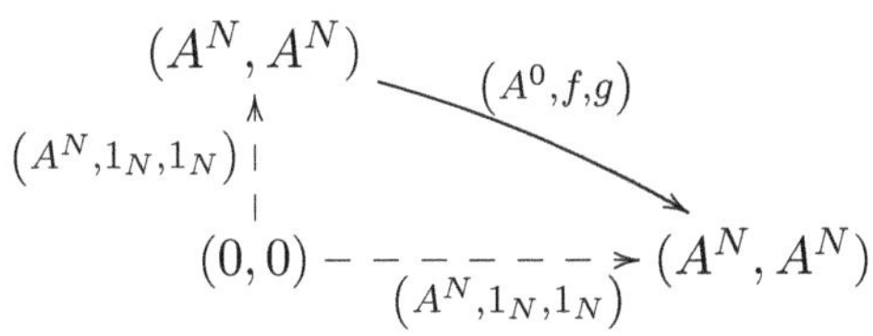

$f, g \in GL(A^N)$

We denote this loop by $\ell(f,g)$. With notation in (4.7.4), note $\ell(f, 1_N) = \eta(f)$. Also note that the arrow $\left(A^0, f, g\right)$ is a composition of $\left(A^0, f, 1_N\right)$ and $\left(A^0, 1_N, g\right)$.

Note that the vertical canonical arrow $(A^N, 1_N, 1_N) : (0,0) \longrightarrow (A^N, A^N)$ is identified with the $(A^N, \alpha, \alpha) : (0,0) \longrightarrow (A^N, A^N)$

for any $\alpha \in GL_N(A)$. So, $\ell(f,g))$ is equivalent to $\ell(f\alpha, g\alpha))$. In particular, $\ell(f,g) = \ell(fg^{-1}, 1_N)$. This establishes that η is surjective. The proof is complete. ■

Theorem 4.7.6. *Let A be a ring. With the notation as in* (4.7.4), *the map η induces an isomorphism $\overline{\eta} : K_1^c(R) \xrightarrow{\sim} \pi_1(\mathbb{B}[(0,0)]_S,$ $(0,0))$.*

Proof. As before, $K_1^c(A) := \frac{GL(A)}{EL(A)}$, and $\underline{K_1^c(A)}$ denote the one object $(\star)$ category, corresponding to the group $K_1^c(A)$. We also have $EL(A) = [GL(A), GL(A)] = [EL(A), EL(A)]$, the commutator subgroup (4.7.2). Recall $\mathbb{B}[(0,0)]_S$ is an H_c-space (see Section 4.8). Therefore, by (4.9.1) $\pi_1(\mathbb{B}[(0,0)]_S, (0,0))$ is an abelian group. Therefore, η factors through a surjective map $\overline{\eta} : K_1^c(A) \twoheadrightarrow \pi_1(\mathbb{B}[(0,0)]_S, (0,0))$, (4.7.5). To prove that $\overline{\eta}$ is injective, we define a left inverse of $\overline{\eta}$. There is a well-defined functor, defined as follows:

$$\Phi : [(0,0)]_S \longrightarrow \underline{K_1^c(A)} \quad \text{where,}$$

$$\begin{cases} \Phi(A^n, A^n) = \star \ \forall\, n \\ \Phi(\overline{(A^k, f, g)}) = EL(A)(fg^{-1}) \ \forall\, f, g \in GL_{k+n}(A) \end{cases}$$

where $\overline{(A^k, f, g)}$ denotes the arrow $(A^n, A^n) \longrightarrow (A^{k+n}, A^{k+n})$, represented by (A^k, f, g), and $EL(A)(fg^{-1})$ denotes the coset. We prove that $\Phi(\overline{(A^k, F, G)})$ is well defined. Suppose $\overline{(A^k, F, G)} : (A^n, A^n) \longrightarrow (A^{k+n}, A^{k+n})$ represents the same map. Then there is $\alpha \in GL_r(A)$, and with $\alpha_n := \alpha \oplus 1_n$, such that $F = \alpha_n f$ and $G = \alpha_n g$. So,

$$\begin{aligned} FG^{-1} &= \alpha_n f(\alpha_n g)^{-1} = \alpha_n (fg^{-1})\alpha_n^{-1} = \alpha_n (fg^{-1})\alpha_n^{-1}(fg^{-1})^{-1} fg^{-1} \\ &= [\alpha_n, fg^{-1}] fg^{-1} \end{aligned}$$

where $[\alpha_n, fg^{-1}]$ denotes the commutator. Therefore, $EL(A)(fg^{-1}) = EL(A)(FG^{-1})$, and it follows that $\Phi(\overline{(A^k, F, G)})$ is well defined. So, Φ defines the map of the classifying spaces $\mathbb{B}\Phi : \mathbb{B}[(0,0)]_S \longrightarrow \mathbb{B}\underline{K_1^c(R)}$, and a map $\pi_1(\Phi) : \pi_1(\mathbb{B}[(0,0)]_S, (0,0)) \longrightarrow \pi_1\left(\mathbb{B}\overline{K_1^c(A)}, \star\right) = K_1^c(A)$. Now,

$$\mathbb{B}\Phi\overline{\rho}(EL(A)\alpha) = \mathbb{B}\Phi(A^0, \alpha, 1_n) = EL(A)\alpha.$$

The proof is complete. ■

4.7.3. *The Agreement of $K_2^c(A)$ and $K_2(A)$*

As before, let A denote a ring, and let $S = \{A^n : n \geq 0\} = \mathscr{F}(A)$ and other notation be as in (4.7.4). Thc proof of coincidcncc of classical $K_2^c(A)$ and Quillen $K_2(A)$, at this stage, is more algebraic than topological. The proof is due to Gersten [G2].

As usual, given a group G, let $\underline{G}$ denote the category that has exactly one object $\star$ and $Mor(\star, \star) = G$. We also use the notation $\mathbb{B}G := \mathbb{B}\underline{G}$ to denote the classifying space.

Before we define $K_2^c(A)$, we define a natural map $\mathbb{B}GL(A) \longrightarrow \mathbb{B}(S^{-1}S)$, as follows.

Lemma 4.7.7. *There is a natural map $\rho : \mathbb{B}GL(A) \longrightarrow \mathbb{B}S^{-1}S$, such that ρ induces the following familiar exact sequence:*

$$1 \longrightarrow EL(A) \longrightarrow GL(A) \xrightarrow{\pi_1(\rho)} K_1(A) \longrightarrow 1$$

$$with \begin{cases} \pi_1(\mathbb{B}GL(A)) = GL(A) \\ \pi_1(\mathbb{B}S^{-1}S) = K_1(A). \end{cases} \tag{4.29}$$

Proof. For completeness, we include the proof (see [Wc2, p. 353]). We have a commutative diagram

$$\begin{array}{ccc} GL_n(A) & \xrightarrow{\beta_n} & Aut_{S^{-1}S}(A^n, A^n) \\ \downarrow{\scriptstyle \iota_n} & & \downarrow{\scriptstyle \odot A} \\ GL_{n+1}(A) & \xrightarrow[\beta_{n+1}]{} & Aut_{S^{-1}S}(A^{n+1}, A^{n+1}) \end{array} \quad \text{where} \quad \begin{cases} \beta(g) = (g, 1_n) \\ \iota_n(g) = g \oplus 1_A. \end{cases}$$

We treat β_n as functors $\underline{GL_n(A)} \longrightarrow S^{-1}S$, sending $\star \mapsto (A^n, A^n)$. This commutative diagram yields a natural equivalence $\beta_n \mapsto \beta_{n+1}\iota_n$. So, it defines a homotopy

$$H_n : \mathbb{B}GL_n(A) \times I \longrightarrow \mathbb{B}S^{-1}S.$$

Now, by mapping telescope, we have a description (refer to Exercise 4.9.2),

$$\mathbb{B}GL(A) = \frac{\coprod_{n=1}^{\infty} \mathbb{B}GL_n(A) \times [n, n+1]}{(x_n, n+1) \sim (\iota_n(x_n), n+1)\ \forall x_n \in \mathbb{B}GL_n(A), n = 1, 2 \ldots} \tag{4.30}$$

(*Note, mapping telescope construction is an iteration of mapping cylinder construction (C.4.11).*) So, the sequence β_n defines a map

$$\rho : \mathbb{B}GL(A) \longrightarrow \mathbb{B}S^{-1}S.$$

The proof is complete. ∎

The following is an immediate consequence of (4.7.7).

Lemma 4.7.8. *Let* $\rho : \mathbb{B}GL(A) \longrightarrow \mathbb{B}S^{-1}S$ *be the map as in* (4.7.7) *and* $\mathscr{F} := \mathscr{F}(\rho, [(0,0)])$ *be the homotopy fiber. Then the sequence* (4.29) *extends to an exact sequence*

$$1 \longrightarrow \pi_2\left(\mathbb{B}S^{-1}S, [(0,0)]\right) \longrightarrow \pi_1(\mathscr{F}) \longrightarrow GL(A) \longrightarrow K_1(A) \longrightarrow 1 \tag{4.31}$$

Proof. It follows from the fact that $\pi_2(\mathbb{B}GL(A)) = 0$. The proof is complete. ∎

For the convenience of our discussions, we sketch the definitions of the Steinberg group, $K_2(A)$, and some other results from [Mi]. We refer to [Mi] for details.

Definition 4.7.9. Let A be a ring. Fix an integer $n \geq 3$. The **Steinberg group** $St_n(A)$ is defined to be the group generated by the set of generators

$$\left\{x_{ij}^{\lambda} : 1 \leq i,j \leq n, i \neq j, \lambda \in A\right\}$$

subject to the relations,

$$\begin{cases} x_{ij}^{\lambda} x_{ij}^{\mu} = x_{ij}^{\lambda+\mu} & \forall 1 \leq i,j \leq n, i \neq j, \text{ and } \lambda, \mu \in A \\ \left[x_{ij}^{\lambda}, x_{jk}^{\mu}\right] = x_{ik}^{\lambda\mu} & \forall 1 \leq i,j,k \leq n, i \neq j, j \neq k, i \neq k, \text{ and } \lambda, \mu \in A \\ \left[x_{ij}^{\lambda}, x_{kp}^{\mu}\right] = 1_n & \forall 1 \leq i,j,k,p \leq n, i \neq j, k \neq p, j \neq k, i \neq p, \\ & \quad \text{and } \lambda, \mu \in A \end{cases} \tag{4.32}$$

So, $St_n(A) = \frac{\mathscr{F}}{\mathscr{R}}$ where

$$\begin{cases} \mathscr{F} = \text{Free Group generated by} \\ \{x_{ij}^{\lambda} : 1 \leq i,j \leq n, i \neq j, \text{ and } \lambda \in A\} \\ \mathscr{R} \subseteq \mathscr{F} \quad \text{is the normal subgroup generated} \\ \qquad\qquad \text{by the relations (4.32)} \end{cases}$$

There is an obvious map $St_n(A) \longrightarrow St_{n+1}(A)$. Define the **infinite Steinberg group** by $St(A) := \lim_n St_n(A)$.

As before, for $1 \leq i, j \leq n$, $i \neq j$ and $\lambda \in A$, let e_{ij}^{λ} denote the $n \times n$ matrix with λ in (i, j) position and zero elsewhere. So, $1_n + e_{ij}^{\lambda} \in EL_n(A)$ is a generator of the group $EL_n(A)$ of elementary matrices. The matrices $\varepsilon_{ij}^{\lambda} := 1_n + e_{ij}^{\lambda}$ satisfy the relations (4.32). So, there is a surjective map

$$\varphi_n : St_n(A) \twoheadrightarrow EL_n(A) \qquad \text{sending} \quad x_{ij}^{\lambda} \mapsto .\varepsilon_{ij}^{\lambda} \tag{4.33}$$

Taking the limit, we have a surjective map

$$\varphi : St(A) \twoheadrightarrow EL(A). \tag{4.34}$$

The classical group $K_2^c(A)$ was defined in [Mi] as

$$K_2^c(A) := \ker(\varphi). \tag{4.35}$$

In what follows, we state some results and definitions from [Mi].

Theorem 4.7.10. *With notation as in (4.7.9), we have*

$$K_2^c(A) = \ker(\varphi) = \text{center}(St(A)).$$

Proof. See [Mi, p. 40]. ■

Remark 4.7.11. Consequently, $K_2^c(A)$ is an abelian group. Further, the sequence (4.29) extends to an exact sequence

$$1 \longrightarrow K_2^c(A) \longrightarrow St(A) \longrightarrow GL(A) \longrightarrow K_1(a) \longrightarrow 1. \tag{4.36}$$

This is yet another extension of the sequence (4.29), along with (4.31).

Proof. Obvious from definition (4.35). ■

Now we define central extensions and universal central extensions of groups.

Definition 4.7.12. Consider an exact sequence

$$1 \longrightarrow \ker(f) \longrightarrow X \xrightarrow{f} G \longrightarrow 1 \qquad \text{of groups.} \tag{4.37}$$

Such a sequence is denoted by (X, f). We say that (4.37) splits if there is a homomorphism, $\epsilon : G \longrightarrow X$ such that $f\epsilon = 1_G$. Also recall that a group G is said to be perfect if the commutator $[G, G] = G$. In our context, by (4.7.2), the Elementary subgroup $EL(A)$ is a perfect subgroup of $GL(A)$.

The sequence (4.37) is defined to be a **Central extension** of G if $\ker(f) \subseteq \text{Center}(f)$. We also say that X is a central extension of G.

A central extension $(U, \mathfrak{v})$ of G is called a **Universal central extension** of G if it has the universal property, among the central extensions of G. This means, given any other central extension (X, f), as in (4.37), there is a unique homomorphism $h : U \longrightarrow X$ such that $\mathfrak{v} = fh$. In other words, there is a unique h such that the diagram

$$\begin{array}{ccccccccc} 1 & \longrightarrow & \ker(g) & \longrightarrow & U & \xrightarrow{\mathfrak{v}} & G & \longrightarrow & 1 \\ & & & & \downarrow h & & \| & & \\ 1 & \longrightarrow & \ker(f) & \longrightarrow & X & \xrightarrow[f]{} & G & \longrightarrow & 1 \end{array} \qquad \text{commutes.} \qquad (4.38)$$

As usual, if G has a universal central extension $(U, \mathfrak{v})$, then it is unique up to natural isomorphisms.

The following is a characterization of universal extensions.

Theorem 4.7.13. *Let G be a group. A central extension $(U, \mathfrak{v})$ of G is universal if and only if* (1) *U is perfect and* (2) *every central extension of U splits.*

Proof. See [Mi, p. 43]. ■

Theorem 4.7.14. *A group G admits a universal central extension if and only if G is perfect.*

Proof. See [Mi, p. 45]. ■

Corollary 4.7.15. *Let A be a ring. Then the infinite elementary group $EL(A)$ admits a universal central extension.*

Proof. In fact, $E(A)$ is perfect, by (4.7.2). Now the corollary follows from (4.7.14). ■

The following is the final theorem from [Mi].

Theorem 4.7.16. *Let A be a ring. Consider the following extension.*

$$1 \longrightarrow K_2^c(A) \longrightarrow St(A) \longrightarrow EL(A) \longrightarrow 1 \tag{4.39}$$

which is the left-hand side of (4.36). *Then* (4.39) *is a universal central extension of $EL(A)$.*

Proof. See [Mi, p. 47], and also [Sw3, p. 208]. ■

The tail-end of the argument comes from [G2], as follows.

Theorem 4.7.17. *Let A be a ring. Consider the following extension:*

$$1 \longrightarrow \pi_2\left(\mathbb{B}S^{-1}S, [(0,0)]\right) \longrightarrow \pi_1(\mathscr{F}) \longrightarrow EL(A) \longrightarrow 1 \tag{4.40}$$

which is the left-hand side of (4.31). Then (4.40) is a universal central extension of $EL(R)$.

Proof. Recall $\pi_2(X, x)$ is commutative, for any $(X, x) \in$ **Top**$_\bullet$. Further, it is not difficult to see that for a homotopy fibration $F \longrightarrow X \longrightarrow B$, the map $\pi_2(B) \longrightarrow \pi_1(X)$ is central. So, (4.40) is central. It is a result of Kervaire [Km] that (4.40) is a universal extension. Refer to Exercise 4.9.3, [Wc2, p. 289], [Sv, Prop. 2.5]) for more complete proofs. ■

Finally, we state the coincidence theorem.

Theorem 4.7.18. *Let A be a ring. Then $K_2^c(A) \simeq K_2(A)$ are isomorphic.*

Proof. Recall, Quillen $K_2(A) = \pi_2\left(\mathbb{B}S^{-1}S, [(0,0)]\right)$. Now the theorem follows by comparing two universal central extensions (4.39) and (4.40). The proof is complete. ■

4.8. H_c-Spaces

For completeness, we include this section on H-spaces. First, we recall the definition of H-spaces [S, p. 36], [W, pp. 116–120].

Definition 4.8.1. Let $(Y, \star) \in \mathbf{Top}_\bullet$. Let $p_1, p_2 : Y \times Y \longrightarrow Y$ denote the projections, and $\iota_1, \iota_2 : Y \longrightarrow Y \times Y$ be the maps defined by $\iota_1(y) = (y, \star)$, $\iota_2(y) = (\star, y)$. Then Y is defined to be a H**-space** if there is a map $\mu : k(Y \times Y) \longrightarrow Y$ in $\mathbf{Top}_\bullet$ such that $\mu\iota_1 = \mu\iota_2 = 1_Y$.

There is a subtle difference between the literature on H-spaces [S, W] and the way it is used in homotopy theory, in particular in K-Theory, literature (e.g. 4.7.6). To iron out this difference, we define H_c-Spaces.

Definition 4.8.2. Let $(Y, \star) \in \mathbf{Top}_\bullet$. Let $p_1, p_2, \iota_1, \iota_2$, be as in (4.8.1), while we consider them as maps $p_1, p_2 : k(Y \times Y) \longrightarrow Y$ and $\iota_1, \iota_2 : Y \longrightarrow k(Y \times Y)$. Then p_1, p_2, ι_1, and ι_2 are continuous. Further, $p_1\iota_1 = p_2\iota_2 = 1_Y$. A topological space Y is said to be an H_c**-space**, if there is a map $\mu : k(Y \times Y) \longrightarrow Y$, such that $\mu\iota_1 = \mu\iota_2 = 1_Y$:

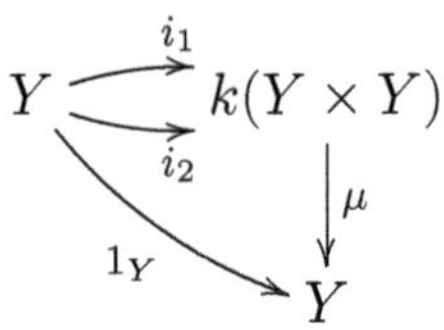

Fix an H_c-Space (Y, μ) as above, we further define the following and add comments:

1. Given $X, Y \in \mathbf{Top}_\bullet$, we denote $[X, Y] := Hom_{\mathcal{H}om\mathbf{Top}_\bullet}(X, Y)$, as in Section B.2.
2. For $\alpha, \beta \in Hom_{\mathbf{Top}_\bullet}(X, Y)$, define the product $\alpha \cdot \beta$ by the commutative diagram

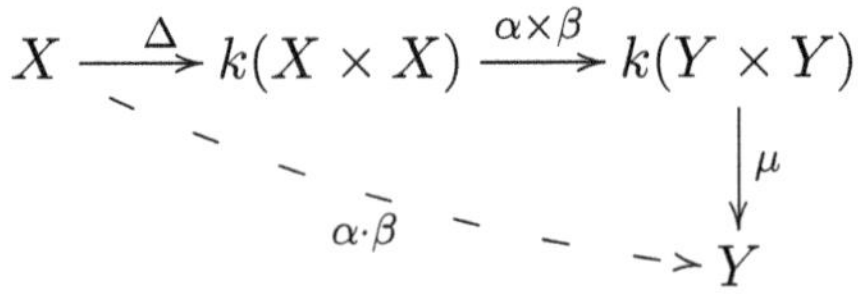

For such $\alpha, \beta \in Hom_{\mathbf{Top}_\bullet}(X, Y)$, further define $[\alpha] \cdot [\beta] := [\alpha \cdot \beta] \in [X, Y]$.

3. Let $e_X : X \longrightarrow Y$ denote the constant map in **Top$_\bullet$**. Then $\alpha \cdot e_X = \alpha = e_X \cdot \alpha$, which follows from the commutative diagram

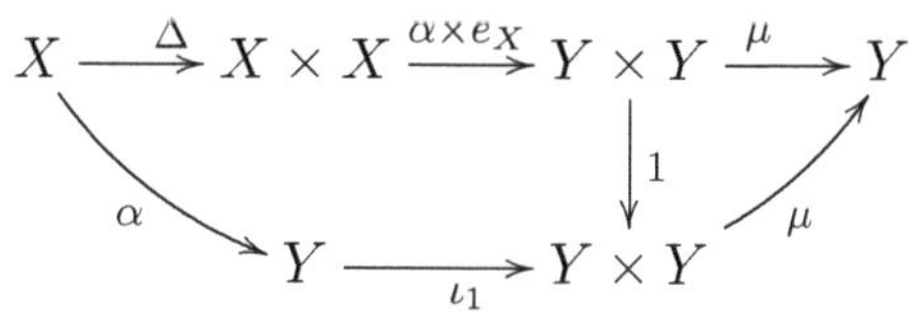

Consequently, $[\alpha] \cdot [e_X] = [\alpha] = [e_X] \cdot [\alpha]$. In other words, under this product $\cdot$ in $[X, Y]$, $[e_X]$ acts as the identity.
4. In some literature, the conditions above that $\mu\iota_1 = \mu\iota_2 = 1_Y$ are loosened and it is assumed only that $\mu\iota_1 \sim \mu\iota_2 \sim 1_Y$ are homotopic, and the product above on $[X, Y]$ makes sense. One further defines (homotopy) associative and symmetric properties of H_c-spaces. (See [W, p. 118].)
Under these conditions, $[X, Y]$ also has a structure of an abelian monoid, with identity, for all $X \in$ **Top$_\bullet$**.
5. In particular, $\forall\, n \geq 0$, the multiplication μ induces a structure of an abelian monoid, with identity on $\pi_n(Y, \star) = [\mathbb{S}^n, Y]$ (*which is not to be mixed up with the homotopy group structures, for* $n \geq 1$).

Definition 4.8.3. Let (Y, μ) be a homotopy associative H_c-space. We say (Y, μ) is **group-like** if there is a map $j : Y \longrightarrow Y$ of pointed spaces, such that $j \cdot 1_Y \simeq 1_Y \cdot j \simeq e_Y$. This means, if $[1_Y]$ has an inverse in the monoid $[Y, Y]$.

In this case, $[X, Y]$ also has a group structure for any $X \in$ **Top$_\bullet$**. This follows from the following commutative diagram, with $\alpha : X \longrightarrow Y$ a map of pointed spaces,

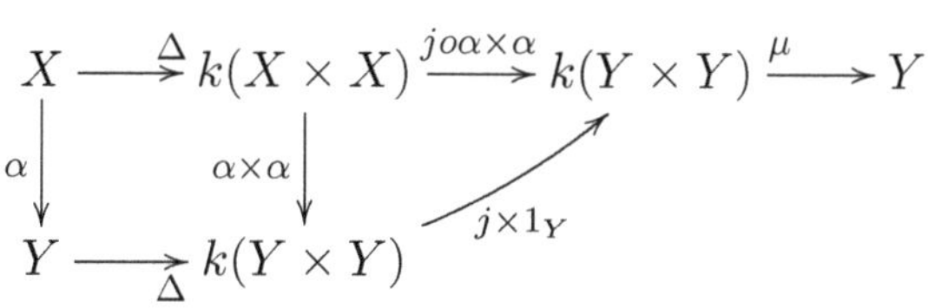

$$\Longrightarrow \begin{cases} [j\alpha \cdot \alpha] = [\mu(jo\alpha \times \alpha)\Delta] \\ = [(j \cdot 1_Y)o\alpha] = [e_Y o\alpha] = [e_X] \end{cases}$$

So, $[\alpha]$ has a left inverse $j\alpha$, for all α. Hence, $[X, Y]$ is a group.

Proposition 4.8.4. *Let (Y, μ) be an associative (symmetric) H_c-space. The map*

$$\Phi : k(Y \times Y) \longrightarrow k(Y \times Y) \quad \text{by} \quad \varphi(x, y) = (x, xy) := (x, \mu(x, y))$$

is continuous.

This map is called the **shear map**. *The shear map is a homotopy equivalence if and only if Y is group-like.* (Note that the shear map is not necessarily continuous on $Y \times Y$.)

Proof. It is clear from the following diagram of product topology:

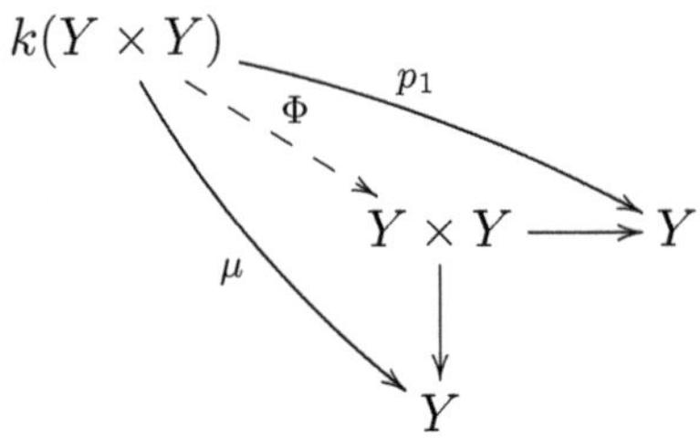

that the map $\Phi : k(Y \times Y) \longrightarrow Y \times Y$ is continuous. By application of the functor $k : \mathbf{Top} \longrightarrow \mathbf{K}$, it follows $\Phi : k(Y \times Y) \longrightarrow k(Y \times Y)$ is continuous. Assume φ is a homotopy equivalence, and $\Psi : k(Y \times Y) \longrightarrow k(Y \times Y)$ is the homotopy inverse of φ. Let $j = p_2 \Psi \iota_1$, as in the diagram

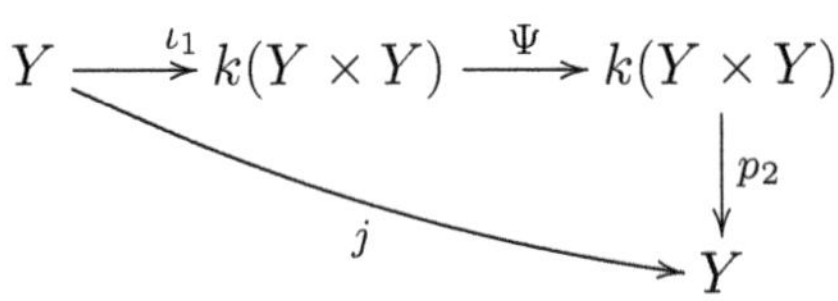

We prove that j is a homotopy inverse of 1_Y. We establish a series of identities. First, $p_1 \varphi = p_1$, $p_2 \varphi = \mu$. Therefore,

$$\begin{cases} p_1 \simeq p_1(\Phi\Psi) = (p_1\Phi)\Psi = p_1\Psi \\ p_1\Psi\iota_1 = p_1\iota_1 = 1_Y \\ p_2 \simeq p_2(\Phi\Psi) = (p_2\Phi)\Psi = \mu\Psi \end{cases}$$

We have

$$\begin{cases} \mu(1_Y \times j)\Delta = \mu(p_1\Psi\iota_1 \times p_2\Psi\iota_1)\Delta \\ = \mu(p_1 \times p_2)(\Psi\iota_1 \times \Psi\iota_1)\Delta \\ = \mu(p_1 \times p_2)\Delta_{Y\times Y}\Psi\iota_1 \\ = \mu\Psi\iota_1 \simeq p_2\iota_1 = e_Y \end{cases} \quad \text{and} \quad \begin{cases} \mu(j \times 1_Y)\Delta = \mu(p_2\Psi i_1 \times p_1\Psi\iota_1)\Delta \\ = \mu(p_2 \times p_1)(\Psi i_1 \times \Psi\iota_1)\Delta \\ = \mu(p_2 \times p_1)\Delta_{Y\times Y}\Psi\iota_1 \\ = \mu\Psi\iota_1 \simeq p_2\iota_1 = e_Y \end{cases}$$

So, $1_Y \cdot j \simeq e_Y$ and $j \cdot 1_Y \simeq e_Y$. This completes the proof that Y is group-like.

Conversely, suppose Y is group-like. Define $\Psi : k(Y \times Y) \longrightarrow k(Y \times Y)$ by $\Psi(u,v) = (u, j(u)v)$ as follows:

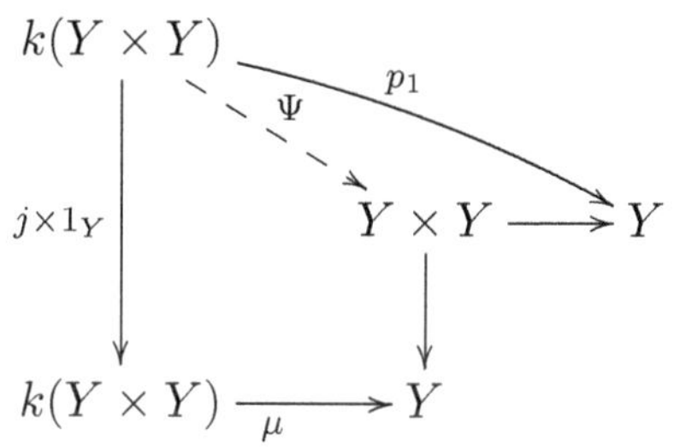

We check,

$$\begin{cases} \Phi\Psi(u,v) = (u, u(j(u)v)) \simeq (u,(uj(u))v) \simeq (u,(e_Y v)) \simeq (u,v) \\ \Psi\Phi(u,v) = \Psi(u,uv) = (u, j(u)(uv)) \simeq (u(j(u)u)v) \\ \quad \simeq (u, e_Y v) \simeq (u,v). \end{cases}$$

The proof is complete. ∎

Lemma 4.8.5. *Suppose (Y,μ) is a connected H_c-space, and is homotopy associative. Then the shear map $\Phi : k(Y\times Y) \longrightarrow k(Y\times Y)$ is a weak equivalence.*

In particular, if $k(Y \times Y)$ has a CW structure, then Φ is a homotopy equivalence (C.3.8). Also, see (C.2.2).

Proof. We fix a base point $\star \in Y$. We need to prove $\Phi_* : \pi_n(k(Y \times Y), (\star,\star)) \xrightarrow{\sim} \pi_n(k(Y \times Y), (\star,\star))$ is an isomorphism, $\forall\ n \geq 1$, since Y is connected. Since the homotopy groups $\pi_n(Y,\star)$

are abelian (see (4.9.1), when $n = 1$), we have

$$\begin{aligned}\pi_n(k(Y\times Y),(\star,\star)) &\cong \pi_n(Y\times Y,(\star,\star)) \cong \pi_n(Y,\star)\times\pi_n(Y,\star)\\ &\cong \pi_n(Y,\star)\oplus\pi_n(Y,\star) \quad \forall\, n\geq 1.\end{aligned}$$

We have maps

$$\begin{cases} p_{1*},p_{2*}:\pi_n(Y,\star)\oplus\pi_n(Y,\star)\longrightarrow\pi_n(Y,\star) \\ i_{1*},i_{2*}:\pi_n(Y,\star)\longrightarrow\pi_n(Y,\star)\oplus\pi_n(Y,\star)\end{cases}, \quad \text{with} \quad \begin{cases}\mu_* i_{1*} = 1\\ \mu_* i_{2*} = 1\end{cases}$$

The following diagram commutes, with either choice of the horizontal arrow:

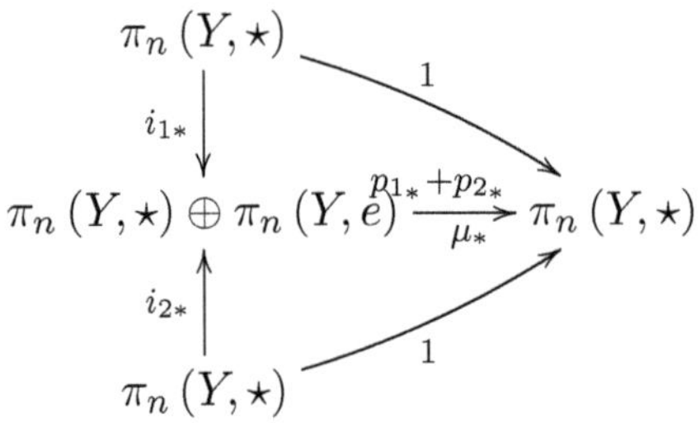

Therefore, $\mu_* = p_{1_*} + p_{2_*}$. Since $p_1\Phi = p_1$, $p_2\Phi = \mu$, we have

$$\begin{cases} p_{1*}\Phi_* = p_{1*} & \text{1st-coordinate of } \Phi_* \\ p_{2*}\Phi_* = \mu_* = p_{1_*} + p_{2_*} & \text{2nd-coordinate of } \Phi_*\end{cases} \Longrightarrow \Phi_*(x,y) = (x, x+y).$$

All the maps are group homomorphisms. This translates to

$$\Phi_* = \begin{pmatrix}1 & 0\\ 1 & 1\end{pmatrix} : \pi_n(Y,\star)\oplus\pi_n(Y,\star)\longrightarrow\pi_n(Y,\star)\oplus\pi_n(Y,\star)$$

which is an isomorphism. The proof is complete. ■

Finally, we have the information we need.

Theorem 4.8.6. *Let (Y,μ) be a connected H_c-space. Further, assume that, $k(Y\times Y)$ has a CW complex structure. Then Y is group-like.*

Proof. Since $k(Y\times Y)$ has a CW structure, the shear map $\Phi : k(Y\times Y)\longrightarrow k(Y\times Y)$ is a homotopy equivalence, by (4.8.5). Now it follows from (4.8.4) that Y is group-like. The proof is complete. ■

4.9. Exercises

Exercise 4.9.1. Let (Y, μ) be an H_c-space, and $\star \in Y$ be a base point. Then $\pi_1(Y, \star)$ is abelian. In fact, for loops α, β, at $\star$, the product $[\alpha] * [\beta] = [\alpha \cdot \beta] \in \pi_1(Y, \star)$, where $\alpha \cdot \beta(t) = \mu(\alpha(t), \beta(t))$ for all $t \in I$. (*To avoid confusion, denote the product in* $\pi_1(Y, \star)$ *by* $*$, *and the product induced by* μ *by* $\cdot$.)

Proof. For simplicity, assume that the constant map $e_{\mathbb{S}^1}$ is the identity, for μ (*not just a homotopy identity*). We prove $[\alpha] * [\beta] = [\beta] * [\alpha] \in \pi_1(Y, \star)$, for loops $\alpha, \beta : I \longrightarrow Y$. Define, $F : I \times I \longrightarrow Y$ by $F(s,t) = \alpha(t) \cdot \beta(t) := \mu(\alpha(t), \beta(t))$, which is given by the commutative diagram

$$\begin{array}{ccc} I \times I & \longrightarrow & k(Y \times Y) \\ & \searrow^{F} & \downarrow{\scriptstyle \mu} \\ & & Y \end{array}$$

Then $F(0,s) = e_{\mathbb{S}^1}\beta(t) = \beta(t)$, $F(t,1) = \alpha(t)e_{\mathbb{S}^1} = \alpha(t)$. So, $F_{|0 \times I \cup I \times 1} = \beta \cup \alpha$, where $\cup$ means concatenation. Likewise, $F(t,0) = \alpha(t)$ and $F(1,s) = \beta(t)$. So, $F_{|I \times 0 \cup I \times 1} = \alpha \cup \beta$. Now, F gives a homotopy from $F_{|I \times 0 \cup I \times 1}$ to $F_{|0 \times I \cup I \times 1}$ by drawing a homotopy along the lines of slope -1. So, $[\alpha] * [\beta] = [\beta] * [\alpha] \in \pi_1(Y, \star)$, and $\pi_1(Y, \star)$ is abelian. Further, note that $\alpha \cdot \beta = F_{|\Delta}$, where $\Delta \subseteq I \times I$ is the diagonal. We can draw a homotopy form $F_{|I \times 0 \cup I \times 1}$ to $F_{|\Delta}$ in the same way, and we have $[\alpha] * [\beta] = [\alpha \cdot \beta]$. This completes the proof when $e_{\mathbb{S}^1}$ is an actual identity for μ. A little further fine-tuning of this argument settles the proof general case. ■

Exercise 4.9.2. Let A be a ring. Prove that the telescopic construction (4.30) is, indeed, equal (or homotopic) to $\mathbb{B}GL(A)$. Complete the proof of Lemma 4.7.7.

Exercise 4.9.3. Let A be a ring. Refer to Theorem 4.7.17. Prove that the sequence (4.40) is a universal central extension of $EL(A)$.
Hint: See [Wc2, p. 289]. ■

Chapter 5

K-Theory of Rings

Among the goals of this chapter is to prove the homotopy invariance theorem (5.2.1) that for Noetherian rings A, the K-theory (or G-theory) of A and of the polynomial rings $A[T]$ coincide. While some readers may not have much interest in the noncommutative situations, the methods do not simplify significantly in the case of commutative rings. For this reason, we do not assume that rings are necessarily commutative. Polynomial rings being graded, the methods from graded rings provide a significant support, in this chapter.

5.1. K-Theory of Graded Rings

Before we proceed, we establish some notation associated to graded rings.

Notation 5.1.1. Let B be a positively graded ring (not necessarily commutative). Except when stated otherwise, by a module we mean a left module. We use the following notation.

$$\begin{cases} B = \bigoplus_{n=0}^{\infty} B_n \quad \text{with } A := B_0, \quad B_n \\ \quad \text{denote the degree } n \text{ component} \\ B_+ := \bigoplus_{n=1}^{\infty} B_n, \quad \text{and we write} \quad B_n = 0 \ \forall\ n \leq -1 \\ Coh(A) \text{ denotes the category of finitely generated} \\ \quad \text{left } A\text{-modules.} \\ Coh_{gr}^{\pm}(B) \text{ denotes the category of finitely generated graded} \\ \quad \text{left } B\text{-modules.} \end{cases}$$

$$\begin{cases} Coh^+_{gr}(B) \text{ denotes the category of finitely generated positively} \\ \quad \text{graded left } B\text{-modules.} \\ \text{For } N = \bigoplus_{n=0}^{\infty} N_n \in Coh^+_{gr}(B) \quad \text{denote} \quad \overline{N} = \frac{N}{B_+N}, \\ \quad \text{and, let } \mathcal{F}_{-1}N = 0, \\ \forall\, p \geq 0,\ \mathcal{F}_pN = \sum_{k=0}^{p} BN_k \in Coh^+_{gr}(B) \ \text{ the submodule} \\ \quad \text{generated by } N_0, N_1, \ldots, N_p \\ \text{For } N \in Coh^{\pm}_{gr}(B), k \in \mathbb{Z}, \text{let } N(k) \in Coh^{\pm}_{gr}(B) \text{ be defined by} \\ \quad N(k)_r = N(k+r). \end{cases}$$

Note that $A \cong \frac{B}{B_+}$ can be thought of as a graded ring, concentrated at degree zero. Accordingly, $\overline{N} = \oplus_{k=0}^{\infty}\overline{N}_k$ can be regarded as a graded A-module. Also note that $\overline{N} = Tor_0^B(A, N)$.

We use the notation $G_n(A) := K_n(Coh(A))$, even though A may not be commutative.

Further, a projective object P in $Coh^+_{gr}(B)$ is called a **graded projective module** over B. It follows that P is a graded projective B-module if and only if $P \oplus Q \cong \oplus_{k=0}^{n} B(-k)^{r_k}$ for some $Q \in Coh^+_{gr}(B)$. Since $\oplus_{k=0}^{n} B(-k)^{r_k}$ are free ungraded B-modules, it also follows that a graded projective B-module is also B-projective.

Some readers may be well advised to review the definition of Tor dimension or flat dimension of a module (5.4.1).

There is a natural $\mathbb{Z}[t]$-action on the K-groups of $Coh^+_{gr}(B)$, defined as follows.

Definition 5.1.2. Let $B = \bigoplus_{i=0}^{\infty} B_i$ and other notation be as in (5.1.1). The shift functor

$$\mathcal{S} : Coh^+_{gr}(B) \longrightarrow Coh^+_{gr}(B) \ \text{ is defined by } \ \mathcal{S}(N) := N(-1).$$

$$\text{So,} \quad \mathcal{S}(N)_k = N_{k-1}.$$

Since $\mathcal{S}$ is an exact functor, it induces

$$\begin{cases} \mathbb{BQ}\mathcal{S} : \mathbb{BQ}Coh^+_{gr}(B) \longrightarrow \mathbb{BQ}Coh^+_{gr}(B) & \text{a continuous function,} \\ \mathbf{K}\mathcal{S} : \mathbf{K}(Coh^+_{gr}(B)) \longrightarrow \mathbf{K}(Coh^+_{gr}(B)) & \text{a continuous function,} \\ t : K_n(Coh^+_{gr}(B)) \longrightarrow K_n(Coh^+_{gr}(B)) & \text{are group homomorphisms,} \\ & \forall\, n \geq 0, \text{ with } t := \pi_n(\mathbf{K}\mathcal{S}) \end{cases}$$

So, $\forall n \geq 0$, the groups $K_n(Coh^+_{gr}(B))$ are $\mathbb{Z}[t]$-modules. The multiplication t^k is induced by the exact functor $N \mapsto N(-k)$.

The following lemma on polynomial rings lists some of the obvious observations.

Lemma 5.1.3. *Suppose A is a Noetherian ring and $B = A[T]$ is the polynomial ring. Use the notation as in (5.1.1). For $N \in Coh^+_{gr}(B)$, we have the following:*

$$\begin{cases} \overline{N} = N_0 \oplus \frac{N_1}{TN_0} \oplus \cdots \oplus \frac{N_k}{TN_{k-1}} \oplus \cdots \\ \mathcal{F}_pN = N_0 \oplus N_1 \oplus \cdots \oplus N_{p-1} \oplus \bigoplus_{r=0}^{\infty} T^r N_p \ . \\ \frac{\mathcal{F}_pN}{\mathcal{F}_{p-1}N} = \bigoplus_{r=0}^{\infty} \frac{T^r N_p}{T^{r+1}N_{p-1}}. \end{cases}$$

There is a natural surjective map

$$\varphi_p : B(-p) \otimes \frac{N_p}{TN_{p-1}} \longrightarrow \frac{\mathcal{F}_pN}{\mathcal{F}_{p-1}N} \quad \textit{defined by}$$

$$\varphi_p(T^r \otimes [n_p]) = [T^r n_p] \ \forall \ n_p \in N_p.$$

Further, if T is a nonzero divisor on N, then the maps φ_p are isomorphisms, $\forall p \geq 0$.

Proof. Obvious! ∎

We would like to extend this lemma (5.1.3), at least to the case of polynomial rings $B = A[T_1, \ldots, T_n]$, in several variables. However, proofs do not appear simpler than that in the case of further generalities, as follows.

Lemma 5.1.4. *Suppose B is a graded ring, with $A := B_0$ and $N \in Coh^+_{gr}(B)$. Use all the notation, as in (5.1.1), and $N_k = 0 \ \forall k \leq -1$. Then*

$$\begin{cases} \overline{N}_n = \frac{N_n}{B_1N_{n-1} + \cdots + B_nN_0} \\ (\mathcal{F}_pN)_n = \sum_{k=0}^{p} B_{n-k}N_k \subseteq N_n \end{cases},$$

$$\left(\overline{\mathcal{F}_pN}\right)_n = \begin{cases} \overline{N}_n & \textit{if } n \leq p \\ 0 & \textit{if } n \geq p+1 \end{cases}$$

$$\frac{\mathcal{F}_pN}{\mathcal{F}_{p-1}N} = \bigoplus_{n=0}^{\infty} \frac{B_{n-p}N_p}{(\mathcal{F}_{p-1}N)_n \cap B_{n-p}N_p}$$

$$= \overline{N}_p \oplus \left(\bigoplus_{n=p+1}^{\infty} \frac{B_{n-p}N_p}{(\mathcal{F}_{p-1}N)_n \cap B_{n-p}N_p} \right). \tag{5.1}$$

With this, in the category $Coh^+_{gr}(B)$, there is a surjective map

$$\varphi_p : B(-p) \otimes_A \overline{N}_p \twoheadrightarrow \frac{\mathcal{F}_pN}{\mathcal{F}_{p-1}N}. \tag{5.2}$$

Suppose $Tor^B_1(A, N) = 0$ and $Tor^A_i\left(B, \overline{N}\right) = 0\ \forall i \geq 1$. Then the map (5.2) *is an isomorphism.* (In particular, if B is flat over A, then the latter condition is always valid.)

Proof. We only prove that the map (5.2) is an isomorphism, while the others are routine checking. Since $A \otimes_B B = A$, by routine algebra argument, it follows from the condition $Tor^A_i\left(B, \overline{N}\right) = 0\ \forall i \geq 1$ that

$$Tor^B_i(A, B \otimes_A \overline{N}) = 0 \quad \forall\, i \geq 1. \tag{5.3}$$

We denote $K_p = \ker(\varphi_p)$, and prove that $K_p = 0$. For the rest of the proof, we write $T_i(M) := Tor^B_i(A, M)$ for any B-module M. In particular, $T_0(N) = \overline{N}$. Consider the exact sequence

$$T_1\left(B(-p) \otimes_A \overline{N}_p\right) \longrightarrow T_1\left(\tfrac{\mathcal{F}_pN}{\mathcal{F}_{p-1}N}\right) \xrightarrow{\partial} T_0(K_p) \longrightarrow T_0\left(B(-p) \otimes_A \overline{N}_p\right) \xrightarrow{\overline{\varphi}_p} T_0\left(\tfrac{\mathcal{F}_pN}{\mathcal{F}_{p-1}N}\right). \tag{5.4}$$

By (5.1, 5.2), $\overline{\varphi}_p$ are isomorphisms, and by (5.3), $T_1\left(B(-p) \otimes_A T_0(N)_p\right) = 0$.

$$\text{Therefore,} \quad \partial : T_1\left(\frac{\mathcal{F}_pN}{\mathcal{F}_{p-1}N}\right) \xrightarrow{\sim} T_0(K_p) \quad \text{are isomorphisms.} \tag{5.5}$$

For $0 \leq n \leq p$, we have $\left(\mathcal{F}_p(N)\right)_n = N_n$. So, $T_1\left(\mathcal{F}_p(N)\right)_n = T_1(N)_n = 0$, by hypothesis. In fact, $(\varphi_p)_n$ are isomorphisms for $0 \leq n \leq p$. This is because $(\varphi_p)_p$ is clearly an isomorphism. And, for $n = 0, 1, \ldots, p-1$, both sides of (5.2) are zero.

Fix $\mathfrak{q}$. We will prove that $\forall p$, $\forall\, n \leq \mathfrak{q}$, $(\varphi_p)_n$ is an isomorphism and $T_1(\mathcal{F}_p(N))_n = 0$. This is valid if $\mathfrak{q} \leq p$. To prove this by downward induction on p, consider the exact sequence

$$T_1\left(\mathcal{F}_{p-1}N\right)_n \longrightarrow T_1\left(\mathcal{F}_pN\right)_n \longrightarrow T_1\left(\frac{\mathcal{F}_pN}{\mathcal{F}_{p-1}N}\right)_n \xrightarrow{d} T_0\left(\mathcal{F}_{p-1}N\right)_n \xrightarrow{\iota} T_0\left(\mathcal{F}_pN\right)_n. \tag{5.6}$$

Assume $T_1\left(\mathcal{F}_pN\right)_n = 0\ \forall\ n \leq \mathfrak{q}$. Then the map d is injective. Further, for $n \geq p$, $T_0\left(\mathcal{F}_{p-1}N\right)_n = 0$. For $n \leq p-1$, $T_0\left(\mathcal{F}_{p-1}N\right)_n = T_0\left(\mathcal{F}_pN\right)_n = T_0(N)_n$, and ι is an isomorphism. Therefore (5.5), $T_0(K_p)_n \cong T_1\left(\frac{\mathcal{F}_pN}{\mathcal{F}_{p-1}N}\right)_n = 0\ \forall\ n \leq \mathfrak{q}$.

With $T_0(K_p)_n = \frac{(K_p)_n}{\sum_{j=1}^{n} B_j(K_p)_{n-j}}$, inductively, $(K_p)_0 = (K_p)_1 = \cdots = (K_p)_{\mathfrak{q}} = 0$. So, $(\varphi_p)_n$ is isomorphism $\forall n \leq \mathfrak{q}$. Now, since $(K_p)_n = 0\ \forall\ n \leq \mathfrak{q}$, from the hidden part of the sequence (5.4), we obtain

$$T_2\left(\frac{\mathcal{F}_pN}{\mathcal{F}_{p-1}N}\right)_n \cong T_2\left(B(-p)\otimes T_0(N)_p\right)_n = 0 \quad n \leq \mathfrak{q}.$$

Now, it follows from the sequence (5.6) $T_1\left(\mathcal{F}_{p-1}N\right)_n = 0\ \forall\ n \leq \mathfrak{q}$. So, we established that, for any arbitrarily fixed integer $\mathfrak{q}$, $\forall n \leq \mathfrak{q}$ $(\varphi_p)_n$ are isomorphisms and $T_1(\mathcal{F}_p(N)_n = 0$. Since $\mathfrak{q}$ is arbitrary, the proof is complete. ∎

The following theorem applies to polynomial rings $B = A[T_1, \ldots, T_n]$ over Noetherian rings A.

Theorem 5.1.5. *Let $B = \oplus_{k=0}^{\infty} B_k$ be a graded (left) Noetherian ring, with $B_0 = A$. Assume B is flat over A as a right A-module, and A has finite Tor dimension over B, as a right B-module. Then there are natural isomorphisms, as follows:*

$$\mathbb{Z}[t] \otimes G_n(A) \xrightarrow{\sim} K_n(Coh^+_{gr}(B)) \qquad \forall\ n \geq 0,$$

where t is an indeterminate.

Proof. Since B is flat over A, the functor $\iota : Coh(A) \longrightarrow Coh^+_{gr}(B)$, sending $M \mapsto B \otimes M$, is exact. So, ι induces a map

$$\Phi_0^B : G_n(A) \longrightarrow K_n(Coh^+_{gr}(B)) \quad \forall\ n \geq 0.$$

As defined in (5.1.2), $K_n(Coh^+_{gr}(B))$ has $\mathbb{Z}[t]$-module structure. Therefore, Φ^B_0 extends to a map

$$\Phi^B : \mathbb{Z}[t] \otimes G_n(A) \xrightarrow{\sim} K_n(Coh^+_{gr}(B)) \quad \forall\, n \geq 0, \quad \text{defined by}$$
$$\Phi^B(t^k \otimes x) = t^k \Phi^B_0(x).$$

Let $\mathscr{E}$ denote the full subcategory of objects $N \in Coh^+_{gr}(B)$, as follows:

$$\mathscr{E} = \left\{ N \in Coh_{gr}(B) : Tor^B_i(A, N) = 0 \ \forall\, i \geq 1 \right\}.$$

Since A has finite Tor dimension over B, by (5.4.1 or 3.5.13), objects $N \in Coh^+_{gr}(B)$ have finite resolutions by objects in $\mathscr{E}$. So, $K_n(\mathscr{E}) \cong K_n(Coh^+_{gr}(B))$, $\forall n$, by resolution theorem (3.5.3).

As in (5.3) (*in this case because B is flat*), $\iota(M) = B \otimes M \in \mathscr{E}$ for all $M \in Obj(Coh(A))$. So, we consider $\iota : Coh(A) \longrightarrow \mathscr{E}$ as an exact functor. Further, the functor

$$\tau : \mathscr{E} \longrightarrow Coh^+_{gr}(A) \quad \text{sending} \quad N \longrightarrow \overline{N} \quad \text{is exact because}$$
$$Tor^B_1(A, N) = 0. \tag{5.7}$$

For integers $p \geq 0$, let $E_p \subseteq \mathscr{E}$ denote the full subcategory of objects N such that $\mathcal{F}_p(N) = N$. For $k \geq 0$, define functors

$$\begin{cases} \beta_k : Coh(A)^{k+1} \longrightarrow E_k & \text{by } \beta(M_0, M_1, \ldots, M_k) = \oplus_{j=0}^k B(-j) \otimes_A M_j \in \mathscr{E} \\ \tau_k : E_k \longrightarrow Coh(A)^{k+1} & \text{by } \tau(N) = \left(\overline{N}_0, \overline{N}_1, \cdots, \overline{N}_k\right). \end{cases}$$

Here β_k is exact because $B(-j)$ is flat over A, and τ_k is exact, by (5.7). It is clear $\tau_k\beta_k = 1$. So, map $\mathbf{K}(Coh(A))^{k+1}) \longrightarrow \mathbf{K}(E_k)$ of the K-theory spaces, induced by β_k, has a homotopy left inverse. Hence, the maps

$$K_n\left(Coh(A)^{k+1}\right) \longrightarrow K_n\left(E_k\right) \quad \text{have left inverse } \forall\, n \geq 0, k \geq 0. \tag{5.8}$$

We have

$$\beta_k\tau_k(N) = \bigoplus_{p=0}^{k} B(-p) \oplus \overline{N}_p$$

For $N \in E_k$, there is a filtration $0 \subseteq \mathcal{F}_0 N \subseteq \mathcal{F}_1 N \subseteq \cdots \subseteq \mathcal{F}_k N = N$. This provides a filtration of the identity functor $1_{E_k} = \mathcal{F}_k$.

The maps $\varphi_p : B(-p) \otimes \overline{N}_p \xrightarrow{\sim} \frac{\mathcal{F}_pN}{\mathcal{F}_{p-1}N}$ are isomorphisms, by (5.1.4), for $p = 0, \ldots, k$. Hence, $\frac{\mathcal{F}_pN}{\mathcal{F}_{p-1}N}$ are exact functors on E_k, for $p = 0, \ldots, k$. By an application of (3.4.6), to the filtration $\mathcal{F}_0 \hookrightarrow \mathcal{F}_1 \hookrightarrow \cdots \mathcal{F}_{k-1} \hookrightarrow \mathcal{F}_k = 1_{E_k}$, we have

$$1 = \sum_{p=0}^{k} \left(\frac{\mathcal{F}_p}{\mathcal{F}_{p-1}}\right)_* = (\beta_k \tau_k)_* : K_n(E_k) \longrightarrow K_n(E_k) \quad \forall\, n \geq 0.$$

This completes the proof that the map (5.8) is an isomorphism, $\forall\, k \geq 0$ and $\forall n \geq 0$. More precisely, β_k induces isomorphisms

$$\left(\bigoplus_{p=0}^{k} \mathbb{Z}t^p\right) \otimes G_n(A) \xrightarrow{\sim} K_n(E_k) \quad \forall\, n \geq 0,\ \forall\, k \geq 0.$$

Since $\mathscr{E} = \lim_k E_k$, passing to the limit (3.3.3), we have

$$\mathbb{Z}[t] \otimes G_n(A) \xrightarrow{\sim} K_n(\mathscr{E}) \quad \text{are isomorphisms, } \forall\, n \geq 0.$$

The proof is complete. ■

The following result on polynomial rings is a consequence of (5.1.5).

Corollary 5.1.6. *Let $B = A[T_1, \ldots, T_n]$ be a polynomial ring over a Noetherian ring A. Then there are natural isomorphisms, as follows:*

$$\mathbb{Z}[t] \otimes G_n(A) \xrightarrow{\sim} K_n(Coh^+_{gr}(B)) \qquad \forall\, n \geq 0,$$

where t is an indeterminate.

Proof. Follows from (5.1.5). ■

The flatness condition on B in the statement of Theorem 5.1.5 can be further relaxed, as follows.

Corollary 5.1.7. *Let $B = \bigoplus_{i=0}^{\infty} B_i$ be a graded (left) Noetherian ring, with $A = B_0$. Assume B has finite Tor dimension over A as a right A-module and A has finite Tor dimension over B as a right B-module. Then there are natural isomorphisms of K-groups as follows:*

$$G_n(A)[t] \xrightarrow{\sim} K_n(Coh^+_{gr}(B)) \qquad \forall\, n \geq 0. \tag{5.9}$$

Proof. First, we explain the natural maps (5.9). Consider the functor

$$\iota : Coh(A) \longrightarrow Coh^+_{gr}(B) \quad \text{defined by} \quad \iota(M) = B \otimes_A M.$$

This is not necessarily an exact functor. Consider the full subcategory

$$\mathscr{P}(\iota) = \{M \in Obj(Coh(A)) : Tor_i^A(B, M) = 0, \ \forall\, i \geq 1\} \hookrightarrow Coh(A). \tag{5.10}$$

By (5.4.1 or 3.5.13), any object $M \in Coh(A)$ has a finite resolution by objects in $\mathscr{P}(\iota)$. Therefore, by resolution theorem (3.5.3), $K_n(\mathscr{P}(\iota)) \xrightarrow{\sim} G_n(A)$ are isomorphic, $\forall\, n \geq 0$. It is also clear that the restriction $\iota_0 : \mathscr{P}(\iota) \longrightarrow Coh^+_{gr}(B)$ of ι is an exact functor. So, ι_0 induces maps $K_n(\mathscr{P}(\iota)) \longrightarrow K_n(Coh^+_{gr}(B))$, for all $n \geq 0$. Thus, we obtain the natural maps (5.9), by composition:

$$\begin{array}{ccc} G_n(A) & \xrightarrow{\Psi_0} & \\ \wr\uparrow & & \searrow \\ K_n(\mathscr{P}(\iota)) & \xrightarrow[\Phi_0]{} & K_n\left(Coh^+_{gr}(B)\right) \end{array} \qquad \text{which extends to}$$

$$\begin{array}{ccc} G_n(A)[t] & \xrightarrow{\Psi} & \\ \wr\uparrow & & \searrow \\ K_n(\mathscr{P}(\iota))\,[t] & \xrightarrow[\Phi]{} & K_n\left(Coh^+_{gr}(B)\right). \end{array}$$

The proof of the corollary is completed by establishing that Φ is an isomorphism $\forall\, n \geq 0$.

Define two more full subcategories:

$$\begin{cases} \mathscr{E} = \left\{N \in Coh^+_{gr}(B) : Tor_i^B(A, N) = 0 \ \forall\, i \geq 1\right\} \hookrightarrow Coh^+_{gr}(B) \\ \mathscr{F} = \left\{N \in \mathscr{E} : \overline{N} := Tor_0^B(A, N) \in \mathscr{P}(\iota)\right\} \hookrightarrow \mathscr{E}. \end{cases}$$

By (5.4.1 or 3.5.13), it follows that objects $N \in Coh^+_{gr}(B)$ have finite resolution by objects in $\mathscr{E}$. So, $K_n(\mathscr{E}) \cong K_n\left(Coh^+_{gr}(B)\right)\ \forall n \geq 0$. Likewise, objects $N \in \mathscr{E}$ have finite resolution by objects in $\mathscr{F}$. So, $K_n(\mathscr{F}) \cong K_n\left(\mathscr{E}\right)\ \forall n \geq 0$.

One checks for $M \in Obj(\mathscr{P}(\iota))$, $\iota_0(M) \in \mathscr{F}$. To complete the proof, we prove that the maps

$$K_n\left(\mathscr{P}(\iota)\right)[t] \xrightarrow{\sim} K_n\left(\mathscr{F}\right) \quad \text{are isomorphisms } \forall\, n \geq 0.$$

This is done by repeating the proof of (5.1.5), which is left for the readers to complete. The proof is complete. ■

Theorem 5.1.7 is formulated in terms of K-theory spaces, as follows.

Corollary 5.1.8. *Let B and A be as in* (5.1.7) *and $\mathscr{P}(\iota)$ as above. Then there are homotopy equivalences of K-theory spaces, as follows:*

$$\coprod_{n=0}^{\infty} \mathbf{G}(A)t^n \xleftarrow{\sim} \coprod_{n=0}^{\infty} \mathbf{K}(\mathscr{P}(\iota))t^n \xrightarrow[\sim]{\beta} \mathbf{K}(Coh_{gr}^{+}(B))$$

where t^n are to be treated as symbols, and $\coprod$ denotes disjoint (disconnected) union. The first term represents the K-theory space of the exact category $\coprod Coh(A)t^n = \lim_n Coh(A)^n$ (and likewise the second term). *The map β is induced by the exact functor*

$$\coprod_n \mathscr{P}(\iota)t^n \longrightarrow Coh_{gr}^{+}(B) \quad \text{sending} \quad \oplus_{j=0}^{\infty} M_j t^j \mapsto \oplus_{j=0}^{\infty} B(-j) \otimes_A M_j.$$

In these direct sums, $M_j = 0$ for all but finitely many j.

Proof. In the proof of (5.1.7), it was proven that the map β is a weak equivalence. Therefore, by Whitehead's theorem (C.3.8), β is a homotopy equivalence. ■

5.2. Homotopy Invariance

The following is the statement of the homotopy invariance theorem.

Theorem 5.2.1. *Let A be a Noetherian ring and $B = A[T]$ be the polynomial ring. Then the natural maps $G_n(A) \xrightarrow{\sim} G_n(B)$ are isomorphisms $n \geq 0$. Consequently, the map of the K-theory spaces $\mathbf{G}(A) \longrightarrow \mathbf{G}(B)$ is a homotopy equivalence.*

Proof. We only prove that the maps $G_n(A) \xrightarrow{\sim} G_n(B)$ are isomorphisms $\forall n \geq 0$, and the latter statement follows from Whitehead's theorem (C.3.8). Let $\mathcal{R} = A[T, Z]$ be the polynomial ring in two variables. By (5.1.6), the following maps are isomorphisms:

$$\begin{cases} \mathbb{Z}[t] \otimes G_n(A) \xrightarrow{\sim} K_n(Coh^+_{gr}(B)) \\ \mathbb{Z}[t] \otimes G_n(A) \xrightarrow{\sim} K_n(Coh^+_{gr}(\mathcal{R})) \end{cases} \quad \forall\, n \geq 0. \tag{5.11}$$

The maps $t^k G_n(A) \longrightarrow K_k(Coh^+_{gr}(B))$, $t^k G_n(A) \longrightarrow K_n(Coh^+_{gr}(\mathcal{R}))$ are, respectively, induced by the functors

$$\begin{cases} \beta^B_k : Coh(A) \longrightarrow Coh^+_{gr}(B) \quad \text{sending } M \mapsto B(-k) \otimes_A M \\ \beta^{\mathcal{R}}_k : Coh(A) \longrightarrow Coh^+_{gr}(\mathcal{R}) \quad \text{sending } M \mapsto \mathcal{R}(-k) \otimes_A M. \end{cases}$$

Since $B \cong \frac{\mathcal{R}}{Z\mathcal{R}}$, we have an inclusion functor

$$\rho : Coh^+_{gr}(B) \hookrightarrow Coh^+_{gr}(\mathcal{R}).$$

We want to compute ρ_*, with respect to these identifications (5.11). For fixed integers k, consider three functors:

$$Coh(A) \longrightarrow Coh^+_{gr}\mathcal{R} \qquad \begin{cases} M \mapsto \beta^{\mathcal{R}}_k(M) = \mathcal{R}(-k) \otimes_A M \\ M \mapsto \beta^{\mathcal{R}}_k(M)(-1) = \mathcal{R}(-(k+1)) \otimes_A M \\ M \mapsto \rho\beta^B_k(M) = B(-k) \otimes_A M. \end{cases}$$

We have an exact sequence

$$0 \longrightarrow \beta^{\mathcal{R}}_k(M)(-1) \xrightarrow{Z} \beta^{\mathcal{R}}_k(M) \longrightarrow \rho\beta^B_k(M) \longrightarrow 0.$$

By the additivity theorem (3.4.4), it follows that

$$\forall n \geq 0, \quad t^k x \in t^k \otimes G_n(A) \quad \rho_*(t^k x) = t^k x - t^{k+1} x \in K_n(\mathcal{R}).$$

So, we have the commutative diagram

$$\begin{array}{ccc} \mathbb{Z}[t] \otimes G_n(A) & \xrightarrow{1-t} & \mathbb{Z}[t] \otimes G_n(A) \\ \wr\downarrow & & \downarrow\wr \\ K_n\left(Coh^+_{gr}(B)\right) & \xrightarrow[\rho_*]{} & K_n\left(Coh^+_{gr}(\mathcal{R})\right). \end{array} \tag{5.12}$$

Let

$$\mathscr{B} = \{N \in Coh^+_{gr}(\mathcal{R}) : Z^n N = 0 \text{ for some } n\} \subseteq Coh^+_{gr}(\mathcal{R})$$

be the full subcategory. In fact, $N \in \mathscr{B} \iff N_Z = 0$. Note $\mathcal{R}_Z = A[T, Z, Z^{-1}]$. It follows from (A.6.9) that $\mathscr{B}$ is a Serre subcategory and $\frac{Coh^+_{gr}(\mathcal{R})}{\mathscr{B}} \cong Coh^{\pm}_{gr}(\mathcal{R}_Z) = Coh^{\pm}_{gr}(A[T, Z, Z^{-1}])$. For graded modules $N = \oplus N_n \in Coh^{\pm}_{gr}(A[T, Z, Z^{-1}])$, it is clear $N = \oplus_n N_0 Z^n$. It follows from this that $Coh(A[T]) \longrightarrow Coh^{\pm}_{gr}(A[T, Z, Z^{-1}])$ is an equivalence of categories, while its inverse is given by the association $N \mapsto \frac{N}{(Z-1)N}$. Therefore, the functor

$$F_1 : Coh^+_{gr}(\mathcal{R}) \longrightarrow Coh(A[T]) \quad \text{defined by } F_1(N) = \frac{N}{(Z-1)N}$$

represents the quotient category $\frac{Coh^+_{gr}(\mathcal{R})}{\mathscr{B}}$ (*which is defined only up to equivalence of categories*). The inclusion ρ identifies $Coh^+_{gr}(B)$ with the full subcategory of objects $N \in Coh^+_{gr}(\mathcal{R})$, such that $ZN = 0$. It follows from dévissage (3.6.2) that the $K_n\left(Coh^+_{gr}(B)\right) \cong K_n(\mathscr{B})\ \forall n$. Consider the following diagram (with $B = A[T]$):

$$\begin{array}{ccccccc}
 & & \mathbb{Z}[t] \otimes G_n(A) & \xrightarrow{1-t} & \mathbb{Z}[t] \otimes G_n(A) & \overset{t=1}{\dashrightarrow} & G_n(A) & \\
 & & \wr\downarrow & & \downarrow\wr & & \downarrow & \\
\cdots \longrightarrow & & K_n\left(Coh^+_{gr}(B)\right) & \xrightarrow[\rho_*]{} & K_n\left(Coh^+_{gr}\mathcal{R}\right) & \xrightarrow[(F_1)_*]{} & G_n(A[T]) & \longrightarrow \cdots
\end{array}$$

In this diagram the second line is the localization exact sequence (3.6.9) for F_1 and the left side rectangle is as in (5.12). Since map $1-t$ is injective, so is ρ_*, for all n. Hence, $(F_1)_*$ is surjective. Again, $co\ker(1-t) = G_n(A)$. Hence, $G_n(A) \longrightarrow G_n(A[T])$ is an isomorphism. The proof is complete. ■

The following are two important corollaries to the homotopy theorem (5.2.1).

Corollary 5.2.2. *Let A be a Noetherian ring, $A[T]$ be the polynomial ring, and $A[T, T^{-1}]$ be the Laurent polynomial ring. Then*

$$G_n(A[T, T^{-1}]) \cong G_n(A) \oplus G_{n-1}(A) \quad \forall n \geq 0,$$

$$\textit{with} \quad G_{-1}(A) = 0.$$

Proof. Using the homomorphism $A[T] \longrightarrow A$ sending, $f(T) \mapsto f(0)$, we identify $Coh(A) \subseteq Coh(A[T])$ with the full subcategory of $N \in Coh(A[T])$ such that $TN = 0$. Let $\mathscr{B} \subseteq Coh(A[T])$ be the full subcategory of objects $N \in Coh(A[T])$ such that $N_T = 0$ (equivalently, $T^k N = 0$ for some $k \geq 0$). We have $\frac{Coh(A[T])}{\mathscr{B}} \cong Coh(A[T, T^{-1}])$ (A.6.9).

By dévissage (3.6.2), $G_n(Coh(A)) \cong K_n(\mathscr{B})$. Combining this with the localization sequence (3.6.9), we have an exact sequence

$$\cdots \longrightarrow G_n(A) \longrightarrow G_n(A[T]) \xrightarrow{\mathfrak{q}_n} G_n(A[T, T^{-1}]) \longrightarrow \cdots$$

By homotopy invariance (5.2.1), we identify $G_n(A) \xrightarrow{\sim} G_n(A[T])$ $\forall\, n \geq 0$. The proof is completed by proving that the map $\mathfrak{q}_n$ splits (has a left inverse) $\forall n$. Let $\varepsilon_1 : A[T, T^{-1}] \longrightarrow A$ be the evaluation map, sending $f(T) \mapsto f(1)$, and let $\varepsilon_1 : Coh(A[T, T^{-1}]) \longrightarrow Coh(A)$ also denote the induced functor. The ring homomorphism ε_1 has finite Tor dimension one. By (3.5.13), this induces a well-defined map $(\varepsilon_1)_* : G_n(A[T, T^{-1}]) \longrightarrow G_n(A)$. The map $(\varepsilon_1)_*$ factors through the full subcategory $\mathscr{P} \subseteq Coh(A[T, T^{-1}])$ of objects $N \in Coh(A[T, T^{-1}])$ such that $T - 1$ is a nonzero divisor on N. Let $\iota : Coh(A) \longrightarrow Coh(A[T])$ be the inclusion. We have two commutative diagrams, where the second one is induced by the first one:

$$\begin{array}{ccccc} Coh(A) & & \xleftarrow{\quad 1 \quad} & & Coh(A) \\ \downarrow & & \searrow & & \uparrow{\scriptstyle \varepsilon_1} \\ Coh(A[T]) & \longrightarrow & Coh(A[T, T^{-1}]) & \hookleftarrow & \mathscr{P} \end{array} \qquad \begin{array}{ccc} G_n(A) & \xleftarrow{\quad 1 \quad} & G_n(A) \\ \wr\downarrow & & \uparrow{\scriptstyle (\varepsilon_1)_*} \\ G_n(A[T]) & \xrightarrow[\mathfrak{q}_n]{} & G_n(A[T, T^{-1}]) \end{array}$$

By homotopy invariance (5.2.1), the first vertical map in the second diagram is an isomorphism. So, $\mathfrak{q}_n$ has a left inverse. The proof is complete. ■

Before we introduce our final result in this section, we recall the following definition.

Definition 5.2.3. Suppose A is a Noetherian ring. Then A is said to be (left) **regular** if every finitely generated left A-module has a finite resolution by (left) projective A-modules [B, p. 122]. As before, $Coh(A)$ denote the category of finitely generated left A-modules. Also, let $\mathscr{P}(A)$ denote the category of finitely generated left projective A-modules. We denote $K_n(A) := K_n(\mathscr{P}(A))$ for all $n \geq 0$.

The following is called the **Fundamental theorem for regular rings**.

Corollary 5.2.4. *Let A be a (left) regular ring. Then*

$$K_n(A) \xrightarrow{\sim} K_n(A[T]), \qquad \textit{and}$$
$$K_n(A[T, T^{-1}]) \cong K_n(A) \oplus K_{n-1}(A) \qquad \forall\, n \geq 0.$$

Proof. Since A is regular, it is fairly standard that $A[T]$ and $A[T, T^{-1}]$ are regular, in case A is commutative and $\dim A < \infty$ [Mh, p. 126]. In the general case also, $A[T]$ and $A[T, T^{-1}]$ are regular, which we prove subsequently (5.3.6). Now the corollary is an immediate consequence of the resolution theorem (3.5.3), and of (5.2.1), (5.2.2). The proof is complete. ∎

5.3. Filtered Rings

The main reason to include this section is to prove that, if A is a regular ring, then so is the polynomial ring $A[T]$. In case A is commutative with $\dim A < \infty$, this is a fairly standard result in commutative algebra [Mh, p. 126]. For the readers interested in further generalities, we prove this result (5.3.6). In this regard, before we get into the discussions on filtered rings, we prove the following on graded rings.

Lemma 5.3.1. *Let B be a Noetherian graded ring, with $A := B_0$ (left) regular. Assume A has finite Tor dimension as a right B-module. Then any $N \in Coh^+_{gr}(B)$ has a finite resolution, by finitely generated projective graded (left) B-modules.*

Proof. There is surjective map $P_0 \twoheadrightarrow N$, where $P_0 = \oplus_{i=1}^r B(-r_i)$ is a graded projective (*free*) B-module. Recursively, we obtain an exact sequence (possibly of infinite length)

$$\cdots \longrightarrow P_n \xrightarrow{\partial_n} \cdots \longrightarrow P_1 \longrightarrow P_0 \longrightarrow N \longrightarrow 0,$$

where $P_n \in Coh^+_{gr}(B)$ are projective (*free*) graded B-modules, and maps are homogeneous. Write $N_n = \ker(\partial_{n-1}) = Im(\partial_n)$.

We have $Tor_k^B(A, N_r) \cong Tor_{k+1}^B(A, N_{r-1}) \cong Tor_{k+r}^B(A, N_0)$. Since $d = Tor\dim_B A$, we have $\forall\ r \geq d, k \geq 1$, $Tor_k^B(A, N_r) = 0$. So, for $r \geq d+1$, the sequences

$$0 \longrightarrow \overline{(N_r)} \longrightarrow \overline{(P_r)} \longrightarrow \overline{(N_{r-1})} \longrightarrow 0 \quad \text{are exact,}$$

with our notation, $\overline{(N_r)} := Tor_0(A, N_r) = \frac{N_r}{B_+N_r}$, and so on. Since A is regular, $\overline{(N_d)}$ has a finite projective dimension $\leq s$. In fact, we have an exact sequence

$$0 \longrightarrow \overline{(N_{d+s})} \longrightarrow \overline{(P_{d+s-1})} \longrightarrow \cdots \longrightarrow \overline{(P_{d+1})} \longrightarrow \overline{(P_d)} \longrightarrow \overline{(N_d)} \longrightarrow 0.$$

So, $\overline{(N_{d+s})}$ is projective [HS, p. 142]. With $M = N_{d+s}$, we have the exact sequences

$$0 \longrightarrow \mathcal{F}_{p-1}M \longrightarrow \mathcal{F}_pM \longrightarrow \frac{\mathcal{F}_pM}{\mathcal{F}_{p-1}M} \longrightarrow 0.$$

By (5.1.4), along with induction, $\mathcal{F}_pM$ is projective for all p. Since $\mathcal{F}_pM = M\ \forall p \gg 0$, we have $M = N_{d+s}$ is projective. The proof is complete. ■

We define increasing filtration of rings and associated other definitions and notation.

Definition 5.3.2. Let R be a ring and $0 = F_{-1}R \subseteq F_0R \subseteq F_1R \subseteq \cdots R$ be an increasing sequence of subgroups of R,

$$\ni \quad \begin{cases} 1 \in F_0R, \quad R = \bigcup_n F_nR, \\ F_nR \cdot F_mR \subseteq F_{m+n}R \quad \forall m, n \end{cases} \qquad \text{Denote } A := F_0R.$$

In this case, we say that $\{F_nR\}$ is an increasing filtration of R. A ring R equipped with such a filtration is called a **filtered ring**. Given such a filtered ring R, the associated graded ring is defined as

$$B := gr(R) = \bigoplus_{n=0}^{\infty} \frac{F_nR}{F_{n-1}R} \qquad \text{Denote}$$

$$B_n := \frac{F_nR}{F_{n-1}R}, \quad B_0 := F_0R = A.$$

Given a filtered ring R, an R-module M is said to have a filtration if there is an increasing sequence of subgroups

$$0 = F_{-1}M \subseteq F_0M \subseteq F_1M \subseteq \cdots \subseteq M$$

$$\ni \quad \begin{cases} M = \bigcup_n F_nM, \\ F_nR \cdot F_mM \subseteq F_{m+n}M \ \forall m, n. \end{cases}$$

In particular, F_nM are A-modules. Given such a filtered R-module, define the associated graded module as follows:

$$gr(M) = \bigoplus_{n=0}^{\infty} \frac{F_nM}{F_{n-1}M}, \quad \text{which is a graded } B\text{-Module.}$$

$$\text{Denote} \quad gr(M)_n := \frac{F_nM}{F_{n-1}M}.$$

We caution that this notion of filtering rings, by increasing subgroups, does not coincide with the usual and well-studied notion of filtering rings by decreasing ideals [Mh, p. 63]. Further, note that F_nR are not required to be ideals. We only talk about increasing filtration in this book. The following would be a typical example, that we may have in our mind.

Example 5.3.3. Let A be a commutative ring and $R = A[T_1, \ldots, T_n]$ be the polynomial ring. Let F_nR be the set of all polynomials $f \in R$ with $\deg f \leq n$. Then $\{F_nR\}$ is a filtration of R. In this case, $gr(R) = R$.

Example 5.3.4. Suppose $(R, \{F_nR\})$ is a filtered ring. Then any free R-module $\mathscr{F} = \oplus_{i\in I} Re_i$ has a filtration, given by $F_r\mathscr{F} = \oplus_{i\in I}(F_rR)e_i$. Further, any submodule $M \subseteq \mathscr{F}$ has a filtration, given by $F_rM = M \cap F_r\mathscr{F}$. In particular, any ideal $I \subseteq R$ has a natural filtration.

The following lemma responds to the immediate questions on filtered rings and modules.

Lemma 5.3.5. *Suppose R is a filtered ring and M is a filtered R-module.* (Unless stated otherwise, by module, we mean left module.) *Use all notation as in* (5.3.2). *Then*

1. *If M is finitely generated, then so is $gr(M)$.*
2. *If $gr(M)$ is a finitely generated B-module, then M is a finitely generated R-module. In particular, if B is Noetherian, then R is (left) Noetherian.*
3. *Assume $gr(M) \in Coh^+_{gr}(B)$. If $gr(M)$ is a projective B-module, then M is a projective R-module.*
4. *Assume $gr(R)$ is Noetherian, and $gr(M) \in Coh^+_{gr}(B)$. If $gr(M)$ has a resolution by finitely generated graded B-modules, of length $\leq n$, then M has a resolution of length $\leq n$, by finitely generated projective R-modules.*

Proof. It is worth observing that, for $x \in F_nM - F_{n-1}M$, the image of x in gr_kM is zero $\forall k \geq n+1$, and x has no representation in gr_kM for $k \leq n-1$. So, x has only a unique nonzero representation $[x] \in gr_nM$. This settles (1). Now assume $gr(M) = B[x_1] + \cdots + B[x_m]$, with $x_i \in F_{d_i}M - F_{d_i-1}M$. Let $x \in F_nM - F_{n-1}M$. If $n = 0$, then it is clear $x \in \sum_{i=1}^m F_0Rx_i$. For $n \geq 1$, we have $[x] = \sum_{i=1}^m [r_i][x_i]$, with $r_i \in F_{n-d_i}R$ (interpret $F_kR = 0 \ \forall \ k \leq -1$). It follows that $x - \sum r_ix_i \in F_{n-1}M$. By induction, it follows that $x \in \sum Bx_i$. This completes the proof of the first part of (2). Note, any (left) ideal $I \subseteq R$ has a filtration (5.3.4). Now the latter part of (2) follows, by the usual characterization of Noetherian rings.

We proceed to construct certain maps, before we prove the rest. Consider $\overline{gr(M)} = Tor_0^B(A, M) = \frac{gr(M)}{B_+gr(M)}$. So,

$$\begin{aligned}\left(\overline{gr(M)}\right)_n &= \frac{gr_n(M)}{B_+gr_n(M)} \\ &\cong \frac{F_nM}{F_{n-1}M + (F_1R \cdot F_{n-1}M + F_2R \cdot F_{n-2}M \cdots F_nR \cdot F_0M)}\end{aligned}$$

which are A-modules, and are direct summands of $\overline{gr(M)}$. If $gr(M)$ is finitely generated, $\left(\overline{gr(M)}\right)_n = 0$ for all but finitely many n. For integers $n \geq 0$, let L_n be A-modules (possibly projective or free) such that we have the following diagram of A-module homomorphisms:

$$\begin{array}{ccccc} L_n & \xrightarrow{\varphi_n} & F_nM & \longrightarrow & gr_n(M) \\ & \searrow^{\psi_n} & & & \downarrow \\ & & & & \left(\overline{gr(M)}\right)_n \end{array} \quad \text{and} \quad \begin{cases} \psi_n \quad \text{is surjective.} \\ L_n = 0 \ \text{ if } \ \left(\overline{gr(M)}\right)_n = 0. \end{cases}$$

Define

$$\begin{cases} F_nP := \bigoplus_j F_{n-j}R \otimes_A L_j = \\ \qquad\qquad F_nR \otimes_A L_0 \oplus F_{n-1}R \otimes_A L_1 \oplus \cdots \oplus F_0R \otimes_A L_n \\ F_{n-1}P := \bigoplus_j F_{n-1-j}R \otimes_A L_j = \\ \qquad\qquad F_{n-1}R \otimes_A L_0 \oplus F_{n-2}R \otimes_A L_1 \oplus \cdots \oplus F_0R \otimes_A L_{n-1} \\ P = \bigoplus_j R \otimes_A L_j = R \otimes_A L_0 \oplus R \otimes_A L_1 \oplus \cdots \end{cases}$$

If $gr(M)$ is finitely generated, then these are all finite direct sums. Assume L_j are projective. Then $F_{n-1}P \subseteq F_nP$. It follows, P together with $\{F_nP\}$ is a filtered R-module. Further, P is a projective R-module. We have

$$\frac{F_nP}{F_{n-1}P} = B_n \otimes_A L_0 \oplus B_{n-1} \otimes L_1 \oplus \cdots \oplus B_0 \otimes L_n.$$

$$\begin{aligned} \text{So,}\quad gr(P) &= B_0 \otimes L_0 \oplus (B_1 \otimes L_0 \oplus B_0 \otimes L_1) \\ &\oplus (B_2 \otimes L_0 \oplus B_1 \otimes L_1 \oplus B_0 \otimes L_2) \cdots \end{aligned}$$

Therefore, $\overline{gr(P)} = B_0 \otimes L_0 \oplus B_0 \otimes L_1 \oplus B_0 \otimes L_2 \cdots \cong L_0 \oplus L_1 \oplus L_2 \oplus \cdots$

Define

$$\begin{array}{ccc} R \otimes_A L_n & \xrightarrow{1 \otimes \varphi_n} & R \otimes F_nM \\ & \searrow^{\Phi_n} & \downarrow \\ & & F_nM. \end{array} \quad \text{and} \quad \Phi = \oplus_n \Phi_n : P \longrightarrow M.$$

Since ψ_n are surjective, it follows $\overline{gr(\Phi)} : \overline{gr(P)} \longrightarrow \overline{gr(M)}$ is surjective. So, $gr(\Phi)$ is surjective. If $gr(M)$ is finitely generated, by Nakayama's lemma (3.4.8), $gr(\Phi)_n$ is surjective $\forall$ n. Inductively, $F_n\Phi : F_nP \twoheadrightarrow F_nM$ are surjective. Hence, $\Phi : P \twoheadrightarrow M$ is surjective.

Let $K := \ker(\Phi)$, and write $F_nK = K \cap P_n$. We have the exact sequences:

$$\begin{cases} 0 \longrightarrow K \longrightarrow P \longrightarrow M \longrightarrow 0 \\ 0 \longrightarrow K_n \longrightarrow P_n \longrightarrow M_n \longrightarrow 0 \quad \forall n \geq 0 \\ 0 \longrightarrow gr(K) \longrightarrow gr(P) \longrightarrow gr(M) \longrightarrow 0. \end{cases} \tag{5.13}$$

To prove (3), take $L_n = \overline{gr(M)}_n$. Note $\overline{gr(M)}$ is a projective A-module, and hence so is L_n. With this choice, $\overline{gr(P)} \xrightarrow{\sim} \overline{gr(M)}$ is an isomorphism. Further, $gr(K) \oplus gr(M) \cong gr(P)$. Hence, $gr(K)$ and K are finitiely generated. Consider the exact sequence

$$Tor_1^B(A, gr(M))) \longrightarrow \overline{gr(K)} \longrightarrow \overline{gr(P)} \longrightarrow \overline{gr(M)}.$$

Since $Tor_1^B(A, gr(M))) = 0$ and the last arrow is an isomorphism, we have $\overline{gr(K)} = 0$. Since $gr(K)$ is finitely generated, by Nakayama's lemma (3.4.8), $gr(K) = 0$, hence $K = 0$. Therefore, Φ is an isomorphism. Since P is projective, so is M.

Now we prove (4). If $n = 0$, then it follows from (3) that M is projective. So, assume $n \geq 1$. Suppose $gr(M)$ has a resolution by finitely generated projective graded modules, of length $\leq n$. With choices of free modules L_n, consider the exact sequences (5.13). We have P is projective. Since $gr(R)$ is Noetherian, $gr(K)$ is finitely generated. It also follows from generalities [HS, p. 142] or from (3.5.4), $gr(K)$ has projective dimension $\leq n - 1$. By induction, K has projective dimension $\leq n - 1$. Hence, M projective dimension $\leq n$. The proof is complete. ■

Lemma 5.3.6. *Let R be a filtered ring and $B = gr(R)$ be the associated graded ring. Write $A = F_0R$. Assume* (1) *B is Noetherian,* (2) *A is regular,* (3) *A has finite Tor dimension as right B-module. Then R is regular.*

In particular, for regular rings A, the polynomial rings $R = A[T]$ are regular. Consequently, so is $A[T, T^{-1}]$.

Proof. From (5.3.5), it follows that R is Noetherian. Suppose $M \in Coh(R)$. There is a surjective map $\partial_0 : P_0 \twoheadrightarrow M$, where P_0 is a finitely generated projective R-module. Let $M_1 = \ker(\partial_0) \subseteq P_0$. Then P_0 is a filtered module, and M_1 inherits a filtered R-module structure from P_0. By (5.3.1), $gr(M_1)$ has finite resolution by projective graded B-modules (5.3.4). Hence, by (5.3.5), M_1 has a finite resolution by projective R-modules, and M has the same. So, R is regular.

Now suppose A is a regular ring. Since A is Noetherian, the polynomial ring $R = A[T]$ is also Noetherian. Now, R is a filtered ring (5.3.3), while its associated graded ring $B = R$. Now, it follows from above that $A[T]$ is regular. The proof is complete. ∎

While it appears technical, the following is an extension of the homotopy invariance theorem (5.2.1), proved by the same methods.

Theorem 5.3.7. *Let $(R, \{F_nR\})$ be a filtered ring, $A := F_0R$, and $B = gr(R)$ be the associated graded ring. Assume B is Noetherian, B has finite Tor dimension over A as a right A-module, and A has finite Tor dimension as a right B-module. Then the maps $G_n(A) \xrightarrow{\sim} G_n(R)$ are isomorphisms, $\forall\, n \geq 0$.*

Further, if A is regular, then so is R, and $K_n(A) \xrightarrow{\sim} K_n(R)$ is an isomorphism, $\forall\, n \geq 0$.

Proof. By (5.3.5), R is Noetherian. Let $d := Tor\dim_A B$. Being a direct summand of B, $Tor\dim_A \frac{F_nR}{F_{n-1}R} \leq d$. So, $Tor\dim_A F_nR \leq d$, for all n. It follows from this, one checks, $Tor\dim_A R \leq d$. So, the maps $G_n(A) \longrightarrow G_n(R)$ are defined, as in (3.5.13).

Consider the graded subring of the polynomial ring $R[Z]$:

$$\mathcal{R} = \bigoplus_{n=0}^{\infty} (F_nR)Z^n \subseteq R[Z], \quad \text{we have} \quad \mathcal{R}_0 = A. \tag{5.14}$$

1. For any A-module M, $Tor_k^A(\mathcal{R}, M) \cong \oplus_{n=0}^{\infty} Tor_k^A(F_nR, M)Z^n = 0$ for $k \geq d+1$. So, $Tor\dim_A \mathcal{R} \leq d$.
2. Then $\frac{\mathcal{R}}{Z\mathcal{R}} \cong \bigoplus_{n=0}^{\infty} \frac{(F_nR)Z^n}{(F_{n-1}R)Z^n} \cong B$. So, we have an exact sequence

$$0 \longrightarrow \mathcal{R} \xrightarrow{Z} \mathcal{R} \longrightarrow B \longrightarrow 0. \tag{5.15}$$

Since Z is a central nonzero divisor on $\mathcal{R}$, one knows $Tor\dim_{\mathcal{R}} A = Tor\dim_B A + 1$ (e.g., *imitate the proof in*

[W, p. 100]). So, A has finite Tor dimension over B, A has finite Tor dimension over $\mathcal{R}$.

3. Let $F_p\mathcal{R} \subseteq \mathcal{R}$ be the set of all $f \in \mathcal{R}$ with all coefficients in F_pR. Then

$$F_p\mathcal{R} = \Big\{ f = \sum a_k Z^k \in \mathcal{R} : \forall\, i = 0, 1, \ldots, p-1;\ a_i \in F_iR, \quad \text{and} \quad \forall\, i \geq p\ a_i \in F_p \Big\}.$$

Then $F_p\mathcal{R}$ is a filtration of $\mathcal{R}$. Further,

$$\frac{F_p\mathcal{R}}{F_{p-1}\mathcal{R}} = \frac{F_pR}{F_{p-1}R} Z^p \oplus \frac{F_pR}{F_{p-1}R} Z^{p+1} \oplus \cdots = \bigoplus_{k=p}^{\infty} gr_p(R) Z^k.$$

So,

$$gr(\mathcal{R}) \cong gr(R)[Z] \qquad \text{is Noetherian.}$$

Hence, $\mathcal{R}$ is Noetherian, by (5.3.5).

So, both the extensions $\iota : A \longrightarrow B$ and $\zeta : A \longrightarrow \mathcal{R}$ satisfy the hypotheses of (5.1.7). Use the same notation $\iota : Coh(A) \longrightarrow Coh^+_{gr}(B)$, $\zeta : Coh(A) \longrightarrow Coh^+_{gr}(\mathcal{R})$ for the induced functors. Let $\mathscr{P}(\iota) \subseteq Coh(A)$, $\mathscr{P}(\zeta) \subseteq Coh(A)$ be subcategories, as defined above (5.10). By (5.1.7), we have the following isomorphisms:

$$\forall\, n \geq 0 \quad \begin{cases} \mathbb{Z}[t] \otimes G_n(A) \xleftarrow{\sim} \mathbb{Z}[t] \otimes K_n(\mathscr{P}(\iota)) \xrightarrow{\sim} K_n(Coh^+_{gr}(B)) \\ \mathbb{Z}[t] \otimes G_n(A) \xleftarrow{\sim} \mathbb{Z}[t] \otimes K_n(\mathscr{P}(\zeta)) \xrightarrow{\sim} K_n(Coh^+_{gr}(\mathcal{R})). \end{cases} \tag{5.16}$$

We check, $\mathscr{P}(\iota) = \mathscr{P}(\zeta) =: \mathscr{P}$. In other words, $\forall\, M \in Coh(A)$,

Claim: $\quad Tor^A_k(B, M) = 0 \ \ \forall\, k \geq 1 \Longleftrightarrow Tor^A_k(\mathcal{R}, M) = 0 \ \ \forall\, k \geq 1.$

Corresponding to the sequence (5.15), we have the exact sequence

$$\cdots \longrightarrow Tor^A_1(\mathcal{R}, M) \xrightarrow{Z} Tor^A_1(\mathcal{R}, M) \longrightarrow Tor^A_1(B, M) \longrightarrow Tor^A_0(\mathcal{R}, M) \xrightarrow{Z} Tor^A_0(\mathcal{R}, M).$$

Assume $Tor^A_k(B, M) = 0 \forall\ k \geq 1$. Then $Tor^A_k(\mathcal{R}, F) = ZTor^A_k(\mathcal{R}, F)$, $\forall\ k \geq 1$. By graded Nakayama's lemma (3.4.8), $Tor^A_k(\mathcal{R}, F) = 0\ \forall\ k \geq 1$. Conversely, assume $Tor^A_k(\mathcal{R}, M) = 0\ \forall\ k \geq 1$. It follows immediately from the exact sequence that

$Tor_k^A(B, M) = 0 \forall\ k \geq 2$. Further, $Tor_1^A(B, M) = 0$ because the multiplication map $Z : \mathcal{R} \otimes_A M \longrightarrow \mathcal{R} \otimes_A M$ is an injective map. So, the claim is established. So, the isomorphisms (5.16), can be consolidated to the following two isomorphisms:

$$\forall\ n \geq 0 \quad \begin{cases} \mathbb{Z}[t] \otimes K_n(\mathscr{P}) \xrightarrow{\sim} K_n(Coh_{gr}^+(B)) \\ \mathbb{Z}[t] \otimes K_n(\mathscr{P}) \xrightarrow{\sim} K_n(Coh_{gr}^+(\mathcal{R})). \end{cases} \tag{5.17}$$

Note, $\mathcal{R}_z = R[Z, Z^{-1}]$. The rest of the proof is a repetition of the arguments of the proof of the Homotopy invariance theorem (5.2.1), by replacing $Coh(A)$ by $\mathscr{P}$, and $B = A[T]$ by R. We leave this for interested readers to complete. The proof is complete. ■

5.4. Exercises

Exercise 5.4.1. Let A be a ring, and N be a (right) A-module. Then **Tor dimension** $\dim_{Tor}^A(N)$ of N is defined to be

$$Tor \dim^A(N) = \sup\left\{k : Tor_k^A(N, M) \neq 0 \text{ for some left } A\text{-module } M\right\}.$$

So, $Tor \dim_A(N)$ need not be finite. Further, $Tor \dim_A(N) = 0$ if and only if N is flat. Tor dimension is also called **Flat dimension** [Wc1, p. 91]. Prove that $Tor \dim_A(N) \leq d$ if and only if there is a resolution

$$0 \longrightarrow F_d \longrightarrow \cdots \longrightarrow F_1 \longrightarrow F_0 \longrightarrow N \longrightarrow 0 \quad \text{with}$$

$$F_i \quad \text{flat (right) } A\text{-modules.}$$

(For left A-modules, we define Tor dimension likewise. In case N is a two-sided A-module, we may have to specify whether we are considering Tor dimension as a left or a right A-module.)

Now let $A \longrightarrow B$ be a homomorphism of Noetherian rings, such that $Tor \dim_A(B)$ **is finite, as a right A-module.** Let

$$\mathscr{P} = \left\{M \in Coh(A) : Tor_k^A(B, M) = 0 \forall\ k \geq 1\right\} \subseteq Coh(A)$$

be the full subcategory. Then

1. Prove that every $M \in Coh(A)$ has a finite resolution by objects in $\mathscr{P}$.

2. Now assume B is graded with $B_0 = A$. Let

$$\mathscr{F} = \{N \in Coh_{gr}(B) : \overline{N} \in \mathscr{P}, \ \ Tor_i^B(A, N) = 0 \ \forall \ i \geq 1\}$$

be the full subcategory of $Coh^+_{gr}(B)$. Let $E \in Coh^+_{gr}(B)$ be such that $Tor_i^B(A, E) = 0 \ \forall \ i \geq 1$. Then E has a finite resolution by objects in $\mathscr{F}$.

Exercise 5.4.2. Let $A \xrightarrow{f} B \xrightarrow{g} C$ be homomorphisms of Noetherian rings. Assume B has finite Tor dimension over A, as a right module, and C has finite Tor dimension over B, as a right module. First, prove that C has finite Tor dimension over A.

Consequently, all the maps f_*, g_*, $(gf)_*$ of G-groups are defined (3.5.13). Prove that the diagram

$$\begin{array}{ccc} G_n(A) & \xrightarrow{f_*} & G_n(B) \\ & \searrow^{(gf)_*} & \downarrow g_* \\ & & G_n(C) \end{array} \qquad \text{commutes } \forall \ n \geq 0.$$

Exercise 5.4.3. Let A be a commutative Noetherian ring. Give an elementary proof that $G_0(A) \xrightarrow{\sim} G_0(A[T])$ is an isomorphism. (**Hint:** *One can possibly imitate the proof of* (5.2.1).)

Exercise 5.4.4. Let $B = \bigoplus_{i=0}^{\infty} B_i$ be a graded Noetherian ring, with $A = B_0$. Assume B has finite Tor dimension over A as a right A-module and A has finite Tor dimension over B, as a right B-module. Complete the proof of (5.1.7) that $\mathbb{Z}[t] \otimes G_n(A) \cong K_n(Coh^+_{gr}(B))$ for all $n \geq 0$.

Exercise 5.4.5. Let $(R, \{F_nR\})$ be a filtered ring, with $A := F_0R$, and $B = gr(R)$ be the associated graded ring. Assume B is Noetherian, B has finite Tor dimension over A as a right A-module, and A has finite Tor dimension as a right B-module. Then complete the proof of (5.3.7) that $G_n(A) \cong G_n(R)$ are isomorphisms, $\forall \ n \geq 0$.

Exercise 5.4.6 ([Q]). Let A be a Noetherian ring and, $\varphi : A \xrightarrow{\sim} A$ be an automorphism. Let $A_\varphi[T]$, and $A_\varphi[T, T^{-1}]$, denote the twisted polynomial ring where $T \cdot a = \varphi(a)T$. (*So, $A_\varphi[T]$ can be a non-commutative ring, even if A is commutative.*) Then prove that

$G_n(A) \cong G_n(A_\varphi[T])$ for all n. Note $\varphi : Coh(A) \longrightarrow Coh(A)$ defines an exact functor. Hence, it defines a map $\varphi_* : G_n(A) \longrightarrow G_n(A)$. Also prove that there is a long exact sequence

$$\cdots \longrightarrow G_n(A) \xrightarrow{1-\varphi_*} G_n(A) \longrightarrow G_n(A_\varphi[T, T^{-1}] \longrightarrow G_{n-1}(A) \longrightarrow \cdots$$

(**Hint:** Apply (5.3.7, and use Localization (3.6.9).)

Exercise 5.4.7. Let A be a commutative regular ring, essentially finite type over a field k. It is known, in this case, every projective $A[T]$-module P is extended from A ([L]). Give an elementary proof that $K_0(A) \cong K_0(A[T])$ and $K_1(A) \cong K_1(A[T])$. (**Hint:** Use [L, V].) Extend the same, for any regular ring A, containing a field, using [Pd].

Problem 5.4.8. Let A be as in (5.4.7). Explore whether results in [L, V] can be used or improved to give an easier proof that $K_n(A) \cong K_n(A[T])$, for all n. What is needed?

Chapter 6

G-Theory of Schemes

In this chapter, we discuss the G-theory of schemes. For a scheme X, by G-theory, we mean the K-theory of the category $Coh(X)$. We made some comments on this previously, mainly on affine schemes (Chapter 5). Before we proceed, we establish some of the notation.

Notation 6.0.1. The structure sheaf of a scheme X will always be denoted by $\mathcal{O}_X$. All schemes X considered subsequently are noetherian, and $\dim X < \infty$. All our definitions on scheme theory are consistent with that in [Hr].

For a Noetherian scheme X, $Coh(X)$ denotes the category of coherent sheaves over X, and $QCoh(X)$ denotes the category of quasi-coherent sheaves over X. The full subcategory of locally free sheaves on X is denoted by $\mathscr{P}(X)$. The category $\mathbb{H}(X)$ is defined to be the full subcategory of objects $M \in Coh(X)$ with finite $\mathscr{P}(X)$-resolution.

For an affine scheme $X = \operatorname{Spec}(A)$, the category of A-modules will, sometimes, be denoted by $QCoh(A)$, and the category of finitely generated A-modules will be denoted by $Coh(A)$ (*instead of* $Mod(A)$ *or* $fMod(A)$).

The K-theory of the category $Coh(X)$ of coherent schemes is referred to as G-theory. The K-theory space (see Section 3.7) of $Coh(X)$ is denoted by $\mathbf{G}(X)$ and K-groups are denoted by $G_n(X)$. The K-theory space of the category $\mathscr{P}(X)$ is denoted by $\mathbf{K}(X)$, and the K-groups of $\mathscr{P}(X)$ are denoted by $K_n(X)$. For affine schemes $X = \operatorname{Spec}(A)$, we use both sets of notation $\mathbf{K}(A) := \mathbf{K}(X)$, $\mathbf{G}(A) := \mathbf{G}(X)$, $K_n(A) := K_n(X)$, $G_n(A) := G_n(X)$. For a Noetherian scheme

X, as a consequence of the resolution theorem (3.5.3), it follows (3.5.9) that

$$K_n(X) \xrightarrow{\sim} K_n(\mathbb{H}(X)) \qquad \forall\, n \geq 0. \tag{6.1}$$

We assume that the readers have some familiarity with the language of scheme theory [Hr, Chap. II, III]. For basic definitions and constructions in scheme theory, we refer to [Hr]. However, for the convenience of some of the readers, we will recall some of the frequently used basic definitions, from [Hr]. First, we record the following.

Lemma 6.0.2. *Let $X = \operatorname{Spec}(A)$ be an affine scheme. Then the functors*

$$\begin{cases} Coh(X) \xrightarrow{\sim} Coh(A) & \mathcal{F} \mapsto \Gamma(X, \mathcal{F}) \\ QCoh(X) \xrightarrow{\sim} QCoh(A) & \mathcal{F} \mapsto \Gamma(X, \mathcal{F}) \end{cases} \quad \textit{are equivalences of categories,}$$

where $\Gamma(X, \mathcal{F})$ denotes the A-module of global sections of $\mathcal{F}$. The inverse of the functor sends an A-module M to its sheafification $\widetilde{M}$. See [Hr] *for details.*

6.1. Preliminary Results

In this section, we list some of the immediate consequences of the results in previous chapters. As an immediate application of the above, we have the following.

Remark 6.1.1. Let X be a Noetherian scheme. Then the functor $\mathscr{P}(X) \longrightarrow Coh(X)$ induces maps

$$\begin{cases} K_n(X) \longrightarrow G_n(X) & \text{of the } K\text{-groups } \forall\, n \geq 0 \\ \mathbf{K}(X) \longrightarrow \mathbf{G}(X) & \text{of the } K\text{-theory spaces.} \end{cases} \tag{6.2}$$

The second map is a result of reformulation (3.7.2).

We define a $K_0(X)$-action on $G_n(X)$, as follows.

Definition 6.1.2. Let X be a Noetherian scheme. Given $\mathcal{E} \in \mathscr{P}(X)$, the tensor product functor

$$\tau_{\mathcal{E}} : Coh(X) \longrightarrow Coh(X) \quad \text{sending} \quad \mathcal{F} \mapsto \mathcal{E} \otimes \mathcal{F} \qquad \text{is exact.}$$

So, $\tau_{\mathcal{E}}$ induces a multiplication map

$$[\mathcal{E}] : G_n(X) \longrightarrow G_n(X) \qquad \forall\, n \geq 0.$$

Exact sequences $0 \longrightarrow \mathcal{K} \longrightarrow \mathcal{E} \longrightarrow \mathcal{Q} \longrightarrow 0$ in $\mathscr{P}(X)$ induce exact sequences

$$0 \longrightarrow \tau_{\mathcal{K}} \longrightarrow \tau_{\mathcal{E}} \longrightarrow \tau_{\mathcal{Q}} \longrightarrow 0 \quad \text{of exact functors.}$$

By Additivity Theorem (3.4.4), we have $[\mathcal{E}] = [\mathcal{K}] + [\mathcal{Q}] \in End(G_n(X))$, where $[\mathcal{E}]$, $[\mathcal{K}]$, $[\mathcal{Q}]$ denote the homomorphisms, respectively, induced by $\tau_{\mathcal{E}}$, $\tau_{\mathcal{K}}$, $\tau_{\mathcal{Q}}$. By the elementary description (3.2.1) of $K_0(X)$, we obtain a group homomorphism, $K_0(X) \longrightarrow End(G_n(X))$. This is also written as a map

$$K_0(X) \otimes G_n(X) \stackrel{\cdot}{\longrightarrow} G_n(X) \qquad \forall\, n \geq 0. \tag{6.3}$$

For $x \in K_0(X)$ and $y \in G_n(X)$, the product is denoted by $x \cdot y \in G_n(X)$.

Perhaps some readers would be content working within the universe of quasi-projective schemes, over affine schemes, which we define.

Definition 6.1.3. Let $Y = \operatorname{Spec}(A)$ be an affine scheme. Let $A[X_0, X_1, \ldots, X_n]$ denote the polynomial ring and $S = A[x_0, \ldots, x_n] = \oplus_{i=0}^{\infty} S_n$ be a finitely generated graded ring over S_0, with $\deg(x_i) = 1$. So, $S = \frac{A[X_0, X_1, \ldots, X_n]}{I}$, where I is a homogeneous ideal. A **projective scheme** over $Y = \operatorname{Spec}(A)$ (*or simply over* A) is defined to be a scheme $X = \operatorname{Proj}(S)$ for some such graded ring S. Such a projective scheme X comes with the structure map $X \longrightarrow Y$, induced by the ring homomorphism $A \longrightarrow S$. The **projective n-space** over $Y = \operatorname{Spec}(A)$ is defined as $\mathbb{P}^n_A = \operatorname{Proj}(A[X_0, X_1, \ldots, X_n])$. So, a projective scheme X is a closed subset of a projective n-space $\mathbb{P}^n_A$ for some n.

For such a projective scheme $X = \operatorname{Proj}(S)$, and a graded S-module M, one associates a sheaf of $\mathcal{O}_X$-module $\widetilde{M}$ [Hr, p. 116]. If M is finitely generated, then $\widetilde{M} \in Coh(X)$. In particular, for $n \in \mathbb{Z}$, we denote $\mathcal{O}(n) := \mathcal{O}_X(p) := \widetilde{S(p)}$. Sometimes $\mathcal{O}(1)$ is called the **twisting sheaf**. Note, $\mathcal{O}(p) \in \mathscr{P}(X)$ is a locally free sheaf of rank one.

A **quasi-projective scheme** X over Y is defined to be an open subset of a projective scheme $Z := \mathrm{Proj}\,(S)$ over Y. So, we have an open embedding $\iota : X \hookrightarrow \mathrm{Proj}\,(S)$. We denote $\mathcal{O}_X(1) = \iota^*\mathcal{O}_Z(1)$.

For future references, we extend the definition of projective n-space $\mathbb{P}^n_A$. For a (Noetherian) scheme Y, the scheme $\mathbb{P}^n_Y = \mathbb{P}^n_{\mathbb{Z}} \times_{\mathrm{Spec}(\mathbb{Z})} Y$ is defined to be the projective n-space over Y. This comes with the projection map $p : \mathbb{P}^n_Y \longrightarrow Y$. A **projective morphism** $f : X \longrightarrow Y$ is defined to be a composition as follows:

$$\begin{array}{ccc} X & \overset{\iota}{\hookrightarrow} & \mathbb{P}^n_Y \\ & {\scriptstyle f}\searrow & \downarrow \\ & & Y \end{array} \qquad \text{where } \iota \text{ is a closed immersion.} \qquad (6.4)$$

So, if $Y = \mathrm{Spec}\,(A)$, then a projective map is the structure map $\mathrm{Proj}\,(S) \longrightarrow \mathrm{Spec}\,(A)$, where S is a graded ring over A.

Projective schemes over affine schemes $\mathrm{Spec}\,(A)$ behave fairly closely like the affine schemes, due to the following.

Proposition 6.1.4. *Let $X = \mathrm{Proj}\,(S)$ be a projective scheme over a Noetherian affine scheme $\mathrm{Spec}\,(A)$, with $S = A[x_0, x_1, \ldots, x_n]$. Let $\mathcal{F}$ be a quasi-coherent sheaf on X. Then there is a graded S-module M such that $\mathcal{F} \cong \widetilde{M}$. If $\mathcal{F} \in Coh(Y)$, then the choice of M is a finitely generated S-module.*

Proof. Given $\mathcal{F}$, write $M = \bigoplus_{n \in \mathbb{Z}} \Gamma\,(Y, \mathcal{F} \otimes \mathcal{O}(n))$. One proves that M has a natural structure of a graded S-module, and there is a natural isomorphism $\widetilde{M} \xrightarrow{\sim} \mathcal{F}$. For details, see [Hr, Prop. 5.15]. The proof is complete. ■

Due to this, results and methods in affine schemes are often improvised to graded modules to obtain results on such projective schemes. And, quasi-projective schemes behave fairly closely like open subschemes of affine schemes (*which are not necessarily affine*). In the context of the resolution theorem (3.5.3), the following is a desirable property of the quasi-projective schemes.

Proposition 6.1.5. *Suppose X is a quasi-projective scheme over a Noetherian affine scheme $Y = \mathrm{Spec}\,(A)$. Then any $\mathcal{F} \in Coh(X)$ is an image of a locally free sheaf $\mathscr{E} \in \mathscr{P}(X)$. More precisely, there is* *an integer n_0 such that $\forall n \geq n_0$, the sheaf $\mathcal{F}(n) := \mathcal{F} \otimes \mathcal{O}(n)$ is an image of a finite direct sum of $\mathcal{O}_X$.* (We say that $\mathcal{F}(n)$ is generated by global sections.)

Consequently, if X is regular, any $\mathcal{F} \in Coh(X)$ has a finite resolution by objects in $\mathscr{P}(X)$.

Proof. Assume $X = \text{Proj}\,(S)$ is projective. So, we can assume $\mathcal{F} = \widetilde{M}$ for some finitely generated graded S-module. It follows that there is a surjective map $\varphi : \oplus_{i=1}^{k} S(-n_i) \twoheadrightarrow M$. Sheafifying, we get a surjective map $\oplus_{i=1}^{k} \mathcal{O}(-n_i) \twoheadrightarrow \widetilde{M} = \mathcal{F}$. It is easy to see that $\mathcal{O}(n)$ is globally generated $\forall n \geq 0$. So, with $n_0 = \max\{n_1, n_2, \ldots, n_k\}$, we have that $\mathcal{F}(n)$ is globally generated, $\forall n \geq n_0$.

In case X is quasi-projective, let $\iota : X \hookrightarrow Z := \text{Proj}\,(S)$ be an embedding of X as an open subset, in a projective scheme Z. For $\mathcal{F} \in Coh(X)$, the pushforward $\iota_*\mathcal{F}$ is a quasi-coherent sheaf on Z. By (6.1.4), $\iota_*\mathcal{F} \cong \widetilde{M}$, for some graded S-module M, which may not be finitely generated. Write $M = \bigoplus_n M_n$. Any set of elements $m_1, m_2, \ldots, m_k$, with $m_i \in M_{n_i}$, defines a map $\bigoplus_{i=1}^{n} S(-n_i) \longrightarrow M$. We get a map $\mathscr{E} := \bigoplus_{i=1}^{n} \mathcal{O}_Z(-n_i) \longrightarrow \widetilde{M}$ by sheafifying. By restriction, we obtain a map $\varphi(m_1, m_2, \ldots, m_k) : \mathscr{E}_{|Y} \longrightarrow \widetilde{M}_{|Y} = \mathcal{F}$. Under the hypotheses, we have choices of $m_1, m_2, \ldots, m_k$ so that this map $\varphi(m_1, m_2, \ldots, m_k)$ is surjective. As before, with $n_0 = \max\{n_1, n_2, \ldots, n_k\}$, we have that $\mathcal{F}(n)$ is globally generated $\forall n \geq n_0$. The proof is complete. ■

To extend (6.1.5) to other types of schemes, we define separated schemes. At this stage, the reason for introduction of separated schemes is to state (6.1.9).

Definition 6.1.6. Let $f : X \longrightarrow Y$ be a morphism of Noetherian schemes. Then f is defined as **separated** if the diagonal map $\Delta : X \longrightarrow X \times_Y X$ is a closed immersion. It follows [Hr, Cor. 4.2] that f is separated if and only if the image of Δ is closed in $X \times_Y X$.

A scheme X is defined as **separated** if $X \longrightarrow \text{Spec}\,(\mathbb{Z})$ is separated.

For the convenience of manipulating separated maps, we include the following.

Lemma 6.1.7. *All the following schemes are assumed to be Noetherian.*

1. *The closed immersions $C \hookrightarrow X$ and open immersions $U \hookrightarrow X$ are separated maps.*
2. *Let* $X \xrightarrow{f} Y \xrightarrow{g} Z$ *be morphisms of schemes. Then* (1) *if* gf *is separated, then* f *is separated, and* (2) *if both* f, g *are separated, then so is* gf.
3. *Let* $f : X \longrightarrow Y$ *be separated, and* $p : Z \longrightarrow Y$ *be a morphism of schemes. Then the base change map* $\mathfrak{f} : X \times_Y Z \longrightarrow Z$ *is separated.* (We say that separated morphisms are closed under base change.)
4. *A morphism* $f : X \longrightarrow Y$ *of schemes is separated if and only if* Y *has an open cover* $\{V_i : i \in I\}$ *such that* $f^{-1}V_i \longrightarrow V_i$ *is separated* $\forall i \in I$. (We say that being separated is a local property, on the base Y. This property allows us to assume that Y is affine for many proofs.)
5. *A map of affine spaces* $\text{Spec}(B) \longrightarrow \text{Spec}(A)$ *is separated. Likewise, the structure maps* $\text{Proj}(S) \longrightarrow \text{Spec}(A)$ *of a projective scheme are separated.*

   Consequently, affine schemes $\text{Spec}(A)$, *projective schemes* $\text{Proj}(S)$, *and the quasi-projective schemes* X *over* A *are separated.*

Proof. For a complete proof, see [Hr, Cor. 4.6], which follows fairly easily from the Valuative criterion for separatedness, established in [Hr, Thm. 4.3]. However, some of the proofs are basic and can be done purely formally.

For example, for an open or closed immersion $X \hookrightarrow Y$, $X \times_Y X = X$. This settles (1). To prove (3), note $(X \times_Y Z) \times_Z (X \times_Y Z) \cong (X \times_Y X) \times_Y Z$. With this identification, the diagonal maps are related as $\Delta(\mathfrak{f}) = \Delta(f) \times 1_Z : (X \times_Y X) \times_Y Z \longrightarrow X \times Z$. Since $\Delta(f)$ is closed, so is $\Delta(\mathfrak{f})$. See [Hr, Cor. 4.6] for the proof of (2, 4), which may be more specific to the category of schemes.

For (5), $\text{Spec}(B) \longrightarrow \text{Spec}(A)$ is separated because the diagonal map Δ is induced by the map $B \otimes_A B \twoheadrightarrow B$, which is surjective. Likewise, $\text{Proj}(S) \times_{\text{Spec}(A)} \text{Proj}(S) \cong \text{Proj}(S \otimes_A S)$ [Hr, Ex. II.5.11], and the diagonal is given by the maps $S \otimes_A S \longrightarrow S$. The proof is complete. ■

While the proof of (6.1.5) was fairly basic, with a little more work it extends to certain separated schemes.

Proposition 6.1.8. *Suppose* X *is a Noetherian, separated, locally factorial scheme. Then any* $\mathcal{F} \in Coh(X)$ *is the image of a locally* *free sheaf $\mathscr{E} \in \mathscr{P}(X)$. Consequently, if X is separated and regular, any $\mathcal{F} \in Coh(X)$ has a finite resolution by objects in $\mathscr{P}(X)$.*

Proof. See [Hr, Ex. III.6.8]. ∎

By an application of the resolution theorem (3.5.3), we obtain the following theorem.

Theorem 6.1.9. *If X is regular and separated, then maps (6.2), $K_n(X) \longrightarrow G_n(X)$ are isomorphisms of groups $\forall n \geq 0$, and the map $\mathbf{K}(X) \longrightarrow \mathbf{G}(X)$ is a homotopy equivalence.*

Proof. Since X is regular and separated, by (6.1.5), $Coh(X) = \mathbb{H}(X)$. By the resolution theorem (3.5.3), the maps $K_n(X) \longrightarrow G_n(X)$ are isomorphisms. Now, by Whitehead's theorem (C.3.8), the map $\mathbf{K}(X) \longrightarrow \mathbf{G}(X)$ is a homotopy equivalence. The proof is complete. ∎

6.1.1. *Closed subschemes and the localization sequence*

An immediate consequence of the Dévissage theorem is the following.

Proposition 6.1.10. *Let X be a Noetherian scheme and X_{red} be the corresponding reduced scheme. Then the maps $G_n(X_{red}) \longrightarrow G_n(X)$ are isomorphisms, $\forall n \geq 0$. Equivalently, the map $\mathbf{G}(X_{red}) \longrightarrow \mathbf{G}(X)$ is a homotopy equivalence.*

Proof. Let $\mathscr{I} \subseteq \mathcal{O}_X$ be the ideal sheaf of the nilpotent elements. Then $\mathscr{I}^k = 0$ for some k. Given $\mathcal{F} \in Coh(X)$, we have a filtration $0 \subseteq \mathscr{I}^{n-1}\mathcal{F} \hookrightarrow \cdots \hookrightarrow \mathscr{I}\mathcal{F} \hookrightarrow \mathcal{F}$. So, the Dévissage theorem (3.6.2) applies to the functor $Coh(X_{red}) \longrightarrow Coh(X)$. Therefore, the maps $G_n(X_{red}) \longrightarrow G_n(X)$ are isomorphisms $\forall n \geq 0$. By an application of Whitehead's theorem (C.3.8), the map $\mathbf{G}(X_{red}) \longrightarrow \mathbf{G}(X)$ is a homotopy equivalence. The proof is complete. ∎

The following is the statement of the Localization theorem of G-theory.

Theorem 6.1.11. *Let X be a Noetherian scheme, and $U \subseteq X$ be an open subset. Let $Z = X - U$. Recall that Z has a default reduced induced closed subscheme structure* [Hr, p. 86]. *Let $\iota : Z \hookrightarrow X$, $\zeta :$*

$U \hookrightarrow X$ denote the inclusion. Then the sequence of maps of G-theory spaces

$$\mathbf{G}(Z) \longrightarrow \mathbf{G}(X) \longrightarrow \mathbf{G}(U) \quad \textit{is a homotopy fibration.} \tag{6.5}$$

Consequently, there is a long exact sequence of G-groups

$$\cdots \longrightarrow G_1(Z) \longrightarrow G_1(X) \longrightarrow G_1(U) \xrightarrow{\partial_1} G_0(Z) \longrightarrow G_0(X) \longrightarrow G_0(U) \longrightarrow 0. \tag{6.6}$$

Proof. Let $Coh_Z(X) \subseteq Coh(X)$ be the full subcategory of objects $\mathcal{F} \in Coh(X)$, supported on Z (meaning $\mathcal{F}_x = 0 \ \forall \ x \in U$). By (A.6.10), $Coh_Z(X)$ is a Serre subcategory of $Coh(X)$, and $\frac{Coh(X)}{Coh_Z(X)} \xrightarrow{\sim} Coh(U)$. By Localization theorem (3.6.9, 3.7.5),

$$\begin{cases} \mathbb{BQ}Coh_Z(X) \longrightarrow \mathbb{QB}Coh(X) \longrightarrow \mathbb{BQ}Coh(U) \\ \mathbf{K}(Coh_Z(X)) \longrightarrow \mathbf{G}(X) \longrightarrow \mathbf{G}(U) \end{cases} \quad \text{are homotopy fibrations,} \tag{6.7}$$

The natural inclusion functor $\iota_* : Coh(Z) \hookrightarrow Coh(X)$ is in fact a functor $\iota_* : Coh(Z) \hookrightarrow Coh_Z(X)$. Objects $\mathcal{F} \in Coh_Z(X)$ have filtration by objects in $Coh(Z)$. So, by Dévissage theorem (3.6.2), $\mathbf{G}(Z) \longrightarrow \mathbf{K}(Coh_Z(X))$ is a homotopy equivalence. This establishes the first part (6.5) of the theorem. The Localization sequence (6.6) follows from the homotopy fibration (6.5). The proof is complete. ■

6.2. Pullback and Pushforward

For suitable morphisms $f : X \longrightarrow Y$ of Noetherian schemes, in this section, we develop pullback $f^* : G_n(Y) \longrightarrow G_n(X)$ and pushforward $f_* : G_n(X) \longrightarrow G_n(Y)$ maps. Then we study the relationship between these maps, which culminates in the projection formulas.

6.2.1. *Pullback maps*

First, we define pullback maps in K-theory of schemes.

Definition 6.2.1. Let $f : X \longrightarrow Y$ be a map of schemes. Then f induces a functor

$$f^* : Coh(Y) \longrightarrow Coh(X) \quad \text{sending} \quad \mathcal{F} \mapsto f^*\mathcal{F}. \tag{6.8}$$

Locally, this functor is given by tensor product [Hr, Prop. 5.2].

1. The restriction $f^* : \mathscr{P}(Y) \longrightarrow \mathscr{P}(X)$ is an exact functor. So, it induces

$$\begin{cases} f^* : \mathbf{K}(Y) \longrightarrow \mathbf{K}(X) & \text{the map of the } K\text{-theory spaces,} \\ f^* : K_n(Y) \longrightarrow K_n(X) & \text{homomorphisms } K\text{-groups } \forall\, n \geq 0. \end{cases}$$

   In fact, $\forall\, n \geq 0$, the association $X \mapsto K_n(X)$ is a contra variant functor on the category of Noetherian schemes.
2. Assume f is flat. Then (6.8) is an exact functor. So, it induces

$$\begin{cases} f^* : \mathbf{G}(Y) \longrightarrow \mathbf{G}(X) & \text{the map of the } G\text{-theory spaces,} \\ f^* : G_n(Y) \longrightarrow G_n(X) & \text{homomorphisms } G\text{-groups } \forall\, n \geq 0. \end{cases} \tag{6.9}$$

   As above, $\forall\, n \geq 0$, the association $X \mapsto G_n(X)$ is a contra variant functor on the category of Noetherian schemes, and flat maps.

We use the trick used in (3.5.13) to relax the flatness condition on f, and define pullback (6.9) for more general maps f. As in (5.4.1), we define Tor dimension:

Definition 6.2.2. Let X be a Noetherian scheme. Assume that given $\mathcal{F} \in Coh(X)$, $\mathcal{F}$ is an image of a locally free sheaf (as in 6.1.5, 6.1.8, 6.2.13). (*In this case, we say that X has enough locally free sheaves.*) Then $Tor_k^X(\mathcal{F}, \mathcal{G})$ is defined $\forall\, k \geq 0$, $\forall \mathcal{F} \in Coh(X)$, and all quasi-coherent sheaves $\mathcal{G}$ on X. For quasi-coherent sheaves $\mathcal{G}$ on X, the **Tor dimension** $Tor \dim_X(\mathcal{G})$ of $\mathcal{G}$ is defined as

$$Tor \dim_X(\mathcal{G}) = \sup \left\{k : Tor_k^X(\mathcal{F}, \mathcal{G}) \neq 0 \text{ for some } \quad \mathcal{F} \in QCoh(X)\right\}$$

It follows

$$Tor \dim_X(\mathcal{G}) = \sup \{Tor \dim_{\mathcal{O}_x}(\mathcal{G}_x) : x \in X\}. \tag{6.10}$$

Let $f : X \longrightarrow Y$ be a morphism of Noetherian schemes, and assume Y has enough locally free sheaves. Then f is said to have **finite Tor dimension**, if $Tor \dim_Y(\mathcal{O}_X) < \infty$. Note that if f is a flat extension, then $Tor \dim_Y(\mathcal{O}_X) = 0$.

Lemma 6.2.3. *Let $f : X \longrightarrow Y$ be a morphism of Noetherian schemes. Assume Y has enough locally free sheaves and f has finite Tor dimension. Let*

$$\mathfrak{Coh}(f,Y) = \{\mathcal{F} \in Coh(Y) : Tor_k^Y(\mathcal{F}, \mathcal{O}_X) = 0 \ \forall\ k \geq 1\} \subseteq Coh(Y) \tag{6.11}$$

be the full subcategory. Then the restriction of the functor f^* (6.8),

$$f^* : \mathfrak{Coh}(f,Y) \longrightarrow Coh(X) \qquad \text{is an exact functor.} \tag{6.12}$$

This is checked locally, and there is nothing new beyond the affine case (3.5.13). So, this induces

$$\begin{cases} \mathbf{K}(\mathfrak{Coh}(f,Y)) \longrightarrow \mathbf{G}(X) & \text{a natural map of } K\text{-theory spaces} \\ K_n(\mathfrak{Coh}(f,Y)) \longrightarrow G_n(X) & \text{homomorphisms of } K\text{-groups } \forall\ n \geq 0. \end{cases} \tag{6.13}$$

Further, every $\mathcal{F} \in Coh(Y)$ has a finite resolution by objects in $\mathfrak{Coh}(f,Y)$. Therefore, by Theorem (3.5.3), we have

$$\begin{cases} \mathbf{K}(\mathfrak{Coh}(f,Y)) \xrightarrow{\sim} \mathbf{G}(Y) & \text{homotopy equivalence of } K\text{-theory spaces} \\ K_n(\mathfrak{Coh}(f,Y)) \xrightarrow{\sim} G_n(Y) & \text{isomorphisms of } K\text{-groups } \forall\ n \geq 0. \end{cases} \tag{6.14}$$

Combining (6.13, 6.14), we obtain the maps

$$f^* : \begin{cases} \mathbf{G}(Y) \xleftarrow{\sim} \mathbf{K}(\mathfrak{Coh}(f,Y)) \longrightarrow \mathbf{G}(X) & \text{of } K\text{-theory spaces} \\ G_n(Y) \xleftarrow{\sim} K_n(\mathfrak{Coh}(f,Y)) \longrightarrow G_n(X) & \text{homomorphisms of} \\ & K\text{-groups } \forall n \geq 0. \end{cases}$$

Proof. That the functor (6.12) is exact can be checked locally, and it follows from the exact sequence of Tor. It remains to be shown that any $\mathcal{F} \in Coh(Y)$ has finite resolution by objects in $\mathfrak{Coh}(f,Y)$. This is a frequently used trick. With $d := Tor \dim \mathcal{O}_X$, we take a resolution

$$0 \longrightarrow \mathcal{K} \longrightarrow \mathcal{E}_{d-1} \xrightarrow{\partial_{d-1}} \cdots \longrightarrow \mathcal{E}_0 \longrightarrow \mathcal{F} \longrightarrow 0$$

$$\text{where} \quad \mathcal{E}_k \in \mathscr{P}(Y), \ \mathcal{K} = \ker(\partial_{d-1}).$$

It follows that $\mathcal{K} \in \mathfrak{Coh}(f,Y)$. Now, by resolution theorem (3.5.3), the statements (6.14) are established. The proof is complete. ∎

The following is on the functorial property of the above pullback homomorphisms.

Corollary 6.2.4. *Let* $X \xrightarrow{f} Y \xrightarrow{g} Z$ *be morphisms of schemes. Assume both* Y, Z *have enough locally free sheaves and both* f, g *have finite Tor dimension. Then the homomorphisms in the diagram*

$$\begin{array}{ccc} G_n(Z) & \xrightarrow{g^*} & G_n(Y) \\ & \searrow^{(fg)^*} & \downarrow f^* \\ & & G_n(X) \end{array}$$

are defined, and the diagram commutes.

Proof. The proof is similar to (5.4.2). Refer to Exercise 6.6.4. ■

The following limit theorem will be of some use to us later.

Proposition 6.2.5. *Let* I *be a filtered category, and* $\iota : I^{op} \longrightarrow \underline{Sch}$ *be a filtered inverse system to the category* $\underline{Sch}$ *of Noetherian schemes. Denote* $\iota(i) = X_i$*. Assume that the limit* $\lim_I X_i$ *exists and* $X := \lim_I X_i$ *is Noetherian. Then*

$$\forall n \geq 0 \quad K_n(X) = \lim_{i \in I} K_n(X_i) \quad \textit{are isomorphisms.}$$

Further, if the transition maps $X_i \longrightarrow X_j$ *are flat, then*

$$\forall n \geq 0 \quad G_n(X) = \lim_{i \in I} G_n(X_i) \quad \textit{are isomorphisms.}$$

We comment that a sufficient condition to ensure that the limit $X := \lim_{i \in I} X_i$ *exists is that the transition maps* $X_i \longrightarrow X_j$ *are affine* (6.6.14).

Proof. For map $\theta \in Mor_I(i, j)$, the pullback functors $\iota(\theta)^* : \mathscr{P}(X_i) \longrightarrow \mathscr{P}(X_j)$ are exact. Now, the association $i \mapsto \mathscr{P}(X_i)$ is a filtered system of exact categories and exact functors. It follows $\mathscr{P}(X) = \lim_{i \in I} \mathscr{P}(X_i)$. Likewise, if f is flat, $Coh(X) = \lim_{i \in I} Coh(X_i)$ limit of exact functors. Now the lemma follows from (3.3.3) (*However, we skip the possible set theoretic issues discussed in [Q]. Interested readers may consult [Sv].*) ■

6.2.2. *Pushforward maps*

For a morphism $f: X \longrightarrow Y$ of Noetherian schemes, to define pushforward, we use tricks similar to what we did to define pullback. The pullback functor $f^*: Coh(Y) \longrightarrow Coh(X)$ was not exact. While defining pullback $f^*: G_n(Y) \longrightarrow G_n(X)$, we assumed that f had finite Tor dimension to remedy this issue. Then we identified the subcategory $\mathfrak{C}oh(f, Y) \subseteq Coh(Y)$ on which f^* was exact. To define pushforward maps of K-groups, we will do the same, for which we need to develop further background.

For morphisms $f: X \longrightarrow Y$ of Noetherian schemes, consider the direct image functor

$$f_*: QCoh(X) \longrightarrow QCoh(Y) \quad \text{sending} \quad \mathcal{F} \mapsto f_*\mathcal{F},$$
$$\text{where } f_*\mathcal{F}(V) = \mathcal{F}(f^{-1}V). \tag{6.15}$$

This functor is only left exact. Further, for $\mathcal{F} \in Coh(X)$, the direct image $f_*\mathcal{F}$ need not be a coherent sheaf on Y. Therefore, (1) to remedy the issue of exactness, we define higher direct images $R^k f_*\mathcal{F}$, and (2) to remedy the issue of coherence property of $f_*\mathcal{F}$, we define proper morphisms $f: X \longrightarrow Y$. For perspective, we comment on the affine case.

Remark 6.2.6. Let $f: X \longrightarrow Y$ be a morphism of Noetherian schemes, with $Y = \operatorname{Spec}(A)$ affine. With the identification (6.0.2) $QCoh(Y) \cong QCoh(A)$, the direct image functor (6.15) translates to

$$f_*: QCoh(X) \longrightarrow QCoh(A) \quad \text{sending} \quad \mathcal{F} \mapsto \Gamma(X, \mathcal{F}).$$

Now, let $f: \operatorname{Spec}(B) \longrightarrow \operatorname{Spec}(A)$ be a morphism of Noetherian affine schemes. Then with the identifications (6.0.2), the direct image functor (6.15)

$$f_*: QCoh(B) \longrightarrow QCoh(A) \quad \text{is defined by} \quad f_*M = M$$
$$\text{considered as an } A\text{-module.}$$

Consequently, f_* is exact, in this case. However, for $M \in Coh(B)$, it is not necessary that $f_*M = M \in Coh(A)$.

We define the sheaf cohomology and the higher direct images next.

Definition 6.2.7. Let $f : X \longrightarrow Y$ be a morphism of Noetherian schemes. Then the category $QCoh(X)$ has enough injectives [BH, p. 91], [Hr2, Sec. II.1]. Given $\mathcal{F} \in QCoh(X)$, consider a resolution by injective $\mathcal{O}_X$-modules in $QCoh(X)$:

$$0 \longrightarrow \mathcal{F} \longrightarrow \mathscr{I}_0 \longrightarrow \mathscr{I}_1 \longrightarrow \mathscr{I}_2 \longrightarrow \cdots$$

$$\text{to be denoted by} \quad \mathscr{I}_\bullet. \tag{6.16}$$

We apply two functors, namely (1) the global section $\Gamma(X, -)$ and (2) direct image functor f_*, to this complex $\mathscr{I}_\bullet$, and obtain two complexes:

$$\begin{cases} 0 \longrightarrow \Gamma(X,\mathcal{F}) \longrightarrow \Gamma(X,\mathscr{I}_0) \longrightarrow \Gamma(X,\mathscr{I}_1) \longrightarrow \cdots & \text{denoted by } \Gamma(X,\mathscr{I}_\bullet) \\ 0 \longrightarrow f_*\mathcal{F} \longrightarrow f_*\mathscr{I}_0 \longrightarrow f_*\mathscr{I}_1 \longrightarrow f_*\mathscr{I}_2 \longrightarrow \cdots & \text{denoted by } f_*\mathscr{I}_\bullet. \end{cases} \tag{6.17}$$

For integers $k \geq 0$, define the following:

1. Define the **sheaf cohomology**,

$$H^k(X,\mathcal{F}) = H^k(\Gamma(X,\mathscr{I}_\bullet))$$
$$\text{cohomology of } \Gamma(X,\mathscr{I}_\bullet) \text{ as a } \Gamma(X,\mathcal{O}_X)\text{-module.} \tag{6.18}$$

2. Define the **higher direct images**,

$$R^k f_*\mathcal{F} = \mathcal{H}^k(f_*\mathscr{I}_\bullet) \quad \text{cohomology of } f_*\mathscr{I}_\bullet \text{ as an } \mathcal{O}_Y\text{-module.} \tag{6.19}$$

We use the notation $\mathcal{H}^k(-)$ to denote the cohomology as a sheaf to distinguish it from the sheaf cohomology $H^k(X, -)$. Up to a natural isomorphism, both $H^k(X,\mathcal{F})$ and $R^k f_*\mathcal{F}$ are independent of the choice of the resolution $\mathscr{I}_\bullet$ of $\mathcal{F}$. Since homologies of complexes in the category $QCoh(Y)$ are in $QCoh(Y)$, it follows that $R^k f_*\mathcal{F} \in QCoh(Y)$. Therefore,

$$R^k f_* : QCoh(X) \longrightarrow QCoh(Y) \quad \text{are covariant functors } \forall\ k \geq 0. \tag{6.20}$$

The following are some of the natural properties of these two functors:

1. Since the direct image functor (6.15) is left exact, there is a natural isomorphism $f_*\mathcal{F} \cong R^0 f_*\mathcal{F}$. For the same reason, $\Gamma(X, \mathcal{F}) \cong H^0(X, \mathcal{F})$.
2. Given an exact sequence

$$0 \longrightarrow \mathcal{K} \longrightarrow \mathcal{F} \longrightarrow \mathcal{G} \longrightarrow 0 \quad \text{in} \quad QCoh(X) \quad (6.21)$$

   there is a connecting morphism $\partial^k : R^k f_*\mathcal{G} \longrightarrow R^{k+1} f_*\mathcal{K}$ such that we obtain a long exact sequence

$$\cdots \longrightarrow R^k f_*\mathcal{K} \longrightarrow R^k f_*\mathcal{F} \longrightarrow R^k f_*\mathcal{G} \xrightarrow{\partial^k} R^{k+1} f_*\mathcal{K} \longrightarrow \cdots \quad (6.22)$$

   starting at degree zero. Likewise, there is a long exact sequence of the cohomologies $H^k(X, -)$.
3. The sequence (6.22) is natural with respect to maps between two short exact sequences (6.21).
4. Since restrictions of injective resolutions to open subsets are the same, it follows immediately, for open subsets $V \subseteq Y$, we have isomorphisms

$$\left(R^k f_*\mathcal{F}\right)_{|V} \cong R^k f_*^V \left(\mathcal{F}_{|f^{-1}V}\right)$$
$$\text{where} \quad f^V := f_{|f^{-1}V} : f^{-1}V \longrightarrow V.$$

See [Hr, §III.1, III.8] for further details. This procedure is known as **Right derived functor of** the left exact functor f_*, and is part of a more general methodology. There is a more workable way to compute the higher direct images $R^k f_*\mathcal{F}$, using Čech cohomology. We give an exposition of the same in Section 6.5.

The following are two observations.

Proposition 6.2.8. *Let $f : X \longrightarrow Y$ be a morphism of Noetherian schemes. Then*

1. *For $\mathcal{F} \in QCoh(X)$, the higher direct images $R^k f_*\mathcal{F}$ are the sheaves associated to the presheaves*

$$V \mapsto H^k\left(f^{-1}V, \mathcal{F}\right).$$

2. *If* $Y = \mathrm{Spec}\,(A)$ *is affine, then*

$$R^k f_* \mathcal{F} = \widetilde{H^k(X, \mathcal{F})} \quad \textit{the sheaf associated to the } A\textit{–module } H^k(X, \mathcal{F}). \tag{6.23}$$

Proof. The first one (1) follows from definition (6.19) of $R^k f_* \mathcal{F}$ and the definition of $H^k\left(f^{-1}V, \mathcal{F}\right)$. To prove (2), we use the identifications $QCoh(Y) \cong QCoh(A)$. With this identification, the complex (6.17) is equivalent to

$$0 \longrightarrow \Gamma(X, \mathcal{F}) \longrightarrow \Gamma(X, \mathscr{I}_0) \longrightarrow \Gamma(X, \mathscr{I}_1) \longrightarrow \Gamma(X, \mathscr{I}_2) \longrightarrow \cdots$$

By definition, $H^k(X, \mathcal{F})$ is the kth-cohomology of this complex, while $R^k f_* \mathcal{F}$ is the sheafification (6.17) of this one. So, $R^k f_* \mathcal{F} \cong \widetilde{H^k(X, \mathcal{F})}$. The proof is complete. ∎

As a possible remedy for failure of exactness of the direct image functor f_*, we defined higher direct images. Next, we try to identify a subcategory of the category of schemes, meaning a type of morphisms $f : X \longrightarrow Y$ such that $\forall\ \mathcal{F} \in Coh(X)$, the direct image $f_* \mathcal{F} \in Coh(Y)$. The so-called **proper morphisms** $f : X \longrightarrow Y$ will foot the bill. However, before we define proper morphisms, we give examples of some familiar families of morphisms f that have this property, as follows.

Theorem 6.2.9. *Let* $f : X \longrightarrow Y$ *be a morphism of Noetherian schemes, of finite dimensions. Assume that* f *is either* (1) *a closed immersion or* (2) *a projective morphism* (6.1.3). *Then for* $\mathcal{F} \in Coh(X)$,

1. $R^k f_* \mathcal{F} = 0 \quad \forall k \gg 0$,
2. *the higher direct images* $R^k f_* \mathcal{F} \in Coh(Y)\ \forall k \geq 0$,
3. *there is an integer* n_0 *such that* $R^k f_* \mathcal{F}(n) = 0\ \forall\ n \geq n_0,\ \forall k \geq 1$.

Proof. This is fairly close to [Hr, Thm II.5.2]. All the questions are local on Y. So, we can assume $Y = \mathrm{Spec}\,(A)$. To prove (1), we use (6.2.8(1)). Since X can be covered by finitely many affine subsets, (1) follows from Čech cohomology (6.5.3). Now, we prove (2, 3).

Let $f : X \longrightarrow Y$ be a closed immersion. Then for $\mathcal{F} \in QCoh(X)$, let $f_*(\mathcal{F}) = \mathcal{F}$ be considered an $\mathcal{O}_Y$-module. So, if $\mathcal{F} \in Coh(X)$, then

$f_*\mathcal{F} \in Coh(Y)$. It is also clear from definition (6.19) that $R^k f_*(\mathcal{F}) = 0\ \forall \mathcal{F},\ \forall k \geq 1$.

Now assume $f : X \longrightarrow Y$ is a projective morphism, with $Y = \mathrm{Spec}\,(A)$. Then as commented on in (6.1.3), $X = \mathrm{Proj}\,(S)$ for some finitely generated graded ring S over A, and f is the structure map $f : \mathrm{Proj}\,(S) \longrightarrow \mathrm{Spec}\,(A)$. So, $X = \mathrm{Proj}\,(S) \subseteq \mathbb{P}^d_A$ is a closed subscheme for some d. Since $Coh(X) \subseteq Coh\left(\mathbb{P}^d_A\right)$, we can assume $X = \mathbb{P}^d_A$, and $S = A[X_0, \ldots, X_d] = \bigoplus S_n$ is the polynomial ring. Using Čech cohomology tools $H^k\left(X, \mathcal{O}(n)\right)$ was explicitly computed in (6.5.4). By equation (6.23), it follows from this description (6.5.4) of $H^k\left(X, \mathcal{O}(n)\right)$ that both the assertions (2, 3) are valid for the direct sum modules $\mathcal{F} = \oplus_{k=1}^m \mathcal{O}(n_k) \in Coh(X)$. Given $\mathcal{F} \in Coh(X)$, there is an exact sequence

$$0 \longrightarrow \mathcal{K} \longrightarrow \mathcal{E} \longrightarrow \mathcal{F} \longrightarrow 0 \quad \text{with} \quad \mathcal{E} = \oplus_{k=1}^m \mathcal{O}(n_k).$$

This leads to the cohomology exact sequence

$$\cdots \longrightarrow H^k\left(X, \mathcal{K}\right) \longrightarrow H^k\left(X, \mathcal{E}\right) \longrightarrow H^k\left(X, \mathcal{F}\right) \longrightarrow H^{k+1}\left(X, \mathcal{K}\right) \longrightarrow \cdots$$

From Čech cohomology (6.5.3), there is an integer d such that $H^k\left(X, -\right) = 0$ for all $k \geq d+1$. Further, by (6.5.4), $H^k\left(X, \mathcal{E}\right)$ are finitely generated (free) A-modules. So, by decreasing induction, $H^k\left(X, \mathcal{F}\right)$ are finitely generated A-modules. Therefore, by equation (6.23), $R^k f_*\mathcal{F} \in Coh(\mathrm{Spec}\,(A))$. By (6.5.4), there is an integer n_0 such that $H^k\left(X, \mathcal{E}(n)\right) = 0\ \forall n \geq n_0, \forall k \geq 1$. It follows that $H^k\left(X, \mathcal{F}(n)\right) = 0\ \forall n \geq n_0, \forall k \geq 1$, by decreasing induction. The proof is complete. ■

For completeness, we define proper morphisms and state that proper morphisms satisfy the properties in (6.2.9).

Definition 6.2.10. Let $f : X \longrightarrow Y$ be a morphism of Noetherian schemes of finite type, and f be separated. We say f is **proper** if f is a universally closed map in the sense that, given any morphism p

of schemes, and the pullback diagram

$$\begin{array}{ccc} X \times_Y W & \overset{\mathfrak{p}}{\dashrightarrow} & X \\ {\scriptstyle \mathfrak{f}} \downarrow & & \downarrow {\scriptstyle f} \\ W & \underset{p}{\longrightarrow} & Y \end{array} \qquad \text{then } \mathfrak{f} \text{ is a closed map.} \tag{6.24}$$

By a closed map, we mean that the image of closed sets are closed.

Properness seems to be a fairly restrictive property. However, it is immediate that the closed immersions $f : X \hookrightarrow Y$ are proper. We will see that projective morphisms $f : X \longrightarrow Y$ are proper. The following is a list of conditions to identify proper maps [Hr, Cor. 4.8].

Lemma 6.2.11. *In the following, all schemes are assumed to be Noetherian.*

1. *Closed immersions $X \hookrightarrow Y$ are proper.*
2. *Let $X \xrightarrow{f} Y \xrightarrow{g} Z$ be morphisms of schemes. If both f and g are proper, then gf is proper. Also, if gf is proper and g separated, then f is proper.*
3. *Let $f : X \longrightarrow Y$ be proper, and $g : Z \longrightarrow Y$ be a morphism of schemes. Then the base change map $X \times_Y Z \longrightarrow Z$ is proper.* (We say proper morphisms are closed under base change.)
4. *A morphism $f : X \longrightarrow Y$ is proper if and only if Y has an open cover $\{V_i : i \in I\}$ such that $f^{-1}V_i \longrightarrow V_i$ are proper $\forall i \in I$.* (We say, being proper is a local property, on the base X. This property allows us to assume that X is affine for many proofs.)

Proof. The separable property is checked by (6.1.7). The finiteness property is also checked routinely. It remains to check that these maps in question are universally closed, which is also routine.

Consider the diagram (6.24). A closed immersion f remains closed under base changes by p. This settles (1). There is nothing new to be checked for the proof of (3). For the proof of (2), consider the pullback diagrams of maps of schemes:

$$\begin{array}{ccc} X\times_Z W & \longrightarrow & X \\ {\scriptstyle\mathfrak{f}}\downarrow & & \downarrow{\scriptstyle f} \\ Y\times_Z W & \longrightarrow & Y \\ {\scriptstyle\mathfrak{g}}\downarrow & & \downarrow{\scriptstyle g} \\ W & \xrightarrow[p]{} & Z \end{array}, \qquad \begin{array}{ccc} X\times_Z W & \longrightarrow & X \\ {\scriptstyle\mathfrak{f}}\downarrow & & \downarrow{\scriptstyle f} \\ W & \xrightarrow[p]{} & Y \\ & \searrow & \downarrow{\scriptstyle g} \\ & & Z \end{array}$$

From the universal property of pullback, we have $X\times_Z W \xrightarrow{\sim} X\times_Y (Y\times_Z W)$. If follows from the first diagram that, if f and g are proper, then $\mathfrak{f}$ and $\mathfrak{g}$ are closed maps, and then so is $\mathfrak{g}\mathfrak{f}$. Therefore, gf is proper. To prove the latter part of (2), use the Valuative criterion for properness (6.6.2).

To prove (4), consider the diagram (6.24). The same proof works to establish f or $\mathfrak{f}$ is a closed map. To prove f is a closed map, let $C \subseteq X$ be a closed subset. Now we have the set theoretic identity

$$Y - f(C) = \bigcup_i (V_i - f(C)\cap V_i) = \bigcup_i \left(V_i - f\left(f^{-1}V_i \cap C\right)\right)$$
$$\text{which is open.} \tag{6.25}$$

The proof is complete. ∎

We prove that the projective morphisms (6.1.3) are proper, as follows.

Lemma 6.2.12. *Let $f : X \longrightarrow Y$ be a projective morphism of Noetherian schemes. Then f is proper.*

Proof. Since the composition of proper maps is proper, we prove that the projection maps $f : \mathbb{P}^n_{\mathbb{Z}} \times Y \longrightarrow Y$ are proper. Since properness is a local property (6.2.11) on Y, it can be assumed $Y = \operatorname{Spec}(A)$ is affine and $X = \mathbb{P}^n_A$. Such a map is clearly of finite type, and by (6.1.7(5)), it is also separated. It remains to be proven that $\mathbb{P}^n_A \longrightarrow \operatorname{Spec}(A)$ is universally closed. So, given a morphism $W \longrightarrow Y$, we need to prove that $\mathbb{P}^n_W \longrightarrow W$ is a closed map. However, this is also a local property, as is evident in the equation (6.25). So, we are required to prove $\mathbb{P}^n_A \longrightarrow \operatorname{Spec}(A)$ is a closed map.

In other words, we prove $\mathrm{Proj}\,(A[X_0, X_1, \ldots, X_n]) \longrightarrow \mathrm{Spec}\,(A)$ is a closed map. A closed subset in $\mathrm{Proj}\,(A[X_0, X_1, \ldots, X_n])$ is defined by $C = V(I)$, where I is a homogeneous ideal. By basic commutative algebra arguments, it follows that $f(V(I)) = V(I \cap A)$ is closed. The proof is complete. ■

So far, only tangible examples of proper maps $X \longrightarrow Y$ are the closed embeddings and the projective morphisms. We will see, in some cases (6.2.14), there will be no more. We define ample invertible sheaves.

Definition 6.2.13. Let X be a Noetherian scheme. An invertible sheaf $\mathcal{L}$ is said to be **ample** if $\forall\, \mathcal{F} \in Coh(X)$, there is an integer n_0 such that $\forall\, n \geq n_0$, $\mathcal{F} \otimes \mathcal{L}^n$ is generated by its global sections. When $\mathcal{L}$ is understood, we write $\mathcal{O}(n) := \mathcal{L}^n$ and $\mathcal{F}(n) := \mathcal{F} \otimes \mathcal{L}^n$. With these notation, $\mathcal{O}(1) := \mathcal{L}$ is ample if and only if $\forall\, \mathcal{F} \in Coh(X)$, there is an integer n_0 such that $\forall n \geq n_0$ there is a surjective map $\mathcal{O}_X^m \twoheadrightarrow \mathcal{F}(n)$, where m may depend on n. Equivalently, there is a surjective map $\mathcal{O}_X(-n)^m \twoheadrightarrow \mathcal{F}$. In particular, any $\mathcal{F} \in Coh(X)$ is an image of an object in $\mathscr{P}(X)$.

Ample invertible sheaves are sometimes referred to as **ample bundles**. Note that the twisting sheaf, defined in (6.1.3), is ample.

We quote the following from [Hr].

Theorem 6.2.14. *Let $f : X \longrightarrow Y$ be a proper morphism of Noetherian schemes, with $Y = \mathrm{Spec}\,(A)$ affine. Assume X has an ample invertible sheaf. Then f is a projective morphism.*

Proof. See [Hr, Rem. II.5.16.1], together with [Hr, Thm. II.7.6]. ■

We state next that Theorem 6.2.9 extends to proper morphisms.

Theorem 6.2.15. *Let $f : X \longrightarrow Y$ be a proper morphism of schemes. Let $\mathcal{F} \in Coh(X)$. Then*

1. $R^k f_* \mathcal{F} = 0 \quad \forall k \gg 0$.
2. *The higher direct images $R^k f_* \mathcal{F} \in Coh(Y)$ $\forall k \geq 0$.*
3. *Further, if X has an ample invertible sheaf $\mathcal{L}$, there is an integer n_0 such that $R^k f_* \mathcal{F}(n) = 0$ $\forall\, n \geq n_0$, $\forall k \geq 1$.*

Proof. As in the proof of (6.2.9), the proof of (1) follows from Čech cohomology (6.5.3).

To prove (3), let $U \subseteq X$ be an open subset. Any $\mathcal{G} \in Coh(U)$ extends to a $\tilde{\mathcal{G}} \in Coh(X)$ [Hr, Ex. II.5.15]. Therefore, $\mathcal{L}_{|U}$ is also ample over U. The question in (3) is local on Y. By replacing Y by affine open subsets $V = \mathrm{Spec}\,(A) \subseteq Y$, and X by $f^{-1}V$, we can assume that $Y = \mathrm{Spec}\,(A)$. In this case, by (6.2.14), $f : X \longrightarrow Y$ is a projective morphism. So, by (6.2.9), the assertion (3) follows.

To prove (2) as well, we can assume $Y = \mathrm{Spec}\,(A)$ is affine. We need to prove that $H^k(X, \mathcal{F})$ is a finitely generated A-module. If X has an ample bundle, it follows from (6.2.9). (*We are unlikely to use this result unless X has such an ample bundle.*) However, for the general case, we refer to [Ga]. The proof is complete. ■

Now we proceed to define the pushforward map for the proper morphisms $f : X \longrightarrow Y$.

Definition 6.2.16. Let $f : X \longrightarrow Y$ be a proper morphism of Noetherian schemes. By (6.2.15), the direct image f_* defines a functor

$$f_* : Coh(X) \longrightarrow Coh(Y), \qquad \text{which is not necessarily exact.}$$

Consider the full subcategory

$$\mathfrak{Coh}(X, f) = \left\{\mathcal{F} \in Coh(X) : R^k f_* \mathcal{F} = 0 \ \forall\ k \geq 1\right\} \subseteq Coh(X). \tag{6.26}$$

Then the restriction functor

$$f_* : \mathfrak{Coh}(X, f) \longrightarrow Coh(Y) \quad \text{is exact, by the exact sequence (6.22).}$$

Consequently, there are maps

$$\begin{cases} \mathbf{K}\mathfrak{Coh}(X, f) \longrightarrow \mathbf{G}(Y) & \text{of } K\text{-theory spaces} \\ K_n\left(\mathfrak{Coh}(X, f))\right) \longrightarrow G_n(Y) & \text{of the } K\text{-groups.} \end{cases} \tag{6.27}$$

Now we try to find suitable situations when $K_n\left(\mathfrak{Coh}(X, f)\right)$ and $G_n(X)$ are identifiable.

Assume X has an ample invertible sheaf $\mathcal{L}$. Replacing $\mathcal{L}$ by $\mathcal{O}(1) := \mathcal{L}^k$ with k high enough, assume that $\mathcal{O}(n)$ is globally generated, $\forall\ n \geq 0$. So, $\forall\ n \geq 0$, there are exact sequences

$\mathcal{O}_X^{r_n} \longrightarrow \mathcal{O}(n) \longrightarrow 0$. Dualizing this sequence, and tensoring with $\mathcal{O}(n)$, we obtain exact sequences

$$0 \longrightarrow \mathcal{O}_X \longrightarrow \mathcal{O}_X(n)^{r_n} \longrightarrow \mathcal{E}_n \longrightarrow 0 \quad \text{with} \quad \mathcal{E}_n \in \mathscr{P}(X). \tag{6.28}$$

For $\mathcal{F} \in Coh(X)$, tensoring (6.28) with $\mathcal{F}$, we obtain exact sequences

$$0 \longrightarrow \mathcal{F} \longrightarrow \mathcal{F}(n)^{r_n} \longrightarrow \mathcal{F} \otimes \mathcal{E}_n \longrightarrow 0.$$

By Theorem 6.2.15(3), when n is large enough, $\mathcal{F}(n)^{r_n} \in \mathfrak{Coh}(f, X)$. By the usual argument in homological algebra, with $d = \dim X$, there is an exact sequence of finite length

$$0 \longrightarrow \mathcal{F} \longrightarrow \mathcal{F}_0 \longrightarrow \mathcal{F}_1 \longrightarrow \cdots \longrightarrow \mathcal{F}_d \longrightarrow 0$$
$$\text{with} \quad \mathcal{F}_k \in \mathfrak{Coh}(X, f). \tag{6.29}$$

So, the resolution theorem (3.5.3) applies to the inclusion $\mathfrak{Coh}(f, X)^{op} \hookrightarrow Coh(X)^{op}$, and we have, respectively, homotopy equivalence and isomorphisms:

$$\begin{cases} \mathbf{K}\mathfrak{Coh}(X, f) = \mathbf{K}\mathfrak{Coh}(X, f)^{op} \cong \mathbf{K}(Coh(X)^{op}) = \mathbf{G}(X) & \text{of } K\text{-theory spaces} \\ K_n\mathfrak{Coh}(X, f) = K_n(X, f)^{op} \cong K_n(Coh(X)^{op}) = G_n(X) & \text{of } K\text{-groups } \forall n \geq 0. \end{cases}$$

Combining with (6.27), we have the **pushforward** maps

$$f_* : \begin{cases} \mathbf{G}(X) \xleftarrow{\sim} \mathbf{K}\mathfrak{Coh}(X, f) \longrightarrow \mathbf{G}(Y) & \text{of } K\text{-theory spaces} \\ G_n(X) \xleftarrow{\sim} K_n\left(\mathfrak{Coh}(X, f)\right) \longrightarrow G_n(Y) & \text{of the } K\text{-groups } \forall\ n \geq 0. \end{cases} \tag{6.30}$$

We summarize all these in the following proposition.

Proposition 6.2.17. *Let $f : X \longrightarrow Y$ be a proper morphism of Noetherian schemes. Assume X has an ample bundle. Then there is a pushforward map, f_*, as above (6.30). Further, let $X \xrightarrow{f} Y \xrightarrow{g} Z$ be proper morphisms of Noetherian schemes. Assume that both X, Y possess ample invertible sheaves. Then the* *following diagram:*

$$\begin{array}{ccc} G_n(X) & \xrightarrow{f_*} & G_n(Y) \\ & \searrow^{(gf)_*} & \big\downarrow g_* \\ & & G_n(Z) \end{array} \qquad \textit{commutes } \forall n \geq 0.$$

Proof. Everything regarding the existence of f_* has been justified in Definition 6.27. The commutativity of the diagram is proved as in (5.4.2). Refer to Exercise 6.6.5. ■

6.2.3. *A projection formula*

Recall the definition (6.1.2) of the action of $K_0(X)$ on $G_n(X)$. Similar to the projection formula (3.5.10) in the affine case, we have the following projection formula for G-theory.

Proposition 6.2.18. *Let $f : X \longrightarrow Y$ be a proper morphism of Noetherian schemes. Assume f has finite Tor dimension and both X, Y support ample bundles. So, both the pullback f^* and pushforward f_* maps are defined. Then for $x \in K_0X$, $y \in G_n(Y)$, we have $f_*(x \cdot f^*y) = f_*(x) \cdot y \in G_n(Y)$. So the diagram*

$$\begin{array}{ccccc} K_0(X) \otimes G_n(Y) & \xrightarrow{1 \times f^*} & K_0(X) \otimes G_n(X) & \xrightarrow{\cdot} & G_n(X) \\ \big\downarrow f_* \otimes 1 & & & & \big\downarrow f_* \\ K_0(Y) \otimes G_n(Y) & & \xrightarrow{\qquad\qquad \cdot \qquad\qquad} & & G_n(Y) \end{array} \qquad \textit{commutes.}$$

$$(6.31)$$

Proof. Let $\mathfrak{Coh}(f, Y) \subseteq Coh(Y)$ and $\mathfrak{Coh}(X, f) \subseteq Coh(X)$ be as defined above (6.11, 6.26). Let $\mathscr{P}(X, f)$ be the full subcategory

$$\mathscr{P}(X, f) = \left\{ \mathcal{E} \in \mathscr{P}(X) : R^k f_* \mathcal{E} = 0, \ \forall\, k \geq 1 \right\}$$
$$= \mathscr{P}(X) \cap \mathfrak{Coh}(X, f). \qquad (6.32)$$

By (6.5.6), the pushforward functor f_* restricts to an exact functor

$$f_* : \mathscr{P}(X, f) \longrightarrow \mathbb{H}(Y), \quad \text{defining maps}$$
$$f_* : K_0\mathscr{P}(X, f) \longrightarrow K_0(\mathbb{H}(Y)) \cong K_0(Y). \qquad (6.33)$$

By construction, for $\mathcal{F} \in \mathscr{P}(X)$, the resolution (6.29) is indeed a resolution in $\mathscr{P}(X, f)$. So, by an application of the resolution theorem (3.5.3 or 3.8.3) to $\mathscr{P}(X, f)^{op} \longrightarrow \mathscr{P}(X)^{op}$, we have an isomorphism $K_0\mathscr{P}(X, f) \cong K_0(X)$. Thus, combining with (6.33), we obtain the natural maps $f_* : K_0(X) \longrightarrow K_0(Y)$. We restrict the diagram (6.31) to $K_0(\mathscr{P}(X, f)) \times K_n(\mathfrak{Coh}(f, Y))$ and prove that the diagram

$$\begin{array}{ccc} K_0(\mathscr{P}(X,f)) \times K_n\left(\mathfrak{Coh}(f,Y)\right) & \xrightarrow{1\cdot f^*} & K_n\mathfrak{Coh}(X,f) \\ \Big\downarrow{\scriptstyle f_*\times 1} & & \Big\downarrow{\scriptstyle f_*} \\ K_0(Y) \times G_n(Y) & \xrightarrow{\quad\cdot\quad} & G_n(Y) \end{array} \tag{6.34}$$

commutes. In particular, for $\mathcal{E} \in (\mathscr{P}(X, f))$, $\mathcal{F} \in \mathfrak{Coh}(f, Y)$, as the top horizontal arrow indicates, we claim that $R^k f_*\mathcal{E} \otimes_X f^*\mathcal{F} = 0\ \forall k \geq 1$. This claim is established separately in (6.5.6). The second vertical map f_* is induced by the exact functor $\mathcal{G} \mapsto f_*\mathcal{G}$.

Fix $\mathcal{E} \in \mathscr{P}(X, f)$, and let $x := [\mathcal{E}] \in K_0(\mathscr{P}(X, f))$. Then the map $y \mapsto f_*(x \cdot f^*y)$ is induced by the exact functor

$$\mathfrak{Coh}(f, Y) \longrightarrow Coh(Y) \quad \text{sending} \qquad \mathcal{F} \mapsto f_*(\mathcal{E} \otimes_X f^*\mathcal{F}). \tag{6.35}$$

For such a fixed $\mathcal{E} \in \mathscr{P}(X, f)$, we establish that the other map $y \mapsto f_*x \cdot y$ is also induced by the same functor (6.35). By definition, $f_*([\mathcal{E}]) = [f_*\mathcal{E}] \in K_0(\mathbb{H}(Y))$. So,

$$\forall\, \mathcal{E} \in \mathscr{P}(X, f), \quad f_*([\mathcal{E}]) = \sum_{i=0}^{d} (-1)^i [P_i] \in K_0(Y), \quad \text{where}$$

$$0 \longrightarrow P_d \longrightarrow \cdots \longrightarrow P_1 \longrightarrow P_0 \longrightarrow f_*\mathcal{E} \longrightarrow 0$$

is a $\mathscr{P}(Y)$-resolution of $f_*\mathcal{E}$. Since $\mathscr{E} \in \mathscr{P}(X)$, for any $\mathcal{F} \in Coh(Y)$, we have $Tor_k^Y(f_*\mathcal{E}, \mathcal{F}) = 0 \iff Tor_k^Y(\mathcal{O}_X, \mathcal{F}) = 0$. So,

$$\mathfrak{Coh}(f, Y) = \left\{\mathcal{F} \in Coh(Y) : Tor_k^Y(f_*\mathcal{E}, \mathcal{F}) = 0,\ \forall k \geq 0\right\}, \quad \text{and} \tag{6.36}$$

$G_n(Y) = K_n(\mathfrak{Coh}(f,Y))$. For $\mathcal{F} \in \mathfrak{Coh}(f,Y)$, the sequence

$$0 \longrightarrow P_d \otimes_Y \mathcal{F} \longrightarrow \cdots \longrightarrow P_1 \otimes_Y \mathcal{F} \longrightarrow P_0 \otimes_Y \mathcal{F} \longrightarrow f_*\mathcal{E} \otimes_Y \mathcal{F} \longrightarrow 0 \quad \text{is exact.}$$

Consider the exact sequence of exact functors $\mathfrak{Coh}(f,Y) \longrightarrow Coh(Y)$:

$$0 \longrightarrow P_d \otimes_Y - \longrightarrow \cdots \longrightarrow P_1 \otimes_Y - \longrightarrow P_0 \otimes_Y - \longrightarrow f_*\mathcal{E} \otimes_Y - \longrightarrow 0.$$

With $x = [\mathcal{E}] \in K_0(X)$, by corollary (3.4.5), it follows that the map

$$K_n(\mathfrak{Coh}(F,Y)) \longrightarrow G_n(Y) \quad \text{sending} \quad y \mapsto f_*x \cdot y$$

is induced by the exact functor

$$\mathfrak{Coh}(f,Y) \longrightarrow Coh(Y) \quad \text{sending} \qquad \mathcal{F} \mapsto f_*\mathcal{E} \otimes \mathcal{F}. \qquad (6.37)$$

There is a natural map (consult [Hr, Ex. II.1.18])

$$f_*\mathcal{E} \otimes_Y \mathcal{F} \longrightarrow f_*(\mathcal{E} \otimes_X f^*\mathcal{F}).$$

In this case, this map is an isomorphism, which we establish separately (6.5.6). Therefore, the functors (6.35) and (6.37) are naturally equivalent. The proof is complete. ■

The following is another useful formula.

Proposition 6.2.19. *Consider the following pullback diagram of Noetherian schemes*

$$\begin{array}{ccc} Z \times_Y X & \xrightarrow{\mathfrak{g}} & X \\ {\scriptstyle \mathfrak{f}}\downarrow & & \downarrow{\scriptstyle f} \\ Z & \xrightarrow[g]{} & Y \end{array} \qquad \textit{we write} \quad W := Z \times_Y X.$$

Assume all the schemes support ample bundles. Assume f is proper and g has finite Tor dimension. Also, assume that X, Z are Tor independent in the following sense:

$$Tor_i^{\mathcal{O}_{Y,y}}(\mathcal{O}_{Z,z}, \mathcal{O}_{X,x}) = 0 \quad \forall\, i \geq 1, y = f(x) = g(z)$$

(This condition is satisfied if f or g is flat.) *Then the diagram*

$$\begin{array}{ccc} G_n(X) & \xrightarrow{\mathfrak{g}^*} & G_n(W) \\ {\scriptstyle f_*}\downarrow & & \downarrow{\scriptstyle \mathfrak{f}_*} \\ G_n(Y) & \xrightarrow[g^*]{} & G_n(Z) \end{array} \qquad \textit{commutes.}$$

Proof. Because of the Tor independence condition, given a $\mathscr{P}(Y)$-resolution of $\mathcal{O}_Z$:

$$\begin{cases} \cdots \longrightarrow P_1 \longrightarrow P_0 \longrightarrow \mathcal{O}_Z \longrightarrow 0 & \text{we obtain} \\ \cdots \longrightarrow f^*P_1 \longrightarrow f^*P_0 \longrightarrow \mathcal{O}_W \longrightarrow 0 & \end{cases} \tag{6.38}$$

a $\mathscr{P}(X)$-resolution of $\mathcal{O}_W$. From this it follows that $\mathfrak{g}$ has finite Tor dimension. Therefore, the pullback maps $\mathfrak{g}^*$ are defined $\forall n \geq 0$. Since $\mathfrak{f}$ is proper, the pushforward maps $\mathfrak{f}_*$ are also defined. We use the notation $\mathfrak{Coh}(\mathfrak{g}, X) \subseteq Coh(X)$, $\mathfrak{Coh}(X, f) \subseteq Coh(X)$, as in (6.11, 6.26). For $\mathcal{F} \in \mathfrak{Coh}(\mathfrak{g}, X) \cap \mathfrak{Coh}(X, F)$, we prove separately (6.5.7) that

$$f_*\mathcal{F} \in \mathfrak{Coh}(g, Y), \qquad \mathfrak{g}^*\mathcal{F} \in \mathfrak{Coh}(W, \mathfrak{f}), \quad \text{and} \quad g^*f_*\mathcal{F} \cong \mathfrak{f}_*\mathfrak{g}^*\mathcal{F}. \tag{6.39}$$

It follows from this that the diagram

$$\begin{array}{ccc} K_n\left(\mathfrak{Coh}(\mathfrak{g}, X) \cap \mathfrak{Coh}(f, X)\right) & \xrightarrow{\ \mathfrak{g}^*\ } & K_n\left(\mathfrak{Coh}(W, \mathfrak{f})\right) \\ \downarrow{\scriptstyle f_*} & & \downarrow{\scriptstyle \mathfrak{f}_*} \\ K_n\left(\mathfrak{Coh}(g, Y)\right) & \xrightarrow[\ g^*\ ]{} & G_n(Z) \end{array} \qquad \text{commutes.}$$

By resolution theorem (3.5.3),

$$K_n\left(\mathfrak{Coh}(g, Y)\right) \cong G_n(Y), \quad K_n\left(\mathfrak{Coh}(W, \mathfrak{f})\right) \cong G_n(W),$$
$$\text{and} \quad K_n\left(\mathfrak{Coh}(\mathfrak{g}, X)\right) \cong G_n(X).$$

So, it would be enough to show that $K_n\left(\mathfrak{Coh}(\mathfrak{g}, X) \cap \mathfrak{Coh}(f, X)\right) \cong K_n\left(\mathfrak{Coh}(\mathfrak{g}, X)\right)$. For $\mathcal{F} \in \mathfrak{Coh}(\mathfrak{g}, X)$ in the sequence (6.29), $\mathcal{F}_i \in \mathfrak{Coh}(\mathfrak{g}, X) \cap \mathfrak{Coh}(f, X)$. So, by resolution theorem (3.5.3) applied to opposite categories, $K_n\left(\mathfrak{Coh}(\mathfrak{g}, X) \cap \mathfrak{Coh}(f, X)\right) \cong K_n\left(\mathfrak{Coh}(\mathfrak{g}, X)\right)$. The proof is complete. ∎

6.3. *G*-Theory of Affine and Projective Bundles

The following is an extension of homotopy invariance theorem (5.2.4), which is still known as homotopy invariance.

Proposition 6.3.1. *Let $f : P \longrightarrow X$ be a flat morphism of Noetherian schemes, whose fibers are affine spaces (i.e., $\cong$ Spec $(k[X_1, \ldots, K_n])$ where k is a field). Then the pullback maps $f^* : G_n(P) \longrightarrow G_n(X)$ are isomorphisms $\forall n \geq 0$. Equivalently, the map $\mathbf{G}P \longrightarrow \mathbf{G}X$ is a homotopy equivalence.*

Proof. The equivalence of two statements follows from Whitehead's theorem (C.3.8). Note that the closed and open subschemes of X inherit the hypotheses of the proposition. We will use Noetherian induction. If the proposition fails, there is a smallest closed subscheme $X_0 \subseteq X$ for which it fails. Replacing X_0 by X, we assume that the proposition is valid for all proper closed subschemes of X. Let $Z \subseteq X$ be a closed subset and $U = X - Z$. Write $P_U = f^{-1}U := U \times_X P$ and $P_Z = f^{-1}Z := Z \times_X P$. By localization theorem (6.1.11), and its natural properties (6.6.8), we have the following commutative diagram of exact sequences of K-groups:

$$\begin{array}{ccccccccc}
\cdots \longrightarrow & G_n(Z) & \longrightarrow & G_n(X) & \longrightarrow & G_n(U) & \longrightarrow & G_{n-1}(Z) & \longrightarrow \cdots \\
& \wr\downarrow & & \downarrow f^* & & \downarrow & & \downarrow\wr & \\
\cdots \longrightarrow & G_n(P_Z) & \longrightarrow & G_n(P) & \longrightarrow & G_n(P_U) & \longrightarrow & G_{n-1}(P_Z) & \longrightarrow \cdots
\end{array} \tag{6.40}$$

ending at $n = 0$ terms. By induction hypotheses, the maps on $G_n(Z)$ are isomorphic. We can assume X is irreducible. If not, write $X = Z_1 \cup Z_2$, with $Z_1 \neq X$, $Z_2 \neq X$. Now $U_1 := X - Z_1 = Z_2 - (Z_1 \cap Z_2)$. The proposition holds for Z_1, Z_2, $(Z_1 \cap Z_2)$. Using the diagram (6.40) for the pair $(Z_1, Z_1 \cap Z_2)$, it follows that the maps $G_n(U_1) \xrightarrow{\sim} G_n(P_{U_1})$ are isomorphisms. Now, use the diagram (6.40) for the pair (X, Z_1), and it follows that the proposition holds for X. So, we assume that X is irreducible. By Dévissage (6.1.10), we can also assume that X is reduced.

Let $x \in X$ denote the generic point of X, and $\kappa(x)$ denote the residue field of x. Let S be the directed set of all closed subsets $Z \subseteq X$, with $x \notin Z$, meaning $Z \neq X$. We take the limit of the diagram (6.40), obtaining the commutative diagram

$$\begin{array}{ccccccccc}
\cdots \longrightarrow & \lim_S G_n(Z) & \longrightarrow & G_n(X) & \longrightarrow & \lim_S G_n(U) & \longrightarrow & \lim_S G_{n-1}(Z) & \longrightarrow \cdots \\
& \wr\downarrow & & \downarrow f^* & & \downarrow & & \downarrow\wr & \\
\cdots \longrightarrow & \lim_S G_n(P_Z) & \longrightarrow & G_n(P) & \longrightarrow & \lim_S G_n(P_U) & \longrightarrow & \lim_S G_{n-1}(P_Z) & \longrightarrow \cdots
\end{array}$$

To establish whether f^* is an isomorphism, it suffices to prove that the maps $\lim_S G_n(U) \xrightarrow{\sim} \lim_S G_n(P_U)$ are isomorphisms $\forall n \geq 0$. We have $\lim_S U = \operatorname{Spec}(\kappa(x))$, and $\lim_S P_U = \kappa(x) \times_X P$. By hypothesis, $\lim_S P_U = \kappa(x) \times_X P \cong \operatorname{Spec}(\kappa(x)[T_1, \ldots, T_n])$. By the homotopy invariance theorem (5.2.4), along with (6.2.5), we have

$$\lim_S G_n(U) \cong G_n(\kappa(x)) \xrightarrow{\sim} G_n(\kappa(x)[T_1, \ldots, T_n]) = \lim_S G_n(P_U).$$

So, the proof is complete. ∎

Readers are referred to [Hr, §II.7] for construction of $\mathbb{P}(\mathscr{E}) := \operatorname{Proj}(\mathcal{S}\mathrm{ym}(\mathscr{E}))$, for locally free sheaves $\mathscr{E}$ on a scheme X. Here $\mathcal{S}\mathrm{ym}(\mathscr{E})$ denotes the symmetric algebra of $\mathscr{E}$ over X. The following is the statement of the projective bundle theorem, for G-theory.

Proposition 6.3.2. *Let X be a Noetherian scheme and let $\mathscr{E} \in \mathscr{P}(X)$ be a locally free sheaf with $rank(\mathscr{E}) = r$. Write $\mathbb{P}\mathscr{E} = \operatorname{Proj}(\mathcal{S}\mathrm{ym}(\mathscr{E}))$. Let $f : \mathbb{P}\mathscr{E} \longrightarrow X$ be the structure map. Then there is a $K_0(\mathbb{P}\mathscr{E})$-module isomorphism*

$$K_0(\mathbb{P}\mathscr{E}) \otimes_{K_0(X)} G_n(X) \xrightarrow{\sim} G_n(\mathbb{P}\mathscr{E}) \quad \textit{sending} \quad y \otimes x \mapsto y \cdot f^*x. \tag{6.41}$$

Equivalently, with notation $\zeta = [\mathcal{O}(-1)] \in K_0(\mathbb{P}\mathscr{E})$, there is an exact sequence

$$0 \longrightarrow G_n(X)[t] \xrightarrow{\lambda(-t)} G_n(X)[t] \xrightarrow{\mathfrak{q}_*} G_n(\mathbb{P}\mathscr{E}) \longrightarrow 0, \tag{6.42}$$

where t is an indeterminate, and

$$\lambda(t) := \lambda_{\mathscr{E}}(t) := \sum_{k=0}^{r} \left[\Lambda^k \mathscr{E}\right] t^k \in K_0(X)[t], \quad \textit{and}$$
$$\mathfrak{q}_*(xt^k) = \zeta^k \cdot f^*x \quad \forall x \in G_n(X).$$

Proof. It is obvious that $G_n[t] \cong K_0(X)[t] \otimes_{K_0(X)} G_n(X)$. By (6.3.3),

$$0 \longrightarrow K_0(X)[t] \xrightarrow{\lambda(-t)} K_0(X)[t] \xrightarrow{\mathfrak{q}_*} K_0(\mathbb{P}\mathscr{E}) \longrightarrow 0$$

is exact, where $\mathfrak{q}_*(xt^k) = \zeta^k \cdot f^*x$

This establishes the equivalence of (6.41) and (6.42). Note that $\lambda(-t) \in K_0(X)[t]$ is a monic polynomial. So, the multiplication map $\lambda(-t) : G_n(X)[t] \longrightarrow G_n(X)[t]$ is injective.

As usual, we use Noetherian induction. So, assume that the theorem is valid for all closed subsets $Z \subseteq X$, $Z \neq X$. Let $U \subseteq X$ be any open subset of X such that $\mathscr{E}_{|U}$ is trivial, and let $A = \Gamma(U, \mathcal{O}_X)$. We write

$$\begin{cases} \mathbb{P}\mathscr{E}_Z := \mathbb{P}\mathscr{E} \times_X Z \cong \mathbb{P}\left(\mathscr{E}_{|Z}\right) \\ \mathbb{P}\mathscr{E}_U := \mathbb{P}\mathscr{E} \times_X U \cong \mathbb{P}\left(\mathscr{E}_{|U}\right) \cong \mathrm{Proj}\left(A[T_0, T_1, \ldots, T_{r-1}]\right) = \mathbb{P}_A^{r-1}. \end{cases}$$

Consider the following diagram map of the localization sequences:

$$\begin{array}{ccccccccc} \cdots \longrightarrow & G_n(Z)^r[t] & \longrightarrow & G_n(X)^r[t] & \longrightarrow & G_n(U)^r[t] & \longrightarrow & G_{n-1}(Z)[t] & \longrightarrow \cdots \\ & \Big\downarrow{\scriptstyle \lambda^Z(-t)} & & \Big\downarrow{\scriptstyle \lambda(-t)} & & \Big\downarrow{\scriptstyle \lambda^U(-t)} & & \Big\downarrow & \\ \cdots \longrightarrow & G_n(Z)^r[t] & \longrightarrow & G_n(X)^r[t] & \longrightarrow & G_n(U)^r[t] & \longrightarrow & G_{n-1}(Z)[t] & \longrightarrow \cdots \\ & \Big\downarrow{\scriptstyle \mathfrak{q}_*^Z} & & \Big\downarrow{\scriptstyle \mathfrak{q}_*} & & \Big\downarrow{\scriptstyle \mathfrak{q}_*^U} & & \Big\downarrow & \\ \cdots \longrightarrow & G_n(\mathbb{P}\mathscr{E}_Z) & \longrightarrow & G_n(\mathbb{P}\mathscr{E}) & \longrightarrow & G_n(\mathbb{P}\mathscr{E}_U) & \longrightarrow & G_{n-1}(\mathbb{P}\mathscr{E}_Z) & \longrightarrow \cdots \end{array} \tag{6.43}$$

where superscript Z or U indicates the maps corresponding to Z or U. All the multiplications $\lambda^Z(-t)$, $\lambda(-t)$, $\lambda^U(-t)$ are known to be injective. By the induction hypothesis, we assume that $\mathfrak{q}_*^Z$ is surjective and $\ker(\mathfrak{q}_*^Z) = Im(\lambda^Z(-t))$. All the rows are exact. By the Snake lemma, it suffices to prove the same for the U-sequence. So, we replace U by X and assume $X = \mathrm{Spec}\,(A)$ is affine, and $\mathscr{E}$ is trivial. Further, $\mathbb{P}\mathscr{E} = \mathrm{Proj}\,(A[T_0, \ldots, T_{r-1}]) = \mathbb{P}_A^{r-1}$. We denote $E = \Gamma(X, \mathscr{E}) \cong A^r$.

Write $S = A[T_0, T_1, \ldots, T_{r-1}]$, and use the notation as in (5.1.1). Let $\mathscr{B} \subseteq Coh_{gr}^+(S)$ be the full subcategory of objects $M \in Coh_{gr}^+(S)$ such that $M_n = 0\ \forall n \gg 0$. By (6.6.9), the functor $\mathfrak{q} : Coh_{gr}^+(S) \longrightarrow Coh\left(\mathbb{P}_A^{r-1}\right)$, where $\mathfrak{q}(M) = \widetilde{M}$, establishes an equivalence of the quotient category $\frac{Coh_{gr}^+(S)}{\mathscr{B}} \xrightarrow{\sim} Coh\left(\mathbb{P}_A^{r-1}\right)$. We treat A as a graded ring, concentrated at degree zero. We have an inclusion functor $\iota : Coh_{gr}^+(A) \hookrightarrow \mathscr{B}$, induced by the graded homomorphism $S \longrightarrow A$. The Dévissage theorem (3.6.2) applies to ι.

So, $K_n(Coh^+_{gr}(A)) \cong K_n(\mathscr{B})$. Consider the diagram

$$\begin{array}{ccccccc}
 & & G_n(A)[t] & \xrightarrow{\lambda(-t)} & G_n(A)[t] & & \\
 & & \tau_A \downarrow \wr & & \wr \downarrow \tau_S & & \\
\cdots \longrightarrow & & K_n(Coh^+_{gr}(A)) & \xrightarrow[\iota_*]{} & K_n(Coh^+_{gr}(S)) & \xrightarrow[q_*]{} & G_n\left(\mathbb{P}^{r-1}_A\right) \longrightarrow \cdots
\end{array} \tag{6.44}$$

In this diagram, the second horizontal line is a Localization exact sequence (3.6.9), in which we identified $K_n(Coh^+(A)) \cong K_n(\mathscr{B})$. The two vertical isomorphisms are obtained by (5.1.5). We claim that the rectangle in (6.44) commutes.

Recall that the map $G_n(A)t^k \longrightarrow G_n(S)$ is induced by the functor $Coh(A) \longrightarrow Coh^+_{gr}(S)$ sending $M \mapsto M \otimes_A S(-k)$. Define functors

$$\Phi_k : Coh(A) \longrightarrow Coh^+_{gr}(S) \quad \text{by sending}$$
$$M \mapsto \left(S(-k) \otimes_A \Lambda^k E\right) \otimes_A M.$$

Note that $\forall\ x \in G_n(A)t^0$, we have

$$\tau_S\left(\lambda(-t)x\right) = \sum_{k=0}^{r}(-1)^k \left(S(-k) \otimes_A \Lambda^k E\right)_* x = \sum_{k=0}^{r}(-1)^k \left(\Phi_k\right)_* x. \tag{6.45}$$

At the degree zero $G_n(A)t^0$, the map $\iota_*\tau_A$ is induced by the following functor:

$$\Psi : Coh(A) \longrightarrow Coh^+_{gr}(S) \quad \text{by sending} \quad M \mapsto M. \tag{6.46}$$

Corresponding to the sequence $(T_0, T_1, \ldots, T_{r-1})$, the Koszul complex $\Lambda^\bullet S \otimes_A E$,

$$0 \longrightarrow S(-r) \otimes_A \Lambda^r E \longrightarrow \cdots \longrightarrow S(-1) \otimes_A \Lambda^1 E \longrightarrow S \otimes_A \Lambda^0 E \longrightarrow A \longrightarrow 0$$

is exact. Further, for $M \in Coh(A)$, the complex $(\Lambda^\bullet S \otimes_A E) \otimes_A M$ is also exact. It follows that

$$0 \longrightarrow \Phi_r \longrightarrow \cdots \longrightarrow \Phi_1 \longrightarrow \Phi_0 \longrightarrow \Psi \longrightarrow 0 \quad \text{is exact.}$$

It follows from (3.4.5), $\Psi_* = \sum_{k=0}^{r}(-1)^k \left(\Phi_k\right)_*$. Therefore, the rectangle in (6.44) commutes. In this diagram, the map $\lambda(-t)$ is injective. Therefore, so is ι_*. Consequently, the localization sequence (6.44) splits into short exact sequences, as follows:

$$\begin{array}{ccccccccc} 0 & \longrightarrow & G_n(A)[t] & \xrightarrow{\lambda(-t)} & G_n(A)[t] & \longrightarrow & G_n\left(\mathbb{P}_A^{r-1}\right) & \longrightarrow & 0 \\ & & \tau_A \downarrow \wr & & \wr\downarrow \tau_S & & \| & & \\ 0 & \longrightarrow & K_n(Coh_{gr}^+(A)) & \xrightarrow[\iota_*]{} & K_n(Coh_{gr}^+(S)) & \xrightarrow[\mathfrak{q}_*]{} & G_n\left(\mathbb{P}_A^{r-1}\right) & \longrightarrow & 0. \end{array}$$

The proof is complete. ■

The following is fairly classical, used above (6.3.2). A proof is also available in [FL, Thm. 2.3].

Theorem 6.3.3. *Let X be a Noetherian scheme and $\mathscr{E} \in \mathscr{P}(Y)$ with rank$(\mathscr{E}) = r$. Write $\mathbb{P}\mathscr{E} = \mathrm{Proj}\,(\mathrm{Sym}(\mathscr{E}))$, where $\mathrm{Sym}(\mathscr{E})$ denotes the symmetric algebra of $\mathscr{E}$ over Y. Let $f : \mathbb{P}\mathscr{E} \longrightarrow Y$ be the structure map. As usual, $\mathcal{O}(n) = \mathrm{Sym}(\mathscr{E})(n)$. Let $\zeta = [\mathcal{O}(-1)] \in K_0(\mathbb{P}\mathscr{E})$ be as above. Then as a $K_0(Y)$-module, $K_0(\mathbb{P}\mathscr{E})$ is freely generated by $1, \zeta, \ldots, \zeta^{r-1}$.*

Proof. In fact, this is the K_0-case of (7.2.3). However, since the Grothendieck group is so basic, we give a direct proof. Write $X = \mathbb{P}\mathscr{E}$. Consider the pullback map $f^* : K_0(Y) \longrightarrow K_0(X)$. Recall (6.6.6) that the pushforward map $f_* : K_0(X) \longrightarrow K_0(Y)$ is induced by the exact functor $\mathscr{P}(X, f) \longrightarrow \mathbb{H}(Y)$, where $\mathscr{P}(X, f) = \{\mathcal{F} \in \mathscr{P}(X) : R^k f_* \mathcal{F} = 0,\ \forall k \geq 1\}$.

By (7.1.2) or by Serre duality, $\mathcal{O}(-n) \in \mathscr{P}(X, f)$ $\forall n \leq r-1$. Further, by (6.5.4), we have $f_*(\mathcal{O}_X) \cong \mathcal{O}_Y$ and $f_*(\mathcal{O}(-n)) = 0$, $\forall 1 \leq n \leq r-1$. So, $f_*(\zeta^n) = 0 \in K_0(Y)$, $\forall 1 \leq n \leq r-1$, and $f_* 1 = 1 \in K_0(Y)$. Now, we prove $1, \zeta, \ldots, \zeta^{r-1}$ are linearly independent over $K_0(Y)$. Suppose

$$f^*(a_0)\zeta^{r-1} + f^*(a_1)\zeta^{r-2} + \cdots + f^*(a_{r-1})\zeta + f^*(a_{r-1}) = 0 \in K_0(X)$$
$$a_i \in K_0(Y).$$

By an application of the projection formula (6.6.7(2)), to this equation, we obtain $a_0 = 0$. Since, ζ is a unit in $K_0(X)$, it follows by induction that $a_i = 0, \forall i$. So, $1, \zeta, \ldots, \zeta^{r-1}$ are linearly independent.

Define a map

$$\lambda : K_0(Y)[t] \longrightarrow K_0(X) \quad \text{by} \qquad \lambda(t) = \zeta^{-1} = [\mathcal{O}_X(1)] .$$

Consider the polynomial

$$p_{\mathscr{E}}(t) = \sum_{i=0}^{r} (-1)^i \left[\Lambda^i \mathscr{E}\right] t^{r-i} \in K_0(Y)[t].$$

From the Koszul complex resolution

$$0 \longrightarrow \mathcal{O}_X(-r) \otimes \Lambda^r \mathscr{E} \longrightarrow \cdots \longrightarrow \mathcal{O}_X(-2) \otimes \Lambda^2 \mathscr{E} \longrightarrow \mathcal{O}_X(-1) \otimes \mathscr{E} \longrightarrow \mathcal{O}_X \longrightarrow 0 ,$$

it follows that $p\left(\zeta^{-1}\right) = 0$. So, the map λ factors through

$$\overline{\lambda} : \frac{K_0(Y)[t]}{(p_{\mathscr{E}}(t))} \longrightarrow K_0(X).$$

Since $p_{\mathscr{E}}(t)$ is monic, the left-hand side is freely generated by $1, t, \ldots, t^{r-1}$. We already proved that $1, \zeta, \ldots, \zeta^{r-1}$ are linearly independent, and hence so is $1, \zeta^{-1}, \ldots, \zeta^{-(r-1)}$. So, $\overline{\lambda}$ is injective. Now, it follows, from the canonical resolution (7.1.7, 7.1.10, 7.2.2), that $\overline{\lambda}$ maps onto $K_0(\mathscr{R}_0(X)) \cong K_0(X)$. ■

6.4. Filtration by Support

In this section, we filter the category $Coh(X)$ by support of its objects. For clarity, we define co-dimension and make a few comments on the same.

Notation 6.4.1. Let X be a scheme. Consistent with [Hr, p. 5, 86], dimension is a topological invariant, not scheme theoretic. So, $\dim X = \dim X_{red}$. Let $Z \subseteq X$ be an irreducible closed subset of X (*not to be mixed up with subscheme*). Then the codimension $co\dim(Z, X)$ of Z in X is defined to be the supremum of integers n such that there is a chain of strictly increasing irreducible closed subsets of X:

$$Z = Z_0 \subset Z_1 \subset \cdots \subset Z_n. \qquad \text{So,} \qquad co\dim(Z,X) = \dim \mathcal{O}_{X,z}$$

where $z \in Z$ is the generic point of Z. For any closed subset $Y \subseteq X$, we define

$$co\dim(Y) := co\dim(Y,X) =$$
$$\inf\{co\dim(Z,X) : Z \subseteq Y, \text{ with } Z \text{ irreducible}\}.$$

So, $co\dim(Y) = \min\{\dim \mathcal{O}_{X,y_1}, \ldots, \dim \mathcal{O}_{X,y_k}\}$, where $Y = Y_1 \cup \cdots \cup Y_k$ is an irreducible decomposition of Y, and $y_i \in Y_i$ is the generic point of Y_i.

We provide some clarification on support of a coherent sheaf.

Remark 6.4.2. Let X be a Noetherian scheme. Recall that for $\mathcal{F} \in Coh(X)$, the support of $\mathcal{F}$ is defined to be $\mathrm{Supp}(\mathcal{F}) = \{x \in X; \mathcal{F}_x \neq 0\}$, which is a closed subset of X. Write $Z(\mathcal{F}) := \mathrm{Supp}(\mathcal{F})$. The support $Z(\mathcal{F})$ is to be considered as a closed subset, not a subscheme. However, $Z(\mathcal{F}) \subseteq X$ possesses many possible closed subscheme structures, including the reduced induced subscheme structure $Z(\mathcal{F})_{\mathrm{red}}$ [Hr, pp. 86]. This is given by the largest ideal sheaf $\mathscr{I}_{\mathrm{red}} \subseteq \mathcal{O}_X$ so that $(\mathscr{I}_{\mathrm{red}})_x \neq \mathcal{O}_{X,x} \iff x \in Z(\mathcal{F})$. However, $\mathcal{F}$ may not qualify as an object of $Coh\,(Z(\mathcal{F})_{\mathrm{red}})$.

Another subscheme structure on $Z(\mathcal{F})$ that directly relates to $\mathcal{F}$ is defined as follows. Define the ideal sheaf $\mathscr{I}(\mathcal{F}) \subseteq \mathcal{O}_X$ by $\mathscr{I}(\mathcal{F})(U) = ann(\mathcal{F}(U)) \subseteq \mathcal{O}_X(U)$, the ideal of annihilators of $\mathcal{F}(U)$. Denote the closed subscheme $Z(\mathcal{F})_{ann} := \left(Z(\mathcal{F}), \frac{\mathcal{O}_X}{\mathscr{I}(\mathcal{F})}\right)$. For any ideal sheaf $\mathscr{I} \subseteq \mathscr{I}(\mathcal{F})$, let $\mathscr{I}$ define the closed subscheme $Y(\mathscr{I})$. This means $Y(\mathscr{I}) = \{x \in X : \mathscr{I}_x \neq \mathcal{O}_{x,X}\}$ and $\mathcal{O}_Y = \frac{\mathcal{O}_X}{\mathscr{I}}$. Then $Z(\mathcal{F}) \subseteq Y(\mathscr{I}) \subseteq X$ are closed subschemes, and $\mathcal{F}$ can be considered as a sheaf on $Y(\mathscr{I})$, as well.

However, by the Dévissage theorem (6.1.10), we have $G_n\,(Z(\mathcal{F})_{ann}) \cong G_n\,((Z(\mathcal{F}))_{\mathrm{red}})\ \forall n \geq 0$. Indeed, by the Dévissage theorem (6.1.10), it does not make any difference which subscheme structure we consider on $Z(\mathcal{F})$.

We make a convention that given any closed subset $Z \subseteq X$, unless stated otherwise, it is given the reduced induced subscheme structure.

Now we define a filtration of $Coh(X)$ by codimension of support of the objects.

Definition 6.4.3. Let X be a Noetherian scheme. For integers $k \geq 0$ and closed subsets $Z \subseteq X$, define the full subcategory $Coh_Z(X) \subseteq Coh(X)$, $Coh^k(X) \subseteq Coh(X)$ by

$$\begin{cases} Coh_Z(X) = \{\mathcal{F} \in Coh(X) : \operatorname{Supp}(\mathcal{F}) \subseteq Z\} \\ Coh^k(X) = \{\mathcal{F} \in Coh(X) : co\dim(\operatorname{Supp}(\mathcal{F})) \geq k\}. \end{cases}$$

Then both $Coh^k(X)$ and $Coh_Z(X)$ are Serre subcategories of $Coh(X)$. If $co\dim Z \geq k$, then $Coh_Z(X) \subseteq Coh^k(X)$ is a Serre subcategory. Further,

$$Coh^k(X) = \varinjlim_{Z} Coh_Z(X) \qquad Z \text{ runs over all closed subsets, with } co\dim Z \geq k. \tag{6.47}$$

For closed subsets $Z \subseteq X$, by Dévissage (6.1.10), we have $G_n(Z) \cong G_n(Coh_Z(X))$, $\forall n \geq 0$. With $d = \dim X$, we have a filtration

$$Coh(X) = Coh^0(X) \supseteq Coh^1(X) \supseteq \cdots \supseteq Coh^{d-1}(X) \supseteq Coh^d(X) \supseteq 0. \tag{6.48}$$

For the convenience of subsequent discussions, we set $Coh^k(X) := Coh(X)$, $\forall k \leq 0$ and $Coh^k(X) := 0$ $\forall k \geq d+1$. For an integer k, we also use the following notation:

$$X^{(k)} = \{x \in X : \dim(\mathcal{O}_{X,x}) = k\} \quad \text{the set of points of codimension } k. \tag{6.49}$$

We record the following immediate consequences.

Proposition 6.4.4. *Let X be a Noetherian scheme and $k \geq 0$, $n \geq 0$ be integers. Then*

$$G_n\left(Coh^k(X)\right) = \varinjlim_{Z} G_n(Z) \quad Z \textit{ runs over all closed subsets, with } co\dim Z \geq k.$$

Further,

1. *Let* $f : X \longrightarrow Y$ *be a flat morphism of Noetherian schemes. Then* $\forall k \geq 0$ *the pullback* f^* *induces exact functors*
$$f^* : Coh^k(Y) \longrightarrow Coh^k(X), \quad \text{and homomorphisms}$$
$$G_n\left(Coh^k(Y)\right) \longrightarrow G_n\left(Coh^k(X)\right).$$
2. *Now let* $\iota : I \longrightarrow \underline{Sch}$*, sending* $i \mapsto X_i$*, be an inverse system of flat (affine) maps. Assume* $X = \lim_{i \in I} X_i$ *exists. Then*
$$K_n\left(Coh^k(X)\right) = \lim_i K_n\left(Coh^k(X_i)\right).$$

Proof. The main part follows from (6.47) and (3.3.3). Then (1) follows from the results in commutative algebra that flat maps preserves codimension (height), and of course pullback f^* is exact.

In the case of (2), let $f_{ij} : X_i \longrightarrow X_j$ denote the transition maps, $f_i : X \longrightarrow X_i$ denote the natural maps. Then the pullback induce maps $f_{ij}^* : Coh^k(X_j) \longrightarrow Coh^k(X_i)$. It follows $Coh^k(X) = \lim_{i \in I} Coh^k(X_i)$, while we skip some of the details. Now, the proposition follows from (3.3.3). The proof is complete. ■

Theorem 6.4.5. *Let* X *be a Noetherian scheme, with* $d = \dim X$*. For* $x \in X$*, denote* $X_x = \mathrm{Spec}\,(\mathcal{O}_{X,x})$*. Then for all integers* $n \geq 0$*, there is a complex,*

$$0 \longrightarrow G_n(X) \xrightarrow{q} \bigoplus_{x \in X^{(0)}} K_n(Coh^0(X_x)) \xrightarrow{d_1} \cdots$$
$$\longrightarrow \bigoplus_{x \in X^{(n-1)}} K_1(Coh^{n-1}(X_x)) \xrightarrow[d_n]{} \bigoplus_{x \in X^{(n)}} K_0(Coh^n(X_x)) \longrightarrow 0. \tag{6.50}$$

Identifying, $\bigoplus_{x \in X^{(n-r)}} K_r(Coh^{n-r}(X_x)) \cong \bigoplus_{x \in X^{(n-r)}} K_r(\kappa(x))$*, we obtain the following complex, to be called the* **Gersten** K**-theory complexes** *of* X *:*

$$0 \longrightarrow G_n(X) \xrightarrow{q} \bigoplus_{x \in X^{(0)}} K_n(\kappa(x)) \xrightarrow{d_1} \bigoplus_{x \in X^{(1)}} K_{n-1}(\kappa(x)) \longrightarrow \cdots$$
$$\longrightarrow \bigoplus_{x \in X^{(n-2)}} K_2(\kappa(x)) \longrightarrow \bigoplus_{x \in X^{(n-1)}} K_1(\kappa(x)) \xrightarrow[d_n]{} \bigoplus_{x \in X^{(n)}} K_0(\kappa(x)) \longrightarrow 0. \tag{6.51}$$

These complexes are natural for closed immersions and flat morphisms.

Proof. By (6.6.11), we have

$$\frac{Coh^p(X)}{Coh^{p+1}(X)} \xrightarrow{\sim} \coprod_{x\in X^{(p)}} Coh^p\left(\mathcal{O}_{X,x}\right) = \coprod_{x\in X^{(p)}} Coh^p\left(X_x\right),$$

where $Coh^p(\mathcal{O}_{X,x})$ coincides with the finite length $\mathcal{O}_{X,x}$-modules. By the Dévissage theorem (3.6.2), we identify $\forall\ x \in X^{(p)}$, $K_n(Coh^p(X_x)) \cong K_n(\kappa(x))\ \forall n \geq 0$. Up to these identifications, (6.50) and (6.51) are isomorphic, and we establish the latter. So, for each $p \geq 0$, we have one localization sequence (3.6.9):

$$\begin{array}{l}\cdots \longrightarrow K_q\left(Coh^{p+1}(X)\right) \xrightarrow{\iota} K_q\left(Coh^p(X)\right) \xrightarrow{\mathfrak{q}} \bigoplus_{x\in X^{(p)}} K_q\left(\kappa(x)\right) \xrightarrow{\partial} \\ K_{q-1}\left(Coh^{p+1}(X)\right) \longrightarrow \cdots \longrightarrow K_0\left(Coh^p(X)\right) \longrightarrow \bigoplus_{x\in X^{(p)}} K_0\left(\kappa(x)\right) \longrightarrow 0.\end{array} \tag{6.52}$$

We combine three such localization sequences corresponding to $Coh^{p+1}(X) \hookrightarrow Coh^p(X)$, $Coh^p(X) \hookrightarrow Coh^{p-1}(X)$, $Coh^{p+2}(X) \hookrightarrow Coh^{p+1}(X)$, and obtain the following diagram:

$$\begin{array}{ccccc}
\bigoplus_{x\in X^{(p-1)}} K_{q+1}\left(\kappa(x)\right) & & & & K_{q-1}\left(Coh^{p+2}(X)\right) \\
\downarrow \partial & \searrow^{d_q} & & & \downarrow \iota \\
K_q\left(Coh^p(X)\right) & \xrightarrow{\mathfrak{q}} & \bigoplus_{x\in X^{(p)}} K_q\left(\kappa(x)\right) & \xrightarrow{\partial} & K_{q-1}\left(Coh^{p+1}(X)\right) \\
\downarrow \iota & & & \searrow_{d_{q+1}} & \downarrow \mathfrak{q} \\
K_q\left(Coh^{p-1}(X)\right) & & & & \bigoplus_{x\in X^{(p+1)}} K_{q-1}\left(\kappa(x)\right).
\end{array} \tag{6.53}$$

Denote $d := \mathfrak{q}\partial$. Thus, we obtain the required complex (6.51) given by the diagonal sequence, of broken arrows, with $n = p + q$. The natural properties for closed immersions and flat morphisms are also obvious from the construction. The proof is complete. ■

We consider the question of exactness of the Gersten complexes (6.51). The following provides a criteria for such exactness.

Proposition 6.4.6. *Let X be a Noetherian scheme, with $d = \dim X$. Then the sequence (6.51) is exact $\forall n \geq 0$ if and only if the maps $\iota : K_q\left(Coh^{p+1}(X)\right) \longrightarrow K_q\left(Coh^p(X)\right)$, induced by the functors $Coh^{p+1}(X) \hookrightarrow Coh^p(X)$, are zero, $\forall p \geq 0$.*

Proof. The condition $\iota = 0$ translates to $\mathfrak{q}$ is injective and ∂ is surjective. So, the assertion follows by chasing the diagram (6.53). ■

With notation as in (6.4.5), for future reference, denote,

$$\mathscr{K}_n^p(X) = \bigoplus_{x \in X^{(p)}} K_{n-p}(\kappa(x)), \quad \text{and let}$$

$$\mathscr{K}_n(X) := \{\mathscr{K}_n^p(X)\} \quad \text{denote the complex.} \tag{6.54}$$

Perhaps the following underscores the importance of the question of exactness of the Gersten complexes (6.51). Indeed, under certain regularity conditions, we would like to realize the Chow groups $A^n(X)$, in this manner. So, we have the following.

Proposition 6.4.7. *Let X be a Noetherian scheme, with* $d = \dim X$. *For integers $n \geq 0$, let $\underline{G_n}$ denote the sheaf of abelian groups, associated to the presheaf $U \mapsto G_n(U)$. Assume $\forall x \in X$, the sequences* (6.51), *corresponding to* Spec $(\mathcal{O}_{X,x})$, *are exact for all $n \geq 0$ (as characterized in* (6.4.6)). *Then the sheaf cohomology*

$$H^p(X, \underline{G_n}) \cong H^p(\mathscr{K}_n(X)) \qquad \forall p \geq 0 \quad \forall n \geq 0.$$

Proof. We sheafify the complexes $\mathscr{K}_n^p(X)$ by defining

$$\underline{\mathscr{K}_n^p}(U) = \bigoplus_{x \in U^{(p)}} K_{n-p}(\kappa(x)) \qquad \text{for open subsets } U.$$

It is clear that $\underline{\mathscr{K}_n^p}$ are flasque sheaves. Further, the complexes (6.51) have sheafification:

$$0 \longrightarrow \underline{G_n} \longrightarrow \underline{\mathscr{K}_n^0} \longrightarrow \underline{\mathscr{K}_n^1} \longrightarrow \cdots \longrightarrow \underline{\mathscr{K}_n^n} \longrightarrow 0.$$

In fact, this is a resolution is $\underline{G_n}$ by flasque sheaves $\underline{\mathscr{K}_n^p}$. This is because, for $x \in X$, the stalks of this complex are precisely the Gersten complex (6.51) of Spec $(\mathcal{O}_{X,x})$, which is exact by hypothesis. Since sheaf cohomology can be computed using flasque resolutions (see 6.5.1), we have $H^p(X, \underline{G_n}) \cong H^p(\mathscr{K}_n(X))$. The proof is complete. ■

6.4.1. *Gersten conjecture*

The following is known as the Gersten conjecture.

Conjecture 6.4.8 (Gersten). Suppose $(R, \mathfrak{m})$ is a regular local ring. Then the Gersten complexes (6.51) for $X = \mathrm{Spec}\,(R)$ are exact.

Quillen proved the conjecture as follows.

Theorem 6.4.9. *Let A be an essentially finite-type regular semilocal ring over a field k. Then the Gersten complexes* (6.51) *of* $X = \mathrm{Spec}\,(A)$ *are exact.*

Proof. First, by hypothesis, $A = S^{-1}R$, where $R = k[x_1, \ldots, x_n]$ is finitely generated k-algebra, S=$R - \cup_{i=1}^m \wp_i$, where $\{\wp_1, \ldots, \wp_m\} \subseteq \mathrm{Spec}\,(R)$ and $R_{\wp_i}$ are regular. We want to reduce the proof to the case when k is perfect. Let $k_0 \hookrightarrow k$ be the prime field of k. We write $R = \frac{k[X_1,\ldots,X_n]}{I}$.

1. **Claim:** We claim that there is an intermediate field extension $k_0 \hookrightarrow \kappa \hookrightarrow k$ and a subring $C \hookrightarrow R$ such that $k_0 \hookrightarrow \kappa$ is finitely generated, C is a finitely generated κ-algebra, and $R = C \otimes_\kappa k$. Further, with $P_i = \wp_i \cap C$, we have $\wp_i = P_i R$. Consequently, with $\mathcal{S} = C - \cup_{i=1}^m P_i$, and $B = \mathcal{S}^{-1}C$, we have (1) B is regular, (2) $A = B \otimes_\kappa k$.
   To see this, let $\kappa \subseteq k$ be a subfield, over k_0 generated by the coefficients of some set of generators of I and of preimages of $\wp_1, \ldots, \wp_m$. Let $J = \kappa[X_1, \ldots X_n] \cap I$, and $C = \frac{\kappa[X_1,\ldots X_n]}{J}$. So $C \subseteq R$. Let $P_i = \wp_i \cap C$. It follows $C \otimes_\kappa k = R$, and $\wp_i = P_i R$. Let $\mathcal{S} = C - \cup_{i=1}^m P_i$, and $B = \mathcal{S}^{-1}C$. Now $B_{P_i} \hookrightarrow A_{\wp_i}$ and $B_{P_i} \otimes_\kappa k = A_{\wp_i}$. So, $B_{P_i} \hookrightarrow A_{\wp_i}$ is a flat extension. Now, B_{P_i} is regular, since $A_{\wp_i}$ is regular [Mh, p. 155, Thm 51]. So, the claim is established.
2. Now, let $\mathscr{I}$ be the directed set of all finitely generated extensions $\kappa \hookrightarrow F \hookrightarrow k$. (So, F is finitely generated over k_0). Then $A = \lim_{\mathscr{I}} F \otimes_\kappa B$, where B is as in (1). Here for $F_1 \hookrightarrow F_2$ in $\mathscr{I}$, the maps $F_1 \otimes_\kappa B \longrightarrow F_2 \otimes_\kappa B$ are flat. So, (6.4.4) applies. We have

$$\begin{cases} \forall k \geq 0, \quad Coh^k(A) = \lim_{\mathscr{I}} Coh^k(F \otimes B), & \text{and} \\ K_n(Coh^k(A)) = \lim_{\mathscr{I}} K_n\,(F \otimes_\kappa B) \quad \forall n \geq 0 \end{cases} \tag{6.55}$$

3. By (6.4.6), it suffices to prove that the maps $K_n(Coh^{k+1}(A)) \longrightarrow K_n(Coh^k(A))$ induced by the functors $Coh^{k+1}(A) \hookrightarrow Coh^k(A)$ are zero maps. By (6.55), it is enough to assume $A = F \otimes_\kappa B$. Since $k_0 \hookrightarrow \kappa \hookrightarrow F$ are finitely generated, we can assume that $A = S^{-1}R$, where R is a finitely generated k_0-algebra. Since, A is a finite-type algebra over k_0 and is regular, it follows that $\Omega_{S^{-1}R/k_0}$ is projective [Hr, Ex. II.8.1]. So, there is $f \in S$ such that Ω_{R_f/k_0} is projective. Again, by replacing R by R_f, we can assume that R is regular [Hr, Ex. II.8.1].

So, we are reduced to the case that R is regular over a perfect field k, and $A = S^{-1}R$, where S is as above. Therefore, R is also smooth.

We prove that the maps $K_n\left(Coh^{p+1}(A)\right) \longrightarrow K_n\left(Coh^p(A)\right)$ vanish.

$$\text{Again,} \quad A = \lim_{f \in S} R_f, \quad \text{and} \quad K_n(Coh^{p+1}(A)) = \lim_{f \in S} K_n(Coh^{p+1}(R_f)).$$

So, it is enough to prove that the maps

$$\lim_{f \in S} K_n(Coh^{p+1}(R_f)) \longrightarrow K_n(Coh^p(A)) \qquad \text{vanish} \quad \forall p \geq 0, \forall n \geq 0.$$

Replacing R by R_f, we prove

$$K_n(Coh^{p+1}(R)) \longrightarrow K_n(Coh^p(A)) \qquad \text{vanish} \quad \forall p \geq 0, \forall n \geq 0.$$

Further,

$$Coh^{p+1}(R) = \lim_{t \neq 0} Coh^p\left(\frac{R}{tR}\right) \qquad t \text{ runs through all nonzero divisors.}$$

So, given such a t, we prove that there is an $f \in S$, such that the horizontal functor

$$\begin{array}{ccc} Coh^p\left(\frac{R}{tR}\right) & \longrightarrow & Coh^p(R_f) \\ & \searrow & \downarrow \\ & & Coh^p(A) \end{array} \qquad \left\{ \begin{array}{l} \text{sending} \quad M \mapsto M_f \quad \text{induces zero maps} \\ K_n\left(Coh^p\left(\frac{R}{tR}\right)\right) \longrightarrow K_n\left(Coh^p(R_f)\right) \end{array} \right. \tag{6.56}$$

on the K-groups. To complete the proof, we use the following lemma (6.4.11). Let $B = k[X_1, \ldots, X_{r-1}] \subseteq R$ be as in (6.4.11). Then

$B \longrightarrow S^{-1}R$ is smooth. So, $S^{-1}\Omega_{R/B}$ is a projective $S^{-1}R$-module of rank one. By semilocal consideration, $S^{-1}\Omega_{R/B}$ is free. Smoothness being an open condition [K, Cor. 8.2], there is an $f_1 \in S$ such that $B \longrightarrow R_{f_1}$ is smooth and $\Omega_{R_{f_1}/B}$ is free.

Consider the diagram

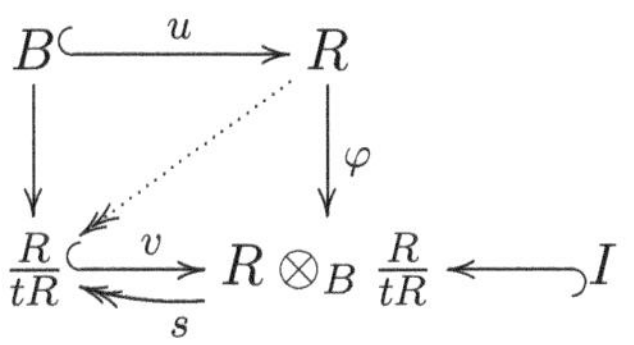

Write, $\overline{R} = \dfrac{R}{tR}$, $\quad \mathcal{R} = R \otimes_B \dfrac{R}{tR}$.

The first vertical map is finite, and hence so is the second one.

1. By base change, $\overline{R} \longrightarrow \mathcal{R}_{f_1}$ is smooth, hence flat. We also have that $\Omega_{\mathcal{R}_{f_1}/\overline{R}} \cong \Omega_{R_{f_1}/B} \otimes_B \overline{R}$ is free.
2. Write $I = \ker(s)$. Since $sv = 1$, we show that, the natural map
$$\varepsilon : \frac{I}{I^2} \xrightarrow{\sim} \frac{\Omega_{\mathcal{R}/\overline{R}}}{I\Omega_{\mathcal{R}/\overline{R}}} \qquad \text{is an isomorphism.}$$
It is clear that for $x \in \mathcal{R}$, we have $dx = d(x-vsx)+d(vsx) = d(x-vsx)$. So, ε is surjective. Since s splits v, the map $D : \mathcal{R} \longrightarrow \frac{I}{I^2}$ sending $x \mapsto [x - vsx]$ is a derivation. By the universal property, there is a map $\overline{s}$, as follows:

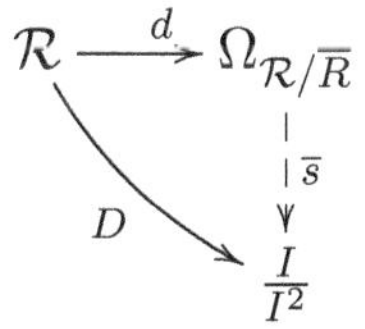

And, $\overline{s}$ induces, a left inverse of ε.

So, ε is an isomorphism.
Since $S^{-1}\overline{R} \cong S^{-1}\frac{\mathcal{R}}{I}$ is semilocal, $S^{-1}\frac{I}{I^2} \cong S^{-1}\frac{\Omega_{\mathcal{R}/\overline{R}}}{I\Omega_{\mathcal{R}/\overline{R}}}$ is free of rank one. Hence $S^{-1}I$ is generated by a nonzero divisor. Hence, there is $f_2 \in S$ such that $I_{f_2} = \mathcal{R}_{f_2}g \cong \mathcal{R}_{f_2}$.
3. With $f = f_1 f_2$, we have that $\overline{R} \longrightarrow \mathcal{R}_f$ is smooth, $\left(\Omega_{\mathcal{R}/\overline{R}}\right)_f$ is free of rank one, and $I_f = \mathcal{R}_f g \cong \mathcal{R}_f$. In particular, $\overline{R} \longrightarrow \mathcal{R}_f$ is flat.

So, we have an exact sequence

$$0 \longrightarrow \mathcal{R}_f \xrightarrow{g} \mathcal{R}_f \xrightarrow{s} \overline{R}_f \longrightarrow 0 \qquad \text{which splits.}$$

For $M \in Coh(\overline{R})$, we have an exact sequence

$$0 \longrightarrow \mathcal{R}_f \otimes_{\overline{R}} M \longrightarrow \mathcal{R}_f \otimes_{\overline{R}} M \xrightarrow{s} M_f \longrightarrow 0. \tag{6.57}$$

Clearly, $M \mapsto M_f$ is an exact functor $Coh^k(\overline{R}) \longrightarrow Coh^k(R_f)$. Since $\overline{R} \longrightarrow \mathcal{R}_f$ is flat, $M \mapsto \mathcal{R}_f \otimes_{\overline{R}} M$ defines an exact functor $Coh^k(\overline{R}) \longrightarrow Coh^k(\mathcal{R}_f)$. Since $R_f \longrightarrow \mathcal{R}_f$ is finite, this is an exact functor $Coh^k(\overline{R}) \longrightarrow Coh^k(R_f)$. We apply the Additivity Theorem (3.4.4) to the functors $Coh^k(\overline{R}) \longrightarrow Coh^k(R_f)$, defined by (6.57), and obtain that, for a given nonzero divisor t, there is $f \in S$ such that the maps

$$K_n\left(Coh^k(\overline{R})\right) \longrightarrow K_n\left(Coh^k(R_f)\right) \qquad \text{vanish} \quad \forall\, n \geq 0.$$

This completes the proof, as was stipulated (6.56). ∎

Before we prove the normalization lemma (6.4.11), used above, we insert the following criteria for smoothness, to suit our purpose.

Lemma 6.4.10. *Let R be an affine algebra over a field k, with $\dim R = r$. Let $\mathfrak{m} \in \operatorname{Spec}(R)$ be a maximal ideal. Assume that R/k is smooth at $\mathfrak{m}$. Let $B = k[X_1, \ldots, X_{r_0}] \hookrightarrow R$ be a polynomial subalgebra, over k, with $r_0 \leq r$. Then R/B is smooth at $\mathfrak{m}$ if and only if $(\Omega_{R/B})_{\mathfrak{m}}$ is free of rank $r - r_0$.*

Proof. Smoothness being a local property [K, Thm. 8.1, 8.2], we can assume R/k is smooth. So, the module of Kähler differentials $\Omega_{R/k}$ is projective of rank r. Let $\wp = \mathfrak{m} \cap B$. Since $k \longrightarrow \frac{R}{\mathfrak{m}}$ is integral, it follows that $\wp$ is also a maximal ideal in B. Suppose R/B is smooth, at $\mathfrak{m}$. By [K, Thm. 8.1(c)], $B_\wp \longrightarrow R_{\mathfrak{m}}$ is flat, and $(\Omega_{R/B})_{\mathfrak{m}}$ is free of rank $\dim \frac{R_{\mathfrak{m}}}{\wp R_{\mathfrak{m}}} = \dim R - \dim B = r - r_0$ [Mh, 21.A, p. 152], as required.

Conversely, assume that $(\Omega_{R/B})_{\mathfrak{m}}$ is free of rank $r - r_0$. Add more variables and consider the exact sequence

$$0 \longrightarrow I \longrightarrow k[X_1, \ldots, X_{r_0}, \ldots, X_N] \longrightarrow R \longrightarrow 0.$$

Since R/k is smooth, we have an exact sequence [K, Thm 8.1(d)]

$$0 \longrightarrow \frac{I}{I^2} \longrightarrow \bigoplus_{i=1}^{N} \frac{k[X_1,\ldots,X_{r_0},\ldots,X_N]}{I} dX_i \longrightarrow \Omega_{R/k} \longrightarrow 0$$

So, $rank\left(\dfrac{I}{I^2}\right) = N - r.$

Further, considering the presentation $R = \frac{B[X_{r_0+1},\ldots,X_N]}{I}$, we obtain the exact sequence

$$\frac{I}{I^2} \longrightarrow \bigoplus_{i=r_0+1}^{N} \frac{B[X_{r_0+1},\ldots,X_N]}{I} dX_i \xrightarrow{\lambda} \Omega_{R/B} \longrightarrow 0.$$

By hypothesis, $(\Omega_{R/B})_{\mathfrak{m}}$ is free, rank $r - r_0$. Therefore, $\lambda_{\mathfrak{m}}$ splits, $(\ker(\lambda))_{\mathfrak{m}}$ is free of rank $(N - r_0) - (r - r_0) = N - r$. By rank consideration, the sequence

$$0 \longrightarrow \left(\frac{I}{I^2}\right)_{\mathfrak{m}} \longrightarrow \bigoplus_{i=r+1}^{N} \left(\frac{B[X_r,\ldots,X_N]}{I}\right)_{\mathfrak{m}} dX_i \xrightarrow{\lambda} (\Omega_{R/B})_{\mathfrak{m}} \longrightarrow 0$$

is exact. Hence, R/B is smooth, at $\mathfrak{m}$, by [K, Thm 8.1(d)]. The proof is complete. ■

The following normalization lemma was used to prove the Gersten conjecture (6.4.9).

Lemma 6.4.11. *Let R be a smooth finitely generated algebra over a field k, with $\dim R = r$, and $\{\wp_1, \ldots, \wp_s\} \subseteq \operatorname{Spec}(R)$. Let $t \in R$ be a nonzero divisor. Then there is a polynomial subalgebra $B = k[X_1, \ldots, X_{r-1}] \subseteq R$ such that*

1. *$B \longrightarrow \frac{R}{tR}$ is finite over B.*
2. *$B \hookrightarrow R$ is smooth at $\wp_i$ for all $i = 1, \ldots, s$.*

Proof. (*When k is infinite, refer to* (6.6.13), *for an alternate, more frequently used, approach to this proof.*) By replacing $\wp_i$ by a maximal ideal containing it, we can assume that $\mathfrak{m}_i := \wp_i$ are maximal ideals in R. Note the module of Käller differentials $\Omega_{R/k}$ is a projective

R-module of rank r. By (6.4.10), it is enough to show the existence $B = k[X_1, \ldots, X_{r-1}] \hookrightarrow R$, of a polynomial subalgebra, such that $(\Omega_{R/B})_{\mathfrak{m}_i}$ are free of rank one, and $B \longrightarrow \frac{R}{tR}$ is finite.

1. We pick a finite-dimensional vector space $V \subseteq R$, such that $R = k[V]$. Further, for $i = 1, \ldots, s$ there are $v_{i1}, \ldots, v_{ir} \in V$, such that images of $dv_{i1}, \ldots, dv_{ir}$ in $\frac{\Omega_{R/k}}{\mathfrak{m}_i \Omega_{R/k}}$ form a basis, and are zero in $\frac{\Omega_{R/k}}{\mathfrak{m}_j \Omega_{R/k}}$ for all $j \neq i$.
2. Consequently, $v_{i1}, \ldots, v_{ir}$ are linearly independent over k.
3. Let $v_1, \ldots, v_m$ be a basis of V. Let $S(V) := \bigoplus S_n(V) = k[Y_1, \ldots, Y_m]$ denote symmetric algebra of V. (*To distinguish elements in $S(V)$ from those in R, we write $Y_i := v_i$*).

We set up some notation:

$$\begin{cases} S_n[V] = \text{Group of homogeneous polynomials of degree } n \\ \mathscr{F}_n[V] = \{f = f_0 + \cdots + f_n \in S(V), \ni f_i \in S_i[V]\} \\ gr_n\,(S(V)) = \frac{\mathscr{F}_n[V]}{\mathscr{F}_{n-1}[V]} \cong S_n(V), \quad gr\,(S(V)) = \bigoplus gr_n\,(S(V)). \end{cases}$$

As indicated before, $\{\mathscr{F}_n[V]\}$ is an increasing filtration of $S(V)$, and $gr\,(S(V))$ is the associated graded ring. (*Recall that as is with any ring with an increasing filtration, there is a well-defined map $S(V) \longrightarrow gr(S(V))$, sending $x \in \mathscr{F}_n[V] - \mathscr{F}_{n-1}[V]$ to $[x] \in \frac{\mathscr{F}_n[V]}{\mathscr{F}_{n-1}[V]}$. However, the map is not very well behaved.*) Write $\overline{R} = \frac{R}{tR}$. Let $\{\mathcal{F}_n\overline{R}\}$ be the increasing filtration of $\overline{R}$, induced by $\{\mathscr{F}_n[V]\}$.

$$\text{So,} \qquad \mathcal{F}_n\overline{R} = \text{Subspace spanned by monomials of } \deg \leq n \text{ in } \overline{R} = \overline{k[V]}.$$

Consider the diagram of surjective maps

$$\begin{array}{ccc} S(V) & \xrightarrow{\mathfrak{q}_0} & R \\ & \searrow^{\mathfrak{q}} & \downarrow \\ & & \overline{R} \end{array} \qquad \text{where} \qquad \mathfrak{q}_0(Y_i) = v_i.$$

So, the map $\mathfrak{q}$ is a map of filtered rings, and $\mathfrak{q}$ induces a map $\mathfrak{g} : gr(S(V)) \longrightarrow gr(\overline{R})$ of graded rings. We establish some more notation and summarize as follows:

$$\begin{cases} \text{Write,} \quad \mathbf{v} = (v_1, v_2, \ldots, v_m) \\ \mathscr{F}_n R(\mathbf{v}) = \{\mathfrak{q}_0(f) = f(\mathbf{v}) \in R : f \in \mathscr{F}_n[V]\} \subseteq R \\ \mathscr{F}_n \overline{R}(\mathbf{v}) = \{\mathfrak{q}(f) = \overline{f(\mathbf{v})} \in \overline{R} : f \in \mathscr{F}_n[V]\} = \frac{\mathscr{F}_n R(\mathbf{v}) + Rt}{Rt} \\ \text{So,} \quad gr_n(\overline{R}) = \frac{\mathscr{F}_n R(\mathbf{v}) + Rt}{\mathscr{F}_{n-1} R(\mathbf{v}) + Rt}. \end{cases}$$

We proceed to compute $\dim gr(\overline{R})$. Consider the diagram

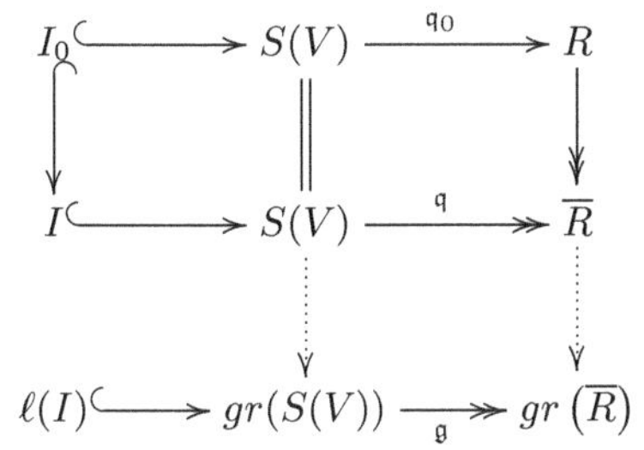

$$\begin{cases} \text{where} \quad I_0 = \ker(\mathfrak{q}_0), \ I = \ker(\mathfrak{q}). \\ \text{With } \mathfrak{q}_0(p) = t \quad \Longrightarrow \quad I = (I_0, p). \\ \textit{The dotted vertical maps are not} \\ \textit{very well behaved.} \end{cases}$$

The map

$$\mathfrak{g} : gr_n(S(V)) \longrightarrow gr_n(\overline{R}) \qquad \text{sends}$$

$$M \in S_n(V) \mapsto [M(\mathbf{v}))] \in \frac{\mathscr{F}_n R(\mathbf{v}) + Rt}{\mathscr{F}_{n-1} R(\mathbf{v}) + Rt}.$$

Let $\ell(I) \subseteq S(V)$ be the ideal generated by the leading components of the elements in I. So,

$$\ell(I) := \langle \{f_n \in S_n(V) : \exists f \in I \ni f = f_n + f_{n-1} + \cdots + f_0, \ f_k \in S_k(V)\}\rangle.$$

Claim: $\ell(I) = \ker(\mathfrak{g})$. To see this, let

$$\begin{cases} f_n \in \ell(I) \quad \text{be a generator of } \ell(I) \\ \text{Then } \exists\ f \in I = (I_0, p) \ni f = f_n + f_{n-1} + \cdots + f_0 \\ f = f_n + f_{n-1} + \cdots + f_0 = g + \lambda p, \text{with } g \in I_0 \\ \text{So,} \quad f_n(\mathbf{v}) + (f_{n-1}(\mathbf{v}) + \cdots + f_0(\mathbf{v})) = 0 + \lambda(\mathbf{v})t \\ \Longrightarrow \mathfrak{g}(f_n) = 0, \quad \text{So,} \quad \ell(I) \subseteq \ker(\mathfrak{g}). \end{cases}$$

Now, let $f_n \in \ker(\mathfrak{g})$ be homogeneous with $\deg(f_n) = n$. So,

$$\begin{cases} f_n(\mathbf{v}) \in \mathscr{F}_n R(\mathbf{v}) + Rt \\ \Longrightarrow f_n(\mathbf{v}) = \beta_{n-1}(\mathbf{v}) + \lambda(\mathbf{v})p(\mathbf{v})\ \beta_{n-1} \in \mathscr{F}_{n-1}[V] \\ \Longrightarrow f_n - \beta_{n-1} - \lambda p =: g \in I_0 \\ \Longrightarrow f_n - \beta_{n-1} \in I \Longrightarrow f_n \in \ell(I) \end{cases}$$

So, the claim $\ell(I) = \ker(\mathfrak{g})$ is established. It is known (6.4.12) that $height(I) = height(\ell(I))$. It follows that

$$\dim gr(\overline{R}) = \dim \overline{R} = r - 1.$$

For any system of parameters $z_1, \ldots, z_{r-1}$ of $gr(\overline{R})$, the extension $k[z_1, \ldots, z_{r-1}] \hookrightarrow gr(\overline{R})$ is finite [BH, p. 37, Thm. 1.5.17]. Choose $z_i \in gr_{p_i}(\overline{R}) = \frac{\mathscr{F}_{p_i}R(\mathbf{v})+tR}{\mathscr{F}_{p_i-1}R(\mathbf{v})+tR}$, with $p_i \geq 2$. Let $Z_i \in \mathscr{F}_{p_i}R(\mathbf{v}) - \mathscr{F}_{p_i-1}R(\mathbf{v})$ be any lift of z_i. So, Z_i can be assumed to be the image of a homogeneous polynomial in $S_{p_i}(V)$. Write

$$\begin{cases} B = k[Z_1, Z_2, \ldots, Z_{r-1}] \subseteq R \\ C = k[z_1, z_2, \ldots, z_{r-1}] \subseteq gr(\overline{R}) \end{cases} \quad \text{We have} \quad gr(\overline{R}) = \sum_{i=1}^{N} C[w_i]$$

where $\deg([w_i]) = d_i$, and for some $w_i \in \mathscr{F}_{d_i}R(\mathbf{v}) - \mathscr{F}_{d_i-1}R(\mathbf{v})$.

$$\textbf{Claim:} \quad \overline{R} = \sum_{i=1}^{N} Bw_i$$

Let $w \in \mathscr{F}_n R(\mathbf{v}) - \mathscr{F}_{n-1}\overline{R}(\mathbf{v})$. If $n = 0$, $w \in k$. So, it works. Now

$$[w] = \sum_{i=1}^{N} [r_i][w_i] \quad r_i \in B_{n-d_i} \quad \text{So,} \quad w - \sum_{i=1}^{n} r_i w_i \in F_{n-1}\overline{R}.$$

By induction $w - \sum_{i=1}^{n} r_i w_i \in \sum_{i=1}^{N} Bw_i$. So, $w \in \sum_{i=1}^{N} Bw_i$. So, $B \longrightarrow \overline{R}$ is finite. Hence, B is also a polynomial ring.

Fix $\mathfrak{m} = \mathfrak{m}_1$. Assume that $\sum_{i=1}^{r-1} \frac{R}{\mathfrak{m}} dZ_i = \sum_{i=1}^{r_0-1} \frac{R}{\mathfrak{m}} dZ_i$, with $r_0 \leq r-1$, and $dZ_1, \ldots, dZ_{r_0-1}$ are linearly independent in $\frac{\Omega_{R/k}}{\mathfrak{m}\Omega_{R/k}}$. There is an integer $1 \leq \iota \leq r$ such that $dv_{1\iota} \notin \sum_{i=0}^{r_0-1} \frac{R}{\mathfrak{m}} dZ_i$. Without loss of generality, write $r_0 := \iota$, and let $X_{r_0} = v_{r_0} + Z_{r_0}$. Then

$Image(X_{r_0}) = Image(Z_{r_0}) = z_{r_0}$. Replacing Z_{r_0} by X_{r_0}, we have that $dZ_1, \ldots, dZ_{r_0-1}, dZ_{r_0}$ are linearly independent in $\frac{\Omega_{R/k}}{\mathfrak{m}\Omega_{R/k}}$. Inductively, we have a choice of $Z_1, \ldots, Z_{r-1}$ such that $dZ_1, \ldots, dZ_{r-1}$ are linearly independent in $\frac{\Omega_{R/k}}{\mathfrak{m}\Omega_{R/k}}$.

From the choices of v_{ij}, by induction, we have a choice of $Z_1, \ldots, Z_{r-1}$, such that $dZ_1, \ldots, dZ_{r-1}$ are linearly independent in $\frac{\Omega_{R/k}}{\mathfrak{m}_i\Omega_{R/k}}$ for all $i = 1, 2, \ldots, s$.

This completes the proof of the lemma. ∎

Lemma 6.4.12. *Let $R = k[X_1, \ldots, X_n]$ be a polynomial ring, over a field k, and $I \subseteq R$ be an ideal. Let $\ell(I)$ be the homogeneous ideal generated by the leading degree terms of the elements in I. Then $height(I) = height(\ell(I))$.*

Proof. Let $A = R[T]$, and consider $\text{Proj}\,(R[T])$. Let $I[T]$ be the ideal generated by polynomials, obtained by homogenizing the polynomials in I. Then $(\ell(I), T) = (I[T], T)$. Note, $T^k \notin I[T]\ \forall k$. So, $height(\ell(I)) = height(I[T])$. Consider $V(I) \subset \text{Spec}\,(R)$ and $V(I[T]) \subseteq \text{Proj}\,(R[T])$. Consider the isomorphism $\varphi : R \xrightarrow{\sim} R\,[T]_{(T)}$, sending $X_i \mapsto \frac{X_i}{T}$. Then $\varphi(I) \cong I[T]_{(T)}$. So, $\dim(V(I)) = \dim V(I[T]_{(T)}) = \dim\,(V(I[T]) \cap D(T)) = \dim V(I[T])$. Therefore, $n - height(I) = n - height(I[T])$, and $height(I) = height(I[T]) = height(\ell(I))$. The proof is complete. ∎

6.4.2. *The Chow groups*

Because of the exactness of Gersten complexes (6.4.9), in the local case, we are able to realize the Chow groups $A^n(X)$, of codimension n-cycles of a scheme X, as a sheaf cohomology of the K-groups $G_n(X)$. Before we proceed, we retrieve the definition [Fu] of the Chow groups $A^p(X)$.

Definition 6.4.13. First, suppose $(R, \mathfrak{m})$ is a local integral domain, with $\dim R = 1$, and $Q(R)$ denotes the field of fractions. Then there is a well-defined group homomorphism

$$Ord : Q(R)^* \longrightarrow \mathbb{Z} \qquad \text{defined by}$$

$$Ord\left(\frac{f}{g}\right) = \ell\left(\frac{A}{(f)}\right) - \ell\left(\frac{A}{(g)}\right) \tag{6.58}$$

where ℓ denotes the length.

Now, let X be a Noetherian scheme, with $\dim X = d < \infty$, and $n \geq 0$ be an integer. As always, for $x \in X$, the residue field of x is denoted by $\kappa(x)$. We identify $\bigoplus_{x \in X^{(n)}} K_0(\kappa(x)) = \mathbb{Z}\left(X^{(n)}\right)$, the free $\mathbb{Z}$-module generated on $X^{(n)}$. Let $x \in X^{(n-1)}$, and $y \in \overline{x} \cap X^{(n)}$. Let $y \in U = \mathrm{Spec}\,(A)$ be an open neighborhood of y. So, $x \in U$. Let $\wp_x \subseteq \wp_y \in \mathrm{Spec}\,(A)$ denote the prime ideals of x, y. Then with $R = \frac{A_{\wp_y}}{\wp_x}$, we have $\dim R = 1$, and $Q(R) \cong \kappa(x)$. According to (6.58), we obtain a group homomorphism

$$Ord_{x2y} : \kappa(x)^* \longrightarrow \mathbb{Z}.$$

If $y \in X^{(n)} - \overline{x}$, let $Ord_{x2y} = 0$. Now, define

$$\begin{cases} Ord_x : \kappa(x)^* \longrightarrow \mathbb{Z}\left(X^{(n)}\right), & \text{by} \quad Ord_x(f) = \sum_{y \in X^{(n)}} Ord_{x2y}(f) y \\ Ord : \bigoplus_{x \in X^{(n-1)}} K_1(\kappa(x)) \longrightarrow \mathbb{Z}\left(X^{(n)}\right), & \text{by} \quad Ord = \bigoplus Ord_x. \end{cases} \tag{6.59}$$

The **Chow group** $A^n(X)$, of codimension n-cycles of X, is defined by the exact sequence

$$\bigoplus_{x \in X^{(n-1)}} K_1(\kappa(x)) \xrightarrow{\quad Ord \quad} \mathbb{Z}\left(X^{(n)}\right) \longrightarrow A^n(X) \longrightarrow 0.$$

We prove that the Chow group $A^n(X)$ is realized by the Gersten complex, as follows.

Theorem 6.4.14. *Let X be a scheme of finite type over a field k, and $n \geq 0$ be an integer. Consider the Gersten complex* (6.51)

$$\cdots \longrightarrow \bigoplus_{x \in X^{(n-1)}} K_1(\kappa(x)) \xrightarrow[d_n]{} \bigoplus_{x \in X^{(n)}} K_0(\kappa(x)) \longrightarrow 0. \tag{6.60}$$

Then $A^n(X) \cong \mathrm{co}\ker(d_n)$. In fact, $d_n = \pm Ord$, where Ord is as defined above (6.59).

Proof. For $x \in X$, we have, $K_1(\kappa(x)) = \kappa(x)^*$ the units of $\kappa(x)$. For $n \geq 0$, we resketch the construction of the Gersten complex (6.60):

$$\begin{array}{ccccc} K_1\left(Coh^{n-1}(X)\right) & \xrightarrow{q} & \bigoplus_{x \in X^{(n-1)}} K_1\left(\kappa(x)\right) & \xrightarrow{\partial} & K_0\left(Coh^{n}(X)\right) \\ \downarrow \iota & & & \searrow d_n & \downarrow q \\ K_1\left(Coh^{n-2}(X)\right) & & & & \bigoplus_{x \in X^{(n)}} K_0\left(\kappa(x)\right). \end{array} \tag{6.61}$$

For this calculation, it is better to keep the equivalent complexes (6.50) in mind. For $y \in X^{(n-1)}$, $x \in X^{(n)}$, let $(d_n)_{y2x} : K_1(\kappa(y)) \longrightarrow K_0(\kappa(x))$ denote the map induced by d_n. We prove $(d_n)_{y2x} = Ord_{y2x}$ for all such x, y.

To prove d_n coincide with Ord, fix $y \in X^{(n-1)}$, and let $Y = \overline{y} \subseteq X$ be the closure. As usual, Y has the reduced induced closed subscheme structure. The inclusion functor $Coh(Y) \hookrightarrow Coh(X)$ induce functors $Coh^j(Y) \hookrightarrow Coh^{j+n-1}(X)$. From the natural properties of the Gersten complexes (6.50), we have the following commutative diagram:

$$\begin{array}{ccccc} K_1\left(\kappa(y)\right) & \xrightarrow{d} & \bigoplus_{x \in Y^{(1)}} K_0\left(\kappa(x)\right) & = & \mathbb{Z}(Y) \\ \downarrow & & \downarrow & & \\ \bigoplus_{x \in X^{(n-1)}} K_1\left(\kappa(x)\right) & \xrightarrow[d_n]{} & \bigoplus_{x \in X^{(n)}} K_0\left(\kappa(x)\right) & = & \mathbb{Z}(X). \end{array}$$

It follows immediately that, if $x \notin Y$, then $(d_n)_{y2x} = Ord_{y2x} = 0$. Also, the vertical maps are natural inclusions.

Now, let $x \in Y^{(1)}$ and $Y_x = \operatorname{Spec}(\mathcal{O}_{Y,x}) = \{y, x\}$. Since the inclusion $Y_x \hookrightarrow Y$ is a flat morphism, from the natural properties, we have a commutative diagram

$$\begin{array}{ccccc} K_1\left(\kappa(y)\right) & \xrightarrow{d} & \bigoplus_{x \in Y^{(1)}} K_0\left(\kappa(x)\right) & = & \mathbb{Z}(Y) \\ \downarrow 1 & & \downarrow & & \\ K_1\left(\kappa(y)\right) & \longrightarrow & K_0\left(\kappa(x)\right) & = & \mathbb{Z}x \end{array}$$

So, it is enough to show that the second line coincides with the order function, which is the case of Spec (R), where $(R, \mathfrak{m})$ is a local integral domain with $\dim R = 1$. This is dealt with in the following (6.4.15). The proof is complete. ■

Lemma 6.4.15. *Let $(R, \mathfrak{m})$ be a local Noetherian domain, with $\dim R = 1$, containing a field. Let $\kappa(\mathfrak{m}) = A/\mathfrak{m}$ and $F = Q(R)$ be the field of fractions. Consider the localization exact sequence* (6.1.11) *for the inclusion $V(\mathfrak{m}) \subseteq \operatorname{Spec}(R)$:*

$$G_1(R) \longrightarrow K_1(F) \xrightarrow{\partial} K_0(\kappa(\mathfrak{m})) \longrightarrow G_0(R) \longrightarrow K_0(F) \longrightarrow 0.$$

Then the map $\partial = \pm Ord$, where the map $Ord : F^* \longrightarrow \mathbb{Z}$ is as above (6.58). *Further, the sign is universal, meaning not dependent on R.*

Proof. We have $K_1(F) = F^*$ and $K_1(R) = R^*$. We also have a commutative diagram

$$\begin{array}{ccc} K_1(R) & & \\ \downarrow & \searrow & \\ G_1(A) & \longrightarrow & K_1(F) \end{array}$$

Suppose $x \in R$, and $x \neq 0$. If x is a unit, then x is in the image of $K_1(R)$ and hence of $G_1(R)$. So, $\partial(x) = 0$, in this case. So, assume $x \in \mathfrak{m}$. Let $k_0 \hookrightarrow R$ be the prime field of R. Since x is a nonunit, x is not algebraic over k_0. So, there is a flat map $\iota : k_0[t] \longrightarrow A$, given by $\iota(t) = x$. Consider the diagram of functors and quotient categories:

$$\begin{array}{ccccc} Coh_{V(t)}(k_0[t]) & \longrightarrow & Coh(k_0[t]) & \longrightarrow & Coh(k_0[t, t^{-1}]) \\ \downarrow & & \downarrow & & \downarrow \\ Coh_{V(\mathfrak{m})}(A) & \longrightarrow & Coh(k_0[t]) & \longrightarrow & Coh(F) \end{array}$$

Use Dévissage for the inclusions $Coh(k_0) \hookrightarrow Coh_{V(t)}(k_0[t])$ and $Coh(\kappa(\mathfrak{m})) \hookrightarrow Coh_{V(\mathfrak{m})}(k_0[t])$. By the natural properties of localization sequences, we obtain the following commutative diagram of exact sequences:

$$\begin{array}{ccccccccccc} K_1(k_0[t]) & \longrightarrow & K_1(k_0[t, t^{-1}]) & \xrightarrow{\partial} & K_0(k_0) & \longrightarrow & K_0(k_0[t]) & \longrightarrow & K_0(k_0[t, t^{-1}]) & \longrightarrow & 0 \\ \downarrow & & \downarrow u & & \downarrow v & & \downarrow & & \downarrow & & \\ G_1(R) & \longrightarrow & K_1(F) & \xrightarrow{\partial} & K_0(\kappa(\mathfrak{m})) & \longrightarrow & G_0(R) & \longrightarrow & K_0(F) & \longrightarrow & 0 \end{array} \tag{6.62}$$

The map v is induced by the functor

$$Coh(k_0) \longrightarrow Coh(k) \quad V \mapsto A \otimes_{k_0[t]} V = \frac{A}{xA} \otimes_{k_0} V \in Coh_{V(\mathfrak{m})}(A).$$

Identifying K_0 with classical K_0, it follows that the map $v : K_0(k_0) = \mathbb{Z} \longrightarrow K_0(\kappa(m)) = \mathbb{Z}$ is given by multiplication by $v = \ell\left(\frac{A}{x}\right)$. Likewise, identifying K_1 with classical K_1, we have $K_1(k_0[t, t^{-1}]) \cong units(k_0[t, t^{-1}]) = k_0^* \cup \{t^n : n \in \mathbb{Z}\}$, and $K_1(F) = F^*$. It follows that the map u is the map induced by $t \mapsto x$.

So, it would suffice to prove that $\partial(t) = \pm 1$. With the identification of K_0, K_1 with classical K_0, K_1, we rewrite the top line of the above diagram:

$$k_0^* \longrightarrow k_0^* \cup \{t^n : n \in \mathbb{Z}\} \xrightarrow{\quad \partial \quad} \mathbb{Z} \longrightarrow \mathbb{Z} \longrightarrow \mathbb{Z} \longrightarrow 0.$$

It is obvious form this exact sequence that $\partial(t) = \pm 1$. Further, note that the sign ± 1 in the second line of the above (6.62) is induced by the sign $\partial(t) = \pm 1$. Consequently, the signs in all the above diagrams are universal and are induced by the sign $\partial(t) = \pm 1$. So, we have proved that, in the Gersten complexes (6.60), the differential $d_n = \pm Ord$. This completes the proof, and also concludes the proof of (6.4.13). ∎

Now we are ready to proclaim that the Chow groups are realized by the sheaf cohomology.

Theorem 6.4.16. *Let X be a regular scheme of finite type over a field. Then there is a canonical isomorphism*

$$A^n(X) = H^n\left(X, \underline{G}_n\right)$$

where $\underline{G}_n$ is the sheaf on X, associated to the presheaf $U \mapsto G_n(U)$.

Proof. By (6.4.9), the hypotheses of (6.4.7) is satisfied. So, $H^n\left(X, \underline{G}_n\right)$ is the sheaf cohomology, which is the cohomology of the Gersten complex (6.60). So, the proof is complete by (6.4.14). ∎

6.5. Čech Cohomology Tools

For a morphism $f : X \longrightarrow Y$, the higher direct images $R^q f_* \mathcal{F}$ are more directly computed, using Čech cohomology. For that reason, we include this section. Our main resource for such information on higher direct images and Čech cohomology remains [Hr]. For convenience and completeness of our discussions, we define sheaf cohomology, for sheaves of abelian groups, on topological spaces.

Definition 6.5.1. Let X be a topological space. Let $\mathbf{Ab}(X)$ denote the category of sheaves of abelian groups on X. Then $\mathbf{Ab}(X)$ has enough injective objects, meaning for any $\mathcal{F} \in \mathbf{Ab}(X)$, there is an exact sequence

$$0 \longrightarrow \mathcal{F} \longrightarrow \mathscr{I}_0 \longrightarrow \mathscr{I}_1 \longrightarrow \cdots \qquad \text{where } \mathscr{I}_i \text{ are injective.} \tag{6.63}$$

Given $\mathcal{F} \in \mathbf{Ab}(X)$ and a resolution (6.63), taking the global sections we get a complex of abelian groups

$$0 \longrightarrow \Gamma(X, \mathscr{I}_0) \longrightarrow \Gamma(X, \mathscr{I}_1) \longrightarrow \cdots \quad \text{to be denoted as} \quad \Gamma(X, \mathscr{I}_\bullet). \tag{6.64}$$

We define (sheaf) cohomology

$$H^n(X, \mathcal{F}) := H^n(\Gamma(X, \mathscr{I}_\bullet)) \quad \forall n \geq 0. \tag{6.65}$$

The following are some immediate facts:

1. $H^n(X, \mathcal{F})$ is well defined, and is independent of the choice of the resolution (6.63) up to an isomorphism.
2. $H^0(X, \mathcal{F}) = \Gamma(X, \mathcal{F})$. Further, for any exact sequence $0 \longrightarrow K \longrightarrow \mathcal{F} \longrightarrow C \longrightarrow 0$, there is a natural long exact sequence
$$0 \longrightarrow H^0(X, K) \longrightarrow H^0(X, \mathcal{F}) \longrightarrow H^0(X, C) \longrightarrow H^1(X, K) \longrightarrow H^1(X, \mathcal{F}) \longrightarrow \cdots \tag{6.66}$$
3. (**Flasque sheaves:**) An object $\mathcal{G} \in \mathbf{Ab}(X)$ is defined to be a **flasque sheaf** if for all pairs of open subsets $V \subseteq U$ of X, the maps $\mathcal{G}(U) \twoheadrightarrow \mathcal{G}(V)$ are surjective. It follows that if $\mathcal{G}$ is flasque, then $H^n(X, \mathcal{G}) = 0 \ \forall n \geq 1$. Consequently, for $\mathcal{F} \in \mathbf{Ab}(X)$, sheaf cohomology, $H^n(X, \mathcal{F})$, can be computed, using a flasque resolution:
$$0 \longrightarrow \mathcal{F} \longrightarrow \mathcal{G}_0 \longrightarrow \mathcal{G}_1 \longrightarrow \cdots \qquad \text{where } \mathcal{G}_i \text{ are flasque.} \tag{6.67}$$

   It is also noteworthy that injective objects are flasque.

4. (**Schemes:**) When X is a scheme, and $\mathcal{F} \in QCoh(X)$, the sheaf cohomology $H^n(X, \mathcal{F})$ defined in (6.2.7) coincides with what is defined here (6.65), treating $\mathcal{F}$ as a sheaf of abelian groups. However, as was observed in (6.2.7), $H^n(X, \mathcal{F})$ had additional structures, which were possible due to the availability of resolutions by injective objects in $QCoh(X)$.

We refer to [Hr, §III.1, III.2] for further details on the above.

Given a topological space X and an open cover $\{U_i : i \in I\}$ of X, for a sheaf $\mathcal{F}$ of abelian groups, Čech cohomology $\check{H}^k(\{U_i\}, \mathcal{F})$ is defined. Since we work with Noetherian schemes, we define it in the following for such finite open covers only.

Definition 6.5.2. Let X be a topological space and $\mathfrak{U} := \{U_0, U_1, \ldots, U_n\}$ be an open cover of X. For $0 \leq i_0 < i_1 < \cdots < i_p \leq n$, let $U_{i_0 i_1 \ldots i_p} := U_{i_0} \cap U_{i_1} \cap \cdots \cap U_{i_p}$. Let

$$\mathcal{F} \in \mathbf{Ab}(X) \quad \text{and let} \qquad C^p(\mathfrak{U}, \mathcal{F}) := \bigoplus_{0 \leq i_0 < i_1 < \cdots < i_p \leq n} \mathcal{F}(U_{i_0 i_1 \ldots i_p}).$$

For $\alpha \in C^p(\mathfrak{U}, \mathcal{F})$, write $\alpha = (\alpha_{i_0 i_1 \ldots i_p})$, with $\alpha_{i_0 i_1 \ldots i_p} \in \mathcal{F}(U_{i_0 i_1 \ldots i_p})$. Define

$$\partial^p : C^p(\mathfrak{U}, \mathcal{F}) \longrightarrow C^{p+1}(\mathfrak{U}, \mathcal{F}) \quad \text{by}$$

$$\partial^p(\alpha)_{i_0 i_1 \cdots i_{p+1}} = \sum_{k=0}^{p+1} (-1)^k \left(\alpha_{i_0 i_1 \ldots i_{k-1} \hat{i}_k i_{k+1} \cdots i_{p+1}} \right)_{|U_{i_0 i_1 \cdots i_{p+1}}}.$$

As usual, the notation $\hat{i}_k$ indicates i_k being dropped. One checks that $\partial^{p+1}\partial^p = 0$, for all $p \geq 0$. So, $(C^p(\mathfrak{U}, \mathcal{F}), \partial^p)$ is a complex of abelian groups, to be denoted by $C^\bullet(\mathfrak{U}, \mathcal{F})$. With respect to the given open cover $\mathfrak{U}$, for integers $p \geq 0$, define p^{th}-Čech cohomology

$$\check{H}^p(\mathfrak{U}, \mathcal{F}) = H^p\left(C^\bullet(\mathfrak{U}, \mathcal{F})\right).$$

Since $C^m(\mathfrak{U}, \mathcal{F}) = 0$, $\forall m \geq n+1$, we have $\check{H}^m(\mathfrak{U}, \mathcal{F}) = 0 \; \forall m \geq n+1$.

The following is a very useful tool for us [Hr, Thm. III.4.5].

Theorem 6.5.3. *Let X be a Noetherian scheme and $\mathfrak{U} := \{U_0, U_1, \ldots, U_n\}$ be an affine open cover of X (i.e., each*

$U_i = \operatorname{Spec}(A_i)$ *is affine). With notation, as in* (6.5.2), *assume* $U_{i_0 i_1 \ldots i_p}$ *is affine, for all* $0 \leq i_0 < i_1 < \cdots < i_p \leq n$ *(e.g., if* X *is separated). Then* $\forall\, \mathcal{F} \in QCoh(X)$, *there is a natural isomorphism*

$$\check{H}^p(\mathfrak{U}, \mathcal{F}) \xrightarrow{\sim} H^p(X, \mathcal{F}) \qquad \forall p \geq 0. \tag{6.68}$$

Proof. By properties of injective resolutions, $\mathcal{F} \hookrightarrow \mathscr{I}_\bullet$, it follows that there is a natural map, as shown (6.68). Further, $\check{H}^0(\mathfrak{U}, \mathcal{F}) \xrightarrow{\sim} H^0(X, \mathcal{F})$. Given an exact sequence,

$$0 \longrightarrow \mathcal{F} \longrightarrow \mathscr{I} \longrightarrow \mathcal{R} \longrightarrow 0, \tag{6.69}$$

the left-hand side of (6.68) leads to a long exact sequence (6.66). In fact, the right-hand side of (6.68) also leads to a long exact sequence. This follows because there is an exact sequence of chain complexes

$$0 \longrightarrow C^\bullet(\mathfrak{U}, \mathcal{F}) \longrightarrow C^\bullet(\mathfrak{U}, \mathscr{I}) \longrightarrow C^\bullet(\mathfrak{U}, \mathcal{R}) \longrightarrow 0$$

Given $\mathcal{F}$, we can obtain an exact sequence (6.69), where $\mathscr{I}$ is an injective X-module. It is clear that $H^p(X, \mathscr{I}) = 0$, $\forall p \geq 1$. One proves that $\check{H}^p(\mathfrak{U}, \mathscr{I}) = 0$ $\forall p \geq 1$. Now the theorem follows by induction. ■

Certain proofs that we may deal with can be reduced to the case of projective spaces $\mathbb{P}^d_A$ over A. So, the following Serre duality is useful.

Theorem 6.5.4. *Let* A *be a Noetherian commutative ring, and* $S = A[X_0, X_1, \ldots, X_d]$ *be the polynomial ring. Let* $X = \mathbb{P}^d_A = \operatorname{Proj}(S)$. *Then*

$$H^k(X, \mathcal{O}(n)) = \begin{cases} 0 & k = 0, \quad n \leq -1 \\ S_n & k = 0, \quad n \geq 0 \\ 0 & 0 < k < d, \quad n \in \mathbb{Z} \\ A & k = d, \quad n = -(d+1) \\ Hom(S_{-n-(d+1)}, A) & k = d \\ 0 & k \geq d + 1 = \dim X. \end{cases}$$

In particular, the nonvanishing terms at degree zero and d *are related, by the natural perfect pairing*

$$\begin{aligned} &H^0(X, \mathcal{O}_X(n)) \times H^d(X, \mathcal{O}_X(-n-d-1)) \\ &\quad \longrightarrow H^d(X, \mathcal{O}_X(-d-1)) = A \end{aligned}$$

of finitely generated free A*-modules.*

Proof. With $S = A[X_0, X_1, \ldots, X_d]$, $\mathfrak{U} = \{D(X_0), D(X_1), \ldots, D(X_d)\}$ is an affine open cover of X. We write $\mathcal{F} = \bigoplus_{n\in\mathbb{Z}} \mathcal{O}(n)$ and compute the Čech cohomology $H^k(\mathfrak{U}, \mathcal{F})$. It is easy to see that

$$C^p(\mathfrak{U}, \mathcal{F}) = \bigoplus_{0\le i_0<i_1<\cdots<i_p\le d} S_{X_{i_0}X_{i_1}\cdots X_{i_p}}. \tag{6.70}$$

The rest of the proof is obtained by explicit computations of the cohomologies of this Čech complex $C^\bullet(\mathfrak{U}, \mathcal{F})$. For details, see [Hr, Thm III.5.1]. ∎

Remark 6.5.5. Subsequently, Quillen uses a very general version (7.1.2) of Serre Duality. In particular, when $X = \mathbb{P}^d_A$ for $N \in QCoh(A)$, formulas for $H^n(X, \mathcal{O}_X(n) \otimes N)$ are given. This case of (7.1.2) is proven by replacing the Čech complex (6.70) by

$$C^p(\mathfrak{U}, \mathcal{F} \otimes N) = \bigoplus_{0\le i_0<i_1<\cdots<i_p\le d} S_{X_{i_0}X_{i_1}\cdots X_{i_p}} \otimes N.$$

We will make a reference to this remark while leaving the details of the proof of (7.1.2), as we did with Serre duality (6.5.4).

6.5.1. *Application of Čech cohomology*

Using Čech cohomology in this subsection, we prove some of the results used before.

Lemma 6.5.6. *Let $f : X \longrightarrow Y$ be a proper morphism of Noetherian schemes, and f has finite Tor dimension. Let $\mathfrak{Coh}(f, Y) \subseteq Coh(Y)$ and $\mathscr{P}(X, f) \subseteq \mathfrak{Coh}(X, f) \subseteq Coh(X)$ be, respectively, as defined above* (6.11, 6.26, 6.32). *Then*

1. *For $\mathcal{E} \in \mathscr{P}(X, f)$, $\mathcal{F} \in \mathfrak{Coh}(f, Y)$, we have $\mathcal{E} \otimes_X f^*\mathcal{F} \in \mathfrak{Coh}(X, f)$, and the natural map*
$$f_*\mathcal{E} \otimes_Y \mathcal{F} \longrightarrow f_*(\mathcal{E} \otimes_X f^*\mathcal{F}) \qquad \textit{is an isomorphism.}$$
2. *For $\mathcal{E} \in \mathscr{P}(X, f)$, we have $f_*\mathcal{E} \in \mathbb{H}(Y)$.*

Proof. All the propositions are local on Y. So, we assume $Y = \operatorname{Spec}(A)$. With $M = H^0(Y, \mathcal{F})$ and $N = H^0(X, \mathcal{E})$, we have $\mathcal{F} = \widetilde{M}$, and $f_*\mathcal{E} = \widetilde{N}$. By (6.36), we have $\widetilde{Tor^A(N, M)} = Tor_i^Y(f_*\mathcal{E}, \mathcal{F}) = 0\ \forall\ i \geq 1$. Further, for open subsets $U = \operatorname{Spec}(B) \subseteq X$, we have, as well, $\widetilde{Tor^A(B, M)} = Tor_i^Y(\mathcal{O}_X, \mathcal{F})_{|U} = 0\ \forall\ i \geq 1$.

Now choose an affine open cover $U_i = \operatorname{Spec}(B_i) \subseteq X$, of X, with $i = 0, 1, \ldots n$, and for $0 \leq i_0 < i_1 < \cdots < i_k \leq n$, let $U_{i_0 i_1 \cdots i_k} := U_{i_0} \cap U_{i_1} \cap \cdots U_{i_k}$. Since f is separated, and Y is affine, X is separated over $\mathbb{Z}$. Therefore, $U_{i_0 i_1 \cdots i_k} \cong \operatorname{Spec}(B_{i_0 i_1 \cdots i_k})$ are affine open subsets of X (see 6.6.3). Write

$$C^k(\mathcal{E}) = \bigoplus_{0 \leq i_0 < i_1 < \cdots < i_k \leq n} \mathcal{E}(U_{i_0 i_1 \cdots i_k}).$$

Consider the Čech complex:

$$C^\bullet(\mathcal{E}):\ 0 \longrightarrow C^0(\mathcal{E}) \longrightarrow C^1(\mathcal{E}) \longrightarrow \cdots \longrightarrow C^n(\mathcal{E}) \longrightarrow 0. \tag{6.71}$$

We have

$$H^k(C^\bullet(\mathcal{E})) = H^k(X, \mathcal{E}). \quad \text{Since } \mathcal{E} \in \mathfrak{Coh}(X, f)$$

$$\begin{cases} H^0(C^\bullet(\mathcal{E})) = H^0(X, \mathcal{E}) \\ H^k(C^\bullet(\mathcal{E})) = 0 \quad \forall\ k \geq 1. \end{cases}$$

Further, assume that $\mathcal{E}_{|U_i}$ are trivial (free). In this case, $\mathcal{E}(U_{i_0 i_1 \cdots i_k})$ is a free $B_{i_0 i_1 \ldots i_k}$-module, and hence $Tor_i^A(B_{i_0 \ldots i_k}, M) = 0,\ i \geq 1$. Consequently, $Tor_i^A(C^p(\mathcal{E}), M) = 0,\ i \geq 1,\ p \geq 0$. From this it follows that $C^\bullet(\mathcal{E}) \otimes_A M$ leads to the following exact sequence:

$$0 \longrightarrow N \otimes_A M \longrightarrow C^0(\mathcal{E}) \otimes_A M \longrightarrow C^1(\mathcal{E}) \otimes_A M \longrightarrow \cdots \longrightarrow C^n(\mathcal{E}) \otimes_A M \longrightarrow 0.$$

It is clear that $C^\bullet \otimes_A M$ is the Čech complex for $\mathcal{E} \otimes f^*\mathcal{F}$. The above sequence being exact, it follows that

$$R^k f_*(\mathcal{E} \otimes f^*\mathcal{F}) = 0, \quad \forall k \geq 1, \qquad \text{and}$$

$$f_*(\mathcal{E} \otimes f^*\mathcal{F}) = R^0 f_*(\mathcal{E} \otimes f^*\mathcal{F}) \cong N \otimes_A M.$$

The first part of this equation means that $\mathcal{E} \otimes f^*\mathcal{F} \in \mathfrak{Coh}(X, f)$, as the first part of (1) requires. For the latter part, note that $H^0(Y, f_*\mathscr{E} \otimes \mathcal{F}) = N \otimes_A M$. This completes the proof of (1).

To prove (2), we can even assume $Y = \mathrm{Spec}\,(A)$, where $(A, \mathfrak{m})$ is a local Noetherian ring. Let $d = Tor\,\dim_Y \mathcal{O}_X$. So,

$$Tor_i^Y(\mathcal{O}_X, \mathcal{K}) = 0 \qquad \forall\, \mathcal{K} \in QCoh(A),\ \forall i \geq d+1.$$

Choose $U_i = \mathrm{Spec}\,(B_i)$, as above, with $\mathcal{E}_{|U_i}$ trivial. As observed above, the Čech complex (6.71) gives an exact sequence

$$0 \longrightarrow N \longrightarrow C^0(\mathcal{E}) \longrightarrow C^1(\mathcal{E}) \longrightarrow \cdots \longrightarrow C^n(\mathcal{E}) \longrightarrow 0$$

where $N := H^0(X, \mathcal{E})$. Arguing as above,

$$Tor_i^A(C^p(\mathcal{E}), K) = 0 \qquad \forall p \geq 0 \quad \forall K \in QCoh(A), \quad \forall i \geq d+1.$$

It follows from all these that

$$Tor_i^A(N, K) = 0,\ \forall K \in QCoh(A), \forall i \geq d+1.$$

In particular, we have

$$Tor_i^A\left(N, \frac{A}{\mathfrak{m}}\right) = 0, \forall i \geq d+1.$$

By (6.2.15), N is a finitely generated A-module. By results in commutative algebra, it follows that $proj\,\dim_A N \leq d$. The proof is complete. ■

The following is another technical lemma we used.

Lemma 6.5.7. *Consider the following pullback diagram of Noetherian schemes*

$$\begin{array}{ccc} Z \times_Y X & \xrightarrow{\mathfrak{g}} & X \\ {\scriptstyle \mathfrak{f}}\downarrow & & \downarrow{\scriptstyle f} \\ Z & \xrightarrow[g]{} & Y \end{array} \qquad \textit{we write} \quad W := Z \times_Y X.$$

Assume that all the schemes support ample bundles. Assume that f is proper and g has finite Tor dimension. Also, assume that X, Z are Tor independent in the following sense:

$$Tor_i^{\mathcal{O}_{Y,y}}(\mathcal{O}_{Z,z}, \mathcal{O}_{X,x}) = 0 \quad \forall\, i \geq 1, y = f(x) = g(z).$$

Then for $\mathcal{F} \in \mathfrak{Coh}(\mathfrak{g}, X) \cap \mathfrak{Coh}(X, f)$, *we have*

$$f_*\mathcal{F} \in \mathfrak{Coh}(g, Y), \qquad \mathfrak{g}^*\mathcal{F} \in \mathfrak{Coh}(W, \mathfrak{f}), \quad \textit{and} \quad g^* f_* \mathcal{F} \cong \mathfrak{f}_* \mathfrak{g}^* \mathcal{F}. \tag{6.72}$$

Proof. For the purpose of this proof, we can assume $Z = \mathrm{Spec}\,(C)$, $Y = \mathrm{Spec}\,(A)$ are affine. In this case, X is projective over Y and W is projective over Z (6.2.14). So, $X = \mathrm{Proj}\,(S)$ for some graded ring $S = S[x_0, x_1, \ldots x_n]$, with $\deg(x_i) = 1$, and $\mathcal{F} = \widetilde{M}$ for some finitely generated graded S-module. While this does not make the rest of the proof any simpler, it makes it much more transparent. Subsequently, for a homogeneous $f \in S$, with $\deg f = p$, we denote $M_{(f)} = \left\{ \frac{m}{f^q} \in M_f : m \in M_{pq} \right\}$.

Let $U_0, U_1, \ldots, U_n$ be any open cover of X. We have a choice of such an affine open cover $U_i = D(x_i) = \mathrm{Spec}\left(S_{(x_i)}\right) \subseteq X$, of X, with $i = 0, 1, \ldots, n$. For $0 \leq i_0 < i_1 < \cdots < i_k \leq n$, let $U_{i_0 i_1 \cdots i_k} := U_{i_0} \cap U_{i_1} \cap \cdots \cap U_{i_k}$. So,

$$U_{i_0 i_1 \cdots i_k} = D(x_{i_0} \cdots x_{i_k}) \cong \mathrm{Spec}\,(B_{i_0 i_1 \cdots i_k}) \quad \text{where}$$
$$B_{i_0 i_1 \cdots i_k} = S_{(x_{i_0} \cdots x_{i_k})}.$$

Further, note that $\mathcal{F}(U_{i_0 i_1 \cdots i_k}) = M_{(x_{i_0} \cdots x_{i_k})}$. (*In the general case,* $U_{i_0 i_1 \cdots i_k}$ *are affine, by (6.6.3).*)

Let $N \in Coh(B_{i_0 i_1 \cdots i_k})$.

$$\textbf{Claim:} \qquad Tor_i^A(N, C) \cong Tor_i^{B_{i_0 i_1 \cdots i_k}}(N, C \otimes_A B_{i_0 i_1 \cdots i_k}) \qquad \forall i \geq 0. \tag{6.73}$$

First, it follows from the Tor independence condition that $Tor_i^A(B_{i_0 i_1 \cdots i_k}, C) = 0$, $\forall i \geq 1$.

$$\text{Let} \quad \cdots \longrightarrow P_1 \longrightarrow P_0 \longrightarrow C \longrightarrow 0$$

be a resolution of C by free $A-$modules,

where P_i are not necessarily finitely generated. By tensor product, we obtain a free resolution

$$\cdots \longrightarrow P_1 \otimes_A B_{i_0 i_1 \cdots i_k} \longrightarrow P_0 \otimes_A B_{i_0 i_1 \cdots i_k} \longrightarrow C \otimes_A B_{i_0 i_1 \cdots i_k} \longrightarrow 0.$$

We also have

$$(P_i \otimes_A B_{i_0 i_1 \cdots i_k}) \otimes_{B_{i_0 i_1 \cdots i_k}} N \cong P_i \otimes_A N \qquad \forall i.$$

So, the claim (6.73) is established. Since $Tor_i^X(\mathcal{F}, \mathcal{O}_W) = 0,\ \forall i \geq 1$, we have

$$\begin{aligned} Tor_i^A(\mathcal{F}(U_{i_0 \dots i_k}), C) &= Tor_i^A(M_{(x_{i_0} \cdots x_{i_k})}, C) = \\ Tor_i^{B_{i_0 i_1 \cdots i_k}} &\left(M_{(x_{i_0} \cdots x_{i_k})}, C \otimes_A B_{i_0 i_1 \cdots i_k} \right) = 0 \end{aligned} \tag{6.74}$$

for all $i \geq 1$. Write

$$C^k(\mathcal{F}) = \bigoplus_{1 \leq i_0 < i_1 < \cdots < i_k \leq n} \mathcal{F}(U_{i_0 i_1 \cdots i_k}).$$

It follows from (6.74)

$$Tor_i^A(C^p(\mathcal{F}), C) = 0 \qquad \forall \quad i \geq 1. \tag{6.75}$$

We have the Čech complex, augmented with $H^0(X, \mathcal{F})$ on the left-hand side

$$0 \longrightarrow H^0(X, \mathcal{F}) \longrightarrow C^0(\mathcal{F}) \longrightarrow C^1(\mathcal{F}) \longrightarrow \cdots \longrightarrow C^n(\mathcal{F}) \longrightarrow 0. \tag{6.76}$$

We have $H^k(C^\bullet(\mathcal{F})) = H^k(X, \mathcal{F})\ \forall k$. Since $\mathcal{F} \in \mathfrak{Coh}(X, f)$, this sequence is exact. This establishes that $f_*\mathcal{F} \in \mathfrak{Coh}(Y, g)$, as required. It also follows from (6.75), the sequence

$$0 \longrightarrow H^0(X, \mathcal{F}) \otimes_A C \longrightarrow C^0(\mathcal{F}) \otimes_A C \longrightarrow C^1(\mathcal{F}) \otimes_A C \longrightarrow \cdots \longrightarrow C^n(\mathcal{F}) \otimes_A C \longrightarrow 0$$

is exact. Write $W_i = Z \times_Y U_i = \operatorname{Spec}\left(C \otimes_A S_{(x_i)}\right)$. Then $\{W_i\}$ is an affine open cover of W, and

$$\begin{aligned} W_{i_0 i_1 \cdots i_k} := W_{i_0} \cap W_{i_1} \cap \cdots \cap W_{i_k} &= \operatorname{Spec}(C \otimes_A B_{i_0 i_1 \cdots i_k}) \\ &= \operatorname{Spec}\left(C \otimes_A S_{(x_{i_0} \cdots x_{i_k})} \right). \end{aligned}$$

Further,

$$\begin{aligned} \mathfrak{g}^* \mathcal{F}(W_{i_0 i_1 \cdots i_k}) &= \mathcal{F}(U_{i_0 i_1 \cdots i_k}) \otimes_{B_{i_0 i_1 \cdots i_k}} (B_{i_0 i_1 \cdots i_k} \otimes_A C) \\ &\cong \mathcal{F}(U_{i_0 i_1 \cdots i_k}) \otimes_A C. \end{aligned} \tag{6.77}$$

By this identity (6.77), corresponding to the cover $\{W_i\}$, the Čech complex, of $\mathfrak{g}^*\mathcal{F}$ augmented with $H^0(X, \mathfrak{g}^*\mathcal{F})$ on the left-hand side is isomorphic, as follows:

$$\begin{array}{ccccccccccccc}
0 & \longrightarrow & H^0(W, \mathfrak{g}^*\mathcal{F}) & \longrightarrow & C^0(\mathfrak{g}^*\mathcal{F}) & \longrightarrow & C^1(\mathfrak{g}^*\mathcal{F}) & \longrightarrow & \cdots & \longrightarrow & C^n(\mathfrak{g}^*\mathcal{F}) & \longrightarrow & 0 \\
 & & \downarrow \wr & & \downarrow \wr & & \downarrow \wr & & & & \downarrow \wr & & \\
0 & \longrightarrow & H^0(X, \mathcal{F}) \otimes_A C & \longrightarrow & C^0(\mathcal{F}) \otimes_A C & \longrightarrow & C^1(\mathcal{F}) \otimes_A C & \longrightarrow & \cdots & \longrightarrow & C^n(\mathcal{F}) \otimes_A C & \longrightarrow & 0
\end{array} \tag{6.78}$$

Since $\mathcal{F} \in \mathfrak{Coh}(\mathfrak{g}, X)$, the first sequence is exact. This completes the proof that $\mathfrak{g}^*\mathcal{F} \in \mathfrak{Coh}(W, \mathfrak{f})$. The first vertical arrow is an isomorphism, because others are. Therefore, $\mathfrak{f}_*\mathfrak{g}^*\mathcal{F} \cong g^* f_* \mathcal{F}$. The proof is complete. ■

6.5.2. *A spectral sequence*

The purpose of this section is to encapsulate some of the above (6.4.5 6.4.6) in a fancier language of spectral sequences, and to alert the readers of the same. Before we proceed, we dissect the proof of (6.4.5), as follows. We use the notation as above:

1. Denote $A_1^{\bullet\bullet} = \{A_1^{pq}\}$ and $E_1^{\bullet\bullet} = \{E_1^{pq}\}$, where
$$E_1^{pq}(X) = \bigoplus_{x \in X^{(p)}} K_{-(p+q)}(\kappa(x)), \quad \text{Also, write}$$
$$A_1^{pq} = K_{-(p+q)}\left(Coh^p(X)\right). \tag{6.79}$$

2. Corresponding to the functors $Coh^{p+1}(X) \hookrightarrow Coh^p(X)$, we have three sets of maps,
$$\iota : A_1^{p+1,q-1} \longrightarrow A_1^{pq}, \quad \mathfrak{q} : A_1^{pq} \longrightarrow E_1^{pq}, \quad \partial : E_1^{pq} \longrightarrow A_1^{p+1,q-1}.$$
In the sense of (6.52), these three sets of maps form an exact triangle:
$$\begin{array}{ccc}
A_1^{\bullet\bullet} & \xrightarrow{\quad \iota \quad} & A_1^{\bullet\bullet} \\
 & {\scriptstyle \partial}\nwarrow \quad \swarrow{\scriptstyle \mathfrak{q}} & \\
 & E^1_{\bullet\bullet} &
\end{array} \tag{6.80}$$
We say **total degree** $\deg(A_1^{pq}) = \deg(E^1_{pq}) = p + q$.

3. Such careful indexing, as above, was designed so that everything falls nicely into the mold of spectral sequences. However, the main pieces of data that we have are the localization sequences (6.52). We resketch the diagram (6.53), as follows:

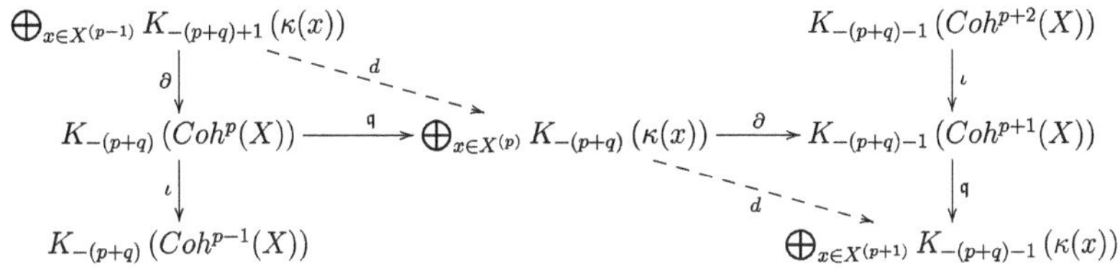

(6.81)

With the above notation, we resketch this diagram, as follows:

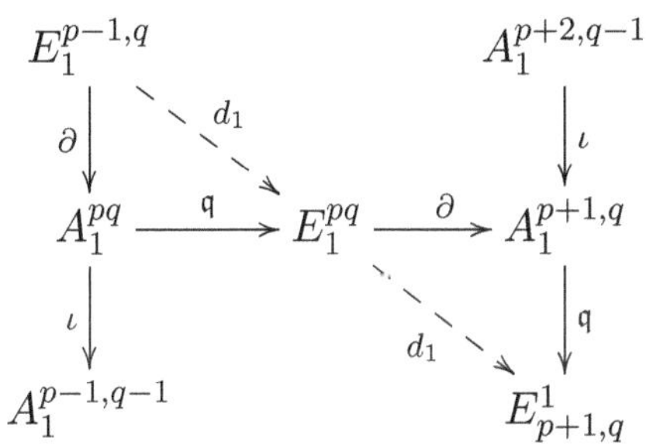

Note, $\begin{cases} d\mathfrak{q} = 0 \\ \partial d = 0 \end{cases}$ (6.82)

Note, for these $E_1^{\bullet\bullet}$, the differentials d_1 run horizontally, left to right, in the pq-plane. In terms of bidegrees, d_1 has bidegree $(1,0)$, ι has bidegree $(-1,1)$, ∂ has bidegree $(1,0)$, and $\mathfrak{q}$ has bidegree $(0,0)$. Subsequently, we want to define similar objects $A_2^{\bullet\bullet}$ and $E_2^{\bullet\bullet}$.

4. Denote

$$A_2^{pq} = \iota(A_1^{p+1,q-1}) \subseteq A_1^{pq}, \quad \text{and} \quad E_2^{pq} = H^p\left(E_1^{\bullet q}\right) =: H^{p,q}\left(E_1^{\bullet\bullet}\right).$$

5. We expand the diagram (6.82):

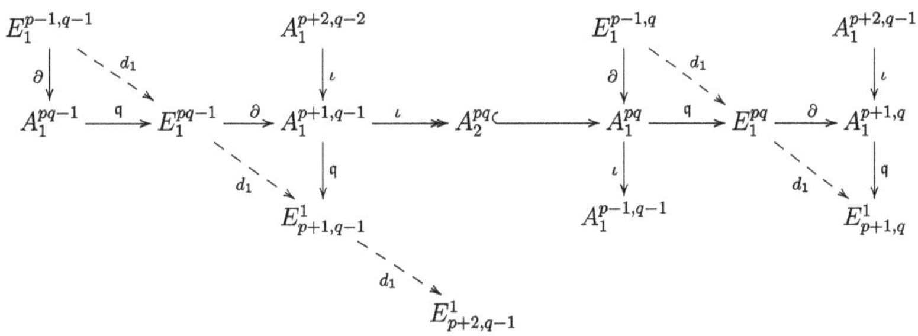

(6.83)

It follows from this diagram that there are maps:

$$\begin{cases} \mathfrak{q}_2 : A_2^{pq} \longrightarrow E_2^{p+1,q-1} \\ \partial_2 : E_2^{pq} \longrightarrow A_2^{p+1,q}, \quad \text{and let} \\ \iota_2 : A_2^{p+1,q-1} \longrightarrow A_2^{pq} \quad \text{be the restriction of } \iota. \end{cases}$$

Like ι, the bidegree of ι_2 is $(-1,1)$, while both preserve the total degree. Both ∂ and ∂_2 have bidegree $(1,0)$. However, $\mathfrak{q}_2$ preserves the total degree, and it has bidegree $(1,-1)$, while $\mathfrak{q}$ had bidegree $(0,0)$.

We summarize the above in the following lemma.

Lemma 6.5.8. *Let X be a Noetherian scheme, with $\dim X < \infty$. With notation, as above, the diagram*

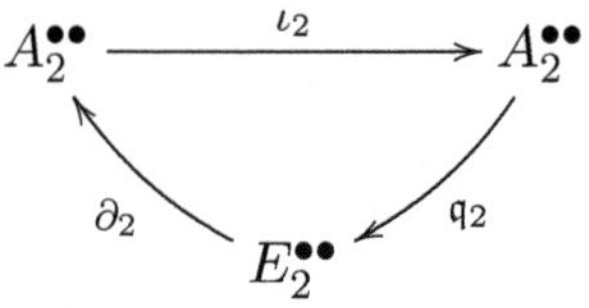

(6.84)

is an exact triangle. Here, the bidegree of ι_2 remains the same as that of ι, which is $(-1,1)$. Likewise, the bidegree of ∂_2 remains the same as that of ∂, which is $(1,0)$. And the bidegree of $\mathfrak{q}_2$ has changed to bidegree $(1,-1)$ from that of $\mathfrak{q}$, which was $(0,0)$.

Iterating this process, for all integers $r \geq 1$, we obtain exact triangles

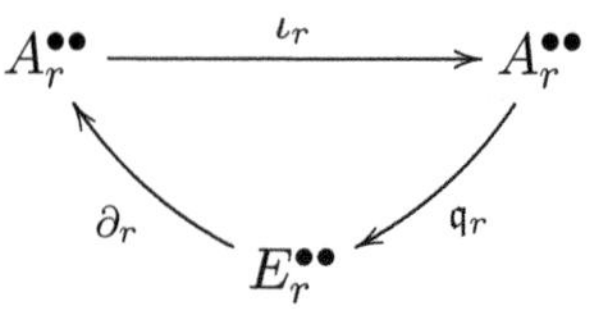

such that the bidegree of ι_r is $(-1,1)$, the bidegree of ∂_r is $(1,0)$, and the bidegree of $\mathfrak{q}_r$ is $(r,-r)$. Further, $E_r^{\bullet\bullet}$ has a differential:

$$d_r : E_r^{pq} \longrightarrow E_r^{p+r,q-(r-1)} \quad \textbf{of bidegree } (r,-r+1), \quad \textit{and}$$
$$E_r^{pq} = H^{pq}\left(E_{r-1}^{\bullet\bullet}\right).$$

Further, by construction, $E_r^{pq} = 0$*, unless* $0 \le p \le d$ *and* $q \le -p$*. Consequently,* $d_r = 0$ *for all* $r \ge d+1$*. So,* $E_r^{pq} = E_{r+1}^{pq}\ \forall\ r \ge d+1$. (In this statement, we have not formalized the structural properties of $E_r^{\bullet\bullet}$ other than providing some data on $E_1^{\bullet\bullet}$, $A_1^{\bullet\bullet}$. All these mean that we have constructed a spectral sequence that converges, which we define next.)

Proof. The exactness of (6.84) follows, by diagram chasing, from the exactness of (6.80). The inductive process of construction of $E_r^{\bullet\bullet}$ and $A_r^{\bullet\bullet}$ is similar to the case of $r = 2$, from $r = 1$. Other comments are self-evident. ■

We define spectral sequences, while interested readers are referred to [Wc1, Chap. 5] for an elaborate exposition of the same.

Definition 6.5.9. Let $\mathscr{A}$ be an abelian category. Fix an integer $a \ge 0$. A **cohomology spectral sequence** in $\mathscr{A}$, starting at E_a, is a family of objects $E_r^{\bullet\bullet} = \{E_r^{pq} : p, q \in \mathbb{Z}\}$, and $r \ge a$, together with differentials $d_r^{\bullet\bullet} : E_r^{\bullet\bullet} \longrightarrow E_r^{\bullet\bullet}$, of bidegree $(r, -(r-1))$ such that

$$\forall\ r \ge a \quad E_{r+1}^{pq} = H^{pq}(E_r^{\bullet\bullet}) \quad \text{the cohomology at bidegree } (p,q).$$

More precisely, $\forall r \ge a$, the differentials are arrows, as follows:

$$d_r^{pq} : E_r^{pq} \longrightarrow E_r^{p+r, q-r+1} \quad \ni \quad d_r^{p+r,q-r+1} d_r^{pq} = 0 \quad \forall p, q.$$

So, the slope of the direction of the differentials d_r^{pq} is downward (negative) in the pq-plane. The following is a list of immediate comments and further definitions:

1. First and foremost, we constructed a spectral sequence in Lemma 6.5.8, staring from the initial data provided by the exact sequences (6.52).
2. We say that the total degree $To\deg(E_r^{pq}) = p+q$ and the bidegree $Bi\deg(E_r^{pq}) = (p,q)$.
3. Such a spectral sequence is said to be **bounded** if $\forall n \in \mathbb{Z}$ the set $\{E_a^{pq} \ne 0 : Tot\deg(E_a^{pq}) = n\}$ is finite.
4. A spectral sequence $E_r^{\bullet\bullet}$ is said to **converge** if

$$\forall (p,q) \quad \exists\ r_{pq} \in \mathbb{N} \ \ni\ \forall\ r \ge r_{pq} \quad E_r^{pq} = E_{r_{pq}}^{pq}.$$

$$\text{Denote} \quad E_\infty^{pq} := E_{r_{pq}}^{pq}.$$

   This is the case whenever the spectral sequence is bounded.

The **Homology spectral sequences** are defined, likewise, by reversing the direction of the arrows. Subsequently, we would comment only on cohomology spectral sequences.

Given a convergent spectral sequence E_r^{pq}, we give a meaning to the often-encountered expression $E_r^{pq} \Longrightarrow H^n$.

Definition 6.5.10. Let $\mathscr{A}$ be an abelian category, and $E_r^{\bullet\bullet}$ be a convergent (cohomology) spectral sequence, starting from E_{r_0}. Suppose that there is a sequence of objects H^n, together with a filtration

$$0 = \mathcal{F}^t H^n \subseteq \cdots \subseteq \mathcal{F}^{p+1} H^n \subseteq \mathcal{F}^p H^n \subseteq \cdots \subseteq \mathcal{F}^s H^n = H^n$$
$$\ni \qquad E_\infty^{pq} \cong \frac{\mathcal{F}^p H^{p+q}}{\mathcal{F}^{p+1} H^{p+q}}.$$

In this case, we write

$$E_{r_0}^{pq} \Longrightarrow H^{p+q}.$$

With Lemma 6.5.8, we have the immediate lemma.

Lemma 6.5.11. *Let X be a Noetherian scheme with* $\dim X = d < \infty$. *As defined in* (6.79), *let*

$$E_1^{pq}(X) = \bigoplus_{x \in X^{(p)}} K_{-(p+q)}(\kappa(x)). \tag{6.85}$$

Then there is a spectral sequence $E_r^{\bullet\bullet}$ such that

$$E_1^{pq}(X) \Longrightarrow G_{-p-q}(X).$$

Note that (6.85) *was constructed from the initial data provided by the exact sequences* (6.52).

Proof. We already checked in (6.5.8) that E_r^{pq} satisfies the conditions of being a spectral sequence. Note, for fixed integer n, $E_1^{d,n-d}, \ldots, E_1^{p,n-p}, \ldots, E_1^{0,n}$ are the only possible nonzero terms of total degree n. So, it is convergent. (*In fact, unless* $0 \le p \le d$, *we have* $E_r^{pq} = 0$. *So,* $d_r^{pq} = 0$ *if* $p + r \ge d + 1$.) To provide a filtration of $G_n(X)$, we consider the filtration

$$0 \subseteq Coh^d(X) \subseteq Coh^{d-1}(X) \subseteq \cdots \subseteq Coh^p(X)$$
$$\subseteq \cdots \subseteq Coh^0(X) = Coh(X).$$

The initial data were created based on localization sequences of the pairs $Coh^{p+1}(X) \subseteq Coh^p(X)$. We define

$$\mathcal{F}^p G_n(X) = Image\left(K_n Coh^p(X) \longrightarrow G_n(X)\right).$$

This gives a filtration

$$0 \subseteq \mathcal{F}^d G_n(X) \subseteq \mathcal{F}^{d-1} G_n(X) \subseteq \cdots \subseteq \mathcal{F}^p G_n(X) \subseteq \cdots \subseteq \mathcal{F}^1 G_n(X)$$
$$\subseteq \mathcal{F}^0 G_n(X) = G_n(X).$$

One checks that

$$\frac{\mathcal{F}^p G_n(X)}{\mathcal{F}^{p+1} G_n(X)} \cong E_\infty^{p,n-p}.$$

For further details, see [Wc1, §5.9]. The proof is complete. ■

Remark 6.5.12. The spectral sequence in (6.5.8) is referred to as the Brown–Gersten–Quillen spectral sequence. As is evident from the construction, starting from the exact sequences (6.52), the process applies to any abelian category. The process is standard and classical. For a more detailed exposition, see [Wc1, §5.9].

To the two equivalent conditions in (6.4.6), we add one more by interpreting them in terms of the Brown-Gersten-Quillen spectral sequence.

Proposition 6.5.13. *Let X be a Noetherian scheme, with $d = \dim X$. Then the following three conditions are equivalent:*

1. *The sequence (6.51) is exact $\forall n \geq 0$*
2. *The maps $\iota : K_q\left(Coh^{p+1}(X)\right) \longrightarrow K_q\left(Coh^1(X)\right)$, induced by the functors $Coh^{p+1}(X) \hookrightarrow Coh^p(X)$, are zero, $\forall p \geq 0$.*
3. *For all q, $E_2^{pq}(X) = 0$ if $p \neq 0$ and the edge map $\mathfrak{q}$ in sequence (6.51) induces isomorphisms $G_n(X) \xrightarrow{\sim} E_2^{0q}$.*

Proof. In (6.4.6), we already proved that (1) $\iff$ (2). Since E_1^{pq} is given by the homologies of the sequences (6.51) (and the $\ker(\mathfrak{q})$ in the case $q = 0$), it follows that (2) $\iff$ (3). The proof is complete. ■

6.6. Exercises

Exercise 6.6.1. Suppose $f : X \longrightarrow Y$ is a finite morphism of Noetherian schemes. Prove that for all $\mathcal{F} \in QCoh(X)$, we have $R^k f_* \mathcal{F} = 0 \ \forall\ k \geq 1$. Then conclude that the push forward maps $f_* : G_n(X) \longrightarrow G_n(Y)$ are well defined, $\forall n \geq 0$.

Exercise 6.6.2. (Optional) Let $f : X \longrightarrow Y$ be a map of Noetherian schemes. Prove the Valuative criteria for separatedness and properness, as follows. Let $(R, \mathfrak{m})$ be any valuation ring and $K = R/\mathfrak{m}$ be its quotient field. Consider the diagram

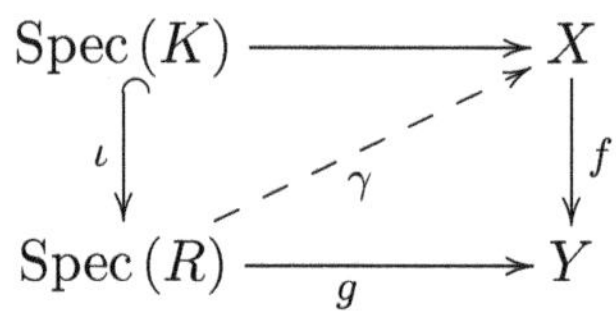

Suppose the outer commutative diagram is given. Prove that

1. f is separated if and only if there is at most one map γ so that the diagram commutes.
2. f is proper if and only if there is exactly one map γ so that the diagram commutes.

See [Hr, §4]

Exercise 6.6.3. (Optional) Let X be a Noetherian separated scheme. Let U, V be two affine open subsets of X. Prove that $U \cap V$ is affine.

Exercise 6.6.4. Under the setup of (6.2.4), prove that the diagram

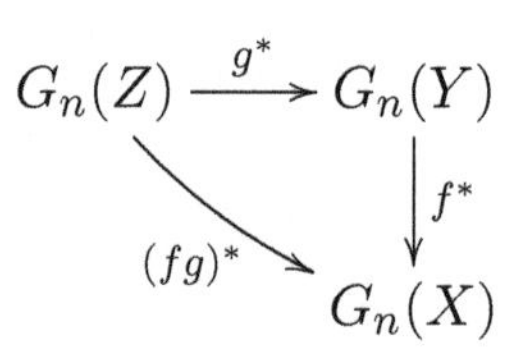

commutes.

Hint: Similar to the proof of (5.4.2).

Exercise 6.6.5. Under the setup of (6.2.17), prove that the diagram

$$\begin{array}{ccc} G_n(X) & \xrightarrow{f_*} & G_n(Y) \\ & \searrow^{(gf)_*} & \downarrow g_* \\ & & G_n(Z) \end{array} \qquad \text{commutes, } \forall \text{n} \geq 0.$$

Hint: Similar to the proof of (5.4.2).

Exercise 6.6.6. Let $f : X \longrightarrow Y$ be a proper map of Noetherian schemes. Assume X has an ample line bundle. Then there are pushforward maps

$$f_* : K_n(X) \longrightarrow K_n(Y) \qquad \forall n \geq 0.$$

Hint: Follows as in (6.2.17). Use (6.5.6), and the argument as in (6.33).

Exercise 6.6.7. Let $f : X \longrightarrow Y$ be a proper morphism of Noetherian schemes, and f have finite Tor dimension. Assume both X, Y support ample bundles. In this case, the pushforward $f_* : G_n(X) \longrightarrow G_n(Y)$ and the pullback $f^* : G_n(Y) \longrightarrow G_n(X)$ are well defined. Prove the following:

1. The pushforward $f_* : K_n(X) \longrightarrow K_n(Y)$ and the pullback $f^* : K_n(Y) \longrightarrow K_n(X)$ maps are well defined. Further, prove that the diagram

$$\begin{array}{ccc} K_n(X) & \longrightarrow & G_n(X) \\ f_* \downarrow & & \downarrow f_* \\ K_n(Y) & \longrightarrow & G_n(Y) \end{array} \qquad \text{commutes } \forall \text{n}.$$

   Hint: Use (6.5.6(2)).
2. There is a $K_0(X)$-action on $K_n(X)$, and likewise for Y. Prove for $x \in K_0(X)$, $y \in K_n(Y)$, we have the same formula $f_*(x) \cdot y = f_*(x \cdot f^* y) \in K_n(Y)$.
3. For $x \in G_n(X)$, $y \in K_0(Y)$ we have the formula $y \cdot f_*(x) = f_*(f^*(y) \cdot x)$.

Hint: Use the projection formula,

$$\forall \mathcal{F} \in Coh(X), \mathscr{E} \in \mathscr{P}(Y), \qquad f_*\mathcal{F} \otimes_Y \mathscr{E} \cong f_*\left(\mathcal{F} \otimes_X f^*\mathcal{E}\right)$$

[Hr, Ex. II.5.1]

Exercise 6.6.8. Let $f : P \longrightarrow X$ be a flat morphism of Noetherian schemes. Let $Z \subseteq X$ be closed and $U = X - Z$. Write $U = X - Z$. Write $P_U = f^{-1}U := U \times_X P$ and $P_Z = f^{-1}Z := Z \times_X P$. Prove that the following diagram:

$$\begin{array}{ccccc} \mathbf{G}(P_Z) & \longrightarrow & \mathbf{G}(P_X) & \longrightarrow & \mathbf{G}(P_U) \\ \downarrow & & \downarrow & & \downarrow \\ \mathbf{G}(Z) & \longrightarrow & \mathbf{G}(X) & \longrightarrow & \mathbf{G}(U) \end{array} \qquad \text{commutes.} \qquad (6.86)$$

Consequently, f induces a map from the localization sequence of the G-groups for the pair (P_Z, P) to the same of (Z, X).

Hint: Use (6.1.11), and work with the Classifying spaces of $\mathbb{Q}Coh(X)$ and others.

Exercise 6.6.9. Let S be a graded commutative ring, with $S_0 = A$ and $S = A[x_0, \cdots, x_n]$, with $\deg(x_i) = 1$. Let $\mathscr{B} \subseteq Coh^+(S)$ be the full subcategory of objects $M \in Coh^+(S)$ such that $M_n = 0\ \forall n \gg 0$. Prove that $\mathscr{B}$ is a Serre subcategory of $Coh^+(S)$. Further, prove that the quotient $\frac{Coh^+(S)}{\mathscr{B}} \xrightarrow{\sim} Coh(\text{Proj}\,(S))$, and that the functor $Coh^+(S) \longrightarrow Coh(\text{Proj}\,(S))$ has the universal property of this quotient.

Hint: This is essentially the same as [Hr, Ex. II.5.9].

Exercise 6.6.10. Let $f : \kappa \longrightarrow F$ be a purely inseparable finite extension of fields, with $[F : \kappa] = p^d$, where $p = char(\kappa)$. Prove that

$$\begin{cases} f_*f^* : K_n(\kappa) \longrightarrow K_n(\kappa) \\ f^*f_* : K_n(F) \longrightarrow K_n(F) \end{cases} \quad \text{are multiplication by} \quad p^d$$

Hint: See [Q, p. 130]

Exercise 6.6.11. Let X be a Noetherian scheme. With notation, as in (6.4.3), prove that $\frac{Coh^k(X)}{Coh^{k+1}(X)} \cong \coprod_{x \in X^{(k)}} Coh^k\left(\mathcal{O}_{X,x}\right)$. (*Note* *that* $Coh^k(\mathcal{O}_{X,x})$ *coincides with the category of finite length* $\mathcal{O}_{X,x}$*-modules.)*

Hint: The natural functor $\iota : Coh^k(X) \longrightarrow \coprod_{x \in X^{(k)}} Coh^k(\mathcal{O}_{X,x})$ vanishes on $Coh^{k+1}(X)$. We need to prove that the induced functor from the quotient category is essentially surjective and fully faithful. Use [Hr, Ex. II.5.15].

Exercise 6.6.12. Let $R = k[[X_1, \ldots, X_d]]$ be the ring of formal power series, over a field k. Prove that $\forall k, n \geq 0$, functors $Coh^{k+1}(R) \hookrightarrow Coh^k(R)$ induce zero maps of K-groups $K_n(Coh^{k+1}(R)) \hookrightarrow K_n(Coh^k(R))$. So, the Gersten complexes (6.51) of $X = \operatorname{Spec}(A)$ are exact.

Hint: Use the proof of (6.4.9) in conjunction with the Weierstrass preparation theorem.

Exercise 6.6.13. Assume k is an infinite field. Prove the normalization lemma (6.4.11), using the general position argument. **Hint:** See [Gd3].

Exercise 6.6.14. (Optional) Let I be a filtered category, and $\iota : I^{op} \longrightarrow \underline{Sch}$ sending $i \mapsto X_i$ be a filtered inverse system to the category $\underline{Sch}$ of Noetherian schemes. Assume that the transition maps $X_i \longrightarrow X_j$ are affine. Then the limit $X := \lim_I X_i$ exists in $\underline{Sch}$. (For the definition of affine morphisms, see [Hr, Ex. II.5.17].)

Chapter 7

K-Theory of Projective Bundles

In Chapter 6, we discussed the projective bundle theorem for G-theory (6.3.2). Let S be a scheme and $\mathscr{E}$ be a locally free sheaf on S. In this chapter, we do K-theory of $\mathbb{P}\mathscr{E} := \text{Proj}\,(\mathcal{S}\text{ym}\,(\mathscr{E}))$, where $\mathcal{S}\text{ym}\,(\mathscr{E})$ denotes the sheaf of symmetric algebras of $\mathscr{E}$.

7.1. The Canonical Resolution of Regular Sheaves on $\mathbb{P}\mathscr{E}$

We set up some notation on the construction of the projective spaces $\mathbb{P}\mathscr{E}$.

Notation 7.1.1. For details regarding the following constructions, readers are referred to [Hr, §II.7]. Let S be a scheme and $\mathscr{E}$ be a locally free sheaf on S, with $rank(\mathscr{E}) = r$. Let $\mathcal{S}\text{ym}\,(\mathscr{E}) = \bigoplus_{n\geq 0} \mathcal{S}\text{ym}\,(\mathscr{E})_n$ denote the sheaf of symmetric algebras of $\mathscr{E}$. The **projective space bundle** of $\mathscr{E}$ is defined as $X := \mathbb{P}\mathscr{E} := \text{Proj}\,(\mathcal{S}\text{ym}\,(\mathscr{E}))$. This comes with a structure map $f : \mathbb{P}\mathscr{E} \longrightarrow S$. As usual, $\mathcal{O}(1) := \mathcal{O}_X(1) = \widetilde{\mathcal{S}\text{ym}\,(\mathscr{E})}\,(1)$ is called the **twisting sheaf** or the **canonical line bundle** on X. As before, for $F \in QCoh(X)$, and $n \in \mathbb{Z}$, denote $F(n) := F \otimes \mathcal{O}_X(n)$.

Perhaps, the most useful fact is, if $U = \text{Spec}\,(A) \subseteq X$ is an affine open subset such that $\mathscr{E}_{|U}$ is trivial, then $f^{-1}U \cong \mathbb{P}^{r-1}_A$. Indeed, $f^{-1}U := \text{Proj}\,(\Gamma(U, \mathscr{E}))$, and $X := \mathbb{P}\mathscr{E} := \text{Proj}\,(\mathcal{S}\text{ym}\,(\mathscr{E}))$ is constructed by gluing such $f^{-1}U \cong \mathbb{P}^{r-1}_A$. The line bundle $\mathcal{O}_X(1)$ is also constructed similarly by gluing. Consult [Hr, §II.7, pp. 160–162] for details. For $F \in QCoh(X)$ and $N \in QCoh(S)$, use the notation

$$F \otimes_S N := F \otimes_X f^* N$$

All schemes we consider are assumed to be Noetherian.

The following are some of the basic and important facts.

Lemma 7.1.2. *Use all the notation as in (7.1.1), and $X := \mathbb{P}\mathscr{E}$. Then we have the following:*

1. *For $F \in QCoh(X)$, the direct images $R^q f_* F \in QCoh(S)$ and $R^q f_* F = 0\ \forall q \geq r$.*
2. *For $F \in QCoh(X)$, $P \in \mathscr{P}(S)$, we have*

$$R^q f_* F \otimes_S P = R^q f_* (F \otimes_S P)$$

3. *For $N \in QCoh(S)$, we have*

$$R^q f_* \left(\mathcal{O}_x(n) \otimes_S N\right) =$$
$$\begin{cases} 0 & q \neq 0,\ r-1 \\ \mathcal{S}\mathrm{ym}\,(\mathscr{E})_n \otimes_S N & q = 0 \\ (\mathcal{S}\mathrm{ym}\,(\mathscr{E})_{-r-n})^* \otimes_S \Lambda^r \mathscr{E} \otimes_S N & q = r-1 \end{cases}$$

4. *If $S = \mathrm{Spec}\,(A)$ is affine, and $F \in Coh(X)$, then F is the quotient of a finite direct sum of $\mathcal{O}_X(-n)$. In fact, there is an integer n_0, such that $\forall n \geq n_0$, $F(n)$ is globally generated.*

Proof. The assertion (1) is a local property on S. So, we can assume that $S = \mathrm{Spec}\,(A)$ is affine and $\mathscr{E}$ is trivial. So, $X = \mathbb{P}_A^{r-1}$. So, $R^q f_* F \cong \widetilde{H^q(X, F)}$. So, $R^q f_* F \in QCoh(S)$. Since $X = \mathbb{P}_A^{r-1}$ has an affine cover of r open sets, $H^q(X, F) = 0\ \forall q \geq r$, by (6.5.2). This settles (1). The assertion (2) is the projection formula [Hr, Ex. III.8.3]. To prove (2), note that if $\mathscr{I}_\bullet$ is an injective resolution of F, then $\mathscr{I}_\bullet \otimes_S P$ is an injective resolution of $F \otimes_S P$. This settles (2).

When $S = \mathrm{Spec}\,(A)$ is affine, assertion (3) is an improved version of the Serre duality (6.5.4) and is proved in the same manner (see Remark 6.5.5). In the general case, when $q \neq 0, r-1$, the vanishing of $R^q f_* \left(\mathcal{O}_x(n) \otimes_S N\right)$ is a local property and follows immediately from the affine case. To deal with the cases $q = 0, r-1$, take a finite open affine cover $U_1, U_2, \ldots, U_m$ of S such that $\mathscr{E}_{|U_i}$ are trivial. With $U_i = \operatorname{Spec}(A_i)$, $f^{-1}U_i = \mathbb{P}^{r-1}_{A_i} = \operatorname{Proj}\left(A_i[X_{i0}, \ldots, X_{i,(r-1)}]\right)$. From the affine case, it follows there are natural isomorphisms:

$$\begin{cases} R^0 f_* \left(\mathcal{O}_x(n) \otimes_S N\right)_{|U_i} \xrightarrow{\sim} \left(\mathcal{S}\mathrm{ym}\,(\mathscr{E})_n \otimes_S N\right)_{|U_i} \\ R^{r-1} f_* \left(\mathcal{O}_x(n) \otimes_S N\right)_{|U_i} \xrightarrow{\sim} \left((\mathcal{S}\mathrm{ym}\,(\mathscr{E})^*_{-r-n} \otimes_S \Lambda^r \mathscr{E} \otimes_S N\right)_{|U_i} \end{cases}$$

Restrictions of these maps on the intersection $U_i \cap U_j$, amount to a change of basis on $\mathscr{E}_{|U_i \cap U_j}$. So, these maps patch to give the desired isomorphism on S. This establishes (3).

With $S = \operatorname{Spec}(A)$, the proof of (4) is similar to (6.1.5), using Čech complexes. We will identify $QCoh(A) \cong QCoh(S)$. First, $\mathcal{S}\mathrm{ym}\,(\mathscr{E}) \cong \bigoplus_{n \in \mathbb{Z}} f_* \mathcal{O}_X(n)$ as (sheaf of) graded A-algebras. Note that $\mathcal{S}\mathrm{ym}\,(\mathscr{E})$ is the symmetric algebra of the projective A-module $\mathscr{E} := \Gamma(X, \mathscr{E})$. In the general case, write $\mathfrak{M}(F) = \bigoplus_{n \in \mathbb{Z}} f_* \mathcal{F}(n)$, which is a graded $\mathcal{S}\mathrm{ym}\,(\mathscr{E})$-module. There is a natural map, $\beta : \widetilde{\mathfrak{M}(F)} \longrightarrow F$ (see [Hr, §II.7]). The restriction of this map on $X_i = f\ {}^1 U_i = \mathbb{P}^{r-1}_{A_i}$ is an isomorphism (6.1.4). Hence, β is an isomorphism. Further, $\mathfrak{M}(F)_{f_i}$ is a surjective image of a finite direct sum of $\mathcal{S}\mathrm{ym}\,(\mathscr{E})_{f_i}(-n)$. Clearing the denominators, for $i = 1, \ldots, m$, there are maps $\varphi_i : \bigoplus_{j=1}^{m_j} \mathcal{O}(-n_{ij}) \longrightarrow \mathfrak{M}(F)$ such that $(\varphi_i)_{f_i}$ is surjective. Adding all these maps, there is a surjective map $\bigoplus_{i=1}^{N} \mathcal{S}\mathscr{E}(-n_i) \twoheadrightarrow \mathfrak{M}(F)$. Sheafifying, we obtain the diagram of surjective maps

$$\begin{array}{ccc} \bigoplus_{i=1}^{N} \mathcal{O}_X(-n_i) & \twoheadrightarrow & \widetilde{\mathfrak{M}(F)} \\ & \searrow\!\!\!\!\searrow & \downarrow \beta \\ & & F \end{array}$$

With $n_0 = \max\{n_1, \ldots, n_N\}$, as in the proof of (6.1.5), $F(n)$ is globally generated. This completes the proof of (4). For further details, see [Hr, II.5.13, II.7.11]. ■

The concept of regularity (Mumford) is a key ingredient in this chapter as follows.

Definition 7.1.3. Use all the notation as in (7.1.1), and $X := \mathbb{P}\mathscr{E}$. For $F \in QCoh(X)$, we say that F is **regular** if

$$R^q f_* F(-q) = 0 \quad \forall\, q \geq 1$$

The full subcategory of regular objects $F \in QCoh(X)$ is denoted by $\mathfrak{R}(X) \subseteq QCoh(X)$. By Lemma 7.1.2(3), we have $\mathcal{O}_X(n) \otimes_S N \in \mathfrak{R}(X)$ for all $N \in QCoh(S)$, and $n \geq 0$.

The following are some lemmas on regularity.

Lemma 7.1.4. *Use all the notation as in* (7.1.1), *and* $X := \mathbb{P}\mathscr{E}$. *Let*

$$0 \longrightarrow K \longrightarrow F \longrightarrow C \longrightarrow 0 \quad \text{be exact in} \quad QCoh(X).$$

Then

1. *If* $K(n)$ *and* $C(n)$ *are regular, then* $F(n)$ *is regular.*
2. *If* $K(n+1)$ *and* $F(n)$ *are regular, then* $C(n)$ *is regular.*
3. *If* $F(n+1)$ *and* $C(n)$ *are regular, and if* $f_*F(n) \longrightarrow f_*C(n)$ *is onto, then* $K(n+1)$ *is regular.*

Proof. Twisting the exact sequence by $\mathcal{O}(n-q)$, we have the exact sequence

$$\cdots \longrightarrow R^q f_* K(n-q) \longrightarrow R^q f_* F(n-q) \longrightarrow R^q f_* C(n-q) \longrightarrow R^{q+1} f_* K(n-q)$$

The rest of the proof follows by chasing this exact sequence. ■

Lemma 7.1.5. *Use all the notation as in* (7.1.1), *and* $X := \mathbb{P}\mathscr{E}$. *For* $F \in QCoh(X)$, *if* F *is regular, then* $F(n)$ *is regular* $\forall n \geq 0$.

Proof. We only need to prove that $F(1)$ is regular. The canonical epimorphism $f^*\mathcal{E} = \mathcal{O}_X \otimes_S \mathcal{E} \twoheadrightarrow \mathcal{O}_X(1)$ induces an epimorphism $\mathcal{O}_X(-1) \otimes_S \mathcal{E} \twoheadrightarrow \mathcal{O}_X$. This leads to the Koszul complex of $f^*\mathscr{E}$:

$$0 \longrightarrow \mathcal{O}_X(-r) \otimes_S \Lambda^r \mathcal{E} \longrightarrow \cdots \longrightarrow \mathcal{O}_X(-1) \otimes_S \mathcal{E} \longrightarrow \mathcal{O}_X \longrightarrow 0 \tag{7.1}$$

which is an exact sequence in $\mathscr{P}(X)$. By tensor product, we obtain the exact sequence

$$0 \longrightarrow F(-r) \otimes_S \Lambda^r \mathcal{E} \longrightarrow \cdots \longrightarrow F(-p) \otimes_S \Lambda^p \mathcal{E} \longrightarrow \cdots \longrightarrow F(-1) \otimes_S \mathcal{E} \longrightarrow F \longrightarrow 0$$

We split this exact sequence as follows:

$$0 \longrightarrow Z_p \longrightarrow F(-p) \otimes_S \Lambda^p \mathcal{E} \longrightarrow Z_{p-1} \longrightarrow 0$$

$$\text{So,} \quad \begin{cases} Z_0 = F \\ Z_{r-1} = F(-r) \otimes_S \Lambda^r \mathcal{E} \end{cases}$$

Since F is regular, by (7.1.2(2)),

$$R^q f_* \left(F(-p) \otimes_S \Lambda^p \mathscr{E}(p-q)\right) = \left(R^q f_* F(-q)\right) \otimes_S \Lambda^p \mathscr{E} = 0 \quad \forall q \geq 1$$

So, $F(-p) \otimes_S \Lambda^p \mathscr{E}(p) = F \otimes_S \Lambda^p \mathscr{E}$ is regular $\forall p$. By (7.1.4(2)), $Z_{r-1}(r)$ is regular. By induction, $Z_p(p+1)$ is regular. Finally, $F(1) = Z_0(1)$ is regular. The proof is complete. ■

Lemma 7.1.6. *Use all the notation as in* (7.1.1)*, and* $X := \mathbb{P}\mathscr{E}$*. Let* $F \in QCoh(X)$ *be regular. Then the canonical map*

$$f^* f_* F =: \mathcal{O}_X \otimes_S f_* F \twoheadrightarrow F \quad \textit{is surjective.}$$

Proof. First, $F(n)$ is regular $\forall n \geq 0$. From the above proof, we have the exact sequence

$$0 \longrightarrow Z_1 \longrightarrow F(-1) \otimes_S \mathcal{E} \longrightarrow F \longrightarrow 0$$

where $Z_1(2)$ is regular.

By (7.1.4), $Z_1(n)$ is regular $\forall n \geq 2$. So, $R^1 f_* Z(n) = R^1 f_* Z(n+1-1) = 0$, $\forall n \geq 1$. So, we obtain the exact sequences

$$0 \longrightarrow f_* Z_1(n) \longrightarrow f_* F(n-1) \otimes_S \mathcal{E} \longrightarrow f_* F(n) \longrightarrow 0$$

$$\forall n \geq 1$$

So, the canonical map

$$\mathcal{S}\mathrm{ym}\,(\mathscr{E}) \otimes_S f_* F \twoheadrightarrow \bigoplus_{n \geq 0} f_* F(n) \quad \text{is surjective.}$$

We have the sheafification $\widetilde{\bigoplus_{n \geq 0} f_* F(n)} \cong F$, which can be checked, locally, by the proof of (6.1.5). So, the proof is complete by sheafification of this surjective map. ■

The following is the celebrated **canonical resolution theorem**.

Theorem 7.1.7. *Use all the notation as in* (7.1.1), *and* $X := \mathbb{P}\mathscr{E}$. *Let* $F \in QCoh(X)$ *be regular. Then* F *has a resolution of the form*

$$0 \longrightarrow f^*T_{r-1}(F)(-(r-1)) \longrightarrow \cdots \longrightarrow f^*T_1(F)(-1)$$
$$\longrightarrow f^*T_0(F)) \longrightarrow F \longrightarrow 0 \tag{7.2}$$

$$\textit{where}\quad T_n : \mathfrak{R}(X) \longrightarrow QCoh(S) \quad \textit{are exact functors.}$$

Further, given a choice of an exact functor T_0F, *and surjective maps,* $T_0F \twoheadrightarrow f_*F$ *(e.g.,* $T_0F = f_*F$, *as in* 7.1.6*), the functors* $T_n(F)$ *are uniquely determined up to isomorphisms. Conversely, if there is such an exact sequence, then* F *is regular. (Recall that* $r = rank(\mathscr{E})$*.)*

Proof. For the convenience of the proof, denote the sequence (7.2) by $C_\bullet$, and $C_{-1} = F$, $C_n = (f^*T_n(F))(-n)$. First, we prove the converse. Suppose such an exact sequence $C_\bullet$ exists. As observed before, $C_n(m) = (f^*T_n(F))(m-n)$ is regular $\forall m \geq n$. So, $C_{r-1}(r-1)$, $C_{r-2}(r-2)$ are regular. Inductively, by 7.1.4(2), it follows that F is regular.

Now, assume F is regular and that there is such an exact sequence $C_\bullet$. We prove that it is uniquely determined by F up to an isomorphism. For $0 \leq n \leq r-1$ and $m \geq 0$, we have $C_n(m) = f^*T_n(F)$(m-n). It follows from (7.1.2(3)) that $R^q\left(C_n(m)\right) = 0 \ \forall q \geq 1$. To see this, we only have to check this for $q = r-1$, and $R^q\left(C_n(m)\right) = \mathcal{S}\text{ym}\left(\mathscr{E}\right)^*_{-r-m+n} \otimes T_n(F) = 0$, because $-r-m+n \leq -1$. So, $f_*(C_\bullet(m)$ is exact $\forall m \geq 0$. Again, using (7.1.2(3)), we obtain exact sequences

$$\cdots \longrightarrow f_*C_n(n) \longrightarrow f_*C_{n-1}(n) \longrightarrow \cdots$$
$$\longrightarrow f_*C_0(n) \longrightarrow f_*F(n) \longrightarrow 0$$

Note that $f_*C_q(n) = f_*\left(\mathcal{O}(n-q) \otimes T_q(F)\right) = \mathcal{S}\text{ym}\left(\mathscr{E}\right)_{n-q} \otimes_S T_qF$. So, $f_*C_n(n) = T_n(F)$, and the terms to the left of $f_*C_q(n)$ are zero. So, the above exact sequence is rewritten as

$$0 \longrightarrow T_n(F) \longrightarrow \mathscr{E} \otimes_S T_{n-1}(F) \longrightarrow \cdots$$
$$\longrightarrow \mathcal{S}\text{ym}\left(\mathscr{E}\right)_n \otimes_S T_0(F) \longrightarrow f_*F(n) \longrightarrow 0 \tag{7.3}$$

So, it is established, inductively, that $T_n(F)$ is uniquely determined by the choice of the map $T_0(F) \twoheadrightarrow f_*F$.

Now, we construct $C_\bullet$. Let $C_{-1} := Z_{-1} := F$, and $T_0(F) := f_*Z_{-1} = f_*F$. For $F \in \mathfrak{R}(X)$, we have $F(1) \in \mathfrak{R}(X)$, and hence $R^1 f_*F = 0$. It follows from this that

$$T_0 : \mathfrak{R}(X) \longrightarrow QCoh(X) \quad \text{is an exact functor.}$$

By (7.1.6), we obtain a surjective map

$$f^*T_0(F) \xrightarrow{\varphi} F.$$

$$\text{By choice} \quad f_*f^*T_0(F) \xrightarrow{\sim} f_*F \quad \text{is an isomorphism.} \tag{7.4}$$

Let $Z_0(F) = \ker(\varphi)$. So, we have an exact sequence

$$0 \longrightarrow Z_0F \longrightarrow \mathcal{O}_X \otimes_S T_0F \xrightarrow{\varphi} F \longrightarrow 0 \tag{7.5}$$

We have that T_0 is exact on $\mathfrak{R}(X)$, and $X \longrightarrow S$ is flat. So, the functor $F \mapsto \mathcal{O}_X \otimes_S T_0F$ is exact on $\mathfrak{R}(X)$. Hence, $Z_0 : \mathfrak{R}(X) \longrightarrow QCoh(X)$ is exact. By (7.1.4(3)), due to the choice (7.4), and because $\mathcal{O}_X \otimes T_0(F)$, F are regular, it follows that $Z_0(F)(1)$ is regular. So, we have defined exact functors

$$\begin{cases} T_0 : \mathfrak{R}(X) \longrightarrow QCoh(S) \\ Z_0 : \mathfrak{R}(X) \longrightarrow QCoh(X) \end{cases} \ni \begin{cases} f_*Z_0(F) = 0 \quad \text{by (7.4)} \\ \text{The sequence (7.5) is exact} \\ Z_0(F)(1) \in \mathfrak{R}(X) \end{cases}$$

Suppose $\forall k \leq n-1$, we have exact functors

$$\begin{cases} T_k : \mathfrak{R}(X) \longrightarrow QCoh(S) \\ Z_k : \mathfrak{R}(X) \longrightarrow QCoh(X) \end{cases}$$

$$\ni \begin{cases} \forall\, 0 \leq k \leq n-1, \quad Z_k(F)(k+1) \in \mathfrak{R}(X), \\ 0 \longrightarrow Z_k(F)(k) \longrightarrow f^*T_kF \xrightarrow{\varphi} Z_{k-1}(k) \longrightarrow 0 \text{ is exact} \\ T_kF = f_*Z_{k-1}(F)(k), \quad f_*Z_k(F)(k) = 0 \end{cases}$$

Define $T_n(F) = f_*\left(Z_{n-1}(F)(n)\right)$. Since $Z_{n-1}(F)(n)$ is regular $\forall F \in \mathfrak{R}(X)$ by the same reasoning as above:

$$T_n : \mathfrak{R}(X) \longrightarrow QCoh(S) \quad \text{is an exact functor.}$$

By (7.1.6), we obtain a surjective map

$$f^*T_n(F) \overset{\varphi}{\twoheadrightarrow} Z_{n-1}(F)(n)\,.$$

By choice $f_*f^*T_n(F) \overset{\sim}{\longrightarrow} f_*Z_{n-1}(F)(n)$ is an isomorphism. (7.6)

Define $Z_n(F)$ by the exact sequence

$$0 \longrightarrow Z_n(F) \longrightarrow f^*T_nF(-n) \xrightarrow{\varphi(-n)} Z_{n-1}(F) \longrightarrow 0 \tag{7.7}$$

Since T_n is exact on $\mathfrak{R}(X)$, and $X \longrightarrow S$ is flat, $F \mapsto f^*T_nF(-n)$ is exact on $\mathfrak{R}(X)$. Further, since $F \mapsto Z_{n-1}(F)$ is also exact on $\mathfrak{R}(X)$, it follows $F \mapsto Z_n(F)$ is exact. So, we have defined two exact functors:

$$\begin{cases} T_n : \mathfrak{R}(X) \longrightarrow QCoh(S) \\ Z_n : \mathfrak{R}(X) \longrightarrow QCoh(X) \end{cases}$$

Twisting (7.7), we have the exact sequence

$$0 \longrightarrow Z_n(F)(n) \longrightarrow f^*T_nF \xrightarrow{\varphi} Z_{n-1}(F)(n) \longrightarrow 0 \tag{7.8}$$

As $\mathcal{O}_X \otimes T_n(F)$, $Z_{n-1}(F)(n)$ are regular, and due to the choice (7.6), $Z_n(F)(n+1)$ is regular, by (7.1.4(3)). It also follows from this sequence and the choice (7.6) that $f_*Z_n(F)(n) = 0$. Now, define $T_{n+1}(F) = f_*Z_n(F)(n)$, and the process continues. In summary, we constructed two sequences of exact functors:

$$\begin{cases} T_n : \mathfrak{R}(X) \longrightarrow QCoh(S) \\ Z_n : \mathfrak{R}(X) \longrightarrow QCoh(X) \end{cases}$$

$$\ni \begin{cases} \forall n \geq 0 \quad Z_n(F)(n+1) \in \mathfrak{R}(X) \\ \text{There is an exact sequence, as shown in (7.8)} \\ T_nF = f_*Z_{n-1}(F)(n), \quad f_*Z_n(F)(n) = 0 \end{cases} \tag{7.9}$$

By concatenating the sequences (7.7), we obtain the required exact sequence (7.2). It remains to be shown that $Z_{r-1}(F) = 0$.

We prove, by induction on q, that $R^qf_*\left(Z_{n+q}(n)\right) = 0$ for all $q \geq 0, n \geq 0$. For $q = 0$, this is valid because $f_*Z_n(n) = 0$, $\forall n \geq 0$.

So, assume that $q \geq 1$. Writing the sequence (7.8) for $n := n+q$ and twisting by $\mathcal{O}_X(-q)$, we have the exact sequence

$$0 \longrightarrow Z_{n+q}(F)(n) \longrightarrow (f^*T_{n+q}F)(-q) \xrightarrow{\varphi} Z_{n+q-1}(F)(n) \longrightarrow 0$$

Consider the exact sequence of $R^i f_*$, corresponding to this exact sequence:

$$R^{q-1}f_*\left(Z_{n+q-1}(F)(n)\right) \longrightarrow R^q f_*\left(Z_{n+q}(F)(n)\right) \longrightarrow R^q f_*\left((f^*T_{n+q}F)(-q)\right)$$

By induction, the left-hand term is zero. The right-hand term is also zero by (7.1.2(3)). So, it is established that $R^q f_*\left(Z_{n+q}(F)(n)\right) = 0\ \forall n \geq 0, q \geq 0$.

It follows from above that $R^q f_*\left(Z_{r-1}(F)(r-1-q)\right) = 0\ \forall q \leq r-1$. However, $R^q f_* = 0,\ \forall\ q \geq r$ by (7.1.2(1)). So, $Z_{r-1}(F)(r-1)$ is regular. Therefore, by (7.1.6), there are surjective maps $f^*f_*(Z_{r-1}(F)(r-1)) \twoheadrightarrow Z_{r-1}(F)(r-1)$. However, $f_*(Z_{r-1}(F)(r-1)) = 0$. Therefore, $Z_{r-1}(F)(r-1) = 0$. So, $Z_{r-1}(F) = 0$. The proof is complete. ■

Now, we consider the regularity properties for locally free sheaves on X.

Lemma 7.1.8. *Use all the notation as in (7.1.1), and $X := \mathbb{P}\mathscr{E}$. Assume S is Noetherian. Let $F \in \mathscr{P}(X)$. Then there is an integer n_0 such that $\forall n \geq n_0$ and $N \in QCoh(S)$, we have the following:*

1. $R^q f_*\left(F(n) \otimes_S N\right) = 0$ *for all* $q \geq 1$.
2. $f_*\left(F(n)\right) \otimes_S N \xrightarrow{\sim} f_*\left(F(n) \otimes_S f^*N\right)$.
3. $f_*F(n) \in \mathscr{P}(S)$.

Proof. The questions are local for the existence of n_0. So, we can assume that $S = \operatorname{Spec}(A)$, and $\mathscr{E}$ is trivial. Hence, $X = \mathbb{P}^{r-1}_A$. If $N \in Coh(S)$, then (1) is, almost, the immediate consequence of (6.2.9), or of [Hr, Thm III.5.2]. We give a complete proof. By (7.1.2), $R^q f_* = 0,\ \forall q \geq r$. So, we use decreasing induction. Let $q_0 \geq 1$, and assume (1) holds $\forall q \geq q_0+1$, $F \in \mathscr{P}(X)$, $\forall N \in QCoh(S)$.

Let $F \in \mathscr{P}(X)$. By (6.1.5), there is an integer m such that $F(m)$ is globally generated. With $L = \mathcal{O}_X(-m)^k$, for some $k \geq 1$, there is an exact sequence

$$0 \longrightarrow K \longrightarrow L \longrightarrow F \longrightarrow 0 \quad \text{in } \mathscr{P}(X) \tag{7.10}$$

This induces the exact sequences

$$0 \longrightarrow K(n) \otimes_A N \longrightarrow L(n) \otimes_A N \longrightarrow F(n) \otimes_A N \longrightarrow 0 \qquad \forall n \tag{7.11}$$

Consider the exact sequence

$$R^{q_0} f_* \left(L(n) \otimes_S N\right) \longrightarrow R^{q_0} f_* \left(F(n) \otimes_S N\right) \longrightarrow R^{q_0+1} f_* \left(K(n) \otimes_S N\right)$$

For $n \gg 0$, the last term vanishes by the induction hypothesis, and first term vanishes by (7.1.2(3)). So, $R^{q_0} f_* \left(F(n) \otimes_S N\right) = 0$, $\forall n \gg 0$. This establishes (1).

To prove (2), consider the exact sequences (7.10, 7.11). We have $\forall n \gg 0$, by (1), the diagrams of exact sequences

$$\begin{array}{ccccccccc} & & f_*(K(n)) \otimes_S N & \longrightarrow & f_*(L(n)) \otimes_S N & \longrightarrow & f_*(F(n)) \otimes_S N & \longrightarrow & 0 \\ & & \downarrow{\scriptstyle u_K} & & \wr\downarrow{\scriptstyle u_L} & & \downarrow{\scriptstyle u_F} & & \\ 0 & \longrightarrow & f_*(K(n) \otimes_S N) & \longrightarrow & f_*(L(n) \otimes_S N) & \longrightarrow & f_*(F(n) \otimes_S N) & \longrightarrow & 0 \end{array}$$

The upper line remains exact because the tensor product is right exact (here $S = \operatorname{Spec}(A)$). The vertical arrow u_L is an isomorphism by (7.1.2(3)). By the Snake Lemma, u_F is surjective. Applying the same to K, we have u_K is surjective. Again, by the Snake Lemma u_F is injective and hence an isomorphism. This settles (2).

To see (3), we want to prove $f_* F(n)$ is flat. Let

$$0 \longrightarrow M \longrightarrow N \longrightarrow C \longrightarrow 0$$

be exact in $QCoh(S)$ or in $Coh(S)$.

Since $F(n) \in \mathscr{P}(X)$, we have

$$0 \longrightarrow F(n) \otimes M \longrightarrow F(n) \otimes N \longrightarrow F(n) \otimes C \longrightarrow 0$$

is exact. By (1), for $n \gg 0$, the sequence

$$0 \longrightarrow f_*(F(n) \otimes M) \longrightarrow f_*(F(n) \otimes N)$$
$$\longrightarrow f_*(F(n) \otimes C) \longrightarrow 0 \quad \text{is exact in } QCoh(S).$$

By (2), $f_*(F(n) \otimes -) \cong f_*(F(n)) \otimes -$. So, it follows that $f_*(F(n))$ is a flat A-module. In fact, $f_*(F(n)) \in Coh(A)$, by (6.2.9) or by the second line of the above exact sequence. So, $f_*(F(n))$ is a projective A-module [Mh, p. 21]. So, (3) is established. The proof is complete. ∎

Lemma 7.1.9. *Use all the notation as in* (7.1.1), *and* $X := \mathbb{P}\mathscr{E}$. *Assume* S *is Noetherian. Let* $F \in \mathscr{P}(X)$ *be such that* $R^q f_* F(n) = 0\ \forall q \geq 1, n \geq 0$. *Then* $f_* F(n) \in \mathscr{P}(S)\ \forall n \geq 0$.

Proof. The question is again local on Y. So, we assume $S = \mathrm{Spec}\,(A)$, $\mathscr{E}$ is trivial and $X = \mathbb{P}^{r-1}_A$. First, $f_*F(n) \in \mathscr{P}(X)$, $\forall n \gg 0$, by (7.1.8(3)). Dualize the Koszul complex (7.1) and tensor it with $F(n)$. We obtain the exact sequence

$$0 \longrightarrow F(n) \longrightarrow F(n+1) \otimes_S \mathscr{E}^* \longrightarrow \cdots$$
$$\longrightarrow F(n+r) \otimes_S \Lambda^r \mathscr{E}^* \longrightarrow 0$$

By hypothesis, with $n \geq 0$, for all the terms of this sequence, $R^q f_* = 0$, $\forall q \geq 1$. So, the application f_* to this sequence leads to the exact sequence

$$0 \longrightarrow f_*F(n) \longrightarrow f_*F(n+1) \otimes_S \mathscr{E}^* \longrightarrow \cdots$$
$$\longrightarrow f_*\left(F(n+r)\right) \otimes_S \Lambda^r \mathscr{E}^* \longrightarrow 0$$

So, by decreasing induction, $f_*F(n) \in \mathscr{P}(S)$. The proof is complete. ∎

In our context of regularity, the following is noteworthy.

Lemma 7.1.10. *Use all the notation as in* (7.1.1), *and* $X := \mathbb{P}\mathscr{E}$. *Assume* S *is Noetherian. Let* $F \in \mathscr{P}(X) \cap \mathfrak{R}(X)$. *Then* $T_i(F) \in \mathscr{P}(S)\ \ \forall\ i \geq 0$, *where* $T_i(F)$ *was defined in the canonical resolution theorem* (7.1).

Proof. Since $\mathcal{F} \in \mathfrak{R}(X)$, by (7.1.5), $\mathcal{F}(n) \in \mathfrak{R}(X)\ \forall n \geq 0$. So, for $n \geq 0, q \geq 1$, we have $R^q f_* F(n) = R^q f_* F(n+q-q) = 0$. By (7.1.9), $f_*F(n) \in \mathscr{P}(S) \forall n \geq 0$. The lemma now follows from the exact sequence (7.3). The proof is complete. ∎

7.2. The Projective Bundle Theorem

Before we state and prove the projective bundle theorem (7.2.3), we establish some notation and preparation.

Notation 7.2.1. Use all the notation as in (7.1.1). We define a sequence of full subcategories of $QCoh(X)$ as follows:

$$\begin{cases} \mathscr{R}_n(X) = \{F \in \mathscr{P}(X) : F(n) \in \mathfrak{R}(X)\} \\ \mathscr{P}_n(X) = \{F \in \mathscr{P}(X) : R^q f_*(F(k)) = 0, \ \forall k \geq n, q \geq 1\} \end{cases}$$

Clearly, $\mathscr{P}_n(X) \subseteq \mathscr{P}_{n+1}(X)$, and by (7.1.5), $\mathscr{R}_n(X) \subseteq \mathscr{R}_{n+1}(X)$. It is also obvious that $\mathscr{R}_n(X) \subseteq \mathscr{P}_n(X)$.

The following is an immediate lemma on these categories.

Lemma 7.2.2. *Use all the notation as in* (7.1.1, 7.2.1). *Then*

$$\forall n \geq 0, \qquad \mathscr{R}_n(X) \subseteq \mathscr{P}_n(X) \subseteq \mathscr{P}(X)$$

are extension closed subcategories of $\mathscr{P}(X)$. *So,* $\mathscr{R}_n(X)$, $\mathscr{P}_n(X)$ *are exact subcategories, and* $K_q(\mathscr{R}_n(X))$, $K_q(\mathscr{P}_n(X))$ *are defined. Further,*

$$K_q(\mathscr{R}_n(X)) \xrightarrow{\sim} K_q(\mathscr{P}_n(X)) \xrightarrow{\sim} K_q(\mathscr{P}(X))$$

are isomorphisms $\forall q \geq 0, n \geq 0$.

Proof. It is immediate from (7.1.8(1)) that $\mathscr{R}_n(X) \subseteq \mathscr{P}(X)$ are extension closed subcategories. It follows from the long exact sequence (6.22) that $\mathscr{P}_n(X) \subseteq \mathscr{P}(X)$ are extension closed subcategories. Consider the exact sequence

$$0 \longrightarrow F \longrightarrow F(1) \otimes_S \mathscr{E}^* \longrightarrow \cdots \longrightarrow F(r) \otimes_S \Lambda^r \mathscr{E}^* \longrightarrow 0 \tag{7.12}$$

obtained by dualizing (7.1) and taking tensor product with F. Fix $n \geq 1$, and let $\iota : \mathscr{P}_{n-1}(X) \hookrightarrow \mathscr{P}_n(X)$ be the inclusion functor. For $p \geq 1$, define functors

$$\begin{cases} \Phi_p : \mathscr{P}_n(X) \longrightarrow \mathscr{P}_{n-1}(X) \\ \Psi_p : \mathscr{R}_n(X) \longrightarrow \mathscr{R}_{n-1}(X) \end{cases} \quad \text{both by sending} \quad F \mapsto F(p) \otimes_S \Lambda^p \mathscr{E}^*$$

It is easy to check that these are well-defined exact functors. From the exact sequences (7.12), we obtain two exact sequences of exact functors:

$$\begin{cases} 0 \longrightarrow 1_{\mathscr{P}_n(X)} \longrightarrow \iota\Phi_1 \longrightarrow \cdots \longrightarrow \iota\Phi_r \longrightarrow 0 \\ 0 \longrightarrow 1_{\mathscr{P}_{n-1}(X)} \longrightarrow \Phi_1\iota \longrightarrow \cdots \longrightarrow \Phi_r\iota \longrightarrow 0 \end{cases}$$

By Corollary 3.4.5, to the additivity theorem, it follows that the map

$$\iota_* : K_q(\mathscr{P}_{n-1}(X) \xrightarrow{\sim} K_q(\mathscr{P}_n(X)) \quad \text{has an inverse} \quad \sum_{p=1}^{r}(-1)^p(\Phi_p)_*$$

So, ι_* is an isomorphism. By (7.1.8(1)), $\mathscr{P}(X)$ is the union of $\mathscr{P}_n(X)$. Therefore, by (3.3.3), $K_q(\mathscr{P}_n(X)) \cong K_q(\mathscr{P}(X))$ for all $q \geq 0$.

Similarly, we prove that the maps $K_q(\mathscr{R}_{n-1}(X)) \xrightarrow{\sim} K_q(\mathscr{R}_n(X)) \xrightarrow{\sim} K_q(\mathscr{P}(X))$ are isomorphisms. This completes the proof of the lemma. (*Indeed, by Whitehead's theorem* (C.3.8), *the maps of K-theory spaces* $\mathbf{K}(\mathscr{P}_n(X)) \longrightarrow \mathbf{K}(\mathscr{P}(X))$, $\mathbf{K}(\mathscr{R}_n(X)) \longrightarrow \mathbf{K}(\mathscr{P}(X))$ *are homotopy equivalences.*) ∎

For a scheme X, recall (6.1.2) the action of $K_0(X)$ on $K_n(X)$. As before, $K_n(X)$ denotes the group $K_n(\mathscr{P}(X))$ $\forall\, n \geq 0$. The following is the statement of the celebrated **projective bundle theorem**.

Theorem 7.2.3. *Use all the notation as in* (7.1.1). *So, S is a Noetherian scheme, $\mathscr{E} \in \mathscr{P}(S)$, with rank$(\mathscr{E}) = r$, $X =$* Proj (*S*ym ($\mathscr{E}$)), *and $f : X \longrightarrow S$ is the structure map. Then for integers $q \geq 0$, there is an isomorphism*

$$K_q(S)^r \xrightarrow{\sim} K_q(X) \quad \textit{sending } (a_0, a_1, \ldots, a_{r-1}) \mapsto \sum_{i=0}^{r-1} z^i \cdot f^* a_i$$

where $z \in K_0(X)$ denotes the class $[\mathcal{O}_X(-1)] \in K_0(X)$.

Proof. We will be working with $\mathscr{P}_0(X) = \{F \in \mathscr{P}(X) : R^q f_*(F(k)) = 0,\ \forall k \geq 0, \forall q \geq 1\}$. For all $0 \leq n \leq r-1$, define exact functors

$$U_n : \mathscr{P}(S) \longrightarrow \mathscr{P}_0(X) \qquad \text{by}$$
$$U_n(N) = \mathcal{O}_X(-n) \otimes_S N = f^*(N)(-n)$$

It follows from (7.1.2(3)) that $U_n(N) \in \mathscr{P}_0(X)$. Since f is flat, U_n is exact. So, U_n induce the maps

$$u_n : K_q(\mathscr{P}(S)) \longrightarrow K_q(\mathscr{P}_0(X)) \qquad \forall q \geq 0$$

Due to Lemma 7.2.2, to complete the proof of the theorem, it suffices to prove that the maps

$$K_q(S)^r \xrightarrow{\sim} K_q(\mathscr{P}_0(X)) \quad \text{sending } (a_0, a_1, \ldots, a_{r-1}) \mapsto \sum_{i=0}^{r-1} u_n(a_n)$$

are isomorphisms. Further, define the functors

$$V_n : \mathscr{P}_0(X) \longrightarrow \mathscr{P}(S) \quad \text{by } V_n(F) = f_*(F(n))$$

By Lemma 7.1.9, indeed, $V_n(F) = f_*(F(n)) \in \mathscr{P}(S)$, and they are exact functors. So, V_n induce maps

$$v_n : K_q\left(\mathscr{P}_0(X)\right) \longrightarrow K_q\left(\mathscr{P}(S)\right)$$

Define

$$v : K_q\left(\mathscr{P}_0(X)\right) \longrightarrow K_q\left(\mathscr{P}(S)\right)^r \quad \text{sending}$$
$$x \mapsto (v_0(x), v_1(x), \ldots, v_{r-1}(x))$$

It follows from (7.1.2(3)) that the composition

$$V_nU_m(N) = f_*(\mathcal{O}_X(n-m) \otimes_S N) = \mathcal{S}\mathrm{ym}\left(\mathscr{E}\right)_{n-m} \otimes_S N$$

So, $v_nu_m = 0 \ \forall m < n$, and $v_nu_n = 1$. So, vu is represented by a triangular matrix, with ones in the diagonal. It follows that vu is an isomorphism and u is injective.

Now, we consider $\mathcal{R}_0(X) = \mathfrak{R}(X) \cap \mathscr{P}(X)$, which consists of all locally free sheaves, which are regular. Further, note that $\mathscr{R}_0(X) \subseteq \mathscr{P}_0(X)$. So, for $F \in \mathscr{R}_0(X)$, by (7.1.10), $T_n(F) \in \mathscr{P}(S)$ for all $n \geq 0$. By construction (7.1.7), $T_n : \mathfrak{R}(X) \longrightarrow QCoh(X)$

is exact. So, $T_n : \mathscr{R}_0(X) \longrightarrow \mathscr{P}(S)$ is an exact functor $\forall n \geq 0$. Let $t_n : K_q(\mathscr{R}_0(X)) \longrightarrow K_q(\mathscr{P}(S))$ be the maps induced by T_n, and define

$$t : K_q(\mathscr{R}_0(X)) \longrightarrow K_q(\mathscr{P}(S))^r \quad \text{by}$$
$$t(x) = \big(t_0(x), -t_1(x), \ldots, (-1)^{r-1} t_{r-1}(x)\big)$$

$$\text{So,} \qquad ut(x) = \sum_{n=0}^{r-1} (-1)^n u_n t_n(x) \quad K_q\left(\mathscr{R}_0(X)\right) \longrightarrow K_q\left(\mathscr{P}_0(X)\right)$$

We claim that ut coincide with the map induced by the inclusion $\zeta : \mathscr{R}_0(X) \hookrightarrow \mathscr{P}_0(X)$. The maps $u_n t_n$ are induced by the compositions

$$\begin{array}{ccc} \mathscr{R}_0(X) & \xrightarrow{T_n} & \mathscr{P}(S) \\ & \searrow^{U_n T_n} & \downarrow U_n \\ & & \mathscr{P}_0(X) \end{array} \qquad \text{sending } F \mapsto T_n(F) \mapsto f^* T_n(F)(-n)$$

The canonical resolution (7.2) can be written as

$$0 \longrightarrow U_{r-1}T_{r-1} \longrightarrow \cdots \longrightarrow U_1 T_1 \longrightarrow U_0 T_0 \longrightarrow \zeta \longrightarrow 0 \tag{7.13}$$

By Corollary 3.4.5, to the additivity theorem, it follows

$$ut = \sum_{n=0}^{r-1} (-1)^n u_n t_n = \zeta_*.$$

By (7.2.2), ζ_* is an isomorphism. So, ut is an isomorphism. So, u is surjective. This completes the proof that u is an isomorphism. ■

7.3. Exercises

Exercise 7.3.1. Let A be a commutative Noetherian ring, $R = A[X_0, X_1, \ldots, X_r]$, and $X = \mathbb{P}^r_A = \text{Proj}\,(R)$. Consider the open affine cover $\mathfrak{U} = \{D(X_0), D(X_1), \ldots, D(X_r)\}$ of X. Let M be finitely generated graded R-module and $\mathcal{F} = \widetilde{M} \in Coh(X)$. Write down the

Čech complex $C^\bullet(\mathfrak{U}, \mathcal{F})$ and compute the cohomologies $H^n(X, \mathcal{F})$. Further, give a necessary condition for $\mathcal{F}$ to be regular.

Hint: This is an open-ended problem designed to analyze the concept of regularity.

Exercise 7.3.2. Give an elementary proof of the projection formula (7.2.3) in the case $q = 0$. So, prove that there is an isomorphism

$$K_0(S)^r \xrightarrow{\sim} K_0(X) \quad \text{sending } (a_0, a_1, \ldots, a_{r-1}) \mapsto \sum_{i=0}^{r-1} z^i \cdot f^* a_i$$

where $z \in K_0(X)$ denotes the class $[\mathcal{O}_X(-1)] \in K_0(X)$.

Chapter 8

Work of Swan on Quadric Hypersurfaces

This chapter provides an exposition of the work of Swan on K-theory of quadric hypersurfaces [Sw1], leading up to the projection formula, and the K-theory of affine spheres. One can view it as a fairly involved applications of the methods in Chapter 7, on the projective bundle theorem (7.2.3). For simplicity, we deviate in two ways from [Sw1]. First, throughout this chapter, we assume 2 is invertible, which was not part of the hypotheses in [Sw1]. For simplicity, we only compute the K-theory of the category of locally free sheaves, while in [Sw1], Swan computed the K-theory of the category of the locally free sheaves with an action of an Azymaya algebra.

8.1. Hypersurfaces in Projective Spaces

We begin our preparatory work on hypersurfaces of projective spaces $\mathbb{P}_R^N$. We establish some notation, as follows.

Notation 8.1.1. Let R be a commutative Noetherian ring, and $B = R[X_0, X_1, \ldots, X_{d+1}] = \bigoplus_{n \geq 0} B_n$ be a polynomial ring. Let $\mathbb{P} := \mathbb{P}_R^{d+1} = \operatorname{Proj}(B)$ be the projective space. Fix a homogeneous polynomial $f \in B_k$, with $\deg(f) = k$. Let $A = \frac{B}{(f)}$, and $X = \operatorname{Proj}(A) \subseteq \mathbb{P}_R^{d+1}$ be a closed subscheme. So, $\dim X = d$. With $S = \operatorname{Spec}(R)$, the structure map is denoted by $\mathfrak{f} : X \longrightarrow S$.

The twisting sheaf on X is denoted by $\mathcal{O}_X(1)$. The higher direct images $R^q\mathfrak{f}_*F$ may be identified with $H^q(X, F)$.

The content $C(f)$ of f is defined to be the ideal of R generated by the coefficients of the monomials in f. (*It is easy to check that the ideal $C(f)$ is invariant under linear change of variables* (8.1.5).) We will assume throughout that $C(f) = R$.

The following is a preliminary lemma.

Lemma 8.1.2. *Use the notation from* (8.1.1). *Assume $C(f) = R$. Then $\forall n \geq 0$, the following*

$$0 \longrightarrow B_n \xrightarrow{f} B_{n+k} \longrightarrow \frac{B_{n+k}}{fB_n} \longrightarrow 0 \tag{8.1}$$

is a split exact sequence. In particular, $\mathrm{Proj}\left(\frac{B}{(f)}\right) \longrightarrow \mathrm{Spec}\,(R)$ *is flat.*

Proof. First, since $C(f) = R$, f is a nonzero divisor in B (see [E, Ex. 3.4, p. 109]). So, the sequence (8.1) is exact. If R is a field, it follows from vector space arguments that the multiplication map f splits. Therefore, the proof is settled when R is a field. Now we deal with the general case of R. We can assume $(R, \mathfrak{m})$ is local. Consider the sequence

$$Hom_R(B_{n+k}, B_n) \longrightarrow Hom_R(B_n, B_n) \longrightarrow C \longrightarrow 0$$

where C is the cokernel.

Tensor this sequence by $\frac{R}{\mathfrak{m}} =: \kappa$. We obtain the exact sequence

$$Hom_R(B_{n+k} \otimes_R \kappa, B_n \otimes_R \kappa) \longrightarrow$$

$$Hom_R(B_n \otimes_R \kappa, B_n \otimes_R \kappa) \longrightarrow \frac{C}{\mathfrak{m}C} \longrightarrow 0.$$

By the case of fields, the first arrow is surjective, and hence $\frac{C}{\mathfrak{m}C} = 0$. By Nakayama's lemma $C = 0$. The proof is complete. ∎

The following is an analogue of Lemma 7.1.2.

Lemma 8.1.3. *Use the notation from* (8.1.1). *So,* $A = \frac{B}{(f)}$, $\deg(f) = k$. *We assume that* $C(f) = R$. *Let* $X = \mathrm{Proj}\,(A) \subseteq \mathbb{P}_R^{d+1}$, $S = \mathrm{Spec}\,(R)$. *Then*

1. *For* $F \in QCoh(X)$, $H^q(X, F) = 0\ \forall q \geq d+2$.
2. *For* $F \in QCoh(X)$, $P \in \mathscr{P}(R)$, *we have*
$$H^q(X, F) \otimes_R P = H^q(X, F \otimes_S P) \qquad \forall q > 0.$$
3. *For* $N \in QCoh(S)$, *we have*
$$H^q(X, \mathcal{O}_X(n) \otimes_R N) = \begin{cases} 0 & q \neq 0,\ d \\ A_n \otimes N & q = 0,\ d \geq 1 \\ A^*_{k-2-d-n} \otimes_R N & q = d,\ d \geq 1. \end{cases}$$
When $d = 0$, *there is an exact sequence*
$$0 \longrightarrow A_n \otimes N \longrightarrow H^0(X, \mathcal{O}_X(n) \otimes N) \longrightarrow$$
$$N \otimes A^*_{k-2-n} \longrightarrow 0.$$
4. *For* $F \in Coh(X)$, *there is an integer* n_0 *such that* $\forall n \geq n_0$, $F(n)$ *is globally generated.*
5. *In particular,* $\mathcal{O}_X(n)$ *is regular* $\forall n \geq k-1$.
6. **Remark:** In (7.1.2), we considered the case of $k = 1$. In this chapter, we are mainly interested in the case of quadrics ($k = 2$). In this case of quadrics, $\mathcal{O}_X(n)$ is regular $\forall n \geq 1$, while $\mathcal{O}_X$ is not. This is a crucial difference in this chapter from the last chapter.

Proof. Since $\mathbb{P}^{d+1}_R$ can be covered by $d+2$ open affine sets, (1) follows from Čech cohomology. Assertion (2) is the projection formula [Hr, Ex. III.8.3]. Since $\mathcal{O}_X(1)$ is ample, (4) follows. Also, (5) follows from (3). We prove (3). From the split exact sequence (8.1), we obtain the exact sequence
$$0 \longrightarrow B(n-k) \otimes N \xrightarrow{\ f\ } B_n \otimes N \longrightarrow A(n) \otimes N \longrightarrow 0.$$
With $\mathbb{P} := \mathbb{P}^{d+1}_R$, this induces the exact sequence
$$0 \longrightarrow \mathcal{O}_{\mathbb{P}}(n-k) \otimes N \longrightarrow \mathcal{O}_{\mathbb{P}}(n) \otimes N \longrightarrow \mathcal{O}_X(n) \otimes N \longrightarrow 0.$$
This leads to the long exact sequence of cohomologies
$$\cdots \longrightarrow H^q(\mathbb{P}, \mathcal{O}_{\mathbb{P}}(n-k) \otimes N) \longrightarrow$$
$$H^q(\mathbb{P}, \mathcal{O}_{\mathbb{P}}(n) \otimes N) \longrightarrow H^q(\mathbb{P}, \mathcal{O}_X(n) \otimes N) \longrightarrow \cdots$$
Now, (3) follows from (7.1.2). The proof is complete. ■

The following is in line with the statement of (7.1.2(1)).

Lemma 8.1.4. *Use the notation from* (8.1.1). *Then* $H^q(X,F) = 0\ \forall q \geq d+1$ *for all* $F \in Coh(X)$.

Proof. As is routine (7.1.8), it follows from (8.1.3(4)) that there is an exact sequence

$$0 \longrightarrow K \longrightarrow L \longrightarrow F \longrightarrow 0 \quad \text{with} \quad L = \mathcal{O}_X(-n)^k, \quad n \gg 0.$$

By (8.1.3(1)), $H^q(X,-) = 0$, $\forall q \geq d+2$, and the sequence

$$H^{d+1}(X,L) \longrightarrow H^{d+1}(X,F) \longrightarrow 0 \quad \text{is exact.}$$

Further, by (8.1.3(3)), $H^{d+1}(X,L) = 0$. The proof is complete. ■

We finish this section with the following exercise.

Exercise 8.1.5. Let $A = R[X_0, X_1, \ldots, X_n]$ be a polynomial ring over a commutative ring R and $f \in A$ be a homogeneous polynomial. Prove that content $C(f)$ is invariant under linear change of variables. Further, prove that if $C(f) = R$, then f is a nonzero divisor in A.

8.2. Canonical Resolution for Projective Schemes

With proper formulation, the canonical resolution theorem (7.1.7) works for projective schemes. We define regularity and establish some notation for projective schemes.

Definition 8.2.1. Let $B = R[X_0, X_1, \ldots, X_r] = \bigoplus_{n\geq 0} B_n$ be a polynomial ring, over a commutative Noetherian ring R. Let $\mathbb{P} := \mathbb{P}^r_R = \text{Proj}\,(B)$ be the projective space, and $S = \text{Spec}\,(R)$. Let $X = \text{Proj}\,(A) \subseteq \mathbb{P}^r_R$ be a closed subscheme. The structure map is denoted by $\mathfrak{f} : X \longrightarrow S$, and $\mathcal{O}_X(1)$ denotes the twisting sheaf. We recapitulate the definition of regularity and some notation from Chapter 7. Since $S = \text{Spec}\,(R)$, we substitute the higher direct images $R^q\mathfrak{f}_*(F)$ by $H^q(X,F)$.

1. As in (7.1.3), for $F \in QCoh(X)$, F is defined to be **regular** if
$$H^q(X, F(-q)) = 0 \quad \forall\, q \geq 1.$$
So, $F \in QCoh(X)$ is regular if and only if it is regular as an object $F \in QCoh\,(\mathbb{P}^r_R)$. The full subcategory of regular objects $F \in QCoh(X)$ is denoted by $\mathfrak{R}(X)$.

2. We are mainly interested in locally free sheaves. Accordingly, as in (7.2.1), for $n \in \mathbb{Z}$, define the following full subcategories of $QCoh(X)$:

$$\begin{cases} \mathscr{R}_n(X) = \{F \in \mathscr{P}(X) : F(n) \in \mathfrak{R}(X)\} \\ \mathscr{P}_n(X) = \{F \in \mathscr{P}(X) : H^q(X, (F(k)) = 0, \ \forall k \geq n, q \geq 1\} \end{cases}$$

Clearly, $\mathscr{P}_n(X) \subseteq \mathscr{P}_{n+1}(X)$, and by (7.1.5), $\mathscr{R}_n(X) \subseteq \mathscr{R}_{n+1}(X)$. It is also obvious that $\mathscr{R}_n(X) \subseteq \mathscr{P}_n(X)$. Further, $\mathscr{P}_0(X) \subseteq \mathfrak{R}(X)$.

In what follows, we write down analogues of some of the lemmas from Chapter 7, for X as in (8.2.1). As in [Sw1], we leave some of the proofs as exercises.

Exercise 8.2.2. Use the notation from (8.2.1). Let

$$0 \longrightarrow K \longrightarrow F \longrightarrow C \longrightarrow 0 \quad \text{be exact in} \quad QCoh(X).$$

Then

1. If $K(n)$ and $C(n)$ are regular, then $F(n)$ is regular.
2. If $K(n+1)$ and $F(n)$ are regular, then $C(n)$ is regular.
3. If $F(n+1)$ and $C(n)$ are regular, and if $\mathfrak{f}_*F(n) \longrightarrow \mathfrak{f}_*C(n)$ is onto, then $K(n+1)$ is regular.

Hint: Use (7.1.4).

Exercise 8.2.3. Use the notation from (8.2.1). For $F \in QCoh(X)$, if F is regular, then $F(n)$ is regular $\forall n \geq 0$. **Hint:** Use (7.1.5).

Exercise 8.2.4. Use the notation from (8.2.1). Let $F \in QCoh(X)$ be regular. Then the canonical map

$$\mathfrak{f}^*\mathfrak{f}_*F =: \mathcal{O}_X \otimes_S \mathfrak{f}_*F \twoheadrightarrow F \qquad \text{is surjective.} \qquad \textbf{Hint: } \text{Use (7.1.6).}$$

Exercise 8.2.5. Use the notation from (8.2.1). Then

$$\forall n \geq 0, \qquad \mathscr{R}_n(X) \subseteq \mathscr{P}_n(X) \subseteq \mathscr{P}(X)$$

are extension closed subcategories of $\mathscr{P}(X)$. So, $\mathscr{R}_n(X)$, $\mathscr{P}_n(X)$ are exact subcategories, and $K_q(\mathscr{R}_n(X))$, $K_q(\mathscr{P}_n(X))$ are defined. Further,

$$K_q(\mathscr{R}_n(X)) \xrightarrow{\sim} K_q(\mathscr{P}_n(X)) \xrightarrow{\sim} K_q(\mathscr{P}(X))$$

are isomorphisms $\forall q \geq 0, n \geq 0$.

Hint: Imitate the proof of (7.2.2) and borrow the exact sequence (7.12) from $\mathbb{P}^r_R$.

By restricting $N \in Coh(R)$, we give a version of (7.1.8) for projective schemes.

Lemma 8.2.6. *Use the notation from* (8.2.1). *Let* $F \in \mathscr{P}(X)$. *Then there is an integer* n_0 *such that* $\forall n \geq n_0$ *and* $N \in Coh(R)$, *we have*

1. $H^q(X, F(n) \otimes_R N) = 0$ *for all* $q \geq 1$.
2. $\Gamma(X, F(n)) \otimes_R N \xrightarrow{\sim} \Gamma(X, F(n) \otimes_R N)$.
3. $\Gamma(X, F(n)) \in \mathscr{P}(R)$.

Proof. Barring the possibility of dependence of n_0 on N, (1) is a direct consequence of (6.2.9), or [Hr, Thm III.5.2]. First, we have

$$H^q(X, -) = 0 \quad \forall q \geq r+1. \tag{8.2}$$

By (6.2.9), there is an integer n_0 such that $H^q(X, \mathcal{O}(n)) = 0$ for all $q \geq 1$ for all $n \geq n_0$. Hence, $H^q\left(X, \mathcal{O}(n) \otimes R^k\right) = 0$ for all $q \geq 1$ for all $n \geq n_0$ and $\forall k \geq 0$. Given $N \in Coh(S)$, there is an exact sequence

$$0 \longrightarrow M_0 \longrightarrow R^k \longrightarrow N \longrightarrow 0. \tag{8.3}$$

By decreasing induction, we have $H^q(X, \mathcal{O}(n) \otimes N) = 0$ for all $n \geq n_0$. So, (1) is established for $F = \mathcal{O}_X$, and hence for $F = \mathcal{O}_X(m)$ for all $m \in \mathbb{Z}$.

Now, let $F \in \mathscr{P}(X)$. By (6.1.5), there is an integer m such that $F(m)$ is globally generated. With $L = \mathcal{O}_X(-m)^k$, for some $k \geq 1$, there is an exact sequence

$$0 \longrightarrow K \longrightarrow L \longrightarrow F \longrightarrow 0 \qquad \text{in } \mathscr{P}(X). \tag{8.4}$$

Since $F \in \mathscr{P}(X)$, this induces the exact sequences

$$0 \longrightarrow K(n) \otimes N \longrightarrow L(n) \otimes N \longrightarrow F(n) \otimes N \longrightarrow 0 \qquad \forall n. \tag{8.5}$$

Consider the exact sequence

$$H^{q_0}(X, L(n) \otimes N) \longrightarrow H^{q_0}(X, F(n) \otimes N)$$
$$\longrightarrow H^{q_0+1}(X, K(n) \otimes N).$$

From the case $F = \mathcal{O}_X \otimes R^k$, it follows that the first term vanishes for $\forall n \gg 0$, and also the last term vanishes $\forall n \gg 0$, by the induction hypothesis. So, $H^{q_0}(X, F(n) \otimes_S N) = 0\ \forall n \gg 0$. This establishes (1).

Now, first we establish (2) for $F = \mathcal{O}_X(n)$. Fix n_0 such that (1) holds for $F = \mathcal{O}_X$. Corresponding to the exact sequence (8.3), we have the commutative diagram of exact sequences:

$$\begin{array}{ccccccccc}
 & & \Gamma(X, \mathcal{O}_X(n)) \otimes M_0 & \longrightarrow & \Gamma(X, \mathcal{O}_X(n)) \otimes R^k & \longrightarrow & \Gamma(X, \mathcal{O}_X(n)) \otimes N & \longrightarrow & 0 \\
 & & \downarrow{\scriptstyle v_{M_0}} & & \wr\downarrow{\scriptstyle v_{R^k}} & & \downarrow{\scriptstyle v_N} & & \\
0 & \longrightarrow & \Gamma(X, \mathcal{O}_X(n) \otimes M_0) & \longrightarrow & \Gamma\left(X, \mathcal{O}_X(n) \otimes R^k\right) & \longrightarrow & \Gamma(X, \mathcal{O}_X(n) \otimes N) & \longrightarrow & 0
\end{array}$$

We get an exact sequence when we tensor the exact sequence (8.3), by $\mathcal{O}_X(n)$. Thus, we get the second line, which is exact $\forall n \geq n_0$ and $\forall N \in Coh(S)$. The top line is exact by right exactness of the tensor product. It is obvious that v_{R^k} is an isomorphism. So, v_N is surjective for all $N \in Coh(R)$. So, v_{M_0} is surjective. By the Snake lemma, it follows that v_N is injective, and hence an isomorphism. So, (2) is established for $\mathcal{O}_X$ and hence for $\mathcal{O}_X(n)$ for all $n \in \mathbb{Z}$.

To establish (2) in the general case, let $F \in \mathscr{P}(X)$, and consider the exact sequences (8.4, 8.5). By (1), we have the diagrams of exact sequences $\forall n \gg 0$,

$$\begin{array}{ccccccccc}
 & & \Gamma(X, K(n)) \otimes N & \longrightarrow & \Gamma(X, L(n)) \otimes N & \longrightarrow & \Gamma(X, F(n)) \otimes N & \longrightarrow & 0 \\
 & & \downarrow{\scriptstyle u_K} & & \wr\downarrow{\scriptstyle u_L} & & \downarrow{\scriptstyle u_F} & & \\
0 & \longrightarrow & \Gamma(X, K(n) \otimes_S N) & \longrightarrow & \Gamma(X, L(n) \otimes N) & \longrightarrow & \Gamma(X, F(n) \otimes N) & \longrightarrow & 0
\end{array}$$

The upper line remains exact because the tensor product is right exact. The vertical arrow u_L is an isomorphism because we just proved the case of $\mathcal{O}(m)$. By the Snake Lemma, u_F is surjective. Applying the same to K, we have u_K is surjective. Again, by the Snake Lemma, u_F is injective, and hence an isomorphism. This settles (2).

To see (3), we want to prove $\Gamma(X, F(n))$ is flat $\forall n \gg 0$. Let

$$0 \longrightarrow M \longrightarrow N \longrightarrow C \longrightarrow 0 \quad \text{be exact in } QCoh(R).$$

Since $F(n) \in \mathscr{P}(X)$, we have

$$0 \longrightarrow F(n) \otimes M \longrightarrow F(n) \otimes N \longrightarrow F(n) \otimes C$$
$$\longrightarrow 0 \quad \text{is exact in } Coh(X).$$

By (1), for $n \gg 0$, the sequence

$$0 \longrightarrow \Gamma(X, F(n) \otimes M) \longrightarrow \Gamma(X, F(n) \otimes N) \longrightarrow$$
$$\Gamma(X, F(n) \otimes C) \longrightarrow 0$$

is exact in $Coh(S)$. By (2), $\Gamma(X, F(n) \otimes -) \cong \Gamma(X, F(n)) \otimes -$. So, it follows that $\Gamma(X, F(n))$ is a flat R-module. In fact, $f_*(F(n)) \in Coh(R)$, by (6.2.9), or by the above exact sequence. So, $\Gamma(X, F(n))$ is a projective R-module [Mh, p. 21]. So, (3) is established. The proof is complete. ■

Exercise 8.2.7. Use the notation from (8.2.1). Let $F \in \mathscr{P}(X)$ be such that $H^q(X, F(n)) = 0$, $\forall q \geq 1, n \geq 0$. Then $\Gamma(X, F(n)) \in \mathscr{P}(R)$ $\forall n \geq 0$. In other words, if $F \in \mathscr{P}_0(X)$, then $\Gamma(X, F(n)) \in \mathscr{P}(R)$ $\forall n \geq 0$.

Proof. First, $f_*F(n) \in \mathscr{P}(X)$ $\forall n \gg 0$, by (8.2.6(3)). The trivial version of the exact sequence (7.1) is induced by the Koszul complex of $(X_0, X_1, \ldots, X_r)$. With $\mathcal{E} = R^{r+1}$, the sequence is as follows:

$$0 \longrightarrow \mathcal{O}_{\mathbb{P}^r_R}(-(r+1)) \otimes_S \Lambda^{r+1}\mathcal{E} \longrightarrow \cdots \longrightarrow$$
$$\mathcal{O}_{\mathbb{P}^r_R}(-1) \otimes_S \mathcal{E} \longrightarrow \mathcal{O}_{\mathbb{P}^r_R} \longrightarrow 0. \tag{8.6}$$

This is a locally split exact sequence. Dualize and tensor it with $F(n)$. We obtain the exact sequences:

$$0 \longrightarrow F(n) \longrightarrow F(n+1) \otimes_S \mathscr{E}^* \longrightarrow$$
$$\cdots \longrightarrow F(n+r+1) \otimes_S \Lambda^{r+1}\mathscr{E}^* \longrightarrow 0.$$

These are exact sequences in $\mathscr{P}(X)$. For convenience, we use $R^q\mathfrak{f}_*$ notation instead of $H^q(X, -)$. By hypothesis, for all the terms of these sequences, $R^q\mathfrak{f}_* = 0$ $\forall q \geq 1$. So, the application of $\mathfrak{f}_*$ to this sequence leads to the exact sequences:

$$0 \longrightarrow \mathfrak{f}_*F(n) \longrightarrow \mathfrak{f}_*F(n+1) \otimes_S \mathscr{E}^* \longrightarrow \cdots \longrightarrow$$
$$\mathfrak{f}_*\left(F(n+r+1)\right) \otimes_S \Lambda^{r+1}\mathscr{E}^* \longrightarrow 0.$$

So, by decreasing induction, $f_*F(n) \in \mathscr{P}(S)$. The proof is complete. ■

For the construction of the canonical resolution (7.1.7), it was sufficient to have $\mathcal{O}_{\mathbb{P}(\mathscr{E})}(1) \in \mathfrak{R}(\mathbb{P}(\mathscr{E}))$. We use this leverage and obtain a canonical resolution theorem, as follows.

Theorem 8.2.8. *Use the notation from* (8.2.1). *So,* $X \hookrightarrow \mathbb{P}^r_R$ *is a closed subscheme. Assume that the structure map* $\mathfrak{f} : X \longrightarrow \mathrm{Spec}\,(R)$ *is flat, and* $\mathcal{O}_X(1)$ *is regular. Inductively,* $\forall p \geq 0$ *two exact functors*

$$\begin{cases} T_p : \mathfrak{R}(X) \longrightarrow QCoh(R) \\ Z_p : \mathfrak{R}(X) \longrightarrow QCoh(X) \end{cases} \quad with \quad \begin{cases} Z_{-1}(F) = F, \\ T_0(F) = \Gamma(X, F) \end{cases}$$

are constructed, together with an exact sequence

$$\cdots \longrightarrow \mathfrak{f}^* T_p(F)(-p)) \xrightarrow{\partial_p} \mathfrak{f}^* T_{p-1}(F)(-(p-1)) \longrightarrow \cdots \longrightarrow$$
$$\mathfrak{f}^* T_0(F)) \longrightarrow F \longrightarrow 0 \tag{8.7}$$

as follows:

1. $T_p(F) := \Gamma(X, Z_{p-1}(F)(p))$*, and* $Z_p(F)$ *is defined by the exact sequence*

$$0 \longrightarrow Z_p(F) \longrightarrow \mathfrak{f}^* T_p(F)(-p) \xrightarrow{\partial_p} Z_{p-1}(F) \longrightarrow 0 \tag{8.8}$$

   such that, $Z_p(F)(p+1) \in \mathfrak{R}(X)$*. Consequently,*

$$\Gamma(X, Z_p(F)(p)) = 0.$$

2. *Since* $Z_{p-1}(p) \in \mathfrak{R}(X)$*, by the inductive hypothesis, it follows from* (8.1.3) *that the canonical map*

$$\partial_p(p) : \mathfrak{f}^* T_p(F) \twoheadrightarrow Z_{p-1}(F)(p) \quad \textit{is surjective.}$$

   which defines ∂_p*.*
3. We remark that, in contrast to (8.2.5), the sequence (8.7) need not terminate.

Proof. The proof is similar to the proof of (7.1.7). Since $F \mapsto Z_{p-1}(F)(p)$ is exact on $\mathfrak{R}(X)$, it follows that $F \mapsto T_p(F)$ is exact.

By twisting (8.8), the sequence

$$0 \longrightarrow Z_p(F)(p) \longrightarrow \mathfrak{f}^*T_p(F) \longrightarrow Z_{p-1}(F)(p) \longrightarrow 0$$

is exact. Since $F \mapsto T_p(F)$ is exact and $\mathfrak{f}$ is flat, the functor $F \mapsto \mathfrak{f}^*T_p(F)$ is exact. Further, the functor $F \mapsto Z_{p-1}(F)(p)$ is exact. It follows that $F \mapsto Z_p(F)(p)$ is exact. The proof is complete. ■

The flatness condition (8.2.8) can be dropped if we define the resolution only on $\mathscr{R}_0(X)$.

Corollary 8.2.9. *Use the notation from* (8.2.1). *So,* $X \hookrightarrow \mathbb{P}^r_R$ *is a closed subscheme. Assume* $\mathcal{O}_X(1)$ *is regular. Inductively,* $\forall p \geq 0$ *two exact functors*

$$\begin{cases} T_p : \mathscr{R}_0(X) \longrightarrow \mathscr{P}(R) \\ Z_p : \mathscr{R}_0(X) \longrightarrow \mathscr{P}(X) \end{cases} \quad \text{with} \quad \begin{cases} Z_{-1}(F) = F, \\ T_0(F) = \Gamma(X, F) \end{cases}$$

are constructed, together with an exact sequence

$$\cdots \longrightarrow \mathfrak{f}^*T_p(F)(-p)) \xrightarrow{\partial_p} \mathfrak{f}^*T_{p-1}(F)(-(p-1)) \longrightarrow \cdots \longrightarrow$$
$$\mathfrak{f}^*T_0(F)) \longrightarrow F \longrightarrow 0 \tag{8.9}$$

as follows:

1. $T_p(F) := \Gamma(X, Z_{p-1}(F)(p))$, *and* $Z_p(F)$ *is defined by the exact sequence*

$$0 \longrightarrow Z_p(F) \longrightarrow \mathfrak{f}^*T_p(F)(-p) \xrightarrow{\partial_p} Z_{p-1}(F) \longrightarrow 0 \tag{8.10}$$

such that $Z_p(F)(p+1) \in \mathscr{R}_0(X)$. *Consequently,*

$$\Gamma(X, Z_p(F)(p)) = 0.$$

2. *Since* $Z_{p-1}(p) \in \mathfrak{R}(X)$, *by the inductive hypothesis, it follows from* (8.1.3) *that the canonical map*

$$\partial_p(p) : \mathfrak{f}^*T_p(F) \twoheadrightarrow Z_{p-1}(F)(p) \quad \textit{is surjective},$$

which defines ∂_p.

Proof. Since $F \in \mathscr{R}_0(X)$, it follows from (8.2.7), $T_0(F) \in \mathscr{P}(R)$. So, $T_0 : \mathscr{R}_0(X) \longrightarrow \mathscr{P}(R)$ is exact. Consequently, $F \mapsto \mathfrak{f}^*T_0F$ is an exact functor $\mathscr{R}_0(X) \longrightarrow \mathscr{P}(X)$. It follows that $Z_0 : \mathscr{R}_0(X) \longrightarrow \mathscr{P}(X)$ is exact. The inductive step works similarly. The proof is complete. ■

8.2.1. *Truncation on* $\mathscr{R}_{-1}(X)$

The canonical resolution (8.2.9) is defined on $\mathscr{R}_0(X)$, while it need not terminate. The goal of this subsection is to extract a finite resolution out of (8.2.9) by restricting them to $\mathscr{R}_{-1}(X)$. The following approach was devised in [Sw1].

Proposition 8.2.10. *Let* $X \hookrightarrow \mathbb{P}^r_R$ *be a projective subscheme, as in* (8.2.1, 8.2.9). *Assume that for some integer* $\mathfrak{d} \in \mathbb{N}$, *we have* $H^q(X, F) = 0$ *for all* $F \in \mathscr{P}(X)$, $\forall q \geq \mathfrak{d}+1$. *Then*

$$\forall\ G \in \mathscr{R}_{\mathfrak{d}}(X),\ \ F \in \mathscr{R}_0(X),\ \ q \geq 1 \quad Ext^q\left(Z_{\mathfrak{d}-1}(F), G\right) = 0.$$

Proof. Consider the exact sequence, corresponding to (8.10):

$$\begin{aligned} &Ext^q\left(Z_{p-1}(F), G\right) \longrightarrow Ext^q\left(T_p(F) \otimes \mathcal{O}(-p), G\right) \longrightarrow \\ &Ext^q\left(Z_p(F), G\right) \longrightarrow Ext^{q+1}\left(Z_{p-1}(F), G\right) \longrightarrow \\ &Ext^{q+1}\left(T_p(F) \otimes \mathcal{O}(-p), G\right) \longrightarrow Ext^{q+1}\left(Z_p(F), G\right). \end{aligned}$$

We have, $T_p(F) \in \mathscr{P}(R)$. Therefore, $H^q\left(X, T_p(F)^* \otimes G(p)\right) = T_p(F)^* \otimes H^q\left(X, G(p)\right)$ because such equalities work for free modules. By [Hr, Prop. III.6.7, 6.3],

$$\begin{cases} Ext^q\left(T_p(F) \otimes \mathcal{O}(-p), G\right) = Ext^q\left(\mathcal{O}_X, T_p(F)^* \otimes G(p)\right) \\ = H^q\left(X, T_p(F)^* \otimes G(p)\right) \\ = T_p(F)^* \otimes H^q\left(X, G(p)\right) = 0 \quad q \geq 1,\ p+q \geq \mathfrak{d}, \\ \text{because} \quad G \in \mathscr{R}_{\mathfrak{d}}(X). \end{cases}$$

So, it follows from the exact sequence that

$$Ext^q\left(Z_p(F), G\right) \cong Ext^{q+1}\left(Z_{p-1}(F), G\right) \qquad \forall q \geq 1, p+q \geq \mathfrak{d}$$

$$\begin{cases} \text{So, } \forall q \geq 1, \quad Ext^q\left(Z_{\mathfrak{d}-1}(F), G\right) \cong Ext^{q+1}\left(Z_{\mathfrak{d}-2}(F), G\right) \\ \cong \cdots \cong Ext^{q+\mathfrak{d}}\left(Z_{-1}(F), G\right) \\ = Ext^{q+\mathfrak{d}}(F, G) = Ext^{q+\mathfrak{d}}\left(\mathcal{O}_X, F^* \otimes G\right) = H^{q+\mathfrak{d}}\left(X, F^* \otimes G\right) = 0 \end{cases}$$

where two equalities in the third line follow from [Hr, Prop. III.6.7, 6.3], which is zero by hypothesis. The proof is complete. ■

Corollary 8.2.11. *Let* $X \hookrightarrow \mathbb{P}^r_R$ *be a projective subscheme, as in* (8.2.1, 8.2.9). *As in* (8.2.10), *assume that* $H^q(X, F) = 0$ *for all* $F \in$

$\mathscr{P}(X)$ $\forall q \geq \mathfrak{d}+1$. *Let*

$$0 \longrightarrow K \longrightarrow F \longrightarrow C \longrightarrow 0$$

be an exact sequence in $\mathscr{R}_0(X)$.

Then the sequence

$$0 \longrightarrow Z_{\mathfrak{d}-1}(K) \longrightarrow Z_{\mathfrak{d}-1}(F) \longrightarrow Z_{\mathfrak{d}-1}(C) \longrightarrow 0$$

is split exact.

Proof. By construction (8.2.9), $Z_{\mathfrak{d}-1}(K) \in \mathscr{R}_{\mathfrak{d}}(X)$. Therefore, by (8.2.10) the sequence

$$Hom\,(Z_{\mathfrak{d}-1}(C), Z_{\mathfrak{d}-1}(F)) \longrightarrow Hom\,(Z_{\mathfrak{d}-1}(C), Z_{\mathfrak{d}-1}(C)) \longrightarrow 0$$

is exact. The proof is complete. ■

Definition 8.2.12. Let $X \hookrightarrow \mathbb{P}^r_R$ be a projective subscheme, as in (8.2.9, 8.2.10). So, there is $\mathfrak{d} \in \mathbb{N}$ such that $H^q(X, F) = 0$ for all $F \in \mathscr{P}(X)$ $\forall q \geq \mathfrak{d}+1$.

Let $F \in \mathscr{R}_{-1}(X)$.

1. We claim that

$$Z_{\mathfrak{d}-1}(\mathcal{O}_X(1))^m \cong Z_{\mathfrak{d}-1}(F) \oplus \mathcal{G} \quad \text{for some } \mathcal{G},\ m.$$

By (8.2.4), the canonical map

$$\mathcal{O}_X \otimes \Gamma(X, F(-1)) \twoheadrightarrow F(-1) \qquad \text{is surjective.}$$

Since $\Gamma(X, F(-1))$ is finitely generated, there are exact sequences

$$0 \longrightarrow \ker(j) \longrightarrow \mathcal{O}_X^m \xrightarrow{\ j\ } F(-1) \longrightarrow 0 \qquad \forall m \gg 0.$$

By (8.2.2(3)), choices can be made, such that $\ker(j)(1)$ is regular. So, we have exact sequences

$$0 \longrightarrow G \longrightarrow \mathcal{O}_X(1)^m \longrightarrow F \longrightarrow 0$$

$$\forall m \gg 0, \quad \text{with} \quad G := ker(j)(1) \text{ regular.}$$

By (8.2.11)

$$Z_{\mathfrak{d}-1}(\mathcal{O}_X(1))^m \cong Z_{\mathfrak{d}-1}(F) \oplus Z_{\mathfrak{d}-1}(G)$$

So, the claim is established.

2. Let $\mathfrak{U} := \mathfrak{U}(X) := Z_{\mathfrak{d}-1}(\mathcal{O}_X(1)) \in \mathscr{R}_{\mathfrak{d}}(X)$, and $\mathfrak{R} = End_X(\mathfrak{U})$. Then $\mathfrak{R}$ is a ring (*possibly noncommutative*). For $W \in \mathscr{P}(X)$, consider $Hom_X(\mathfrak{U}, W)$. Recall that

$$\varphi \in Hom_X(\mathfrak{U}, W) \quad \text{is given by maps} \quad \{\varphi_U : \mathfrak{U}(U) \longrightarrow W(U)\}.$$

Now, for $g \in \mathfrak{R}$, $\varphi \in Hom_X(\mathfrak{U}, W)$, there is an obvious action, by composition:

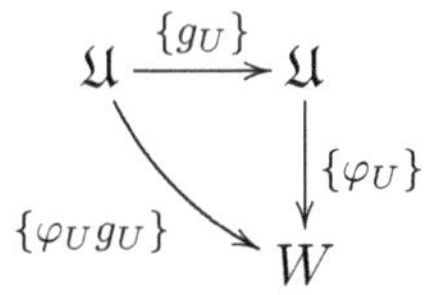

Define $\quad \varphi \star g := \varphi g.$

This gives $Hom_X(\mathfrak{U}, W)$ a structure of a right $\mathfrak{R}$-module.

3. **Tensor Product:** Given a right $\mathfrak{R}$-module M, and a $\mathcal{F} \in QCoh(X)$, on which $\mathfrak{R}$ has a left action, the tensor product $M \otimes_{\mathfrak{R}} \mathcal{F}$ is a sheaf of abelian groups $M \otimes_{\mathfrak{R}} \mathcal{F} \in \mathbf{Ab}(X)$, together with a balanced map $\beta : M \times \mathcal{F} \longrightarrow M \otimes_{\mathfrak{R}} \mathcal{F}$, with the following universal property:

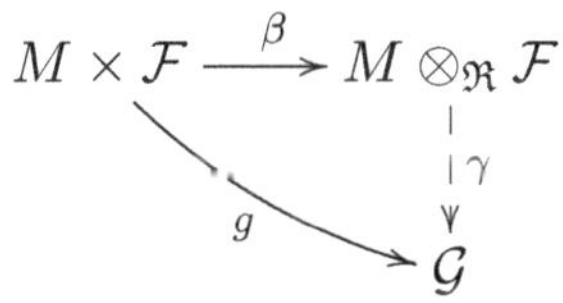

This means, given any other balanced map g, with $\mathcal{G} \in \mathbf{Ab}(X)$, there is a unique map γ in $\mathbf{Ab}(X)$, making the diagram commute. It is clear that $M \otimes_{\mathfrak{R}} \mathcal{F}$ is the sheaf associated to the presheaf

$$V \mapsto M \otimes_{\mathfrak{R}} \mathcal{F}(V).$$

4. Note, $\mathfrak{R}$ acts on $\mathfrak{U}(X)$ from left. So, $Hom_X(\mathfrak{U}, W) \otimes_{\mathfrak{R}} \mathfrak{U}$ is defined. In fact, there is a natural map

$$Hom_X(\mathfrak{U}, W) \otimes_{\mathfrak{R}} \mathfrak{U} \longrightarrow W \tag{8.11}$$

induced by the balanced map

$$Hom_X(\mathfrak{U}, W) \times \mathfrak{U}(V) \longrightarrow W(V) \qquad \text{sending} \quad (g, s) \mapsto g(s).$$

Lemma 8.2.13. *Let $W \in \mathscr{P}(X)$ be such that $W \oplus W_1 = \mathfrak{U}^m$. Then*

$$\begin{cases} Hom_X(\mathfrak{U}, W) \oplus Hom_X(\mathfrak{U}, W_1) \cong \mathfrak{R}^m, & \textit{and, the map} \\ Hom_X(\mathfrak{U}, W) \otimes_{\mathfrak{R}} \mathfrak{U} \xrightarrow{\sim} W & \textit{is an isomorphism.} \end{cases} \quad (8.11)$$

Proof. The first equality is clear because $\mathfrak{R}^m \cong Hom_X(\mathfrak{U}, \mathfrak{U}^m)$. It is also clear that the latter map is an isomorphism for $W = \mathfrak{U}^m$. The property is inherited by the direct summands. ■

We truncate the canonical resolution (8.9), as follows.

Proposition 8.2.14. *Let $X \hookrightarrow \mathbb{P}^r_R$ be a projective subscheme, as in (8.2.9, 8.2.10). So, there is $\mathfrak{d} \in \mathbb{N}$ such that $H^q(X, F) = 0$ for all $F \in \mathscr{P}(X)$ $\forall q \geq \mathfrak{d} + 1$.*

1. *For $F \in \mathscr{R}_{-1}(X)$, define*

$$T : \mathscr{R}_{-1}(X) \longrightarrow \mathscr{P}_r(\mathfrak{R}) \qquad \textit{by} \quad T(F) = Hom_X(\mathfrak{U}, Z_{\mathfrak{d}-1}(F)). \quad (8.12)$$

   Then T is an exact functor, where $\mathscr{P}_r(\mathfrak{R})$ denotes the category of finitely generated projective right $\mathfrak{R}$-modules. (We remark, eventually, we will prove that $T(F) \in \mathscr{P}(R)$, in the case of our interest.)

2. *The resolution (8.9), truncated as follows,*

$$\begin{array}{ccccc} 0 \longrightarrow & T(F) \otimes_{\mathfrak{R}} \mathfrak{U} & \longrightarrow & \mathcal{O}(-(\mathfrak{d}-1)) \otimes_R T_{\mathfrak{d}-1}(F) \\ & \wr \downarrow & \nearrow & \\ & Z_{\mathfrak{d}-1}(F) & & \end{array}$$

$$\longrightarrow \mathcal{O}(-(\mathfrak{d}-2)) \otimes_R T_{\mathfrak{d}-2}(F) \longrightarrow \cdots \quad (8.13)$$

   is exact.

3. *Of course, $F \mapsto T_p(F)$ are exact on $\mathscr{R}_{-1}(X) \subseteq \mathscr{R}_0(X)$.*

Proof. By (8.2.11), T is exact. It is clear from (8.2.13), that $T(F) \in \mathscr{P}_r(\mathfrak{R})$, and the vertical map in (8.13) is an isomorphism. Hence, the sequence (8.13) is exact, as is (8.9). The proof is complete. ■

Almost everything above applies to the quadratic hyper surfaces $X \subseteq \mathbb{P}^{d+1}_R$, which remains our main interest in this chapter.

Theorem 8.2.15. *Let $X \subseteq \mathbb{P}_R^{d+1}$ be a quadric hyper surface. So, with notation as in* (8.1.1), $A = \frac{B}{(f)}$, $\deg(f) = 2$, $X = \operatorname{Proj}(A) \subseteq \mathbb{P}_R^{d+1}$*, and we assume $C(f) = R$. Then all three versions* (8.2.8, 8.2.9, 8.2.14) *of the resolution theorems are valid for X.*

Proof. By (8.1.3), $\mathcal{O}_X(1) \in \mathfrak{R}(X)$. So, (8.2.9) is valid. By (8.1.2), the structure map $\mathfrak{f} : X \longrightarrow \operatorname{Spec}(R)$ is flat. So, (8.2.8) is valid. By (8.1.4), $H^q(X, F) = 0\ \forall q \geq d+1$ for all $F \in \mathscr{P}(X)$. So, (8.2.14) is valid. The proof is complete. ■

Remark 8.2.16. There seems to be a dichotomy between [Sw1, p. 126] and us, whether $\mathfrak{R} = End_X(\mathfrak{U})$ acts on $\mathfrak{U}$, $Hom(\mathfrak{U}, W)$ from left or right (8.2.12). However, given $f \in \mathfrak{R}$ and $x \in Hom(\mathfrak{U}, W)$, $y \in \mathfrak{U}$, there is no ambiguity how f acts on x or y. Whether it is a left or right action depends on how the product is defined in $\mathfrak{R}$. For $f, g \in \mathfrak{R}$, we define $f \star g(x) = f(g(x))$ by the usual composition. This seems to be the norm in the modern literature. However, one can also consider the $\mathfrak{R}^{op}$ where the product is defined, the way the arrows run. Some readers may be more comfortable working with left modules, while we view $T(F)$ as a right $\mathfrak{R}$-module (8.12). On the other hand, some readers may not be comfortable working with $\mathfrak{R}^{op}$. So, we made our choice, which runs contrary to [Sw1].

8.3. Quadratic Spaces

As the chapter heading conveys, we define quadratic spaces.

Definition 8.3.1. Let R be a commutative Noetherian ring. (*For simplicity, whenever we discuss quadratic R-modules, we assume $1/2 \in R$. This is not assumed in [Sw1], and all the results are valid without this restriction. However, this assumption $1/2 \in R$ simplifies the methods significantly.*) A quadratic module over R is a pair $(M, \mathfrak{q})$ such that $M \in QCoh(R)$ and $\mathfrak{q} : M \longrightarrow R$ is a map such that

$$\begin{cases} \mathfrak{q}(rm) = r^2\mathfrak{q}(m) & \forall r \in R, m \in M \\ B : M \times M \longrightarrow R & \text{defined by} \\ B(x, y) = \mathfrak{q}(x + y) - \mathfrak{q}(x) - \mathfrak{q}(y) & \text{is bilinear} \end{cases}$$

The inner product notation $\langle u, v\rangle := B(u, v)$ is also used. So, B induces a map $M \longrightarrow M^*$, and $2\mathfrak{q}(m) = B(m, m)$. So, $\mathfrak{q}$ is completely determined by the symmetric bilinear map B (since $1/2 \in R$). However, it has been customary to define a quadratic module as a pair $(M, \mathfrak{q})$, as above.

If $(M, \mathfrak{q})$ is a quadratic R-module and $R \longrightarrow A$ is a homomorphism of commutative rings, then there is a unique quadratic A-module $(M \otimes A, \mathfrak{q}_1)$ such that the diagram

$$\begin{array}{ccc} M & \longrightarrow & M \otimes A \\ {\scriptstyle \mathfrak{q}}\downarrow & & \downarrow{\scriptstyle \mathfrak{q}_1} \\ R & \longrightarrow & A \end{array} \qquad \text{commutes.}$$

In particular, localization of quadratic modules makes sense. The case of free R-modules has a special place in terms of techniques. Suppose $M = \oplus_{i=1}^{n} Re_i$ is free, with basis $e_1, \ldots, e_n$. We have the following:

1. Use the notation for the matrix $\mathbf{B}(\{e_i\}) := (B(e_i, e_j))$ to indicate its dependence on the basis. So, $\mathbf{B}(\{e_i\})$ is a symmetric matrix. The bilinear map B is given by this matrix $\mathbf{B}(\{e_i\})$.
2. For $u = \sum_{i=1}^{n} x_i e_i$, $v = \sum_{j=1}^{n} y_i e_j$, we have

$$\begin{cases} \langle u, v\rangle = \sum_{i,j=1}^{n} x_i y_j B(e_i, e_j) = \mathbf{x}^t \mathbf{B}(\{e_i\})\mathbf{y} \\ \mathfrak{q}(u) = \frac{1}{2}\mathbf{x}^t \mathbf{B}(\{e_i\})\mathbf{x} \end{cases}$$

$$\text{where} \quad \begin{cases} \mathbf{x}^t = (x_1, \ldots, x_n) \\ \mathbf{y}^t = (y_1, \ldots, y_n) \end{cases}$$

   With dual basis $\{e_i^*\}$ of M^*, the linear map $M \longrightarrow M^*$ is given by the matrix $\mathbf{B}(\{e_i\})$.
3. (**Change of basis**): Suppose $\{\mathfrak{e}_1, \ldots, \mathfrak{e}_n\}$ is another basis of M, and

$$\begin{pmatrix} \mathfrak{e}_1 \\ \mathfrak{e}_2 \\ \cdots \\ \mathfrak{e}_n \end{pmatrix} = C \begin{pmatrix} e_1 \\ e_2 \\ \cdots \\ e_n \end{pmatrix} \qquad \text{then} \quad \mathbf{B}(\{\mathfrak{e}_i\}) = C\mathbf{B}(\{e_i\})C^t.$$

4. For a fixed integer $n \geq 0$ let $M \cong R^n$ be as above. Fix a basis $\{e_1, \ldots, e_n\}$ of M. Then any quadratic R-module (M, q) can be

considered a homogeneous quadratic polynomial

$$\mathfrak{q} = \sum_{i \le j} a_{ij} x_i x_j \quad \text{with}$$
$$a_{ij} = \frac{B(e_i, e_j) + B(e_j, e_i)}{2} = B(e_i, e_j) = B(e_j, e_i). \tag{8.14}$$

Therefore, quadratic R-modules $(M, \mathfrak{q})$ are in one-to-one correspondence with the homogeneous quadratic polynomials in $R[X_1, \ldots, X_n]$.

5. We are mostly concerned when M is projective. In this case, the bilinear map B can, locally, be described by the matrix $\mathbf{B}(\{e_i\})$, for any choice of local basis $\{e_i\}$.

Subsequently, we mainly consider quadratic R-modules $(P, \mathfrak{q})$, where $P \in \mathscr{P}(R)$. In analogy to the case of free modules (the following comment 8.14), the quadratic R-modules $(P, \mathfrak{q})$ can be identified with elements of degree two in the symmetric algebra $\mathcal{S}\text{ym}(P^*)$.

Lemma 8.3.2. *Let R be a commutative Noetherian ring, and $P \in \mathscr{P}(R)$. Let $\mathcal{S}\text{ym}(P^*) = \bigoplus_{n \ge 0} \mathcal{S}\text{ym}(P^*)_n$ be the symmetric algebra. Let $Quad(P)$ denote the R-module of all quadratic R-modules $(P, \mathfrak{q})$. Then there is a natural isomorphism:*

$$\mu : \mathcal{S}\text{ym}(P^*)_2 \xrightarrow{\sim} Quad(P). \tag{8.15}$$

Proof. There is a natural map $\mu_0 : P^* \otimes P^* \longrightarrow Quad(P)$ defined by $\mu_0(f \otimes g)(x) = f(x)g(x)$. This map factors through a map

$$\mu : \mathcal{S}\text{ym}(P^*)_2 \longrightarrow Quad(P).$$

Now, it is enough to check locally that μ is an isomorphism. In the local case, by (8.14), it is obvious that μ is an isomorphism. The proof is complete. ∎

We introduce the following notation.

Notation 8.3.3. Let R be a commutative Noetherian ring and $(P, \mathfrak{q})$ be a quadratic R-module, with $P \in \mathscr{P}(R)$. Use the same notation, $\mathfrak{q} \in \mathcal{S}\text{ym}(P^*)_2$, for the preimage of $\mathfrak{q} \in Quad(P)$. Denote

$$\begin{cases} B = \mathcal{S}\text{ym}(P^*) \\ A = S(\mathfrak{q}) = \frac{\mathcal{S}\text{ym}(P^*)}{(\mathfrak{q})} \end{cases} \quad \text{and} \quad \begin{cases} \mathbb{B}(P^*) = \text{Proj}(B) \\ X(\mathfrak{q}) = \text{Proj}(S(\mathfrak{q})) \subseteq \mathbb{B}(P^*). \end{cases}$$

If $P \oplus Q \cong R^{N+1}$, then $X(\mathfrak{q}) \subseteq \mathbb{P}^N_R = \operatorname{Proj}\left(\mathcal{S}\mathrm{ym}\left(R^{N+1}\right)^*\right)$ is a closed subscheme.

We define nondegenerate quadratic R-modules.

Definition 8.3.4. Let R be a commutative Noetherian ring. A **quadratic space** is a quadratic R-module $(P, \mathfrak{q})$ such that $P \in \mathscr{P}(R)$, and the symmetric map $\varphi_{\mathfrak{q}} : P \longrightarrow P^*$, induced by B, is an isomorphism. In other words, the following is a commutative diagram of isomorphisms:

$$\begin{array}{ccc} P & \xrightarrow[\sim]{\varphi_{\mathfrak{q}}} & P^* \\ \wr\downarrow & \nearrow_{\varphi_{\mathfrak{q}}^*} & \\ P^{**} & & \end{array}$$

A quadratic space is also referred to as nondegenerate or nonsingular. So, $(P, \mathfrak{q})$ is nondegenerate if and only if, locally, the matrix $(B(e_i, e_j))$ is invertible, which is obvious.

The projective hyper surface $\operatorname{Proj}\left(\frac{\kappa[X_0, X_1]}{(X_0^2+X_1^2)}\right)$ is smooth, while $\operatorname{Proj}\left(\frac{\kappa[X_0, X_1, X_2]}{(X_0^2+X_1^2)}\right)$ is not. We characterize the smoothness property of $X(\mathfrak{q})$ as follows.

Proposition 8.3.5. *Use the notation from* (8.3.3). *As usual, we assume* 2 *is invertible. Then*

$$C(\mathfrak{q}) = R \Longleftarrow (P, \mathfrak{q}) \text{ is nondegenerate} \iff X(\mathfrak{q}) \text{ is smooth over } R.$$

Here $C(\mathfrak{q})$ *denotes the content of* $\mathfrak{q}$, *as defined in* (8.2.1), *considering* $\mathfrak{q} \in R[X_0, X_1, \ldots, X_N]$. *Refer to* [Hr, p. 268] *or [K] for definition of smoothness.*

Proof. The questions are local. So, we assume $(R, \mathfrak{m})$ is local, and $P = \oplus_{i=0}^n Re_i$ is free of rank $n+1$ (say). With the evaluation maps, $X_i = e_i^* \in P^*$, we have $\mathcal{S}(P^*) = R[X_0, \ldots, X_n]$. So, $\mathfrak{q}$ is given by a polynomial, as in (8.14).

It is clear that $(P, \mathfrak{q})$ is nondegenerate if and only if the symmetric matrix $\mathbf{B}(\{e_i\})$ is invertible. The content $C(\mathfrak{q})$ is generated by $B(e_i, e_j)$. So, if $(P, \mathfrak{q})$ is nondegenerate, $C(\mathfrak{q}) = R$. Therefore, by (8.1.2), $\mathrm{Proj}\,(S(\mathfrak{q})) \longrightarrow \mathrm{Spec}\,(R)$ is flat. Since $1/2 \in R$, there is an invertible matrix C such that $C\mathbf{B}(\{e_i\})C^t$ is a diagonal matrix (8.3.9). With the change basis

$$\begin{pmatrix} \mathfrak{e}_1 \\ \mathfrak{e}_2 \\ \cdots \\ \mathfrak{e}_n \end{pmatrix} = C \begin{pmatrix} e_1 \\ e_2 \\ \cdots \\ e_n \end{pmatrix}, \quad \text{we have}$$

$$\mathbf{B}(\{\mathfrak{e}_i\}) = \begin{pmatrix} \lambda_0 & 0 & \cdots 0 \\ 0 & \lambda_1 & \cdots 0 \\ \cdots & \cdots & \cdots 0 \\ 0 & \cdots 0 & \lambda_n \end{pmatrix} \quad \text{with} \quad \lambda_i \notin \mathfrak{m}.$$

So, with this new basis, we have $\mathfrak{q} = \lambda_0 X_0^2 + \lambda_1 X_1^2 + \cdots + \lambda_n X_n^2$. So, the Jacobian matrix is given by

$$\left(2\lambda_0 X_0 \; 2\lambda_1 X_1 \; \cdots \; 2\lambda_n X_n \right).$$

Therefore, by the Jacobian criterion, $X(\mathfrak{q})$ is smooth over R.

Conversely, suppose $X(\mathfrak{q})$ is smooth over R. We prove that the map $\varphi : P \longrightarrow P^*$ is an isomorphism. From rank consideration, it suffices to show φ is surjective. By Nakayama's lemma, it is enough to show that $\varphi \otimes \kappa$ is an isomorphism, where $\kappa = \frac{R}{\mathfrak{m}}$. Since $X(\mathfrak{q} \otimes \kappa) = X(\mathfrak{q}) \otimes \kappa$ is smooth over κ, we can assume that $R = \kappa$ is a field, and P is a vector space. To prove $\varphi : P \longrightarrow P^*$ is surjective, it is enough to prove that $\varphi \otimes \overline{\kappa}$ is surjective, where $\kappa \longrightarrow \overline{\kappa}$ is the algebraic closure. So, we can assume that $R = \kappa$ is algebraically closed. In this case of fields, a change of basis as above is possible. Further, since $\kappa = \overline{\kappa}$, a change of basis as above is possible such that $\mathfrak{q} = \sum_{i=0}^{r} X_i^2 \in R[X_0, \ldots, X_r, X_{r+1}, \ldots, X_n]$ with $r \le n$ [Lt, p. 12]. The Jacobian matrix is given by

$$\left(2X_0 \; 2X_1 \; \cdots \; 2X_r, 0, \ldots, 0 \right).$$

By the Jacobian criterion, $n - r = 0$. So, $\mathfrak{q} = \sum_{i=0}^{n} X_i^2$, which is nondegenerate. The proof is complete. ■

As was the case with the quadratic hyper surfaces of $\mathbb{P}_R^{d+1}$, almost everything in the previous sections remains valid for $X(\mathfrak{q})$, where $(P, \mathfrak{q})$ is nondegenerate.

Theorem 8.3.6. *Let* $(P, \mathfrak{q})$ *be a nondegenerate quadratic* R*-module and* $X(\mathfrak{q}) = \operatorname{Proj}(S(\mathfrak{q}))$. *Then all three versions* (8.2.8, 8.2.9, 8.2.14) *of the resolution theorems are valid for* $X(\mathfrak{q})$. (We remark, eventually, we will prove that $T(F) \in \mathscr{P}(R)$, in this case.)

Proof. Write $X = X(\mathfrak{q})$. Assume $rank(P) = d + 2$, and hence $\dim X(\mathfrak{q}) = d$. First, $\mathcal{O}_X(1)$ is regular. All the hypotheses are checked locally. Locally, we can assume $P \cong R^{d+2}$ is free. In this case, $X(q) \cong \operatorname{Proj}\left(\frac{R[X_0,\ldots,X_d]}{(\mathfrak{q})}\right)$. Locally, $\mathcal{O}_X(1)$ is regular, by (8.1.3), and hence $\mathcal{O}_X(1)$ is regular. So, (8.2.9) is valid. Since $X \longrightarrow \operatorname{Spec}(R)$ is flat, (8.2.8) is valid. Finally, it is locally checked, by (8.1.4), that $\forall q \geq d+1$, $H^q(X, F) = 0$ for all $F \in \mathscr{P}(X)$. So, (8.2.14) is valid. The proof is complete. ∎

8.3.1. *Clifford algebra*

In this section, we define the Clifford algebra of a quadratic R-module (M, q). Among the original sources of these constructions are [ABS], [BR, p. 116].

Definition 8.3.7. Let R be a commutative Noetherian ring, and $(M, \mathfrak{q})$ be a quadratic R-module. The **Clifford algebra** $C(\mathfrak{q}) := C(M, \mathfrak{q})$ of $\mathfrak{q}$ is defined to be

$$C(\mathfrak{q}) = \frac{T(M)}{I(\mathfrak{q})} \quad \text{where} \quad \begin{cases} T(M) \text{ is the tensor algebra of } M \\ I(\mathfrak{q}) \text{ is the two-sided ideal generated by} \\ \{x \otimes x - \mathfrak{q}(x) : x \in M\}. \end{cases}$$

The following are some of the immediate comments and definitions:

1. The map $M \longrightarrow C(\mathfrak{q})$ has the usual universal property. This means, given any R-algebra A, and a linear map $\rho : P \longrightarrow A$, such that $\rho(x)^2 = \mathfrak{q}(x)$ for all $x \in P$, then there is a unique map $\tilde{\rho}$ such that the diagram

$$\begin{array}{ccc} P & \longrightarrow & C(\mathfrak{q}) \\ & \searrow^{\rho} & \downarrow{\scriptstyle \tilde{\rho}} \\ & & A \end{array} \qquad \text{commutes.}$$

   It is interesting to note that the universal property does not distinguish A and its opposite ring A^{op}.

2. If $\mathfrak{q} = 0$, then $C(\mathfrak{q}) = \Lambda^\bullet M$ is the exterior algebra of M.
3. As in exterior algebra, the image of $x_1 \otimes x_2 \otimes \cdots \otimes x_n \in T^n(M)$ in $C(\mathfrak{q})$ is denoted by $x_1 \wedge x_2 \wedge \cdots \wedge x_n \in C(\mathfrak{q})$. It follows that
$$\mathfrak{q}(x+y) = \mathfrak{q}(x) + x \wedge y + y \wedge x + \mathfrak{q}(y).$$
4. There is $\mathbb{Z}_2$-grading $C(\mathfrak{q}) = C_0(\mathfrak{q}) \oplus C_1(\mathfrak{q})$ into the even part $C_0(\mathfrak{q})$ and the odd part $C_1(\mathfrak{q})$, induced by the grading on $T(M)$.
5. If $M = \oplus_{i=0}^d Re_i$ is free, with basis $e_0, e_1, \ldots, e_n$, then $C(\mathfrak{q})$ has a basis [ABS, BR]
$$\{e_{i_0} \wedge e_{i_1} \wedge \cdots \wedge e_{i_r} : 0 \le i_0 < i_1 < \cdots < i_r \le d; r = 0, 1, \ldots, d\}.$$
6. For two quadratic R-modules $(M_1, \mathfrak{q}_1)$ and $(M_2, \mathfrak{q}_2)$, it follows [BR, pp. 97, 116] that
$$C(\mathfrak{q}_1 \perp q_2) = C(\mathfrak{q}_1) \hat{\otimes}_R C(\mathfrak{q}_2) \qquad \text{is the tensor product of}$$
$\mathbb{Z}_2$-graded algebras,

   which is the usual tensor product, with multiplication given by
$$(x_1 \otimes y_1) \cdot (x_2 \otimes y_2) := (-1)^{\deg(x_2)\deg(y_1)} (x_1 x_2) \otimes (y_1 y_2).$$
7. **Notation:** Because of remark (8.2.16), we will mostly be working with right $C(\mathfrak{q})$-modules. Notation will have a subscript l, r, to indicate whether we are talking about left or right modules. However, we will sometimes drop the subscript l, r, if it is clear from the context. The category of $C_0(\mathfrak{q})$-modules (left or right) is denoted by $QCoh(C_0(\mathfrak{q}))$. The category of $\mathbb{Z}_2$-graded $C(\mathfrak{q})$-modules (left or right) are denoted by $QCoh_{\mathbb{Z}_2}(C(\mathfrak{q}))$. Further, $Coh(C_0(\mathfrak{q})) \subseteq QCoh(C_0(\mathfrak{q}))$, $Coh_{\mathbb{Z}_2}(C(\mathfrak{q})) \subseteq QCoh_{\mathbb{Z}_2}(C(\mathfrak{q}))$ denote the respective full subcategories of finitely generated modules. There are two natural functors
$$\begin{cases} \Phi_l : QCoh_{\mathbb{Z}_2}(C(\mathfrak{q})) \longrightarrow QCoh(C_0(\mathfrak{q})) \\ \quad \text{defined by} \quad \Phi(M_0 \oplus M_1) = M_0 \\ \Psi_l : QCoh(C_0(\mathfrak{q})) \longrightarrow Qcoh_{\mathbb{Z}_2}(C(\mathfrak{q})) \\ \quad \text{defined by} \quad \Psi(N) = C(\mathfrak{q}) \otimes_{C_0(\mathfrak{q})} N. \end{cases} \tag{8.16}$$

Similarly, denote the functors Φ_r, Ψ_r functors of categories of right modules. Note that $\Psi_r(N) = N \otimes_{C_0(\mathfrak{q})} C(\mathfrak{q})$, differs in order of tensor product.

Lemma 8.3.8. *Let R be a commutative Noetherian ring and $(P, \mathfrak{q})$ be a nondegenerate quadratic R-module. Then the functors Φ_l and Ψ_l, defined in* (8.16)*, are inverses of each other. Likewise, the functors Φ_r and Ψ_r are inverses of each other.* (Consult (8.3.9) to relax the condition that $\mathfrak{q}$ is nondegenerate.)

Proof. We prove the first one and drop the subscript l. It is clear that $\Phi\Psi = 1$. For $M = M_0 \oplus M_1 \in QCoh_{\mathbb{Z}_2}(C(\mathfrak{q}))$, we have $\Psi\Phi(M_0 \oplus M_1) = C(\mathfrak{q}) \otimes M_0$. Further, there is a natural map

$$\tau : C(\mathfrak{q}) \otimes_{C_0(\mathfrak{q})} M_0 \longrightarrow M_0 \oplus M_1.$$

We prove τ is an isomorphism. Let $K = \ker(\tau)$ and $C = co\ker(\tau)$. Since, τ is an isomorphism on M_0 (even part), K and C have even part zero. To prove $K = C = 0$, we can assume $(R, \mathfrak{m})$ is local and P is free with basis $e_1, \ldots, e_n$. By a change of basis (8.3.9), we can assume that $\mathfrak{q} = \lambda_1 X_1^2 + \lambda_2 X_2^2 + \cdots + \lambda_n X_n^2$, where $\lambda_i \notin \mathfrak{m}$ are units, and $X_i = e_i^* \in P^*$. For $m \in K = K_1$, we have $e_1 \cdot m = 0$ and hence $e_1^2 \cdot m = \lambda_1 m = 0$, and $m = 0$. So, $K = 0$. Similarly, $C = 0$. So, τ is an isomorphism. The proof is complete. ∎

Exercise 8.3.9. Let $(R, \mathfrak{m})$ be a local Noetherian ring, $1/2 \in R$. Let F be a free R-module with $rank(F) = n$, and $(F, \mathfrak{q})$ be a quadratic R-module. Then there is a choice of basis $\{\mathfrak{e}_1, \mathfrak{e}_2, \ldots, \mathfrak{e}_n\}$ of F such that the symmetric matrix for $\mathfrak{q}$ is given by

$$\mathbf{B}(\{\mathfrak{e}_i\}) = \begin{pmatrix} \Delta & \mathbf{0}_{12} \\ \mathbf{0}_{21} & M \end{pmatrix} \quad \text{where}$$

$$\Delta = \begin{pmatrix} \lambda_1 & 0 & \cdots & 0 \\ 0 & \lambda_1 & \cdots & 0 \\ \cdots & \cdots & \cdots & \cdots \\ 0 & 0 & \cdots & \lambda_r \end{pmatrix}, \quad \text{with } \lambda_i \notin \mathfrak{m} \text{ are units,}$$

and M is a square matrix of order $n - r$, with all its entries in $\mathfrak{m}$, and $\mathbf{0}_{12}$, $\mathbf{0}_{21}$ are zero matrices of appropriate size.

8.4. Canonical Resolution and Minimal Resolution

The goal of this section is to describe the canonical resolution (8.2.8, 8.2.9) in a more tangible manner. We set up some notation.

Notation 8.4.1. Let R be a commutative Noetherian ring, $I \subseteq R[X_0, X_1, \ldots, X_N]$ be a homogenous ideal. Let $A = \frac{R[X_0, X_1, \ldots, X_N]}{I}$, $X = \text{Proj}(A) \subseteq \mathbb{P}_R^N$. Let $F \in Coh(X)$ and $F = \widetilde{M}$ for some graded A-module M. We assume

$$\begin{cases} \mathcal{O}(1), \quad F \in \mathscr{R}_0(X) \quad \text{are regular} \\ A_n \xrightarrow{\sim} \Gamma(X, \mathcal{O}_X(n)) \quad \text{is isomorphism } \forall n \geq 1 \\ M_n = 0 \quad \forall n \leq -1, \text{ and } \quad M_n \xrightarrow{\sim} \Gamma(X, F(n)) \\ \quad \text{is isomorphism } \forall n \geq 0. \end{cases} \tag{8.17}$$

By (8.2.6), $M_0 = \Gamma(X, \mathcal{F}) \in \mathscr{P}(R)$. Further, by (8.2.4) or its proof, M_0 generates M, and we have an exact sequence

$$0 \longrightarrow N \longrightarrow A \otimes M_0 \longrightarrow M \longrightarrow 0. \tag{8.18}$$

Sheafifying, we have $\widetilde{N} \cong Z_0(F)$, and we obtain the exact sequence

$$0 \longrightarrow \widetilde{N} \longrightarrow \mathcal{O}_X \otimes M_0 \longrightarrow F \longrightarrow 0$$

$$\text{by} \quad (8.2.2(3)) \quad \widetilde{N}(1) = Z_0(F)(1) \in \mathscr{R}_0(X).$$

For convenience, in analogy to previous notation, we denote the category of such graded finitely generated A-modules M, as in (8.17), by $\mathscr{R}_0(A)$.

Lemma 8.4.2. *Use all the notation from* (8.4.1). *Then* $\widetilde{N}(1)$ *satisfies the conditions in* (8.17). *So,* (1) $\widetilde{N}(1) \in \mathscr{R}_0(X)$, (2) $N(1)_n = N(n+1) = 0 \ \forall n \leq -1$, *(3)* $N(1)_n = N_{n+1} \xrightarrow{\sim} \Gamma(X, \widetilde{N}(1))$ *is an isomorphism* $\forall n \geq 0$. *So,* $N(1) \in \mathscr{R}_0(A)$.

Proof. We already observed that $\widetilde{N}(1) = Z_0(F)(1) \in \mathscr{R}_0(X)$. Since $M_0 \xrightarrow{\sim} \Gamma(X, F)$ is an isomorphism, $N_0 = 0$. So, $N(1)_n = 0 \ \forall n \leq -1$. We also have the diagram of exact sequences

$$\begin{array}{ccccccccc} 0 & \longrightarrow & N_n & \longrightarrow & A_n \otimes M_0 & \longrightarrow & M_n & \longrightarrow & 0 \\ & & \downarrow & & \downarrow \wr & & \downarrow \wr & & \\ 0 & \longrightarrow & \Gamma(X, \widetilde{N}(n)) & \longrightarrow & \Gamma(X, \mathcal{O}_X(n)) \otimes M_0 & \longrightarrow & \Gamma(X, F(n)) & \longrightarrow & 0 \end{array}$$

By hypothesis, the last two vertical arrows are isomorphisms $\forall n \geq 0$. So, the first vertical arrow is an isomorphism $\forall n \geq 0$, while $N_0 = 0$. So, it follows $N(1)_n \xrightarrow{\sim} \Gamma(X, \widetilde{N}(1)(n))$ is isomorphism $\forall n \geq 0$. ■

We define minimal resolution.

Definition 8.4.3. Use all the notation from (8.4.1). We define a sequence of functors

$$\begin{cases} T_p : \mathscr{R}_0(A) \longrightarrow \mathscr{P}(R) \\ Z_p : \mathscr{R}_0(A) \longrightarrow Coh^+_{gr}(A) \end{cases} \quad \ni \quad Z_{p-1}(M)(p) \in \mathscr{R}_0(A)$$

together with an exact sequence

$$\cdots \longrightarrow A(-p) \otimes T_p(M) \longrightarrow \cdots \longrightarrow A \otimes T_0(M) \longrightarrow M \longrightarrow 0 \tag{8.19}$$

where $T_0(M) = M_0$, $Z_0(M) := N = \ker(A_n \otimes M_0 \longrightarrow M)$, as in (8.18). Inductively, when T_{p-1} and Z_{p-1} are defined, define T_p and Z_p, as follows: $T_p(M) = (Z_{p-1}(p))_0$, and

$$0 \longrightarrow Z_p(M) \longrightarrow A(-p) \otimes T_p(M) \longrightarrow Z_{p-1}(M) \longrightarrow 0$$

where the sequence is exact. The process continues by (8.4.2). Such a resolution is necessarily minimal because the degree-zero map $A_0 \otimes N_0 \longrightarrow M_0$ is an isomorphism.

By sheafification of (8.19), we obtain the canonical resolution (8.2.9) of F.

Example 8.4.4. Let $A = R[x_0, \ldots, x_N] = \oplus_{n \geq 0} A_n$ be a graded commutative ring, and $X = \operatorname{Proj}(A) \subseteq \mathbb{P}^N_R$. Let $M = \oplus_{n \geq 0} A_{n+1} \subseteq A(1)$. It follows that $\widetilde{M} = \widetilde{A(1)} = \mathcal{O}_X(1)$ [Hr, Ex. II.5.9]. One can describe M by the exact sequence

$$0 \longrightarrow M(-1) \longrightarrow A \longrightarrow R \longrightarrow 0.$$

Now, further assume that $\mathcal{O}(1) \in \mathscr{R}_0(X)$. So, (8.4.1, 8.4.3) are applicable. Thus, we obtain the minimal resolution (8.19) of M, which leads to the canonical resolution (8.2.9) of $\mathcal{O}_X(1)$. This, in particular, applies to $X = X(\mathfrak{q})$, where $(P, \mathfrak{q})$ is a nondegenerate quadratic R-module.

The following is an application of the Koszul complex constructions.

Example 8.4.5. Let $R \longrightarrow A$ be a homomorphism of commutative rings (not necessarily graded), and $f : M \longrightarrow A$ be an R-module homomorphism, with $M \in Coh(R)$. The exterior algebra of M is denoted by $\Lambda^\bullet M$. Then there is a unique derivation, in the graded sense [T]

$$d : \Lambda^\bullet M \otimes_R A \longrightarrow \Lambda^\bullet M \otimes_R A \qquad \ni \quad d_1(m \otimes 1) = f(m) \quad \forall m \in M$$

of degree -1, and d_n is necessarily defined as follows:

$$\begin{aligned} & d_n\left((m_1 \wedge \cdots \wedge m_n) \otimes a\right) \\ & \quad = \sum (-1)^r \left(m_1 \wedge \cdots \wedge \hat{m}_r \wedge \cdots \wedge m_n\right) \otimes af(m_r) \end{aligned}$$

where the $\hat{m}_r$-notation is used to indicate that m_r is being dropped. We denote this complex by $K_\bullet(f) := \Lambda^\bullet M \otimes A$, and $K_n(f) = \Lambda^n M \otimes A$. Also, note that $\Lambda^\bullet M \otimes_R A$ has a skew symmetric multiplicative structure:

$$\begin{cases} xy = (-1)^{mn} yx & x \in \Lambda^m M, y \in \Lambda^n M \\ x^2 = 0 & x \in \Lambda^{2n+1} M \otimes A, 1/2 \in R. \end{cases} \tag{8.20}$$

Now we use the derivation (8.4.5), as follows.

Example 8.4.6. Let R be a commutative Noetherian ring, and $(P, \mathfrak{q})$ be a quadratic R-module, with $P \in \mathscr{P}(R)$. As in (8.3.3), let $B = \mathcal{S}\mathrm{ym}\,(P^*) = \bigoplus_{n \geq 0} \mathcal{S}\mathrm{ym}\,(P^*)_n$, and $A = S(\mathfrak{q}) = \bigoplus_{n \geq 0} S_n(\mathfrak{q})$. Consider the maps $\iota_0 : P^* = \mathcal{S}\mathrm{ym}\,(P^*)_1 \hookrightarrow \mathcal{S}\mathrm{ym}\,(P^*)$, $\iota : P^* = S_1(\mathfrak{q}) \hookrightarrow S(\mathfrak{q})$. By (8.4.5), we obtain two Koszul complexes,

$$\begin{cases} K_\bullet(\iota_0) = \Lambda^\bullet P^* \otimes \mathcal{S}\mathrm{ym}\,(P^*) \\ K_\bullet(\iota) = \Lambda^\bullet P^* \otimes S(\mathfrak{q}). \end{cases}$$

Here the former is the more familiar Koszul complex, with $H_0(K_\bullet(\iota_0)) = R$ and $H_n(K_\bullet(\iota_0)) = 0,\ \forall n \neq 0$. It is easy to compute $H_n(K_\bullet(\iota))$, as follows. Recall the identification $\mathcal{S}\mathrm{ym}\,(P^*)_2 \xrightarrow{\sim} \mathcal{Q}uad(P)$, and we use the same notation $\mathfrak{q}$ for its preimage. There is an exact sequence of complexes

$$0 \longrightarrow K_\bullet(\iota_0) \xrightarrow{\mathfrak{q}} K_\bullet(\iota_0) \longrightarrow K_\bullet(\iota) \longrightarrow 0$$

$$\text{So,} \quad \begin{cases} H_0\left(K_\bullet(\iota)\right) = R, \\ H_1\left(K_\bullet(\iota)\right) = R, \\ H_n\left(K_\bullet(\iota)\right) = 0 \quad \forall n \geq 2. \end{cases} \tag{8.21}$$

The differential: $d_n : \Lambda^n P^* \otimes S_k(\mathfrak{q}) \longrightarrow \Lambda^{n-1} P^* \otimes S_{k+1}(q)$ sends

$$\begin{aligned} &(f_1 \wedge \cdots \wedge f_n) \otimes \overline{g_1 \otimes \cdots \otimes g_k} \\ &\quad \mapsto \sum \pm (f_1 \wedge \cdots \wedge \hat{f_r} \wedge \cdots f_n) \otimes \overline{f_r \otimes g_1 \otimes \cdots \otimes g_k} \end{aligned} \tag{8.22}$$

where $\hat{f_r}$-notation indicates f_r is dropped. So, with $n = 1$, $k = 1$, we have

$$f_1 \otimes g_1 \mapsto \overline{f_1 \otimes g_1} \quad A_1 \otimes P^* = P^* \otimes P^* \longrightarrow S_2(\mathfrak{q}) = \overline{\mathcal{S}\mathrm{ym}\,(P^*)_2} \tag{8.23}$$

which is the canonical quotient map to the symmetric algebra. Let $\tilde{\mathfrak{q}} \in P^* \otimes P^*$ be a lift of $\mathfrak{q} \in \mathcal{S}\mathrm{ym}\,(P^*)_2$. Then $\tilde{\mathfrak{q}}$ represents a degree one element in $\Lambda^1 P^* \otimes P^*$ in the complexes:

$$\mathfrak{t} := \tilde{\mathfrak{q}} \in (K_\bullet(\iota_0))_1 = P^* \otimes \mathcal{S}\mathrm{ym}\,(P^*), \qquad \text{and}$$
$$\mathfrak{t} := \tilde{\mathfrak{q}} \in (K_\bullet(\iota_1))_1 = P^* \otimes S(\mathfrak{q})$$

where we write $\mathfrak{t} := \tilde{\mathfrak{q}} \in K_1(\iota)$ to distinguish it from being an element of $\mathcal{S}\mathrm{ym}\,(P^*)$ or $S(\mathfrak{q})$. So, we have proved that

$$d_1(\mathfrak{t}) = 0 \in (K_\bullet(\iota))_0 = S(\mathfrak{q}). \quad \text{Further,} \quad H_1(K(\iota)) = R\bar{\mathfrak{t}} \tag{8.24}$$

which follows from the definition of the map $H_1(K(\iota_0)) \xrightarrow{\sim} H_0(K(\iota))$, as in (8.21).

The trick of the divided power algebra was used in the paper of Tate [T], to kill $\bar{\mathfrak{t}}$ to get a resolution of R. The situation here is much simpler. Write

$$\Gamma[T] = \bigoplus \Gamma_{2n}[T] = \bigoplus_{n \geq 0} RT^{(n)} \quad \text{a free } R\text{-module, graded with } \deg T^{(n)} = 2n.$$

We write $\Gamma_{2n+1}[T] = 0$, and hence there is only a trivial differential of degree -1. Define a new differentially graded algebra, as follows:

$$\mathcal{K}_n := \mathcal{K}_n(\iota, \mathfrak{t}) : = \left(\bigoplus_{q=0}^{n} \Lambda^{n-q} P^* \otimes \Gamma^q[T] \right) \otimes S(\mathfrak{q})$$
$$= \left(\bigoplus_{q=0}^{2q \le n} \Lambda^{n-2q} P^* T^{(q)} \right) \otimes S(\mathfrak{q}).$$

Refer to [T] for basic facts on differentials of graded algebras. Define the differential

$$\partial_n : \mathcal{K}_n(\iota, \mathfrak{t}) \longrightarrow \mathcal{K}_{n-1}(\iota, \mathfrak{t}) \quad \text{by}$$
$$\partial_n \left(\omega T^{(q)} \right) = (d\omega) T^{(q)} + (-1)^k \omega \mathfrak{t} T^{(q-1)}, \quad \forall \omega \in \Lambda^k P^*.$$

It follows from (8.20, 8.24) that ∂ defines a complex (differential). For clarity, we have

$$\begin{cases} \mathcal{K}_{2n} = \big(\Lambda^{2n} P^* \oplus \Lambda^{2(n-1)} P^* T^{(1)} \oplus \cdots \oplus \\ \quad \Lambda^{2(n-p)} P^* T^{(p)} \oplus \cdots \oplus \Lambda^0 P^* T^{(n)} \big) \otimes S(\mathfrak{q}) \\ \mathcal{K}_{2n-1} = \big(\Lambda^{2n-1} P^* \oplus \Lambda^{2n-3} P^* T^{(1)} \oplus \cdots \oplus \\ \quad \Lambda^{2(n-p)-1} P^* T^{(p)} \oplus \cdots \oplus \Lambda^1 P^* T^{(n-1)} \big) \otimes S(\mathfrak{q}) \\ \mathcal{K}_2 = \big(\Lambda^2 P^* \oplus \Lambda^0 P^* T^{(1)} \big) \otimes S(\mathfrak{q}) \\ \mathcal{K}_1 = \Lambda^1 P^* \otimes S(\mathfrak{q}) \\ \mathcal{K}_0 = S(\mathfrak{q}). \end{cases} \tag{8.25}$$

The following is noteworthy:

1. if n is even, $\mathcal{K}_n$ starts with $\Lambda^0 P^* \otimes T^{(n/2)}$.
2. if n is odd, $\mathcal{K}_n$ starts with $\Lambda^1 P^* \otimes T^{((n+1)/2)}$.

Consider the tail of the complex:

$$\begin{array}{ccccc} \mathcal{K}_2 & \longrightarrow & \mathcal{K}_1 & \longrightarrow & \mathcal{K}_0 \\ \| & & \| & & \| \\ \big(\Lambda^2 P^* \oplus R T^{(1)} \big) \otimes S(\mathfrak{q}) & \longrightarrow & P^* \otimes S(\mathfrak{q}) & \longrightarrow & S(\mathfrak{q}) \end{array}$$

It follows from (8.24) that

$$H_1(\mathcal{K}_\bullet) = 0, \qquad \text{and} \qquad H_0(\mathcal{K}_\bullet) = R.$$

So, we have the complex $\mathcal{K}_\bullet := \mathcal{K}_\bullet(\iota, \mathfrak{t})$, as follows:

$$\cdots \longrightarrow \mathcal{K}_n \longrightarrow \cdots \longrightarrow \mathcal{K}_3 \longrightarrow \mathcal{K}_2 \longrightarrow P^* \otimes S(\mathfrak{q})$$

$$\longrightarrow S(\mathfrak{q}) \longrightarrow R \longrightarrow 0. \tag{8.26}$$

Twisting appropriately, to match degrees, we obtain a complex

$$\cdots \longrightarrow \mathcal{K}_n(-(n-1)) \longrightarrow \cdots \longrightarrow \mathcal{K}_2(-1) \longrightarrow P^* \otimes S(\mathfrak{q})$$

$$\longrightarrow S(\mathfrak{q})(1) \longrightarrow R \longrightarrow 0.$$

With $\mathcal{L}_k = \mathcal{K}_{k+1}(-k)$, and $M = S(\mathfrak{q})_+(1) = P^* S(\mathfrak{q})(1)$, we obtain another complex:

$$\cdots \longrightarrow \mathcal{L}_k \longrightarrow \cdots \longrightarrow \mathcal{L}_2 \longrightarrow \mathcal{L}_1 \longrightarrow \mathcal{L}_0 \longrightarrow M \longrightarrow 0. \tag{8.27}$$

Then both the complexes (8.26, 8.27) are exact. The complex (8.27) is a minimal resolution.

Proof. We only need to prove that (8.26) is exact. For notational convenience, write $C_n = K_n(\iota) = S(\mathfrak{q}) \otimes \Lambda^n P^*$. Consider the exact sequence

$$0 \longrightarrow C_n \xrightarrow{\ i\ } \mathcal{K}_n \xrightarrow{\ j\ } \mathcal{K}_{n-2} \longrightarrow 0$$

where i is the inclusion $x \mapsto xT^{(0)}$, and

$$j\left(x_0 + x_1T^{(1)} + x_2T^{(2)} \cdots\right) = \left(x_1 + x_2T^{(1)} + x_3T^{(2)} \cdots\right).$$

This defines an exact sequence of the complexes, with shifted degrees

$$0 \longrightarrow C_\bullet \xrightarrow{\ i\ } \mathcal{K}_\bullet \xrightarrow{\ j\ } \mathcal{K}_\bullet(-2) \longrightarrow 0$$

Accordingly, we obtain the exact sequence

$$\cdots \longrightarrow H_n(C_\bullet) \xrightarrow{\ i_*\ } H_n(\mathcal{K}_\bullet) \xrightarrow{\ j_*\ } H_n(\mathcal{K}_\bullet(-2)) \longrightarrow \cdots$$

which is, more explicitly,

$$H_{n+1}(C_\bullet) \xrightarrow{\ i_*\ } H_{n+1}(\mathcal{K}_\bullet) \xrightarrow{\ j_*\ } H_{n-1}(\mathcal{K}_\bullet) \xrightarrow{\ d_*\ } H_n(C_\bullet)$$

$$\xrightarrow{\ i_*\ } H_n(\mathcal{K}_\bullet) \xrightarrow{\ j_*\ } H_{n-2}(\mathcal{K}_\bullet). \tag{8.28}$$

From (8.21), it follows that

$$\begin{cases} \forall n \geq 1, \ H_{2n+1}(\mathcal{K}_\bullet) \cong H_1(\mathcal{K}_\bullet) = 0 \\ \forall n \geq 1, \ H_{2n}(\mathcal{K}_\bullet) \cong H_2(\mathcal{K}_\bullet) \end{cases} \quad \text{given by} \quad j_*$$

It remains to be shown that $H_2(\mathcal{K}_\bullet) = 0$. From (8.28), we obtain the exact sequence

$$0 \xrightarrow{i_*} H_2(\mathcal{K}_\bullet) \xrightarrow{j_*} H_0(\mathcal{K}_\bullet) \xrightarrow{d_*} H_1(C_\bullet) \xrightarrow{i_*} 0.$$

We also have $H_1(C_\bullet) = R\mathfrak{t}$, and $H_0(\mathcal{K}_\bullet) \cong R$. By definition of the map d_*, it follows that $d_*(1) = \mathfrak{t}$. So, d_* is an isomorphism. Consequently, $H_2(\mathcal{K}_\bullet) = 0$, as required. The proof is complete. ■

The following is the final result in this section.

Corollary 8.4.7. *With notation as above, the sheafification of* (8.27)

$$\cdots \longrightarrow \widetilde{\mathcal{L}_k} \longrightarrow \cdots \longrightarrow \widetilde{\mathcal{L}_2} \longrightarrow \widetilde{\mathcal{L}_1} \longrightarrow \widetilde{\mathcal{L}_0} \longrightarrow M \longrightarrow 0 \tag{8.29}$$

is the minimal resolution of $\mathcal{O}_X(1)$, *where* $X = \mathrm{Proj}\,(S(\mathfrak{q}))$, *as in* (8.2.8, 8.2.9, 8.2.14) *(assume* $(P, \mathfrak{q})$ *is nondegenerate).*

8.5. The Clifford Sequence

We set up some notation.

Notation 8.5.1. Let R be a commutative Noetherian ring and $(P, \mathfrak{q})$ be a nondegenerate quadratic R-module, where P is a projective R-module, with $rank(P) = d + 2$. Let

$$\begin{cases} B := \mathcal{S}\mathrm{ym}\,(P^*) = \bigoplus_{n\geq 0} \mathcal{S}\mathrm{ym}\,(P^*)_n \\ \quad \text{be the symmetric algebra, and} \\ A := S(\mathfrak{q}) = \frac{B}{(\mathfrak{q})} = \bigoplus_{n\geq 0} S_n(\mathfrak{q}), \quad \text{and write } A_n := S_n(\mathfrak{q}) = \frac{B_n}{\mathfrak{q}B_{n-2}}. \end{cases}$$

Denote $\mathbb{P} := \mathbb{P}(P^*) := \mathrm{Proj}\,(B) = \mathrm{Proj}\,(\mathcal{S}\mathrm{ym}\,(P^*))$ and $X = \mathrm{Proj}\,(S(\mathfrak{q}))$. So, $\dim X = d$. We have $B_0 = A_0 = R$ and $B_1 =$

$A_1 = P^*$. For $n \in \mathbb{Z}$, the maps $B_1 \otimes B(n) \longrightarrow B(n+1)$, and $A_1 \otimes A(n) \longrightarrow A(n+1)$ induce two sets of equivalent maps:

$$\begin{cases} P^* \otimes \mathcal{O}_{\mathbb{P}}(n) \longrightarrow \mathcal{O}_{\mathbb{P}}(n+1) \\ \mathcal{O}_{\mathbb{P}}(n) \longrightarrow Hom(P^*, \mathcal{O}_{\mathbb{P}}(n+1)) \\ \mathcal{O}_{\mathbb{P}}(n) \longrightarrow P \otimes \mathcal{O}_{\mathbb{P}}(n+1) \end{cases} \qquad \begin{cases} P^* \otimes \mathcal{O}_X(n) \longrightarrow \mathcal{O}_X(n+1) \\ \mathcal{O}_X(n) \longrightarrow Hom(P^*, \mathcal{O}_X(n+1)) \\ \mathcal{O}_X(n) \longrightarrow P \otimes \mathcal{O}_X(n+1) \end{cases}$$

The third maps are instrumental in our subsequent constructions, which we represent by the following commutative diagrams:

$$\begin{array}{ccc} \mathcal{O}_{\mathbb{P}}(n) & \longrightarrow & Hom(P^*, \mathcal{O}_{\mathbb{P}}(n+1)) \\ & {\scriptstyle \varpi_n^{\mathbb{P}}} \searrow & \uparrow \wr \\ & & P \otimes \mathcal{O}_{\mathbb{P}}(n+1) \end{array} \qquad \begin{array}{ccc} \mathcal{O}_X(n) & \longrightarrow & Hom(P^*, \mathcal{O}_X(n+1)) \\ & {\scriptstyle \varpi_n} \searrow & \uparrow \wr \\ & & P \otimes \mathcal{O}_X(n+1) \end{array} \tag{8.30}$$

The vertical maps send $p \otimes x \mapsto (f \mapsto f(p)x)$. If $P = \oplus Re_i$ is free, the maps in (8.30) are described, respectively, in these two cases, as follows:

$$\begin{cases} \varpi_n^{\mathbb{P}}(f_1 \dots f_k) = \sum e_i \otimes (e_i^* f_1 \dots f_k) \\ \varpi_n(f_1 \dots f_k) = \sum e_i \otimes \overline{(e_i^* f_1 \dots f_k)} \end{cases} \qquad \text{with} \quad f_i \in P^* \tag{8.31}$$

Use the notation from Section 8.3.1 on Clifford algebras $C(\mathfrak{q})$. Consider the canonical map $\chi : P \hookrightarrow C_0(\mathfrak{q}) \hookrightarrow C(\mathfrak{q})$. For $M = M_0 \oplus M_1 \in Coh_{r,\mathbb{Z}_2}(C(\mathfrak{q}))$, there are maps

$$\begin{cases} M_0 \otimes_R P \longrightarrow M_1 \; m \otimes e \mapsto m \cdot e \\ M_1 \otimes_R P \longrightarrow M_0 \; m \otimes e \mapsto m \cdot e \end{cases} \qquad \text{which we denote by } \cdot .$$

Combining with the above (8.30), we define maps

$$\begin{cases} \Phi_n^{\mathbb{P}} := (\cdot \otimes 1)(1_{M_0} \otimes \varpi_n^{\mathbb{P}}) : M_0 \otimes \mathcal{O}_{\mathbb{P}}(n) \longrightarrow M_1 \otimes \mathcal{O}_{\mathbb{P}}(n+1) \\ \Phi_n^{\mathbb{P}} := (\cdot \otimes 1)(1_{M_1} \otimes \varpi_n^{\mathbb{P}}) : M_1 \otimes \mathcal{O}_{\mathbb{P}}(n) \longrightarrow M_0 \otimes \mathcal{O}_{\mathbb{P}}(n+1) \end{cases} \quad \text{and,} \tag{8.32}$$

$$\begin{cases} \varphi_n := (\cdot \otimes 1)(1_{M_0} \otimes \varpi_n) : M_0 \otimes \mathcal{O}_X(n) \longrightarrow M_1 \otimes \mathcal{O}_X(n+1) \\ \varphi_n := (\cdot \otimes 1)(1_{M_1} \otimes \varpi_n) : M_1 \otimes \mathcal{O}_X(n) \longrightarrow M_0 \otimes \mathcal{O}_X(n+1) \end{cases} \tag{8.33}$$

We display the first map in (8.33), as follows:

$$\begin{array}{ccc} M_0 \otimes \mathcal{O}_X(n) & \xrightarrow{1_{M_0} \otimes \varpi} & M_0 \otimes P \otimes \mathcal{O}_X(n+1) \\ & {\scriptstyle \varphi_n} \searrow & \downarrow {\scriptstyle \cdot \otimes 1} \\ & & M_1 \otimes \mathcal{O}_X(n+1) \end{array}$$

We sometimes ignore the subscripts or superscripts. Module theoretically, if $P = \oplus Re_i$ is free, these maps are defined by

$$\varphi(m \otimes \overline{f_1 \cdots f_k}) = \sum (m \cdot e_i) \otimes \overline{e_i^* \otimes f_1 \cdots f_k}. \tag{8.34}$$

We define the Clifford sequences, as follows.

Definition 8.5.2. Let $M = M_0 \oplus M_1 \in Coh_{r,\mathbb{Z}_2}(C(\mathfrak{q}))$. Use the notation, as above (8.30, 8.32, 8.33). Define two **Clifford sequences**, as follows. First, define $\mathbf{C}l^{\mathbb{P}}_{\bullet}(M)$, by

$$\cdots \longrightarrow M_{n+d+1} \otimes \mathcal{O}_{\mathbb{P}}(-n) \xrightarrow{\Phi} M_{n+d} \otimes \mathcal{O}_{\mathbb{P}}(-(n-1))$$
$$\xrightarrow{\Phi} M_{n+d-1} \otimes \mathcal{O}_{\mathbb{P}}(-(n-2)) \longrightarrow \cdots \tag{8.35}$$

where the subscript in M_i is taken mod 2. Subsequently, we will see (8.5.3) that this sequence is not necessarily a complex. We are primarily interested in the following **Clifford sequence**, for X, to be denoted by $\mathbf{C}l_{\bullet}(M)$ (*without any superscript*), as follows:

$$\cdots \longrightarrow M_{n+d+1} \otimes \mathcal{O}_X(-n) \xrightarrow{\varphi} M_{n+d} \otimes \mathcal{O}_X(-(n-1))$$
$$\xrightarrow{\varphi} M_{n+d-1} \otimes \mathcal{O}_X(-(n-2)) \longrightarrow \cdots \tag{8.36}$$

and denote $\mathbf{C}l_n(M) := M_{n+d+1} \otimes \mathcal{O}_X(-n)$. (*Note that the choice of subscript $n+d+1$ is a minor question of management of parity mod* 2.) We also use the notation $Cl_n(M) = M_{n+d+1} \otimes S(\mathfrak{q})(-n)$ so that $\widetilde{Cl_n(M)} = \mathbf{C}l_n(M)$.

Define $\mathfrak{U}_n(M) := co\ker(\varphi : \mathbf{C}l_{n+2}(M) \longrightarrow \mathbf{C}l_{n+1}(M))$. So, we have the exact sequence

$$\mathbf{C}l_{n+2}(M) \xrightarrow{\varphi} \mathbf{C}l_{n+1}(M) \twoheadrightarrow \mathfrak{U}_n(M) \longrightarrow 0.$$

For convenience, we have the following notation:

$$\begin{cases} \mathbf{C}l_{\bullet} := \mathbf{C}l_{\bullet}(C(\mathfrak{q})) \\ \mathfrak{U}_n := \mathfrak{U}_n(C(\mathfrak{q})) \\ \text{Also, recall (8.2.12)} \quad \mathfrak{U} = Z_{d-1}\mathcal{O}_X(1) \end{cases} \tag{8.37}$$

Exercise 8.5.3. Use the notation from (8.5.2). Prove

1. In the sequence (8.35), the composition $\Phi^2 = \mathfrak{q}$ is multiplication by $\mathfrak{q} \in B$.
2. The Clifford sequence (8.36) is a complex.

Hint: We can assume $(R, \mathfrak{m})$ is local. In fact, by a proper choice of basis (8.3.9), we can assume that $\mathfrak{q} = \sum_{i=0}^{d+1} \lambda_i X_i^2$, with $\lambda_i \notin \mathfrak{m}$ units.

Lemma 8.5.4. *Use the notation as above* (8.5.2). *Then*

$$M_0 \otimes_{C_0(\mathfrak{q})} \mathbf{Cl}_n(C(\mathfrak{q})) \xrightarrow{\sim} \mathbf{Cl}_n(M) \qquad \textit{is an isomorphism.}$$

In particular,

$$\mathfrak{U}_n(M) \cong M_0 \otimes_{C_0(\mathfrak{q})} \mathfrak{U}_n(C(\mathfrak{q})).$$

Proof. The first part follows from (8.3.8) or from its proof that $M_0 \otimes_{C_0(\mathfrak{q})} C(\mathfrak{q}) \cong M$. The latter follows from the commutative diagram

$$\begin{array}{ccccc} M_0 \otimes_{C_0(\mathfrak{q})} \mathbf{Cl}_{n+2}(C(\mathfrak{q})) & \xrightarrow{\varphi} & M_0 \otimes_{C_0(\mathfrak{q})} \mathbf{Cl}_{n+1}(C(\mathfrak{q})) & \twoheadrightarrow & M_0 \otimes_{C_0(\mathfrak{q})} \mathfrak{U}_n(C(\mathfrak{q})) \\ \downarrow \wr & & \downarrow \wr & & \downarrow \wr \\ \mathbf{Cl}_{n+2}(M) & \xrightarrow{\varphi} & \mathbf{Cl}_{n+1}(M) & \twoheadrightarrow & \mathfrak{U}_n(M). \end{array}$$

The proof is complete. ■

8.5.1. *$C(\mathfrak{q})$-action on the Exterior Algebra*

We define a $C(\mathfrak{q})$-action on the Exterior Algebra in this section.

Definition 8.5.5. Continue to use the notation above (8.5.1). Let $\Lambda = \Lambda^\bullet P^*$ be the exterior algebra. Write $\Lambda = \Lambda_0 \oplus \Lambda_1$, where $\Lambda_0 = \bigoplus \Lambda^{2n} P^*$ is the even part, and $\Lambda_1 = \bigoplus \Lambda^{2n+1} P^*$ is the odd part. We proceed to define a $C(\mathfrak{q})$-action on Λ.

For $x \in P \cong P^{**}$, as in (8.4.5), there is a unique derivation $\partial_x : \Lambda \longrightarrow \Lambda$ determined by $\partial_x(f) = f(x)$, for all $f \in P^*$. It is necessarily given by the formula [T]

$$\partial_x(f_0 \wedge f_1 \wedge \cdots \wedge f_n) = \sum_{r=0}^{n} (-1)^r f_r(x) f_0 \wedge f_1 \wedge \cdots \wedge \hat{f_r} \wedge \cdots \wedge f_n. \quad (8.38)$$

As before, let $\mathfrak{t} = \tilde{\mathfrak{q}} \in P^* \otimes P^*$ be a lift of $\mathfrak{q} \in \mathcal{S}\mathrm{ym}\,(P^*)_2$. Let $\mathfrak{t}_x := \partial_x \otimes 1_{P^*}(\mathfrak{t}) \in P^* = \Lambda^1 P^*$. Define

$$L_x : \Lambda \longrightarrow \Lambda \quad L_x(\omega) = \mathfrak{t}_x \wedge \omega. \quad \text{Clearly,} \quad L_x^2(\omega) = \mathfrak{t}_x \wedge \mathfrak{t}_x \wedge \omega = 0. \tag{8.39}$$

Further, we have $\partial_x(\mathfrak{t}_x) = \partial_x(\partial_x \otimes 1(\mathfrak{t})) = \mathfrak{t}(x,x) = \mathfrak{q}(x)$. So,

$$(L_x\partial_x + \partial_x L_x)(\omega) = \mathfrak{t}_x \wedge \partial_x(\omega) + \partial_x(\mathfrak{t}_x \wedge \omega) = \partial_x(\mathfrak{t}_x) \wedge \omega = \mathfrak{q}(x)\omega.$$

So, it follows

$$\text{with} \quad \rho_x = L_x + \partial_x, \quad \text{we have} \quad (\rho_x)^2 = \mathfrak{q}(x).$$

So, by universal property, there are two extensions ρ, as follows:

$$\begin{array}{ccc} P & \longrightarrow & C(\mathfrak{q}) \\ & \searrow^{\rho} & \downarrow{\scriptstyle \rho} \\ & & End(\Lambda) \end{array}, \qquad \text{and} \qquad \begin{array}{ccc} P & \longrightarrow & C(\mathfrak{q}) \\ & \searrow^{\rho} & \downarrow{\scriptstyle \rho} \\ & & End(\Lambda)^{op} \end{array}$$

The first diagram gives a left action of $C(\mathfrak{q})$ on Λ and the second diagram gives a right action of $C(\mathfrak{q})$ on Λ.

Lemma 8.5.6. *Continue to use the notation above* (8.5.1), *with* $rank(P) = d + 2 = r$. *Then there are isomorphisms of* $\mathbb{Z}_2$*-graded* $C(\mathfrak{q})$*-modules:*

$$\begin{cases} C(\mathfrak{q}) \otimes_R \det(P^*) \xrightarrow{\sim} \Lambda(r) = \Lambda(d) \\ \quad \textit{induced by left action of } C(\mathfrak{q}) \textit{ on } \det(P^*) \subseteq \Lambda \\ \det(P^*) \otimes_R C(\mathfrak{q}) \xrightarrow{\sim} \Lambda(r) = \Lambda(d) \\ \quad \textit{induced by right action of } C(\mathfrak{q}) \textit{ on } \det(P^*) \subseteq \Lambda \end{cases}$$

where $\Lambda(k)_n := \Lambda_{k+n}$, *while subscript* $k + n$ *is taken mod* 2.

Proof. We only establish the first isomorphism. As usual, we check this locally. So, assume $(R, \mathfrak{m})$ is local and P is free with basis $e_1, \ldots, e_r$, and dual basis $X_1, \ldots, X_r$. By a proper choice of basis, we can assume $\mathfrak{q} = \lambda_1 X_1^2 + \cdots + \lambda_r X_r^2$, with $\lambda_i \notin \mathfrak{m}$ units. With this choice, the preimage of $\mathfrak{q}$ in $P^* \otimes P^*$ is $\mathfrak{t} := \sum \lambda_i X_i \otimes X_i$. Note that

$$\mathfrak{t}_{e_i} = \sum_{j=1}^{r} (\partial_{e_i} \otimes 1)(\lambda_j X_j \otimes X_j) = \lambda_i X_i, \quad \forall i.$$

Recall that we use $\wedge$-notation in $C(\mathfrak{q})$. So, $C(\mathfrak{q}) \otimes \det P^*$ has a basis $(e_{i_1} \wedge \cdots \wedge e_{i_s})\omega$, with $\{1 \le i_1 < i_2 < \cdots < i_s \le r\}$, while Λ has a basis $X_{j_1} \wedge X_{j_2} \wedge \cdots \wedge X_{j_t}$ with $\{1 \leq j_1 < j_2 < \cdots < j_t \leq r\}$. Let $\omega_r := X_1 \wedge X_2 \wedge \cdots \wedge X_r \in \det P^*$ denote the generator of $\det P^*$. We prove that

$$(e_{i_1} \wedge \cdots \wedge e_{i_s}) \cdot \omega_r = \pm X_{j_1} \wedge X_{j_2} \wedge \cdots \wedge X_{j_t}$$

where $\{i_1 < i_2 < \cdots < i_s\}$ and $\{j_1 < i_2 < \cdots < j_t\}$ are disjoint partitions of $\{1, 2, \ldots, n\}$. This will establish the lemma. The statement is obviously valid for $s = 0$. We prove by induction on s. Assume that the statement is valid for s and we prove it for $s + 1$. Without loss of generality, let $i_1 = 1, \ldots i_s = s, i_{s+1} = s + 1, j_1 = s + 2, \ldots, j_t = n$. By induction, we have

$$(e_1 \wedge \cdots \wedge e_s) \cdot \omega_r = \pm X_{s+1} \wedge X_{s+2} \wedge \cdots \wedge X_r$$

So,

$$\begin{aligned}
&(e_1 \wedge \cdots \wedge e_s \wedge e_{s+1}) \cdot \omega_r \\
&= \pm\, e_{s+1} \cdot (X_{s+1} \wedge X_{s+2} \wedge \cdots \wedge X_r) \\
&= \pm\, (L_{e_{s+1}} + \partial_{e_{s+1}}) (X_{s+1} \wedge X_{s+2} \wedge \cdots \wedge X_r) \\
&= \pm\, \left(\mathfrak{t}_{e_{s+1}} \wedge (X_{s+1} \wedge X_{s+2} \wedge \cdots \wedge X_r) + (X_{s+2} \wedge \cdots \wedge X_r)\right) \\
&= \pm\, (\lambda_{j+1} X_{s+1} \wedge (X_{s+1} \wedge X_{s+2} \wedge \cdots \wedge X_r) + (X_{s+2} \wedge \cdots \wedge X_r)) \\
&= \pm\, (X_{s+2} \wedge \cdots \wedge X_r)\,.
\end{aligned}$$

The proof is complete. ■

Exercise 8.5.7. With notation as above, define the right action of $C(\mathfrak{q})$ on Λ more directly by repeating the above proof. **Hint:** Consult [T] for basic facts about differentials of graded algebras. Let $\tau_x = (1_{P^*} \otimes \partial_x)(\mathfrak{t})$. Define $L_x(\omega) = \pm \omega \wedge \tau_x$.

Exercise 8.5.8. Establish the second isomorphism in (8.5.6). Also, prove that the diagram

$$\begin{array}{ccc}
C(\mathfrak{q}) \otimes \det P^* & \longrightarrow & \Lambda(d) \\
\downarrow & \nearrow & \\
\det P^* \otimes C(\mathfrak{q}) & &
\end{array}$$

commutes.

8.5.2. *Comparison of resolutions*

Now, we compare the canonical resolutions and the Clifford sequence. For integers k and sequences $C_\bullet := \mathbf{Cl}_\bullet(\Lambda)$ or $C_\bullet := \mathcal{L}_\bullet$ and others, we denote $C_\bullet^{\geq k}$, the sequence (a complex or not)

$$\cdots \longrightarrow C_{k+1} \longrightarrow C_k \longrightarrow 0.$$

Lemma 8.5.9. *Continue to use the notation above* (8.5.1), *with* $rank(P) = d+2$. *Then the sequences*

$$Can_\bullet^{\geq d}(\mathcal{O}_X(1)) \cong \widetilde{\mathcal{L}}_\bullet^{\geq d} \cong \mathbf{Cl}_\bullet^{\geq d}(\Lambda(d))$$

are naturally isomorphic. Here, $\mathcal{L}_\bullet$ *is as defined in* (8.27), $Can_\bullet(\mathcal{O}_X(1))$ *is as in* (8.3.6).

Proof. We demonstrated (8.4.7) that $Can_\bullet(\mathcal{O}_X(1)) \cong \widetilde{\mathcal{L}}_\bullet$. We prove $\widetilde{\mathcal{L}}_\bullet^{\geq d} \cong \mathbf{Cl}_\bullet^{\geq d}(\Lambda(d))$. Recall that $\mathbf{Cl}_\bullet^{\geq d}(\Lambda(d))$ is defined by using the right action on $\Lambda(d)$. We have

$$\mathbf{Cl}_n(\Lambda(d)) = \Lambda(d)_{n+d+1} \otimes \mathcal{O}(-n) = \Lambda_{n+1} \otimes \mathcal{O}(-n) = \widetilde{Cl_n(\Lambda(d))}.$$

It follows from description (8.25) that

$$\mathcal{K}_n = \begin{cases} \left(\cdots \oplus \Lambda^2 P^* \otimes T^{\left(\frac{n}{2}-1\right)} \oplus \Lambda^0 P^* \otimes T^{\left(\frac{n}{2}\right)}\right) \otimes S(\mathfrak{q}) & n \text{ even} \\ \left(\cdots \oplus \Lambda^3 P^* \otimes T^{\left(\frac{n+1}{2}-1\right)} \oplus \Lambda^1 P^* \otimes T^{\left(\frac{n+1}{2}\right)}\right) \otimes S(\mathfrak{q}) & n \text{ odd}. \end{cases}$$

By definition, $\mathcal{L}_n = \mathcal{K}_{n+1}(-n)$. It follows from this, $\forall n \geq d$, there is an identification

$$\Lambda_{n+1} \xrightarrow{\sim} \mathcal{L}_n \quad \text{induced by} \quad \Lambda^p P^* \xrightarrow{\sim} \Lambda^p P^* \otimes T^{\left(\frac{n+1-p}{2}\right)}$$
$$\text{with} \quad n+1-p \in 2\mathbb{Z}.$$

So, $\forall n \geq d$, as modules, $Cl_n(\Lambda(d)) \cong \mathcal{L}_n$. It remains to be checked that the differentials on $\mathcal{L}_\bullet$ coincide with $\varphi : Cl_n(\Lambda) \longrightarrow Cl_{n-1}(\Lambda)$. By definition, for the differential

$$\partial_n : \mathcal{K}_n(\iota, \mathfrak{t}) \longrightarrow \mathcal{K}_{n-1}(\iota, \mathfrak{t}),$$
$$\partial_n\left(\omega T^{(q)}\right) = (d\omega)T^{(q)} + (-1)^{k+1}\omega \mathfrak{t} T^{(q-1)}, \quad \forall \omega \in \Lambda^{k+1} P^*.$$

Writing $\mathfrak{t} = \sum \zeta_i \otimes \eta_i$, we have

$$\begin{aligned} &\partial_n \left(\omega T^{(q)}\right) = (d\omega)T^{(q)} + \mathfrak{t}\omega T^{(q-1)} = \\ &d\omega T^{(q)} + (-1)^{k+1} \sum (\omega \wedge \zeta_i) \otimes \eta_i T^{(q-1)}. \end{aligned} \tag{8.40}$$

Now, we describe the maps

$$\varphi : Cl_n(\Lambda(d)) \longrightarrow Cl_{n-1}(\Lambda(d)).$$

We adapt (8.33) to define φ, as follows:

$$\begin{array}{ccc} \Lambda_{n+1} \otimes \mathcal{O}_X(-n) & \xrightarrow{1_\Lambda \otimes \theta(1)} & \Lambda_{n+1} \otimes P \otimes \mathcal{O}_X(-n+1) \\ & \searrow^{\varphi} & \downarrow {\scriptstyle (\mathfrak{d}+L)\otimes 1} \\ & & \Lambda_n \otimes \mathcal{O}_X(-n+1) \end{array} \tag{8.41}$$

where, following (8.38, 8.39, Exercise 8.5.7), the maps $\mathfrak{d}$ and L above are defined by

$$\begin{aligned} &\mathfrak{d}(\omega, x) = \partial_x(\omega), \\ &L(\omega, x) = L_x(\omega) = (-1)^{n+1}\omega \wedge \tau_x = (-1)^{n+1}\omega \wedge (1 \otimes \partial_x)(\mathfrak{t}). \end{aligned}$$

So, we can write

$$\varphi = (\mathfrak{d} \otimes 1)(1 \otimes \theta(1)) + (L \otimes 1)(1 \otimes \theta(1)) =: \varphi_1 + \varphi_2.$$

By comparison with (8.40), we show that $\varphi_1(\omega) = d(\omega)$ and φ_2 agrees with the latter part of (8.40). The map $\theta(1) : \mathcal{O}_X(n) \longrightarrow P \otimes \mathcal{O}_X$ $(n+1)$ is characterized by

$$\theta(1)(z) = \sum p_i \otimes \alpha_i \quad \text{where} \quad \sum u(p_i)\alpha_i = uz \quad \forall u \in P^*,$$

while α_i, p_i depend on z. With $\omega = u_{i_0} \wedge u_{i_1} \wedge \cdots \wedge u_{i_k} \in \Lambda^{k+1}P^*$, and $z \in \mathcal{O}_X(n)$ (locally), we have

$$\varphi_1(\omega \otimes z) = (\mathfrak{d} \otimes 1)(1 \otimes \theta(1))(\omega \otimes z) = (\mathfrak{d} \otimes 1)\left(\omega \otimes \sum_i p_i \otimes \alpha_i\right)$$

$$= \sum_i \left(\sum_j (-1)^j u_j(p_i) u_{i_0} \wedge u_{i_1} \wedge \cdots \hat{u}_j \wedge \cdots \wedge u_{i_k}\right) \otimes \alpha_i$$

$$= \sum_j (-1)^j \left(u_{i_0} \wedge u_{i_1} \wedge \cdots \hat{u}_j \wedge \cdots \wedge u_{i_k} \otimes \left(\sum_i u_j(p_i)\alpha_i\right)\right)$$

$$= \sum_j (-1)^j \left(u_{i_0} \wedge u_{i_1} \wedge \cdots \hat{u}_j \wedge \cdots \wedge u_{i_k} \otimes (u_j z)\right)$$

$$= d(\omega \otimes z).$$

For the last equality, see (8.22). So, φ_1 agrees with the first part of (8.40). Then

$$\varphi_2(\omega \otimes z) = (L \otimes 1)(1 \otimes \theta(1))(\omega \otimes z) = (L \otimes 1)\left(\omega \otimes \sum_i p_i \otimes \alpha_i\right)$$

$$= (-1)^{n+1} \sum_i \omega \wedge (1 \otimes \partial_{p_i})(\mathfrak{t}) \otimes \alpha_i$$

$$= (-1)^{n+1} \omega \wedge \sum_i \sum_j \zeta_j \eta_j(p_i) \otimes \alpha_i$$

$$= (-1)^{n+1} \omega \wedge \sum_j \zeta_j \otimes \eta_j z.$$

This also agrees with the latter part of (8.40). This completes the proof that the differential on $\mathcal{L}_n$ agrees with that on $Cl_n(\Lambda)$ for $n \geq d$. The proof is complete. ■

Corollary 8.5.10. *Continue to use the notation above* (8.5.1), *with* $rank(P) = d + 2$. *Then* $Can_\bullet^{\geq d}(\mathcal{O}_X(1)) \cong \det P^* \otimes \mathbf{Cl}_\bullet^{\geq d}(C(\mathfrak{q}))$.

Proof. We have from (8.5.9), in conjunction with (8.5.6),

$$Can_\bullet^{\geq d}(\mathcal{O}_X(1)) \cong \mathbf{Cl}_\bullet^{\geq d}(\Lambda(d)) \cong \mathbf{Cl}_\bullet^{\geq d}(\det(P) \otimes C(\mathfrak{q}))$$
$$\cong \det(P) \otimes \mathbf{Cl}_\bullet(C(\mathfrak{q}))^{\geq d}.$$

The last isomorphism can be seen directly, or use (8.5.4). ■

Corollary 8.5.11. *Continue to use the notation above* (8.5.1). *Let* $r := rank(P) = d + 2$. *Recall other notation from* (8.37). *Then*

$$\mathfrak{U} := Z_{d-1}(\mathcal{O}_X(1)) \cong \det(P) \otimes \mathfrak{U}_{d-1}(C(\mathfrak{q})).$$

Proof. Note that $Can_\bullet$ is exact. So, we have the diagram of exact sequences:

$$\begin{array}{ccccccc}
Can_{d+1}(\mathcal{O}_X(1)) & \longrightarrow & Can_d(\mathcal{O}_X(1)) & \longrightarrow & Z_{d-1}(\mathcal{O}_X(1)) & \longrightarrow & 0 \\
\downarrow \wr & & \downarrow \wr & & \downarrow \wr & & \\
\det(P) \otimes \mathbf{Cl}_{d+1}(C(\mathfrak{q})) & \longrightarrow & \det(P) \otimes \mathbf{Cl}_d(C(\mathfrak{q})) & \longrightarrow & \det(P) \otimes \mathfrak{U}_{d-1}(C(\mathfrak{q})) & \longrightarrow & 0.
\end{array}$$

By (8.5.10), the first two vertical arrows are isomorphisms. So, the third induced vertical arrow is also an isomorphism. ■

The following are some comments on the Clifford sequence (8.36).

Lemma 8.5.12. *Use the notation as above, and* $M = M_0 \oplus M_1 \in Coh_{\mathbb{Z}_2}(C(\mathfrak{q}))$. *Then*

1. *The Clifford sequence* (8.36) $\mathbf{Cl}_\bullet(M)$ *is locally split exact.*
2. *Suppose* $M \in \mathscr{P}(R)$. *Then* $\mathfrak{U}_n(M) \in \mathscr{P}(X)$.

Proof. First, we prove (1). By (8.5.4), it is enough to prove it for $M = C(\mathfrak{q})$. In fact, it would be enough to prove it for $\mathbf{Cl}_\bullet^{\geq d}(C(\mathfrak{q}))$. By (8.5.10), $\det P^* \otimes \mathbf{Cl}_\bullet^{\geq d}(C(\mathfrak{q})) \cong Can_\bullet^{\geq d}(\mathcal{O}_X(1))$, which is a resolution of $Z_{d-1}(\mathcal{O}(1)) \in \mathscr{P}(X)$. This establishes (1).

By (8.5.4), $\mathfrak{U}_n(M) \cong \mathfrak{U}_n(C(\mathfrak{q})) \otimes_{C_0(\mathfrak{q})} M_0$. So, to prove (2), it is enough to prove $\mathfrak{U}_n(C(\mathfrak{q})) \in \mathscr{P}(X)$ and, equivalently, to prove $\mathfrak{U}_n(C(\mathfrak{q}))(-2d) \in \mathscr{P}(X)$. By twisting the sequence (8.36), by $\mathcal{O}_X(-2d)$ it follows that $\mathfrak{U}_n(C(\mathfrak{q}))(-2d) \cong \mathfrak{U}_{n+2d}(C(\mathfrak{q})) \cong Z_{n+2d}(\mathcal{O}(1)) \in \mathscr{P}(X)$. This establishes (2). The proof is complete. ■

Lemma 8.5.13. *Continue to use the notation above (8.5.1), with rank$(P) = d+2$. Suppose $M, N \in Coh_{\mathbb{Z}_2}(C(\mathfrak{q}))$ are finitely generated projective (right) $C(\mathfrak{q})$-modules. Then the natural map*

$$\iota : Hom_{C(\mathfrak{q})}(M,N)_0 \xrightarrow{\sim} Hom_X(\mathfrak{U}_n(M), \mathfrak{U}_n(N)) \quad \text{is an isomorphism } \forall n,$$

where $Hom_{C(\mathfrak{q})}(M,N)_0$ denotes the set of degree-zero $C(\mathfrak{q})$-homomorphisms. In particular,

$$C_0(\mathfrak{q}) \xrightarrow{\sim} End_X(\mathfrak{U}_n(C(\mathfrak{q}))) \text{ is an isomorphism } \forall n.$$

Proof. For $g \in Hom_{C(\mathfrak{q})}(M,N)_0$, the map $\iota(g)$ is defined by the diagram

$$\begin{array}{ccccc}
M_{n+2+d+1} \otimes \mathcal{O}_X(-(n+2)) & \longrightarrow & M_{n+1+d+1} \otimes \mathcal{O}_X(-(n+1)) & \twoheadrightarrow & \mathfrak{U}_n(M) \\
\downarrow g\otimes 1 & & \downarrow g\otimes 1 & & \downarrow \iota(g) \\
N_{n+2+d+1} \otimes \mathcal{O}_X(-(n+2)) & \longrightarrow & N_{n+1+d+1} \otimes \mathcal{O}_X(-(n+1)) & \twoheadrightarrow & \mathfrak{U}_n(N).
\end{array}$$

Twisting by $\mathcal{O}_X(n)$, and replacing M, N by $M(1), N(1)$ if necessary, it suffices to prove it for $n = 0$. So, the above diagram reduces to

$$\begin{array}{ccccccc}
M_0 \otimes \mathcal{O}_X(-2) & \longrightarrow & M_1 \otimes \mathcal{O}_X(-1) & \twoheadrightarrow & \mathfrak{U}_0(M) & \hookrightarrow & M_0 \otimes \mathcal{O}_X(0) \\
\downarrow g_0\otimes 1 & & \downarrow g_1\otimes 1 & & \downarrow \iota(g) & & \downarrow g_0\otimes 1 \\
N_0 \otimes \mathcal{O}_X(-2) & \longrightarrow & N_1 \otimes \mathcal{O}_X(-1) & \twoheadrightarrow & \mathfrak{U}_0(N) & \hookrightarrow & N_0 \otimes \mathcal{O}_X(0).
\end{array} \tag{8.42}$$

By (8.5.12(1)), Clifford sequences are (locally split) exact. So, we have exact sequences

$$\begin{array}{ccccccccc}
M_0 \otimes \mathcal{O}_X(-2) & \xrightarrow{\varphi^M} & M_1 \otimes \mathcal{O}_X(-1) & \xrightarrow{j} & \mathfrak{U}_0(M) & \hookrightarrow & M_0 \otimes \mathcal{O}_X & & \\
 & & & & \downarrow f & & & & \\
 & & N_1 \otimes \mathcal{O}_X(-1) & \twoheadrightarrow & \mathfrak{U}_0(N) & \xrightarrow{i} & N_0 \otimes \mathcal{O}_X & \xrightarrow{\varphi^N} & N_1 \otimes \mathcal{O}_X(1).
\end{array}$$

It follows from this diagram that the association $f \mapsto ifj$ establishes an isomorphism:

$$\begin{aligned} &\mathcal{H}om_X\left(\mathfrak{U}_0(M), \mathfrak{U}_0(N)\right) \cong \\ &\left\{f \in \mathcal{H}om_X\left(M_1 \otimes \mathcal{O}_X(-1),\ N_0 \otimes \mathcal{O}_X\right) : f\varphi^M = \varphi^N f = 0\right\}. \end{aligned} \tag{8.43}$$

To check that ι is an isomorphism, we can assume $(R, \mathfrak{m})$ is local, and P is free with basis $e_1, \ldots, e_r$, with $r = d+2$. Let $X_i = e_i^* \in P^*$ be the dual basis. So,

$$B = R[X_1, \ldots, X_r], \qquad S(\mathfrak{q}) = \frac{B}{(\mathfrak{q})} =: A, \quad \text{and}$$

$$A_1 = B_1 = P^* = \bigoplus_{i=1}^{r} RX_i.$$

We may denote the image of X_i in x_i, as it makes a difference, only when we take the product. Since $M \in \mathscr{P}(R)$, by [Hr, Ex. II.5.1], (8.1.3), we have

$$\begin{aligned} \mathcal{H}om_X\left(M_1 \otimes \mathcal{O}_X(-1), N_0 \otimes \mathcal{O}_X\right) &\cong \left(\bigoplus_{i=1}^{r} RX_i\right) \otimes M_1^* \otimes N_0 \\ &\cong \left(\bigoplus_{i=1}^{r} RX_i\right) \otimes \mathcal{H}om_R(M_1, N_0). \end{aligned}$$

Let f be an element, as in (8.43). We proceed to show that f is in the image of ι. So,

$$f \in \mathcal{H}om_X\left(M_1 \otimes \mathcal{O}_X(-1), N_0 \otimes \mathcal{O}_X\right). \quad \text{Write}$$

$$f = \sum f_i \otimes x_i \quad \text{with } f_i \in \mathcal{H}om_R(M_1, N_0).$$

For such an $f = \sum f_i \otimes x_i$, write

$$F := \sum f_i \otimes X_i \in \mathcal{H}om_R(M_1, N_0) \otimes B_1 = \mathcal{H}om_R(M_1, N_0) \otimes A_1.$$

With $\mathbb{P} = \operatorname{Proj}(B)$, we compare f and F, as in the diagram

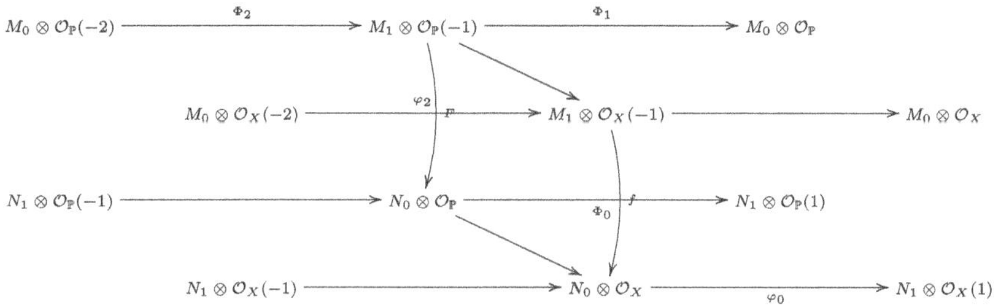

By the comment (8.34), φ_ϵ is given by $m \mapsto \sum m \cdot e_i \otimes x_i$, and Φ_ϵ is similarly defined by $m \mapsto \sum m \cdot e_i \otimes X_i$. It follows from $\varphi_0 f = f\varphi_1 = 0$ that

$$\begin{aligned}\Phi_0 F \in \ker\big(&Hom_{\mathbb{P}}\left(M_1 \otimes \mathcal{O}_{\mathbb{P}}(-1), N_1 \otimes \mathcal{O}_{\mathbb{P}}(1)\right)\\ &\longrightarrow Hom_X\left(M_1 \otimes \mathcal{O}_X(-1), N_1 \otimes \mathcal{O}_X(1)\right)\big).\end{aligned}$$

We have the commutative diagram

$$\begin{array}{ccc} Hom_{\mathbb{P}}\left(M_1 \otimes \mathcal{O}_{\mathbb{P}}(-1), N_1 \otimes \mathcal{O}_{\mathbb{P}}(1)\right) & \longrightarrow & Hom_X\left(M_1 \otimes \mathcal{O}_X(-1), N_1 \otimes \mathcal{O}_X(1)\right) \\ \wr\downarrow & & \downarrow\wr \\ Hom(M_1, N_1) \otimes B_2 & \longrightarrow & Hom(M_1, N_1) \otimes A_2. \end{array}$$

Since $Hom(M_1, N_1)$ is free and $\deg(\mathfrak{q}) = 2$, it follows that $\Phi_0 F = g \otimes \mathfrak{q}$ for some $g \in Hom(M_1, N_1)$. Likewise,

$$\begin{array}{ccc} Hom\left(M_0 \otimes \mathcal{O}_{\mathbb{P}}(-2), N_0 \otimes \mathcal{O}_{\mathbb{P}}\right) & \longrightarrow & Hom\left(M_1 \otimes \mathcal{O}_X(-2), N_0 \otimes \mathcal{O}_X\right) \\ \wr\downarrow & & \downarrow\wr \\ Hom(M_0, N_0) \otimes B_2 & \longrightarrow & Hom(M_0, N_0) \otimes A_2. \end{array}$$

By the same reasoning, $F\Phi_2 = h \otimes \mathfrak{q}$ for some $h \in Hom(M_0, N_0)$. Note that $\Phi^2 = \mathfrak{q}(X)$. We have

$$\begin{cases} \Phi_0(F - \Phi_1(g \otimes 1)) = 0 \\ (F - (h \otimes 1)\Phi_1)\Phi_2 = 0. \end{cases}$$

Composing, one more time with Φ, we have $\mathfrak{q}(X)(F - \Phi_1(1 \otimes g)) = (F - (1 \otimes h)\Phi_1)\mathfrak{q}(X) = 0$. Since $\mathfrak{q}(X)$ is a nonzero divisor on B, it follows that $F = \Phi_1(g \otimes 1) = (h \otimes 1)\Phi_1$. Expanding these, we have

$$\sum_i f_i \otimes X_i = \sum_i e_i g \otimes X_i = \sum_i he_i \otimes X_i. \quad \text{Hence} \quad e_i g = he_i \quad \forall i.$$

It follows from this, the map

$$(g, h) : M_0 \oplus M_1 \longrightarrow N_0, \oplus N_1 \quad \text{is a } C(\mathfrak{q})\text{-linear map.}$$

It also follows from definition (8.42) that $\iota(g, h) = f$. So, ι is surjective.

To prove ι is injective, let $\iota(g) = 0$. Then $(g_0 \otimes 1)\varphi_1 = 0$. Since $B_1 = A_1$, we have $(g_0 \otimes 1)\Phi_1 = 0$. So, $(g_0 \otimes 1)\mathfrak{q}(X) = (g_0 \otimes 1)\Phi^2 = 0 \in B_1 \otimes Hom(M_0, N_0)$. Since $Hom(M_0, N_0)$ is free and $\mathfrak{q}(X)$ is a nonzero divisor, $(g_0 \otimes a) = 0$, and hence $g_0 = 0$. By diagonalizing $\mathfrak{q}$, we see that $g = 0$. The proof is complete. ■

The following is the final result in this section.

Corollary 8.5.14. *Continue to use the notation above* (8.5.1), *with* $rank(P) = d + 2$. *Then*

$$C_0(\mathfrak{q}) \cong End_X\left(Z_{d-1}(\mathcal{O}_X(1))\right).$$

Proof. By (8.5.11), $Z_{d-1}(\mathcal{O}_X(1)) \cong \det(P) \otimes \mathfrak{U}_{d-1}(C(\mathfrak{q}))$. By (8.5.13),

$$\begin{aligned} End_X\left(Z_{d-1}(\mathcal{O}_X(1))\right) &\cong End_X\left(\det(P)\right) \otimes \mathfrak{U}_{d-1}(C(\mathfrak{q}))) \\ &\cong End_X\left(\mathfrak{U}_{d-1}(C(\mathfrak{q}))\right) \cong C_0(\mathfrak{q}) \end{aligned}$$

The proof is complete. ■

8.6. K-Theory of Quadric Hypersurfaces

In analogy to the projective bundle theorem (7.2.3), in this section we prove the projection formula for quadratic hypersurfaces. We continue to use the notation as in (8.5.1), with $rank(P) = d + 2$.

Recall the notation (8.2.12), $\mathfrak{U} := Z_{d-1}(\mathcal{O}_X(1))$, and $\mathfrak{R} := End_X(\mathfrak{U}) \cong C_0(\mathfrak{q})$, by (8.5.14). Define

$$\begin{cases} U_n : \mathscr{P}(R) \longrightarrow \mathscr{P}(X) \quad \text{by} \quad U_n(M) = M \otimes_R \mathcal{O}_X(-n) \\ \quad n = 0, 1, \ldots, d-1 \\ U : \mathscr{P}_r(C_0(\mathfrak{q})) \longrightarrow \mathscr{P}(X) \quad \text{by} \quad U(M) = M \otimes_{C_0(\mathfrak{q})} \mathfrak{U}. \end{cases} \tag{8.44}$$

As was observed in (8.2.12), $\mathfrak{U}$ has left $C_0(\mathfrak{q})$-action. Therefore, the functor U is defined on the category $\mathscr{P}_r(C_0(\mathfrak{q}))$ of projective right $C_0(\mathfrak{q})$-modules. We have that $K \oplus M \cong C_0(\mathfrak{q})^m$ is free as right $C_0(\mathfrak{q})$-modules for some K. So, for $M \in \mathscr{P}_r(C_0(\mathfrak{q}))$, $U(M) \in \mathscr{P}(X)$. It is clear that U_i, U are exact, because short exact sequences in $\mathscr{P}(R)$ and $\mathscr{P}_r(C_0(\mathfrak{q}))$ split. These exact functors induce maps $u_0, u_1, \ldots, u_{d-1}, u$ of K-groups. Define

$$\begin{aligned} & v := (u_0, u_1, \ldots, u_{d-1}; u) : \\ & K_q(R)^d \bigoplus K_q(\mathscr{P}_r(C_0(\mathfrak{q})) \longrightarrow K_q(X) \qquad \forall q \geq 0. \end{aligned} \tag{8.45}$$

Having established all the machineries, we have the projection formula, as follows.

Theorem 8.6.1. *Use the notation as in* (8.5.1), *with* $rank(P) = d+2$. *Then the maps* v, *defined in* (8.45), *are isomorphisms* $\forall q \geq 0$.

Proof. As expected, we follow the proof of (7.2.3), fairly closely. Recall the extension closed subcategories

$$\begin{cases} \mathscr{R}_n(X) = \{F \in \mathscr{P}(X) : F(n) \in \mathfrak{R}(X)\} \\ \mathscr{P}_n(X) = \{F \in \mathscr{P}(X) : H^q(X, (F(k)) = 0, \ \forall k \geq n, q \geq 1\}. \end{cases}$$

We have that, by (8.2.5),

$$K_q\left(\mathscr{R}_n(X)\right) \xrightarrow{\sim} K_q\left(\mathscr{P}_n(X)\right) \xrightarrow{\sim} K_q\left(\mathscr{P}(X)\right)$$
$$\text{are isomorphisms } \forall q \geq 0, n \geq 0.$$

Consider the functors (8.12), as in the truncated canonical resolution (8.13)

$$\begin{cases} T_n : \mathscr{R}_{-1}(X) \longrightarrow \mathscr{P}(R) & n = 0, 1, \ldots, d-1 \\ T : \mathscr{R}_{-1}(X) \longrightarrow \mathscr{P}_r(C_0(\mathfrak{q})) & \text{by} \quad T(F) = Hom_X(\mathfrak{U}, Z_{\mathfrak{d}-1}(F)). \end{cases}$$

Denote the induced maps of the K-groups, as follows:

$$\begin{cases} t_n : K_q(X) = K_q(\mathscr{R}_{-1}(X)) \longrightarrow K_q(\mathscr{P}(R)) & n = 0, 1, \ldots, d-1 \\ t : K_q(X) = K_q(\mathscr{R}_{-1}(X)) \longrightarrow K_q(\mathscr{P}_r(C_0(\mathfrak{q}))) \end{cases}$$

Let

$$\begin{aligned} \mathfrak{t} = (t_0, -t_1, \ldots, (-1)^{d-1} t_{d-1}; (-1)^d t) &: K_q(X) \\ = K_q(\mathscr{R}_{-1}(X)) &\longrightarrow K_n(R)^d \bigoplus K_n(\mathscr{P}_r(C_0(\mathfrak{q})). \end{aligned}$$

We have

$$v\mathfrak{t} = \sum_{n=0}^{d-1} (-1)^n u_n t_n + (-1)^d ut = \sum_{n=0}^{d-1} (-1)^n [U_n T_n]_* + (-1)^d [UT]_*.$$

Note that $U_n T_n(F) = \mathcal{O}_X(-n) \otimes T_n(F)$, $UT(F) = T(F) \otimes_{C_0(\mathfrak{q})} \mathfrak{U}$. So, the truncated canonical resolution (8.13) coincides with the exact sequence

$$\begin{aligned} 0 \longrightarrow UT \longrightarrow U_{d-1}T_{d-1} \longrightarrow \cdots \longrightarrow U_1 T_1 \\ \longrightarrow U_0 T_0 \longrightarrow \zeta \longrightarrow 0 \end{aligned}$$

where $\zeta : \mathscr{R}_{-1}(X) \longrightarrow \mathscr{P}(X)$ is the inclusion functor. By Additivity (3.4.5), $v\mathfrak{t} = \zeta_*$, which is an isomorphism. So, v is surjective.

We prove that v is injective. Define functors

$$\begin{cases} W_n : \mathscr{P}_0(X) \longrightarrow \mathscr{P}(R) & \text{by} \\ \quad W_n(F) = \Gamma(X, F(n)) \\ W : \mathscr{P}_0(X) \longrightarrow \mathscr{P}_r(C_0(\mathfrak{q})) & \text{by} \\ \quad W(F) = Hom_X(\mathfrak{U}, F) = \Gamma(X, F \otimes \mathfrak{U}^*). \end{cases}$$

By (8.2.7), W_n are exact functors, as defined. Further, by (8.6.2, 8.2.7), $W(F) \in \mathscr{P}(R)$, and by (8.6.3), $W(F) \in \mathscr{P}(C_0(\mathfrak{q}))$. Let

$$w = (w_0, w_1, \ldots, w_{d-1}; w) : K_n(X) \longrightarrow K_n(R)^d \bigoplus K_n(\mathscr{P}_r(C_0(\mathfrak{q})))$$

be the maps induced by $(W_0, \ldots, W_{d-1}; W)$. Note that, by (8.6.4), $U_0, \ldots, U_{d-1}, U$ take values in $\mathscr{P}_0(X)$. As in the proof of (7.2.3), we prove wv is a triangular matrix, with ones on the diagonal.

1. By (8.1.3), we have

$$\begin{aligned} W_iU_j(M) &= \Gamma\left(X, M \otimes \mathcal{O}(i-j)\right) \\ &= \Gamma\left(X, \mathcal{O}(i-j)\right) \otimes M = \begin{cases} M & \text{if } i = j \\ 0 & \text{if } i < j \end{cases} \end{aligned}$$

2. For $M \in \mathscr{P}_r(C_0(\mathfrak{q}))$, we have

$$WU(M) = Hom(\mathfrak{U}, M \otimes_{C_0(\mathfrak{q})} \mathfrak{U}) \cong Hom(\mathfrak{U}, \mathfrak{U}) \otimes_{C_0(\mathfrak{q})} M = M$$

   where the last equality is obtained by (8.5.13).
3. Finally, claim that

$$W_iU(M) = \Gamma\left(X, M \otimes_{C_0(\mathfrak{q})} \mathfrak{U}(i)\right) = \Gamma(X, \mathfrak{U}(i)) \otimes M = 0.$$

   By construction (8.2.8), we have $\Gamma(X, \mathfrak{U}(d-1)) = \Gamma(X, Z_{d-1}(\mathcal{O}_X(1))(d-1)) = 0$. For $i < d-1$, consider the exact sequence

$$0 \longrightarrow Z_{d-1}(\mathcal{O}(1))(i) \longrightarrow T_{d-1} \otimes \mathcal{O}_X(i-(d-1)) \longrightarrow Z_{d-2}(\mathcal{O}_X(1))(n) \longrightarrow 0$$

   So,

$$0 \longrightarrow \Gamma(X, Z_{d-1}(\mathcal{O}(1))(i)) \longrightarrow \Gamma(X, T_{d-1} \otimes \mathcal{O}_X(i-(d-1))) \quad \text{is exact.}$$

   The second term is zero $\forall i \leq d-2$. So, the claim is established.

This completes the proof that wv is a triangular matrix, with ones on the diagonal. Hence, wv is an isomorphism. Therefore, v is injective. This completes the proof of the theorem. ∎

Lemma 8.6.2. *Use the notation as in* (8.5.1), *with* $rank(P) = d+2$. *If* $\mathcal{F} \in \mathscr{P}_0(X)$, *then* $\mathcal{F} \otimes \mathfrak{U}^* \in \mathscr{P}_0(X)$, *where* $\mathfrak{U}^* = Hom_X(\mathfrak{U}, R)$.

Proof. We need to prove that $H^q(X, \mathcal{F} \otimes \mathfrak{U}^*(k)) = 0$ for all $k \geq 0$, $q \geq 1$. By [Hr, Prop. III.6.3, III.6.7], $H^q(X, \mathcal{F} \otimes \mathfrak{U}^*(k)) = Ext^q(\mathfrak{U}, \mathcal{F}(k))$. Since $H^q(X, -) = 0 \forall q \geq d+1$, we have $\mathscr{P}_0(X) \subseteq \mathscr{R}_d(X)$. So, it follows that $Ext^q(\mathfrak{U}, \mathcal{F}(k)) = 0$, by (8.2.10). The proof is complete. ■

Lemma 8.6.3. *Use the notation as in* (8.5.1), *with* $rank(P) = d+2$. *Let* P *be a finitely generated (right)* $C_0(\mathfrak{q})$*-module such that* $P \in \mathscr{P}(R)$. *Then* $P \in \mathscr{P}_r(C_0(\mathfrak{q}))$.

Proof. Write $\mathfrak{R} = C_0(\mathfrak{q})$. Let

$$0 \longrightarrow K \longrightarrow \mathfrak{R}^m \longrightarrow P \longrightarrow 0 \quad \text{be an exact sequence}$$

of $\mathfrak{R}$-modules. Note that K is a finitely generated R-module and hence a finitely generated $\mathfrak{R}$-module. This leads to an exact sequence

$$0 \longrightarrow Hom_{\mathfrak{R}}(P, K) \longrightarrow Hom_{\mathfrak{R}}(P, \mathfrak{R}^m)$$
$$\longrightarrow Hom_{\mathfrak{R}}(P, P) \longrightarrow Ext^1_{\mathfrak{R}}(P, K).$$

We prove that $Ext^1_{\mathfrak{R}}(P, K) = 0$. Since P is a finitely presented R-module, and $R \longrightarrow \mathfrak{R}$ is flat, we have

$$0 = Ext^1_R(P, K) \otimes_R \mathfrak{R} \cong Ext^1_{\mathfrak{R}}(P \otimes_R \mathfrak{R}, K \otimes_R \mathfrak{R}) = Ext^1_{\mathfrak{R}}(P, K).$$

The proof is complete. ■

Lemma 8.6.4. *Use the notation as in* (8.5.1), *with* $rank(P) = d+2$. *Then* $U_n(M) \in \mathscr{P}_0(X)$ *for all* $M \in \mathscr{P}(R)$. *Further,* $U(M) \in \mathscr{P}_0(X)$, *for all* $M \in \mathscr{P}_r(C_0(\mathfrak{q}))$.

Proof. We can assume that $(R, \mathfrak{m})$ is local. We have $U_n(M) = M \otimes_R \mathcal{O}_X(-n)$. By (8.1.2), $H^q(X, U_n(M)(k)) = A^*_{-d+n-k} \otimes M = 0$, for all $n = 0, 1, \ldots, d-1$, $k \geq 0$. So, $U_n(M) \in \mathscr{P}_0(X)$.

For $M \in \mathscr{P}_r(C_0(\mathfrak{q}))$, $U(M) = M \otimes_{C_0(\mathfrak{q})} \mathfrak{U}$. Since M is projective (free), we only need to prove $\mathfrak{U} \in \mathscr{P}_0(X)$. With notation as in (8.37), we have $\mathfrak{U} \cong \det(P) \otimes \mathfrak{U}_{d-1} \cong \mathfrak{U}_{d-1}$. So, it suffices to prove the statement for $\mathfrak{U}_{d-1}$. With $M = C(\mathfrak{q}) = M_0 \oplus M_1$, using the notation and some information from the latter part of the proof of (8.5.13), the following diagram emerges:

$$\begin{array}{ccccccccc} 0 \longrightarrow & M_0 \otimes \mathcal{O}_{\mathbb{P}}(-(d+1)) & \xrightarrow{\Phi} & M_1 \otimes \mathcal{O}_{\mathbb{P}}(-d) & \longrightarrow & \mathfrak{U}_{d-1} & \longrightarrow 0 \\ & \downarrow & & \downarrow & & \| & \\ & M_0 \otimes \mathcal{O}_X & \xrightarrow[\varphi]{} & M_1 \otimes \mathcal{O}_X & \longrightarrow & \mathfrak{U}_{d-1} & \longrightarrow 0 \end{array}$$

Here the second line represents the definition of $\mathfrak{U}_{d-1}$. Since $\Phi^2 = \mathfrak{q}$ is a nonzero divisor, we have that Φ is injective. Since the kernels of the first two vertical maps are also a multiple of $\mathfrak{q}$, it follows that the first line is an exact sequence. Twist this sequence by $\mathcal{O}_{\mathbb{P}}(k)$ and consider the exact sequence of cohomologies:

$$H^q\left(\mathbb{P}, M_0 \otimes \mathcal{O}_{\mathbb{P}}(-(d+1)+k)\right) \longrightarrow H^q\left(X, \mathfrak{U}_{d-1}(k)\right)$$
$$\longrightarrow H^{q+1}\left(\mathbb{P}, M_1 \otimes \mathcal{O}_{\mathbb{P}}(-d+k)\right).$$

By (7.1.2), it follows that $H^d\left(X, \mathfrak{U}_{d-1}(k)\right) = 0$ for all $k \geq 0$. Hence, $H^q\left(X, \mathfrak{U}_{d-1}(k)\right) = 0$ for all $k \geq 0, q \geq 1$, by (8.1.3). The proof is complete. ■

8.6.1. *Graded interpretation*

We give a graded interpretation of u induced by the functor U in (8.44). Such graded interpretations allow us to use the Clifford sequence description of the canonical resolution.

Lemma 8.6.5. *Use the notation above* (8.5.1). *Then there is an equivalence of categories*

$$\Phi : \mathscr{P}_{r,\mathbb{Z}_2}(C(\mathfrak{q})) \xrightarrow{\sim} \mathscr{P}_r(C_0(\mathfrak{q})) \quad \text{sending} \quad M_0 \oplus M_1 \mapsto M_0,$$

where $\mathscr{P}_{r,\mathbb{Z}_2}(C(\mathfrak{q}))$ *denotes the* $\mathbb{Z}_2$*-graded right finitely generated projective* $C(\mathfrak{q})$*-modules.*

Proof. Similar to (8.3.8). ■

Corollary 8.6.6. *Use the notation above* (8.5.1). *Then the map* $u :$ $K_n\left(\mathscr{P}_r(C_0(\mathfrak{q}))\right) \longrightarrow K_n(X)$ *induced by the functor* U *in* (8.44) *is* *equivalent to the map*

$$u_g : K_n\left(\mathscr{P}_{r,\mathbb{Z}_2}(C(\mathfrak{q}))\right) \longrightarrow K_n(X)$$

which is induced by the functor

$$U_g : \mathscr{P}_{r,\mathbb{Z}_2}(C(\mathfrak{q})) \longrightarrow \mathscr{P}(X), \quad \textit{sending} \quad M \mapsto \det P^* \otimes_R \mathfrak{U}_{d-1}(M).$$

We also remark that, since $\det P^*$ *is invertible, this multiplier can be dropped from the definition of* U_g.

Proof. Consider the diagram

$$\begin{array}{ccc} \mathscr{P}_{r,\mathbb{Z}_2}(C(\mathfrak{q})) & \xrightarrow{\Phi} & \mathscr{P}_r(C_0(\mathfrak{q})) \\ & \searrow^{U_g} & \downarrow U \\ & & \mathscr{P}(X) \end{array}$$

$$\text{So,} \quad \begin{cases} U\Phi(M) = M_0 \otimes_{C_0(\mathfrak{q})} \mathfrak{U} \\ U_g(M) = \det P^* \otimes_R \mathfrak{U}_{d-1}(M). \end{cases} \tag{8.46}$$

It follows from (8.5.4, 8.5.11) that $U\Phi(M) \cong U_g(M)$ are naturally isomorphic. The proof is complete. ∎

8.7. The Affine Case

The goal of this section is to apply the projection formula (8.6.1) and obtain some results in the affine case. At a basic level, we are interested in the affine case $\sum_i a_i X_i^2 = 1$ over a field, which includes the affine spheres. More generally, for nondegenerate quadratic R-modules $(P, \mathfrak{q})$, the case of $(P \oplus R, q \perp \pm 1)$ is of some interest. We do some preparatory work. We start with a lemma regarding affine subspaces.

Lemma 8.7.1. *Let* R *be a commutative Noetherian ring. Suppose that* $P = P_1 \oplus P_2$ *is a finitely generated projective* R*-module and* $rank(P_2) = 1$.

1. *Write* $\mathbb{P}(P) = \operatorname{Proj}(\mathcal{S}\mathrm{ym}(P^*))$. *Then* $V = \mathbb{P}(P) - \mathbb{P}(P_1)$ *is affine.*

2. *Let* $(P, \mathfrak{q}) = (P_1, \mathfrak{q}_1) \perp (P_2, \mathfrak{q}_2)$ *be the orthogonal sum of nondegenerate quadratic R-modules* $(P_1, \mathfrak{q}_1)$, $(P_2, \mathfrak{q}_2)$. *Write* $X := X(\mathfrak{q})$, $X_1 := X(\mathfrak{q}_1)$. *Then* $U(\mathfrak{q}_1, \mathfrak{q}_2) := U := X(\mathfrak{q}) - X(\mathfrak{q}_1)$ *is also affine.*

Proof. It is enough to prove when $(R, \mathfrak{m})$ is local [Hr, Ex. II.2.17]. So, we assume P_1 and P_2 are free. Let $X_0, \ldots, X_{n-1}$ be a basis of P_1^*, and X_n be a basis of P_2^*. So, $\mathbb{P}(P_1) = (X_n = 0)$. In this case, $V = \mathbb{P}^r - \mathbb{P}^{r-1} = (X_n \neq 0) \cong \mathbb{A}^r$. Now, note that U is the closed subset of V defined by $\mathfrak{q} = 0$. So, U is affine. ■

Lemma 8.7.2. *Let R be a commutative Noetherian ring. Let* $(P_1, \mathfrak{q}_1)$, $(P_2, \mathfrak{q}_2)$ *be nondegenerate R-modules, and* $rank(P_2) = 1$. *Let* $(P, \mathfrak{q}) = (P_1, \mathfrak{q}_1) \perp (P_2, \mathfrak{q}_2)$. *Write* $X := X(\mathfrak{q})$, $X_1 := X(\mathfrak{q}_1)$. *Let* $\mathscr{I}$ *be the sheaf of ideals of* $X_1 \subseteq X$. *Then*

$$\mathscr{I} = \mathcal{O}_X(-1) \otimes P_2^*.$$

With $U = U(\mathfrak{q}_1, \mathfrak{q}_2) = X(\mathfrak{q}) - X(\mathfrak{q}_1)$ *and* $\mathcal{O}_U(n) = \mathcal{O}_X(n)_{|U}$, *we have*

$$\mathcal{O}_U(n-1) \otimes P_2^* \approx \mathcal{O}_U(n) \quad \text{and} \quad \mathcal{O}_U(n) \approx \mathcal{O}_U \otimes (P_2^*)^n.$$

Proof. Write

$$A = \frac{\mathcal{S}\text{ym}\,(P^*)}{(\mathfrak{q})}, \quad C = \frac{\mathcal{S}\text{ym}\,(P_1^*)}{(\mathfrak{q}_1)}. \qquad \text{So,}$$
$$X_1 = \text{Proj}\,(C) \subseteq X = \text{Proj}\,(A).$$

Note that $\mathcal{S}\text{ym}\,(P^*) \cong \mathcal{S}\text{ym}\,(P_1^*) \otimes \mathcal{S}\text{ym}\,(P_2^*)$. It follows that $\frac{\mathcal{S}\text{ym}(P^*)}{P_2^* \mathcal{S}\text{ym}(P^*)} \cong \mathcal{S}\text{ym}\,(P_1^*)$, with $\mathfrak{q} \mapsto \mathfrak{q}_1$. This induces an exact sequence

$$0 \longrightarrow P_2^* \otimes A(-1) \longrightarrow A \longrightarrow C \longrightarrow 0.$$

We check this locally. So $P_2^* = RX_0$ and $P_1^* = \sum_{i=1}^n RX_i$. Suppose $X_0 f = 0$ for some $f \in A$. So, in $R[X_0, \ldots, X_n]$, we have

$$X_0 f = \mathfrak{q}h = \mathfrak{q}(X_0 h_1 + h_2(X_1, \ldots, X_n)). \quad \textit{Hence} \quad X_0(f - h_1\mathfrak{q}) = \mathfrak{q}h_2.$$

Substituting $X_0 = 0$, we have $\mathfrak{q}_1 h_2 = 0$, and so $h_2 = 0$. (Note that $\mathfrak{q}_1$ is a nonzero divisor, because it is nondegenerate.) Hence $f = 0$.

So, the exactness of the sequence is established. It follows now that the sequence

$$0 \longrightarrow P_2^* \otimes \mathcal{O}_X(-1) \longrightarrow \mathcal{O}_X \longrightarrow \mathcal{O}_{X_1} \longrightarrow 0 \quad \text{is exact.} \tag{8.47}$$

This settles the first part. The first statement regarding U follows by twisting this sequence by n and restricting it to the open set U. The second part follows by induction. So, the proof is complete. ■

Before we proceed, we set up some notation.

Notation 8.7.3. Let R denote a commutative Noetherian ring. Let $(P_1, \mathfrak{q}_1), (P_2, \mathfrak{q}_2)$ be nondegenerate R-modules, with $rank(P_1) = d+1$, $rank(P_2) = 1$. Let $(P, \mathfrak{q}) = (P_1, \mathfrak{q}_1) \perp (P_2, \mathfrak{q}_2)$. Write $X := X(\mathfrak{q})$, $X_1 = X(\mathfrak{q}_1)$. So, X has relative dimension d, and X_1 has relative dimension $d-1$ over $\mathrm{Spec}\,(R)$. As before, $\mathbb{H}(X)$ denotes the full subcategory of $F \in Coh(X)$ that has a finite locally free resolution. So, the functor

$$\iota : \mathscr{P}(X_1) \longrightarrow \mathbb{H}(X) \quad \text{is well defined. And} \quad K_q(\mathbb{H}(X)) \cong K_q(X)\, \forall\, q.$$

Corresponding to (8.44), for X_1, we have functors

$$\begin{cases} U_n^1 : \mathscr{P}(R) \longrightarrow \mathscr{P}(X_1) \quad \text{defined by} \\ \quad U_n^1(M) = M \otimes \mathcal{O}_{X_1}(-n) \qquad n = 0, 1, \ldots, d-2 \\ U_g^1 : \mathscr{P}_{r,\mathbb{Z}_2}(C(\mathfrak{q}_1)) \longrightarrow \mathscr{P}(X_1) \quad \text{defined by} \\ \quad U_g^1(M) = \det(P_1^*) \otimes \mathfrak{U}_{d-2}(M). \end{cases}$$

The maps of K-groups induced by these functors will be denoted by u_n^1 and u_g^1. Corresponding maps for X will be denoted by u_i and u, as in (8.44). We use the notation $K_q^{gr}(C(\mathfrak{q}_1)) := K_q\left(\mathscr{P}_{r,\mathbb{Z}_2}(C(\mathfrak{q}_1))\right)$ and $K_q(C_0(\mathfrak{q})) = K_q\left(\mathscr{P}_r(C_0(\mathfrak{q})\right)$. By the projection formula (8.6.1), ι_* is computed as follows:

$$\begin{array}{ccc} \bigoplus_{n=0}^{d-2} K_q(R) \bigoplus K_q^{gr}(C(\mathfrak{q}_1)) & & \bigoplus_{n=0}^{d-1} K_q(R) \bigoplus K_q^{gr}(C(\mathfrak{q})) \\ \downarrow \wr & & \downarrow \wr \\ K_q(X_1) & \xrightarrow{\quad \iota_* \quad} & K_q(\mathbb{H}(X)) \end{array} \tag{8.48}$$

The composition $\iota U_n^1(M) = \iota_*(M \otimes \mathcal{O}_{X_1}(-n))$. From (8.47), we obtain the exact sequence

$$0 \longrightarrow M \otimes \mathcal{O}_X(-n-1) \otimes P_2^* \longrightarrow M \otimes \mathcal{O}_X(-n) \longrightarrow$$
$$M \otimes \mathcal{O}_{X_1}(-n) \longrightarrow 0.$$

So, by the Additivity theorem (3.4.5)

$$\iota_* u_n^1 = \left(\iota U_n^1\right)_* = u_n - u_{n+1} o \alpha_* \tag{8.49}$$

where $\alpha : \mathscr{P}(R) \longrightarrow \mathscr{P}(R)$ is the functor $M \mapsto \det P^* \otimes M$. To compute $\iota_* u_g^1$, define the functors

$$\begin{cases} \varepsilon_0 : \mathscr{P}_r(C_0(\mathfrak{q}_1)) \longrightarrow \mathscr{P}_r(C_0(\mathfrak{q})) & \text{by} \quad \varepsilon_0(M) = M \otimes_{C_0(\mathfrak{q}_1)} C_0(\mathfrak{q}) \\ \varepsilon : \mathscr{P}_{r,\mathbb{Z}_2}(C(\mathfrak{q}_1)) \longrightarrow \mathscr{P}_{r,\mathbb{Z}_2}(C(\mathfrak{q})) & \text{by} \quad \varepsilon(M) = M \otimes_{C(\mathfrak{q}_1)} C(\mathfrak{q}). \end{cases} \tag{8.50}$$

These two functors are compatible in the sense of the commutative diagram

$$\begin{array}{ccc} \mathscr{P}_r(C_0(\mathfrak{q}_1)) & \longrightarrow & \mathscr{P}_r(C_0(\mathfrak{q})) \\ \wr\downarrow & & \downarrow\wr \\ \mathscr{P}_{r,\mathbb{Z}_2}(C(\mathfrak{q}_1)) & \longrightarrow & \mathscr{P}_{r,\mathbb{Z}_2}(C(\mathfrak{q})) \end{array}$$

while the vertical functors are equivalences (8.6.5).

Lemma 8.7.4. *Use the notation as above* (8.7.3). *Then*

$$C(\mathfrak{q}) = C(\mathfrak{q}_1) \bigoplus C(\mathfrak{q}_1) \otimes P_2, \quad More, precisely,$$

$$\begin{cases} C_0(\mathfrak{q}) = C_0(\mathfrak{q}_1) \oplus C_1(\mathfrak{q}_1) \otimes P_2 \\ C_1(\mathfrak{q}) = C_1(\mathfrak{q}_1) \oplus C_0(\mathfrak{q}_1) \otimes P_2. \end{cases}$$

Consequently, for $M = M_0 \oplus M_1 \in \mathscr{P}_{r,\mathbb{Z}_2}(C(\mathfrak{q}_1))$, *we have*

$$\varepsilon(M) = N_0 \oplus N_1 \quad where \quad \begin{cases} N_0 = M_0 \oplus (M_1 \otimes P_2) \\ N_1 = M_1 \oplus (M_0 \otimes P_2). \end{cases} \tag{8.51}$$

Proof. From the generalities of Clifford algebras (8.3.7),

$$C(\mathfrak{q}) = C(\mathfrak{q}_1)\hat{\otimes}C(\mathfrak{q}_2).$$

Since $rank(P_2) = 1$, the assertion follows. ■

With notation as in (8.51), $\mathfrak{U}_{d-1}\varepsilon(M) = \mathfrak{U}_{d-1}(N)$ is given by the exact sequence:

$$(M_0 \oplus (M_1 \otimes P_2)) \otimes \mathcal{O}_X(-d-1) \xrightarrow{\partial} (M_1 \oplus (M_0 \otimes P_2))$$

$$\otimes \mathcal{O}_X(-d) \xrightarrow{\partial} \mathfrak{U}_{d-1}(N) \longrightarrow 0.$$

This sequence is displayed in the following diagram:

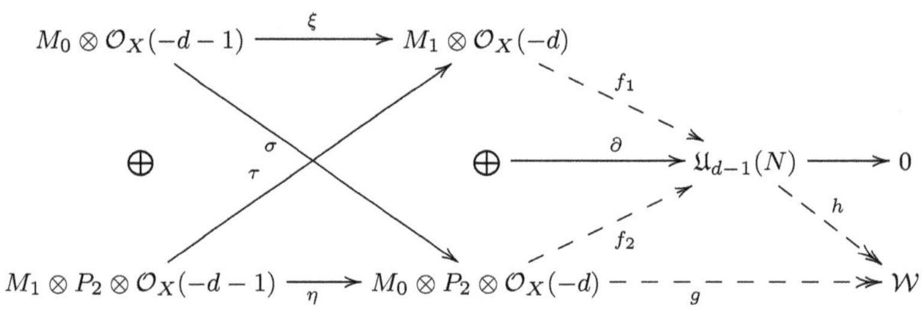

(8.52)

where $\mathcal{W} := co\ker(f_1)$. It also follows that the composition $g = hf_2$ is surjective. So, we obtain an exact sequence

$$M_1 \otimes \mathcal{O}_X(-d) \xrightarrow{f_1} \mathfrak{U}_{d-1}(N) \xrightarrow{h} \mathcal{W} \longrightarrow 0.$$

Lemma 8.7.5. *We use the notation* $\mathfrak{U}_n^1(M)$*, with the superscript, to denote the object over* X_1*, defined in* (8.5.2).

1. *Then the sequence*

$$M_1 \otimes P_2 \otimes \mathcal{O}_X(-d-1) \xrightarrow{\eta_0} M_0 \otimes P_2 \otimes \mathcal{O}_{X_1}(-d) \longrightarrow$$

$$P_2 \otimes \mathfrak{U}_{d-1}^1(M) \longrightarrow 0$$

factors through a natural isomorphism $\gamma : \mathcal{W} \xrightarrow{\sim} P_2 \otimes \mathfrak{U}_{d-1}^1(M)$.

2. *The sequence*

$$0 \longrightarrow M_1 \otimes \mathcal{O}_X(-d) \xrightarrow{f_1} \mathfrak{U}_{d-1}(N) \xrightarrow{h} \mathcal{W} \longrightarrow 0$$

is exact. Consequently, we have an exact sequence

$$0 \longrightarrow M_1 \otimes \mathcal{O}_X(-d) \xrightarrow{f_1} \mathfrak{U}_{d-1}(N) \xrightarrow{\gamma h}$$

$$P_2 \otimes \mathfrak{U}^1_{d-1}(M) \longrightarrow 0. \tag{8.53}$$

Proof. Refer to (8.52) for the notation. Checking locally, it follows from the description (8.31), that $co\ker(\sigma) = M_0 \otimes P_2 \otimes \mathcal{O}_{X_1}(n)$. Since, $h = co\ker(f_1)$, it is not difficult to see that

$$N_0 \otimes \mathcal{O}_X(-d-1) \xrightarrow{(\sigma,\eta)} M_0 \otimes P_2 \otimes \mathcal{O}_X(-d) \xrightarrow{g} \mathcal{W} \longrightarrow 0$$

is exact (this can also be seen by chasing the diagram (8.31)). This leads to the diagram of exact sequences, inducing the natural map γ:

$$\begin{array}{ccccccc}
N_0 \otimes \mathcal{O}_X(-d-1) & \xrightarrow{(\sigma,\eta)} & M_0 \otimes P_2 \otimes \mathcal{O}_X(-d) & \xrightarrow{g} & \mathcal{W} & \longrightarrow & 0 \\
\downarrow & & \downarrow p & & \downarrow \gamma & & \\
M_1 \otimes \mathcal{O}_{X_1}(-d-1) & \xrightarrow[\eta_0]{} & M_0 \otimes P_2 \otimes \mathcal{O}_{X_1}(-d) & \longrightarrow & P_2 \otimes \mathfrak{U}^1_{d-1}(M) & \longrightarrow & 0.
\end{array}$$

Since the solid vertical arrows are surjective, γ is also surjective. To prove the injectivity of γ, locally, let $\gamma(w) = 0$. Let $g(x) = w$. Then $p(x)$ has a preimage, which has a preimage $y \in \mathcal{O}_X(-d-1) \otimes M_1 \otimes P_2$. Hence, $\eta(y) - x \in \ker(p) = image(\sigma)$. So, $\sigma(z) = \eta(y) - x$ for some z. Therefore, $w = g(x) = g\eta(y) - g\sigma(z) = 0$. This establishes (1).

To establish (2), we prove f_1 is injective. We assume that $(R, \mathfrak{m})$ is local. Let $e_0, e_1, \ldots, e_d$ be a basis of P_1 and e be a basis of P_2. Let $X_i = e_i^*$, and $Y = e^*$. By (8.34), the map

$$\varphi := \begin{pmatrix} \xi & \sigma \\ \tau & \eta \end{pmatrix} = \sum_{i=0}^{d} e_i X_i + eY \qquad \text{both sides acting from the right.}$$

Note that, for $m \in M_i$, by definition, $m \cdot e := m \otimes e$. Further, since $\mathfrak{q} = q_1 \perp q_2$, we have $e_i \cdot e = -e \cdot e_i$. It follows that

$$\forall\, m_0 \otimes a \in M_0 \otimes \mathcal{O}_X(-d-1) \qquad \begin{cases} \xi(m_0 \otimes a) = \sum_{i=0}^d m_0 \cdot e_i \otimes X_i a \\ \sigma(m_0 \otimes a) = m_0 \otimes e \otimes Y a. \end{cases}$$

Further, $\forall m_1 \otimes e \otimes a \in M_1 \otimes P_2 \otimes \mathcal{O}_X(-d-1)$, we have

$$\varphi(m_1 \otimes e \otimes a) = \sum_{i=0}^{d} (m_1 \otimes e \otimes a) e_i X_i + (m_1 \otimes e \otimes a) e Y$$

$$= \sum_{i=0}^{d} (m_1 \otimes e \otimes a) e_i X_i + m_1 \otimes a \mathfrak{q}(e) Y$$

$$\text{So,} \qquad \begin{cases} \eta = -\sum_{i=0}^d e_i X_i \\ \tau = \mathfrak{q}(e) Y \end{cases} \qquad \text{right multiplications.}$$

Now suppose that $f_1(w) = 0$. Then $\partial(w, 0) = 0$. So, for some u, v, we have

$$\begin{cases} \xi(u) + \tau(v) = w \\ \sigma(u) + \eta(v) = 0 \end{cases} \quad \text{Hence,} \quad \begin{cases} u \sum e_i X_i + v\mathfrak{q}(e) Y = w \\ uY = v \sum e_i X_i. \end{cases}$$

Apply $\sum e_i X_i$, from right, to the second equation

$$uY \sum e_i X_i = v \left(\sum X_i e_i\right)^2 = v\mathfrak{q}_1 = v \sum \mathfrak{q}_1(e_i) X_i^2 \Longrightarrow$$

$$wY = \left(u \sum e_i X_i\right) Y + v\mathfrak{q}(e) Y^2$$

$$= v \sum \mathfrak{q}_1(e_i) X_i^2 + v\mathfrak{q}(e) Y^2 = v\mathfrak{q}$$

Therefore, $wY = \mathfrak{q}v = 0$, since $\mathfrak{q} = 0$ on X. By (8.7.2), Y is regular on $\mathcal{O}_X$. Therefore, $w = 0$. The proof is complete. ∎

Lemma 8.7.6. *Use the notation as above* (8.7.3). *Consider the functors*

$$\begin{cases} U_g : \mathscr{P}_{r,\mathbb{Z}_2}(C(\mathfrak{q})) \longrightarrow \mathscr{P}(X) \\ U_g^1 : \mathscr{P}_{r,\mathbb{Z}_2}(C(\mathfrak{q}_1)) \longrightarrow \mathscr{P}(X_1) \\ \varepsilon : \mathscr{P}_{r,\mathbb{Z}_2}(C(\mathfrak{q}_1)) \longrightarrow \mathscr{P}_{r,\mathbb{Z}_2}(C(\mathfrak{q})) \end{cases}$$

and others. As before, denote the maps of the K-groups, induced by these functors, by lowercase letters. We have

$$\iota_* u_g^1 = u_{d-1}\beta_* - u_g\varepsilon_* : K_q^{gr}(C(\mathfrak{q}_1)) \longrightarrow K_q(X)$$

where β is the functor, as follows:

$$\beta : \mathscr{P}_{r,\mathbb{Z}_2}(C(\mathfrak{q}_1)) \longrightarrow \mathscr{P}(R) \quad \textit{sending} \quad M \mapsto \det(P_1^*) \otimes M_1.$$

Proof. From (8.6.6), we have,

$$U_g(N) = \mathfrak{U}_{d-1}(N) \otimes \det(P^*) = \det(P_1^*)P_2^* \otimes \mathfrak{U}_{d-1}(N).$$

So, the exact sequence (8.53) is equivalent to

$$\begin{aligned} 0 \longrightarrow M_1 \otimes \det(P^*) \otimes \mathcal{O}_X(-d) \longrightarrow U_g(N) \\ \longrightarrow P_1^* \otimes \mathfrak{U}_{d-1}^1(M) \longrightarrow 0. \end{aligned} \tag{8.54}$$

The Clifford sequence being locally split exact (8.5.12), we have the exact sequence

$$0 \longrightarrow \mathfrak{U}_{d-1}^1(M) \longrightarrow M_1 \otimes \mathcal{O}_{X_1}(-d+1) \longrightarrow \mathfrak{U}_{d-2}^1(M) \longrightarrow 0.$$

Since $U_g^1(M) = \det(P_1^*) \otimes \mathfrak{U}_{d-2}^1(M)$, combining with (8.54), we have the exact sequence

$$\begin{aligned} 0 \longrightarrow M_1 \otimes \det(P^*) \otimes \mathcal{O}_X(-d) \longrightarrow U_g(N) \longrightarrow \\ M_1 \otimes \det(P_1^*) \otimes \mathcal{O}_{X_1}(-d+1) \longrightarrow U_g^1(M) \longrightarrow 0. \end{aligned} \tag{8.55}$$

The following exact sequence is obtained by $M_1 \otimes \det(P_1) \otimes -$, with the known exact sequence

$$\begin{aligned} 0 \longrightarrow M_1 \otimes \det(P^*) \otimes \mathcal{O}_X(-d) \longrightarrow \\ M_1 \otimes \det(P_1^*) \otimes \mathcal{O}_X(-d+1) \longrightarrow M_1 \otimes \det(P_1^*) \otimes \mathcal{O}_{X_1}(-d+1) \longrightarrow 0. \end{aligned} \tag{8.56}$$

By the Additivity Theorem (3.4.5), applied to (8.55, 8.56), we have

$$\iota_* u_g^1 = u_{d-1}\beta_* - u_g\varepsilon_*.$$

The proof is complete. ■

Having computed $\iota_* u_g^1$ and $\iota_* u_n^1$ (8.49), we describe the maps $\iota_* : K_q(X_1) \longrightarrow K_q(X)$ in terms of the identifications, as in diagram (8.48):

$$\begin{array}{ccccccccc}
K_q(R)\oplus & & K_q(R)\oplus\cdots & & K_q(R)\oplus & & K_q^{gr}(C(\mathfrak{q}_1)) & & \\
\downarrow{\scriptstyle 1} & \overset{-\alpha_*}{\searrow} & \downarrow{\scriptstyle 1} & \overset{-\alpha_*}{\searrow} & \downarrow{\scriptstyle 1} & \overset{-\alpha_*}{\searrow} & \downarrow{\scriptstyle \beta_*} & \overset{-\varepsilon_*}{\searrow} & \\
K_q(R)\oplus & & K_q(R)\oplus\cdots & & K_n(R)\oplus & & K_q(R)\oplus & & K_q^{gr}(C(\mathfrak{q}))
\end{array} \tag{8.57}$$

From this description of ι_*, we have the following.

Lemma 8.7.7. *Use the notation as above* (8.7.3). *Consider the maps*

$$\chi(\mathfrak{q}_1,\mathfrak{q}) := (\beta_*, -\varepsilon_*) : K_q^{gr}(C(\mathfrak{q}_1)) \longrightarrow K_q(R) \oplus K_q^{gr}(C(\mathfrak{q})).$$

Then $\chi(\mathfrak{q}_1,\mathfrak{q})$ *and* ι_* *have the same kernel and cokernel. The same is described in the following diagram:*

$$\begin{array}{ccccccc}
\ker(\chi(\mathfrak{q}_1,\mathfrak{q})) & \hookrightarrow & K_q^{gr}(C(\mathfrak{q}_1)) & \xrightarrow{\chi(\mathfrak{q}_1,\mathfrak{q})} & K_q(R)\oplus K_q^{gr}(C(\mathfrak{q})) & \twoheadrightarrow & \operatorname{co}\ker(\chi(\mathfrak{q}_1,\mathfrak{q})) \\
\downarrow\wr & & \downarrow{\scriptstyle u_g^1} & & \downarrow{\scriptstyle (u_{d-1},u_g)} & & \downarrow\wr \\
\ker(\iota_*) & \hookrightarrow & K_q(X_1) & \xrightarrow[\iota_*]{} & K_q(X) & \twoheadrightarrow & \operatorname{co}\ker(\iota_*).
\end{array}$$

Proof. Obvious from the description (8.57) of ι_*. ■

Now we assume that R is regular, and use the localization sequence (6.1.11).

Theorem 8.7.8. *Use the notation as above* (8.7.3). *Further, assume that R is regular. As before, $X = X(\mathfrak{q})$, $X_1 = X(\mathfrak{q}_1)$, and $U = X - X_1$. Then there is an exact sequence*

$$\cdots \longrightarrow K_{n+1}(U) \xrightarrow{\partial} K_n^{gr}(C(\mathfrak{q}_1)) \xrightarrow{\chi(\mathfrak{q}_1,\mathfrak{q})}$$

$$K_n(R) \oplus K_n^{gr}(C(\mathfrak{q})) \xrightarrow{(\rho_0,\rho)_*} K_n(U) \longrightarrow \cdots \tag{8.58}$$

where $(\rho_0, \rho)_*(x) = (u_{d-1}x, u_g(x))_{|U}$. *So,* ρ_0, ρ *are the functors*

$$\begin{cases} \rho_0 : \mathscr{P}(R) \longrightarrow \mathscr{P}(U) \\ \quad \rho_0(M) = M \otimes \mathcal{O}_U(-(d-1)) = M \otimes P_2^{d-1} \otimes \mathcal{O}_U \\ \rho : \mathscr{P}_{r,\mathbb{Z}_2}(C(\mathfrak{q})) \longrightarrow \mathscr{P}(U) \\ \quad \rho(M) = U_g(M)_{|U} = \mathfrak{U}(M)_{|U} = \mathfrak{U}_{d-1}(M) \otimes \det(P^*)_{|U}. \end{cases} \tag{8.59}$$

Proof. Since R is regular, G-theory and K-theory of X, X_1 and U coincide. By (6.1.11), we have an exact sequence

$$\cdots \longrightarrow K_{n+1}(U) \longrightarrow K_n(X_1) \xrightarrow{\iota_*} K_n(X) \longrightarrow K_n(U) \longrightarrow \cdots$$

The desired exact sequence is obtained from this, in conjunction with (8.7.7).

8.7.1. *Special case of* $\mathfrak{q} = \mathfrak{q}_1 - T^2$

Now, we consider the case of spheres, or of the slightly more general case, when $P_2 = R$ and $\mathfrak{q}_2 = -T^2$. As before, $\mathfrak{q}_1$ is nondegenerate. We set up our notation, as follows.

Notation 8.7.9. Let R denote a regular ring, with $1/2 \in R$ (*as always in this chapter*). Let $(P_1, \mathfrak{q}_1)$ denote a nondegenerate quadratic R-module. Let $P = P_1 \oplus Re \cong P_1 \oplus R$ and $\mathfrak{q} = \mathfrak{q}_1 \perp -T^2 =: \mathfrak{q}_1 \perp \langle -1 \rangle$. In this case,

$$\begin{cases} S(\mathfrak{q}) = \frac{\mathcal{S}ym(P_1^*)\mathcal{S}[T]}{(\mathfrak{q}_1 - T^2)}, \\ X(\mathfrak{q}) = \operatorname{Proj}(S(\mathfrak{q})), \end{cases} \qquad \begin{cases} U := X - X(\mathfrak{q}_1) \cong \operatorname{Spec}(R(\mathfrak{q}_1)), \text{ where} \\ R(\mathfrak{q}_1) := \frac{\mathcal{S}ym(P_1^*)}{(\mathfrak{q}_1 - 1)}, \qquad \mathcal{O}_U(n) \cong \mathcal{O}_U \ \forall n. \end{cases}$$

Lemma 8.7.10. *With notation as in* (8.7.9),

$$C_0(\mathfrak{q}) = C_0(\mathfrak{q}_1 \perp \langle -1 \rangle) \cong C(\mathfrak{q}_1).$$

Proof. First, $C(\langle -1 \rangle) \cong R \oplus Re$, with $e^2 = -1$. So,

$$C(\mathfrak{q}) = C(\mathfrak{q}_1)\hat{\oplus}C(\langle -1 \rangle) \cong (C_0(\mathfrak{q}_1) \oplus C_1(\mathfrak{q}_1)e) \bigoplus (C_1(\mathfrak{q}_1) \oplus C_0(\mathfrak{q}_1)e)$$

with $e^2 = -1$ and $ea = (-1)^{\deg(a)}ae$ for $a \in C(\mathfrak{q}_1)$, homogeneous. So, $C(\mathfrak{q}_1) \longrightarrow C_0(\mathfrak{q})$ is an isomorphism, sending $a_0 + a_1 \longleftrightarrow a_0 + a_1 e$, with $a_i \in C_i(\mathfrak{q}_1)$. The proof is complete. ∎

By (8.6.5),

$$K_q^{gr}\left(C(\mathfrak{q})\right) \cong K_q\left(C_0(\mathfrak{q})\right) \cong K_q\left(C(\mathfrak{q}_1)\right)$$

given by the functor

$$\begin{array}{ccc} \mathscr{P}_{r,\mathbb{Z}_2}(C(\mathfrak{q})) & \longrightarrow & \mathscr{P}_r(C_0(\mathfrak{q})) \\ & \searrow & \downarrow \\ & & \mathscr{P}_r(C(\mathfrak{q}_1)) \end{array} \qquad \text{sending} \quad N_0 \oplus N_1 \mapsto N_0.$$

So, $C(\mathfrak{q}_1)$ acts on N_0, by $m \cdot e_i := m \cdot e_i e$.

Lemma 8.7.11. *Use notation as in* (8.7.9). *Let* $P_1 = \oplus_{i=0}^d Re_i$ *be free, with basis* $\{e_i\}$. *Then* $\rho(N) = \mathfrak{U}_{d-1}(N) \otimes \det(P^*)_{|U}$ *is, equivalently, given by the cokernel as follows:*

$$N_0 \otimes \mathcal{O}_U \xrightarrow{\sum e_i X_i - 1} N_0 \otimes \mathcal{O}_U \longrightarrow \rho(N) \longrightarrow 0.$$

Proof. Since $\det(P^*)$ is invertible, we identify $\rho(N) = \mathfrak{U}_{d-1}(N)_{|U}$. Using the definition, we have the exact sequence, together with the commutative diagram:

$$\begin{array}{ccccccc} N_0 \otimes \mathcal{O}_U(-d-1) & \xrightarrow{\sum e_i X_i + eT} & N_1 \otimes \mathcal{O}_U(-d) & \longrightarrow & \rho(N) & \longrightarrow & 0 \\ {\scriptstyle T=1}\downarrow & & {\scriptstyle T=1}\downarrow{\scriptstyle z \mapsto ze} & & & & \\ N_0 \otimes \mathcal{O}_U & \xrightarrow[\sum e_i X_i - 1]{} & N_0 \otimes \mathcal{O}_U & & & & \end{array}$$

Since the vertical maps are isomorphisms, the proof is complete. ∎

Rewriting (8.7.8), we obtain the following corollary (writing $\mathfrak{q}$ for $\mathfrak{q}_1$).

Corollary 8.7.12. *Use notation as in* (8.7.9). *Assume that* R *is regular. Then there is an exact sequence,*

$$\cdots \longrightarrow K_{n+1}(R(\mathfrak{q})) \xrightarrow{\partial} K_n^{gr}(C(\mathfrak{q})) \xrightarrow{(-\beta_*, \varepsilon_*)}$$

$$K_n(R) \oplus K_n(C(\mathfrak{q})) \xrightarrow{(\rho_0, \rho)_*} K_n(R(\mathfrak{q})) \longrightarrow \cdots \qquad (8.60)$$

where $K_n(C(\mathfrak{q})) = K_n(\mathscr{P}_r(C(\mathfrak{q})))$, *the* K*-groups of the category of right projective* $C(\mathfrak{q})$*-modules (ungraded). We reinterpret the functors:*

$$\begin{cases} \beta : \mathscr{P}_{r,\mathbb{Z}_2}(C(\mathfrak{q})) \longrightarrow \mathscr{P}(R) & \beta(M) = M_1 \\ \varepsilon : \mathscr{P}_{r,\mathbb{Z}_2}(C(\mathfrak{q})) \longrightarrow \mathscr{P}_r(C(\mathfrak{q})) & \varepsilon(M) = M \text{ (ungraded)} \\ \rho_0 : \mathscr{P}(R) \longrightarrow \mathscr{P}(R(\mathfrak{q})) & \rho_0(M) = M \otimes R(\mathfrak{q}) \\ \rho : \mathscr{P}_r(C(\mathfrak{q})) \longrightarrow \mathscr{P}(R(\mathfrak{q})) & \rho(N) = \Gamma(U, \mathfrak{U}_{d-1}(N)). \end{cases}$$

(So, ε is obtained by forgetting the grading.) *In particular,*

$$K_0^{gr}(C(\mathfrak{q})) \xrightarrow{(-\beta_*, \varepsilon_*)} K_0(R) \oplus K_0(C(\mathfrak{q})) \xrightarrow{(\rho_0, \rho)_*}$$

$$K_0(R(\mathfrak{q})) \longrightarrow 0 \quad \text{is exact.} \tag{8.61}$$

If P_1 *is free, then* ρ *is, equivalently, given by the cokernel of* $\sum e_i X_i - 1$, *as in* (8.7.11).

Swan defined $ABS(\mathfrak{q})$ as follows [Sw2].

Definition 8.7.13. Use notation as in (8.7.9). Define $ABS(\mathfrak{q}) = co\ker(\varepsilon)$. So, we have an exact sequence

$$K_0^{gr}(C(\mathfrak{q})) \xrightarrow{\varepsilon} K_0(C(\mathfrak{q})) \longrightarrow ABS(\mathfrak{q}) \longrightarrow 0.$$

Now, $ABS(\mathfrak{q})$ coincides with the Grothendieck group of the $R(\mathfrak{q})$ (*the sphere*), in the following sense.

Corollary 8.7.14. *Under the hypotheses of* (8.7.12), *we have an isomorphism*

$$\overline{\rho} : ABS(\mathfrak{q}) \xrightarrow{\sim} \widetilde{K}_0(R(\mathfrak{q})) \quad \textit{where} \quad \widetilde{K}_0(R(\mathfrak{q})) := \frac{K_0(R(\mathfrak{q}))}{Im(\rho_0)}.$$

The map is induced by ρ. *So, the sequence*

$$K_0(R) \longrightarrow K_0(R(\mathfrak{q})) \longrightarrow ABS(\mathfrak{q}) \longrightarrow 0 \quad \textit{is exact.}$$

Proof. Write $\widetilde{K}_0(R(\mathfrak{q})) = \frac{K_0(R(\mathfrak{q}))}{Im(\rho_0)}$. From (8.61), ρ induces the map $\overline{\rho}$, as follows:

$$\begin{array}{ccccccc} K_0^{gr}(C(\mathfrak{q})) & \xrightarrow{\varepsilon} & K_0(C(\mathfrak{q})) & \longrightarrow & ABS(\mathfrak{q}) & \longrightarrow & 0 \\ & & \| & & \downarrow{\scriptstyle \overline{\rho}} & & \\ K_0(R) & \xrightarrow[\rho_0]{} & K_0(C(\mathfrak{q})) & \longrightarrow & \widetilde{K}_0(R(\mathfrak{q})). & & \end{array}$$

It follows from (8.61) that $\overline{\rho}$ is an isomorphism. ■

When $R = k$ is a field, then the map ρ_0 splits by the rank map. In general, there is no reason to expect the map ρ_0 to be injective.

Example 8.7.15. Let $R = \frac{\mathbb{R}[X,Y]}{(x^2+y^2-1)}$ be the algebraic coordinate ring of the circle. Let $\mathfrak{q} = -t^2$. Then $R(\mathfrak{q}) = \frac{\mathbb{C}[X,Y]}{(x^2+y^2-1)}$. In this case, by (8.8.7), $K_0(R) = \mathbb{Z} \oplus \mathbb{Z}_2$, and $K_0(R(\mathfrak{q})) = \mathbb{Z}$ by (8.8.8). So, ρ_0 is not injective.

Remark 8.7.16. The result (8.7.14) has certain special significance. With $R = \mathbb{R}$, this result amounts to saying that the algebraic K-theory (i.e. Grothendieck group) of $R(\mathfrak{q})$, coincides with the topological K-theory. Likewise for $R = \mathbb{C}$. All these culminated in the work of Swan [Sw2, Thm. 7.8], following [ABS, F]. See Section 8.8, for further details.

8.8. Algebraic and Topological K-Theory of Spheres

The goal of this section is to elaborate on the remark (8.7.16) that the algebraic and topological K-theory (i.e. Grothendieck group) coincide. This is a result of Swan [Sw2]. Before we proceed, we establish the following equivalences of categories, related to Clifford algebras.

Lemma 8.8.1. *Let R be a commutative Noetherian ring (with $1/2 \in R$), and $(P, \mathfrak{q})$ be a quadratic R-module, with P projective. As usual, denote $\langle 1 \rangle := (Re, T^2)$, where $Re \cong R$. Then we have*

$$\begin{cases} Coh_l(C(\mathfrak{q} \perp \langle 1 \rangle)) \approx Coh_{l,\mathbb{Z}_2}(C(\mathfrak{q})) \approx Coh_{l,\mathbb{Z}_2}(C(-\mathfrak{q})) \\ Coh_r(C(\mathfrak{q} \perp \langle 1 \rangle)) \approx Coh_{r,\mathbb{Z}_2}(C(\mathfrak{q})) \approx Coh_{r,\mathbb{Z}_2}(C(-\mathfrak{q})). \end{cases}$$

Subscript l indicates categories of left modules, and subscript r indicates categories of right modules.

Proof. We only prove the statement about categories of left modules. Note that $C(\mathfrak{q}) \subseteq C(\mathfrak{q} \perp \langle 1 \rangle)$ is a subring. Also, $e^2 = 1 \in C(\mathfrak{q} \perp \langle 1 \rangle)$. So, $e^2 - 1 = (e-1)(e+1) = 0$. Given $M \in Coh_l(C(q \perp \langle 1 \rangle))$, M is also a $C(\mathfrak{q})$-module. Given $m \in M$, we can write $m = \frac{(e+1)m}{2} + \frac{(e-1)m}{2}$. It follows that

$$M = M_0 \oplus M_1 \qquad \text{with} \qquad M_0 := \ker(e-1),\ M_1 = \ker(e+1).$$

For $p \in P$, we have $pe = -ep \in C(\mathfrak{q} \perp \langle 1 \rangle)$. It follows from this that the decomposition $M = M_0 \oplus M_1$ gives a $\mathbb{Z}_2$ graded $C(\mathfrak{q})$-module structure on M. Conversely, suppose $N = N_0 \oplus N_1 \in Coh_{l,\mathbb{Z}_2}(C(\mathfrak{q}))$. Let e act on N_0 as 1 and on N_1 as -1. This gives a $C(q \perp \langle 1 \rangle)$-module structure on N. This establishes $Coh_l(C(\mathfrak{q} \perp \langle 1 \rangle)) \approx Coh_{l,\mathbb{Z}_2}(C(\mathfrak{q}))$.

Now, we prove $Coh_{l,\mathbb{Z}_2}(C(\mathfrak{q})) \approx Coh_{l,\mathbb{Z}_2}(C(-\mathfrak{q}))$. Let $M = M_0 \oplus M_1 \in Coh_{l,\mathbb{Z}_2}(C(-\mathfrak{q}))$. For $p \in P$, define a new action of p on M, by $p \star m := pm\ \forall m \in M_0$, $p \star m := -pm\ \forall m \in M_1$. We have the commutative diagrams

$$\begin{array}{ccc} M_0 & \xrightarrow{p} & M_1 \\ & {\scriptstyle -\mathfrak{q}(m)} \searrow & \downarrow {\scriptstyle p} \\ & & M_0 \end{array} \quad \begin{array}{ccc} M_1 & \xrightarrow{p} & M_0 \\ & {\scriptstyle -\mathfrak{q}(m)} \searrow & \downarrow {\scriptstyle p} \\ & & M_1 \end{array} \implies \begin{array}{ccc} M_0 & \xrightarrow{p\star} & M_1 \\ & {\scriptstyle \mathfrak{q}(m)} \searrow & \downarrow {\scriptstyle p\star} \\ & & M_0 \end{array} \quad \begin{array}{ccc} M_1 & \xrightarrow{p\star} & M_0 \\ & {\scriptstyle \mathfrak{q}(m)} \searrow & \downarrow {\scriptstyle p\star} \\ & & M_1 \end{array}$$

By the universal property of $P \longrightarrow C(\mathfrak{q})$, this extends to a $C(\mathfrak{q})$-action on M. By the reversibility of the process, it follows that $Coh_{l,\mathbb{Z}_2}(C(\mathfrak{q})) \approx Coh_{l,\mathbb{Z}_2}(C(-\mathfrak{q}))$. The proof is complete. ■

As a consequence, we have the following commutative diagram of functors:

$$\begin{array}{ccc} \mathscr{P}_{r,\mathbb{Z}_2}(C(\mathfrak{q})) & \longrightarrow & \mathscr{P}_r(C(\mathfrak{q})) \\ \uparrow \wr & \nearrow & \\ \mathscr{P}_r(C(\mathfrak{q} \perp \langle 1 \rangle)) & & \end{array}$$

where the vertical arrow is an equivalence, and the diagonal arrow sends $M \mapsto M$, considering M as a $C(\mathfrak{q})$-module via the inclusion $C(\mathfrak{q}) \subseteq C(\mathfrak{q} \perp \langle 1 \rangle)$. Therefore, under the hypotheses of (8.7.12), the exact sequence in (8.7.12) can be rewritten as

$$\begin{array}{l} \cdots \longrightarrow K_{n+1}(R(\mathfrak{q})) \xrightarrow{\partial} K_n(C(\mathfrak{q} \perp \langle 1 \rangle)) \xrightarrow{(-\beta_*, \varepsilon_*)} \\ K_n(R) \oplus K_n(C(\mathfrak{q})) \xrightarrow{(\rho_0, \rho)_*} K_n(R(\mathfrak{q})) \longrightarrow \cdots \end{array} \tag{8.62}$$

where

$$\begin{cases} \beta : \mathscr{P}_r(C(\mathfrak{q} \perp \langle 1 \rangle) \longrightarrow \mathscr{P}(R) & \beta(M) = \ker(e+1) \\ \varepsilon : \mathscr{P}_r(C(\mathfrak{q} \perp \langle 1 \rangle)) \longrightarrow \mathscr{P}_r(C(\mathfrak{q})) & \varepsilon(M) = M \\ \rho_0 : \mathscr{P}(R) \longrightarrow \mathscr{P}(R(\mathfrak{q})) & \rho_1(M) = M \otimes R(\mathfrak{q}) \\ \rho : \mathscr{P}_r(C(\mathfrak{q})) \longrightarrow \mathscr{P}(R(\mathfrak{q})) & \rho_2(N) = \Gamma(U, \mathfrak{U}_{d-1}(N)) \\ & \quad = co\ker\left(\sum e_i X_i - 1\right). \end{cases}$$

In particular, of our particular interest in this section, is the exact sequence

$$K_0(C(\mathfrak{q} \perp \langle 1 \rangle)) \longrightarrow K_0(R) \oplus K_0(C(\mathfrak{q})) \xrightarrow{(\rho_0,\rho)_*} K_0(R(\mathfrak{q})) \longrightarrow 0.$$

Because of all these identifications (8.6.5, 8.7.10, 8.8.1), the same exact sequence has a variety of descriptions (8.58, 8.60, 8.62). Accordingly, $ABS(\mathfrak{q})$ defined in (8.7.13) has a variety of equivalent formulations. The following is one.

Corollary 8.8.2. *Suppose R is a regular Noetherian ring. Let $(P, \mathfrak{q})$ be a nondegenerate quadratic R-module. The definition of $ABS(\mathfrak{q})$ is equivalent to the exact sequence*

$$K_0(C(\mathfrak{q} \perp \langle 1 \rangle)) \longrightarrow K_0(C(\mathfrak{q})) \longrightarrow ABS(\mathfrak{q}) \longrightarrow 0.$$

Part of this endeavor, mainly due to Swan [Sw3, Sw4], is to bring these modern approaches [Sw2, Q] to the mold of [ABS], in which the topological K-theory of spheres was computed. The following is from [ABS].

Theorem 8.8.3. *Let $k = \mathbb{R}, \mathbb{C}$. Let $\mathbb{S}^n := \mathbb{S}^n(k)$ denote the real or complex sphere $\mathfrak{q}_n = \sum_{i=0}^n x_i^2 = 1$, and $ABS(k, \mathfrak{q}_n)$ denote $ABS(\mathfrak{q}_n)$, corresponding to k. Also, write*

$$A_n(k) = \frac{k[X_0, X_1, \dots, X_n]}{(\mathfrak{q}_n - 1)} \quad \textit{the algebraic coordinate ring of} \quad \mathbb{S}^n(k)$$

Then the following:

$$\begin{array}{ccc} ABS(\mathbb{R}, \mathfrak{q}_n) & \xrightarrow{\sim} & \widetilde{K}_0(A_n(\mathbb{R})) \\ & \searrow{\scriptstyle\sim} & \downarrow{\scriptstyle\wr} \\ & & \widetilde{KO}\,(\mathbb{S}^n(\mathbb{R})) \end{array} \qquad \begin{array}{ccc} ABS(\mathbb{C}, \mathfrak{q}_n) & \xrightarrow{\sim} & \widetilde{K}_0(A_n(\mathbb{C})) \\ & \searrow{\scriptstyle\sim} & \downarrow{\scriptstyle\wr} \\ & & \widetilde{KU}\,(\mathbb{S}^n(\mathbb{C})) \end{array}$$

are commutative diagrams of isomorphisms. As has been standard $KO(\mathbb{S}^n(\mathbb{R}))$, $KU(\mathbb{S}^n(\mathbb{C}))$ denote the Grothendieck groups of, respectively, the categories $Vect(\mathbb{S}^n(\mathbb{R}))$, $Vect(\mathbb{S}^n(\mathbb{C}))$, of real and complex vector bundles. And, $\widetilde{KO}(\mathbb{S}^n(\mathbb{R}))$ and $\widetilde{KU}(\mathbb{S}^n(\mathbb{C}))$ denote the kernel of the rank map. Likewise, $\widetilde{K}_0(A_n(k))$ denotes the kernel of the rank map, which coincides with the notation in (8.7.14).

Proof. The horizontal isomorphisms follow from (8.7.14). The diagonal maps are known to be isomorphisms [ABS, F], [Sw2, Thm. 7.8]. Hence, the vertical maps are also isomorphisms. ■

Therefore, to compute the algebraic invariant $\widetilde{K}_0(A_n(k))$ for $k = \mathbb{R}$, $\mathbb{C}$, we need to compute $ABS(k, \mathfrak{q}_n)$, which was done in [ABS]. Subsequently, for $k = \mathbb{R}$, $\mathbb{C}$, the notation $C(k, \mathfrak{q}_n)$ will denote the Clifford algebra of $\mathfrak{q}_n$ over k. Originally, in [ABS], the graded version of the description of the $ABS(k, \mathfrak{q}_n)$ was used, as follows:

$$K_0^{gr}(C(k, \mathfrak{p}_{n+1})) \longrightarrow K_0^{gr}(C(k, \mathfrak{p}_n)) \longrightarrow ABS(k, \mathfrak{q}_n) \longrightarrow 0$$

where $\mathfrak{p}_n = X_1^2 + X_2^2 + \cdots + X_n^2$, and hence $\mathfrak{q}_n = X_0^2 + \mathfrak{p}_n$. For the remaining part of the calculations, we will work with categories of left modules because computationally, choice of left or right modules would not make any difference. By (8.8.1), we have $C(k, \mathfrak{p}_n) \cong C(k, -\mathfrak{p}_n)$, and consequently, $K_0^{gr}(C(k, \mathfrak{p}_n)) \cong K_0^{gr}(C(k, -\mathfrak{p}_n))$. In a series of exercises, we compute $K_0^{gr}(C(k, \mathfrak{p}_n))$ and $ABS(k, \mathfrak{q}_n)$, as follows.

Exercise 8.8.4. Let $\mathbb{H}$ denote the quaternion algebra. Establish the following chart, for Clifford algebras:

n	$C(\mathbb{R}, -\mathfrak{p}_n)$	$C(\mathbb{R}, \mathfrak{p}_n)$
1	$\mathbb{C}$	$\mathbb{R} \oplus \mathbb{R}$
2	$\mathbb{H}$	$\mathbb{M}_2(\mathbb{R})$
3	$\mathbb{H} \oplus \mathbb{H}$	$\mathbb{M}_2(\mathbb{C})$
4	$\mathbb{M}_2(\mathbb{H})$	$\mathbb{M}_2(\mathbb{H})$
5	$\mathbb{M}_4(\mathbb{C})$	$\mathbb{M}_2(\mathbb{H}) \oplus \mathbb{M}_2(\mathbb{H})$
6	$\mathbb{M}_8(\mathbb{R})$	$\mathbb{M}_4(\mathbb{H})$
7	$\mathbb{M}_8(\mathbb{R}) \oplus \mathbb{M}_8(\mathbb{R})$	$\mathbb{M}_8(\mathbb{C})$
8	$\mathbb{M}_{16}(\mathbb{R})$	$\mathbb{M}_{16}(\mathbb{R})$

While computing these, keep track of the generators and grading. Further, for all $n \geq 1$, we have

$$C(\mathbb{R}, -\mathfrak{p}_{n+8}) \cong \mathbb{M}_8(C(\mathbb{R}, -\mathfrak{p}_n)), \qquad C(\mathbb{R}, \mathfrak{p}_{n+8}) \cong \mathbb{M}_8(C(\mathbb{R}, \mathfrak{p}_n)).$$

Exercise 8.8.5. Identify the maps $C(\mathbb{R}, -\mathfrak{p}_n) \hookrightarrow C(\mathbb{R}, -\mathfrak{p}_{n+1})$, in (8.8.4), as follows. We write $C_n = C(\mathbb{R}, -\mathfrak{p}_n)$. Then

$$\begin{array}{ccccccccccccccc}
C_1 & \to & C_2 & \to & C_3 & \to & C_4 & \to & C_5 & \to & C_6 & \to & C_7 & \to & C_8 \\
\| & & \| & & \| & & \| & & \| & & \| & & \| & & \| \\
\mathbb{C} & \to & \mathbb{H} & \xrightarrow[\Delta]{} & \mathbb{H} \oplus \mathbb{H} & \xrightarrow[\Delta]{} & \mathbb{M}_2(\mathbb{H}) & \xrightarrow[\iota_4]{} & \mathbb{M}_4(\mathbb{C}) & \xrightarrow[\iota_5]{} & \mathbb{M}_8(\mathbb{R}) & \xrightarrow[\Delta]{} & \mathbb{M}_8(\mathbb{R}) \oplus \mathbb{M}_8(\mathbb{R}) & \xrightarrow[\Delta]{} & \mathbb{M}_{16}(\mathbb{R})
\end{array}$$

Since these are maps of graded algebras, some of the maps are obvious. The maps Δ in the diagram are diagonal maps. The map ι_4 is induced by $\mathbb{H} \hookrightarrow \mathbb{M}_2(\mathbb{C})$, and ι_5 is induced by $\mathbb{C} \hookrightarrow \mathbb{M}_2(\mathbb{R})$.

Exercise 8.8.6. Let R be a ring (not necessarily commutative) and $S = \mathbb{M}_n(R)$ the ring of square matrices of order n. Let $\mathbf{R}(n)$, $\mathbf{C}(n)$, respectively, denote the groups of row, column matrices of length n. Then S acts on $\mathbf{C}(n)$ from left, and on $\mathbf{R}(n)$ from right. Define the functors

$$\begin{cases} Coh(R) \longrightarrow Coh(S) \ M \mapsto \mathbf{C}(n) \otimes_R M \\ Coh(S) \longrightarrow Coh(S) \ M \mapsto \mathbf{R}(n) \otimes_S M. \end{cases}$$

Prove that these two functors are the inverse of each other. This is known as the **Morita equivalence**.

Exercise 8.8.7. Identify the simple submodules of $C(\mathbb{R}, \mathfrak{p}_n)$ and establish the following table.

n	$K_0^{gr}(C(\mathbb{R}, -\mathfrak{p}_n))$	$ABS(\mathbb{R}, \mathfrak{q}_n)$
1	$\mathbb{Z}$	$\mathbb{Z}_2$
2	$\mathbb{Z}$	$\mathbb{Z}_2$
3	$\mathbb{Z}$	0
4	$\mathbb{Z} \oplus \mathbb{Z}$	$\mathbb{Z}$
5	$\mathbb{Z}$	0
6	$\mathbb{Z}$	0
7	$\mathbb{Z}$	0
8	$\mathbb{Z} \oplus \mathbb{Z}$	$\mathbb{Z}$

$$\text{and} \quad \begin{cases} K_0^{gr}\left(C(\mathbb{R}, -\mathfrak{p}_n)\right) \cong K_0^{gr}\left(C(\mathbb{R}, -\mathfrak{p}_{n+8})\right) \\ ABS\left(\mathbb{R}, \mathfrak{q}_n\right) = ABS\left(\mathbb{R}, \mathfrak{q}_{n+8}\right). \end{cases} \quad \forall n$$

Hint: Use (8.8.4). Note that all the matrix algebras are simple algebras. So, they have a unique irreducible isomorphism class of irreducible (ungraded) modules. Thus, the first column of $K_0^{gr}\left(C(\mathbb{R}, -\mathfrak{p}_n)\right)$ is computed. To compute $ABS(\mathbb{R}, \mathfrak{q}_n)$, use (8.8.5).

Exercise 8.8.8. Compute $C(\mathbb{C}, -\mathfrak{p}_n)$ and $C(\mathbb{C}, \mathfrak{p}_n)$ for all $n \geq 1$, as in (8.8.4). Further, establish the table

n	$K_0^{gr}\left(C(\mathbb{C}, -\mathfrak{p}_n)\right)$	$ABS\left(\mathbb{C}, \mathfrak{q}_n\right)$
1	$\mathbb{Z}$	$\mathbb{Z}_2$
2	$\mathbb{Z} \oplus \mathbb{Z}$	$\mathbb{Z}$

$$\text{and} \quad \begin{cases} K_0^{gr}\left(C(\mathbb{C}, -\mathfrak{p}_n)\right) \cong K_0^{gr}\left(C(\mathbb{C}, -\mathfrak{p}_{n+2})\right) \\ ABS\left(\mathbb{C}, \mathfrak{q}_n\right) = ABS\left(\mathbb{C}, \mathfrak{q}_{n+2}\right). \end{cases} \quad \forall n$$

This completes the computation of $\widetilde{K}_0\left(A_n(k)\right) \cong ABS\left(k, \mathfrak{q}_n\right)$ for $k = \mathbb{R}, \mathbb{C}$.

Chapter 9

Epilogue: K-Theory

9.1. Introduction to the Epilogues

At the beginning of this project, it was neither envisioned nor was it part of the plan, that anything beyond Quillen K-theory would be part of this book. The plan was to produce something faithful to the pedagogic approach, suitable for second-year graduate students in midlevel universities. However, it has been about half a century since the work of Quillen [Q] appeared. While the work of Quillen [Q] provided the foundation, the research in K-theory flourished during this half a century. For completeness, in this chapter and subsequently, we give an overview of the progress made during this period. The approach in these chapters differs from the preceding chapters in that this is more of an overview, and we refrain from any attempt to give complete details, faithful to the pedagogic style. My excuse for this decision is that the book has already grown long enough. Among what we tried to achieve in these chapters is to draw the attention of the readers to how all these are built on the foundation and analogies to the theory, and on methodologies already developed in the preceding and appendix chapters. This is all that is needed, in most cases, and a motivated reader will be able to work out a complete proof without having to go back to the original sources. Everything in K-theory, to this date, remains a part of a single long story.

G-theoretically, Quillen's work [Q] was fairly complete and satisfactory. That was not the case with K-theory of schemes. There was no localization sequence for K-theory, and Dévissage did not work. Waldhausen [Wh] vastly generalized the work of Quillen [Q] to address these issues regarding K-theory of schemes, together with [TT]. Scheme theoretically, perhaps, the most beautiful application of [Q] is the work of Swan on quadrics, which we covered in Chapter 8. Negative K-theory is another area of interest. We provide some overview of all these in this chapter.

The other important application of methods of K-theory is hermitian K-theory, the work of Schlichting [Sm3, Sm4], and others. We give an overview of the same in Chapter 10. Triangulated categories became an integral part of the parameters in the subsequent theory. In that, the theory on Triangulated Witt groups is a central piece [Bp, BW]. We give an overview of these in Chapter 11.

As outlined above, we discuss K-theory in the rest of this chapter.

9.2. Waldhausen K-Theory

Waldhausen expands Quillen K-theory in two directions.

1. First, it defines K-theory of categories with cofibrations, which encapsulates exact categories.
2. Further, it incorporates weak equivalences and defines K-theory of categories with cofibrations and weak equivalences. The K-theory of categories of bounded chain complexes together with quasi-isomorphisms are studied in this manner.

Relaxing the definition of exact categories, we define categories with cofibrations.

Definition 9.2.1. A category $\mathscr{C}$ with a distinguished zero object 0 is called a pointed category. A **category with cofibration** is a pointed category $\mathscr{C}$, together with a subcategory $co(\mathscr{C})$, satisfying three conditions (1, 2, 3), listed in what follows. The arrows in $co(\mathscr{C})$ are called **cofibrations** and are denoted by $\hookrightarrow$.

1. Any isomorphism in $\mathscr{C}$ is in $co(\mathscr{C})$.
2. For all objects $A \in \mathscr{C}$, the arrow $0 \longrightarrow A$ is in $co(\mathscr{C})$.
3. Any pair of arrows $C \xleftarrow{f} A \overset{\iota}{\hookrightarrow} B$, with $\iota \in co(\mathscr{C})$, $f \in \mathscr{C}$ can be completed to a pushforward diagram in $\mathscr{C}$:

$$\begin{array}{ccc} A & \overset{\iota}{\hookrightarrow} & B \\ {\scriptstyle f}\downarrow & & \downarrow{\scriptstyle \varphi} \\ C & \underset{\zeta}{\hookrightarrow} & B \coprod_A C \end{array} \qquad \ni \quad \zeta \in co(\mathscr{C}).$$

We say that cofibrations admit cobase change or closed under cobase change. In particular, $\mathscr{C}$ has finite coproducts.

Further, given any cofibration $\iota : A \hookrightarrow B$, we denote $B/A := B \coprod_A 0$. So, we have the pushforward diagram

$$\begin{array}{ccc} A & \overset{\iota}{\hookrightarrow} & B \\ \downarrow & & \downarrow{\scriptstyle \mathfrak{q}} \\ 0 & \underset{\zeta}{\hookrightarrow} & B/A \end{array} \qquad \ni \quad \zeta \in co(\mathscr{C}).$$

The map $\mathfrak{q} : B \twoheadrightarrow B/A$ is thought of as the quotient map. We also use this double-headed arrow $\twoheadrightarrow$ to denote this quotient map. For any choice of object $C \cong B/A$, we say that $A \overset{\iota}{\hookrightarrow} B \overset{\mathfrak{q}}{\twoheadrightarrow} C$ is a **cofibration sequence**.

The category of small categories with cofibrations will be denoted by co**Cat** (maps are functors).

Definition 9.2.2. Let $\mathscr{C}$ be a category with cofibrations $co(\mathscr{C})$. A subcategory of **weak equivalences** in $\mathscr{C}$ is defined as a subcategory $\mathbf{w} := w(\mathscr{C})$ of $\mathscr{C}$ such that the following conditions are satisfied:

1. Any isomorphism in $\mathscr{C}$ is in $\mathbf{w}$.
2. (Gluing lemma) Arrows in $\mathbf{w}$ are closed under pushforward, in the sense of the following pushforward diagram:

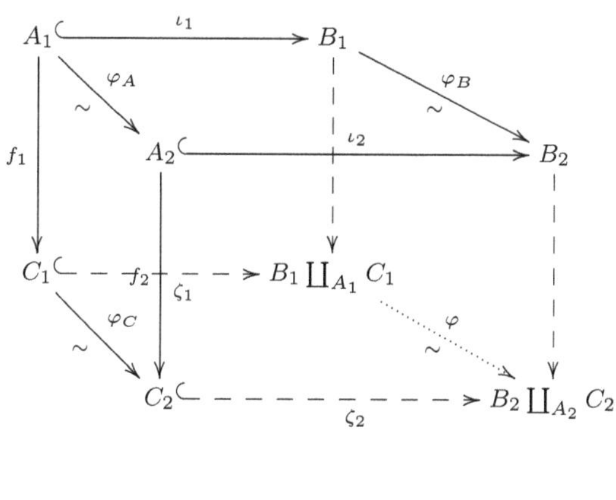

Here ι_1 and ι_2 are cofibrations, and other notation are obvious, with front and back walls being pushforward diagrams. If the diagonal arrows $\varphi_A, \varphi_B, \varphi_C$ are in $\mathbf{w}$, then so is the induced morphism φ.

Such a triple $\mathscr{C} := (\mathscr{C}, co(\mathscr{C}), \mathbf{w})$ is called a category with cofibrations and weak equivalences. We simply say that $\mathscr{C}$ is a **category with cofibrations and weak equivalences.** We also call them **Waldhausen categories.** The category of small Waldhausen categories will be denoted by $\mathbf{w}co\mathbf{Cat}$ (maps are functors).

Further, we introduce two more definitions for future reference:

1. The category $\mathbf{w}$ is said to satisfy **saturation axiom** if for $A \xrightarrow{f} B \xrightarrow{g} C$ composable arrows in $\mathscr{C}$, when two of f, g, gf are in $\mathbf{w}$, then so is the third one.
2. The category $\mathbf{w}$ is said to satisfy **extension axiom** if given a commutative diagram

$$\begin{array}{ccccc} A_1 & \longrightarrow & B_1 & \twoheadrightarrow & C_1 \\ {\scriptstyle f_A}\downarrow & & \downarrow{\scriptstyle f_B} & & \downarrow{\scriptstyle f_C} \\ A_2 & \longrightarrow & B_2 & \twoheadrightarrow & C_2 \end{array}$$

where the rows are cofibration sequences, if $f_A, f_C \in \mathbf{w}$, then $f_B \in \mathbf{w}$.

We proceed to define the K-theory of Waldhausen categories.

Definition 9.2.3. Let $\mathscr{C} := (\mathscr{C}, co(\mathscr{C}))$ be a small category with cofibrations. We describe Waldhausen's $S_\bullet(\mathscr{C})$ construction as a simplicial category

$$\Delta^{op} \longrightarrow \mathbf{Cat}, \quad \text{the category of small categories.}$$

For integers $n \geq 0$, let $Ar[n] = \{(i,j) \in [n] \times [n] : 0 \leq i \leq j \leq n\}$ denote the partially ordered set. Let $Fun(Ar[n], \mathscr{C})$ denote the category of functors $Ar[n] \longrightarrow \mathscr{C}$ and natural transformations. The category $S_n(\mathscr{C}) \subseteq Fun(Ar[n], \mathscr{C})$ is defined to be the subcategory, whose objects as functors,

$$A : Ar[n] \longrightarrow \mathscr{C} \quad \ni$$
$$\begin{cases} A_{ii} = 0 \quad \forall\, i \\ A_{ij} \hookrightarrow A_{ik} \twoheadrightarrow A_{jk} \quad \text{is a cofibration sequence } \forall i \le j \le k. \end{cases}$$

Therefore, for such an object $A \in S_n(\mathscr{C})$, there is a pushforward diagram

$$\begin{array}{ccccc} & & A_{ij} & \hookrightarrow & A_{ik} \\ & & \downarrow & & \downarrow \\ A_{jj} & = & 0 & \longrightarrow & A_{jk} \end{array} \qquad \forall i \le j \le k$$

Often, an object $A \in S_n(\mathscr{C})$ is displayed by a diagram, as follows:

$$\begin{array}{ccccccccccccc}
0 & \hookrightarrow & A_{01} & \hookrightarrow & A_{02} & \hookrightarrow & A_{03} & \hookrightarrow & \cdots & \hookrightarrow & A_{0,n-1} & \hookrightarrow & A_{0n} \\
 & & & & \downarrow & & \downarrow & & \downarrow & & \downarrow & & \downarrow \\
 & & 0 & \hookrightarrow & A_{12} & \hookrightarrow & A_{13} & \hookrightarrow & \cdots & \hookrightarrow & A_{1,n-1} & \hookrightarrow & A_{1n} \\
 & & & & & & \downarrow & & \downarrow & & \downarrow & & \downarrow \\
 & & & & 0 & \hookrightarrow & A_{23} & \hookrightarrow & \cdots & \hookrightarrow & A_{2,n-1} & \hookrightarrow & A_{2n} \\
 & & & & & & & & \downarrow & & \downarrow & & \downarrow \\
 & & & & & & 0 & \hookrightarrow & \cdots & \hookrightarrow & A_{3,n-1} & \hookrightarrow & A_{3n} \\
 & & & & & & & & & & & & \downarrow \\
 & & & & & & & & & & & & \vdots \\
 & & & & & & & & & & & & \downarrow \\
 & & & & & & & & & & 0 & \hookrightarrow & A_{n-1,n} \\
 & & & & & & & & & & & & \downarrow \\
 & & & & & & & & & & & & 0
\end{array} \tag{9.1}$$

As usual, given $A, B \in S_n(\mathscr{C})$, a morphism is a natural transformation $A \longrightarrow B$, which can be displayed by extending the above to a cubical diagram. We record the following two comments:

1. This category $S_n(\mathscr{C})$ is equivalent to the category $S_n^{0\bullet}(\mathscr{E})$, whose objects are the first line of (9.1).
2. The category $S_n(\mathscr{C})$ inherits a structure of a category with cofibrations $co(S_\bullet(\mathscr{E}))$ by defining an arrow $f : A \longrightarrow B$ as a cofibration if the maps $f_{ij} : A_{ij} \longrightarrow B_{ij}$ are in $co(\mathscr{E})$ for all $(i, j) \in Ar[n]$.

It is clear that

$$S_\bullet(\mathscr{E}) : \Delta^{op} \longrightarrow \mathbf{Cat} \quad \text{sending} \quad [n] \mapsto S_n(\mathscr{E})$$

is a simplicial category. To define face and degeneracies, note that the first line of (9.1) dictates the rest of these maps. The face map $d_i : S_n(\mathscr{E}) \longrightarrow S_{n-1}(\mathscr{E})$ for $1 \leq i \leq n-1$ is obtained by composing maps in (9.1), for $i = 0$ by forgetting the first line in (9.1), and for $i = n$ by dropping the last vertical line in (9.1). The degeneracies $s_i : S_n(\mathscr{E}) \longrightarrow S_{n+1}(\mathscr{E})$ are obtained by inserting identities in (9.1). By (2), we can consider

$$S_\bullet(\mathscr{C}) : \Delta^{op} \longrightarrow co\mathbf{Cat}$$
$$\text{as a simplicial object in the category} \quad co\mathbf{Cat}. \tag{9.2}$$

We incorporate weak equivalences into this construction of $S_\bullet(\mathscr{E})$, and define the **K**-theory space of categories with cofibrations and weak equivalences.

Definition 9.2.4. Let $\mathscr{C} = (\mathscr{C}, co\mathscr{C}, \mathbf{w})$ be a category with cofibrations and weak equivalences. Then for integers $n \geq 0$, a morphism $f : A \longrightarrow B$ in $S_n(\mathscr{C})$ is defined to be a weak equivalence if the arrows $f_{ij} : A_{ij} \longrightarrow B_{ij}$ are in $\mathbf{w}$ for all $(i, j) \in Ar[n]$. This gives $S_n(\mathscr{C})$ the structure of a Waldhausen category. Following (9.2), the association $n \mapsto S_n\mathscr{C}$ defines the functor

$$S_\bullet(\mathscr{C}) : \Delta^{op} \longrightarrow \mathbf{w}co\mathbf{Cat}$$
$$\text{as a simplicial object in the category} \quad \mathbf{w}co\mathbf{Cat}. \tag{9.3}$$

The subcategory of weak equivalences is denoted by $\mathbf{w}S_n(\mathscr{C}) \subseteq S_n(\mathscr{C})$. For such a category $\mathscr{C} = (\mathscr{C}, co\mathscr{C}, \mathbf{w})$ with cofibrations and weak equivalences, define the following:

1. In analogy to the definition of classifying spaces of the $\mathbb{Q}$-categories (Section 3.2), define

$$\mathbb{B}(\mathscr{C}, \mathbf{w}) = |\mathbf{w}S_\bullet(\mathscr{C})|$$

the geometric realization of the simplicial category $\mathbf{w}S_\bullet(\mathscr{C}) \subseteq S_\bullet(\mathscr{C})$. This is the geometric realization of the simplicial topological space

$$n \mapsto |\mathbf{w}S_n(\mathscr{C})| = \mathbb{B}\left(\mathbf{w}S_n(\mathscr{C})\right) = \left|N_\bullet\left(\mathbf{w}S_n(\mathscr{C})\right)\right|.$$

Here $N_\bullet(-)$ denotes the nerve of the category. By (1.3.2, 9.2.6), $\mathbb{B}(\mathscr{C}, \mathbf{w})$ is the geometric realization of the bisimplicial set

$$(m, n) \mapsto N_m\left(\mathbf{w}S_n(\mathscr{C})\right).$$

2. The $\mathbf{K}$-theory space of $\mathscr{C} = (\mathscr{C}, co\mathscr{C}, \mathbf{w})$ is defined to be the pointed space,

$$\mathbf{K}\left(\mathscr{C}, \mathbf{w}\right) := \Omega\left|\mathbf{w}S_\bullet(\mathscr{C})\right| = \Omega\mathbb{B}(\mathscr{C}, \mathbf{w}) \qquad \text{the loop space.}$$

3. Further, for integers $n \geq 0$, define

$$K_n\left(\mathscr{C}, \mathbf{w}\right) = \pi_n\left(\mathbf{K}\left(\mathscr{C}, \mathbf{w}\right), 0\right).$$

4. For future reference, we define the simplicial set

$$s_\bullet(\mathscr{C}) : \Delta^{op} \longrightarrow \mathbf{Sets} \quad \text{by} \quad s_n(\mathscr{C}) = Obj\left(S_n(\mathscr{C})\right). \tag{9.4}$$

Note that $s_\bullet(\mathscr{C}) = N_0\left(\mathbf{w}S_\bullet(\mathscr{E})\right)$ is the set of vertices in the $\mathbf{w}$-direction. So, there is a natural map

$$|s_\bullet(\mathscr{C})| \longrightarrow \left|k \mapsto \left|N_k\left(\mathbf{w}S_\bullet(\mathscr{E})\right|\right|\right. = \left|\mathbf{w}S_\bullet\mathscr{E}\right|. \tag{9.5}$$

The following is an immediate and important observation.

Lemma 9.2.5. *Let $\mathscr{C} = (\mathscr{C}, co\mathscr{C}, \mathbf{w})$ be a category with cofibrations and weak equivalences. Then there is a natural map*

$$|\mathbf{w}\mathscr{C}| \longrightarrow \Omega\mathbb{B}(\mathscr{C}, \mathbf{w}) = \Omega\left|\mathbf{w}S_\bullet(\mathscr{C})\right| = \mathbf{K}(\mathscr{C}, \mathbf{w}). \tag{9.6}$$

Proof. In fact,

$$\mathbb{B}(\mathscr{C}, \mathbf{w}) = \frac{\coprod_n |\mathbf{w}S_n(\mathscr{E})| \times \Sigma^n}{\sim} \quad \text{where } \Sigma^n$$

is the standard n-simplex.

In the numerator, the first factor $|\mathbf{w}S_0(\mathscr{C})| \times \Sigma^0 = 0$ is a singleton, and the second factor is $|\mathbf{w}S_1(\mathscr{E})| \times \Sigma^1 = |\mathbf{w}\mathscr{C}| \times \Sigma^1$. So, for $x \in |\mathbf{w}\mathscr{C}|$, the association $x \mapsto x \times \Sigma^1 \in |\mathbf{w}S_1(\mathscr{E})| \times \Sigma^1$ defines the required map (9.6). The proof is complete. ■

Now we iterate the $S_\bullet$-construction. Before that, we record the following definition.

Definition 9.2.6. An n-fold simplicial object in a category $\mathscr{C}$ is a functor

$$K_{\bullet_1\bullet_2\cdots\bullet_n} : (\Delta^{op})^n \longrightarrow \mathscr{C}.$$

Given such a simplicial object, define the diagonal object

$$\mathfrak{D}K_\bullet : \Delta^{op} \longrightarrow \mathscr{C} \qquad \text{by} \quad \mathfrak{D}K_r = K_{r,r,\ldots,r}.$$

Assume $\mathscr{C} = \mathbf{Top}$ or $\mathscr{C} = \mathbf{SSet}$. With the help of the Bisimplicial Lemma 1.3.2, the geometric realization $|K_{\bullet_1\bullet_2\cdots\bullet_n}|$ can be defined and computed inductively, or by the formula

$$|K_{\bullet_1\bullet_2\cdots\bullet_n}| = |\mathfrak{D}K_\bullet|.$$

Definition 9.2.7. Let $\mathscr{C} = (\mathscr{C}, co\mathscr{C}, \mathbf{w})$ be a small category with cofibrations and weak equivalences. By (9.2.4), $S_n\mathscr{C} \in \mathbf{w}co\mathbf{Cat}$. So, the process of $S_\bullet$-construction can be iterated. For integers $n \geq 0$, inductively define the n-fold simplicial object in $\mathbf{w}co\mathbf{Cat}$:

$$S^{(n)}(\mathscr{C}) := S^{(n)}_{\bullet_1\bullet_2\cdots\bullet_n}(\mathscr{C}) : (\Delta^{op})^n \longrightarrow \mathbf{w}co\mathbf{Cat} \ \text{ by}$$

$$S^{(n)}_{m_1 m_2\cdots m_n}(\mathscr{C}) = S_{m_1}\left(S^{(n-1)}_{m_2\cdots m_n}(\mathscr{C})\right).$$

Let $\mathbf{w}_n := \mathbf{w}S_{\bullet_1\bullet_2\cdots\bullet_n}(\mathscr{C}) \subseteq S_{\bullet_1\bullet_2\cdots\bullet_n}(\mathscr{C})$ denote the simplicial subcategory of weak equivalences. The diagonal of $S^{(n)}_{\bullet_1\bullet_2\cdots\bullet_n}(\mathscr{C})$ is

$$\mathfrak{D}\left(S^{(n)}(\mathscr{C})\right)_{\bullet} : \Delta^{op} \longrightarrow \mathbf{w}co\mathbf{Cat} \quad \text{defined by}$$
$$\mathfrak{D}\left(S^{(n)}(\mathscr{C})\right)_{m} = S_m\left(S^{(n-1)}_{m\cdots m}(\mathscr{C})\right).$$

Let $\mathbf{w}_n\mathfrak{D}\left(S^{(n)}(\mathscr{C})\right)_{\bullet} \subseteq \mathfrak{D}\left(S^{(n)}(\mathscr{C})\right)_{\bullet}$ denote the subcategory of weak equivalences. As before, we denote

$$\mathbb{B}(\mathscr{C}, \mathbf{w_n}) := \left|\mathbf{w}S^{(n)}_{\bullet_1\bullet_2\cdots\bullet_n}(\mathscr{C})\right| = \left|\mathbf{w}\mathfrak{D}\left(S^{(n)}(\mathscr{C})\right)_{\bullet}\right|$$

where the last equality follows from (1.3.2, 9.2.6). Thus, we obtain a sequence to CW complexes

$$\left\{|\mathbf{w}\mathscr{C}|, |\mathbf{w}S_{\bullet}\mathscr{C}|, \ldots, \left|\mathbf{w}S^{(n)}_{\bullet_1\bullet_2\cdots\bullet_n}\mathscr{C}\right|, \cdots\right\} \tag{9.7}$$

where the K-theory space is $\mathbf{K}(\mathscr{C}, \mathbf{w}) = \Omega\,|\mathbf{w}S_{\bullet}\mathscr{C}|$.

We summarize the above, as follows.

Lemma 9.2.8. *Let $\mathscr{C} = (\mathscr{C}, co\mathscr{C}, \mathbf{w})$ be a small category with cofibrations and weak equivalences. Then there are continuous natural maps*

$$\begin{cases} \sigma_0 : |\mathbf{w}\mathscr{C}| \longrightarrow \Omega\mathbb{B}(\mathscr{C}, \mathbf{w_1}) \\ \sigma_n : \mathbb{B}(\mathscr{C}, \mathbf{w_n}) \longrightarrow \Omega\mathbb{B}(\mathscr{C}, \mathbf{w_{n+1}}) \quad \forall n \geq 1. \end{cases} \tag{9.8}$$

Further, the maps in the second line are homotopy equivalences.

Proof. The map in the first line was obtained in (9.6), and maps in the second line are obtained by the repeated applications of (9.6). It follows from (9.8), that the maps $\mathbb{B}(\mathscr{C}, \mathbf{w_n}) \longrightarrow \Omega\mathbb{B}(\mathscr{C}, \mathbf{w_{n+1}})$ are homotopy equivalences $\forall n \geq 1$. The proof is complete. ∎

We would like to encapsulate information in the above (9.2.8) into the mold of Ω-spectrum.

Definition 9.2.9. A **spectrum** is a sequence of pointed topological spaces $\mathbf{E} := \{E_0, E_1, \ldots\}$, together with maps $\sigma_n : E_n \longrightarrow \Omega E_{n+1}$, to be called the bonding maps. Given such a spectrum, $\forall n \in \mathbb{Z}$, define the homotopy groups

$$\pi_n(\mathbf{E}) = co \lim_{k\geq 0, n+l\geq 0, k-l\geq 0} \pi_{n+l}\left(\Omega^{k-l}E_k\right). \tag{9.9}$$

One checks that, for fixed k, l, with $k-l \geq 0$, $n+l \geq 0$, we have

$$\pi_n(\mathbf{E}) = co\lim\left(\pi_{n+l}\left(\Omega^{k-l}E_k\right) \longrightarrow \pi_{n+l}\left(\Omega^{k-l+1}E_{k+1}\right) \longrightarrow \right.$$
$$\left.\pi_{n+l}\left(\Omega^{k-l+2}E_{k+2}\right) \longrightarrow \cdots\right). \tag{9.10}$$

Such a spectrum $\mathbf{E} := \{E_i : i \geq 0\}$ is called an **Ω-spectrum** if the bonding maps σ_n are homotopy equivalences. If $\mathbf{E}$ is an Ω-spectrum, then for integers $n \in \mathbb{Z}$,

$$\pi_n(\mathbf{E}) := \pi_{n+k}(E_k) = \pi_{n+k+1}(E_{k+1}) \qquad \forall n+k \geq 0, \; k \geq 0.$$

So, for $n \geq 0$, $\pi_n(\mathbf{E}) = \pi_n(E_0)$, and for $n \leq -1$, we have $\pi_n(\mathbf{E}) = \pi_0(E_{-n})$.

1. A map of spectra $\mathbf{f} : \mathbf{E} = \{E_n\} \longrightarrow \mathbf{F} = \{F_n\}$ is a sequence of maps $f_i : E_i \longrightarrow F_i$ such that the diagram

$$\begin{array}{ccc} E_n & \longrightarrow & \Omega E_{n+1} \\ {\scriptstyle f_n}\downarrow & & \downarrow{\scriptstyle \Omega f_n} \\ F_n & \longrightarrow & \Omega F_{n+1} \end{array} \quad \text{commutes } \forall n.$$

2. A map $\mathbf{f} : \mathbf{E} = \{(E_n, x_n)\} \longrightarrow \mathbf{F} = \{(F_n, y_n)\}$ of spectra is defined as an **equivalence** of spectra if it induces isomorphisms $\pi_n(\mathbf{E}) \xrightarrow{\sim} \pi_n(\mathbf{F})$ $\forall n \in \mathbb{Z}$. In this case, we also (*loosely*) say that $\mathbf{f}$ is a homotopy equivalence of spectra.
3. Given a map $\mathbf{f} : \mathbf{E} = \{(E_n, x_n)\} \longrightarrow \mathbf{F} = \{(F_n, y_n)\}$ of spectra, its homotopy fiber is defined as $\mathscr{F}(\mathbf{f}) := \{\mathscr{F}(f_n, y_n)\}$, together with the induced maps $\mathscr{F}(f_n, y_n) \longrightarrow \Omega\mathscr{F}(f_{n+1}, y_{n+1})$.
4. A sequence of maps $\mathbf{X} \xrightarrow{\iota} \mathbf{E} \xrightarrow{\mathbf{f}} \mathbf{F}$ of spectra is defined as a **homotopy fibration** of spectra if (1) $\mathbf{f}\iota$ is homotopic to the trivial map in $\mathbf{Top}_\bullet$ and (2) the induced map $\mathbf{X} \longrightarrow \mathscr{F}(\mathbf{f})$ is an equivalence of spectra.

Remark 9.2.10. In order to cast (9.2.8) into this mold of Ω-spectrum, write

$$\begin{cases} E_n = \Omega \left| \mathbf{w}S^{(n+1)}(\mathscr{C}) \right|, \\ \sigma_n : E_n \longrightarrow \Omega E_{n+1} \text{ are the bonding maps induced by (9.8)} \\ \mathrm{Sp}(\mathscr{C}, \mathbf{w}) = \{E_n : n \geq 0\} \quad \text{So, } E_0 = \Omega \left| \mathbf{w}S^{(1)}(\mathscr{C}) \right| \\ \quad \text{is the degree 0 term.} \end{cases} \tag{9.11}$$

So, $E_0 = \Omega \left| \mathbf{w}S_\bullet(\mathscr{C}) \right| = \mathbf{K}(\mathscr{C}, \mathbf{w})$ is the K-theory space. By (9.2.13), $\mathbf{E} = \{E_n\}$ is an Ω-spectrum, to be called the K-theory spectrum of $\mathscr{C}$. Accordingly, $K_n(\mathscr{C}, \mathbf{w}) = \pi_n(\mathbf{E})$ is defined for all $n \in \mathbb{Z}$. However, it follows from (9.2.13) that $K_n(\mathscr{E}, \mathbf{w}) = 0, \forall n \leq -1$. Such a spectrum, with trivial negative homotopy groups, is referred to as a **connective spectrum**. So, we do not get anything new (or negative K-groups) in this manner. However, this discussion on spectrum will be instructive later.

Exercise 9.2.11. Let $(\mathscr{C}, co(\mathscr{C}), \mathbf{w})$ be a small Waldhausen category. Then $K_0(\mathscr{C}, \mathbf{w}) = \pi_0 \left(\mathbf{K}(\mathscr{C}, \mathbf{w}) \right) = \pi_1 \left(\mathbf{K}(\left| \mathbf{w}S_\bullet(\mathscr{C}) \right|) \right)$ is the free group (or free abelian group) generated by symbols $[A]$, where A runs through all objects of $\mathscr{C}$ subject to the following two relations:

1. $[A] = [B]$ if there is a morphism $A \xrightarrow{\sim} B$ in $\mathbf{w}$.
2. If $A \hookrightarrow B \twoheadrightarrow C$ is cofibration sequence, then $[A][C] = [B]$.

The group defined by these relationships is commutative because $A \hookrightarrow A \coprod_0 B \twoheadrightarrow B$ is a cofibration sequence. **Hint:** Similar to (3.2.3).

For completeness, we insert the following from [Wh], which we used above.

Lemma 9.2.12. *Let $X_\bullet \in \mathbf{SSet}$ be a simplicial set. Define the path space $\mathbf{P}X_\bullet \in \mathbf{SSet}$ by $\mathbf{P}X_n = X_{n+1}$. Also, define the constant simplicial set $\mathbf{C}X_\bullet$ by $\mathbf{C}X_n = X_0$. Then $|\mathbf{P}X_\bullet| \cong |\mathbf{C}X_\bullet|$. In particular, if $X_0 = \star$ is singleton, then $|\mathbf{P}X_\bullet|$ is contractible.*

Lemma 9.2.13. *Let $(\mathscr{C}, co(\mathscr{C}), \mathbf{w})$ be a Waldhausen category. By identifying $\mathbf{w}\mathscr{C} \cong \mathbf{w}S_1\mathscr{C}$, we obtain a sequence $\mathbf{w}\mathscr{C} \longrightarrow \mathbf{P}(\mathbf{w}S_\bullet\mathscr{C}) \longrightarrow \mathbf{w}S_\bullet\mathscr{C}$. The composite map is constant. Since the middle term is contractible, we obtain a map $|\mathbf{w}\mathscr{C}| \longrightarrow$*

$\Omega\,|\mathbf{w}S_\bullet\mathscr{C}| = \mathbf{K}(\mathscr{C},\mathbf{w})$. Up to homotopy, this map coincides with the map (9.6). Applying the same to $\mathscr{C} := S_\bullet\mathscr{C}$, we obtain the diagram

$$\begin{array}{ccccc} |\mathbf{w}S_\bullet\mathscr{C}| & \longrightarrow & \left|\mathbf{P}(\mathbf{w}S_\bullet^{(2)}\mathscr{C})\right| & \longrightarrow & \left|\mathbf{w}S_\bullet^{(2)}\mathscr{C}\right| \\ \downarrow{\scriptstyle \sigma_1} & & & & \\ \Omega\left|\mathbf{w}S_\bullet^{(2)}\mathscr{C}\right|. & & & & \end{array}$$

Then

1. *The horizontal sequence is a homotopy fibration, while $\left|\mathbf{P}(\mathbf{w}S_\bullet^{(2)}\mathscr{C})\right|$ is contractible.*
2. *Consequently, by (C.3.8), the vertical map σ_1 is a homotopy equivalence. The notation σ_1 was used to match the notation in (9.8).*
3. *Further, $\pi_1\left(\left|\mathbf{w}S_\bullet^{(2)}\mathscr{C}\right|\right) = \pi_0\left(\left|\mathbf{w}S_\bullet^{(2)}\mathscr{C}\right|\right) = 0$.*
4. *Inductively, we have that $\sigma_n : \left|\mathbf{w}S_\bullet^{(n)}\mathscr{C}\right| \longrightarrow \left|\mathbf{w}S_\bullet^{(n+1)}\mathscr{C}\right|$ is a homotopy equivalence, and $\pi_i\left(\left|\mathbf{w}S_\bullet^{(n)}\mathscr{C}\right|\right) = 0\ \forall i \le n-1$.*

Proof. See [Wh, Sec. 1.5] for further details. ■

9.2.1. *Exactness and Functorial properties*

Definition 9.2.14. Let $(\mathscr{C}, co(\mathscr{C}), \mathbf{w}\mathscr{C})$, $(\mathscr{D}, co(\mathscr{D}), \mathbf{w}\mathscr{D})$ be two Waldhausen categories. Subsequently, we may use the notation $\mathbf{w_C} := \mathbf{w}(\mathscr{C})$, etc. A functor $F : \mathscr{C} \longrightarrow \mathscr{D}$ is called an **exact functor** if F respects the structure of Waldhausen categories. This means that $F(co(\mathscr{C})) \subseteq co(\mathscr{D})$, $F(\mathbf{w_C}) \subseteq \mathbf{w_D}$, and F preserves pushforward, along cofibrations. The last condition means that, for any pair of arrows $C \longleftarrow A \overset{\iota}{\hookrightarrow} B$ in $\mathscr{C}$ with $\iota \in co(\mathscr{C})$, the induced map $FB \coprod_{FA} FC \xrightarrow{\sim} F(B \coprod_A C)$ is an isomorphism.

Let $F, G : \mathscr{C} \longrightarrow \mathscr{D}$ be exact functors of Waldhausen categories. We say that F and G are **weakly equivalent** if there is a natural transformation $\eta : F \longrightarrow G$ such that $\eta_A : FA \longrightarrow GA$ is in $\mathbf{w_D}$ for all $A \in Obj(\mathscr{C})$.

The following is on functorial properties.

Lemma 9.2.15. *Let $F : \mathscr{C} \longrightarrow \mathscr{D}$ be an exact functor of Waldhausen categories. Then F induces the maps,*

$$\begin{cases} |S_\bullet F| : |\mathbf{w}S_\bullet(\mathscr{C})| \longrightarrow |\mathbf{w}S_\bullet(\mathscr{D})| & \textit{in } \mathbf{Top} \\ KF : \mathbf{K}(\mathscr{C}, \mathbf{w_C}) \longrightarrow \mathbf{K}(\mathscr{C}, \mathbf{w_D}) & \textit{in } \mathbf{Top} \\ \mathrm{Sp}F : \mathrm{Sp}(\mathscr{C}, \mathbf{w_C})) \longrightarrow \mathrm{Sp}(\mathscr{D}, \mathbf{w_D})) & \textit{of spectra.} \end{cases}$$

These associations are functorial, from the category of Waldhausen categories and exact functors, to **Top** *or to the category of spectra, as the case may be.*

Let $F, G : \mathscr{C} \longrightarrow \mathscr{D}$ be exact functors of Waldhausen categories. Assume there is a weak equivalence $\eta : F \longrightarrow G$. Then η induces homotopies $|S_\bullet F| \sim |S_\bullet G|$, $KF \sim KG$, $\mathrm{Sp}F \sim \mathrm{Sp}G$.

Proof. Similar to (2.1.6, 2.1.9). Also see [Wh, Prop. 1.3.1], [TT, 1.5.4]. ■

Before we proceed, recall the versions Theorem AW (2.3.9), and Theorem BW (2.3.10) in [Wh], of Theorem A, B (2.2.2, 2.2.3).

9.2.2. *Agreement with Quillen K-theory*

In this section, we would like to comment on the relationship between Quillen K-theory and Waldhausen K-theory. Before we state (9.2.18) and establish the relationship, we do some preparatory work.

Suppose that $\mathscr{B}$ is a category, and $\mathscr{A} \subseteq \mathscr{B}$ is a subcategory. Assume $Obj(\mathscr{A}) = Obj(\mathscr{B})$. For integers $m, n \geq 0$, let N_{mn} be the set of all commutative diagrams

$$\begin{array}{ccccccccccc} A_{00} & \longrightarrow & A_{01} & \longrightarrow & A_{02} & \longrightarrow & A_{03} & \longrightarrow & \cdots & \longrightarrow & A_{0n} \\ \downarrow & & \downarrow & & \downarrow & & \downarrow & & & & \downarrow \\ A_{10} & \longrightarrow & A_{11} & \longrightarrow & A_{12} & \longrightarrow & A_{13} & \longrightarrow & \cdots & \longrightarrow & A_{1n} \\ \downarrow & & \downarrow & & \downarrow & & \downarrow & & & & \downarrow \\ A_{20} & \longrightarrow & A_{21} & \longrightarrow & A_{22} & \longrightarrow & A_{23} & \longrightarrow & \cdots & \longrightarrow & A_{2n} \\ \downarrow & & \downarrow & & \downarrow & & \downarrow & & & & \downarrow \\ \cdots & \longrightarrow & \cdots & \longrightarrow & \cdots & \longrightarrow & \cdots & \longrightarrow & \cdots & \longrightarrow & \cdots \\ \downarrow & & \downarrow & & \downarrow & & \downarrow & & & & \downarrow \\ A_{m0} & \longrightarrow & A_{m1} & \longrightarrow & A_{m2} & \longrightarrow & A_{m3} & \longrightarrow & \cdots & \longrightarrow & A_{mn} \end{array} \qquad (9.12)$$

where the horizontal arrows are in $\mathscr{B}$ and the vertical arrows are in $\mathscr{A}$. This defines a bisimplicial set, $\mathscr{A}\mathscr{B} := N_{\bullet\bullet} : \Delta^{op} \times \Delta^{op} \longrightarrow$ **SSet**. So, for fixed m, $N_{m\bullet} \in$ **SSet**. An n-simplex in $N_{m\bullet}$ consists of m composable arrows in $\mathscr{N}_n\mathscr{B}$. Likewise, for fixed n, $N_{\bullet n} \in$ **SSet**, and an m-simplex consists of n arrows in $\mathscr{N}_m\mathscr{A}$.

The nerve $\mathscr{N}_\bullet\mathscr{B}$ can be considered a bisimplicial set by defining $B_{mn} = \mathscr{N}_n\mathscr{B}$, with the vertical maps being identities. So, there is a natural inclusion $\iota : B_{\bullet\bullet} \hookrightarrow \mathscr{A}\mathscr{B}$. Since $\mathbb{B}\mathscr{B} = |B_{\bullet\bullet}|$, we obtain a natural map $\mathbb{B}\mathscr{B} \longrightarrow |\mathscr{A}\mathscr{B}|$.

Lemma 9.2.16 (Swallowing Lemma). *With the notation, as above, the map $\mathbb{B}\mathscr{B} \longrightarrow |\mathscr{A}\mathscr{B}|$ is a homotopy equivalence.*

Proof. Fix $m \geq 0$. It is clear that $\mathscr{N}_\bullet\mathscr{B} \cong B_{m\bullet}$ as simplicial sets. Consider the inclusion $\iota_m : B_{m\bullet} \subseteq N_{m\bullet}$ of simplicial sets. There is a map $p_m : N_{m\bullet} \longrightarrow \mathscr{N}_\bullet\mathscr{B}$ sending an object (9.12) to the first row. The diagram

$$\begin{array}{ccc} B_{m\bullet} & \xrightarrow{\iota_m} & N_{m\bullet} \\ & {\scriptstyle 1}\searrow & \downarrow{\scriptstyle p_m} \\ & & \mathscr{N}_\bullet\mathscr{B} \end{array}$$

gives a left inverse of the map $|\iota_m| : \mathbb{B}\mathscr{B} \longrightarrow |N_{m\bullet}|$. In fact, $\iota_m p_m$ takes the object (9.12) to m copies of the first row, connected by identity. Rotating ninety degrees and redrawing (9.12), we obtain a natural transformation $\iota_m p_m \mapsto 1$, as follows:

$$\begin{array}{ccccccccc} A_{0\bullet} & \xrightarrow{1} & A_{0\bullet} & \xrightarrow{1} & \cdots & \longrightarrow & A_{0\bullet} & \xrightarrow{1} & A_{0\bullet} \\ {\scriptstyle 1}\downarrow & & \downarrow{\scriptstyle f_{0\bullet}} & & & & \downarrow & & \downarrow{\scriptstyle f_{m-1\bullet}\cdots f_{0\bullet}} \\ A_{0\bullet} & \xrightarrow[f_{0\bullet}]{} & A_{1\bullet} & \xrightarrow[f_{1\bullet}]{} & \cdots & \longrightarrow & A_{m-1\bullet} & \xrightarrow[f_{m-1\bullet}]{} & A_{m\bullet} \end{array}$$

Therefore, $|\iota_m|\,|p_m|$ is homotopic to identity. So, $|\iota_m|$ is a homotopy equivalence, for all m. By Bisimplicial Lemma 1.3.2, $\mathbb{B}\mathscr{B} \longrightarrow |\mathscr{A}\mathscr{B}|$ is a homotopy equivalence. ■

We also insert the following [Sg, Prop. A.1], on the so-called edgewise subdivision.

Exercise 9.2.17. Let $K \in \mathbf{SimTop}$ be a simplicial space. Use the notation from (1.5.6), and define the functor $\eth : \Delta \longrightarrow \Delta$ given by $\eth([n]) = [n][n] = [2n+1]$. Then the composition $K^e = K\eth \in \mathbf{SimTop}$. Further, the geometric realizations $|K| \cong |K^e|$ are naturally homeomorphic. This is often referred to as the **edgewise subdivision** of K. **Hint:** Use the definition of geometric realization. See [Sg, Prop. A.1] for details. ∎

The following reconciles Quillen and Waldhausen K-theory.

Proposition 9.2.18. *Suppose that $\mathscr{E}$ is an exact category. Fix a zero object 0 in $\mathscr{E}$. Let $co(\mathscr{E})$ be the subcategory of inflations in $\mathscr{E}$, and $\mathbf{i} := \mathbf{i}\mathscr{E}$ be the subcategory of isomorphisms in $\mathscr{E}$. Then the triple $(\mathscr{E}, co(\mathscr{E}), \mathbf{i})$ forms a category with cofibrations and weak equivalences. Then there is a natural homotopy equivalence $\mathbb{B}\mathbb{Q}\mathscr{E} \cong |\mathbf{i}S_\bullet(\mathscr{E})|$.*

Proof. By (9.2.16), $\mathcal{N}_\bullet\mathbb{Q}\mathscr{E} \longrightarrow (\mathbf{i}\mathbb{Q}\mathscr{E})(\mathbb{Q}\mathscr{E})$ is a homotopy equivalence. Since isomorphisms in $\mathscr{E}$ and $\mathbb{Q}\mathscr{E}$ coincide (3.8.1), the objects in $(\mathbf{i}\mathbb{Q}\mathscr{E})(\mathbb{Q}\mathscr{E})$ are as in diagram (9.12), where the horizontal arrows are in $\mathbb{Q}\mathscr{E}$ and vertical arrows are isomorphisms in $\mathscr{E}$.

By (9.2.17), the geometric realizations $|n \mapsto |\mathbf{i}S_n\mathscr{E}||$ and $|n \mapsto |(\mathbf{i}S_n\mathscr{E})^e||$ are naturally homeomorphic. By definition, $(\mathbf{i}S_n\mathscr{E})^e = \mathbf{i}S_{2n+1}\mathscr{E}$. We proceed to define a map

$$(\mathbf{i}S_n\mathscr{E})^e = \mathbf{i}S_{2n+1}\mathscr{E} \longrightarrow \mathscr{N}_n\mathbb{Q}\mathscr{E}. \tag{9.13}$$

To do this, we treat $[2n+1]$ as a totally ordered set:

$$[2n+1] = \{n' < \cdots < 2' < 1' < 0' < 0 < 1 < 2 < \cdots < n\}.$$

An object $\mathbf{A} \in (\mathbf{i}S_n\mathscr{E})^e = \mathbf{i}S_{2n+1}\mathscr{E}$ is given by

$$0 \hookrightarrow A_{n'(n-1)'} \hookrightarrow \cdots \hookrightarrow A_{n'1'} \hookrightarrow A_{n'0'} \hookrightarrow$$

$$A_{n'0} \hookrightarrow A_{n'1} \hookrightarrow \cdots \hookrightarrow A_{n'n}$$

at the top row, together with choices of $A_{ij} \cong \frac{A_{n'j}}{A_{n'i}}$, in the lower rows. Given such an object $\mathbf{A}$, we obtain n arrows

$$A_{0'0} \longrightarrow A_{1'1} \longrightarrow A_{2'2} \longrightarrow \cdots \longrightarrow A_{r'r} \longrightarrow \cdots$$

$$\longrightarrow A_{n'n} \quad \text{in} \quad \mathbb{Q}\mathscr{E}.$$

When $n = 3$, it is given as follows:

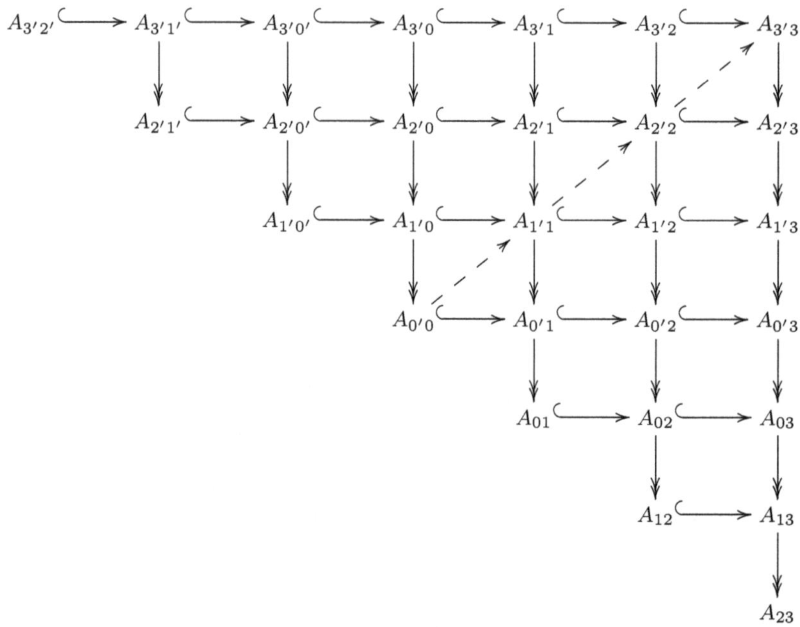

Three broken arrows represent an element of $N_3\mathbb{Q}\mathscr{E}$. Let $s_\bullet\mathscr{E} \in \mathbf{SSet}$ be as in (9.4). One checks that the above defines

$$\begin{cases} (s_\bullet\mathscr{E})^e \longrightarrow \mathscr{N}_\bullet\mathbb{Q}\mathscr{E} & \text{a map of simplicial sets, to the nerve} \\ (\mathbf{i}S_\bullet\mathscr{E})^e \longrightarrow (\mathbf{i}\mathbb{Q}\mathscr{E})(\mathbb{Q}\mathscr{E}) & \text{a map of bisimplicial sets.} \end{cases}$$

By further inspection, it follows that the second map is an equivalence in each degree (in fact, bijection). Hence, it induces a homotopy equivalence. Now, we have a commutative diagram

$$\begin{array}{ccc} (s_\bullet\mathscr{E})^e & \longrightarrow & \mathscr{N}_\bullet\mathbb{Q}\mathscr{E} \\ {\scriptstyle \zeta_0}\downarrow & & \downarrow \\ (\mathbf{i}S_\bullet\mathscr{E})^e & \longrightarrow & (\mathbf{i}\mathbb{Q}\mathscr{E})(\mathbb{Q}\mathscr{E}) \end{array}$$

Here, the first vertical map ζ_0 is as in (9.5). Note that $|(s_\bullet\mathscr{E})^e|$ is the 0-skeleton in $|(\mathbf{i}S_\bullet\mathscr{E})^e|$. Further, the face and degeneracy maps in $\{|N_n(\mathbf{w}S_\bullet(\mathscr{E})^e|\}$ are homotopy equivalences. So, it follows that ζ_0 is a homotopy equivalence.

The second vertical map is a homotopy equivalence by (9.2.16), and so is the second horizontal arrow. So, the first horizontal arrow is also a homotopy equivalence. It follows that, $|\mathbf{i}S_\bullet\mathscr{E}| \cong |\mathbf{i}S_\bullet\mathscr{E}^e| \cong |\mathscr{N}_\bullet\mathbb{Q}\mathscr{E}|$ are homotopy equivalences. The proof is complete. ∎

9.2.3. *The chain complex categories*

The main point of Waldhausen K-theory is, for most readers, to extend Quillen K-theory of exact categories to the K-theory of various subcategories of the chain complex category $Ch(QCoh(X))$ of quasi-coherent modules over Noetherian schemes X. Most importantly, to extend to the K-theory of the category $Ch^b(\mathscr{P}(X))$, $Ch^b(Coh(X))$ of bounded chain complexes, respectively, of locally free sheaves and coherent X-modules. We establish some notation and basic definitions on these.

Notation 9.2.19. Let $\mathscr{A}$ be an abelian category, and $Ch(\mathscr{A})$ denote the category of chain complexes. We use homology chain complex notation. So (contrary to [TT]), an object $M_\bullet \in Ch(\mathscr{A})$ is described by sequences

$$\cdots \longrightarrow M_{n+1} \xrightarrow{\partial_{n+1}} M_n \xrightarrow{\partial_n} M_{n-1} \longrightarrow \cdots \quad \text{with} \quad \partial\partial = 0.$$

So, $Ch(\mathscr{A})$, together with chain complex maps $M_\bullet \longrightarrow N_\bullet$, is an abelian category. The full subcategory $Ch^b(\mathscr{A}) \subseteq Ch(\mathscr{A})$ consists of all $M_\bullet \in Ch(\mathscr{A})$ such that $M_n = 0$ for all but finitely many n. The full subcategory $Ch^L(\mathscr{A}) \subseteq Ch(\mathscr{A})$ (resp. $Ch^R(\mathscr{A}) \subseteq Ch(\mathscr{A})$) consists of all $M_\bullet \in Ch(\mathscr{A})$ such that $M_n = 0$ for all $n \ll 0$ (resp. $n \gg 0$). Superscript L (resp. R) indicates that $M_\bullet$ is supported on the left-hand side (resp. right-hand side).

1. The homotopy category $\mathcal{K}(\mathscr{A})$ of $\mathscr{A}$ has the same objects as $Ch(\mathscr{A})$, and maps are homotopy equivalence classes of maps in $Ch(\mathscr{A})$.
2. A map $f : M_\bullet \longrightarrow N_\bullet$ in $Ch(\mathscr{A})$ is called a **quasi-isomorphism** if f induces isomorphisms $f_* : H_n(M_\bullet) \xrightarrow{\sim} H_n(N_\bullet)$ of homologies $\forall n$.
3. The quasi-isomorphisms in $\mathcal{K}(\mathscr{A})$ satisfy the calculus of fractions (see Section A.5.1). The derived category $\mathbf{D}(\mathscr{A})$ is obtained from $\mathcal{K}(\mathscr{A})$ by inverting the quasi-isomorphisms. The derived

categories $\mathbf{D}^b(\mathscr{A})$, $\mathbf{D}^L(\mathscr{A})$, $\mathbf{D}^R(\mathscr{A})$, corresponding, respectively, to $Ch^b(\mathscr{A}), Ch^L(\mathscr{A}), Ch^R(\mathscr{A})$, are defined likewise. These definitions of derived categories, coincide with the more elaborate definitions given in (11.2.22).

4. Similar definitions are given for other subcategories of $Ch(\mathscr{A})$. Of particular interest to us is when $\mathscr{A} = QCoh(X)$ for some Noetherian scheme X. In this case, we consider $Ch^b(\mathscr{P}(X)), Ch^b(Coh(X)) \subseteq Ch(QCoh(X)$, and their derived categories.

   More generally, let $\mathscr{E}$ be an exact category embedded in an abelian category $\mathscr{E} \hookrightarrow \mathscr{A}$. Then $Ch^b(\mathscr{E}), Ch^L(\mathscr{E}), Ch^R(\mathscr{E})$, and their derived categories are defined, likewise. The derived categories are, respectively, denoted by $\mathbf{D}^b(\mathscr{E})$, $\mathbf{D}^L(\mathscr{E})$, $\mathbf{D}^R(\mathscr{E})$. See (11.2.22), for a more precise definition.
5. (**Structure of Waldhausen category:**) The category $Ch(\mathscr{A})$ is given a (default) structure of Waldhausen category, as follows:
   (a) An injective map $\iota : K_\bullet \longrightarrow M_\bullet$ is defined to be a **cofibration** if there is a degreewise split exact sequence

$$0 \longrightarrow K_\bullet \xrightarrow{\iota} M_\bullet \longrightarrow C_\bullet \longrightarrow 0 \tag{9.14}$$

   We denote the subcategory of cofibrations, by $co(Ch(\mathscr{E}))$.
   (b) The subcategory $qIso(Ch(\mathscr{A}))$ of quasi-isomorphisms in $Ch(\mathscr{A})$ is defined to be the subcategory of weak equivalences.

   Then the triple $(Ch(\mathscr{A}), co(Ch(\mathscr{A})), qIso(Ch(\mathscr{A})))$ forms a Waldhausen category. Other subcategories $\mathscr{C} \subseteq Ch(\mathscr{A})$ mentioned above, including those of our particular interest (4), inherit this structure of Waldhausen category by requiring the exact sequence (9.14) to be in $\mathscr{C}$. For example, in the case of $\mathscr{C} = Ch^b(\mathscr{P}(X))$, we require the sequence (9.14) to be in $Ch^b(\mathscr{P}(X))$, and in the case of $\mathscr{C} = Ch^b(\mathscr{E})$, we require the sequence (9.14) to be in $Ch^b(\mathscr{E})$.
6. Consequently, for these subcategories $(\mathscr{C}, \mathbf{w_C}) \subseteq (Ch(\mathscr{A}), co(Ch(\mathscr{A})), qIso(Ch(\mathscr{A})))$, the Waldhausen K-theory space $\mathbf{K}(\mathscr{C}, \mathbf{w_C})$ and the K-theory spectra $\mathrm{Sp}(\mathscr{C}, \mathbf{w_C})$ are defined. Also note that, for some of the examples above, the Quillen K-theory

space $\mathbf{K}(\mathscr{A})$, $\mathbf{K}(\mathscr{P}(X))$, $\mathbf{K}(Coh(X))$, $\mathbf{K}(\mathscr{E})$, and the K-groups are defined. Subsequently, we will prove that, for such subcategories, Waldhausen K-theory and Quillen K-theory coincide.

We define the cylinder and the cone of a morphism $f_\bullet : B_\bullet \longrightarrow C_\bullet$ of chain complexes.

Definition 9.2.20. Let $\mathscr{A}$ be an abelian category and $f_\bullet : B_\bullet \longrightarrow C_\bullet$ be a map of chain complexes. For the differentials, use the notation $d_n^B : B_n \longrightarrow B_{n-1}$, $d_n^C : C_n \longrightarrow C_{n-1}$, etc. We define two complexes, the cylinder $\mathbf{cyl}(f_\bullet), \mathbf{cone}(f_\bullet) \in Ch(\mathscr{A})$.

Let $\mathbf{cyl}(f_\bullet)_n = B_n \oplus B_{n-1} \oplus C_n$. Define the differential

$$\partial_n : \mathbf{cyl}(f_\bullet)_n \longrightarrow \mathbf{cyl}(f_\bullet)_{n-1} \quad \text{by} \quad \partial_n = \begin{pmatrix} d_n^B & 1_{B_n} & 0 \\ 0 & -d_{n-1}^B & 0 \\ 0 & -f_n & d_n^C \end{pmatrix}.$$

Then $(\mathbf{cyl}(f_\bullet), \partial) \in Ch(\mathscr{A})$, to be called the **mapping cylinder** of $f_\bullet$.

Now, let $\mathbf{cone}(f_\bullet)_n = B_{n-1} \oplus C_n$, and define the differential

$$\partial_n : \mathbf{cone}(f_\bullet)_n \longrightarrow \mathbf{cone}(f_\bullet)_{n-1} \quad \text{by} \quad \partial_n = \begin{pmatrix} -d_{n-1}^B & 0 \\ -f_{n-1} & d_n^C \end{pmatrix}.$$

The following exercise is of some future use.

Exercise 9.2.21. Let $\mathscr{A}$ be an abelian category, and $f_\bullet : B_\bullet \longrightarrow C_\bullet$ be a map in $Ch(\mathscr{A})$. Then the natural maps

$$\begin{cases} j : C_\bullet \longrightarrow \mathbf{Cyl}(f_\bullet) \\ T : \mathbf{Cyl}(f_\bullet) \longrightarrow C_\bullet \end{cases} \quad \text{are quasi-isomorphisms.}$$

9.2.4. *Expected results*

Almost all of Chapter 3, on K-theory of exact categories, extends to the K-theory of Waldhausen categories of $\mathscr{E}$. In preparation for the additivity theorem, we define the following.

Definition 9.2.22. Let $(\mathscr{C}, co(\mathscr{C}), \mathbf{w_C})$, $(\mathscr{A}, co(\mathscr{A}), \mathbf{w_A})$, $(\mathscr{B}, co(\mathscr{B}), \mathbf{w_B})$ be Waldhausen categories, and $(\mathscr{A}, co(\mathscr{A}), \mathbf{w_A}) \hookrightarrow (\mathscr{C}, co(\mathscr{C}), \mathbf{w_C})$, $(\mathscr{B}, co(\mathscr{B}), \mathbf{w_B}) \hookrightarrow (\mathscr{C}, co(\mathscr{C}), \mathbf{w_C})$ be inclusion exact functors. In analogy to Section 3.4, define the category $\varepsilon(\mathscr{A}, \mathscr{C}, \mathscr{B})$, as follows:

1. Objects in $\varepsilon(\mathscr{A}, \mathscr{C}, \mathscr{B})$ are cofibration sequences $A \hookrightarrow C \twoheadrightarrow B$, with $A \in Obj(\mathscr{A})$, $C \in Obj(\mathscr{C})$, $B \in Obj(\mathscr{B})$.
2. Morphisms in $\varepsilon(\mathscr{A}, \mathscr{C}, \mathscr{B})$ are given by commutative diagrams

$$\begin{array}{ccccc} A & \hookrightarrow & C & \twoheadrightarrow & B \\ {\scriptstyle f_A}\downarrow & & \downarrow{\scriptstyle f_C} & & \downarrow{\scriptstyle f_B} \\ A_1 & \hookrightarrow & C_1 & \twoheadrightarrow & B_1 \end{array} \quad \text{with} \quad \begin{cases} f_A \in Mor_{\mathscr{A}}(A, A_1) \\ f_C \in Mor_{\mathscr{C}}(C, C_1) \\ f_B \in Mor_{\mathscr{B}}(B, B_1). \end{cases} \tag{9.15}$$

3. A morphism (9.15) is defined as a cofibration in $\varepsilon(\mathscr{A}, \mathscr{C}, \mathscr{B})$, if $f_A \in co(\mathscr{A})$, $f_B \in co(\mathscr{B})$ and the map $C \coprod_A A_1 \longrightarrow C$ is in $co(\mathscr{C})$. In this case, $f_C \in co(\mathscr{C})$. The subcategory of cofibrations will be denoted by $co\,(\varepsilon(\mathscr{A}, \mathscr{C}, \mathscr{B}))$.
4. A morphism (9.15) is defined to be a weak equivalence in $\varepsilon(\mathscr{A}, \mathscr{C}, \mathscr{B})$, if $f_A \in \mathbf{w_A}$, $f_C \in \mathbf{w_C}$, $f_B \in \mathbf{w_B}$. The category of weak equivalences in $\varepsilon(\mathscr{A}, \mathscr{C}, \mathscr{B})$ will be denoted by $\mathbf{w}_\varepsilon$.

Then the triple $(\varepsilon(\mathscr{A}, \mathscr{C}, \mathscr{B}), co\,(\varepsilon(\mathscr{A}, \mathscr{C}, \mathscr{B})), \mathbf{w}_\varepsilon)$ forms a Waldhausen category. Further, there are three exact functors (9.16), sending $E := \left(A \hookrightarrow C \twoheadrightarrow B \right) \in \varepsilon(\mathscr{A}, \mathscr{C}, \mathscr{B})$, as follows:

$$\begin{cases} s : \varepsilon(\mathscr{A}, \mathscr{C}, \mathscr{B}) \longrightarrow \mathscr{A} \;\; E \mapsto A \\ t : \varepsilon(\mathscr{A}, \mathscr{C}, \mathscr{B}) \longrightarrow \mathscr{A} \;\; E \mapsto C \\ q : \varepsilon(\mathscr{A}, \mathscr{C}, \mathscr{B}) \longrightarrow \mathscr{A} \;\; E \mapsto B. \end{cases} \tag{9.16}$$

There is also another exact functor

$$\odot : \mathscr{A} \times \mathscr{B} \longrightarrow \varepsilon(\mathscr{A}, \mathscr{C}, \mathscr{B}) \quad \text{defined by} \quad A \odot B = \left(A \hookrightarrow A \coprod\nolimits_0 B \twoheadrightarrow B \right).$$

One checks that $A \odot B$ is a cofibration sequence. (We prefer this notation $\odot$ because it reminds us of the direct sum $\oplus$, while [W,TT], respectively, used $\vee, \cup$. However, we hesitate to use $\oplus$ for $\coprod_0$, because $\mathscr{C}$ is not even assumed to be additive.)

The following analogue of (3.4.1) is also referred to as the additivity theorem.

Theorem 9.2.23 (Additivity Theorem). *Use the notation in (9.2.22). Then*

$$\text{the functor} \qquad (s,q) : \varepsilon(\mathscr{A},\mathscr{C},\mathscr{B}) \longrightarrow \mathscr{A} \times \mathscr{B}$$

induces homotopy equivalences,

$$\begin{cases} |\mathbf{w}(s,q)| : |\mathbf{w}S_\bullet\varepsilon(\mathscr{A},\mathscr{C},\mathscr{B})| \longrightarrow |\mathbf{w}S_\bullet\mathscr{A}| \times |\mathbf{w}S_\bullet\mathscr{B}| \\ \quad \text{of spaces} \\ K(s,q) : \mathbf{K}\left(\varepsilon(\mathscr{A},\mathscr{C},\mathscr{B}), \mathbf{w}_\varepsilon\right) \longrightarrow \mathbf{K}\left(\mathscr{A}, \mathbf{w_A}\right) \times \mathbf{K}\left(\mathscr{B}, \mathbf{w_B}\right) \\ \quad \text{of spaces} \\ \mathrm{Sp}(s,q) : \mathrm{Sp}\left(\varepsilon(\mathscr{A},\mathscr{C},\mathscr{B}), \mathbf{w}_\varepsilon\right) \longrightarrow \mathrm{Sp}(\mathscr{A}, \mathbf{w_A}) \times \mathrm{Sp}(\mathscr{B}, \mathbf{w_B}) \\ \quad \text{of spectra.} \end{cases}$$

Further, the functor $\odot$ is a natural homotopy inverse of (s,q).

Outline of the Proof. We only need to establish the first homotopy equivalence. The proof follows the line of arguments in the proof of (3.4.1), and uses the versions Theorem AW (2.3.9) and Theorem BW (2.3.10). ■

The following is the analogue of the additivity theorem (3.4.4).

Corollary 9.2.24. *Let $(\mathscr{C}, co(\mathscr{C}), \mathbf{w_C}), (\mathscr{D}, co(\mathscr{D}), \mathbf{w_D})$ be two Waldhausen categories, and $G, F, H : \mathscr{C} \longrightarrow \mathscr{D}$ be three exact functors. Let*

$$\begin{array}{l} 0 \longrightarrow G \longrightarrow F \longrightarrow H \longrightarrow 0 \\ \qquad \text{be a sequence of natural transformations, } \ni. \end{array} \tag{9.17}$$

1. *$\forall A \in Obj(\mathscr{C})$, the sequence $G(A) \hookrightarrow F(A) \twoheadrightarrow H(A)$ is a cofibration sequence.*
2. *For all cofibrations $A \hookrightarrow B$ in $\mathscr{C}$, the induced arrow $F(A) \coprod_{G(A)} G(B) \longrightarrow F(B)$ is a cofibration.*

Then there is a natural homotopy between the maps

$$\begin{cases} \mathbf{K}F \simeq \mathbf{K}G + \mathbf{K}H : \mathbf{K}(\mathscr{C}, \mathbf{w_C}) \longrightarrow \mathbf{K}(\mathscr{D}, \mathbf{w_D}) \quad \text{of spaces} \\ \mathrm{Sp}F \simeq \mathrm{Sp}G + \mathrm{Sp}H : \mathrm{Sp}(\mathscr{C}, \mathbf{w_C}) \longrightarrow \mathrm{Sp}(\mathscr{D}, \mathbf{w_D}) \text{ of spectra} \end{cases}$$

where $+$ denotes the maps induced by $\odot$.

Proof. Exactly similar to the proof of (3.4.4). ■

We define the cylinder functor, which is technically important [TT, Def. 1.3.1].

Definition 9.2.25. Let $(\mathscr{C}, co(\mathscr{C}), \mathbf{w_C})$ be a Waldhausen category. Denote the category of arrows of $\mathscr{C}$ by $Ar(\mathscr{C}) := Fun([1], \mathscr{C})$ (where $[1] := \{0 < 1\}$). A **cylinder functor** on $\mathscr{C}$ is a functor $T : Ar(\mathscr{C}) \longrightarrow \mathscr{C}$ that associates arrows $f : A \longrightarrow B$ in $\mathscr{C}$ to the commutative diagram

$$\begin{array}{ccccc} A & \xrightarrow{j_1} & T(f) & \xleftarrow{j_2} & B \\ & {\scriptstyle f}\searrow & \downarrow{\scriptstyle p} & \swarrow{\scriptstyle 1_B} & \\ & & B & & \end{array} \quad \text{where} \begin{cases} j_1 : A \longrightarrow T(f) \\ j_2 : B \longrightarrow T(f) \\ p : T(f) \longrightarrow B \end{cases} \text{are natural transformations,}$$

$$\ni \ \forall \text{ morphisms } (g_1, g_2) : f_1 \longrightarrow f_2 \text{ given by commuting diagrams } \begin{array}{ccc} A_1 & \xrightarrow{f_1} & B_1 \\ {\scriptstyle g_1}\downarrow & & \downarrow{\scriptstyle g_2} \\ A_2 & \xrightarrow[f_2]{} & B_2 \end{array},$$

$$\text{the diagram} \quad \begin{array}{ccccc} A_1 \coprod_0 B_1 & \xrightarrow{j_1 \coprod_0 j_2} & T(f_1) & \xrightarrow{p} & B_1 \\ {\scriptstyle g_1 \oplus g_2}\downarrow & & \downarrow{\scriptstyle T(g_1, g_2)} & & \downarrow{\scriptstyle g_2} \\ A_2 \coprod_0 B_2 & \xrightarrow[j_1 \coprod_0 j_2]{} & T(f_2) & \xrightarrow[p]{} & B_2 \end{array} \quad \text{commutes.}$$

Further,

1. $A \odot B \hookrightarrow T(f)$ is a cofibration.
2. If g_1, g_2 are in $\mathbf{w_C}$, then $T(g_1, g_2)$ is in $\mathbf{w_C}$.
3. If g_1, g_2 are in $co(\mathscr{C})$, then $T(f_1) \coprod_{A_1 \odot B_1} (A_2 \coprod B_2) \longrightarrow T(f_2)$ is in $co(\mathscr{C})$. Consequently, $T(g_1, g_2)$ is in $co(\mathscr{C})$.
4. $T(0 \hookrightarrow A) = A$ and both $j_1 = p = 1_A$.

A cylinder functor T is said to **satisfy the cylinder axiom**, if

$$(p : T(f) \longrightarrow B) \in \mathbf{w_C} \qquad \forall f. \tag{9.18}$$

We remark that for an abelian category $\mathscr{A}$, the category $Ch(\mathscr{A})$ has a cylinder functor, as described in (9.2.20). In fact, the cylinder functor $\mathbf{cyl}(f)$ satisfies the cylinder axiom (9.2.21).

Lemma 9.2.26. *Let $(\mathscr{C}, co(\mathscr{C}))$ be a small category with cofibrations. Suppose that there are two subcategories $\mathbf{v} \subseteq \mathbf{w} \subseteq \mathscr{C}$, such that*

$$\begin{cases} (\mathscr{C}, \mathbf{v}) :- (\mathscr{C}, co(\mathscr{C}), \mathbf{v}) \\ (\mathscr{C}, \mathbf{w}) := (\mathscr{C}, co(\mathscr{C}), \mathbf{w}) \end{cases} \quad \text{are Waldhausen categories.}$$

Assume $(\mathscr{C}, \mathbf{w})$ satisfies the extension and saturation axiom (see 9.2.2).

Let $\mathscr{C}^{\mathbf{w}} = \{A \in Obj(\mathscr{C}) : (0 \hookrightarrow A) \in \mathbf{w}\}$ be the full subcategory of $\mathscr{C}$. (We say that $\mathscr{C}^{\mathbf{w}}$ is the subcategory of $\mathbf{w}$-acyclic objects in $\mathscr{C}$.) Then

$$\begin{cases} (\mathscr{C}^{\mathbf{w}}, \mathbf{v}) := (\mathscr{C}^{\mathbf{w}}, co(\mathscr{C}) \cap \mathscr{C}^{\mathbf{w}}, \mathbf{v} \cap \mathscr{C}^{\mathbf{w}}) \\ (\mathscr{C}^{\mathbf{w}}, \mathbf{w}) := (\mathscr{C}^{\mathbf{w}}, co(\mathscr{C}) \cap \mathscr{C}^{\mathbf{w}}, \mathbf{w} \cap \mathscr{C}^{\mathbf{w}}) \end{cases} \quad \text{are Waldhausen categories.}$$

Further, if $\mathscr{C}$ has a functor that is a cylinder functor on both $(\mathscr{C}, \mathbf{v})$ and $(\mathscr{C}, \mathbf{w})$, then T is a cylinder functor on $(\mathscr{C}^{\mathbf{w}}, \mathbf{v})$.

Theorem 9.2.27 (Localization Theorem). *Use all the notation from (9.2.26). Under all the hypotheses in (9.2.26), further assume that $(\mathscr{C}, \mathbf{w})$ has a cylinder functor T, which is also a cylinder functor on $(\mathscr{C}, \mathbf{v})$. Assume that T satisfies the cylinder axiom (9.18) on $(\mathscr{C}, \mathbf{w})$. Then the diagrams*

$$\begin{array}{ccc} |S_\bullet(\mathscr{C}^w, \mathbf{v})| & \longrightarrow & |S_\bullet(\mathscr{C}, \mathbf{v})| \\ \downarrow & & \downarrow \\ |S_\bullet(\mathscr{C}^w, \mathbf{w})| & \longrightarrow & |S_\bullet(\mathscr{C}, \mathbf{w})| \end{array} \qquad \begin{array}{ccc} \mathbf{K}(\mathscr{C}^w, \mathbf{v}) & \longrightarrow & \mathbf{K}(\mathscr{C}, \mathbf{v}) \\ \downarrow & & \downarrow \\ \mathbf{K}(\mathscr{C}^w, \mathbf{w}) & \longrightarrow & \mathbf{K}(\mathscr{C}, \mathbf{w}) \end{array} \qquad \begin{array}{ccc} \mathrm{Sp}(\mathscr{C}^w, \mathbf{v}) & \longrightarrow & \mathrm{Sp}(\mathscr{C}, \mathbf{v}) \\ \downarrow & & \downarrow \\ \mathrm{Sp}(\mathscr{C}^w, \mathbf{w}) & \longrightarrow & \mathrm{Sp}(\mathscr{C}, \mathbf{w}) \end{array}$$

are homotopy cartesian in the respective categories. (In the first diagram, we use the notation $|S_\bullet(\mathscr{C}, \mathbf{w})| := |\mathbf{w}S_\bullet(\mathscr{C})|$ etc.) Note, the lower left corner terms are contractible. Consequently, we obtain sequences of homotopy fibrations, and there is a long exact sequence of K-groups.

Proof. See [Wh, Thm. 1.6.4] and [TT, Thm. 1.8.2]. ■

The following has a flavor of the resolution theorem (3.5.2).

Theorem 9.2.28 (Approximation Theorem). *Let $(\mathscr{C}, co(\mathscr{C}), \mathbf{w}_{\mathbf{C}})$ and $(\mathscr{D}, co(\mathscr{D}), \mathbf{w}_{\mathbf{D}})$ be two small Waldhausen categories. Assume that both satisfy the saturation axiom, and that $\mathscr{C}$ has a cylinder functor, satisfying the cylinder axiom. Let $F : \mathscr{C} \longrightarrow \mathscr{D}$ be an exact functor, satisfying the following two conditions:*

1. *For morphisms f in $\mathscr{C}$, $f \in \mathbf{w_C} \Longleftrightarrow F(f) \in \mathbf{w_D}$.*
2. *Given $A \in Obj(\mathscr{C})$ and a morphism $f : FA \longrightarrow B$ in $\mathscr{D}$, there are a morphism $g : A \longrightarrow A_1$ in $\mathscr{C}$ and a weak equivalence $(h : FA_1 \longrightarrow B) \in \mathbf{w_D}$ such that the diagram*

$$\begin{array}{ccc} FA & \xrightarrow{f} & B \\ & \searrow^{F(g)} & \uparrow \wr\, h \\ & & FA_1 \end{array} \quad \textit{commutes.}$$

Then F induces homotopy equivalences $\mathbf{K}F : \mathbf{K}(\mathscr{C}, \mathbf{w_C}) \xrightarrow{\sim} \mathbf{K}(\mathscr{D}, \mathbf{w_D})$ of spaces and $\mathrm{Sp}F : \mathrm{Sp}(\mathscr{C}, \mathbf{w_C}) \xrightarrow{\sim} \mathrm{Sp}(\mathscr{D}, \mathbf{w_D})$ of spectra.

Proof. See [Wh, Thm. 1.6.7], [TT, 1.9.1]. ∎

9.3. *K*-Theory of Complicial Exact Categories

For an abelian category $\mathscr{A}$, a complicial biWaldhausen category [TT, 1.2.11] was defined as consisting of certain subcategories $\mathscr{E} \subseteq \mathscr{A}$ satisfying certain conditions. Evidently, this was to encapsulate the examples of the common chain complex categories discussed in Section 9.2.3. Schlichting [Sm1] introduced the complicial exact category for the same purpose and implements Waldhausen K-theory. We give an overview from [Sm1].

Definition 9.3.1. As before, let $Ch^b(\mathscr{P}(\mathbb{Z}))$ denote the category of bounded chain complexes of finitely generated free $\mathbb{Z}$-modules. For $M_\bullet, N_\bullet \in Ch^b(\mathscr{P}(\mathbb{Z}))$, define the tensor product

$$(M_\bullet \otimes N_\bullet)_n = \bigoplus M_k \otimes N_{n-k}, \quad \text{and} \quad d(x \otimes y) = dx \otimes y + (-1)^{\deg x} x \otimes dy.$$

With this product, $Ch^b(\mathscr{P}(\mathbb{Z}))$ has a structure of a symmetric monoidal category (see Section 4.1). The unit of this product is the complex $\mathbb{1}$, which is $\mathbb{Z}\varepsilon_0 \cong \mathbb{Z}$ at degree zero and is zero elsewhere. There are two other distinguished objects on $Ch^b(\mathscr{P}(\mathbb{Z}))$. First, the complex $C_\bullet$ such that $C_1 = \mathbb{Z}\eta_1 \cong \mathbb{Z}$, $C_0 = \mathbb{Z}\eta_0 \cong \mathbb{Z}$, $C_i = 0$, $\forall i \neq 0, 1$, and $d(\eta_1) = \eta_0$. Then we have $T_\bullet$, where $T_1 = \mathbb{Z}\tau_1 \cong \mathbb{Z}$, and $T_i = 0$, $\forall i \neq 1$. We have an exact sequence

$$0 \longrightarrow \mathbb{1} \longrightarrow C_\bullet \longrightarrow T_\bullet \longrightarrow 0 \quad \text{and it splits in each degree.} \tag{9.19}$$

Definition 9.3.2. An exact category $\mathscr{E}$, together with a bi-exact (tensor) product

$$\otimes : Ch^b(\mathscr{P}(\mathbb{Z})) \times \mathscr{E} \longrightarrow \mathscr{E} \tag{9.20}$$

is said to be a **complicial exact category** if it is a $Ch^b(\mathscr{P}(\mathbb{Z}))$-category (see definition 4.2.1). In particular, there are natural isomorphisms $\alpha : (A \otimes B) \otimes X \cong A \otimes (B \otimes X)$, $\lambda : \mathbb{1} \otimes X \cong X$, together with other natural properties. We have the following:

1. For $X \in \mathscr{E}$, denote $CX := C_\bullet \otimes X$ and $TX = T_\bullet \otimes X$.
2. For $X \in \mathscr{E}$, by the tensor product with (9.19), we obtain a conflation

$$0 \longrightarrow X \longrightarrow CX \longrightarrow TX \longrightarrow 0\,, \quad \text{which is functorial.} \tag{9.21}$$

3. For arrows $f : X \longrightarrow Y$ in $\mathscr{E}$, the **cone** $C(f)$ of f is defined by the pushforward diagram

$$\begin{array}{ccc} X & \hookrightarrow & CX \\ {\scriptstyle f}\downarrow & & \downarrow \\ Y & \hookrightarrow & C(f) \end{array} \tag{9.22}$$

4. It is worth mentioning that, combining (9.21, 9.22) for any arrow $f : X \longrightarrow Y$, we obtain a complex

$$X \xrightarrow{f} Y \longrightarrow C(f) \longrightarrow TX \tag{9.23}$$

to be called a triangle.

The definition of exact categories with weak equivalence is a little more stringent than that of Waldhausen categories because exact categories have both inflations and deflations.

Definition 9.3.3. Suppose that $\mathscr{E}$ is an exact category. Then a subcategory of weak equivalences is a subcategory $\mathbf{w} \subseteq \mathscr{E}$ such that (1) all isomorphisms in $\mathscr{E}$ are in $\mathbf{w}$, (2) $\mathbf{w}$ is closed under pushforward along inflations in $\mathscr{E}$, (3) $\mathbf{w}$ is closed under pullback along deflations in $\mathscr{E}$, (4) for composable arrows $X \xrightarrow{f} Y \xrightarrow{g} Z$ in $\mathscr{E}$, if two of f, g, gf are in $\mathbf{w}$, then so is the third one. In this case, we say that $(\mathscr{E}, \mathbf{w})$ is an exact category with weak equivalences.

Now suppose that $(\mathscr{E}, \mathbf{w})$ is an exact category with weak equivalences, where $\mathscr{E}$ is a complicial exact category. We say that $(\mathscr{E}, \mathbf{w})$ is a **complicial exact category with weak equivalences** if the product (9.20) preserves weak equivalences. That means that, for quasi-isomorphisms f in $Ch^b(\mathscr{P}(\mathbb{Z}))$, and $g \in \mathbf{w}$, the product $f \otimes g \in \mathbf{w}$.

For completeness, we include the following definition.

Definition 9.3.4. Let $(\mathscr{E}, \mathbf{w_E})$, $(\mathscr{D}, \mathbf{w_D})$ be two complicial exact categories with weak equivalences. An exact functor $F : \mathscr{E} \longrightarrow \mathscr{D}$ is said to be a **functor of complicial exact categories with weak equivalences** if F is a functor of $Ch^b(\mathscr{P}(\mathbb{Z}))$-categories (4.2.4), and $F(\mathbf{w_E}) \subseteq \mathbf{w_D}$.

The following is the example of our main interest.

Exercise 9.3.5. Let $\mathscr{E}$ be an exact category. Then the category $Ch^b(\mathscr{E})$ can be made into a complicial exact category. Let $F(\mathbb{Z})$ be the category of based finitely generated free $\mathbb{Z}$-modules. Then there is an equivalence of categories $Ch^b(F(\mathbb{Z})) \cong Ch^b(\mathscr{P}(\mathbb{Z}))$. Define the (tensor) product $F(\mathbb{Z}) \times \mathscr{E}$ by defining $(\oplus \mathbb{Z}e_i) \otimes X := \oplus X e_i$, where Xe_i stands for a copy of X. Likewise, for matrices $(a_{ij}) \in \mathbb{M}_{m\times n}(\mathbb{Z})$, and $f : X \longrightarrow Y$ in $\mathscr{E}$, define $((a_{ij}) \otimes f) := (a_{ij} f)$. This extends to a (tensor) product

$$Ch^b(\mathscr{P}(\mathbb{Z})) \times Ch^b(\mathscr{E}) \longrightarrow Ch^b(\mathscr{E}).$$

Now, let $qIso := qIso(Ch^b(\mathscr{E}))$ denote the subcategories of quasi-isomorphisms in $Ch^b(\mathscr{E})$. Then $\left(Ch^b(\mathscr{E}), qIso\right)$ forms a complical exact category with weak equivalences. As was defined in (9.2.20), $\left(Ch^b(\mathscr{E}), qIso\right)$ has a cylinder functor $\mathbf{cyl} : Ar(Ch^b(\mathscr{E})) \longrightarrow Ch^b(\mathscr{E})$, which satisfies cylinder axiom (9.2.21).

Lemma 9.3.6. *Let $(\mathscr{E}, \mathbf{w})$ be a complicial exact category with weak equivalences.*

1. *Let* $Inf(\mathscr{E})$ *be the subcategory of inflations in* $\mathscr{E}$. *Then* $(\mathscr{E}, Inf(\mathscr{E}), \mathbf{w})$ *is a Waldhausen category. Therefore, the Waldhausen* K*-theory space* $\mathbf{K}(\mathscr{E}, \mathbf{w})$ *and* K*-theory spectra* $\mathrm{Sp}(\mathscr{E}, \mathbf{w})$ *are defined.*
2. *Further, the cylinder functor on* $Ch^b(\mathscr{P}(\mathbb{Z}))$ *induces a cylinder functor* $T : Ar(\mathscr{E}) \longrightarrow \mathscr{E}$, *by* $T(f) = \mathbf{cyl}\,(1_{\mathbb{1}}) \otimes f$. *Clearly,* T *satisfies the cylinder axiom.*

The following is an agreement between K-theory of exact categories $\mathscr{E}$ and of the complicial exact category $(Ch^b(\mathscr{E}), qIso)$ with weak equivalences.

Theorem 9.3.7. *Let* $\mathscr{E}$ *be an exact category. Consider the natural inclusion functor* $\mathscr{E} \longrightarrow Ch^b(\mathscr{E})$, *which sends* $M \in Obj(\mathscr{E})$ *to the complex* M, *concentrated at degree zero. This functor induces a homotopy equivalence*

$$\mathbf{K}(\mathscr{E}) \xrightarrow{\sim} \mathbf{K}(Ch^b(\mathscr{E}), qIso).$$

Proof. See [TT, 1.11.7]. ∎

Proposition 9.3.8. *Let* $(\mathscr{E}, \mathbf{w})$ *be a small complicial exact category with weak equivalences. Let* $\mathscr{E}^w = \{X \in Obj(\mathscr{E}) : (0 \longrightarrow X) \in \mathbf{w}\}$ *be the full exact subcategory.* (The objects $X \in \mathscr{E}^w$ are referred to as **acyclic objects**.) *Then*

$$\mathbf{K}(\mathscr{E}, \mathbf{w}) \longrightarrow \mathbb{BQ}\mathscr{E}^w \longrightarrow \mathbb{BQ}\mathscr{E} \quad \textit{is a homotopy fibration.} \tag{9.24}$$

Proof. The subcategory $\mathbf{i} \subseteq \mathbf{w}$ of isomorphisms forms a weak equivalence in $\mathscr{E}$. By the localization theorem (9.2.27), applied to $\mathbf{i} \subseteq \mathbf{w}$, the sequence

$$|S_\bullet(\mathscr{E}^w, \mathbf{i})| \longrightarrow |S_\bullet(\mathscr{E}, \mathbf{i})| \longrightarrow |S_\bullet(\mathscr{E}, \mathbf{w})| \text{ is a homotopy fibration.}$$

By (9.2.18), $|S_\bullet(\mathscr{E}^w, \mathbf{i})| \cong \mathbb{BQ}\mathscr{E}^w$, and $|S_\bullet(\mathscr{E}, \mathbf{i})| \cong \mathbb{BQ}\mathscr{E}$. Rotating (C.3.10) this sequence, we obtain the desired homotopy fibration (9.24). The proof is complete. ∎

To state the localization theorem, we need to develop some more background. For a definition of a Triangulated category, see (11.1.1).

Definition 9.3.9. Let $(\mathscr{E}, \mathbf{w})$ be a complicial exact category. We associate $\mathscr{E} \mapsto \mathfrak{T}\mathscr{E}$ with a triangulated category in a functorial manner.

1. We directly define
$$\mathfrak{T}(\mathscr{E}, \mathbf{w}) := \mathbf{w}^{-1}\mathscr{E} \qquad \text{obtained by calculus of fractions.} \tag{9.25}$$
Indeed, $\mathfrak{T}(\mathscr{E}, \mathbf{w})$ is a triangulated category, where the distinguished triangles are defined as those isomorphic, in $\mathfrak{T}(\mathscr{E}, \mathbf{w})$, to the triangles (9.23). Note that the subcategory $\mathscr{E}^w = \{X \in Obj(\mathscr{E}) : (0 \longrightarrow X) \in \mathbf{w}\} \subseteq \mathscr{E}$ of acyclic objects is killed in $\mathfrak{T}(\mathscr{E}, \mathbf{w})$.
2. We sketch a proof that $\mathfrak{T}(\mathscr{E}, \mathbf{w})$ is a triangulated category.
   (a) For a complicial exact category $\mathscr{E}$ (with or without weak equivalences), one defines a so-called **stable category** $\underline{\mathscr{E}}$, whose objects are the same as the objects of $\mathscr{E}$. Morphisms in $\underline{\mathscr{E}}$ are homotopy equivalence classes of maps in $\mathscr{E}$, while we refer to [Sm1, 3.2.5, A.2.14] for a definition of homotopy of maps in $\mathscr{E}$. This is similar to the homotopy category of the chain complex category of any exact category.
   It turns out that the stable category $\underline{\mathscr{E}}$ is a triangulated category, while the exact triangles are those isomorphic to the triangles (9.23).
   (b) In our case of complicial exact category $(\mathscr{E}, \mathbf{w})$ with weak equivalences, the inclusion of the stable categories $\underline{\mathscr{E}^w} \subseteq \underline{\mathscr{E}}$ is an inclusion of triangulated categories.
   (c) Define
$$\mathfrak{T}(\mathscr{E}, \mathbf{w}) := \frac{\underline{\mathscr{E}}}{\underline{\mathscr{E}^w}}, \qquad \text{the Verdier quotient (11.2.15).}$$
From the construction of the Verdier quotient (11.2.15), it follows that $\mathfrak{T}(\mathscr{E}, \mathbf{w}) := \mathbf{w}^{-1}\mathscr{E}$. ■
3. We say that $\mathfrak{T}(\mathscr{E}, \mathbf{w})$ is the associated triangulated category of $(\mathscr{E}, \mathbf{w})$. This association $(\mathscr{E}, \mathbf{w}) \mapsto \mathfrak{T}(\mathscr{E}, \mathbf{w})$ is functorial from the category of small complicial exact categories with weak equivalence and functors of the same (9.3.4) to the category of small triangulated categories and triangle functors.

4. Let $\mathscr{A}$ be an exact category, and $\mathscr{E} = (Ch^b(\mathscr{A}), qIso)$ be the category of bounded chain complexes and quasi-isomorphisms. Then $\mathfrak{T}\left((Ch^b(\mathscr{A}), qIso)\right) \cong \mathbf{D}^b(\mathscr{A})$, the bounded derived category (9.2.19).

Further details on these are available in [Sm1, 3.2.5, 3.2.15, 3.2.17, A.2.14], [Sm5].

We give the following **connective version** of the Thomason–Waldhausen Localization theorem for complicial exact categories with weak equivalences.

Theorem 9.3.10 (Localization Theorem). *Let* $(\mathscr{C}, \mathbf{w_C}) \longrightarrow (\mathscr{D}, \mathbf{w_D}) \longrightarrow (\mathscr{E}, \mathbf{w_E})$ *be a sequence of functors of complicial exact categories with weak equivalences. Assume that the induced sequence* $\mathfrak{T}(\mathscr{C}, \mathbf{w_C}) \xrightarrow{\iota} \mathfrak{T}(\mathscr{D}, \mathbf{w_D}) \xrightarrow{\mathfrak{q}} \mathfrak{T}(\mathscr{E}, \mathbf{w_E})$ *of the associated triangulated categories is exact, as defined in* (11.2.15)*. Then the induced maps of the K-theory spaces*

$$\mathbf{K}(\mathscr{C}, \mathbf{w_C}) \longrightarrow \mathbf{K}(\mathscr{D}, \mathbf{w_D}) \longrightarrow \mathbf{K}(\mathscr{E}, \mathbf{w_E})$$

is a homotopy fibration.

Consequently, there is a long exact sequence of K-groups ending at degree zero.

Proof. The proof is based on (9.2.27). Refer to [Sm1, Thm 3.2.23]. ■

The following on invariance [Sm1, Thm. 3.2.24] is a reformulation of [TT, Thm. 1.9.8].

Theorem 9.3.11 (Triangular Invariance). *Let* $F : (\mathscr{E}, \mathbf{w_E}) \longrightarrow (\mathscr{D}, \mathbf{w_D})$ *be an exact functor of complicial exact categories with weak equivalences* (9.3.4)*. Suppose the induced functor* $\mathfrak{T}(\mathscr{E}, \mathbf{w_E}) \longrightarrow \mathfrak{T}(\mathscr{D}, \mathbf{w_D})$ *of the associated triangulated categories is an equivalence of triangulated categories. Then F induces a homotopy equivalence*

$$\mathbf{K}(\mathscr{E}, \mathbf{w_E}) \longrightarrow \mathbf{K}(\mathscr{D}, \mathbf{w_D}) \text{ of spaces.}$$

Proof. Follows from (9.3.10). ■

Before we close this section, we define K_0 of triangulated categories and make a basic connection. This definition is analogous to the definition of the Grothendieck group $K_0(\mathscr{E})$ of an exact category (3.2.1).

Definition 9.3.12. Let $\mathcal{T}$ be a small triangulated category. Let $\mathbb{Z}(Obj(\mathcal{T}))$ denote the free abelian group generated by the set $Obj(\mathcal{T})$. Let $\mathscr{R}(\mathcal{T}) \subseteq \mathbb{Z}(Obj(\mathcal{T}))$ be the subgroup generated by the set

$$\{Y - X - Z : \exists \text{ a distinguished triangle}$$
$$X \longrightarrow Y \longrightarrow Z \longrightarrow TX \ \}.$$

The Grothendieck group $K_0(\mathcal{T})$ is defined by

$$K_0(\mathcal{T}) = \frac{\mathbb{Z}(Obj(\mathcal{T}))}{\mathscr{R}(\mathcal{T})}.$$

This is also referred to as the Grothendieck group of $\mathcal{T}$.

The following is a fundamental link between the K-theory of complicial exact categories and the Grothendieck group (9.3.12).

Proposition 9.3.13. *Let $(\mathscr{E}, \mathbf{w})$ be a complicial exact category with weak equivalences. Let $\mathfrak{T}(\mathscr{E}, \mathbf{w})$ be the triangulated category associated to $(\mathscr{E}, \mathbf{w})$. Then there is a natural isomorphism*

$$K_0(\mathscr{E}, \mathbf{w}) \cong K_0(\mathfrak{T}(\mathscr{E}, \mathbf{w})).$$

Proof. Refer to [Sm1, 3.2.22], while we provide some outline. By (9.3.8), we have an exact sequence

$$K_0(\mathscr{E}^w) \longrightarrow K_0(\mathscr{E}) \longrightarrow K_0(\mathscr{E}, \mathbf{w}) \longrightarrow 0$$

where $\mathscr{E}^w \subseteq \mathscr{E}$ is the subcategory of the acyclic objects in $\mathscr{E}$. From definitions of Grothendieck groups, there is a natural map $K_0(\mathscr{E}) \longrightarrow K_0(\mathfrak{T}(\mathscr{E}, \mathbf{w}))$, which is surjective. This factors through a surjective map $K_0(\mathscr{E}, \mathbf{w}) \twoheadrightarrow K_0(\mathfrak{T}(\mathscr{E}, \mathbf{w}))$. Since both groups are defined by generators and relations, one defines the inverse map $K_0(\mathfrak{T}(\mathscr{E}, \mathbf{w})) \longrightarrow K_0(\mathscr{E}, \mathbf{w})$ by elementary methods. ∎

9.4. Negative $\mathbb{K}$-Theory

Negative K-theory appears naturally. For example, for commutative ring R, we can define $K_{-1}(R)$ by the following exact sequence:

$$K_0(R[T]) \oplus K_0\left(R[T^{-1}]\right) \longrightarrow K_0\left(R[T,T^{-1}]\right) \longrightarrow K_{-1}(R) \longrightarrow 0.$$

We follow the functorial approach from [Sm1, Sm2]. Much of this hinges on the work of Schlichting on suspension and delooping [Sm2].

9.4.1. *Cofinality and idempotent completion*

Among the sources for this section are [BS, Sm1]. In analogy to cofinality for symmetric monoidal categories (Section 4.6), we define cofinal extension for exact categories.

Definition 9.4.1. Let $\mathscr{A}$ be an additive category and $\mathscr{B} \subseteq \mathscr{A}$ be a full subcategory.

1. We say that $\mathscr{B} \subseteq \mathscr{A}$ is **cofinal** if every object in $\mathscr{A}$ is a direct factor (i.e summand) of an object in $\mathscr{B}$. We also say that $\mathscr{B}$ is cofinal in $\mathscr{A}$. In this case, we also say that the inclusion $\mathscr{B} \subseteq \mathscr{A}$ is an **equivalence up to factor** (i.e. direct summand).
   For a commutative ring A, the category $F(A)$ of finitely generated free modules is cofinal in $\mathscr{P}(A)$ (see 4.6.3).
2. If $\mathscr{A}$ is an exact category, an exact subcategory $\mathscr{B} \subseteq \mathscr{A}$ is said to be **cofinal** if it is cofinal in the above sense and if $\mathscr{B}$ is a fully exact subcategory, meaning that (1) $\mathscr{B}$ is extension closed and (2) it preserves and detects conflations.
3. An additive category $\mathscr{A}$ is called **idempotent complete** if for all idempotent arrows $p = p^2 : A \longrightarrow A$ there is an isomorphism $\varphi : A \xrightarrow{\sim} X \oplus Y$ such that the diagram

$$\begin{array}{ccc} A & \xrightarrow{p} & A \\ \varphi\downarrow\wr & & \wr\downarrow\varphi \\ X\oplus Y & \xrightarrow{\begin{pmatrix}1_X & 0\\ 0 & 0\end{pmatrix}} & X\oplus Y \end{array} \qquad \text{commutes.}$$

   This means that, $A \cong \ker(p) \oplus Image(p)$.

4. Given an additive category $\mathscr{A}$, there is a largest additive category $\widetilde{\mathscr{A}}$ such that $\mathscr{A} \subseteq \widetilde{\mathscr{A}}$ is cofinal. In fact, $\widetilde{\mathscr{A}}$ is idempotent complete. Such an $\widetilde{\mathscr{A}}$ would be called the **idempotent completion** of $\mathscr{A}$. Existence of such a completion is proved in what follows (9.4.2).

The following is from [BS], [Sm1, 2.4.1].

Lemma 9.4.2. *Let $\mathscr{A}$ be an additive category. Then there is a largest additive category $\widetilde{\mathscr{A}}$ such that the fully faithful functor $\iota : \mathscr{A} \subseteq \widetilde{\mathscr{A}}$ is cofinal. In fact, $\widetilde{\mathscr{A}}$ is idempotent complete. Further,*

1. *The canonical functor $\iota : \mathscr{A} \longrightarrow \widetilde{\mathscr{A}}$ has the universal property in the sense that, given any additive functor $\zeta : \mathscr{A} \longrightarrow \mathscr{B}$ such that $\mathscr{B}$ is idempotent complete, then there is a natural additive functor $F : \widetilde{\mathscr{A}} \longrightarrow \mathscr{B}$, unique up to natural equivalence such that the diagram*

$$\begin{array}{ccc} \mathscr{A} & \xrightarrow{\iota} & \widetilde{\mathscr{A}} \\ & \searrow & \downarrow F \\ & & \mathscr{B} \end{array} \qquad \textit{commutes.}$$

2. *If $\mathscr{A}$ is an exact category, then $\widetilde{\mathscr{A}}$ has a natural structure of an exact category, and ι is a fully exact functor.*
3. *If $\mathscr{A}$ is a triangulated category (see* 11.1.1*), then $\widetilde{\mathscr{A}}$ has a natural structure of a triangulated category, and ι is a full and faithful triangle functor.*

Proof. First, we construct the category $\widetilde{\mathscr{A}}$. Let the objects of $\widetilde{\mathscr{A}}$ be the pairs (A, p) where $A \in Obj(\mathscr{A})$ and $p = p^2 : A \longrightarrow A$ is an idempotent arrow in $\mathscr{A}$. A morphism $(A, p) \longrightarrow (B, q)$ is an arrow $f : A \longrightarrow B$ in $\mathscr{A}$ such that the diagram

$$\begin{array}{ccc} A & \xrightarrow{p} & A \\ {\scriptstyle f}\downarrow & \overset{f}{\searrow} & \downarrow{\scriptstyle f} \\ B & \xrightarrow[q]{} & B \end{array} \qquad \text{commutes.}$$

The composition in $\widetilde{\mathscr{A}}$ is inherited from the compositions in $\mathscr{A}$. The identities in $\widetilde{\mathscr{A}}$ are defined to be $1_{(A,p)} := p$. The group structure on

$Mor_{\widetilde{\mathscr{A}}}((A,p),(B,q))$ is also inherited from that in $Mor_{\mathscr{A}}(A,B)$. The direct product $(A,p)\times(B,q)=(A\times B, p\times q)$ exists. So, $\widetilde{\mathscr{A}}$ is also an additive category. Therefore, $(A,p)\times(B,q)=(A\times B,p\times q)=(A,p)\oplus(B,q)$. An arrow $f:(A,p)\longrightarrow(A,p)$ in $\widetilde{\mathscr{A}}$ is an idempotent if and only if f is an idempotent in $\mathscr{A}$ (as well as $fp=f=pf$).

Suppose that $f:(A,p)\longrightarrow(A,p)$ is idempotent in $\widetilde{\mathscr{A}}$. Then the arrows

$$\begin{cases}(f\ p-f):(A,f)\oplus(A,p-f)\longrightarrow(A,p),\\ \begin{pmatrix}f\\ p-f\end{pmatrix}:(A,p):\longrightarrow(A,f)\oplus(A,p-f)\end{cases}$$

are inverses of each other. This follows from direct matrix multiplications:

$$(f\ p-f)\begin{pmatrix}f\\ p-f\end{pmatrix}=p=1_{(A,p)},$$

$$\begin{pmatrix}f\\ p-f\end{pmatrix}(f\ p-f)=\begin{pmatrix}f & 0\\ 0 & p-f\end{pmatrix}.$$

Now, the diagram,

$$\begin{array}{ccc}(A,p) & \xrightarrow{\quad f\quad} & (A,p)\\ {\scriptstyle\begin{pmatrix}f\\ p-f\end{pmatrix}}\Big\downarrow\wr & & \wr\Big\downarrow{\scriptstyle\begin{pmatrix}f\\ p-f\end{pmatrix}}\\ (A,f)\oplus(A,p-f) & \xrightarrow[\begin{pmatrix}f & 0\\ 0 & 0\end{pmatrix}]{} & (A,f)\oplus(A,p-f)\end{array}\quad\text{commutes.}$$

However, $\begin{pmatrix}f & 0\\ 0 & 0\end{pmatrix}:=\begin{pmatrix}1_{(A,f)} & 0\\ 0 & 0\end{pmatrix}.$

This establishes that $\widetilde{\mathscr{A}}$ is an idempotent complete additive category.

Define the canonical inclusion functor $\iota:\mathscr{A}\longrightarrow\widetilde{\mathscr{A}}$ by sending $X\mapsto(X,1_X)$. Given any object $(X,p)\in Obj(\widetilde{\mathscr{A}})$, we can write $(X,1)\cong(X,p)\oplus(X,1-p)$. So, $\mathscr{A}\subseteq\widetilde{\mathscr{A}}$ is cofinal. To establish that $\widetilde{\mathscr{A}}$ is the largest cofinal category to $\mathscr{A}$, let $\zeta:\mathscr{A}\longrightarrow\mathscr{B}$ be a cofinal inclusion of additive categories. We extend ζ to $\tilde{\zeta}:\mathscr{B}\hookrightarrow\widetilde{\mathscr{A}}$. For

$Y \in Obj(\mathscr{B})$, $Y \oplus X \cong A$ for some $X \in Obj(\mathscr{B})$, $A \in Obj(\mathscr{A})$. Define p_A as follows:

$$\begin{array}{ccc} A & \xrightarrow{q_A} & Y \\ & \searrow^{p_A} & \downarrow i_A \\ & & A \end{array} \qquad \text{Define} \quad \tilde{\zeta}(Y) = (A, p_A).$$

Note that p_A is a morphism in $\mathscr{A}$, because ζ is full and faithful. One checks that $\tilde{\zeta}$ is well defined (always up to isomorphism).

To prove (2), assume that $\mathscr{A}$ is an exact category. Declare a sequence

$$(K, w) \xrightarrow{f} (A, p) \xrightarrow{g} (C, q) \text{ exact in } \widetilde{\mathscr{A}} \text{ if}$$

$$K \xrightarrow{f} A \xrightarrow{g} C \text{ is exact in } \mathscr{A}.$$

This gives a structure of an exact category on $\widetilde{\mathscr{A}}$, and $\iota : \mathscr{A} \longrightarrow \widetilde{\mathscr{A}}$ is fully exact. This settles (2). If $\mathscr{A}$ is a triangulated category, $\widetilde{\mathscr{A}}$ is given the structure of a triangulated category, similarly. This settles (3).

Proof of (1) is routine. For more details, refer to [BS]. The proof is complete. ■

There are many versions of cofinality theorems [Wh, Prop. 1.5.9], [TT, Thm. 1.10.1], [Gd2, Thm. 1.1]. The following version suits our purpose.

Theorem 9.4.3 (Cofinality). *Suppose that $\mathscr{E} \subseteq \mathscr{D}$ is a fully exact cofinal extension of exact categories. Then the maps $K_n(\mathscr{E}) \xrightarrow{\sim} K_n(\mathscr{D})$ are isomorphisms for all $n \geq 1$, and $K_0(\mathscr{E}) \hookrightarrow K_0(\mathscr{D})$ is injective. In particular, the same applies for the idempotent completion $\mathscr{E} \subseteq \widetilde{\mathscr{E}}$.*

Proof. See [Gd2, Thm. 1.1], [Gd4, Thm. 6.1]. ■

9.4.2. *Negative $\mathbb{K}$-theory spectrum of exact categories*

For exact categories $\mathscr{E}$, we define their suspension $\mathbb{S}\mathscr{E}$ as an exact category.

Definition 9.4.4. Let $\mathscr{E}$ be an exact category. The **filter category** $\mathscr{F}\ell(\mathscr{E})$ of $\mathscr{E}$ is an exact category, defined as follows: The objects of $\mathscr{F}\ell(\mathscr{E})$ are sequences

$$A_0 \overset{i_0}{\hookrightarrow} A_1 \overset{i_1}{\hookrightarrow} A_2 \overset{i_2}{\hookrightarrow} A_3 \hookrightarrow \cdots, \text{ where } i_r \text{ are inflations.} \tag{9.26}$$

Such an object is sometimes denoted by $A_\bullet$. Given another object

$$B_\bullet : \quad B_0 \overset{j_0}{\hookrightarrow} B_1 \overset{j_1}{\hookrightarrow} B_2 \overset{j_2}{\hookrightarrow} B_3 \hookrightarrow \cdots$$

in $\mathscr{F}\ell(\mathscr{E})$, a morphism $f_\bullet : A_\bullet \longrightarrow B_\bullet$ is a sequence of morphism $f_n : A_n \longrightarrow B_n$ in $\mathscr{E}$ such that the obvious diagram

$$\begin{array}{ccccccccc} A_0 & \overset{i_0}{\hookrightarrow} & A_1 & \overset{i_1}{\hookrightarrow} & A_2 & \overset{i_2}{\hookrightarrow} & A_3 & \hookrightarrow & \cdots \\ \downarrow f_0 & & \downarrow f_1 & & \downarrow f_2 & & \downarrow f_3 & & \\ B_0 & \underset{j_0}{\hookrightarrow} & B_1 & \underset{j_1}{\hookrightarrow} & B_2 & \underset{j_2}{\hookrightarrow} & B_3 & \hookrightarrow & \cdots \end{array}$$

commutes. A pair of composable morphisms

$$K_\bullet \xrightarrow{f_\bullet} M_\bullet \xrightarrow{g_\bullet} C_\bullet \quad \text{in} \quad \mathscr{F}\ell(\mathscr{E})$$

is declared a conflation in $\mathscr{F}\ell(\mathscr{E})$ if

$$K_n \xrightarrow{f_n} M_n \xrightarrow{g_n} C_n \quad \text{are conflations in } \mathscr{E} \quad \forall\, n \geq 0.$$

This defines the structure of an exact category on $\mathscr{F}\ell(\mathscr{E})$. Consult [K, Section 5] for a proof.

Now, we define the countable envelop of $\mathscr{E}$.

Definition 9.4.5. Let $\mathscr{E}$ be an exact category. The **countable envelop** $\mathfrak{E}(\mathscr{E})$ of $\mathscr{E}$ is an exact category, defined as follows:

The objects of $\mathfrak{E}(\mathscr{E})$ are the same (9.26) as that of the filtering category $\mathscr{F}\ell(\mathscr{E})$. Given objects $A_\bullet, B_\bullet$ in $\mathfrak{E}(\mathscr{E})$, the set of morphisms is defined as

$$Mor_{\mathfrak{E}(\mathscr{E})}(A_\bullet, B_\bullet) := \lim_i co\lim_j Mor_{\mathscr{E}}(A_i, B_j).$$

Morphisms in $Mor_{\mathfrak{E}(\mathscr{E})}(A_\bullet, B_\bullet)$ are represented by $f_{ij} \in Mor_{\mathscr{E}}(A_i, B_j)$, to be denoted by $[f_{ij}]$. So, $Mor_{\mathfrak{E}(\mathscr{E})}(A_\bullet, B_\bullet)$ can be viewed as the limit of the diagrams:

$$\begin{array}{ccc} & Mor_{\mathscr{E}}(A_i, B_j) & \\ \swarrow & & \searrow \\ Mor_{\mathscr{E}}(A_{i-1}, B_j) & & Mor_{\mathscr{E}}(A_i, B_{j+1}) \\ \searrow & & \swarrow \\ & Mor_{\mathscr{E}}(A_{i-1} B_{j+1}) & \end{array}$$

There is a natural functor $\mathscr{Fl}(\mathscr{E}) \longrightarrow \mathfrak{E}(\mathscr{E})$. A pair of composable morphisms

$$K_\bullet \xrightarrow{[f_{ij}]} M_\bullet \xrightarrow{[g_{ij}]} C_\bullet \quad \text{in} \quad \mathfrak{E}(\mathscr{E})$$

is declared a conflation if it is isomorphic to the image of a conflation in $\mathscr{Fl}(\mathscr{E})$. Further, there is a fully faithful exact functor $\iota_0 : \mathscr{E} \longrightarrow \mathscr{Fl}(\mathscr{E})$, sending

$$X \mapsto \left(X \xrightarrow[1_X]{} X \xrightarrow[1_X]{} X \xrightarrow[1_X]{} X \longrightarrow \cdots \right). \quad \text{Composing,} \quad \begin{array}{ccc} \mathscr{E} & \xrightarrow{\iota_0} & \mathscr{Fl}(\mathscr{E}) \\ & \underset{\iota}{\searrow} & \downarrow \\ & & \mathfrak{E}(\mathscr{E}) \end{array}$$

we obtain a fully exact functor $\iota : \mathscr{E} \longrightarrow \mathfrak{E}(\mathscr{E})$. (It is not claimed that the vertical functor is full.) See [K, Appendix B], [Sm2, §3.1], [Sm1] for further details.

Definition 9.4.6. Use the notation as in (9.4.5). We revisit the construction (A.5.9). Consider the fully exact functor $\iota : \mathscr{E} \longrightarrow \mathfrak{E}(\mathscr{E})$. Let Σ be the class of all weak isomorphisms in $\mathfrak{E}(\mathscr{E})$, as defined in (A.5.9). Then

1. It follows from the proofs of [Sm2, Lem. 3.2] and [Sm2, Lem. 1.13] that Σ satisfies the calculus of left fractions. We define

$$\frac{\mathfrak{E}(\mathscr{E})}{\mathscr{E}} := \Sigma^{-1}\left(\mathfrak{E}(\mathscr{E})\right) \quad \text{obtained by calculus of fractions.}$$

At this stage, it is not claimed that $\frac{\mathfrak{E}(\mathscr{E})}{\mathscr{E}}$ is an exact category.

2. Now suppose that $\mathscr{E}$ is idempotent complete. Then it follows from [Sm2, Lem. 3.2] that $\frac{\mathfrak{C}(\mathscr{E})}{\mathscr{E}}$ has a structure of an exact category, and the functor $\mathscr{E} \longrightarrow \frac{\mathfrak{C}(\mathscr{E})}{\mathscr{E}}$ satisfies the universal property for the quotient category (A.5.8).
3. Finally, we define [Sm2, Def. 3.3] the **suspension** of $\mathscr{E}$ by

$$\mathbb{S}\mathscr{E} := \frac{\mathfrak{C}(\widetilde{\mathscr{E}})}{\widetilde{\mathscr{E}}} \qquad \text{where}$$

$$\widetilde{\mathscr{E}} \text{ denotes the idempotent completion of } \mathscr{E}. \tag{9.27}$$

The following is a key lemma for us.

Lemma 9.4.7. *Let $\mathscr{E}$ be a small idempotent complete exact category. Then the K-theory space $\mathbf{K}(\mathfrak{C}(\mathscr{E}))$ is contractible. Consequently, there is a homotopy equivalence $\mathbf{K}\mathscr{E} \xrightarrow{\sim} \Omega\mathbf{K}\mathbb{S}\mathscr{E}$.*

Proof. See [Sm2, Lem. 3.2, Thm 3.4]. To outline the proof, we have $\widetilde{\mathscr{E}} = \mathscr{E}$ and $\mathbb{S}\mathscr{E} = \frac{\mathfrak{C}(\mathscr{E})}{\mathscr{E}}$. Now, the sequence

$$\mathbf{K}\mathscr{E} \longrightarrow \mathbf{K}(\mathfrak{C}(\mathscr{E})) \longrightarrow \mathbf{K}\frac{\mathfrak{C}(\mathscr{E})}{\mathscr{E}} \quad \text{is a homotopy fibration[Sm2, Thm. 2.1].}$$

Further, the middle term is contractible [Sm2, Lem. 3.2]. ■

The following is an immediate consequence of (9.4.7) and the definition (9.27).

Corollary 9.4.8. *Suppose that $\mathscr{E}$ is a small exact category. Then there is a homotopy equivalence*

$$\mathbf{K}\widetilde{\mathscr{E}} \xrightarrow{\sim} \Omega\mathbf{K}\mathbb{S}\mathscr{E}.$$

We define the K-theory spectrum, as follows.

Definition 9.4.9. Let $\mathscr{E}$ be a small exact category. Inductively, define

$$\mathbb{S}^{n+1}\mathscr{E} = \mathbb{S}\left(\mathbb{S}^n\mathscr{E}\right) = \frac{\mathfrak{C}\left(\widetilde{\mathbb{S}^n\mathscr{E}}\right)}{\widetilde{\mathbb{S}^n\mathscr{E}}}.$$

By (9.4.8), we have maps of K-theory spaces

$$\begin{array}{ccc} \mathbf{K}\mathbb{S}^n\mathscr{E} & \xrightarrow{\sigma_n} & \Omega\mathbf{K}\mathbb{S}^{n+1}\mathscr{E} \\ \downarrow & \nearrow_{\sim} & \\ \mathbf{K}\widetilde{\mathbb{S}^n\mathscr{E}} & & \end{array}$$

The diagonal map is a homotopy equivalence, by (9.4.8). With $\mathbb{S}^0\mathscr{E} = \mathscr{E}$, define the $\mathbb{K}$-theory spectrum as follows:

$$\mathbb{K}\mathscr{E} = \{(\mathbf{K}\mathbb{S}^n\mathscr{E}, \sigma_n) : n \geq 0\}. \tag{9.28}$$

Use definitions (9.9, 9.10), and define the $\mathbb{K}$-groups

$$\mathbb{K}_n(\mathscr{E}) := \pi_n\left(\mathbb{K}\mathscr{E}\right) \qquad \forall\, n \in \mathbb{Z} \tag{9.29}$$

The following is an immediate consequence of the definition (9.4.9).

Lemma 9.4.10. *Let $\mathscr{E}$ be a small exact category. Then*

$$\mathbb{K}_n(\mathscr{E}) = \begin{cases} K_n(\mathscr{E}) & \textit{if } n \geq 1 \\ K_0(\widetilde{\mathscr{E}}) & \textit{if } n = 0 \\ K_0(\widetilde{\mathbb{S}^{-n}\mathscr{E}}) & \textit{if } n \leq -1. \end{cases}$$

Proof. For $n \geq 0$, we use the formula (9.10), with $k = l = 0$, and consider the diagram

$$\begin{array}{ccccccccc} \pi_n(\mathbf{K}\mathscr{E}) & \longrightarrow & \pi_n\left(\Omega\mathbf{K}\mathbb{S}\mathscr{E}\right) & \longrightarrow & \pi_n\left(\Omega^2\mathbf{K}\mathbb{S}^2\mathscr{E}\right) & \longrightarrow & \pi_n\left(\Omega^3\mathbf{K}\mathbb{S}^3\mathscr{E}\right) & \longrightarrow & \\ \downarrow & \nearrow_{\sim} & \downarrow & \nearrow_{\sim} & \downarrow & \nearrow_{\sim} & \downarrow & & \\ \pi_n(\mathbf{K}\widetilde{\mathscr{E}}) & & \pi_n\left(\Omega\mathbf{K}\widetilde{\mathbb{S}\mathscr{E}}\right) & & \pi_n\left(\Omega^2\mathbf{K}\widetilde{\mathbb{S}^2\mathscr{E}}\right) & & \pi_n\left(\Omega^3\mathbf{K}\mathbb{S}^3\mathscr{E}\right) & & \end{array}$$

The diagonal maps are isomorphisms because they are induced by homotopy equivalences. In fact, for $n \geq 1$, all the vertical maps are isomorphisms by cofinality (9.4.3). This establishes the case of $n \geq 1$ of the lemma. For $n = 0$, except for the first vertical arrow, all the vertical arrows are isomorphisms by cofinality (9.4.3), which settles the $n = 0$ case.

For $n \leq -1$, we use the formula (9.10), with $k = l = -n$, and consider the diagram

$$\begin{array}{ccccccccc}
\pi_0(\mathbf{K}\mathbb{S}^k\mathscr{E}) & \longrightarrow & \pi_0\left(\Omega\mathbf{K}\mathbb{S}^{k+1}\mathscr{E}\right) & \longrightarrow & \pi_0\left(\Omega^2\mathbf{K}\mathbb{S}^{k+2}\mathscr{E}\right) & \longrightarrow & \pi_0\left(\Omega^3\mathbf{K}\mathbf{S}^3\mathscr{E}\right) & \longrightarrow & \\
\downarrow & \overset{\sim}{\nearrow} & \downarrow & \overset{\sim}{\nearrow} & \downarrow & \overset{\sim}{\nearrow} & \downarrow & & \\
\pi_0(\mathbf{K}\widetilde{\mathbb{S}^k}\mathscr{E}) & & \pi_0\left(\Omega\mathbf{K}\widetilde{\mathbb{S}^{k+1}}\mathscr{E}\right) & & \pi_0\left(\Omega^2\mathbf{K}\widetilde{\mathbb{S}^{k+2}}\mathscr{E}\right) & & \pi_0\left(\Omega^3\mathbf{K}\mathbb{S}^{k+3}\mathscr{E}\right) & &
\end{array}$$

By the same reasoning, as in the case of $n = 0$, except for the first one, all the vertical arrows are isomorphisms. This settles the case of $n \leq -1$. The proof is complete. ■

Remark 9.4.11. It is worth drawing the attention to the fact that the negative $\mathbb{K}$-theory discussed above works purely within the setup of Quillen K-theory. In particular, this was without any reference to weak equivalences. Further, the $\mathbb{K}$-theory spectrum (9.28) is not necessarily an Ω-spectrum. This (9.28) is referred to as the nonconnective $\mathbb{K}$-theory spectrum because negative $\mathbb{K}$-groups (homotopy groups) are possible.

Perhaps, the most celebrated result from the recent past would be the following from [CHSW], which settles a long-standing conjecture.

Theorem 9.4.12. *Let k be a field, with char$(k) = 0$. Let X be a scheme over k, essentially of finite type, and* $\dim X = d$. *Then* $\mathbb{K}_n(\mathscr{P}(X)) = 0$, $\forall n \leq -(d+1)$.

Proof. See [CHSW, Thm. 6.2]. ■

9.4.3. *Negative $\mathbb{K}$-theory of complicial exact categories*

The $\mathbb{K}$-theory spectrum (9.28) of exact categories has an analogue for complicial exact categories $(\mathscr{E}, \mathbf{w})$ with weak equivalences, thanks to Waldhausen K-theory. We combine the arguments of Sections 9.3 and 9.4.2.

Definition 9.4.13. Let $(\mathscr{E}, \mathbf{w})$ be a complicial exact category with weak equivalences. The construction of suspension (9.27) is improvised to define the **suspension** of $(\mathscr{E}, \mathbf{w})$,

$$\mathbb{S}(\mathscr{E}, \mathbf{w}) := (\mathbb{S}(\mathscr{E}, \mathbf{w}), \mathbb{S}\mathbf{w}) \quad \text{and the map}$$

$$\sigma_0 : \mathbf{K}(\mathscr{E}, \mathbf{w}) \longrightarrow \Omega\mathbf{K}(\mathbb{S}(\mathscr{E}, \mathbf{w}))$$

where $\mathbb{S}(\mathscr{E}, \mathbf{w})$ is a complicial exact category with weak equivalences, and σ_0 is a map of spaces. Iteratively, define

$$\mathbb{S}^{n+1}(\mathscr{E}, \mathbf{w}) := \mathbb{S}(\mathbb{S}^n(\mathscr{E}, \mathbf{w})), \quad \text{and the maps}$$

$$\sigma_n : \mathbf{K}(\mathbb{S}^n(\mathscr{E}, \mathbf{w})) \longrightarrow \Omega\mathbf{K}\left(\mathbb{S}^{n+1}(\mathscr{E}, \mathbf{w})\right).$$

Now, define the $\mathbb{K}$-theory spectrum

$$\mathbb{K}(\mathscr{E}, \mathbf{w}) = \{(\mathbf{K}(\mathbb{S}^n(\mathscr{E}, \mathbf{w})), \sigma_n) : n \geq 0\}. \tag{9.30}$$

Define the $\mathbb{K}$-groups

$$\mathbb{K}_n(\mathscr{E}, \mathbf{w}) = \pi_n(\mathbb{K}(\mathscr{E}, \mathbf{w})) \qquad \forall\, n \in \mathbb{Z}. \tag{9.31}$$

Again, this is referred to as the **nonconnective $\mathbb{K}$-theory** because nontrivial negative $\mathbb{K}$-groups are possible.

The following is a more explicit description of these groups.

Lemma 9.4.14. *Let $(\mathscr{E}, \mathbf{w})$ be a small complicial exact category with weak equivalence. Then*

$$\mathbb{K}_n(\mathscr{E}, \mathbf{w}) = \begin{cases} K_n(\mathscr{E}, \mathbf{w}) & \text{if } n \geq 1 \\ K_0(\widetilde{\mathfrak{T}(\mathscr{E}, \mathbf{w})}) & \text{if } n = 0 \\ K_0(\widetilde{\mathfrak{T}\mathbb{S}^{-n}(\mathscr{E}, \mathbf{w})}) & \text{if } n \leq -1. \end{cases}$$

The following is the nonconnective version of the Thomason–Waldhausen Localization theorem.

Theorem 9.4.15 (Localization Theorem). *Let $(\mathscr{C}, \mathbf{w_C}) \longrightarrow (\mathscr{D}, \mathbf{w_D}) \longrightarrow (\mathscr{E}, \mathbf{w_E})$ be a sequence of functors of complicial exact categories with weak equivalences. Assume that the induced sequence $\mathfrak{T}(\mathscr{C}, \mathbf{w_C}) \xrightarrow{\iota} \mathfrak{T}(\mathscr{D}, \mathbf{w_D}) \xrightarrow{\mathfrak{q}} \mathfrak{T}(\mathscr{E}, \mathbf{w_E})$ of the associated triangulated categories is* **exact up to factor** (11.2.15). *Then the induced maps of the $\mathbb{K}$-theory spectra*

$$\mathbb{K}(\mathscr{C}, \mathbf{w_C}) \longrightarrow \mathbb{K}(\mathscr{D}, \mathbf{w_D}) \longrightarrow \mathbb{K}(\mathscr{E}, \mathbf{w_E}) \quad \text{is a homotopy fibration.}$$

Consequently, there is a long exact sequence of $\mathbb{K}$-groups not necessarily ending on either side.

Proof. Refer to [Sm1, 3.2.27], [Sm5, Thm. 9]. ∎

The following is the nonconnective version of Triangular invariance (9.3.11).

Theorem 9.4.16 (Triangular Invariance). *Let* $F : (\mathscr{E}, \mathbf{w_E}) \longrightarrow (\mathscr{D}, \mathbf{w_D})$ *be an exact functor of complicial exact categories with weak equivalences* (9.3.4)*. Suppose the induced functor* $\mathfrak{T}(\mathscr{E}, \mathbf{w_E}) \longrightarrow \mathfrak{T}(\mathscr{D}, \mathbf{w_D})$ *of the associated triangulated categories is an equivalence up to factor. Then* F *induces an equivalence*

$$\mathbb{K}(\mathscr{E}, \mathbf{w_E}) \longrightarrow \mathbb{K}(\mathscr{D}, \mathbf{w_D}) \textit{ of spectrum.}$$

Proof. Follows from (9.4.14). ∎

The following is an agreement theorem.

Theorem 9.4.17. *Let* $\mathscr{E}$ *be an exact category, and* $(Ch^b(\mathscr{E}), qIso)$ *be the associated category of bounded chain complexes, which is a complicial exact category with weak equivalences. Then*

$$\mathbb{K}_n(\mathscr{E}) = \mathbb{K}_n\left(Ch^b(\mathscr{E}), qIso\right) \qquad \forall n \in \mathbb{Z}.$$

In other words, the natural map

$$\mathbb{K}\mathscr{E} \xrightarrow{\sim} \mathbb{K}\left(Ch^b(\mathscr{E}), qIso\right) \quad \textit{is an equivalence of spectra.}$$

Proof. For $n \geq 1$, this follows from (9.4.14), and the agreement (9.3.7). Refer to [Sm1, 3.2.30] for the rest of the proof. ∎

Remark 9.4.18. For a scheme X and an open subset $U \subseteq X$ and $Z = X - U$ (with reduced induced scheme structure), we have interest in writing down long exact sequences of K-groups, relating $\mathscr{P}(X)$, $\mathscr{P}(U)$, and $\mathscr{P}(Z)$. Thus, the theory of complicial exact categories with weak equivalences becomes useful.

9.5. K-Theory of Schemes

In Quillen K-theory, we were able to write down long exact sequences of G-groups of schemes (6.1.11). However, that was not the case with

K-theory. That is where the theory of complicial exact categories becomes useful. The literature indicates that the category of perfect complexes works best.

Definition 9.5.1. Let X be a Noetherian separable scheme. We follow [TT, §2]. A complex $M_\bullet \in Ch(QCoh(X))$ is defined to be a **perfect complex** if there is an affine open cover $X = \cup_{i\in I} U_i$ such that, $\forall i \in I$, the restriction $(M_\bullet)_{|U_i}$ is quasi-isomorphic to an object $P_\bullet \in Ch^b(\mathscr{P}(U_i))$. It follows that this definition is independent of the choice of the cover $\{U_i\}$.

1. Let $\mathbf{Perf}(X) \subseteq QCoh(X)$ denote the full subcategory of perfect complexes.
2. For a closed subset $Z \subseteq X$, let $\mathbf{Perf}_Z(X)$ denote the full subcategory of complexes $M_\bullet \in \mathbf{Perf}(X)$, such that $\forall n$, the homologies $\mathcal{H}_n(M_\bullet)$, with support in Z. With $U = X - Z$, this means $\mathcal{H}_n(M_\bullet)_{|U} = 0, \forall n$.
   (a) It follows that $(\mathbf{Perf}_Z(X), qIso)$ is a complicial exact category with weak equivalences.
   (b) It also follows that the associated triangulated category
   $$\mathfrak{T}(\mathbf{Perf}_Z(X), qIso) \cong \mathbf{D}(\mathbf{Perf}_Z(X), qIso)$$
   is the corresponding derived category.
3. **Example.** Let
   $$0 \longrightarrow \mathcal{O}_X \longrightarrow \mathscr{I}_0 \longrightarrow \mathscr{I}_1 \longrightarrow \mathscr{I}_2 \longrightarrow \cdots$$
   be an injective resolution of $\mathcal{O}_X$. Then the complex $\mathscr{I}_\bullet \in \mathbf{Perf}(X)$.

We have the following.

Theorem 9.5.2. *Let X be a Noetherian separated scheme. Assume that X has an ample invertible sheaf (more generally, an ample family of line bundles [TT, Def. 2.1]). Then the inclusion $Ch^b(\mathscr{P}(X)) \subseteq \mathbf{Perf}(X)$ induces an equivalence of the derived categories $\mathbf{D}(Ch^b(\mathscr{P}(X))) \xrightarrow{\sim} \mathbf{D}^b(\mathbf{Perf}(X))$. Therefore,*

$$\mathbb{K}(Ch^b(\mathscr{P}(X))) \xrightarrow{\sim} \mathbb{K}(\mathbf{Perf}(X)) \qquad \textit{is a homotopy equivalence.}$$

Proof. Refer to [TT, Cor 3.9]. ■

Now we proceed to state the localization theorem for schemes.

Theorem 9.5.3 (Localization Theorem). *Let X be a Noetherian separated scheme. Let $U \subseteq X$ be an open subset and $Z = X - U$. Then the sequence of the derived categories*

$$\mathbf{D}(\mathbf{Perf}_Z(X)) \longrightarrow \mathbf{D}(\mathbf{Perf}(X)) \longrightarrow \mathbf{D}(\mathbf{Perf}(U))$$

is exact up to factor, in the sense of (11.2.15). *Consequently, the sequence*

$$\mathbb{K}(\mathbf{Perf}_Z(X)) \longrightarrow \mathbb{K}(\mathbf{Perf}(X)) \longrightarrow \mathbb{K}(\mathbf{Perf}(U))$$

is a homotopy fibration. Therefore, there is a long exact sequence of $\mathbb{K}$-groups.

Proof. For the first part, refer to [Sm1, 3.4.9]. The latter part follows from (9.4.15). ∎

9.5.1. *Quasi-projective schemes*

We proceed to discuss some of the results on quasi-projective schemes from [M].

Notation 9.5.4. Let X be a Noetherian scheme, with $\dim X = d$. Assume that $Coh(X)$ has enough locally free sheaves. To be consistent with the notation in [M], we denote $\mathbb{M}(X) := \mathbb{H}(X) \subseteq Coh(X)$, the full subcategory of objects $\mathcal{F} \in Coh(X)$, with $\dim_{\mathscr{P}(X)} \mathcal{F} < \infty$. As before, for $x \in X$, denote $X_x = \mathrm{Spec}\,(\mathcal{O}_{X,x})$.

1. Similar to the filtration of $Coh(X)$ by codimension of support (6.4.3), we define a filtration of $\mathbb{M}(X)$, while we filter it by grade.
   (a) Recall that, for $\mathcal{F} \in Coh(X)$, $grade(\mathcal{F}) := \min\{t : \mathcal{E}xt^t(\mathcal{F}, \mathcal{O}_X) \neq 0\}$. We remark that, if X is Cohen–Macaulay, then $grade(\mathcal{F}) = co\dim\,(Supp(\mathcal{F}))$ (see [Mh]).
   (b) For integers $k \geq 0$, denote

$$\begin{cases} Coh_g^k(X) := \{\mathcal{F} \in Coh(X) : grade(\mathcal{F}) \geq k\} \\ \mathbb{M}^k(X) := \{\mathcal{F} \in \mathbb{M}(X) : grade(\mathcal{F}) \geq k\} \\ C\mathbb{M}^k(X) := \{\mathcal{F} \in \mathbb{M}(X) : grade(\mathcal{F}) = k = \dim_{\mathscr{P}(X)}(\mathcal{F})\}. \end{cases}$$

   So, we have a filtration, $\mathbb{M}(X) = \mathbb{M}^0(X) \supseteq \mathbb{M}^1(X) \supseteq \cdots \supseteq \mathbb{M}^d(X) \supseteq 0$, by grade. We will strictly be using this filtration by grade, and the notation without the subscript g will be the norm. However, our final applications assume that X is Cohen–Macaulay, in which case this filtration coincides with the filtration by codimension of the support. Note that $\mathbb{M}^k(X)$ is a Serre subcategory of $\mathbb{M}(X)$, and $Coh^k_g(X)$ is a Serre subcategory of $Coh(X)$.

2. Also, for $\mathscr{E} = \mathscr{P}(X), \mathbb{M}(X)$ (and others), the derived category will be denoted by $\mathbf{D}^b(\mathscr{E})$. For integers $k \geq 0$,

$$\begin{cases} \mathscr{C}h^k(\mathscr{E}) := \{\mathcal{F}_\bullet \in Ch^b(\mathscr{E}) : \forall\, i\ \mathcal{H}_i(\mathcal{F}_\bullet) \in Coh^k_g(X)\} \\ \mathscr{D}^k(\mathscr{E}) := \{\mathcal{F}_\bullet \in \mathbf{D}^b(\mathscr{E}) : \forall\, i\ \mathcal{H}_i(\mathcal{F}_\bullet) \in Coh^k_g(X)\} \end{cases}$$

will denote the full subcategory of such objects. (*Note the difference between two fonts* $\mathscr{D}, \mathbf{D}$.)

(a) In fact, $(\mathscr{C}h^k(\mathscr{E}), qIso)$ is a complicial exact category with weak equivalences.

(b) Also, $\mathscr{D}^k(\mathscr{E})$ is a triangulated subcategory of $\mathbf{D}^b(\mathscr{E})$.

We comment on the fact that these categories $C\mathbb{M}^k(X)$, $\mathbb{M}^k(X)$ and others mentioned above are fairly well behaved. Readers are referred to [M] for further details. The following is a key result in [M] that facilitates application of some of the above.

Proposition 9.5.5. *Let X be a Noetherian quasi-projective scheme over a Noetherian affine scheme* Spec(A), *and let $k \geq 0$ be a fixed integer. Consider the commutative diagram of functors of derived categories:*

$$\begin{array}{ccccccc} \mathbf{D}^b(C\mathbb{M}^{k+1}(X)) & \xrightarrow[\sim]{\varsigma} & \mathbf{D}^b(\mathbb{M}^{k+1}(X)) & \xrightarrow[\sim]{\iota} & \mathscr{D}^{k+1}(\mathbb{M}(X)) & \xleftarrow[\sim]{\iota'} & \mathscr{D}^{k+1}(\mathscr{P}(X)) \\ \downarrow{\scriptstyle\alpha} & & \downarrow{\scriptstyle\beta} & & \downarrow{\scriptstyle\gamma} & & \downarrow{\scriptstyle\eta} \\ \mathbf{D}^b(C\mathbb{M}^k(X)) & \xrightarrow[\varsigma]{\sim} & \mathbf{D}^b(\mathbb{M}^k(X)) & \xrightarrow[\iota]{\sim} & \mathscr{D}^k(\mathbb{M}(X)) & \xleftarrow[\iota']{\sim} & \mathscr{D}^k(\mathscr{P}(X)). \end{array} \tag{9.32}$$

Then all the horizontal functors are equivalences of derived categories and all the vertical functors are fully faithful.

For the proof of (9.5.5), we refer to [M]. Because of this result (9.5.5), we can apply the Triangular Equivalences (9.4.16, 9.3.11) to the respective chain complex categories, with quasi-isomorphisms. We can further use the agreement theorems (9.3.7, 9.4.17). In fact, we would also like to combine it with the Localization theorems (9.3.10, 9.4.15). For this purpose, we state the following.

Lemma 9.5.6. *Suppose that X is a Cohen–Macaulay quasi-projective scheme over an affine scheme* Spec (A), *and that $k \geq 0$ is an integer. Then the sequence of derived categories*

$$\mathscr{D}^{k+1}\left(\mathscr{P}(X)\right) \longrightarrow \mathscr{D}^{k}\left(\mathscr{P}(X)\right) \longrightarrow \coprod_{x \in X^{(k)}} \mathscr{D}^{k}\left(\mathscr{P}(X_x)\right)$$

is exact up to factor. If X is regular, this sequence is exact.

Proof. See [Bp2]. ■

Now we apply (9.5.5) and execute the plan stated above.

Theorem 9.5.7. *Suppose that X is a Noetherian quasi-projective scheme over an affine scheme* Spec (A) *and that $k \geq 0$ is an integer. Consider the diagram of $\mathbb{K}$-theory spectra and maps:*

$$\begin{array}{ccccc}
\mathbb{K}\left(C\mathbb{M}^{k+1}(X)\right) & & \mathbb{K}\left(C\mathbb{M}^{k}(X)\right) & & \coprod_{x\in X^{(k)}} \mathbb{K}\left(C\mathbb{M}^{k}(X_x)\right) \\
\downarrow \wr & & \wr \downarrow \Psi & & \downarrow \wr \\
\mathbb{K}\left(\mathscr{C}h^{k+1}(\mathbb{M}(X)), qIso\right) & \longrightarrow & \mathbb{K}\left(\mathscr{C}h^{k}(\mathbb{M}(X)), qIso\right) & \longrightarrow & \coprod_{x\in X^{(k)}} \mathbb{K}\left(\mathscr{C}h^{k}(\mathbb{M}(X_x)), qIso\right) \\
\uparrow \wr & & \wr \uparrow \Phi & & \uparrow \wr \\
\mathbb{K}\left(\mathscr{C}h^{k+1}(\mathscr{V}(X)), qIso\right) & \longrightarrow & \mathbb{K}\left(\mathscr{C}h^{k}(\mathscr{P}(X)), \mathscr{Q}\right) & \longrightarrow & \coprod_{x\in X^{(k)}} \mathbb{K}\left(\mathscr{C}h^{k}(\mathscr{P}(X_x)), qIso\right)
\end{array}$$

Then the vertical maps are equivalences of $\mathbb{K}$-theory spectra. Further, if X is Cohen–Macaulay, then the second line and the third line are homotopy fibrations of $\mathbb{K}$-theory spectra.

Proof. Here the middle upward arrow $\Phi : \mathbb{K}(\mathscr{C}h^k(\mathscr{P}(X), qIso)) \to \mathbb{K}(\mathscr{C}h^k(\mathbb{M}(X), qIso))$ is induced by the functor $\iota' : (\mathscr{C}h^k(\mathscr{P}(X), qIso)) \to (\mathscr{C}h^k(\mathbb{M}(X), qIso))$ of complicial exact categories with weak equivalences. By (9.5.5), ι' induces an equivalence of the associated triangulated categories. Therefore, by Triangular equivalence (9.3.11), Φ is a homotopy equivalence. Likewise, the other two upward arrows are homotopy equivalences.

The middle downward arrow Ψ is a composition of three maps, as follows:

$$\begin{array}{ccc} \mathbb{K}\left(C\mathbb{M}^k(X)\right) & \xrightarrow{\Psi'} & \mathbb{K}\left(Ch^b\left(C\mathbb{M}^k(X), qIso\right)\right) \\ {\scriptstyle\Psi}\downarrow & & \downarrow{\scriptstyle\zeta'} \\ \mathbb{K}\left(\mathscr{C}h^k\left(\mathbb{M}(X), qIso\right)\right) & \xleftarrow[\iota]{} & \mathbb{K}\left(Ch^b\left(\mathbb{M}^k(X), qIso\right)\right). \end{array}$$

Now, ζ' and ι are induced by the corresponding functors of complicial exact categories, with weak equivalences. Again, by (9.3.11) in conjunction with Theorem 9.5.5, ζ' and ι are equivalences of spectra. Now, Ψ' is an equivalence of spectra, by the agreement theorem (9.4.17). Hence, so is Ψ.

It remains to be shown that, when X is Cohen–Macaulay, the third line is a homotopy fibration of $\mathbb{K}$-theory spectra. To do this, consider the sequence of complicial exact categories with weak equivalences (not necessarily exact):

$$\begin{aligned} &\left(\mathscr{C}h^{k+1}\left(\mathscr{P}(X)\right), qIso\right) \longrightarrow \left(\mathscr{C}h^{k}\left(\mathscr{P}(X)\right), qIso\right) \\ &\longrightarrow \textstyle\coprod_{x\in X^{(k)}} \left(\mathscr{C}h^{k}\left(\mathscr{P}(X_x)\right), qIso\right). \end{aligned} \tag{9.33}$$

By (9.5.6), the corresponding sequence of the derived categories is exact up to factor. Therefore, by an application of the nonconnective version of the Thomason–Waldhausen Localization theorem (9.4.15), the third line in the statement of the theorem is a homotopy fibration of $\mathbb{K}$-theory spectra. The proof is complete. ■

Note that, if X is Cohen–Macaulay, then for $x \in X^{(k)}$, $C\mathbb{M}^k(X_x) = \mathbb{M}^k(X_x)$ is the category of finite length $\mathcal{O}_{X,x}$-modules with finite projective dimension. The following is an immediate consequence of (9.5.7).

Corollary 9.5.8. *Let X and k be as in Theorem* 9.5.7. *Assume that X is Cohen–Macaulay. Then for any integer n, there is an exact sequence of $\mathbb{K}$-groups,*

$$\begin{array}{l} \cdots \longrightarrow \mathbb{K}_n\left(C\mathbb{M}^{k+1}(X)\right) \longrightarrow \mathbb{K}_n\left(C\mathbb{M}^{k}(X)\right) \longrightarrow \\ \oplus_{x\in X^{(k)}}\mathbb{K}_n\left(\mathbb{M}^k(X_x)\right) \longrightarrow \mathbb{K}_{n-1}\left(C\mathbb{M}^{k+1}(X)\right) \longrightarrow \cdots \end{array} \tag{9.34}$$

Proof. Follows from Theorem 9.5.7. The proof is complete. ■

Remark 9.5.9. Note that we cannot apply the connective version of Thomason–Waldhausen Localization theorem (9.3.10) in the setup of (9.5.7). This is because the sequence of derived categories, corresponding to (9.33), is not necessarily exact. However, if X is regular, the sequence of derived categories is exact, and we have the version of Theorem 9.5.7 and Corollary 9.5.8, respectively, for **K**-theory spaces and groups.

Proposition 9.5.10. *Let X be as in Theorem 9.5.7. Assume that X is Cohen–Macaulay. Then there is a Gersten complex, similar to* (6.50), *constructed as follows:*

$$\begin{array}{ccccc}
\oplus_{x\in X^{(k-1)}}\mathbb{K}_{n+1}\left(\mathbb{M}^{k-1}(X_x)\right) & & & & \mathbb{K}_{n-1}\left(C\mathbb{M}^{k+2}(X)\right) \\
\downarrow & \searrow & & & \downarrow \\
\mathbb{K}_n\left(C\mathbb{M}^{k}(X)\right) & \longrightarrow & \oplus_{x\in X^{(k)}}\mathbb{K}_n\left(C\mathbb{M}^{k}(X_x)\right) & \longrightarrow & \mathbb{K}_{n-1}\left(C\mathbb{M}^{k+1}(X)\right) \\
\downarrow & & & \searrow & \downarrow \\
\mathbb{K}_n\left(C\mathbb{M}^{k-1}(X)\right) & & & & \oplus_{x\in X^{(k+1)}}\mathbb{K}_{n-1}\left(C\mathbb{M}^{k+1}(X_x)\right)
\end{array}$$

The broken diagonal arrows form the Gersten complex:

$$0 \longrightarrow \mathbb{K}_n(X) \longrightarrow \bigoplus_{x\in X^{(0)}}\mathbb{K}_n(\mathbb{M}^0(X_x)) \longrightarrow \bigoplus_{x\in X^{(1)}} K_{n-1}(\mathbb{M}^1(X_x))$$

$$\cdots \longrightarrow \bigoplus_{x\in X^{(r)}} K_{n-r}(\mathbb{M}^r(X_x)) \longrightarrow \bigoplus_{x\in X^{(r+1)}} \mathbb{K}_{n-r-1}(\mathbb{M}^{r+1}(X_x)) \longrightarrow \cdots$$

$$\longrightarrow \bigoplus_{x\in X^{(d-1)}} \mathbb{K}_{n-d-1}(\mathbb{M}^{d-1}(X_x)) \longrightarrow \bigoplus_{x\in X^{(d)}} \mathbb{K}_{n-d}(\mathbb{M}^{d}(X_x)) \longrightarrow 0$$

In the spirit of the K-theory spectral sequence (6.85), *under the hypotheses as above, we can write down a spectral sequence [Bp3]*

$$\mathbb{E}(X)_1^{pq} = \begin{cases} \bigoplus_{x\in X^{(p)}} \mathbb{K}_{-p-q}(X_x) & 0 \le p \le d \\ 0 & \text{otherwise.} \end{cases} \tag{9.35}$$

The spectral sequence (9.35) *is derived directly from the exact sequences* (9.34) *using standard methods, as is shown here. The method is explained in Section* 6.5.2, *and more elaborately in* [Wc1, Sec. 5.9]. *The line $q = 0$ of this sequence is the Gersten complex above* (9.35).

Remark 9.5.11. Quillen's version of the Gersten complex (6.50) deals with the category $Coh(X)$ and its filtration by codimension of support, where X is a Noetherian scheme. This complex (9.35) is exactly analogous to the G-theory Gersten complex (6.50), written for the subcategories of $\mathbb{M}(X) \subseteq Coh(X)$. Using (9.5.5), $C\mathbb{M}^k(X)$ can be replaced by $\mathbb{M}^k(X)$ in (9.34). Other than that there is no analogue of (9.34), at the level of such exact categories. However, at the level of the category of perfect complexes, there is an analogue, even when X is only a Noetherian separated scheme, as follows (9.5.12).

Exercise 9.5.12. Let X be a Noetherian separated scheme, with $\dim X = d < \infty$. Denote the usual filtering of $Coh(X)$ and $\mathbf{P}erf(X)$, respectively, by

$$\begin{cases} Coh_u^k(X) = \{\mathscr{F} \in Coh(X) : co\dim(Supp(\mathscr{F})) \geq k\} \\ \mathbf{P}erf^k(X) = \\ \{\mathscr{F}_\bullet \in \mathbf{P}erf(X) : \text{the homologies } H_i(\mathscr{F}_\bullet) \in Coh_u^k(X)\ \forall\ i\} \\ \text{Also, denote} \quad X_u^{(k)} = \left\{x \in X : co\dim \overline{\{x\}} = k\right\} \end{cases}$$

where $0 \leq k \leq d$.

Prove that the sequence

$$\mathbb{K}\mathbf{P}erf^{k+1}(X) \longrightarrow \mathbb{K}\mathbf{P}erf^k(X) \longrightarrow \coprod_{x \in X_u^{(k)}} \mathbb{K}\mathbf{P}erf^k(X_x) \tag{9.36}$$

is a homotopy fibration of $\mathbb{K}$-theory spectra. Further, derive a Gersten complex similar to (9.35). **Hint:** Follow the above proofs. These are from [TT, Eqn. 10.3.6], [Bp3].

We remark that, for a Cohen–Macaulay local ring $(A, \mathfrak{m})$ with $\dim A = k$, it follows from (9.5.7) that there is a zigzag equivalence of spectra $\mathbb{K}C\mathbb{M}^k(\mathrm{Spec}\,(A)) \xrightarrow{\sim} \mathbb{K}\mathbf{P}erf^k(\mathrm{Spec}\,(A))$. Therefore, when X is Cohen–Macaulay (not necessarily quasi-projective), the Gersten complex derived from (9.36) takes the form exactly like (9.35), and extends it.

Exercise 9.5.13. Consider the local ring

$$A = \left(\frac{\kappa[X, Y, Z, W]}{(XW - YZ)}\right)_{(X,Y,Z,W)} \qquad \text{where } \kappa \text{ is an infinite field.}$$

Let $X = \mathrm{Spec}\,(A)$. Prove that the Gersten complex (9.35) is not exact for $X = \mathrm{Spec}\,(A)$.

Proof. Dutta, Hochster, and McLaughlin [DHM] gave an example of a module $M \in C\mathbb{M}^3(X)$ with $\chi(M, \frac{A}{\wp}) \neq 0$. Note that there is a fully faithful exact functor $C\mathbb{M}^k(X) \longrightarrow \mathbf{P}erf(X)$, which respects the filtrations. Consider the commutative diagram

$$\begin{array}{ccc} K_0(C\mathbb{M}^3(X)) & \xrightarrow{\iota_1} & K_0(C\mathbb{M}^2(X)) \\ \downarrow & & \downarrow \\ K_0(\mathbf{P}erf^3(X)) & \xrightarrow[\iota_2]{} & K_0(\mathbf{P}erf^2(X)) \end{array}$$

Levine proved [Bp3, Prop. 9] that $\iota_2[M] \neq 0$. Therefore, $\iota_1[M] \neq 0$. The assertion is proved by proving an analogue of (6.4.6). ■

Chapter 10

Epilogue: Hermitian K-Theory

Let X be a scheme. Then the category $\mathscr{P}(X)$ is endowed with an obvious duality, sending $P \mapsto P^* := \mathcal{H}om\,(P, \mathcal{O}_X)$. Incorporating this duality, classically, this gave rise to two groups: (1) The Grothendieck–Witt group $GW_0(\mathscr{P}(X), *)$, and (2) The Witt group $W_0(\mathscr{P}(X), *)$. Not surprisingly, following [Q, W, TT], this led to the definition of (higher) Grothendieck–Witt and (higher) Witt groups of Exact categories with duality. This is generally known as Hermitian K-theory. We will avoid getting into the classical part of the theory, and give an overview of the homotopy approach to Hermitian K-theory. To a great extent, Schlichting is the main architect of this approach to Hermitian K-theory [Sm3, Sm4, Sm5, Sm6]. In a sense, all that culminated in [Sm3], where Hermitian K-theory was developed, in the framework of dg categories, with a close analogy to K-theory [Q, W, TT]. Throughout this chapter, we assume that 2 is invertible. Refer to Chapter 11 for information related to Triangulated categories.

10.1. Hermitian K-Theory of Exact Categories

We start with the definition of a category with duality.

Definition 10.1.1. A **category with duality** is a triple $(\mathscr{C}, *, \eta)$ where $\mathscr{C}$ is a category, $* : \mathscr{C}^{op} \longrightarrow \mathscr{C}$ is a functor, and $\eta : 1_{\mathscr{C}} \longrightarrow *^2$ is a natural transformation; meaning that there are natural arrows $\eta_A : A \longrightarrow A^{**}$ for all $A \in Obj(\mathscr{C})$ such that $\eta_A^* \eta_{A^*} = 1_{A^*}$. Subsequently, we will be using such dualities in a variety of contexts, as follows:

1. For exact categories with duality, η_A will be assumed to be an isomorphism.
2. For exact categories with weak equivalences and duality, we assume that η_A is a weak equivalence (to be defined).
3. Likewise, for dg categories with weak equivalences and duality, we assume that η_A is a weak equivalence (to be defined).

All these will be made precise as we encounter these respective situations. Suppose that $(\mathscr{C}, *, \eta)$ is a category with duality.

1. A **symmetric form** is a pair (P, φ) where $P \in Obj(\mathscr{E})$ and $\varphi : P \longrightarrow P^*$ is a morphism such that the diagram

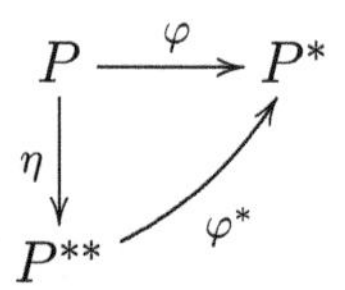

commutes.

   If φ is an isomorphism, we say that (P, φ) is a **symmetric space** or is nonsingular or nondegenerate. These definitions are analogous to quadratic modules (8.3.1) and quadratic spaces (8.3.4).
2. Given two symmetric forms (P_1, φ_1) and (P_2, φ_2), a map $f : P_1 \longrightarrow P_2$ in $\mathscr{C}$ is called a map $f : (P_1, \varphi_1) \longrightarrow (P_2, \varphi_2)$ of symmetric forms if $\varphi_1 = f^*\varphi_1 f$. Such a map is called an **isometry** if f is an isomorphism.
3. Given two symmetric forms (P_1, φ_1) and (P_2, φ_2), the orthogonal sum is defined as $(P_1, \varphi_1) \perp (P_2, \varphi_2) = (P_1 \oplus P_2, \varphi_1 \oplus \varphi_2)$.
4. Let (P, φ) be a symmetric form, and $f : M \longrightarrow P$ be a morphism. Then $(Q, f^*\varphi f)$ defines a symmetric form.

Before we proceed, we introduce the following notation.

Notation 10.1.2. Let $\mathscr{C} := (\mathscr{C}, \vee, \varpi)$ be a category with duality.

1. Then $\mathscr{C}^h$ or $\mathscr{C}_h$ denote the category of symmetric forms in $\mathscr{C}$. So, the objects in $\mathscr{C}_h$ are symmetric forms (X, φ) in $\mathscr{C}$. A map

$(X, \varphi) \longrightarrow (Y, \psi)$ in $\mathscr{C}_h$ is a morphism $f : X \longrightarrow Y$ in $\mathscr{C}$ such that the diagram

$$\begin{array}{ccc} X & \xrightarrow{\varphi} & X^\vee \\ {\scriptstyle f}\downarrow & & \uparrow{\scriptstyle f^\vee} \\ Y & \xrightarrow[\psi]{} & Y^\vee \end{array} \quad \text{commutes.} \tag{10.1}$$

2. Suppose that $\mathbf{w}$ is a subcategory of morphisms in $\mathscr{C}$. Assume that (1) $\varpi_A \in \mathbf{w}$ for all objects A in $\mathscr{C}$ and (2) $f \in \mathbf{w} \Longrightarrow f^\vee \in \mathbf{w}$. Then the notation $\mathbf{w}\mathscr{C} := \mathbf{w}$ is also used to denote the category whose objects are the same as that of $\mathscr{C}$ and morphisms are the morphisms of $\mathbf{w}$. So, an object in $(\mathbf{w}\mathscr{C})_h$ is a symmetric form (X, φ) in $\mathscr{C}$, with $\varphi \in \mathbf{w}$. For objects $(X, \varphi), (Y, \psi)$ in $(\mathbf{w}\mathscr{C})_h$, a morphism $f : (X, \varphi) \longrightarrow (Y, \psi)$ in $(\mathbf{w}\mathscr{C})_h$ is a map $f \in \mathbf{w}$ so that the diagram (10.1) commutes.
3. Therefore, if $\mathbf{i}$ denotes the category of isomorphisms in $\mathscr{C}$, then an object (X, φ) in $(\mathbf{i}\mathscr{C})_h$ is a symmetric space in $\mathscr{C}$.

In this section, we consider exact categories with duality.

Definition 10.1.3. An **exact category with duality** is a category with duality $(\mathscr{E}, *, \eta)$, where $\mathscr{E}$ is an exact category, $* : \mathscr{E}^{op} \longrightarrow \mathscr{E}$ is an exact functor, and $\eta : 1_{\mathscr{E}} \longrightarrow *^2$ is a natural equivalence; meaning that $\eta_A : A \xrightarrow{\sim} A^{**}$ is a natural isomorphism for all $A \in Obj(\mathscr{E})$.

Example 10.1.4. Suppose that X is a Noetherian scheme and $\mathcal{L}$ is an invertible sheaf. For $P \in \mathscr{P}(X)$, let $P \mapsto P^* := \mathcal{H}om\,(P, \mathcal{L})$, and $ev : P \xrightarrow{\sim} P^{**}$ be the evaluation map. Then $(\mathscr{P}(X), *, ev)$ is an exact category with duality. In particular, we can take $\mathcal{L} = \mathcal{O}_X$.

Definition 10.1.5. Suppose that $(\mathscr{E}, *, \eta)$ is an exact category with duality, and (P, φ) is a symmetric space.

1. Suppose that $\iota : L \hookrightarrow P$ is an inflation. Then ι^* is a deflation, and hence so is $\iota^*\varphi$. Define the **orthogonal complement** $L^\perp := \ker(\iota^*\varphi)$. Since $\ker(\iota^*\varphi) \cong \ker(\iota^*)$, we can embed $L^\perp \subseteq P^*$.

2. Let $\iota : L \hookrightarrow P$ be an inflation such that $\iota^* \varphi \iota = 0$. In this case, we have a commutative diagram

$$\begin{array}{ccccc} L & \overset{\iota}{\hookrightarrow} & P & \overset{p}{\twoheadrightarrow} & C \\ {\scriptstyle \zeta_L}\downarrow & \nearrow & \wr\downarrow{\scriptstyle \varphi} & & \downarrow{\scriptstyle \zeta_C} \\ L^{\perp} & \hookrightarrow & P^* & \underset{\iota^*}{\twoheadrightarrow} & L^* \end{array} .$$

with $C = co\ker(\iota)$ for some morphisms ζ_L, ζ_C.

Such an inflation ι is said to be a **totally isotropic** subspace if, further, the induced map $\zeta_L : L \longrightarrow L^{\perp}$ is an inflation.

3. A totally isotropic subspace $\iota : L \hookrightarrow P$ is said to be a **Lagrangian** if $L = L^{\perp}$. So, $\iota : L \hookrightarrow P$ is a Lagrangian if and only if

$$L \hookrightarrow P \underset{\iota_* \varphi}{\twoheadrightarrow} L^* \qquad \text{is a conflation.}$$

In this case, both ζ_L, ζ_C are isomorphisms. And, we have a commutative diagram

$$\begin{array}{ccccc} L & \overset{\iota}{\hookrightarrow} & P & \overset{p}{\twoheadrightarrow} & C \\ \wr\downarrow & & \wr\downarrow{\scriptstyle \varphi} & & \wr\downarrow{\scriptstyle \zeta_C} \\ C^* & \hookrightarrow & P^* & \twoheadrightarrow & L^* \end{array}$$

4. A symmetric space (P, φ) is called **metabolic** if it has a Lagrangian $L \hookrightarrow P$.
5. Let $L \in Obj(\mathscr{E})$. Define

$$H(L) := \left(L \oplus L^*, \begin{pmatrix} 0 & 1 \\ 1 & \eta \end{pmatrix} \right).$$

Then $H(L)$ is a symmetric space, to be called a **hyperbolic space**. It follows that $L \hookrightarrow H(L)$ is a Lagrangian. So, $H(L)$ is metabolic.

We give the definition of the Witt group and the Grothendieck–Witt (GW) group [Sm2].

Definition 10.1.6. Suppose that $(\mathscr{E}, *, \eta)$ is a small exact category with duality. Let $\mathbb{MI}(\mathscr{E})$ denote the set of all isometry classes of symmetric spaces. Note that $\mathbb{MI}(\mathscr{E})$ has a structure of a monoid under orthogonal sum.

1. Let $\mathscr{R}Wi(\mathscr{E}) \subseteq \mathbb{MI}(\mathscr{E})$ be the submonoid generated by metabolic spaces. Define the **Witt group**

$$W(\mathscr{E}) := W_0(\mathscr{E}) := \frac{\mathbb{MI}(\mathscr{E})}{\mathscr{R}Wi(\mathscr{E})}.$$

   In fact, the Witt group $W_0(\mathscr{E})$ turns out to be a group. This is because, for any symmetric space (P, φ), the space $(P, \varphi) \perp (P, -\varphi)$ is metabolic.
2. We define the **Grothendieck group** of a monoid. Let $\mathscr{M}$ be an abelian monoid. The Grothendieck group of $\mathscr{M}$ is defined to be the abelian group $\mathbf{Gr}(\mathscr{M})$, together with homomorphism $\iota : \mathscr{M} \longrightarrow \mathbf{Gr}(\mathscr{M})$, satisfying the following universal property:

$$\begin{array}{ccc} \mathscr{M} & \xrightarrow{\ \iota\ } & \mathbf{Gr}(\mathscr{M}) \\ & \searrow^{g} & \downarrow{\scriptstyle \mathfrak{g}} \\ & & G \end{array}$$

   This means, given any homomorphism g, with G an abelian group, there is a unique group homomorphism $\mathfrak{g}$ such the $\mathfrak{g}\iota = g$.
3. Now consider the abelian group $\mathbf{Gr}(\mathbb{MI}(\mathscr{E}))$. The **Grothendieck–Witt group** $GW(\mathscr{E}) := GW_0(\mathscr{E})$ of $\mathscr{E}$ is defined as the group $\mathbf{Gr}(\mathbb{MI}(\mathscr{E}))$ modulo the relations

$$[(P, \varphi)] = [H(L)] \quad \text{whenever} \quad L \hookrightarrow P \quad \text{is a Lagrangian.}$$

The following is a routine lemma.

Lemma 10.1.7. *Suppose that $(\mathscr{E}, *, \eta)$ is a small exact category with duality. Then there is a natural exact sequence*

$$K_0(\mathscr{E}) \xrightarrow{H} GW_0(\mathscr{E}) \longrightarrow W_0(\mathscr{E}) \longrightarrow 0$$

where H denotes the map induced by the association $L \mapsto H(L)$.

Similar to K-theory spaces $\mathbf{K}(\mathscr{E})$ of exact categories $\mathscr{E}$ (Chapter 3), we associate a GW-space whose homotopy groups are defined as the higher GW-groups of $(\mathscr{E}, *, \eta)$. We do this by implementing Quillen $\mathbb{Q}$-construction, as follows.

Definition 10.1.8. Suppose that $\mathscr{E} := (\mathscr{E}, *, \eta)$ is a small exact category with duality. We define a new category $\mathbb{Q}(\mathscr{E}, h)$, similar to the Quillen $\mathbb{Q}$-category $\mathbb{Q}\mathscr{E}$. The objects of $\mathbb{Q}(\mathscr{E}, h)$ are symmetric spaces (P, φ). A morphism $(P_1, \varphi_1) \longrightarrow (P_2, \varphi_2)$ in $\mathbb{Q}(\mathscr{E}, h)$ is a morphism $P_1 \longrightarrow P_2$ in $\mathbb{Q}\mathscr{E}$, represented by (U, p, ι), as follows, and a commutative diagram

$$\begin{array}{ccccc} P_1 & \overset{p}{\twoheadleftarrow} & U & \overset{\iota}{\hookrightarrow} & P_2 \\ {\scriptstyle \varphi_1}\downarrow\wr & & \downarrow & & \wr\downarrow{\scriptstyle \varphi_2} \\ P_1^* & \underset{p^*}{\hookrightarrow} & U^* & \underset{\iota^*}{\twoheadleftarrow} & P_2^* \end{array}$$

$$\ni \quad \ker(p) \xrightarrow{\sim} \ker(\iota^* \varphi_2) \quad \text{is an isomorphism.}$$

This is equivalent to

$$\begin{array}{ccc} U & \overset{\iota}{\hookrightarrow} & P_2 \\ {\scriptstyle p}\downarrow & & \downarrow{\scriptstyle \iota^*\varphi_2} \\ P_1 & \underset{p^*\varphi_1}{\longrightarrow} & U^* \end{array} \qquad \text{being bicartesian.}$$

There is a forgetful exact functor

$$\mathfrak{f} : \mathscr{E}(h) \longrightarrow \mathscr{E}, \quad \text{inducing a map} \quad |\mathbb{Q}\mathfrak{f}| : \mathbb{B}\mathbb{Q}(\mathscr{E}, h) \longrightarrow \mathbb{B}\mathbb{Q}\mathscr{E} \quad \text{of spaces.}$$

The **Grothendieck–Witt (GW) space** of $(\mathscr{E}, *, \eta)$ is defined as

$$\mathbf{GW}(\mathscr{E}) := \mathbf{GW}(\mathscr{E}, *, \eta) := \mathscr{F}(|\mathbb{Q}\mathfrak{f}|, 0) \qquad \text{the homotopy fiber.} \tag{10.2}$$

So, we have homotopy fibrations

$$\mathbf{GW}(\mathscr{E}) \longrightarrow \mathbb{B}\mathbb{Q}(\mathscr{E}, h) \longrightarrow \mathbb{B}\mathbb{Q}\mathscr{E}, \quad \mathbf{K}(\mathscr{E}) \longrightarrow \mathbf{GW}(\mathscr{E}) \longrightarrow \mathbb{B}\mathbb{Q}(\mathscr{E}, h)$$

where the second one is obtained by rotation of the first one (C.3.10).

Before we define the higher Grothendieck groups, we state the following two lemmas.

Lemma 10.1.9. *Suppose that $(\mathscr{E}, *, \eta)$ is a small exact category with duality. Then there are natural isomorphisms*

$$\begin{cases} W(\mathscr{E}) = W_0(\mathscr{E}) \cong \pi_0\left(\mathbb{BQ}(\mathscr{E}, h), 0\right) \\ GW_0(\mathscr{E}) \cong \pi_0\left(\mathbf{GW}(\mathscr{E}), 0\right). \end{cases}$$

Proof. See [Sm4, Prop. 4.8], [Sm4, Prop. 4.11]. ■

Definition 10.1.10. Suppose that $(\mathscr{E}, *, \eta)$ is a small exact category with duality. For integer $n \geq 0$, define

$$GW_n(\mathscr{E}) := GW_n(\mathscr{E}, *, \eta) := \pi_n\left(\mathbf{GW}(\mathscr{E}), 0\right).$$

Then there is an exact sequence

$$\begin{array}{c} \cdots \longrightarrow K_1(\mathscr{E}) \longrightarrow GW_1(\mathscr{E}) \longrightarrow \pi_1\left(\mathbb{BQ}(\mathscr{E}, h), 0\right) \\ \\ \cdots \longrightarrow K_0(\mathscr{E}) \longrightarrow GW_0(\mathscr{E}) \longrightarrow\!\!\!\!\rightarrow W(\mathscr{E}). \end{array} \tag{10.3}$$

We remark that $\pi_1\left(\mathbb{BQ}(\mathscr{E}, h), 0\right)$ has an explicit description [Sm4, Prop. 4.9], and so does $K_1(\mathscr{E})$ (3.8.7). This provides a description of $GW_1(\mathscr{E})$.

Remark 10.1.11. Suppose that $(\mathscr{E}, *, \eta)$ is a small exact category with duality.

1. Assume that $\mathscr{E}$ is idempotent complete. Construction of suspension (9.27) is improvised to define a suspension $\mathbb{S}(\mathscr{E}) := \mathbb{S}(\mathscr{E}, *, \eta)$, which is an exact category with duality. It is also proved that [Sm4, Thm. 9.11]

$$\mathbf{GW}(\mathscr{E}) \xrightarrow{\sim} \Omega\mathbf{GW}(\mathbb{S}\mathscr{E}).$$

2. In general, let $\widetilde{\mathscr{E}}$ denote the idempotent completion of $\mathscr{E}$, which has a natural structure of an exact category with duality. Define the suspension of $\mathscr{E}$ by

$$\mathbb{S}(\mathscr{E}) := \mathbb{S}(\mathscr{E}, *, \eta) := \mathbb{S}\widetilde{\mathscr{E}}, \qquad \text{Further, let} \qquad \mathbb{S}^{n+1}\mathscr{E} := \mathbb{S}\left(\widetilde{\mathbb{S}^n\mathscr{E}}\right).$$

As in (9.28), with $\mathbb{S}^0\mathscr{E} = \mathscr{E}$, define the $\mathbb{G}W$-theory spectrum:

$$\mathbb{G}W(\mathscr{E}) := \mathbb{G}W(\mathscr{E}, *, \eta) := \{(\mathbf{GW}\mathbb{S}^n\mathscr{E}, \sigma_n) : n \geq 0\} \tag{10.4}$$

and define the $\mathbb{G}W$-groups

$$\mathbb{G}W_n(\mathscr{E}) := \mathbb{G}W_n(\mathscr{E}, *, \eta) := \pi_n(\mathbb{G}W(\mathscr{E})) \qquad \forall\, n \in \mathbb{Z}. \tag{10.5}$$

See [Sm4, Thm. 9.12] for details.

10.1.1. *Exact categories with weak equivalences and duality*

We define exact categories with weak equivalences and duality.

Definition 10.1.12. An exact category with weak equivalences and duality is a quadruple $(\mathscr{E}, *, \eta, \mathbf{w})$ such that (1) $(\mathscr{E}, *, \eta)$ is a category with duality, (2) $\mathscr{E}$ is exact and $* : \mathscr{E}^{op} \longrightarrow \mathscr{E}$ is exact, (3) $(\mathscr{E}, \mathbf{w})$ is an exact category with weak equivalences (9.3.3), (4) for morphisms, $f \in \mathbf{w} \iff f^* \in \mathbf{w}$, (5) $\forall\, A \in Obj(\mathscr{E})$, $\eta_A \in \mathbf{w}$. (See [Sm3, §4.2, p. 1767].)

Imitating Definition 10.1.6, we define Grothendieck groups and Witt groups of category with weak equivalences and duality [Sm6, Def. 1, 2].

Definition 10.1.13. Suppose that $(\mathscr{E}, *, \eta, \mathbf{w})$ is a small exact category with weak equivalences and duality. A symmetric form (X, φ) is defined as a symmetric $\mathbf{w}$-space if $\varphi \in \mathbf{w}$. The Grothendieck–Witt group $GW_0(\mathscr{E}) := GW_0(\mathscr{E}, *, \eta, \mathbf{w})$ is defined as the free abelian group generated by symmetric $\mathbf{w}$-spaces $[(X, \varphi)]$ subject to the following relations:

1. $[(X, \varphi)] + [(Y, \psi)] = [(X, \varphi) \perp (Y, \psi)]$.
2. For all $g : Y \longrightarrow X$ morphisms in $\mathbf{w}$, we have $[(X, \varphi)] = [(Y, g^*\varphi g)]$.

3. Let $E_\bullet = L \hookrightarrow P \twoheadrightarrow C$ be a conflation. Consider a map $\varphi_\bullet : E_\bullet \longrightarrow E_\bullet^*$, as follows:

$$\begin{array}{ccccccc} E_\bullet & & L & \hookrightarrow & P & \twoheadrightarrow & C \\ \varphi_\bullet \downarrow & & \varphi_L \downarrow & & \downarrow \varphi & & \downarrow \varphi_C \\ E_\bullet^* & & C^* & \hookrightarrow & P^* & \twoheadrightarrow & L^* \end{array} \quad \ni \quad \begin{cases} \varphi_L, \varphi_C, \varphi \in \mathbf{w} \\ \varphi^*\eta = \varphi \\ \varphi_L^*\eta = \varphi_C, \varphi_C^*\eta = \varphi_L. \end{cases}$$

Such a map $\varphi_\bullet$ can be considered a symmetric form in the category $\varepsilon(\mathscr{E})$ of conflations in $\mathscr{E}$. Given such a map, note that (P, φ) is a symmetric $\mathbf{w}$-space. We impose a relation

$$[(P, \varphi)] = \left[\left(L \oplus C, \begin{pmatrix} 0 & \varphi_C \\ \varphi_L & 0 \end{pmatrix}\right)\right].$$

Further, we define the Witt group $W_0(\mathscr{E}) := W_0(\mathscr{E}, *, \eta, \mathbf{w})$ as the free abelian group generated by symmetric $\mathbf{w}$-spaces $[(P, \varphi)]$ subject to the relations (1), (2), and given any map $\varphi_\bullet : E_\bullet \longrightarrow E_\bullet^*$, as in (3), impose the relation

$$[(P, \varphi)] = 0.$$

For such exact categories with weak equivalences and duality, the $S_\bullet$-construction (9.2.3) of Waldhausen is implemented, as follows, which is referred to as $R_\bullet$-construction.

Definition 10.1.14. Suppose that $(\mathscr{E}, *, \eta, \mathbf{w})$ is a small exact category with weak equivalences and duality. We fix a zero object $0 \in \mathscr{E}$, and treat $\mathscr{E}$ as a pointed exact category. By edgewise subdivision (9.2.17) of $S_\bullet\mathscr{E}$, we obtain a simplicial category $R_\bullet(\mathscr{E})$. It is enriched to a simplicial pointed exact category with weak equivalences and duality.

For integers $n \geq 0$, let $\underline{n} = [n][n] = [2n+1]$ (use the notation 1.5.6). We relabel and write

$$\underline{n} = \{\mathfrak{n} < \cdots < \mathfrak{2} < \mathfrak{1} < \mathfrak{o} < 0 < 1 < 2 < \cdots < n\}.$$

Taking the mirror image from the middle of the two zeros, we obtain a duality $\vee : \underline{n}^{op} \xrightarrow{\sim} \underline{n}$, given by $i \leftrightarrow \mathfrak{i}$. It follows that $[n] \mapsto \mathfrak{d}([n]) = \underline{n}$ is, in fact, a simplicial category with duality.

As in (9.2.3), $Ar(\underline{n})$ denotes the category of arrows in $\underline{n}$, and $Fun(Ar(\underline{n}), \mathscr{E})$ denotes the category of functors and natural transformations. So, $Fun(Ar(\underline{n}), \mathscr{E}) \cong Fun(Ar([2n+1]), \mathscr{E})$. However, there is a duality on $Fun(Ar(\underline{n}), \mathscr{E})$. An object $A \in Fun(\underline{n}, \mathscr{E})$ is given by arrows $f : \{A_{i_1 j_1} \longrightarrow A_{i_2 j_2} : (i_1, j_1) \le (i_2, j_2)\}$, satisfying the composition laws. Its dual A^* is given by

$$(A^*)_{ij} = (A_{j^\vee i^\vee})^* \quad \text{and arrows} \quad \begin{array}{ccc} (A^*)_{i_1 j_1} & \dashrightarrow & (A^*)_{i_2 j_2} \\ \| & & \| \\ A^*_{j_1^\vee i_1^\vee} & \xrightarrow[f^*]{} & A^*_{j_2^\vee i_2^\vee} \end{array}$$

As in (9.2.3), $R_n(\mathscr{E}) \subseteq Fun(\underline{n}, \mathscr{E})$ is defined to be the full subcategory of objects,

$$A : Ar(\underline{n}) \longrightarrow \mathscr{E} \quad \ni \quad \begin{cases} A_{ii} = 0 \quad \forall\, i \\ A_{ij} \hookrightarrow A_{ik} \twoheadrightarrow A_{jk} \text{ is a exact } \forall i \le j \le k. \end{cases}$$

Then $R_n(\mathscr{E})$ is a pointed exact category with weak equivalences and duality, where

1. The zero object is the constant 0 functor.
2. A sequence $A \longrightarrow B \longrightarrow C$ is defined to be a conflation, if $A_{ij} \hookrightarrow B_{ij} \twoheadrightarrow C_{ij}$ is a conflation $\forall (i, j)$.
3. A map $f : A \longrightarrow B$ is defined to be a weak equivalence if $f_{ij} : A_{ij} \longrightarrow B_{ij}$ is a weak equivalence $\forall (i, j)$.

It follows that

$$[n] \mapsto R_n(\mathscr{E}) \quad \text{is a simplicial pointed exact category with duality and weak equivalences.}$$

This will be denoted by $R_\bullet(\mathscr{E})$, which is a simplicial pointed exact category with weak equivalences and duality.

The inclusions $[n] \hookrightarrow \underline{n} = [n][n]$ to the latter half induces a map of the simplicial categories $R_\bullet(\mathscr{E}) \longrightarrow S_\bullet(\mathscr{E})$. With notation as in

(10.1.2), we have the composition of maps of simplicial categories

$$\begin{array}{ccc} (\mathbf{w}R_\bullet(\mathscr{E}))_h & \longrightarrow & \mathbf{w}R_\bullet(\mathscr{E}) \\ & \searrow^{\mathfrak{f}} & \downarrow \\ & & \mathbf{w}S_\bullet(\mathscr{E}) \end{array}$$

Define the GW-space of $\mathscr{E}$ by $\mathbf{GW}(\mathscr{E}) = \mathscr{F}(\mathfrak{f}, 0)$, the homotopy fiber

$$\mathbf{GW}(\mathscr{E}) \longrightarrow |(\mathbf{w}R_\bullet(\mathscr{E}))_h| \xrightarrow{|\mathfrak{f}|} |\mathbf{w}S_\bullet(\mathscr{E})|\,. \tag{10.6}$$

([Sm6, Def. 3, p. 358], [Sm3, p. 1769].) So, $|(\mathbf{w}R_\bullet(\mathscr{E}))_h|$ is the geometric realization of the simplicial space $\{n \mapsto |(\mathbf{w}R_n(\mathscr{E}))_h|\}$, with $|(\mathbf{w}R_0(\mathscr{E}))_h| = |\mathbf{w}\mathscr{E}|$. This induces a canonical map as follows:

$$\begin{array}{ccccc} |\mathbf{w}\mathscr{E}| & & & & \\ \downarrow & \searrow & & & \\ \mathbf{GW}(\mathscr{E}) & \longrightarrow & |(\mathbf{w}R_\bullet(\mathscr{E}))_h| & \xrightarrow{|\mathfrak{f}|} & |\mathbf{w}S_\bullet(\mathscr{E})| \end{array} \tag{10.7}$$

For Noetherian schemes X, various subcategories of the category $Ch(QCoh(X))$ of chain complexes are of our main interest. Accordingly, we have the following example.

Example 10.1.15. Let $\mathscr{E}$ be an exact category. Let $Ch^b(\mathscr{E})$ be the category of bounded chain complexes. So, the objects in $Ch^b(\mathscr{E})$ are chain complexes,

$$E_\bullet := \left(\cdots \longrightarrow E_{n+1} \xrightarrow{d_{n+1}} E_n \xrightarrow{d_n} E_{n-1} \longrightarrow \cdots \right) \tag{10.8}$$

with only finitely many $E_n \neq 0$.

1. A sequence $K_\bullet \longrightarrow E_\bullet \longrightarrow C_\bullet$ in $Ch^b(\mathscr{E})$ is defined as a conflation if the sequence is degreewise split exact. (So, this is independent of the exact category structure in $\mathscr{E}$.) This gives $Ch^b(\mathscr{E})$ a structure of an exact category.

2. A complex $M_\bullet \in Ch^b(\mathscr{E})$ is defined to be **acyclic** if $\forall n$, the differentials d_n decompose as a composition of a deflation followed by an inflation:

$$\begin{array}{ccccc} M_{n+1} & \xrightarrow{d_{n+1}} & & & M_{n-1} \\ \downarrow & & \searrow & \overset{d_n}{\nearrow} & \uparrow \\ Im(d_{n+1}) & \hookrightarrow & M_n & \twoheadrightarrow & Im(d_n) \end{array} \tag{10.9}$$

and the lower line is a conflation.

3. A morphism $f_\bullet : M_\bullet \longrightarrow N_\bullet$ in $Ch^b(\mathscr{E})$ is defined to be a quasi-isomorphism if $\mathbf{cone}(f_\bullet)$ is acyclic (see 9.2.20). Denote the set of all quasi-isomorphisms by $qIso$.

Then $(Ch^b(\mathscr{E}), qIso)$ is an exact category with weak equivalences.

Now assume that $(\mathscr{E}, *, \eta)$ is an exact category with duality. For $E_\bullet \in Ch^b(\mathscr{E})$, as in (10.8), define the dual

$$E_\bullet^* := \left(\cdots \longrightarrow E^*_{n-1} \xrightarrow{\pm d^*_{-n}} E^*_{-n} \xrightarrow{\pm d^*_{-(n+1)}} E^*_{-(n+1)} \longrightarrow \cdots \right). \tag{10.10}$$

$$\text{So,} \quad \begin{cases} (E_\bullet^*)_n = (E_{-n})^* \\ (d^*)_n = (-1)^{n+1} d^*_n. \end{cases}$$

Also, define $\varpi : E_\bullet \longrightarrow E_\bullet^{**}$ by $\varpi_n := (-1)^n \eta_{E_n}$. For $f = \{f_n\} : E_\bullet \longrightarrow N_\bullet$, define $f^* = \{f_n^*\}$. All these signs are picked to be consistent with (10.2.16), subsequently. Then

1. $(Ch^b(\mathscr{E}), *, \varpi, qIso)$ is an exact category with duality and weak equivalences.
2. Further, note that $(\mathscr{E}, *, \eta, \mathbf{i})$ can be considered as an exact category with weak equivalences and duality, where $\mathbf{i}$ denotes the isomorphisms.
3. The inclusion by degree zero induces a functor

$$(\mathscr{E}, *, \eta, \mathbf{i}) \longrightarrow (Ch^b(\mathscr{E}), *, \varpi, qIso)$$

of categories with weak equivalences and duality.

Exercise 10.1.16. Under the setup of (10.1.15), prove that a morphism $f : E_\bullet \longrightarrow G_\bullet$ is a quasi-isomorphism if and only if $f^* : G_\bullet^* \longrightarrow F_\bullet^*$ is a quasi-isomorphism.

Proof. Let f be quasi-isomorphism. By definition the $cone(f)$ is acyclic. By dualizing the diagram (10.9) it follows that the dual $cone(f)^*$ is acyclic. However, $cone(f)^* = cone(f^*)$. So, f^* is a quasi-isomorphism. The converse follows from double dual identification. The proof is complete. ■

We have the following agreement theorem.

Theorem 10.1.17. *Let $(\mathscr{E}, *, \eta)$ be an exact category with duality. Then we have the following commutative diagram of homotopy equivalences of GW-spaces:*

$$\begin{array}{ccc} \mathbf{GW}(\mathscr{E}, *, \eta) & \xrightarrow{\sim} & \mathbf{GW}(\mathscr{E}, *, \eta, \mathbf{i}) \\ & \searrow^{\sim} & \downarrow \wr \\ & & \mathbf{GW}(Ch^b(\mathscr{E}), *, \varpi, qIso) \end{array}$$

where $\mathbf{GW}(\mathscr{E}, *, \eta)$ was defined in (10.2), and the other two spaces are defined as in (10.6).

Proof. See [Sm6, Prop. 2, 6]. ■

We define higher GW groups.

Lemma 10.1.18. *Suppose that $\mathscr{E} := (\mathscr{E}, *, \eta, \mathbf{w})$ is an exact category with weak equivalences and duality. Then it follows that* [Sm6, Prop. 6]

$$\pi_0(\mathbf{GW}(\mathscr{E}), 0) = GW_0(\mathscr{E}), \quad \textit{as defined in } (10.1.13).$$

Define higher GW-groups

$$GW_n(\mathscr{E}) := \pi_n(\mathbf{GW}(\mathscr{E}), 0) \quad \forall n \geq 0.$$

We iterate the $R_\bullet \mathscr{E}$ construction, as in (9.2.7).

Definition 10.1.19. Let $\mathscr{E} := (\mathscr{E}, *, \eta, \mathbf{w})$ be a small pointed exact category with weak equivalences and duality. Inductively, define multisimplicial exact categories $R^{(n)}(\mathscr{E}) := R^{(n)}_{\bullet_1 \bullet_2 \cdots \bullet_n}(\mathscr{E})$, with weak equivalences and duality. Let $\mathbb{R}^{(n)}_\bullet(\mathscr{E})$ denote the diagonal of $R^{(n)}(\mathscr{E})$.

$$\text{So,} \qquad \begin{cases} R^{(n)}_{m_1 m_2 \cdots m_n}(\mathscr{E}) = R_{m_1}\left(R^{(n-1)}_{m_2 \cdots m_n}(\mathscr{E})\right) \\ \mathbb{R}^{(n)}_m(\mathscr{E}) = R^{(n)}_{m,m,\cdots,m}(\mathscr{E}). \end{cases}$$

Proposition 10.1.20. *Let $\mathscr{E} := (\mathscr{E}, *, \eta, \mathbf{w})$ be a small pointed exact category with weak equivalences and duality. Use the notation as in* (10.1.2). *Then the sequence, obtained as in* (10.7)

$$\left|\left(\mathbf{w}\mathbb{R}_\bullet^{(n)}(\mathscr{E})\right)_h\right| \longrightarrow \left|\left(\mathbf{w}R_\bullet\mathbb{R}_\bullet^{(n)}(\mathscr{E})\right)_h\right| \longrightarrow \left|\left(\mathbf{w}S_\bullet\mathbb{R}_\bullet^{(n)}(\mathscr{E})\right)\right| \tag{10.11}$$

is a homotopy fibration $\forall n \geq 1$. *Consequently, there is a homotopy fibration*

$$\left|\left(\mathbf{w}\mathbb{R}_\bullet^{(n)}(\mathscr{E})\right)_h\right| \longrightarrow \left|\left(\mathbf{w}\mathbb{R}_\bullet^{(n+1)}(\mathscr{E})\right)_h\right| \longrightarrow \left|\left(\mathbf{w}S_\bullet^{(n+1)}(\mathscr{E})\right)\right|. \tag{10.12}$$

Proof. See [Sm3, Prop. 4.1, p. 1770] and [Sm3, Cor. 4.2, p. 1770]. The second homotopy fibration follows from the first one by identifying the respective terms. ■

10.2. dg Categories with Weak Equivalences and Duality

Starting from this section, we give an overview of [Sm3], which was done in the framework of dg categories (differentially graded categories). In fact, dg categories are the abstractions of the categories of various chain complex categories. Contrary to [Sm3], we use homology notation.

Definition 10.2.1. Fix a commutative ring k. A **dg k-module** M is a graded k-module $M = \oplus_{n\in\mathbb{Z}} M_n$, and a k-linear map (differential) $d : M \longrightarrow M$ such that $d(M_n) \subseteq M_{n-1}$, and $d^2 = 0$. So, a dg k-module M is another name for what is known as a (*homology*) chain complex of k-modules. For $m \in M_n$, we say that m is a homogeneous element, with $\deg(m) = n$. A morphism $f : M \longrightarrow N$ of dg k-modules is a k-linear map, with $f(M_n) \subseteq N_n$ for all n, and $fd = df$. We write $Mor(M, N)$ to denote the set of all such morphisms. The category of dg k-modules is denoted by dg$\mathbf{Mod}_k$. According to our notation in other chapters, dg$\mathbf{Mod}_k = Ch(QCoh(k))$.

There are three natural functors

$$\begin{cases} Z_0 : \mathrm{dg}\mathbf{Mod}_k \longrightarrow QCoh(k) & Z_0(M) = \ker(d_0 : M_0 \longrightarrow M_{-1}) \\ B_0 : \mathrm{dg}\mathbf{Mod}_k \longrightarrow QCoh(k) & B_0(M) = Image(d_1 : M_1 \longrightarrow M_0) \\ H_0 : \mathrm{dg}\mathbf{Mod}_k \longrightarrow QCoh(k) & H_0(M) = \frac{Z_0(M)}{B_0(K)}. \end{cases}$$

Indeed, for each n, we can write such functors Z_n, B_n, H_n.

For $M, N \in \mathrm{dg}\mathbf{Mod}_k$, we have the tensor product

$$\begin{cases} (M \otimes N)_n = \bigoplus_{i+j=n} M_i \otimes N_j, \\ x \in M_p, y \in N_q \quad \text{define} \quad \mathrm{d}(x \otimes y) = \mathrm{d}x \otimes y + (-1)^p x \otimes \mathrm{d}y. \end{cases}$$

Also, define the functions $[M, N] \in \mathrm{dg}\mathbf{Mod}_k$ by

$$\begin{cases} [M, N]_n = \prod_r Hom_k(M_r, N_{r+n}) \\ f = \{f_r\} \in [M, N]_n \quad \text{define} \quad d(f) = dof - (-1)^n fod. \end{cases} \tag{10.13}$$

For $f_r \in Hom_k(M_r, N_{r+n})$, we display df_r by the commutative diagram:

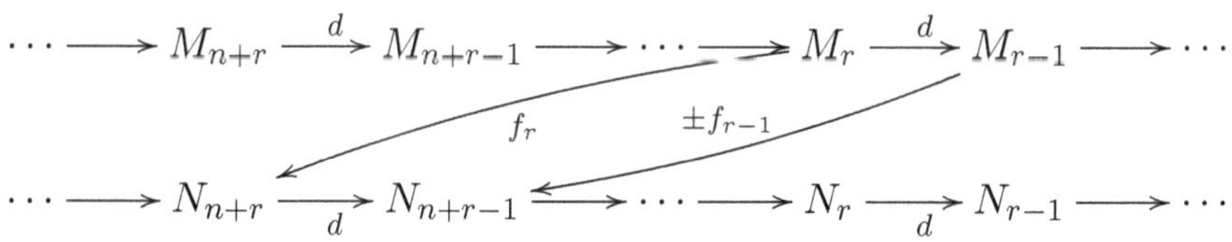

So, for the dg k-module $[M, N] := \bigoplus_{n \in \mathbb{Z}} [M, N]_n$, the k-module

$$Z_0[M, N] = \ker\left(d : [M, N]_0 \longrightarrow [M, N]_{-1}\right) = Hom(M, N)$$

is the k-module of all chain complex maps. For $M, N \subset \mathrm{dg}\mathbf{Mod}_k$, there are three natural maps

$$\begin{cases} M \otimes N \longrightarrow N \otimes M & \forall x \in M_p, y \in N_q \text{ send } x \otimes y \mapsto (-1)^{pq} y \otimes x \\ ev : [M, N] \otimes M \longrightarrow N & f \otimes x \mapsto f(x) \\ \nabla : M \longrightarrow [N, M \otimes N] & x \mapsto (y \mapsto x \otimes y). \end{cases}$$

Further, let $\mathbb{1} = k$ denote the complex concentrated at degree zero. Then the tensor product $\otimes$ endows $\mathrm{dg}\mathbf{Mod}_k$ a structure of a symmetric monoidal category, where the unit is $\mathbb{1} = k$ and the symmetry is given by

$$\chi : M \otimes N \xrightarrow{\sim} N \otimes M \quad by \quad \chi(x \otimes y) = (-1)^{\deg(x)\deg(y)} y \otimes x.$$

In fact, dg$\mathbf{Mod}_k$ is a closed monoidal category, in the sense that the tensor product $\otimes$ and the coevaluation ∇ interact well with each other [H, Def. 4.1.13].

We define dg categories.

Definition 10.2.2. Fix a commutative ring k. A (pointed) dg category $\mathscr{A}$ is given by the following data:

1. A nonempty set $Obj(\mathscr{A})$ of objects.
2. For any two objects $A, B \in Obj(\mathscr{A})$, there is an object $\mathscr{A}(A, B) \in \mathrm{dg}\mathbf{Mod}_k$, to be called the **mapping complex**.
3. For any $A \in Obj(\mathscr{A})$, there is a unit morphism $1_A \in \mathscr{A}(A, A)_0$.
4. For $A, B, C \in Obj(\mathscr{A})$, there is a map of dg k-modules, to be referred to as a composition,
$$\mathscr{A}(B, C) \otimes \mathscr{A}(A, B) \longrightarrow \mathscr{A}(A, C) \quad \text{sending} \quad f \otimes g \mapsto fg,$$
by notation.

   This satisfies all the usual associativity and unit conditions ($1_A g = g$, $f 1_C = f$) of composition.
5. A choice of the zero object $0_{\mathscr{A}}$.

The following are some immediate consequences:

1. $\forall A \in Obj(\mathscr{A})$, we have $d1_A = 0$. So, $1_A \in Z_0\mathscr{A}(A, A)$.
2. For $f \in \mathscr{A}(A, B)$ and $g \in \mathscr{A}(B, C)$, if $df = 0$ and $dg = 0$, then $d(gf) = 0$. This defines a category $Z_0\mathscr{A} \subseteq \mathscr{A}$, whose objects are the same as $Obj(\mathscr{A})$, and the morphisms are $Z_0(\mathscr{A})(A, B) := Z_0(\mathscr{A}(A, B))$.
3. So, we think of $Z_0(\mathscr{A}) \subseteq \mathscr{A}$ as the underlying category, and $\mathscr{A}$ with additional homological structure. We can also associate a so-called **homotopy category** $H_0\mathscr{A}$ whose objects are the same as the objects of $\mathscr{A}$, and $H_0\mathscr{A}(A, B) = H_0(\mathscr{A}(A, B))$.
4. We only consider pointed dg categories. So, henceforth by a dg category, we mean a pointed dg category.

We remark that dg categories are defined over a fixed commutative ring k. Sometimes, we will ignore mentioning k.

A basic example of dg categories is as follows.

Example 10.2.3. Suppose $\mathscr{A}$ is a small additive category. For objects $X, Y \in Obj(\mathscr{A})$, let $\mathscr{A}(X,Y) := Mor_{\mathscr{A}}(X,Y)$ be concentrated at degree zero, and let $d = 0$. This gives $\mathscr{A}$ a structure of a dg $\mathbb{Z}$-category.

In particular, for a ring R, category $QCoh(R)$ of left R-modules can be considered a dg category. Similarly, various other suitable subcategories of $QCoh(R)$ can also be considered dg categories. These subcategories need not be closed under direct sum.

We define morphisms of dg categories and natural transformations.

Definition 10.2.4. Let $\mathscr{A}$ and $\mathscr{B}$ be two dg categories. A morphism $F : \mathscr{A} \longrightarrow \mathscr{B}$ is given by the following:

1. To each $A \in Obj(\mathscr{A})$, it associates an object $FA \in Obj(\mathscr{B})$.
2. For all $A, B \in Obj(\mathscr{A})$, there is a morphism of dg k-modules $F : \mathscr{A}(A,B) \longrightarrow \mathscr{B}(FA, FB)$ such that (1) $\forall A \in Obj(\mathscr{A})$, we have $F(1_A) = 1_{FA}$, (2) for composable f, g, $F(gf) = F(g)F(f)$, (3) $F(0) = 0$.

Such a morphism is also called a **dg functor**. This defines a category, to be denoted by dg$\mathbf{Cat}_k$, whose objects are all small dg categories and whose morphisms are dg functors. The following are some more definitions:

1. Given morphisms $F, G : \mathscr{A} \longrightarrow \mathscr{B}$ of dg categories, a **natural transformation** $\tau : F \longrightarrow G$ is a collection $\{\tau_A \in Z_0\mathscr{B}(FA, GA) : A \in Obj\mathscr{A}\}$ such that $\forall f \in \mathscr{A}(A,B)$, we have $\tau_B F(f) = G(f)\tau_A$. We say that τ is a **natural isomorphism** if τ_A is an isomorphism in $Z_0\mathscr{B}(FA, GA)$ $\forall A$.
2. A morphism $F : \mathscr{A} \longrightarrow \mathscr{B}$ of dg categories is called an equivalence of dg categories if there is a morphism $G : \mathscr{B} \longrightarrow \mathscr{A}$ such that GF and FG are naturally isomorphic to the identity functors. In this case, F induces an equivalence of the underlying categories $Z_0\mathscr{A} \longrightarrow Z_0\mathscr{B}$.

We also define the tensor product of dg categories.

Definition 10.2.5. Let $\mathscr{A}, \mathscr{B} \in$ dg$\mathbf{Cat}_k$. Define $\mathscr{A} \otimes \mathscr{B} \in$ dg$\mathbf{Cat}_k$, as follows:

1. The objects of $\mathscr{A} \otimes \mathscr{B}$ are pairs $(A, B) \in Obj(\mathscr{A}) \times Obj(\mathscr{B})$, while $(0, B)$, $(A, 0)$ are identified as the 0 of $\mathscr{A} \otimes \mathscr{B}$.

2. Now, define

$$(\mathscr{A} \otimes \mathscr{B})((A_1, B_1), (A_2, B_2)) = \mathscr{A}(A_1, A_2) \otimes_k \mathscr{B}(B_1, B_2).$$

3. Define the compositions

$$\begin{aligned}(\mathscr{A}(A_2, A_3) \otimes_k \mathscr{B}(B_2, B_3)) \times (\mathscr{A}(A_1, A_2) \otimes_k \mathscr{B}(B_1, B_2))\\ \longrightarrow (\mathscr{A}(A_1, A_3) \otimes_k \mathscr{B}(B_1, B_3))\end{aligned}$$

by the formula

$$(f_1 \otimes g_1)o(f_2 \otimes g_2) = (-1)^{\deg(g_1)\deg(f_2)}(f_1 f_2 \otimes g_1 g_2).$$

Also, define the identity

$$1_{(A,B)} = 1_A \otimes 1_B \in \mathscr{A}(A, A) \otimes \mathscr{B}(B, B).$$

Definition 10.2.6. Let $\mathscr{A} \in \text{dg}\mathbf{Cat}_k$ and $\mathscr{C}$ be any category. Define the dg category $dgFun(\mathscr{C}, \mathscr{A}) \in \text{dg}\mathbf{Cat}_k$, as follows:

1. Objects of $dgFun(\mathscr{C}, \mathscr{A})$ are functors $F : \mathscr{C} \longrightarrow Z_0\mathscr{A}$.
2. For, $F, G \in Obj(dgFun(\mathscr{C}, \mathscr{A}))$, define the mapping complexes $[F, G] = \bigoplus_{n\in\mathbb{Z}}[F, G]_n$ naturally. An element in $[F, G]_n$ is represented by $\{\alpha_X \in [F(X), G(X)]_n : X \in Obj(\mathscr{C})\}$ such that $\forall f \in Mor_{\mathscr{C}}(X, Y)$, we have $G(f)\alpha_X = \alpha_Y F(f)$.

Subsequently, for our purpose, a useful application of this construction is the case when $\mathscr{C} = Ar([n])$. For $\mathscr{A}, \mathscr{B} \in \text{dg}\mathbf{Cat}_k$, more generally, one defines $dgFun(\mathscr{A}, \mathscr{B})$ in $\text{dg}\mathbf{Cat}_k$.

The following is a fundamental example of a dg category, which is in the spirit of $Ch^b(\mathscr{P}(\mathbb{Z}))$, discussed in (9.3.1).

Example 10.2.7. Let k be a commutative ring. Let $\mathscr{C}_k$ denote the category of bounded complexes $M_\bullet$, with $M_i = k^{n_i}$. So, $M_i = 0$ for all but finitely many i. We view it as the full subcategory of $\text{dg}\mathbf{Mod}_k$. So, it inherits the structure of a dg k-category, where the zero object is the zero complex. As in (9.3.1), $\mathscr{C}_k$ has the structure of a symmetric monoidal category. The unit of $\mathscr{C}_k$ is the complex $\mathbb{1}$, which is $k\varepsilon_0 \cong k$ at degree zero and is zero elsewhere. There are two other distinguished objects in $\mathscr{C}_k$. First, the complex $C_\bullet$ such that $C_1 = k\eta_1 \cong k$, $C_0 = k\eta_0 \cong k$, $C_i = 0$, $\forall i \neq 0, 1$, and $d(\eta_1) = \eta_0$.

Then we have $T_\bullet$, where $T_1 = k\tau_1 \cong k$, and $T_i = 0,\ \forall i \neq 1$. We have an exact sequence

$$0 \longrightarrow \mathbb{1} \longrightarrow C_\bullet \longrightarrow T_\bullet \longrightarrow 0. \tag{10.14}$$

So, $\mathscr{C}_k \in \mathrm{dg}\mathbf{Cat}$. Also, for any $\mathscr{A} \in \mathrm{dg}\mathbf{Cat}$, we denote $\mathscr{C}_k\mathscr{A} := \mathscr{C}_k \otimes \mathscr{A}$. There is a fully faithful dg functor

$$\mathscr{A} \longrightarrow \mathscr{C}_k\mathscr{A} \qquad \text{sending} \quad A \mapsto \mathbb{1} \otimes A. \tag{10.15}$$

We have more definitions to incorporate exactness.

Definition 10.2.8. Let $\mathscr{A}$ be a pointed dg category. A sequence $K \xrightarrow{f} M \xrightarrow{g} C$ of morphisms in $Z_0\mathscr{A}$ is defined as exact if (1) $gf = 0$ and (2) there are $r \in \mathscr{A}(M,K)^0$, $s \in \mathscr{A}(C,M)^0$ such that $1_K = rf$, $1_C = gs$, $1_M = fr + gs$.

1. A pointed dg category $\mathscr{A}$ is defined to be **exact** if the exact sequences in $\mathscr{A}$ induce a structure of an exact category on $Z_0\mathscr{A}$.
2. A pointed dg category $\mathscr{A}$ is defined to be **pretriangulated** if (1) $\mathscr{A}$ is exact and (2) the functor (10.15) is an equivalence.
   In fact, the condition (2) is the same as requiring that the restriction $Z_0\mathscr{A} \longrightarrow Z_0(\mathscr{C}_k\mathscr{A})$ of the functor (10.15) be an equivalence of exact categories.

We expand on our example 10.1.15.

Example 10.2.9. Let $\mathscr{E}$ be a k-linear exact category (meaning that $Hom_{\mathscr{E}}(M,N)$ are k-modules, and compositions are bilinear). Let $Ch^b(\mathscr{E})$ be the category of bounded chain complexes, as in (10.1.15). Then $Ch^b(\mathscr{E})$ has a dg category structure, which we denote by $dg(\mathscr{E})$.

1. The objects of $dg(\mathscr{E})$ are the same as the objects of $Ch^b(\mathscr{E})$.
2. For chain complexes $M_\bullet, N_\bullet \in Ch^b(\mathscr{E})$, the mapping complexes

$$\begin{cases} [M_\bullet, N_\bullet] = \bigoplus_{n\in\mathbb{Z}} [M_\bullet, N_\bullet]_n = \bigoplus_{n\in\mathbb{Z}} \prod_r Mor_{\mathscr{E}}(M_r, N_{r+n}) \\ \forall\ f_{ij} \in Mor_{\mathscr{E}}(M_i, N_j),\ \text{let } \deg(f_{ij}) = j - i,\ d(f_{ij}) = df_{ij} - (-1)^{\deg(f_{ij})} f_{ij} d \end{cases} \tag{10.16}$$

   This is analogous to Definition 10.2.1, and the differential coincides with (10.13).

3. So, $Z_0 dg(\mathscr{E}) = Ch^b(\mathscr{E})$, as defined in (10.1.15). One checks that $H_0 Ch^b \mathscr{E} = \mathcal{K}^b(\mathscr{E})$ is the usual homotopy category.
4. (**Example**) In particular, consider the dg category $dg(\mathscr{F}(k))$ where $\mathscr{F}(k)$ is the exact category of finitely generated free k-modules. Then $Z_0\mathscr{C}_k \longrightarrow Z_0(dg(\mathscr{F}(k)))$ is an equivalence of categories.
5. In fact, $dg(\mathscr{E})$ is a pretriangulated dg category [Sm3, 1.21]. The exact category structure was already defined in (10.1.15). Similar to (9.3.5), the inverse τ of the functor $\iota : dg(\mathscr{E}) \longrightarrow \mathscr{C}_k \otimes dg(\mathscr{E})$ is defined naturally, as follows:

$$\text{For } F_\bullet \in \mathscr{C}_k,\ M_\bullet \in dg(\mathscr{E}) \quad \text{define} \quad \tau(F_\bullet, M_\bullet)_n = \bigoplus_{i+j=n} F_i \otimes M_j$$

where for $F_i = k^{n_i}$, we write $F_i \otimes M_j := M_j^{n_i}$. The restriction of the differential $\tau(F_\bullet, M_\bullet)_n \longrightarrow \tau(F_\bullet, M_\bullet)_{n-1}$ on $F_i \otimes M_j$ is given by

$$d = d_i \otimes 1_{M_j} + (-1)^i 1_{F_i} \otimes d_j : F_i \otimes M_j \longrightarrow (F_{i-1} \otimes M_i) \bigoplus (F_i \otimes E_{j-1}).$$

It is immediate that $\tau\iota = 1_{Ch^b(\mathscr{E})}$ and one checks that $\iota\tau \sim 1_{\mathscr{C}_k \otimes Ch^b(\mathscr{E})}$ are naturally isomorphic.

So, ι and τ are inverses of each other (Def. 10.2.4).

Refer to [Sm3, 1.21, 1.4, 1.27, 1.30, 1.36, 1.38] for further information on the dg category structure of $dg(\mathscr{E})$.

The following is an important observation.

Lemma 10.2.10. *Use the notation as above* (10.2.9). *Then* $M_\bullet \mapsto \tau(T_\bullet, M_\bullet)$ *is given by*

$$\tau(T_\bullet, M_\bullet)_n = M_{n-1} \quad \textit{and the differential}$$
$$\tau(d)_n = -d_{n-1} : M_{n-1} \longrightarrow M_{n-2}.$$

So, this identification via τ *amounts to the usual shift functor in* $dgCh^b(\mathscr{E})$.

Notation 10.2.11. We remark that, for a pretriangulated dg category $\mathscr{A}$, the functor

$$\iota : \mathscr{A} \longrightarrow \mathscr{C}_k \otimes \mathscr{A} \quad \text{has an inverse} \quad \tau : \mathscr{C}_k \otimes \mathscr{A} \longrightarrow \mathscr{A} \tag{10.17}$$

which makes $\mathscr{A}$ look exactly like a complicial exact category (9.3.2). We plan to follow a similar track, as we did in Section 9.3, on complicial exact categories with weak equivalences. As in (9.3.2), for $X \in Obj(\mathscr{A})$, use the notation

$$CX := \tau(C_\bullet \otimes X), \qquad TX := \tau(T_\bullet \otimes X).$$

Consider the exact sequence (10.14), and tensor it with X. We obtain an exact sequence

$$0 \longrightarrow X \longrightarrow CX \longrightarrow TX \longrightarrow 0.$$

Given a morphism $f : X \longrightarrow Y$ in $\mathscr{A}$, we have the following pushforward diagram:

$$\begin{array}{ccccc} X & \hookrightarrow & CX & \twoheadrightarrow & TX \\ {\scriptstyle f}\downarrow & & \downarrow & & \| \\ Y & \underset{g}{\hookrightarrow} & C(f) & \underset{h}{\twoheadrightarrow} & TX \end{array} \tag{10.18}$$

Here $C(f)$ is the pushforward of f and $X \hookrightarrow CX$. It exists because $Z_0(\mathscr{A})$ is an exact category. Thus, we get a sequence

$$X \xrightarrow{f} Y \xrightarrow{g} C(f) \xrightarrow{h} TX \qquad \text{in} \quad Z_0\mathscr{A}. \tag{10.19}$$

This sequence is called a **standard triangle**, and $C(f)$ is called the cone of f.

Remark 10.2.12. For a dg category $\mathscr{A}$, in [Sm3], a pretriangulated dg category $\mathscr{A}^{ptr}$ was constructed with a full inclusion $\mathscr{A} \subseteq \mathscr{A}^{ptr}$. The construction is a little more technical, which we skip. This category $\mathscr{A}^{ptr}$ is called the pretriangulated hull of $\mathscr{A}$. While there was no universal property established, $\mathscr{A}^{ptr}$ seems the smallest such pretriangulated dg category containing $\mathscr{A}$. We leave it to the reader to investigate. All the dg categories we consider will be pretriangulated and hence $\mathscr{A} = .\mathscr{A}^{ptr}$. Most importantly, the functor $\iota : \mathscr{A} \longrightarrow \mathscr{C}_k \otimes \mathscr{A}$ has an inverse $\tau : \mathscr{C}_k \otimes \mathscr{A} \longrightarrow \mathscr{A}$.

For a definition of triangulated categories, see (11.1.1).

Lemma 10.2.13. *Let $\mathscr{A}$ be a pretriangulated dg category. Then $H_0\mathscr{A}$ is a triangulated category.*

Outline of the Proof. See [Sm3, Rem. 1.22]. While we did not elaborate, this entails the same arguments as in (9.3.9(2a)) applied to $Z_0\mathscr{A}$. Everything in (9.3.2) makes sense for $Z_0\mathscr{A}$. It turns out that the stable category $\underline{Z_0\mathscr{A}} \cong H_0\mathscr{A}$. We have

1. The shift functor in $\underline{Z_0\mathscr{A}} \cong H_0\mathscr{A}$ is given by $X \mapsto TX$.
2. Distinguished triangles in $\underline{Z_0\mathscr{A}} \cong H_0\mathscr{A}$ are those that are isomorphic to (10.19) in $H_0(\mathscr{A})$.

■

We proceed to incorporate weak equivalences to dg categories.

Definition 10.2.14. As usual, let k be a commutative ring. A dg category with weak equivalences is a pair $(\mathscr{A}, \mathbf{w})$, where $\mathscr{A}$ is a (pretriangulated) dg k-category and $(Z_0\mathscr{A}, \mathbf{w})$ is an exact category with weak equivalences. We further assume that $\mathbf{w}$ is saturated. (See [Sm3, §1.8, p. 1745].) Note that this is purely a property of the underlying category $Z_0\mathscr{A}$. A dg functor $F : (\mathscr{A}, \mathbf{w_A}) \longrightarrow (\mathscr{B}, \mathbf{w_B})$ between two such categories is a dg functor F, which induces an exact functor $(Z_0\mathscr{A}, \mathbf{w_A}) \longrightarrow (Z_0\mathscr{B}, \mathbf{w_B})$. Such a dg functor is also called an **exact dg functor**.

In this case, let $\mathscr{A}^w = \{A \in Obj(\mathscr{A}) : (0 \longrightarrow A) \in \mathbf{w}\}$ be the dg subcategory of $\mathbf{w}$-acyclic objects. Then $\mathscr{A}^w$ is also a pretriangulated dg subcategory. We define

$$\mathfrak{T}(\mathscr{A}, \mathbf{w}) := \frac{H_0\mathscr{A}}{H_0\mathscr{A}^w} \quad \text{the Verdier quotient of triangulated categories.} \tag{10.20}$$

We say that $\mathfrak{T}(\mathscr{A}) := \mathfrak{T}(\mathscr{A}, \mathbf{w})$ is the **associated triangulated category** of $\mathscr{A}$.

Now we define dg categories with duality.

Definition 10.2.15. Suppose that $\mathscr{A}$ is a dg category. Then the opposite dg category $\mathscr{A}^{op}$ is defined naturally. Objects of $\mathscr{A}^{op}$ are the same as that of $\mathscr{A}$, and $\mathscr{A}^{op}(M, N) = \mathscr{A}(N, M)$. For $f \in \mathscr{A}^{op}(M, N)$, $g \in \mathscr{A}^{op}(N, G)$, define $go^{op}f := (-1)^{\deg(f)\deg(g)} fg$. A dg functor $F : \mathscr{A} \longrightarrow \mathscr{B}$ naturally defines a dg functor $F^{op} : \mathscr{A}^{op} \longrightarrow \mathscr{B}^{op}$. Further, $(\mathscr{A} \otimes \mathscr{B})^{op} \cong \mathscr{A}^{op} \otimes \mathscr{B}^{op}$.

1. A **dg category with duality** is a triple $(\mathscr{A}, *, \eta)$, where $\mathscr{A}$ is a dg category, $* : \mathscr{A}^{op} \longrightarrow \mathscr{A}$ is a dg functor, and $\eta : 1_{\mathscr{A}} \longrightarrow *^2$ is a natural transformation, meaning $\forall A \in Obj(\mathscr{A})$, there is a natural morphism $\eta_A : A \longrightarrow A^{**}$ in $Z_0\mathscr{A}$, and an identification of triple dual (10.1.1), as in the commutative diagram:

$$\begin{array}{ccc} A^* & \xrightarrow{\eta_{A^*}} & (A^*)^{**} \\ & \nwarrow^{(\eta_A)^*} & \downarrow \wr \\ & & (A^{**})^* \end{array}$$

   Given a dg category with duality $(\mathscr{A}, *, \eta)$, we have that $(Z_0\mathscr{A}, *, \eta)$ is a category with duality. In fact, $* : Z_0\mathscr{A}^{op} \longrightarrow Z_0\mathscr{A}$ is exact. Since η is not assumed to be an isomorphism, we do not say that $(Z_0\mathscr{A}, *, \eta)$ is an exact category with duality. We save it till we incorporate weak equivalences. (See [Sm3, Def. 1.29, p. 1747].)
2. A **dg category with weak equivalences and duality** is a quadruple $(\mathscr{A}, *, \eta, \mathbf{w})$, where (1) $(\mathscr{A}, *, \eta)$ is a dg category with duality, (2) $(\mathscr{A}, \mathbf{w})$ is a dg category with weak equivalences, (3) $\forall A \in Obj(\mathscr{A})$, $\eta_A \in \mathbf{w}$, (4) The dg category $A^w = \{A : (0 \longrightarrow A) \in \mathbf{w}\}$ is closed under duality, (5) $f \in \mathbf{w} \iff f^* \in \mathbf{w}$. (See [Sm3, Def. 1.37, p. 1751].)

   A functor of such dg categories with weak equivalences and duality is a dg functor that respects all these structures.

   In this case, $(Z_0\mathscr{A}, *, \eta, \mathbf{w})$ is an exact category with weak equivalences and duality.

We expand on the examples (10.1.15, 10.2.9), of our primary interest.

Example 10.2.16. Let k be a commutative ring. Let $\mathscr{E}$ be a k-linear exact category. Then the pretriangulated dg category $dg(\mathscr{E})$ was defined in (10.2.9), whose underlying exact category is $Ch^b(\mathscr{E})$. It follows immediately that the quasi-isomorphisms induce a structure of an exact category with weak equivalences on $dg(\mathscr{E})$. We use two notation for this $dg(\mathscr{E}) := (dg(\mathscr{E}), qIso)$. It follows that the associated triangulated category (10.20) is

$$\mathfrak{T}((dg(\mathscr{E}), qIso) = \mathbf{D}^b(\mathscr{E}) \quad \text{the bounded derived category}$$

(see Section 11.2.2).

Now assume that $(\mathscr{E}, *, \eta)$ is an exact category with duality. Then $(Ch^b(\mathscr{E}), *, \varpi, qIso)$ is an exact category with duality and weak equivalences (10.1.15). This structure extends to $dg(\mathscr{E})$ as a dg category with duality and weak equivalences. At object level, for $E_\bullet \in Ch^b(\mathscr{E})$, we already defined its dual $E^*_\bullet$ (10.1.15). What remains to be defined is the dual f^* for $f = \{f_{ij}\} \in [M_\bullet, N_\bullet] = \bigoplus_{n\in\mathbb{Z}} \prod_r Mor_{\mathscr{E}}(M_r, N_{r+n})$, where $f_{ij} : M_i \longrightarrow N_j$. Define

$$f^* \in [N^*_\bullet, M^*_\bullet] = \bigoplus_{n\in\mathbb{Z}} \prod_r Hom(N^*_{-r}, M^*_{-r-n}) \quad \text{given by}$$

$$f^* = \{(-1)^{i(j-i)} f^*_{ij}\}.$$

Also, define

$$\eta_{M_\bullet} : M_\bullet \longrightarrow M^{**}_\bullet \quad \text{by} \quad \{(-1)^n \eta_{E_n}\}.$$

This completes the description of $dg(\mathscr{E}) := (Ch^b(\mathscr{E}, *, \varpi, qIso)$ as a dg category with weak equivalences and duality. (See [Sm3, Ex. 1.30, p. 1747].)

Corollary 10.2.17. *As before, let k be a commutative ring and $\mathscr{F}(k)$ be the category of finitely generated free k-modules. Now, $(\mathscr{F}(k), *, ev)$ is an exact category with duality, where for $M \in \mathscr{F}(k)$, $M^* := Hom(M, k)$ and ev is the evaluation map $M \xrightarrow{\sim} M^{**}$. So, $dg(\mathscr{F}(k))$ is a dg category with weak equivalences and duality, as defined in* (10.2.16). *Consequently, $\mathscr{C}_k$ inherits the structure of a dg category with weak equivalences and duality (identifying $(k^n)^* \cong k^n$).*

10.3. Hermitian K-Theory of dg Categories

For dg categories $\mathscr{A}$, the Waldhausen $S_\bullet$-construction (9.2.3) and $R_\bullet$-construction (10.1.14), extend, as dg categories.

Definition 10.3.1. Let $\mathscr{A} = (\mathscr{A}, *, \varpi, \mathbf{w})$ be a pointed dg category with weak equivalences and duality. Now, we define $S_\bullet\mathscr{A}$ as a dg category, exactly as in (9.2.3) (we use the same notation). Note that exact sequences are defined (10.2.8) in $\mathscr{A}$, even when $\mathscr{A}$ is not exact. As a dg category, $S_n(\mathscr{A}) \subseteq dgFun(Ar([n]), \mathscr{A})$ is a dg subcategory. By defintion (10.2.6), an object $A \in S_n(\mathscr{A})$ is a functor $A : Ar([n]) \longrightarrow Z_0\mathscr{A}$, satisfying the exactness conditions, as in (9.2.3). So, $S_n\mathscr{A}$ is a pointed dg category with weak equivalences and duality.

Likewise, define $R_n\mathscr{A} \subseteq dgFun(Ar(\underline{n}), \mathscr{A})$, exactly as in (10.1.14), as a pointed dg category with weak equivalences and duality. Objects in $R_n\mathscr{A}$ are functors $A : Ar(\underline{n}) \longrightarrow Z_0\mathscr{A}$, satisfying the exactness conditions in (10.1.14).

Inductively, define multisimplicial dg categories with weak equivalences and duality $R^{(n)}(\mathscr{A}) := R^{(n)}_{\bullet_1\bullet_2\cdots\bullet_n}(\mathscr{A})$. Let $\mathbb{R}^{(n)}_\bullet(\mathscr{A})$ denote the diagonal of $R^{(n)}(\mathscr{A})$.

So,
$$\begin{cases} R^{(n)}_{m_1m_2\cdots m_n}(\mathscr{A}) = R_{m_1}\left(R^{(n-1)}_{m_2\cdots m_n}(\mathscr{A})\right) \\ \mathbb{R}^{(n)}_m(\mathscr{A}) = R^{(n)}_{m,m,\cdots,m}(\mathscr{A}). \end{cases}$$

Now, assume that $\mathscr{A}$ is exact. Then

$$\mathbf{w}\mathbb{R}^{(n)}_\bullet\mathscr{A} = \mathbf{w}Z_0\mathbb{R}^{(n)}_\bullet\mathscr{A} = \mathbf{w}\mathbb{R}^{(n)}_\bullet Z_0\mathscr{A}.$$

Since $Z_0\mathscr{A}$ is an exact category, nothing new is obtained beyond (10.1.14), when we consider the geometric realizations. However, there are triangulated categories, associated to the dg categories, which provide an extra handle to formulate the theory.

The following is a restatement of the latter part (10.12) of (10.1.20) in the framework of dg categories.

Proposition 10.3.2. *Let $\mathscr{A} := (\mathscr{A}, *, \varpi, \mathbf{w})$ be a small pointed dg category with weak equivalences and duality. Assume that $\mathscr{A}$ is exact. Then there is a homotopy fibration*

$$\left|\left(\mathbf{w}\mathbb{R}^{(n)}_\bullet(\mathscr{A})\right)_h\right| \longrightarrow \left|\left(\mathbf{w}\mathbb{R}^{(n+1)}_\bullet(\mathscr{A})\right)_h\right| \longrightarrow \left|\left(\mathbf{w}S^{(n+1)}_\bullet(\mathscr{A})\right)\right|. \tag{10.21}$$

Here, we use the notation $\left|\left(\mathbf{w}\mathbb{R}^{(n)}_\bullet(\mathscr{A})\right)_h\right| := \left|\left(\mathbf{w}Z_0\mathbb{R}^{(n)}_\bullet(\mathscr{A})\right)_h\right|$ *and likewise.*

Proof. There is nothing new here beyond (10.12) of (10.1.20). See [Sm3, Prop. 4.4]. ■

10.3.1. *Shifted dualities in dg categories*

For dg categories $\mathscr{A}$ with duality and integers $n \in \mathbb{Z}$, we define a natural shifted dg category $\mathscr{A}^{[n]}$ with duality. For any $A_\bullet \in \mathscr{C}_k$, there are natural shifted dg categories in $\mathscr{C}^{[A_\bullet]}$ with duality. We use these to define the shifted dg category with duality $\mathscr{A}^{[n]}$ $\forall\, n \in \mathbb{Z}$. We summarize and list some of the natural dualities on $\mathscr{C}_k$.

Definition 10.3.3. As usual, k denotes a commutative ring. Let $A_\bullet \in \mathscr{C}_k$. Then $A_\bullet$ defines a duality structure

$$\mathscr{C}^{[A_\bullet]} := \left(\mathscr{C}, \vee_{A_\bullet}, \varpi^{A_\bullet}\right) \quad \text{on the dg category} \quad \mathscr{C}_k \quad \text{as follows :} \tag{10.22}$$

1. On the objects define the duality functor
$$\vee_{A_\bullet} : \mathscr{C}^{op} \longrightarrow \mathscr{C} \qquad \text{by} \qquad M_\bullet^{\vee_{A_\bullet}} := [M_\bullet, A_\bullet].$$
Strictly speaking, we identify $Hom(k^n, k) \cong k^n$.
2. For maps $f \in [M_\bullet, N_\bullet] = \bigoplus [M_\bullet, N_\bullet]_n$, define
$$f^{\vee_{A_\bullet}} \in \left[N_\bullet^{\vee_{A_\bullet}}, M_\bullet^{\vee_{A_\bullet}}\right] \quad \text{by} \quad f^{\vee_{A_\bullet}}(g) := (-1)^{\deg(f)\deg(g)} gf$$
$$\forall\, g \in N_\bullet^{\vee_{A_\bullet}}.$$
3. To define the double dual identification $\varpi^{A_\bullet} : 1_{\mathscr{C}} \longrightarrow \vee_{A_\bullet}\vee_{A_\bullet}$, let $M_\bullet \in \mathscr{C}$ and $m \in M$ be homogeneous. Define
$$\varpi^{A_\bullet}(m) \in M_\bullet^{\vee_{A_\bullet}\vee_{A_\bullet}} \quad \text{by}$$
$$\varpi^{A_\bullet}(m)(f) = (-1)^{\deg(m)\deg(f)} f(m) \forall f \in M_\bullet^{\vee_{A_\bullet}}.$$

This completes the definition of the duality structure (10.22).

The following is an obvious lemma.

Lemma 10.3.4. *Let k be a commutative ring and $A_\bullet, B_\bullet \in \mathscr{C}_k$. Then the tensor product*

$$\otimes : \mathscr{C}^{[A_\bullet]} \otimes \mathscr{C}^{[B_\bullet]} \xrightarrow{\sim} \mathscr{C}^{[A_\bullet \otimes B_\bullet]}$$

defines an equivalence of dg categories with duality.

Proof. Routine! ■

We will mainly be interested in the following shifted dualities.

Definition 10.3.5. As usual, let k be a commutative ring, and fix an integer $n_0 \in \mathbb{Z}$. Let $\kappa[n_0] \in \mathscr{C}_k$ be the complex, $\kappa[n_0]_{n_0} = k$ and $\kappa[n_0]_n = 0$, $\forall n \neq n_0$. Following (10.3.3), denote $\vee_{n_0} := \vee_{\kappa[n_0]} : \mathscr{C}_k^{op} \longrightarrow \mathscr{C}_k$ and $\varpi^{n_0} := \varpi^{[\kappa[n_0]]} : 1_{\mathscr{C}} \longrightarrow \vee n_0 \vee_{n_0}$. So,

$$\forall \quad M_\bullet \in \mathscr{C}_k, \quad M_\bullet^{\vee_{n_0}} = [M_\bullet, \kappa[n_0]] = \bigoplus_{n\in\mathbb{Z}} [M_\bullet, \kappa[n_0]]_n = \bigoplus_{n\in\mathbb{Z}} M_{n_0-n}^*.$$

By (10.13), at degree n, the differential is given by

$$\partial_n := (-1)^{n+1} d_{n_0-n+1}^* : M_{n_0-n}^* \longrightarrow M_{n_0-n+1}^*. \tag{10.23}$$

For $F \in [M_\bullet, N_\bullet]_n$, the dual $F^{\vee_{n_0}}$ is defined by (10.13). So,

$$M_\bullet^{\vee_{n_0}\vee_{n_0}} = \bigoplus_{n\in\mathbb{Z}} M_n^{**} \quad \text{with the differential} \qquad \partial = \{d_n^{**}\}.$$

The double dual identification $\varpi^{n_0} : M_\bullet \longrightarrow M_\bullet^{\vee_{n_0}\vee_{n_0}}$ is given by

$$\varpi^{n_0} = (-1)^{n(n_0-n)} ev : M_n \xrightarrow{\sim} M_n^{**} \text{ where } ev \text{ is the evaluation map.} \tag{10.24}$$

Note that $2 \mid n^2 - n$. So, $\varpi_n^{n_0} = (-1)^{nn_0-n} ev$. Finally, we have that

$$\mathscr{C}_k^{[n_0]} := (\mathscr{C}_k, \vee_{n_0}, \varpi^{n_0}) \quad \text{is a dg category with duality.}$$

It is worth pointing out that $\varpi_n^0 = (-1)^n ev$, and $\mathscr{C}_k^{[0]}$ is equivalent to $dgCh^b(\mathscr{F}(k))$, as defined in (10.2.17).

In fact, $(\mathscr{C}_k, \vee_{n_0}, \varpi^{n_0}, qIso)$ is a dg category with duality and weak equivalences. Sometimes, we use the same notation $\mathscr{C}_k^{[n_0]} := (\mathscr{C}_k, \vee_{n_0}, \varpi^{n_0}, qIso)$ when there is no scope of confusion.

The following is a simple lemma that leads to the periodic nature of $\mathscr{C}^{[n]}$.

Lemma 10.3.6. *Let k be a commutative ring. Let $T : \mathscr{C}_k \longrightarrow \mathscr{C}_k$ be the shift functor of dg categories (see Lemma 10.2.10). Then T^2 establishes an isomorphism*

$$T^2 : \mathscr{C}^{[0]} \xrightarrow{\sim} \mathscr{C}^{[4]} \quad \textit{of categories with duality and quasi-isomorphisms.}$$

Proof. It is proved by routine checking that, for objects and maps in $\mathscr{C}$, we have

$$T^2(\star^{\vee_0}) = (T^2(\star))^{\vee_4} \quad \text{and} \quad T^2(\varpi^0) = \varpi^4.$$

Details are left to the readers. ■

The following is the statement on the periodic nature of $\mathscr{C}^{[n]}$.

Lemma 10.3.7. *Let $n, m \in \mathbb{Z}$. Then there is a natural equivalence*

$$\mu_{mn} : \mathscr{C}_k^{[m]} \otimes \mathscr{C}_k^{[n]} \xrightarrow{\sim} \mathscr{C}_k^{m+n}$$

of dg categories with duality and weak equivalences.

Further, the isomorphism T^2 (10.3.6) *leads to an isomorphism*

$$T^2 : \mathscr{C}^{[n]} \xrightarrow{\sim} \mathscr{C}^{[n+4]} \quad \textit{of dg categories with weak equivalence and duality.} \tag{10.25}$$

This (10.25) *isomorphism is obtained by the following commutative diagram of equivalences of dg categories with duality and weak equivalences:*

$$\begin{array}{ccccc}
\mathscr{C}_k^{[n]} & \xrightarrow{\sim} & \mathscr{C}_k^{[n]} \otimes \mathscr{C}_k^{[0]} & \xrightarrow{1 \otimes T^2} & \mathscr{C}_k^{[n]} \otimes \mathscr{C}_k^{[4]} \\
 & & & \searrow^{\sim} & \downarrow \wr \\
 & & & & \mathscr{C}_k^{[n+4]}
\end{array} \qquad \forall n \in \mathbb{Z}.$$

Proof. Follows from (10.3.4) and (10.3.6). Also see [Sm3, §1.12, p. 1749]. ■

Now we define shifted categories of dg categories $\mathscr{A}$ with duality.

Definition 10.3.8. Let $\mathscr{A} := (\mathscr{A}, *, \varpi)$ be a dg category with duality. Then for $n \in \mathbb{Z}$, define the n-shifted dg category, with duality

$$\mathscr{A}^{[n]} := \mathscr{C}_k^{[n]} \otimes \mathscr{A}$$

It follows from (10.3.7) that there are equivalences of dg categories with weak equivalence and duality:

$$\begin{cases} \left(\mathscr{A}^{[n]}\right)^{[m]} \cong \mathscr{A}^{[n+m]} & \forall n, m \in \mathbb{Z} \\ \mathscr{A}^{[n+4]} \cong \mathscr{A}^{[n]} & \forall n \in \mathbb{Z}. \end{cases}$$

1. All these categories $\mathscr{A}^{[n]}$ have the same underlying dg category $\mathscr{C}_k \otimes \mathscr{A}$, with different duality, with periodicity four.
2. If $\mathscr{A}$ is pretriangulated, then the underlying dg category of $\mathscr{A}^{[n]}$ is $\mathscr{C}_k \otimes \mathscr{A} \cong \mathscr{A}$, which are equivalent.
3. If $\mathscr{A} := (\mathscr{A}, *, \varpi, \mathbf{w})$ is a dg category with weak equivalences and duality, then so is $\mathscr{A}^{[n]}$.

We define the cone functor, for subsequent use.

Definition 10.3.9. Let $\mathscr{A} := (\mathscr{A}, *, \varpi, \mathbf{w})$ be a small pointed pretrinagulated dg category with weak equivalences and duality. Consider the dg category $dgFun([1], \mathscr{A})$, whose objects are morphisms $f : X \longrightarrow Y$ in $Z_0\mathscr{A}$. Consider the inverse functor

$$\otimes : \mathscr{C} \times \mathscr{A} \longrightarrow \mathscr{A} \quad \text{and, write } CX := C \otimes X. \tag{10.26}$$

Then for $f : X \longrightarrow Y$ in $Z_0\mathscr{A}$, there is a cone $C(f)$, defined by the pushforward as follows:

$$\begin{array}{ccccc} X & \hookrightarrow & CX & \twoheadrightarrow & TX \\ {\scriptstyle f}\downarrow & & \downarrow & & \Vert \\ Y & \hookrightarrow & C(f) & \twoheadrightarrow & TX \end{array}$$

Proposition 10.3.10. *Use the notation, as in* (10.3.9). *Then the association $f \mapsto C(f)$ defines a duality preserving functor*

$$\mathrm{Cone} : dgFun([1], \mathscr{A}) \longrightarrow \mathscr{A}^{[1]}. \tag{10.27}$$

This functor is referred to as the cone functor.

Proof. The definition of the functor $f \mapsto C(f)$ is clear. It remains to prove that the duality compatibility works in $\mathscr{A}^{[1]}$. Refer to ([Sm3, p. 1753]) for details. ∎

Proposition 10.3.11. *Let $\mathscr{A} := (\mathscr{A}, *, \varpi, \mathbf{w})$ be a small pointed pretrinagulated, dg k-category with weak equivalences and duality.*

Assume that $1/2 \in k$. *Consider the commutative diagram*

$$\begin{array}{ccc} \mathscr{A} & \xhookrightarrow{\iota} & dgFun([1],\mathscr{A}) \\ {\scriptstyle \mathbf{Cone}_{|\mathscr{A}}}\downarrow & & \downarrow{\scriptstyle \mathbf{Cone}} \\ (\mathscr{A}^{[1]})^w & \longrightarrow & \mathscr{A}^{[1]} \end{array} \quad \text{where} \begin{cases} \iota(X) = 1_X \\ \mathbf{Cone}_{|\mathscr{A}}(X) = \mathbf{Cone}(1_X) = CX. \end{cases}$$

For $n \geq 1$, *the application of* $\left|\left(\mathbf{w}\mathbb{R}_\bullet^{(n)}(-)\right)_h\right|$ *yields a homotopy Cartesian square of pointed spaces, while the lower left corner space is contractible. In particular,* $\forall n \geq 1$, *the sequence*

$$\left|\left(\mathbf{w}\mathbb{R}_\bullet^{(n)}(\mathscr{A})\right)_h\right| \longrightarrow \left|\left(\mathbf{w}\mathbb{R}_\bullet^{(n)}(dgFun([1],\mathscr{A}))\right)_h\right| \longrightarrow \left|\left(\mathbf{w}\mathbb{R}_\bullet^{(n)}(\mathscr{A}^{[1]})\right)_h\right|$$

is a homotopy fibration.

Proof. See [Sm3, Prop. 4.9]. The key ingredient for the proof is the localization theorem (9.2.26) of Waldhausen. ■

10.3.2. *The GW spectrum of dg categories*

Definition 10.3.12. Let $\mathscr{A} := (\mathscr{A}, *, \varpi, \mathbf{w})$ be a pretriangulated dg k-category with weak equivalences and duality. Then define

$$GW(\mathscr{A})_n := \left|\left(\mathbf{w}\mathbb{R}_\bullet^{(n)}\mathscr{A}^{[n]}\right)_h\right| := \left|p \mapsto \left(\mathbf{w}\mathbb{R}_p^{(n)}\mathscr{A}^{[n]}\right)_h\right|. \tag{10.28}$$

In particular, $GW(\mathscr{A})_0 = |(\mathbf{w}\mathscr{A})_h| = |p \mapsto (\mathbf{w}\mathscr{A})_h|$. More generally, assuming $1/2 \in k$, even when $\mathscr{A} := (\mathscr{A}, *, \varpi, \mathbf{w})$ is not pretriangulated, we define

$$GW(\mathscr{A})_n := GW(\mathscr{A}^{ptr})_n \qquad \text{(See [Sm3, Prop. 5.6]).}$$

Where $\mathscr{A}^{ptr}$ is the pretriangulated hull of $\mathscr{A}$, defined in [Sm3, Def. 1.16]. The definition of $\mathscr{A}^{ptr}$ is fairly technical. Since most of the dg categories we are interested in are pretriangulated (e.g., 10.2.9), it is not needed for our purpose.

Now define the Grothendieck–Witt spectrum

$$GW(\mathscr{A}) := \{GW(\mathscr{A})_n : n \geq 0\}. \tag{10.29}$$

We define the bonding maps $GW(\mathscr{A})_n \longrightarrow \Omega GW(\mathscr{A})_{n+1}$ in the following (10.3.13), and prove that they are homotopy equivalences for $n \geq 1$. Consequently, the GW-spectrum (10.29) of $\mathscr{A}$ is a positive Ω-spectrum (meaning that for $n \geq 1$ the bonding maps (10.31) are homotopy equivalences). Now define **Grothendieck–Witt groups** of $\mathscr{A}$, by (9.2.9)

$$GW_n(\mathscr{A}) := GW_n(\mathscr{A}, *, \varpi, \mathbf{w}) := \pi_n\left(GW(\mathscr{A})\right) n \in \mathbb{Z}. \tag{10.30}$$

We prove that the spectrum (10.29) is a positive Ω-spectrum.

Lemma 10.3.13. *Under the hypotheses of* (10.3.12), *there are homotopy equivalences,*

$$\begin{cases} GW(\mathscr{A})_n \xrightarrow{\sim} \Omega GW(\mathscr{A})_{n+1} \; n \geq 1 \\ \mathbf{GW}(Z_0\mathscr{A}) \xrightarrow{\sim} \Omega GW(\mathscr{A})_1 \;\text{ see (10.6).} \end{cases} \tag{10.31}$$

Proof. We can assume that $\mathscr{A}$ is pretriangulated. For $n \geq 1$, by (10.3.11) and (10.3.2), we have two homotopy fibrations

$$\left|\left(\mathbf{w}\mathbb{R}_{\bullet}^{(n+1)}(\mathscr{A}^{[n]})\right)_h\right| \longrightarrow \left|\left(\mathbf{w}\mathbb{R}_{\bullet}^{(n+1)}(dgFun([1], \mathscr{A}^{[n]}))\right)_h\right| \longrightarrow \left|\left(\mathbf{w}\mathbb{R}_{\bullet}^{(n+1)}(\mathscr{A}^{[n+1]})\right)_h\right|$$

$$\left|\left(\mathbf{w}\mathbb{R}_{\bullet}^{(n)}(\mathscr{A}^{[n]})\right)_h\right| \longrightarrow \left|\left(\mathbf{w}\mathbb{R}_{\bullet}^{(n+1)}(\mathscr{A}^{[n]})\right)_h\right| \longrightarrow \left|\left(\mathbf{w}S_{\bullet}^{(n+1)}(\mathscr{A}^{[n]})\right)\right|.$$

By rotating the first sequence, we obtain a homotopy fibration (C.3.10):

$$\Omega\left|\left(\mathbf{w}\mathbb{R}_{\bullet}^{(n+1)}(\mathscr{A}^{[n+1]})\right)_h\right| \longrightarrow \left|\left(\mathbf{w}\mathbb{R}_{\bullet}^{(n+1)}(\mathscr{A}^{[n]})\right)_h\right|$$
$$\longrightarrow \left|\left(\mathbf{w}\mathbb{R}_{\bullet}^{(n+1)}(dgFun([1], \mathscr{A}^{[n]}))\right)_h\right|.$$

One proves that the 3rd-terms of the 2nd, 3rd-sequences are homotopically equivalent:

$$\left|\left(\mathbf{w}\mathbb{R}_\bullet^{(n+1)}(dgFun([1],\mathscr{A}^{[n]}))\right)_h\right| \cong \left|\left(\mathbf{w}S_\bullet^{(n+1)}(\mathscr{A}^{[n]})\right)\right|$$

[Sm3, Cor. 4.11].

So, comparing the second and third sequences, we obtain homotopy equivalences:

$$\left|\left(\mathbf{w}\mathbb{R}_\bullet^{(n)}(\mathscr{A}^{[n]})\right)_h\right| \cong \Omega\left|\left(\mathbf{w}\mathbb{R}_\bullet^{(n+1)}(\mathscr{A}^{[n+1]})\right)_h\right|.$$

This establishes the case $n \geq 1$ of (10.31). The second line follows by the same argument, by replacing the second homotopy fibration by (C.3.10). The proof is complete. ∎

Corollary 10.3.14. *Under the hypotheses of* (10.3.12),

$$GW_p(\mathscr{A}) = GW_p(Z_0\mathscr{A}) \qquad \forall p \geq 0.$$

In particular,

$$GW_0(\mathscr{A}) = GW^0(\mathfrak{T}\mathscr{A}) \qquad \textit{as defined in } (11.1.6).$$

Proof. It follows from (10.3.13) that, for $p \geq 0$, we have

$$GW_p(\mathscr{A}) = \pi_p\Omega GW_1(\mathscr{A}) = \pi_p(\mathbf{GW}(Z_0\mathscr{A})) = GW_p(Z_0\mathscr{A}).$$

See [Sm3, Prop. 3.8] for the second statement. The proof is complete. ∎

We have the following on agreement.

Proposition 10.3.15. *Suppose that* $(\mathscr{E},*,\eta)$ *is an exact category with duality. Consider the dg category* $dg(\mathscr{E}) := (Ch^b(\mathscr{E}),*,\varpi,qIso)$ *of bounded chain complexes with weak equivalence and duality* (10.2.16). *Then*

$$GW_n(dg(\mathscr{E})) = \begin{cases} GW_n(\mathscr{E},*,\eta) & n \geq 0 \\ W^{-n}(\mathbf{D}^b(\mathscr{E})) & n \leq -1 \end{cases}$$

where $W^{-n}(\mathbf{D}^b(\mathscr{E}))$ *denotes the triangular Witt groups of Balmer* (11.2.11).

Proof. The case $n \geq 0$ follows from (10.1.17). See [Sm3, p. 1781] for the case $n \leq -1$. ∎

We define the shifted $G\mathcal{W}$-spectrum.

Definition 10.3.16. Let $\mathscr{A} := (\mathscr{A}, *, \varpi, \mathbf{w})$ be a small dg k-category with weak equivalence and duality. Assume that $1/2 \in k$. For integers $n \in \mathbb{Z}$, define the n-shifted $G\mathcal{W}$-spectrum, and GW-groups by

$$G\mathcal{W}^{[n]}(\mathscr{A}) := G\mathcal{W}(\mathscr{A}^{[n]}), \quad \text{and} \quad GW_p^{[n]}(\mathscr{A}) = \pi_p\left(G\mathcal{W}^{[n]}(\mathscr{A})\right) \quad \forall p \in \mathbb{Z}.$$

These are sometimes called stable homotopy spectra or groups. By (10.3.8), these groups have periodicity 4.

So far, to define the $G\mathcal{W}$-spectrum, only the structure of the underlying exact category has been used. The dg category structure on $\mathscr{A}$ is at our disposal to make further statements.

Theorem 10.3.17 (Invariance). *Assume that $1/2 \in k$. Let $F : \mathscr{A} \longrightarrow \mathscr{B}$ be a functor of dg k-categories with weak equivalences and duality. Assume that the induced functor $\mathfrak{T}(\mathscr{A}) \xrightarrow{\sim} \mathfrak{T}(\mathscr{B})$ of the associated triangulated categories is an equivalence. Then F induces equivalence*

$$G\mathcal{W}^{[n]}(\mathscr{A}) \xrightarrow{\sim} G\mathcal{W}^{[n]}(\mathscr{B}) \qquad \text{of the } G\mathcal{W} \text{ spectra} \quad \forall n \in \mathbb{Z}.$$

By equivalence of such positive Ω-spectra, we mean that the maps $G\mathcal{W}^{[n]}(\mathscr{A})_r \xrightarrow{\sim} G\mathcal{W}^{[n]}(\mathscr{B})_r$ are homotopy equivalences $\forall r \geq 1$ [Sm3, p. 1830]. *Consequently, the maps*

$$GW_p^{[n]}(\mathscr{A}) \xrightarrow{\sim} GW_p^{[n]}(\mathscr{A}) \qquad \text{are isomorphisms}, \quad \forall n, p \in \mathbb{Z}.$$

Proof. See [Sm3, Thm. 6.5]. ∎

Before we state the localization theorem, we consider the exactness of dg categories.

Definition 10.3.18. A sequence $\mathscr{A}_0 \longrightarrow \mathscr{A}_1 \longrightarrow \mathscr{A}_2$ of dg categories with weak equivalences is said to be **quasi-exact** if the sequence $\mathfrak{T}\mathscr{A}_0 \longrightarrow \mathfrak{T}\mathscr{A}_1 \longrightarrow \mathfrak{T}\mathscr{A}_2$ of associated triangulated categories is exact (see 10.2.14, 11.2.15).

A functor $(\mathscr{A}, \mathbf{w}_A) \longrightarrow (\mathscr{B}, \mathbf{w}_B)$ of dg categories with weak equivalences is said to be a **quasi-equivalence** if the associated functor $\mathfrak{T}(\mathscr{A}, \mathbf{w}_A) \xrightarrow{\sim} \mathfrak{T}(\mathscr{B}, \mathbf{w}_B)$ is an equivalence.

Theorem 10.3.19 (Localization). *Assume that $1/2 \in k$. Let $\mathscr{A}_0 \longrightarrow \mathscr{A}_1 \longrightarrow \mathscr{A}_2$ be a sequence of dg categories with weak equivalences and duality. Assume that the sequence is quasi-exact. Then $\forall n \in \mathbb{Z}$, there is a homotopy fibration of GW-spectra*

$$GW^{[n]}(\mathscr{A}_0) \longrightarrow GW^{[n]}(\mathscr{A}_1) \longrightarrow GW^{[n]}(\mathscr{A}_2).$$

Proof. See [Sm3, Thm. 6.6]. ■

Remark 10.3.20. For $\mathscr{A} \in dg\mathbf{Cat}WD_k$, the GW-spectrum $GW(\mathscr{E})$ is not, in general, a connective spectrum (9.2.10). In other words, the negative GW-groups $GW_p(\mathscr{A})$ need not vanish for $p \leq -1$ (see 10.3.15). However, this GW-spectrum $GW(\mathscr{E})$ is treated as an analogue of the connective K-theory spectrum (9.2.10).

Because of the nature of exact sequences of derived categories, such as (9.5.6), the localization theorem (10.3.19) may not be as widely useful as we may desire. This leads to a desire to develop a GW-theory, that is analogous to the nonconnective $\mathbb{K}$-theory. We will pursue this in subsequent sections.

10.4. Nonconnective Hermitian $\mathbb{G}W$-Theory

We proceed to develop a nonconnective version of the GW-theory. We only give an overview of [Sm3, §8]. First, we give some background on the category Sp of symmetric spectra and Bi Spectra BiSp, from [Sm3, App. B].

10.4.1. *The category* Sp *of symmetric spectra*

Let k be a commutative ring, with $1/2 \in k$. Let $dg\mathbf{Cat}WD_k$ denote the category of small dg categories with weak equivalence and duality and dg functors of such dg categories. The purpose of this section is to provide a functorial interpretation of the association $\mathscr{A} \mapsto GW(\mathscr{A})$

from the category $dg\mathbf{Cat}WD_k \longrightarrow \mathrm{Sp}$ to a suitable category Sp of symmetric spectra, to be defined.

Definition 10.4.1. Let $I = [0,1]$ be the unit interval, $\mathbb{S}^0 := \{0,1\}$, $\mathbb{S}^1 = \frac{I}{\mathbb{S}^0}$ be the circle, and $\mathbb{S}^n = \mathbb{S}^1 \wedge \cdots \wedge \mathbb{S}^1$ be the n-sphere, obtained by an n-fold smash product. Let Σ denote the category whose objects are $\overline{n} := \{1, 2, \ldots, n\}$, $\overline{0} = \phi$. The morphisms are, $\forall\ m \neq n\ Mor(\overline{m}, \overline{n}) = \phi$, and $\forall\ n\ Mor(\overline{n}, \overline{n}) = \Sigma_n$ is the group of permutation elements in $\overline{n}$. The smash product $\wedge$ provides the category $\mathbf{Top}_\bullet$ a structure of a symmetric monoidal category, with $\mathbb{S}^0$ being the identity.

A **symmetric sequence** is a functor $\Sigma \longrightarrow \mathbf{Top}_\bullet$. So, a symmetric sequence is a sequence $\{X_0, X_1, \ldots\}$ in $\mathbf{Top}_\bullet$, together with base point preserving left action of Σ_n on X_n.

A **symmetric spectrum** is a functor (i.e. a symmetric sequence)

$$X := \{X_n : n \geq 0\} : \Sigma \longrightarrow \mathbf{Top}_\bullet, \quad \text{together with maps}$$
$$e_{n,m} : \mathbb{S}^n \wedge X_m \to X_{n+m}$$

in $\mathbf{Top}_\bullet$, for $\forall\ n, m \in \mathbb{N}$, to be called the bonding maps, such that

1. The bonding maps $e_{n,m}$ are $\Sigma_n \times \Sigma_m$-equivariant maps, via the natural inclusions $\Sigma_n \times \Sigma_m \subseteq \Sigma_{n+m}$.
2. The maps $e_{0,m} : \mathbb{S}^0 \wedge X_m \xrightarrow{\sim} X_m$ are the natural identifications $\forall m$.
3. For $\forall p, n, m$, the diagram

$$\begin{array}{ccc} \mathbb{S}^p \wedge \mathbb{S}^n \wedge X_m & \xrightarrow{1 \wedge e_{n,m}} & \mathbb{S}^p \wedge X_{n+m} \\ {\scriptstyle e_{p,n} \wedge 1} \downarrow & & \downarrow {\scriptstyle e_{p,n+m}} \\ \mathbb{S}^{p+n} \wedge X_m & \xrightarrow[e_{p+n,m}]{} & X_{p+n+m} \end{array} \qquad \text{commutes,}$$

where the first vertical $e_{p,n}$ is the natural identification.

A map $f : X = \{X_n : n \geq 0\} \longrightarrow Y := \{Y_n : n \geq 0\}$ of symmetric spectra, is a sequence of maps $f_n : X_n \longrightarrow Y_n$ that are Σ_n-equivariant, and they commute with the bonding maps $e_{n,m}$.

The category of such symmetric spectrum and such maps is denoted by Sp.

A spectrum $X = \{X_n\}$ is called an **Ω-spectrum** if the bonding maps $e_{1,m} : \mathbb{S}^1 \wedge X_m \longrightarrow X_{1+m}$ are weak equivalences. (Since we work with CW complexes, weak equivalences coincide with homotopy equivalences.)

Example 10.4.2. The following are some important examples of symmetric spectra:

1. $\mathbb{S} := \mathbb{S}_{\mathrm{Sp}} := \{\mathbb{S}^n : n \geq 0\}$, together with the natural identifications $e_{n,m} : \mathbb{S}^n \wedge \mathbb{S}^m \xrightarrow{\sim} \mathbb{S}^{n+m}$, form a symmetric spectrum. The spectrum $\mathbb{S}$ is referred to as the **sphere spectrum**. Sometimes, we use the notation $\mathbb{S}_{\mathrm{Sp}} := \mathbb{S}$ to avoid confusion.
2. Let $X \in \mathbf{Top}_\bullet$. Then $\Sigma X := \{\mathbb{S}^n \wedge X : n \geq 0\}$ is a symmetric spectrum, referred to as the **suspension spectrum**. The bonding maps are induced by those in $\mathbb{S}$. Sometimes we use the notation $\mathbb{S}^n_{\mathrm{Sp}} := \Sigma\mathbb{S}^n$, for $n \geq 0$. So,
$$\Sigma : \mathbf{Top}_\bullet \longrightarrow \mathrm{Sp} \qquad \text{is a functor.}$$

We use the symmetric monoidal structure $(\mathbf{Top}_\bullet, \wedge, \mathbb{S}^0)$ to provide a similar structure on Sp, as follows.

Definition 10.4.3. Fix an integer $n \geq 0$. Let
$$\mathscr{I}_n = \left\{ (A,B) : \begin{cases} A := (1 \leq i_1 < i_2 < \cdots < i_r \leq n) \\ B := (1 \leq j_1 < j_2 < \cdots < j_{n-r} \leq n) \end{cases} \ni A \cap B = \phi \right\}.$$

Note that for $(A,B) \in \mathscr{I}_n$, $A \cup B = \overline{n}$. Now, let $X = \{X_n\}, Y = \{Y_n\}$ be symmetric sequences. For $(A,B) \in \mathscr{I}_n$, denote $X_A := X_r$, $Y_B := Y_{n-r}$. Define
$$(X \wedge Y)_n := \coprod_{(A,B)\in\mathscr{I}_n} X_A \wedge Y_B, \qquad \text{the coproduct in } \mathbf{Top}_\bullet. \quad (10.32)$$

1. So, $(X \wedge Y)_n$ has $\binom{n}{r}$ copies of $X_r \wedge Y_{n-r}$.
2. Let $\sigma \in \Sigma_n$. So, $\sigma : \overline{n} \xrightarrow{\sim} \overline{n}$ is an isomorphism. Let $(A,B) \in \mathscr{I}_n$. We have,
$$\begin{cases} X_A = X_{\sigma(A)} = X_r \\ Y_B = Y_{\sigma(B)} = Y_{n-r} \end{cases}. \text{ Write, } \begin{cases} A = (1 \leq i_1 < i_2 < \cdots < i_r \leq n) \\ \sigma(A) = (1 \leq \iota_1 < \iota_2 < \cdots < \iota_r \leq n) \\ \iota_p \mapsto \sigma(i_p) \text{ is a permutation of } \sigma(A). \end{cases}$$
Identifying $\sigma(A) \cong \overline{r}$, we obtain a permutation $\overline{r} \xrightarrow{\sim} \overline{r}$. Since X is a symmetric sequence, this defines an action, $\sigma_A(X) : X_{\sigma(A)} \xrightarrow{\sim}$

$X_{\sigma(A)}$. Likewise, there is an action $\sigma_B(Y) : Y_{\sigma(B)} \xrightarrow{\sim} Y_{\sigma(B)}$. This defines an action $\sigma(X \wedge Y) : (X \wedge Y)_n \xrightarrow{\sim} (X \wedge Y)_n$, whose restriction on $X_A \wedge X_B$ is given by the composition

$$\begin{array}{ccc} X_A \wedge Y_B & \xrightarrow{1 \wedge 1} & X_{\sigma(A)} \wedge Y_{\sigma(B)} \\ & \searrow & \downarrow X_{\sigma(A)} \wedge \sigma_B(Y) \\ & & X_{\sigma(A)} \wedge Y_{\sigma(B)} \end{array} \tag{10.33}$$

We summarize the above as follows.

Lemma 10.4.4. *Let X, Y be symmetric sequences. Then $X \wedge Y = \{(X \wedge Y)_n : n \geq 0\}$, as defined in (10.32), is a symmetric sequence. The action of Σ_n on $(X \wedge Y)_n$ is defined as in (10.33).*

1. *In particular, the wedge product $\wedge$ defines a symmetric monoidal product on the category of symmetric sequences of pointed spaces, where $\mathbb{1} := \{\mathbb{S}^0, pt, pt, \ldots\}$ acts as the unit. The product $\wedge$ is also referred to as the smash product.*
2. *With $\mathbb{S} = \mathbb{S}_{\mathrm{Sp}}$, there is a map of symmetric sequences $e : \mathbb{S} \wedge \mathbb{S} \longrightarrow \mathbb{S}$, induced by the identifications $\mathbb{S}^n \wedge \mathbb{S}^m \xrightarrow{\sim} \mathbb{S}^{n+m}$. Further, there is a natural map of spectra $\mathbb{1} \longrightarrow \mathbb{S}$. Jargon from commutative algebra is borrowed.*

   (a) *One says that $\mathbb{S}$ is a commutative monoid, in the sense that $\mathbb{S}^n \wedge \mathbb{S}^m \cong \mathbb{S}^m \wedge \mathbb{S}^n$.*

   (b) *One also says that a symmetric spectrum X is a left $\mathbb{S}$-module, in the sense that the following diagrams*

$$\begin{array}{ccc} \mathbb{S} \wedge \mathbb{S} \wedge X & \xrightarrow{e \wedge 1_X} & \mathbb{S} \wedge X \\ {\scriptstyle 1_{\mathbb{S}} \wedge e} \downarrow & & \downarrow {\scriptstyle e} \\ \mathbb{S} \wedge X & \xrightarrow[e]{} & X \end{array}, \qquad \begin{array}{ccc} \mathbb{1} \wedge X & \longrightarrow & \mathbb{S} \wedge X \\ & \searrow & \downarrow {\scriptstyle e} \\ & & X \end{array} \quad \text{commute.}$$

Lemma 10.4.5. *The category Sp of symmetric spectra is a symmetric monoidal category under the smash product $X \wedge_{\mathbb{S}} Y$, defined by* *the pushforward diagram*

$$\begin{array}{ccc} X \wedge \mathbb{S} \wedge Y & \xrightarrow{1_X \wedge e_Y} & X \wedge Y \\ {\scriptstyle (e_X\tau)\wedge 1_Y}\downarrow & & \downarrow \\ X \wedge Y & \dashrightarrow & X \wedge_{\mathbb{S}} Y \end{array}$$

where $\tau : X \wedge S \longrightarrow S \wedge X$ *is the switch. In this product,* $\mathbb{1} := \mathbb{S}$ *acts as the unit. With* $\mathscr{C} := (\mathbf{Top}_\bullet, \wedge, \mathbb{S}^0)$, *we write that* $\mathrm{Sp}(\mathscr{C}, \mathbb{S}^1) := \mathrm{Sp}(\mathbf{Top}_\bullet, \mathbb{S}^1) := (sym\mathrm{Sp}, \wedge_{\mathbb{S}}, \mathbb{S})$ *is a symmetric monoidal category, where sym*Sp *denotes the set of all symmetric spectra. (The first coordinate* $\mathbb{S}^1$ *is viewed as the generator of the sphere spectrum* $\mathbb{S} = \{\wedge^n \mathbb{S}^1\}$.*)*

Lemma 10.4.6. *Let* k *be a commutative ring, with* $1/2 \in k$. *Then the category* $dg\mathbf{Cat}WD_k$ *of small dg categories with weak equivalences and duality is a symmetric monoidal category under* $\otimes$. *Then*

1. *For* $\mathscr{A} \in dg\mathbf{Cat}WD_k$, *the spectrum* $GW(\mathscr{A}) \in \mathrm{Sp}$ *is a symmetric spectrum. The unit* $\mathbb{1}_{dg}$ *of* $\mathscr{A} \in dg\mathbf{Cat}WD_k$ *is given by one object dg category* k, *with duality given by* $k \otimes k \xrightarrow{\sim} k$. *It follows that* $GW(\mathbb{1}_{dg}) = \mathbb{1} \in \mathrm{Sp}$, *where* $\mathbb{1} := \{\mathbb{S}^0, pt, pt, \ldots\}$.
2. *Denote* $GW(k) := GW(\mathscr{C}_k)$. *It follows that* (1) $GW(k)$ *is a commutative symmetric ring spectrum,* (2) $GW(\mathscr{A})$ *is a module spectrum over* $GW(k)$ [Sm3, p. 1779 end].
3. *The association* $\mathscr{A} \mapsto GW(\mathscr{A})$ *establishes a functor*

$$(dg\mathbf{Cat}WD_k, \otimes, \mathbb{1}_{dg}) \longrightarrow (sym\mathrm{Sp}, \wedge_{\mathbb{S}}, \mathbb{S}) \quad \textit{of symmetric}$$

monoidal categories.

Proof. See [Sm3, §5.1, 5.2]. The bonding maps of $GW(\mathscr{A})$ are obtained by iterating (10.3.13). ■

Remark 10.4.7. In the modern literature, homotopy theory has been consumed within the enriched theory of Model categories [H]. The jargon of Model categories seems unavoidable for our discussions on nonconnective version of the Grothendieck–Witt theory while we need it minimally (see 10.4.14). Refer to [H] for definitions and other basics on Model categories and to [HSS, H2], [Sm3, Appendix B] for other related information.

For the convenience of our discussions, we include some outlines. A Model category $\mathscr{C}$ is a category with three subcategories: (1) weak equivalences $\mathbf{w} \subseteq \mathscr{C}$, (2) cofibrations $coF\mathscr{C} \subseteq \mathscr{C}$, (3) fibrations $F\mathscr{C} \subseteq \mathscr{C}$, subject to further conditions [H, §1.1].

The homotopy category $Ho\mathscr{C} := \mathbf{w}^{-1}\mathscr{C}$ of a Model category $\mathscr{C}$ is obtained by inverting the weak equivalences. For objects $X, Y \in \mathscr{C}$, let $[X, Y]_{Ho\mathscr{C}}$ denote the set of all morphisms $X \longrightarrow Y$ in $Ho\mathscr{C}$. This set of morphisms $[X, Y]_{Ho\mathscr{C}}$ is to be treated as a newer avatar of the set of all homotopically equivalent maps.

Consider the category $\mathscr{C} := sys\mathrm{Sp}$ of symmetric spectra and the maps of the same (without any reference to the product $\wedge_{\mathbb{S}}$-structure).

1. There are many variations of model structures on $\mathscr{C} := sys\mathrm{Sp}$. The so-called (positive) **projective stable model structure** is the model structure of our choice. See [Sm3, §B.2-3] for more. We summarize some of the basic facts on this model structure, as follows.

   (a) The model structure endows $\mathrm{Sp} = (sys\mathrm{Sp}, \wedge_{\mathbb{S}}, \mathbb{S})$ with a structure of a symmetric monoidal model category [H2, Def. 6.2]. It only means that the model structure behaves well with the moniodal structure. We denote the homotopy category by $Ho\mathrm{Sp}$.

   (b) The parenthetic qualification "(positive)" only points to the variation of the same, with no conditions imposed, at degree $n = 0$. The $GW(\mathscr{A})$-spectrum of a dg category being a positive Ω-spectra (10.3.13), this variation is useful for us. However, they have the same homotopy category $Ho(sys\mathrm{Sp})$.

   (c) The Fibrant objects are precisely the (positive) Ω-spectra $X \in sym\mathrm{Sp}$.

   (d) A map $X \longrightarrow Y$ of (positive) Ω-spectra is a stable equivalence (i.e. a weak equivalence) in this structure if and only if $X_n \longrightarrow Y_n$ is a weak equivalence $(\forall n \geq 1)$ $\forall n \geq 0$.

2. For a $X \in sym\mathrm{Sp}$ and $n \in \mathbb{Z}$, the homotopy groups $\pi_n(X)$ defined in (9.2.9) are sometimes referred to as the **naive homotopy group** in the model category literature. For our purpose, this suffices.

3. For $n \in \mathbb{Z}$, and $Y \in sym\mathrm{Sp}$, define the **true homotopy group (set)**

$$\pi_n^t(Y) := [\mathbb{S}_{\mathrm{Sp}}^n, Y]_{Ho\mathrm{Sp}}$$

where $\mathbb{S}_{\mathrm{Sp}}^n = \Sigma\mathbb{S}^n$ is the suspension for $n \geq 0$ and it makes sense when $n \leq -1$ in this model structure. The $\mathcal{GW}(\mathscr{A})$-spectrum of a dg category is a (positive) Ω-spectrum (10.3.13). In this case, these two homotopy groups coincide. Readers interested in model structures on Sp are referred to [HSS, H2].

10.4.2. *The Bispetra* BiSp

We plan to iterate the process of the construction of the category $\mathrm{Sp}(\mathbf{Top}_\bullet, \mathbb{S}^1)$ of symmetric spectra from $(\mathbf{Top}_\bullet, \wedge, \mathbb{S}^0)$. In fact, we repeat the process, starting from $\mathrm{Sp}(\mathbf{Top}_\bullet, \mathbb{S}^1) := (sym\mathrm{Sp}, \wedge_{\mathbb{S}}, \mathbb{S})$. The process works for any symmetric monoidal model category $(\mathscr{C}, \otimes, \mathbb{1})$ [H2, Def. 6.2].

Construction 10.4.8. Suppose that $(\mathscr{C}, \wedge, \mathbb{1})$ is a symmetric monoidal category. Assume that finite coproduct $\coprod$ exists in $\mathscr{C}$, and that the product $\wedge$ commutes with the coproduct, in the sense that

$$(A \wedge C) \coprod (B \wedge C) \xrightarrow{\sim} (A \coprod B) \wedge C, \quad \star \xrightarrow{\sim} \star \wedge A$$

are isomorphisms, for all A, B, C.

As in (10.4.1), a **symmetric sequence** in $\mathscr{C}$ is a functor $\Sigma \longrightarrow \mathscr{C}$. The category of symmetric sequences is denoted by $\mathscr{C}^\Sigma$. Then $\mathscr{C}^\Sigma$ has a structure of a monoidal category, where for $X, Y \in \mathscr{C}^\Sigma$, the product $X \wedge Y = \{(X \wedge Y)_n\}$ is defined as in (10.4.3).

Let $K \in \mathscr{C}$. Let $\mathbb{S}_K := \mathbb{S}ym(K) := (\mathbb{1}, K, K^2, \ldots)$, where $K^n = K \wedge K \wedge \cdots \wedge K$ is the n-fold product. Note that $\mathbb{S}ym(K) \in \mathscr{C}^\Sigma$, and $\mathbb{S}ym(K)$ has a commutative monoid structure, in the sense $K^m \wedge K^n \xrightarrow{\sim} K^{m+n}$.

As in (10.4.1), a **symmetric K-spectrum** is a symmetric sequence $X = \{X_n\} \in \mathscr{C}^\Sigma$, together with $\Sigma^n \times \Sigma^m$-equaivariant maps $e_{n,m} : K^n \wedge X_m \longrightarrow X_{n+m}$ such that $e_{0,m} : \mathbb{1} \wedge X_m \xrightarrow{\sim} X_m$ is the natural identification, and $\forall p, n, m$ the diagram

$$\begin{array}{ccc} K^p \wedge K^n \wedge X_m & \xrightarrow{1\wedge e_{n,m}} & K^p \wedge X_{n+m} \\ {\scriptstyle e_{p,n}\wedge 1}\downarrow & & \downarrow{\scriptstyle e_{p,n+m}} \\ K^{p+n} \wedge X_m & \xrightarrow[e_{p+n,m}]{} & X_{p+n+m} \end{array} \qquad \text{commutes.}$$

So, a symmetric K-spectrum X is a left $\mathbb{S}ym(K)$-module. The module structure is given by maps $e : \mathbb{S}ym(K) \wedge X \longrightarrow X$ such that the diagrams

$$\begin{array}{ccc} \mathbb{S}ym(K) \wedge \mathbb{S}ym(K) \wedge X & \xrightarrow{e\wedge 1_X} & \mathbb{S}ym(K) \wedge X \\ {\scriptstyle 1_{\mathbb{S}ym(K)}\wedge e}\downarrow & & \downarrow{\scriptstyle e} \\ \mathbb{S}ym(K) \wedge X & \xrightarrow[e]{} & X \end{array} \quad ,$$

$$\begin{array}{ccc} \mathbb{1} \wedge X & \longrightarrow & \mathbb{S}ym(K) \wedge X \\ & \searrow & \downarrow{\scriptstyle e} \\ & & X \end{array} \qquad \text{commute.}$$

The category $\mathrm{Sp}(\mathscr{C}, K)$ of the symmetric K-spectra is defined as the category of the left $\mathbb{S}ym(K)$-modules $X \in \mathscr{C}^{\Sigma}$, together with $\mathbb{S}ym(K)$-module maps. Further, as in (10.4.5), for $X, Y \in \mathrm{Sp}(\mathscr{C}, K)$, we define a product $X \wedge_{\mathbb{S}_K} Y$ by the pushforward diagram

$$\begin{array}{ccc} X \wedge \mathbb{S}ym(K) \wedge Y & \xrightarrow{1_X\wedge e_Y} & X \wedge Y \\ {\scriptstyle (e_X\tau)\wedge 1_Y}\downarrow & & \downarrow \\ X \wedge Y & \dashrightarrow & X \wedge_{\mathbb{S}_K} Y \end{array} \qquad \text{where} \qquad \mathbb{S}_K = \mathbb{S}ym(K).$$

Before we close this construction, with integers $n \geq 0$, we list a few natural functors:

$$\begin{cases} Ev_n : \mathrm{Sp}(\mathscr{C}, K) \longrightarrow \mathscr{C} & X = \{X_r\} \mapsto X_n \\ F_0 : \mathscr{C} \longrightarrow \mathrm{Sp}(\mathscr{C}, K) & M \mapsto (M, M \wedge K, M \wedge K^2, \ldots) \\ K : \mathrm{Sp}(\mathscr{C}, K) \longrightarrow \mathrm{Sp}(\mathscr{C}, K) & X \mapsto F_0(K) \wedge_{\mathbb{S}_K} X. \end{cases} \tag{10.34}$$

The functor F_0 is a left adjoint to Ev_0. The functor Ev_n is called the evaluation functor. Often, in the literature, $F_0M = M$ is identified. So, $F_0\mathbb{1} = \mathbb{S}_K$, and $F_0K^n = (K^n, K^{n+1}, \ldots) =: K^n$ are identified. Notation for the functor K is also an abuse of notation.

Lemma 10.4.9. *Let* $(\mathscr{C}, \wedge, \mathbb{1})$ *be a symmetric monoidal category, as in* (10.4.8). *Then* $(\mathrm{Sp}(\mathscr{C}, K), \wedge_{\mathbb{S}_K}, \mathbb{S}_K)$ *is a symmetric monoidal category, with* $\mathbb{S}_K$ *as the unit.*

Proof. See [Sm3, §B.9]. ∎

Lemma 10.4.10. *Let* $(\mathscr{C}, \wedge, \mathbb{1})$ *be a symmetric monoidal model category* [H2, Def. 6.2]. *(This means that* $\mathscr{C}$ *has a model category structure that is compatible with the monoidal structure.) Let* K *be cofibrant (meaning that* $\star \longrightarrow K$ *is a cofibration). Then* $\mathrm{Sp}(\mathscr{C}, K)$ *has a model category structure, similar to the* ***projective stable model structure*** *in* (10.4.7). *In this model structure, the functor*

$$K : \mathrm{Sp}(\mathscr{C}, K) \longrightarrow \mathrm{Sp}(\mathscr{C}, K) \quad \textit{is a left Quillen equivalence.}$$

Proof. The model structure is constructed, as was done for $sys\mathrm{Sp}$, in (10.4.7). Refer to [H2, §1, 3], and also [Sm3, §B.9]. ∎

We define the homotopy sets.

Definition 10.4.11. Let $(\mathscr{C}, \wedge, \mathbb{1})$ be a symmetric monoidal model category [H2, Def. 6.2]. Let K be cofibrant. Let $\mathrm{Sp}(\mathscr{C}, K)$ be as in (10.4.10). Let $X \in \mathrm{Sp}(\mathscr{C}, K)$, and $n \in \mathbb{Z}$.

1. Define the (naive) homotopy set
$$\pi_n(X) := \lim_k [K^{k+n}, X_k]_{Ho\mathscr{C}}.$$
2. Define the true homotopy set
$$\pi_n^t(X) := [K^n, X]_{Ho(\mathrm{Sp}(\mathscr{C},K))}.$$
This makes sense $\forall n \in \mathbb{Z}$, by the Quillen equivalence (10.4.10).

The following is an example of our interest.

Proposition 10.4.12. *Let* $\mathrm{Sp} := (\mathbf{Top}_\bullet, \mathbb{S}^1) = (sys\mathrm{Sp}, \wedge_{\mathbb{S}}, \mathbb{S})$ *be as above, which is a symmetric monoidal category. Then* Sp *is a symmetric monoidal model category.*

Proof. Follows from (10.4.10). ∎

Now we define the category of bispectra BiSp.

Definition 10.4.13. Consider the category $\mathrm{Sp} := (\mathbf{Top_\bullet}, \mathbb{S}^1) = (sys\mathrm{Sp}, \wedge_{\mathbb{S}}, \mathbb{S})$, which is a symmetric monoidal model category. The sphere spectrum $\mathbb{S} = (\mathbb{S}^0, \mathbb{S}^1, \mathbb{S}^2, \ldots) \in sys\mathrm{Sp}$ is the unit of the monoidal structure. Denote $\mathbb{S}^1_{\mathrm{Sp}} := (\mathbb{S}^1, \mathbb{S}^2, \mathbb{S}^3, \ldots)$. Let $K := \tilde{\mathbb{S}}^1_{\mathrm{Sp}} \in \mathrm{Sp}$ be the cofibrant replacement of $\mathbb{S}^1_{\mathrm{Sp}}$. (This means that K is cofibrant and $K \longrightarrow \mathbb{S}^1_{\mathrm{Sp}}$ is a fibration and a weak equivalence (also called trivial fibration).) Now, define

$$\mathrm{BiSp}(\mathbb{S}^1\text{–}\tilde{\mathbb{S}}^1_{\mathrm{Sp}}) := \mathrm{Sp}(\mathrm{Sp}, K) := \mathrm{Sp}\left(\mathrm{Sp}, \tilde{\mathbb{S}}^1_{\mathrm{Sp}}\right) \tag{10.35}$$

as in (10.4.8, 10.4.9). We will often abbreviate $\mathrm{BiSp} := \mathrm{BiSp}(\mathbb{S}^1\text{–}\tilde{\mathbb{S}}^1_{\mathrm{Sp}})$. So, BiSp is a symmetric monoidal model category. We say that BiSp is the category of $\mathbb{S}^1\text{–}\tilde{\mathbb{S}}^1_{\mathrm{Sp}}$ **bispectra**. We summarize some of the main points:

1. Denote $K^n := (\tilde{\mathbb{S}}^1_{\mathrm{Sp}})^n := K \wedge_{\mathbb{S}} K \wedge_{\mathbb{S}} \cdots \wedge_{\mathbb{S}} K \in \mathrm{Sp}$, the n-fold product.
2. The unit in BiSp, $\mathbb{S}_K = \mathbb{S}ym(K) = (\mathbb{1}, K, K^2, \ldots) \in \mathrm{Sp}^{\Sigma}$, where $\mathbb{1} := \mathbb{S}$.
3. So, a bispectrum $\mathcal{X} = \{\mathcal{X}_n : n \geq 0\} \in \mathrm{BiSp}$ is a left $\mathbb{S}_K$-module, and $(\mathrm{BiSp}, \wedge_{\mathbb{S}_K}, \mathbb{S}_K)$ is a symmetric monoidal model category (10.4.10, 10.4.11).
4. Two functors (10.34)

$$\begin{cases} \tilde{\mathbb{S}}^1_{\mathrm{Sp}} \wedge_{\mathbb{S}} - : \mathrm{Sp} \longrightarrow \mathrm{Sp} & X \mapsto \tilde{\mathbb{S}}^1_{\mathrm{Sp}} \wedge_{\mathbb{S}} X \\ F_0 : \mathrm{Sp} \longrightarrow \mathrm{BiSp} & F_0(M) = (M, M \wedge_{\mathbb{S}} \tilde{\mathbb{S}}^1_{\mathrm{Sp}}, M \wedge_{\mathbb{S}} \tilde{\mathbb{S}}^2_{\mathrm{Sp}}, \ldots) =: \mathbb{S}(M) \wedge G_0(K) \end{cases}$$

   are Quillen equivalences, where $\tilde{\mathbb{S}}^n_{\mathrm{Sp}}$ is the n-fold product. (See [H2, 9.1].)
5. Let $\mathcal{X} = (\mathcal{X}_0, \mathcal{X}_1, \mathcal{X}_2, \ldots) \in \mathrm{BiSp}$ be a bispectrum. Assume that $\mathcal{X}_n \in \mathrm{Sp}$ are positive Ω-spectrum $\forall n \geq 0$. Under further suitable conditions [Sm3, Lemma B.16], two homotopy groups (10.4.11) are isomorphic:

$$\pi_n^t(\mathcal{X}) \cong \pi_n(\mathcal{X}) = \lim_k [K^{n+k}, X_k]_{Ho\mathrm{Sp}} \qquad \forall n \in \mathbb{Z}. \tag{10.36}$$

Remark 10.4.14. It is worth drawing the attention to (10.3.13) that, for a dg category $\mathscr{A}$, the GW-spectrum $GW(\mathscr{A})$ is a positive Ω-spectrum. Since we would only be considering the bispectrum of GW–spectra, the limit formula (10.36) suffices to compute the nonconnective Grothendieck–Witt groups, subsequently. This remains within the category Sp, while we need a cofibrant replacement K of $\mathbb{S}^1_{\mathrm{Sp}}$.

10.4.3. *The Karoubi $\mathbb{G}W$-spectra*

In Section 9.4.2, we defined the suspension $\mathbb{S}\mathscr{E}$ of an exact category. Likewise, we define the suspension of a dg category $\mathscr{A}$, while we use a different construction.

Construction 10.4.15. Let $\mathbf{C} := \mathbb{M}_\infty(\mathbb{Z})$ denote the ring of all $\mathbb{N} \times \mathbb{N}$-matrices $(a_{ij})_{i,j\in\mathbb{N}}$, with $a_{ij} \in \mathbb{Z}$ such that, in each row and column, only finitely many $a_{ij} \neq 0$. This ring is sometimes called the cone ring. There is a natural map $\mathbb{Z} \hookrightarrow \mathbf{C}$, sending $n \in \mathbb{Z}$ to the diagonal matrix. Since $\mathbf{C}$ is torsion free, it is flat over $\mathbb{Z}$.

Let $\mathscr{I}(\mathbb{Z}) \subseteq \mathbf{C}$ be the two-sided ideal of matrices with only finitely many nonzero entries. Define the **suspension ring**, $\mathfrak{s} := \mathfrak{s}(\mathbb{Z}) = \frac{\mathbf{C}}{\mathscr{I}(\mathbb{Z})}$. The transposition makes both $\mathbf{C}$ and $\mathfrak{s}$ into rings with involution. Let $\varepsilon = (a_{ij}) \in \mathbf{C}$ be the matrix with $a_{00} = 1$ and zero elsewhere. Let $\tilde{\mathbf{C}}$ denote the full subcategory of left $\mathbf{C}$-modules consisting of $\mathbf{C}, \mathbf{C}\varepsilon$. Note that $\varepsilon = \varepsilon^2$ and hence $\mathbf{C}\varepsilon$ is also projective. It follows:

$$\begin{cases} Hom(\mathbf{C}, \mathbf{C}) = \mathbf{C} \\ Hom(\mathbf{C}, \mathbf{C}\varepsilon) = \{a \in \mathbf{C} : \varepsilon a = a\} \\ Hom(\mathbf{C}\varepsilon, \mathbf{C}) = \{a \in \mathbf{C} : a\varepsilon = a\} \\ Hom(\mathbf{C}\varepsilon, \mathbf{C}\varepsilon) = \{a \in \mathbf{C} : a\varepsilon = \varepsilon a\} = \mathbb{Z}. \end{cases}$$

In fact, $\tilde{\mathbf{C}}$ is a category with duality, given by identity on the objects and transposition on the maps. Further, $\tilde{\mathbf{C}}$ is considered a dg category (10.2.3). Further, $\mathbb{Z}$ is treated as one-object dg category and $\mathfrak{s}$ is treated as a dg category with two objects $\{0, \mathfrak{s}\}$. We have a duality-preserving sequence of dg functors

$$\mathbb{Z} \xrightarrow{\iota} \tilde{\mathbf{C}} \xrightarrow{p} \mathfrak{s} \qquad \text{where} \quad \begin{cases} \iota(\star) = \mathbf{C}\varepsilon, \ \iota(n) = n\varepsilon \\ p \text{ sends } \ \mathbf{C} \mapsto \mathfrak{s}, \varepsilon \mapsto 0. \end{cases} \tag{10.37}$$

The composition $p\mathfrak{s}$ is the trivial functor. Given a dg category $\mathscr{A}$, with duality and weak equivalences, write $\tilde{\mathbf{C}}\mathscr{A} := \tilde{\mathbf{C}} \otimes_{\mathbb{Z}} \mathscr{A}$, and

$\mathfrak{s}\mathscr{A} := \mathfrak{s} \otimes_{\mathbb{Z}} \mathscr{A}$. By the tensor product with (10.37), we obtain a sequence of pointed dg categories

$$\mathscr{A} \xrightarrow{\iota_{\mathscr{A}}} \tilde{\mathbf{C}}\mathscr{A} \xrightarrow{p_{\mathscr{A}}} \mathfrak{s}\mathscr{A}\,, \quad \text{with trivial composition.} \tag{10.38}$$

Part of the following arguments go, mutatis mutandis [Sm5, §2.2], as in Section 9.4.2.

We define Morita equivalence and Morita exact sequences.

Definition 10.4.16. A dg functor $\iota : (\mathscr{A}, \mathbf{w_A}) \longrightarrow (\mathscr{B}, \mathbf{w_B})$ with weak equivalences is said to be a **Morita equivalence** if the induced functor $\mathfrak{T}(\mathscr{A}, \mathbf{w_A}) \longrightarrow \mathfrak{T}(\mathscr{B}, \mathbf{w_B})$ of the associated triangulated categories is cofinal (9.4.1).

Let $(\mathscr{A}, \mathbf{w_A}) \longrightarrow (\mathscr{B}, \mathbf{w_B}) \longrightarrow (\mathscr{C}, \mathbf{w_C})$ be a sequence of dg functors with weak equivalence. Such a sequence is said to be **Morita exact** if the sequence of the associated triangulated categories, $\mathfrak{T}(\mathscr{A}, \mathbf{w_A}) \longrightarrow \mathfrak{T}(\mathscr{B}, \mathbf{w_B}) \longrightarrow \mathfrak{T}(\mathscr{C}, \mathbf{w_C})$, is exact up to factor (11.2.15).

Lemma 10.4.17. *Let $dg\mathbf{Cat}W_k$ denote the category of small dg categories with weak equivalences. The functors $\mathbf{C}, \mathfrak{s} : dg\mathbf{Cat}W_k \longrightarrow dg\mathbf{Cat}W_k$ preserve Morita equivalences and Morita exact sequences. Further, for $\mathscr{A} \in dg\mathbf{Cat}W_k$,*

1. *The sequence (10.38)*

$$\mathscr{A} \xrightarrow{\iota_{\mathscr{A}}} \tilde{\mathbf{C}}\mathscr{A} \xrightarrow{p_{\mathscr{A}}} \mathfrak{s}\mathscr{A} \qquad \textit{is Morita exact.}$$

2. *Assume that $1/2 \in k$. Then the spectrum $GW(\tilde{\mathbf{C}}\mathscr{A})$ is contractible.*

Proof. See [Sm3, Lem. 8.2, 8.4]. ■

With the description (10.36) of the homotopy sets in BiSp in mind, we wish to develop a description of the cofibrant replacement $K = \tilde{\mathbb{S}}^1_{\mathrm{Sp}}$ of $\mathbb{S}^1_{\mathrm{Sp}}$ in the category Sp. We use the notation $X_{\mathrm{Sp}} := F_0(X)$ to denote the image of $X \in \mathbf{Top_\bullet}$ in the category Sp. Let $\tilde{\mathbb{S}}^0_{\mathrm{Sp}}$ be a cofibrant replacement of $\mathbb{S}^0_{\mathrm{Sp}} := (\mathbb{S}^0, pt, pt, \ldots) \in \mathrm{Sp}$. We consider the diagram

$$\begin{array}{ccccccc} \star & \longrightarrow & \tilde{\mathbb{S}}^0_{\mathrm{Sp}} & \xrightarrow{c_I} & \tilde{I}_{\mathrm{Sp}} & \xrightarrow{p} & \tilde{\mathbb{S}}^1_{\mathrm{Sp}} \\ & \searrow & \downarrow \wr & & \wr \downarrow w_I & & \downarrow w \\ & & \mathbb{S}^0_{\mathrm{Sp}} & \longrightarrow & I_{\mathrm{Sp}} & \xrightarrow[q]{} & \mathbb{S}^1_{\mathrm{Sp}} \end{array} \qquad \text{Here} \quad \begin{cases} \mathbb{S}^1_{\mathrm{Sp}} = \frac{I_{\mathrm{Sp}}}{\mathbb{S}^0_{\mathrm{Sp}}}, & \text{and,} \\ K = \tilde{\mathbb{S}}^1_{\mathrm{Sp}} := \frac{\tilde{I}_{\mathrm{Sp}}}{\tilde{\mathbb{S}}^0_{\mathrm{Sp}}} \end{cases}$$

Here, p, q are quotient maps, and $w_I c_I$ is a decomposition of the diagonal map, with c_I a cofibration and w_I a trivial fibration. One checks that K is cofibrant, and $w : \tilde{\mathbb{S}}^1_{\mathrm{Sp}} \longrightarrow \mathbb{S}^1_{\mathrm{Sp}}$ is a trivial fibration. So, $\tilde{\mathbb{S}}^1_{\mathrm{Sp}}$ is a cofibbrant replacement of $\mathbb{S}^1_{\mathrm{Sp}}$.

Since $GW(\tilde{\mathbf{C}}(k))$ is contractible, the map $GW(\tilde{\mathbf{C}}(k)) \longrightarrow pt$ is a trivial fibration. Since c_I is a cofibration, by its lifting property, we obtain the map ρ, and the extended commutative diagram:

$$\begin{array}{ccccc} \tilde{\mathbb{S}}^0_{\mathrm{Sp}} & \longrightarrow & \mathbb{S}^0_{\mathrm{Sp}} \xrightarrow{\langle 1 \rangle} & & GW(k) \\ c_I \downarrow & & & & \downarrow \iota_k \\ \tilde{I}_{\mathrm{Sp}} & & \overset{\rho}{\dashrightarrow} & & GW(\tilde{\mathbf{C}}(k)) \\ \downarrow & & & & \downarrow p_k \\ \tilde{\mathbb{S}}^1_{\mathrm{Sp}} & & \overset{\sigma}{\dashrightarrow} & & GW(\mathfrak{s}(k)) \end{array} \tag{10.39}$$

Here, the map ρ of spectra factors through σ because $p_k \iota_k$ is trivial.

Lemma 10.4.18. *Let $\mathscr{A}$ be a dg k-category with weak equivalence and duality. Assume that $1/2 \in k$. Further, assume that the associated triangulated category $\mathfrak{T}\mathscr{A}$ is idempotent complete. Then the composition*

$$\begin{array}{ccc} \tilde{\mathbb{S}}^1_{\mathrm{Sp}} \wedge_{\mathbb{S}} GW(\mathscr{A}) & \xrightarrow{\sigma \wedge 1} & GW(\mathfrak{s}(k)) \wedge_{\mathbb{S}} GW(\mathscr{A}) \\ & \searrow & \downarrow \cup \\ & & GW(\mathfrak{s}\mathscr{A}) \end{array}$$

is a weak equivalence (i.e. stable equivalence).

Proof. See [Sm3, Lem. 8.5]. ∎

We define Karoubi $\mathbb{G}W$-spectra, from [Sm3, §8.2].

Definition 10.4.19. Let k be a commutative ring with $1/2 \in k$, and $\mathscr{A} \in dg\mathbf{Cat}WD_k$ be a dg category with weak equivalence and duality. As usual, $\mathfrak{s}^n\mathscr{A} := \mathfrak{s} \otimes \cdots \mathfrak{s} \otimes \mathscr{A}$ denotes the n-fold tensor product of $\mathfrak{s}$, followed by $\otimes\mathscr{A}$. Then Σ_n acts on $\mathfrak{s}^n\mathscr{A}$. Define

$$\mathbb{G}W(\mathscr{A})_n := GW(\mathfrak{s}^n\mathscr{A}) \qquad \text{and,} \quad \mathbb{G}W(\mathscr{A}) := \{\mathbb{G}W(\mathscr{A})_n : n \geq 0\} \in \mathrm{Sp}^{\Sigma}.$$

So, $\mathbb{G}W(\mathscr{A})$ is a symmetric sequence in Sp.

Lemma 10.4.20. *Use the setup of* (10.4.19). *Note that* $\mathbb{G}W(\mathscr{A})_n$ *are a positive* Ω*-spectrum* (10.3.13) $\forall n \geq 0$. *Then the symmetric sequence*

$$\mathbb{G}W(\mathscr{A}) \in \mathrm{BiSp}(\mathbb{S}^1 – \tilde{\mathbb{S}}^1_{\mathrm{Sp}})$$

is a bi $\mathbb{S}^1 – \tilde{\mathbb{S}}^1_{\mathrm{Sp}}$*-spectrum. Consequently, we have the following:*

1. *The association* $\mathscr{A} \mapsto \mathbb{G}W(\mathscr{A})$ *defines a functor*

   $$dg\mathbf{Cat}WD_k \longrightarrow \mathrm{BiSp}(\mathbb{S}^1 – \tilde{\mathbb{S}}^1_{\mathrm{Sp}}) \quad \textit{of the symmetric monoidal categories.}$$

2. *The bispectrum* $\mathbb{G}W(\mathscr{A})$ *is semisimple, in the sense of* [Sm3, §B.11]. *Consequently, the limit formula* (10.36) *can be used to compute the true homotopy groups* $\pi_r^t(\mathbb{G}W(\mathscr{A}))$.

The bispectrum $\mathbb{G}W(\mathscr{A})$ *is referred to as the* ***Kaurobi*** $\mathbb{G}W$***-spectrum.***

Proof. With $K = \tilde{\mathbb{S}}^1_{\mathrm{Sp}}$ and $\mathbb{S}_K := \mathbb{S}ym(K) := (\mathbb{1}, K, K^2, \ldots) \in \mathrm{Sp}^{\Sigma}$, we need to give a $\mathbb{S}_K$-module structure on $\mathbb{G}W(\mathscr{A})$. By repeated application of (10.4.18) for the dg category $\mathscr{C}_k$, there is a map

$$\mathbb{S}ym(K) \longrightarrow \mathbb{G}W(k) := \{\mathbb{G}W\mathfrak{s}^n\mathscr{C}_k : n \geq 0\}$$

For $\mathscr{A} \in dg\mathbf{Cat}WD_k$, the symmetric sequence $\mathbb{G}W(\mathscr{A})$ has a natural structure of a $\mathbb{G}W(k)$-module. So, $\mathbb{G}W(\mathscr{A})$ is a $\mathbb{S}ym(K)$-module.

This establishes the main assertion. The assertion (10.4.20) is obvious. For the proof of (2), we refer to [Sm3, Lem. B.16], [Sm3, Prop. 8.7]. The proof is complete. ■

Definition 10.4.21. Let k be a commutative ring with $1/2 \in k$, and $\mathscr{A} \in dg\mathbf{Cat}WD_k$. For $r \in \mathbb{Z}$, define **Karoubi** $\mathbb{G}W$-groups

$$\mathbb{G}\mathfrak{W}_r(\mathscr{A}) := \pi_r^t\left(\mathbb{G}W(\mathscr{A})\right) = \lim_p [\tilde{\mathbb{S}}_{\mathrm{Sp}}^{p+r}, \mathbb{G}W(\mathscr{A})_p]_{Ho\mathrm{Sp}}.$$

Further, denote

$$\begin{cases} \mathbb{G}W^{[n]}\mathscr{A} := \mathbb{G}W\left(\mathscr{A}^{[n]}\right) \\ \mathbb{G}\mathfrak{W}_r^{[n]}(\mathscr{A}) := \mathbb{G}\mathfrak{W}_r\left(\mathscr{A}^{[n]}\right). \end{cases}$$

Now we formulate some results.

Proposition 10.4.22. *Let k be a commutative ring with $1/2 \in k$, and $\mathscr{A} \in dg\mathbf{Cat}WD_k$. Then*

$$\mathbb{G}\mathfrak{W}_p^{[n]}(\mathscr{A}) = \begin{cases} GW_p^{[n]}(\mathscr{A}) & if\ p \geq 1 \\ GW_0^{[n]}(\widetilde{\mathfrak{T}\mathscr{A}}) & if\ p \geq 0 \\ GW_0^{[n]}(\widetilde{\mathfrak{T}(\mathfrak{s}^{-p}\mathscr{A})}) & if\ p \leq -1 \end{cases} \qquad \forall n, p \in \mathbb{Z}.$$

Here $\mathfrak{T}\mathscr{A}$ denotes the associated triangulated category of $\mathscr{A}$, and $\widetilde{\mathfrak{T}\mathscr{A}}$ is the idempotent completion. For $p \geq 0$, the groups $GW_p^{[n]}(\mathscr{A})$ are as defined in (10.3.16)*. The groups GW_0 of triangulated categories are defined as in* (11.1.6)*.*

Proof. See [Sm3, Prop. 8.7]. ■

Theorem 10.4.23 (Invariance). *Let k be a commutative ring with $1/2 \in k$, and let $F : \mathscr{A} \longrightarrow \mathscr{B}$ be an exact dg functor $dg\mathbf{Cat}WD_k$. Assume that F is a Morita equivalence. Then the induced map*

$$\mathbb{G}W(\mathscr{A}) \xrightarrow{\sim} \mathbb{G}W(\mathscr{B}) \quad \textit{of the Karoubi } \mathbb{G}W \textit{ spectra}$$

is an equivalence in BiSp*. Consequently, the induced group homomorphism*

$$\mathbb{G}\mathfrak{W}_p(\mathscr{A}) \xrightarrow{\sim} \mathbb{G}\mathfrak{W}_p(\mathscr{B}) \quad \textit{are isomorphisms } \forall p \in \mathbb{Z}.$$

Proof. See [Sm3, Thm. 8.9]. ■

Theorem 10.4.24 (Localization theorem). *Let k be a commutative ring with $1/2 \in k$. Let $\mathscr{A} \longrightarrow \mathscr{B} \longrightarrow \mathscr{C}$ be a Morita exact sequence in $dg\mathbf{Cat}WD_k$. Then $\forall n \in \mathbb{Z}$, the induced sequence*

$$\mathbb{G}W^{[n]}(\mathscr{A}) \longrightarrow \mathbb{G}W^{[n]}(\mathscr{B}) \longrightarrow \mathbb{G}W^{[n]}(\mathscr{C}) \quad in \quad BiSp$$

is a homotopy fibration.

Proof. See [Sm3, Prop. 8.10]. ■

10.4.4. *Karoubi $\mathbb{G}W$ theory for quasi-projective schemes*

We apply the above on $\mathbb{G}W$ theory to the dg category $dg\mathscr{P}(X)$ of bounded chain complexes of locally free sheaves on quasi-projective schemes X over affine schemes. As is pointed out in [Sm3, §9], for Noetherian schemes X, applications of the above to the dg categories of prefect complexes $\mathbf{Perf}(X)$ are possible but avoidable. We will be reusing the notation we developed above, from Sections 9.5 and 9.5.1, Definition 9.5.1, Notation 9.5.4, Example 10.1.15, Example 10.2.9, Example 10.2.16, and others.

Definition 10.4.25. Let X be a Noetherian scheme, over a Noetherian commutative ring A, with $1/2 \in A$. Fix an invertible sheaf $\mathcal{L}$ on X. The duality $P \mapsto P^* := \mathcal{H}om(P, \mathcal{L})$ on $\mathscr{P}(X)$ induces a duality on $dg\mathscr{P}(X)$, as in (10.2.16). For $E_\bullet \in Ch^b(\mathscr{P}(X))$, denote $E_\bullet^\vee := E_\bullet^{\vee_\mathcal{L}}$ its dual. Denote the doubledual identification by $\varpi := \varpi^\mathcal{L} : E_\bullet \xrightarrow{\sim} E_\bullet^{\vee\vee}$. We will often drop the superscript $\mathcal{L}$ when there is no risk of confusion. So, we use two notation

$$dg\mathscr{P}(X) := dg(\mathscr{P}(X), \mathcal{L}) := \left(Ch^b(\mathscr{P}(X), \vee, \varpi, qIso\right),$$

as a dg category with weak equivalence and duality. For $n, p \in \mathbb{Z}$, denote,

$$\begin{cases} GW^{[n]}(X, \mathcal{L}) := GW^{[n]}(dg(\mathscr{P}(X))) \\ \mathbb{G}W^{[n]}(X, \mathcal{L}) := \mathbb{G}W^{[n]}(dg(\mathscr{P}(X))) \end{cases} \begin{cases} GW_p^{[n]}(X, \mathcal{L}) := GW_p^{[n]}(dg(\mathscr{P}(X))) \\ \mathbb{G}\mathfrak{W}_p^{[n]}(X, \mathcal{L}) := \mathbb{G}\mathfrak{W}_p^{[n]}(dg(\mathscr{P}(X))) \end{cases} \tag{10.40}$$

When $n = 0$, we drop the superscript $[n]$, and when $\mathcal{L} = \mathcal{O}_X$ or $\mathcal{L}$ is understood, we drop the second coordinate $\mathcal{L}$. It also follows that

$$(dg(\mathscr{P}(X), \mathcal{L}))^{[n]} = \mathscr{C}_k^{[n]} \otimes dg(\mathscr{P}(X), \mathcal{L}) \cong ((dg(\mathscr{P}(X), \mathcal{L}))^n \qquad \text{are equivalent,}$$

where the right-hand side is defined, exactly the way category $\mathscr{C}_k^{[n]}$ was defined (10.3.5), by replacing the complex $\kappa[n]$ by the complex $\kappa[n, \mathcal{L}] \in Ch^b(\mathscr{P}(X)$, where $\kappa[n, \mathcal{L}]_n = \mathcal{L}$ and $\kappa[n, \mathcal{L}]_m = 0\ \forall m \neq n$. Subsequently, we look at these categories in this latter old-fashioned way. So, for a fixed integer n_0, an object $E_\bullet$ and its dual $E_\bullet^{\vee_\mathcal{L}} \in (dg(\mathscr{P}(X), \mathcal{L}))^{n_0}$ look like the following:

$$E_\bullet := \cdots \longrightarrow M_{n_0} \longrightarrow M_{n_0-1} \longrightarrow \cdots \longrightarrow M_0 \longrightarrow M_{-1} \longrightarrow \cdots$$

$$E_\bullet^{\vee_\mathcal{L}} := \cdots \longrightarrow M_0^* \longrightarrow M_1^* \longrightarrow \cdots \longrightarrow M_{n_0}^* \longrightarrow M_{n_0+1}^* \longrightarrow \cdots$$

There are issues regarding signs of the differentials in $E_\bullet^{\vee_\mathcal{L}}$, exactly similar to (10.3.5). The case $n_0 = 0$ corresponds to (10.1.15).

Theorem 10.4.26. *Let X be a Noetherian scheme over a commutative ring k, with $1/2 \in k$. Fix an invertible sheaf $\mathcal{L}$ on X. Fix $n \in \mathbb{Z}$. Then $\forall p \geq 0$, we have*

$$\mathbb{G}\mathfrak{W}_p^{[n]}(X, \mathcal{L}) = GW_p^{[n]}(X, \mathcal{L}) = GW_p^{[n]}\left(Ch^b(X), \vee^{\mathcal{L}}, \varpi\right), \quad \textit{where}$$

the last term denotes the GW group of the exact categories with quasi-isomorphisms and n-shifted duality (10.1.18).

Proof. Note that the associated triangulated categories $\mathfrak{T}dg\ Ch^b(X) \cong \mathbf{D}^b(\mathscr{P}(X))$ are idempotent complete. So, the first equality follows from (10.4.22). The proof is complete. ■

The nonconnective version of the localization theorem [Sm3, Thm. 9.5] is as follows.

Theorem 10.4.27 (Localization theorem). *Let X be a quasi-projective scheme over a Noetherian commutative ring A, with $1/2 \in A$. Fix an invertible sheaf $\mathcal{L}$ on X. Let $Z \subseteq X$ be a closed subset and $U = X - Z$. Let $dg_Z(\mathscr{P}(X), \mathcal{L}) \subseteq dg(\mathscr{P}(X), \mathcal{L})$ denote the dg* *subcategory of objects $E_\bullet \in dg(\mathscr{P}(X))$ such that $supp(\mathcal{H}_n(E_\bullet)) \subseteq Z$ (meaning $(E_\bullet)_x$ is acyclic $\forall x \in U$). For $n \in \mathbb{Z}$, denote*

$$\mathbb{G}W^{[n]}(X, Z, \mathcal{L}) = \mathbb{G}W^{[n]}(dg_Z(\mathscr{P}(X), \mathcal{L})).$$

Then the sequence

$$\mathbb{G}W^{[n]}(X, Z, \mathcal{L}) \longrightarrow \mathbb{G}W^{[n]}(X, \mathcal{L}) \longrightarrow \mathbb{G}W^{[n]}\left(U, \mathcal{L}_{|U}\right)$$

is a homotopy fibration.

Proof. Consider the sequence

$$dg_Z(\mathscr{P}(X), \mathcal{L}) \longrightarrow dg(\mathscr{P}(X), \mathcal{L}) \longrightarrow dg(\mathscr{P}(U), \mathcal{L}_{|U}).$$

The induced sequence of triangulated categories,

$$\mathbf{D}^b_Z(X) \longrightarrow \mathbf{D}^b(X) \longrightarrow \mathbf{D}^b(U) \tag{10.41}$$

is exact up to factor, by the proof of [Sm1, Thm. 3.4.9]. So, the proof is complete by (10.4.24). ■

10.4.5. *Gersten complex for the Karoubi $\mathbb{G}\mathfrak{W}$-groups*

Now proceed to develop a Gersten complex for Karoubi $\mathbb{G}\mathfrak{W}$-groups, similar to (9.35), for quasi-projective schemes. We reuse the notation from Sections 9.5 and 9.5.1, in particular from (9.5.4), while we update some of the notation.

Notation 10.4.28. We continue with the notation in (9.5.4), (10.4.25). Sometimes we use the prefix *dg* to denote the corresponding dg category. Fix an integer $k \geq 0$. Note that there is no natural duality on $\mathbb{M}^k(X)$. However, there is a duality on $C\mathbb{M}^k(X)$, given by

$$\mathscr{F} \mapsto \mathscr{F}^* = \mathcal{E}xt^k(\mathscr{F}, \mathcal{L}).$$

This induces a duality on $dgC\mathbb{M}^k(X)$ and on $\mathbf{D}^b(C\mathbb{M}^k(X))$. We denote both of them by $(\vee, \varpi)$. For the rest of this chapter, we will be working with such a fixed locally free sheaf $\mathcal{L}$ and dualities induced by it. For the simplicity of notation, we will often ignore mentioning $\mathcal{L}$. We simply denote this dg category of bounded complexes of objects in $C\mathbb{M}^k(X)$ by $dgC\mathbb{M}^k(X)$ as a dg category with weak equivalences and duality.

Proposition 10.4.29. *Let X be a Noetherian quasi-projective scheme over an affine scheme* Spec (A)*, and let $k \geq 0$ be an integer. Then there is a duality-preserving equivalence*

$$\mathbf{D}^b\left(C\mathbb{M}^k(X)\right) \longrightarrow T^k\mathscr{D}^k\left(\mathscr{P}(X)\right)$$

of the triangulated categories, where T denotes the shift, and the duality on $T^k\mathscr{D}^k\left(\mathscr{V}(X)\right)$ is $\# := T^k\mathcal{H}om\left(-, \mathcal{L}\right)$ (see 11.2.7*).*

Proof. It is a standard fact that there is a functor $\mathbb{M}(X) \longrightarrow \mathbf{D}^b\left(\mathscr{P}(X)\right)$, by resolution (e.g. see [Kb1, Thm. 12.1], [M3, 3.3]). The restriction of this functor to $C\mathbb{M}^k(X)$ extends to a functor $\mathbf{D}^b\left(C\mathbb{M}^k(X)\right) \to \mathscr{D}^k\left(\mathscr{P}(X)\right)$. It turns out that this functor represents the composite functor in (9.32). Hence, the functor is an equivalence. Now, routine checking establishes that this functor preserves the duality, as required. The proof is complete. ■

While we obtained an equivalence of derived categories in (10.4.29), there is no natural dg functor from $dg(C\mathbb{M}^k(X))$ to $dg(\mathscr{P}(X))$. This necessitates us to zigzag via the category of perfect complexes. We introduce the following further notation.

Notation 10.4.30. Let X denote a Noetherian scheme.

1. In analogy to previous notation, for integers $k \geq 0$, we denote

$$\begin{cases} \mathbf{P}erf(X) = \text{Category of Perfect complexes of } \mathcal{O}_{\mathrm{X}} - \text{modules} \\ \mathbf{D}(\mathbf{P}erf(X)) = \text{Derived category of } \mathbf{P}erf(X) \\ \mathbf{P}erf^k(X) := \left\{\mathcal{F}_\bullet \in \mathbf{P}erf(X) : \forall\, i\ \mathcal{H}_i\left(\mathcal{F}_\bullet\right) \in Coh^k_g(X)\right\} \\ \mathscr{D}^k\mathbf{P}erf(X) := \left\{\mathcal{F}_\bullet \in \mathcal{D}^b(Perf(X)) : \forall\, i\ \mathcal{H}_i\left(\mathcal{F}_\bullet\right) \in Coh^k_g(X)\right\}. \end{cases}$$

   To avoid confusion, we will use the prefix dg to denote the respective dg categories. So, $dg\mathbf{P}erf(X)$ will denote the dg category whose objects are the same as those of $\mathbf{P}erf(X)$.
2. In fact, $\mathbf{P}erf(X)$ has a duality. Fix a minimal injective resolution $I_\bullet$ of $\mathcal{L}$, as follows:

$$0 \longrightarrow \mathcal{L} \longrightarrow I_0 \longrightarrow I_{-1} \longrightarrow I_{-2} \longrightarrow \cdots. \qquad \text{Clearly, } I_\bullet \in \mathbf{P}erf(X).$$

   For $\mathcal{F}_\bullet \in \mathbf{P}erf(X)$, denote $\mathcal{F}^\vee := \mathcal{H}om(\mathcal{F}_\bullet, I_\bullet)$.

The following proposition on derived equivalences is a reformulation of results in [TT].

Proposition 10.4.31. *Let X be a quasi-projective scheme over a Noetherian affine scheme* Spec(A). *Use all the notation as in* (10.4.30). *Then the association $\mathcal{F}_\bullet \mapsto \mathcal{F}_\bullet^\vee$, for $\mathcal{F} \in \mathbf{P}erf(X)$ provides a duality structure on the dg category $dg\mathbf{P}erf(X)$. So, $dg\mathbf{P}erf(X) := (\mathbf{P}erf(X), \vee, \varpi, qIso)$ is a dg category with weak equivalence and duality. Further, for integers $k \geq 0$, there is a natural equivalence*

$$\mathscr{D}^k(\mathscr{P}(X)) \to \mathscr{D}^k\mathbf{P}erf(X) \qquad \textit{of derived categories.}$$

Proof. See [TT, Lem. 3.8], [Sm1, Prop. 3.4.8]. ∎

We cast the above within dg category structures.

Proposition 10.4.32. *Let X be a quasi-projective scheme over a Noetherian affine scheme* Spec(A). *Use all the notation as in* (10.4.28, 10.4.30). *Let $k \geq 0$ be an integer. We consider two functors of dg categories*

$$\begin{cases} dgC\mathbb{M}^k(X) \longrightarrow T^k\left(dg\mathbf{P}erf^k(X)\right) \\ dg^k\mathscr{P}(X) \hookrightarrow dg\mathbf{P}erf^k(X). \end{cases}$$

Then both are functors of dg categories with weak equivalences and duality. Further, the induced functors of the associated triangulated categories are equivalences.

Proof. Immediate from (10.4.29) and (10.4.31), respectively. ∎

The following useful diagram is analogous to the diagram (9.32) in the context of dg categories with weak equivalences and dualities.

Corollary 10.4.33. *Let X be a quasi-projective scheme over a Noetherian affine scheme* Spec(A). *Use all the notation as in* (10.4.28, 10.4.30). *Let $k \geq 0$ be an integer. Consider the diagram*

$$\begin{array}{ccccc} T^{-1}dgC\mathbb{M}^{k+1}(X) & \longrightarrow & T^k dg\mathbf{P}erf^{k+1}(X) & \longleftarrow & T^k dg^{k+1}\mathscr{P}(X) \\ & & \downarrow & & \downarrow \\ dgC\mathbb{M}^k(X) & \longrightarrow & T^k dg\mathbf{P}erf^k(X) & \longleftarrow & T^k dg^k\mathscr{P}(X). \end{array} \tag{10.42}$$

In this diagram, all the arrows are functors of dg categories with weak equivalences and dualities. Further, the horizontal arrows induce equivalences of the associated triangulated categories, and the right-hand square commutes. Note that there is no natural vertical functor on the left-hand side.

Proof. Follows from Proposition 10.4.32. ∎

Theorem 10.4.34. *Suppose that X is a quasi-projective scheme over an affine scheme* $\mathrm{Spec}\,(A)$, *with* $1/2 \in A$ *and* $k \geq 0, r$ *are integers. In the following, weak equivalences and dualities in the respective categories will be as in* (10.4.32). *Then the maps in the following zigzag sequences:*

$$GW^{[r]}\,(dgC\mathbb{M}^k(X)) \xrightarrow{\zeta} GW^{[k+r]}\,(dgPerf^k(X)) \xleftarrow{\Phi} GW^{[k+r]}\,(dg^k\mathscr{P}(X)) \qquad \text{in Sp}$$

$$\mathbb{G}W^{[r]}\,(dgC\mathbb{M}^k(X)) \xrightarrow[\zeta]{} \mathbb{G}W^{[k+r]}\,(dgPerf^k(X)) \xleftarrow[\Phi]{} \mathbb{G}W^{[k+r]}\,(dg^k(\mathscr{P}(X)) \qquad \text{in BiSp}$$

*are stable homotopy equivalences in the respective symmetric monoidal model categories (*Sp *and* BiSp*). In particular, they induce isomorphisms of GW-groups and $\mathbb{G}\mathfrak{W}$-groups, respectively.*

Proof. Follows from Proposition 10.4.32 and the Invariance theorems (10.3.17, 10.4.23). ∎

The Main Theorem on Karoubi $\mathbb{G}W$-spectra of quasi-projective schemes is as follows.

Theorem 10.4.35. *Suppose that X is a quasi-projective scheme over an affine scheme* $\mathrm{Spec}\,(A)$, *with* $1/2 \in A$, *and* $k \geq 0, n$ *are integers. Assume further that X is a Cohen–Macaulay scheme. Consider the following diagram of $\mathbb{G}W$-bi spectra:*

$$\begin{array}{ccccc}
\mathbb{G}W^{[n-(k+1)]}\,(dgC\mathbb{M}^{k+1}(X)) & & \mathbb{G}W^{[n-k]}\,(dgC\mathbb{M}^k(X)) & & \coprod_{x\in X^{(k)}} \mathbb{G}W^{[n-k]}\,(dgC\mathbb{M}^k\,(X_x)) \\
\downarrow & & \downarrow & & \downarrow \\
\mathbb{G}W^{[n]}\,(dg\mathbf{P}erf^{k+1}(X)) & & \mathbb{G}W^{[n]}\,(dg\mathbf{P}erf^k X)) & & \coprod_{x\in X^{(k)}} \mathbb{G}W^{[n]}\,(\mathbf{P}erf^k(X_x)) \\
\uparrow & & \uparrow & & \uparrow \\
\mathbb{G}W^{[n]}\,(dg^{k+1}\mathscr{P}(X)) & \longrightarrow & \mathbb{G}W^{[n]}\,(dg^k\mathscr{P}(X)) & \longrightarrow & \coprod_{x\in X^{(k)}} \mathbb{G}W^{[n]}\,(dg^k\mathscr{P}(X_x))
\end{array}$$

In this diagram, all the vertical arrows are equivalences of homotopy bispectra and the bottom sequence is a homotopy fibration of bispectra. In particular, there is a sequence of zigzag maps of bispectra

$$\begin{aligned} &\mathbb{G}W^{[n-1]}\left(dgC\mathbb{M}^{k+1}(X)\right) \longrightarrow \mathbb{G}W^{[n]}\left(dgC\mathbb{M}^{k}(X)\right) \\ &\longrightarrow \textstyle\coprod_{x \in X^{(k)}} \mathbb{G}W^{[n]}\left(dgC\mathbb{M}^{k}\left(X_x\right)\right) \end{aligned} \qquad (10.43)$$

that is a homotopy fibration.

Proof. It follows directly from Theorem 10.4.34 that the vertical rows are equivalences. It remains to be shown that the bottom row is a fibration. This follows from the localization theorem (10.4.27) and Lemma 9.5.6. The proof is complete. ∎

We write down a long exact sequence of the $\mathbb{G}\mathfrak{W}$-groups, as follows.

Corollary 10.4.36. *With notation as above* (10.4.35) *and* $p \in \mathbb{Z}$, *there is an exact sequence of* $\mathbb{G}\mathfrak{W}$*-groups:*

$$\begin{array}{l} \mathbb{G}\mathfrak{W}_{p+1}^{[n-1]}\left(dg\left(C\mathbb{M}^{k+1}(X)\right)\right) \longrightarrow \mathbb{G}\mathfrak{W}_{p+1}^{[n]}\left(dg\left(C\mathbb{M}^{k}(X)\right)\right) \longrightarrow \coprod_{x \in X^{(k)}} \mathbb{G}\mathfrak{W}_{p+1}^{[n]}\left(dg\left(C\mathbb{M}^{k}\left(X_x\right)\right)\right) \\ \longrightarrow \mathbb{G}\mathfrak{W}_{p}^{[n-1]}\left(dg\left(C\mathbb{M}^{k+1}(X)\right)\right) \longrightarrow \mathbb{G}\mathfrak{W}_{p}^{[n]}\left(dg\left(C\mathbb{M}^{k}(X)\right)\right) \longrightarrow \coprod_{x \in X^{(k)}} \mathbb{G}\mathfrak{W}_{p}^{[n]}\left(dg\left(C\mathbb{M}^{k}\left(X_x\right)\right)\right) \\ \longrightarrow \mathbb{G}\mathfrak{W}_{p-1}^{[n-1]}\left(dg\left(C\mathbb{M}^{k+1}(X)\right)\right) \longrightarrow \mathbb{G}\mathfrak{W}_{p-1}^{[n]}\left(dg\left(C\mathbb{M}^{k}(X)\right)\right) \longrightarrow \coprod_{x \in X^{(k)}} \mathbb{G}\mathfrak{W}_{p-1}^{[n]}\left(dg\left(C\mathbb{M}^{k}\left(X_x\right)\right)\right). \end{array}$$

A Gersten Complex

Let X be as in Theorem 10.4.35. For integer $k \geq 0$, we combine three of the exact sequences above (10.4.36)

$$\begin{array}{ccccc} \oplus_{x \in X^{(k-1)}} \mathbb{G}\mathfrak{W}_{p+1}^{[n+1]}\left(dgC\mathbb{M}^{k-1}(X_x)\right) & & & & \mathbb{G}\mathfrak{W}_{p-1}^{[n-2]}\left(dgC\mathbb{M}^{k+2}(X)\right) \\ \downarrow & \searrow & & & \downarrow \\ \mathbb{G}\mathfrak{W}_{p}^{[n]}\left(dgC\mathbb{M}^{k}(X)\right) & \longrightarrow & \oplus_{x \in X^{(k)}} \mathbb{G}\mathfrak{W}_{p}^{[n]}\left(dgC\mathbb{M}^{k}(X_x)\right) & \longrightarrow & \mathbb{G}\mathfrak{W}_{p-1}^{[n-1]}\left(dgC\mathbb{M}^{k+1}(X)\right) \\ \downarrow & & & \searrow & \downarrow \\ \mathbb{G}\mathfrak{W}_{p}^{[n+1]}\left(dgC\mathbb{M}^{k-1}(X)\right) & & & & \oplus_{x \in X^{(k+1)}} \mathbb{G}\mathfrak{W}_{p-1}^{[n-1]}\left(dgC\mathbb{M}^{k+1}(X_x)\right) \end{array}$$

With $d = \dim X$, for all $n \in \mathbb{Z}$, we obtain a Gersten complex of $\mathbb{GW}$-groups as follows:

$$
\begin{array}{c}
0 \\
\downarrow \\
\mathbb{GW}_p^{[n]}\,(X, \mathcal{L}) \\
\downarrow \\
\oplus_{x \in X^{(0)}} \mathbb{GW}_p^{[n]}\left(dgC\mathbb{M}^0(X_x)\right) \\
\downarrow \\
\oplus_{x \in X^{(1)}} \mathbb{GW}_{p-1}^{[n-1]}\left(dgC\mathbb{M}^1(X_x)\right) \\
\downarrow \\
\cdots \\
\downarrow \\
\oplus_{x \in X^{(k-1)}} \mathbb{GW}_{p-(k-1)}^{[n-(k-1)]}\left(dgC\mathbb{M}^{k-1}(X_x)\right) \\
\downarrow \\
\oplus_{x \in X^{(k)}} \mathbb{GW}_{p-k}^{[n-k]}\left(dgC\mathbb{M}^k(X_x)\right) \\
\downarrow \\
\oplus_{x \in X^{(k+1)}} \mathbb{GW}_{p-(k+1)}^{[n-(k+1)]}\left(dgC\mathbb{M}^{k+1}(X_x)\right) \\
\downarrow \\
\cdots \\
\downarrow \\
\oplus_{x \in X^{(d)}} \mathbb{GW}_{p-d}^{[n-d]}\left(dgC\mathbb{M}^d(X_x)\right) \\
\downarrow \\
0
\end{array}
\tag{10.44}
$$

A case of our particular interest is when $p = n = k$, as follows:

$$
\begin{array}{c}
0 \\
\downarrow \\
\mathbb{G}\mathfrak{W}_k^{[k]}(X, \mathcal{L})) \\
\downarrow \\
\oplus_{x \in X^{(0)}} \mathbb{G}\mathfrak{W}_k^{[k]}\left(dgC\mathbb{M}^0(X_x)\right) \\
\downarrow \\
\oplus_{x \in X^{(1)}} \mathbb{G}\mathfrak{W}_{k-1}^{[k-1]}\left(dgC\mathbb{M}^1(X_x)\right) \\
\downarrow \\
\cdots \\
\downarrow \\
\oplus_{x \in X^{(k-1)}} \mathbb{G}\mathfrak{W}_1^{[1]}\left(dgC\mathbb{M}^{k-1}(X_x)\right) \\
\downarrow \\
\oplus_{x \in X^{(k)}} \mathbb{G}\mathfrak{W}_0^{[0]}\left(dgC\mathbb{M}^k(X_x)\right) \\
\downarrow \\
\oplus_{x \in X^{(k+1)}} \mathbb{G}\mathfrak{W}_{-1}^{[-1]}\left(dgC\mathbb{M}^{k+1}(X_x)\right) \\
\downarrow \\
\cdots \\
\downarrow \\
\oplus_{x \in X^{(d)}} \mathbb{G}\mathfrak{W}_{k-d}^{[k-d]}\left(dgC\mathbb{M}^d(X_x)\right) \\
\downarrow \\
0
\end{array}
\tag{10.45}
$$

Remark 10.4.37. Assume that X is a Cohen–Macaulay scheme.

1. We point out that, for $x \in X^{(k)}$, the category $C\mathbb{M}^k(X_x)$ is the category of $\mathcal{O}_{X,x}$-modules of finite length and finite projective dimension. This category $C\mathbb{M}^k(X_x)$ is an exact category with duality. For exact categories with duality $\mathscr{E} := (\mathscr{E}, \vee, \varpi)$, the $\mathcal{G}W$-spectrum and $\mathbb{G}W$-bispectrum have not been discussed in the literature. It is desirable that the $\mathbb{G}W$-bispectrum of $C\mathbb{M}^k(X)$ be inserted in the literature and $\mathbb{G}W(dgC\mathbb{M}^k(X))$ be expressed in terms of $\mathbb{G}W$-bispectrum of $C\mathbb{M}^k(X)$. This requires further study.
2. There is no agreement theorem, similar to (10.3.15), for $\mathbb{G}W$-theory. However, by (10.3.8), these groups are 4-periodic. For $r = 0, 2$ and $p \geq 0$, (10.4.26) describes the groups $\mathbb{G}\mathfrak{W}_p^{[r]}(dgC\mathbb{M}^k(X))$, in terms of $GW_p^{[r]}\left(C\mathbb{M}^k(X), \mathcal{E}xt^k(-, \mathcal{L}), \pm\varpi\right)$.
3. Assume that X is quasi-projective and regular. Then

$$G\mathfrak{W}_p^{[n]}(X, \mathcal{L}) = GW_p^{[n]}(X, \mathcal{L}) \qquad n, p \in \mathbb{Z}.$$

Further, we obtain both the complexes (10.44, 10.45), with GW-groups.

Proof. See [Sm3, Prop. 9.3], for the first part. The latter proposition follows, because the corresponding sequence of the derived categories is exact. So, the localization theorem (10.3.19) applies.
4. Some of the above results could be stated for schemes X, with an ample family of line bundles, which is the standard hypothesis in [Sm3]. To keep things simple, we have decided to state all the results for quasi-projective schemes. We make some comments in further generality for regular schemes in what follows.

10.4.6. *Further generality for regular schemes*

For regular schemes, the localization theorem (10.3.19) applies as follows.

Theorem 10.4.38 (Localization theorem). *Let X be a Noetherian separated regular scheme over a commutative ring A, with $1/2 \in A$. Fix an invertible sheaf $\mathcal{L}$ on X. Let $Z \subseteq X$ be a closed subset and $U = X - Z$. Use the other notation as above* (10.4.27), *and for $n \in \mathbb{Z}$, denote*

$$GW^{[n]}(X, Z, \mathcal{L}) = GW^{[n]}(dg_Z(\mathscr{P}(X), \mathcal{L})).$$

Then the sequence

$$GW^{[n]}(X, Z, \mathcal{L}) \longrightarrow GW^{[n]}(X, \mathcal{L}) \longrightarrow GW^{[n]}\left(U, \mathcal{L}_{|U}\right)$$

is a homotopy fibration of spectra. Consequently, there are long exact sequences of GW-groups.

Proof. We follow the same argument, as in the proof of (10.4.27). In this case, the sequence (10.41) of the derived categories, associated to the dg categories in question, is exact. So, the theorem follows from the localization theorem (10.3.19). The proof is complete. ■

The following is a version of (10.4.35), for regular schemes.

Theorem 10.4.39. *Let X be a Noetherian separated regular scheme over a commutative ring A, with $1/2 \in A$. Let $d := \dim X$ and $0 \leq k \leq d$ be an integer. Fix an invertible sheaf $\mathcal{L}$ on X.*

1. *Then as in* (10.4.35), *$\forall n \in \mathbb{Z}$, the sequences*

$$GW^{[n]}\left(dg^{k+1}\mathscr{P}(X)\right) \longrightarrow GW^{[n]}\left(dg^{k}\mathscr{P}(X)\right) \longrightarrow \coprod_{x \in X^{(k)}} GW^{[n]}\left(dg^{k}\mathscr{P}(X_x)\right)$$

$$\mathbb{G}W^{[n]}\left(dg^{k+1}\mathscr{P}(X)\right) \longrightarrow \mathbb{G}W^{[n]}\left(dg^{k}\mathscr{P}(X)\right) \longrightarrow \coprod_{x \in X^{(k)}} \mathbb{G}W^{[n]}\left(dg^{k}\mathscr{P}(X_x)\right)$$

   are homotopy fibrations. Consequently, there are long exact sequences, one with GW-groups and one with $\mathbb{G}\mathfrak{W}$-groups.

2. *Corresponding to two homotopy fibrations* (1), *we obtain two sets of Gersten sequences, similar to* (10.44, 10.45), *one with GW-groups, one with $\mathbb{G}\mathfrak{W}$-groups. Two sets of complexes seem to differ on the negative groups. We have*
   (a) *$\forall p - k \geq 0$ and $x \in X^{(k)}$, we have*

$$\mathbb{G}\mathfrak{W}_{p-k}^{[n-k]}\left(dgC\mathbb{M}^k(X_x)\right) = GW_{p-k}^{[n-k]}\left(dgC\mathbb{M}^k(X_x)\right)$$

   .

   (b) *$\forall p - k \leq -1$ and $x \in X^{(k)}$, we have*

$$\begin{cases} GW_{p-k}^{[n-k]}\left(dgC\mathbb{M}^k(X_x)\right) = W^{n-p}\left(\mathbf{D}^b C\mathbb{M}^k(X_x)\right) \\ \mathbb{G}\mathfrak{W}_{p-k}^{[n-k]}\left(dgC\mathbb{M}^k(X_x)\right) = GW_0^{[n]}\left(\mathfrak{T}\mathfrak{s}^{k-p}\widetilde{dgC\mathbb{M}^k(X_x)}\right). \end{cases}$$

Proof. We have, by Lemma 9.5.6, the sequences of derived categories,

$$\mathscr{D}^{k+1}(\mathscr{P}(X)) \longrightarrow \mathscr{D}^{k}(\mathscr{P}(X)) \longrightarrow \coprod_{x\in X^{(k)}} \mathscr{D}^{k}(\mathscr{P}(X_x))$$

are exact. So, both versions of localization theorems (10.3.19, 10.4.24) apply. Thus, (1) is established.

The existence of two sets of complexes (2) follows similarly, as in (10.44, 10.45). Note that $\mathbf{D}^b C\mathbb{M}^k(X_x)$ is idempotent complete. So, the equalities in (2a) and second line of (2b), follow directly from (10.4.22). The equality in the first line in (2b) follows from agreement (10.3.15). The proof is complete. ■

To be consistent with the literature, we define the following spectral sequences.

Definition 10.4.40. Suppose that X is a Noetherian scheme over a Noetherian ring A with $1/2 \in A$, with $d = \dim X$. Fix an invertible sheaf $\mathcal{L}$. All dualities considered will be induced by $P \mapsto Hom(P, \mathcal{L})$ in $\mathscr{P}(X)$. Fix integer $n \in \mathbb{Z}$.

1. Suppose that X is regular and separated. Define

$$E(X,n)_1^{kq} := EGW(X,n)_1^{kq} = \begin{cases} \bigoplus_{x\in X^{(k)}} GW^{[n]}_{n-k-q}\left(dg^k\mathscr{P}(X_x)\right) & 0 \le k \le d \\ 0 & \text{otherwise} \end{cases}$$

   This is page one of a spectral sequence, to be referred to as the **Gersten GW spectral sequence**. The $q = 0$ line is (isomorphic to) the complex (10.45), rewritten with GW-groups, to be referred to as **Gersten GW-complex**.
2. Suppose that X is either regular and separated or that X is quasi-projective over Spec (A). Define

$$\mathbb{E}(X,n)_1^{kq} := \mathbb{E}\mathbb{G}\mathfrak{W}(X,n)_1^{kq} := \begin{cases} \bigoplus_{x\in X^{(k)}} \mathbb{G}\mathfrak{W}^{[n]}_{n-k-q}\left(dg^k\mathscr{P}(X_x)\right) & 0 \le k \le d \\ 0 & \text{otherwise} \end{cases} \tag{10.46}$$

   This is page one of a spectral sequence, to be referred to as the **Gersten Karoubi $\mathbb{G}\mathfrak{W}$ spectral sequence**. The $q = 0$ line is (isomorphic to) the complex (10.45), to be referred to as **Gersten $\mathbb{G}\mathfrak{W}$-complex**.

Compare these with the Gersten Spectral sequences (6.79, 9.36) for $K, \mathbb{K}$-theory and Lemma 6.5.11. Also see [FS]. Further, compare (10.45) with the Gersten $K, \mathbb{K}$-theory complexes (6.4.5, 9.35) and the Gersten–Witt complex (11.17).

10.5. Nori Homotopy Obstructions

Suppose $X = \mathrm{Spec}\,(A)$ is a regular affine scheme. For projective A-modules P, Nori outlined a homotopy obstruction [M4], for P to split as $P \cong Q \oplus A$. As a reaction to this, Barge and Morel introduced the Chow–Witt groups [BM, Mf], which we define next.

Definition 10.5.1. Consider the setup of (10.4.40). Let $0 \leq k \leq d$ be an integer.

1. Suppose that X is regular and separated. Define the **Chow–Witt group**, of codimension k-cycles, as

$$\widetilde{CH}^k(X, \mathcal{L}) := H^k\left(E_1^{\bullet,0}(X,k)\right), \quad \text{the cohomology at degree } k \tag{10.47}$$

   of the complex (10.45), rewritten with GW-groups. More explicitly, it is the homology of the following sequence:

$$\bigoplus_{x\in X^{(k-1)}} GW_1^{[1]}\left(dgC\mathbb{M}^{k-1}(X_x)\right) \longrightarrow \bigoplus_{x\in X^{(k)}} GW_0^{[0]}\left(dgC\mathbb{M}^{k}(X_x)\right)$$
$$\longrightarrow \bigoplus_{x\in X^{(k+1)}} W^0\left(\mathbf{D}^bC\mathbb{M}^{k+1}(X_x)\right) \tag{10.48}$$

   where the last term is obtained by identification by (10.4.39(2b)). These terms can further be identified, as follows:

$$\begin{cases} \forall x \in X^{(k)} & GW_0^{[0]}\left(dgC\mathbb{M}^k(X_x)\right) \cong GW^0\left(\mathbf{D}^b(C\mathbb{M}^k(X_x))\right) \cong GW\left(C\mathbb{M}^k(X_x)\right) \\ \forall x \in X^{(k+1)} & W^0\left(\mathbf{D}^bC\mathbb{M}^{k+1}(X_x)\right) \cong W\left(C\mathbb{M}^{k+1}(X_x)\right). \end{cases} \tag{10.49}$$

2. Suppose that X is either regular and separated or X is quasi-projective over $\mathrm{Spec}\,(A)$. Define the (nonconnective) **Chow–Witt**

group, of codimension k-cycles, as

$$\widetilde{\mathbb{C}H}^k(X,\mathcal{L}) := H^k\left(\mathbb{E}_1^{\bullet,0}(X,k)\right), \qquad \text{the cohomology at degree } k \tag{10.50}$$

of the complex (10.45) of $\mathbb{GW}$-groups.

The definition (10.47) of Chow–Witt group $\widetilde{CH}^k(X,\mathcal{L})$ differs from the original definition given in [BM]. This equivalent definition was worked out in [FS]. A further elaborate study of $\widetilde{CH}^k(X,\mathcal{L})$ is available in [Fj]. The definition (10.45) of the (nonconnective) Chow–Witt group $\widetilde{\mathbb{C}H}^k(X,\mathcal{L})$ is an improvisation of the same because of the availability of the nonconnected version of the Gerster–Karoubi $\mathbb{GW}$-sequence (10.45). Thus far, it has not been studied well enough to be of any further use.

Before we proceed, we state the following on homotopy invariance of Chow–Witt groups.

Proposition 10.5.2. *Let X be a regular scheme over a field κ, with $1/2 \in \kappa$, $d = \dim X$. Fix an integer $0 \le k \le d$ and an invertible sheaf $\mathcal{L}$ on X. All the dualities considered are induced by the duality $Q \mapsto Hom(Q,\mathcal{L})$ on $\mathscr{P}(X)$. Let $p : X \times \mathbb{A}^1 \longrightarrow X$ be the projection map. Then p induces an isomorphism*

$$p^* : \widetilde{CH}^k(X,\mathcal{L}) \xrightarrow{\sim} \widetilde{CH}^k(X\times\mathbb{A}^1, p^*\mathcal{L}) \quad \text{of the } \textit{Chow–Witt groups.}$$

Proof. See [Fj, Cor. 11.3.3]. ■

For a projective A-module P, with $rank(P) = k$, we define the Nori Homotopy sets $\pi_0(\mathcal{L}O(P))$ and define a natural map $\pi_0(\mathcal{L}O(P)) \longrightarrow \widetilde{CH}^k(X,\Lambda^k P)$.

We give a summary of some of the definitions and results from [MM, M5].

Definition 10.5.3. Let A be a Noetherian commutative ring, with $\dim A = d$, and P be a projective A-module with $rank(P) = k$.

By a **local** P**-orientation**, we mean a pair (I, ω), where I is an ideal of A and $\omega : P \twoheadrightarrow \frac{I}{I^2}$ is a surjective homomorphism. We will use the same notation ω for the map $\frac{P}{IP} \twoheadrightarrow \frac{I}{I^2}$, induced by ω. Denote

$$\begin{cases} \mathcal{LO}(P) = \{(I,\omega) : (I,\omega) \text{ is a local } P \text{ orientation}\} \\ \mathcal{LO}^k(P) = \{(I,\omega) \in \mathcal{LO}(P) : height(I) = k\} \\ \widetilde{\mathcal{LO}}(P) = \{(I,\omega) \in \mathcal{LO}(P) : height(I) \geq k\} \\ \text{Note } \widetilde{\mathcal{LO}}(P) = \mathcal{LO}^k(P) \cup \{(A,0)\}. \end{cases} \tag{10.51}$$

For $(I_0, \omega_0), (I_1, \omega_1) \in \mathcal{LO}(P)$, we write $(I_0, \omega_0) \sim (I_1, \omega_1)$ if there is an $(I, \omega) \in \mathcal{LO}(P[T])$ such that $(I(0), \omega(0)) = (I_0, \omega_0)$ and $(I(1), \omega(1)) = (I_1, \omega_1)$. In this case, we say that (I_0, ω_0) is homotopic to (I_1, ω_1). Now $\sim$ generates a chain equivalence relation on $\mathcal{LO}(P)$, which we call the chain homotopy relation. The (Nori) **Homotopy obstruction** set $\pi_0(\mathcal{LO}(P))$ is defined as the set of equivalence classes of elements in $\mathcal{LO}(P)$.

Similarly, define $\pi_0\left(\widetilde{\mathcal{LO}}(P)\right)$ as the set of all equivalence classes of elements in $\widetilde{\mathcal{LO}}(P)$. By the usual basic element theory [M2], it does not make any difference whether equivalence is defined by using $P[T]$-orientations $(I, \omega) \in \mathcal{LO}(P[T])$ or $(I, \omega) \in \widetilde{\mathcal{LO}}(P[T])$. So, we have a pushforward diagram

$$\begin{array}{ccc} \mathcal{LO}(P[T]) & \xrightarrow{T=0} & \mathcal{LO}(P) \\ {\scriptstyle T=1}\downarrow & & \downarrow \\ \mathcal{LO}(P) & \longrightarrow & \pi_0(\mathcal{LO}(P)) \end{array} \quad \text{in } \underline{Sets}. \tag{10.52}$$

A similar pushforward diagram could not possibly be used to define $\pi_0\left(\widetilde{\mathcal{LO}}(P)\right)$.

1. It follows that there is a natural map

$$\varphi : \pi_0\left(\widetilde{\mathcal{LO}}(P)\right) \longrightarrow \pi_0(\mathcal{LO}(P)). \tag{10.53}$$

2. There are two distinguished elements in $\pi_0(\mathcal{LO}(P))$, namely the equivalence classes

$$\mathbf{e}_1 = [(A,0)] \in \pi_0(\mathcal{LO}(P)) \quad \varepsilon(P) := \mathbf{e}_0 = [(0,0)] \in \pi_0(\mathcal{LO}(P)).$$

The (Nori) **Homotopy obstruction class** of P is defined as $\varepsilon(P)$. Readers should note that the house $\pi_0(\mathcal{LO}(P))$ of the obstruction class $\varepsilon(P)$ depends on P.

Further note that, given any surjective map $f : P \twoheadrightarrow I$, onto an ideal I, f induces an element $(I, \overline{f}) \in \mathcal{LO}(P)$. It is immediate that $\mathbf{e}_0 = [(I, \overline{f})]$. Consequently, $\mathbf{e}_0$ has similar representations in $\pi_0\left(\widetilde{\mathcal{LO}}(P)\right)$.

3. Under certain conditions, we define an addition on $\pi_0(\mathcal{LO}(P))$. In this addition $\mathbf{e}_1$ would be the natural identity.

The following are from [MM, M5].

Theorem 10.5.4. *Let A and P be as in* (10.5.3), *with $rank(P) = k$. Then the map φ is surjective. Assume further that A is a regular ring containing a field κ, with $1/2 \in \kappa$. Then*

1. *The relation $\sim$ is an equivalence relation on $\mathcal{LO}(P)$.*
2. *The natural map* (10.53), $\varphi : \pi_0\left(\widetilde{\mathcal{LO}}(P)\right) \longrightarrow \pi_0(\mathcal{LO}(P))$, *is a bijection.*
3. *If $2k \geq d+2$, then $\pi_0\left(\widetilde{\mathcal{LO}}(P)\right)$ has a structure of an abelian monoid, with $\mathbf{e}_1$ as the additive identity.*
4. *Assume that A is essentially smooth over an infinite perfect field κ, and $2k \geq d+3$. Then*

$$P \cong Q \oplus A \iff \varepsilon(P) = \mathbf{e}_1.$$

Proof. See [MM]. In fact, this is the culmination of the work by many authors, including [MV, BK, BS1, BS2, M4] and others. ∎

Remark 10.5.5. For the rest of this section, we will be working under the following setup. Assume that $X = \mathrm{Spec}\,(A)$ is a regular affine scheme over a field κ, with $1/2 \in \kappa$ and $d = \dim X$. Fix a projective A-module P, with $rank(P) = k \leq d$, and $\mathcal{L} := \Lambda^k P$.

All dualities considered will be induced by the duality $Q \mapsto Q^* := Hom(Q, \mathcal{L})$ on $\mathscr{P}(X)$.

We have the following diagram:

$$\begin{array}{ccccc} GW^0\left(\mathbf{D}^b(C\mathbb{M}^k(X))\right) & \xrightarrow{\iota} & \bigoplus_{x\in X^{(k)}} GW^0(\mathbf{D}^b(C\mathbb{M}^k(X_x))) & \xrightarrow{\mathfrak{q}} & W^0\left(\mathbf{D}^b(C\mathbb{M}^{k+1}(X))\right) \\ & & & \searrow^{\partial} & \downarrow \\ & & & & \bigoplus_{x\in X^{(k+1)}} W^0\left(\mathbf{D}^b(C\mathbb{M}^{k+1}(X_x))\right). \end{array} \tag{10.54}$$

Using the identification, as in (10.49), this can be given an alternate description, as follows:

$$\begin{array}{ccccc} GW\left(C\mathbb{M}^k(X)\right) & \xrightarrow{\iota} & \bigoplus_{x\in X^{(k)}} GW(C\mathbb{M}^k(X_x)) & \xrightarrow{\mathfrak{q}} & W\left(C\mathbb{M}^{k+1}(X)\right) \\ & & & \searrow^{\partial} & \downarrow \\ & & & & \bigoplus_{x\in X^{(k+1)}} W\left(C\mathbb{M}^{k+1}(X_x)\right). \end{array} \tag{10.55}$$

However, the map $\mathfrak{q}$ in (10.55) seems to require the description (10.54) at the level of the Triangulated categories (11.2.18).

Given $(I, \omega) \in \widetilde{\mathcal{L}O}(P)$, with $I \neq A$, we associate $\Phi(I, \omega) : \frac{A}{I} \xrightarrow{\sim} Ext^k\left(\frac{A}{I}, \mathcal{L}\right)$, a symmetric isomorphism in $C\mathbb{M}^k(X)$, unique up to isometry. Consider any lift of ω to a surjective map, as follows:

$$\begin{array}{ccc} P & \overset{f}{\twoheadrightarrow} & I \cap J \\ & \searrow^{\omega} & \downarrow \\ & & \frac{I}{I^2} \end{array} \qquad \ni \qquad I + J = A, \ height(I) \geq k$$

The Koszul complex of f induces a symmetric isomorphism $\mathfrak{f} : \frac{A}{I\cap J} \xrightarrow{\sim} Ext^k\left(\frac{A}{I\cap J}, \mathcal{L}\right)$. We can write $\mathfrak{f} = \mathfrak{f}_I \oplus \mathfrak{f}_J$, where $\mathfrak{f}_I : \frac{A}{I} \xrightarrow{\sim} Ext^k\left(\frac{A}{I}, \mathcal{L}\right)$ is a symmetric isomorphism, and likewise $\mathfrak{f}_J$. Note that $\mathfrak{f}_I$, so defined, is unique up to isometry. As before, let $\mathbb{M}\mathcal{I}(C\mathbb{M}^k(X))$ denote the monoid of isometry classes of symmetric isomorphisms in $C\mathbb{M}^k(X)$. Define,

$$\Phi : \widetilde{\mathcal{L}O}(P) \longrightarrow \mathbb{M}\mathcal{I}(C\mathbb{M}^k(X)) \quad \text{by} \quad \Phi(I, \omega) := \begin{cases} [\mathfrak{f}_I] & \text{if } I \neq A \\ 0 & \text{if } I = A \end{cases}$$

Guided by the diagrams (10.54, 10.55), we obtain the map Ω, as follows:

$$\begin{array}{ccccccc} \widetilde{\mathcal{L}O}(P) & \xrightarrow{\Phi} & \mathbb{MI}(C\mathbb{M}^k(X)) & \xrightarrow{\eta} & GW\left(C\mathbb{M}^k(X)\right) & \xrightarrow{\iota} & \ker(\partial) \\ & & & & & & \downarrow q \\ & & & \overset{\Omega}{\searrow} & & & \widetilde{CH}^k(X, \mathcal{L}) \end{array} \tag{10.56}$$

It is best to keep in mind that the following functor, by degree-zero inclusion,

$$C\mathbb{M}^k(X) \longrightarrow \mathbf{D}^b(C\mathbb{M}^k(X)) \quad \text{induces isomorphisms of } GW\text{-groups.}$$

So, the symmetric isomorphism $\mathfrak{f}_I : \frac{A}{I} \xrightarrow{\sim} Ext^k\left(\frac{A}{I}, \mathcal{L}\right)$, constructed above, can be considered as the same in $\mathbf{D}^b(C\mathbb{M}^k(X))$, concentrated at degree zero. Therefore, the map Φ or the map Ω has an interpretation in the derived categories. Further, we have identifications $\mathbf{D}^b(C\mathbb{M}^k(X)) \xrightarrow{\sim} T^k\mathscr{D}^k\left(\mathscr{P}(X)\right)$.

We proceed to show that Ω factors through a map $\pi_0\left(\widetilde{\mathcal{L}O}(P)\right) \longrightarrow \widetilde{CH}^k(X)$.

Theorem 10.5.6. *Consider the setup of* (10.5.5). *Then the map* Ω *in* (10.56) *factors through a set theoretic map* Θ_P*, as in the following commutative diagram:*

$$\begin{array}{ccc} \widetilde{\mathcal{L}O}(P) & \xrightarrow{\beta} & \pi_0\left(\widetilde{\mathcal{L}O}(P)\right) \\ & \overset{\Omega}{\searrow} & \downarrow \Theta_P \\ & & \widetilde{CH}^k(X, \mathcal{L}) \end{array} \tag{10.57}$$

Proof. Let $(I_0, \omega_0), (I_1, \omega_1) \in \widetilde{\mathcal{L}O}(P)$ be such that $\beta((I_0, \omega_0)) = \beta((I_1, \omega_1)) \in \pi_0\left(\widetilde{\mathcal{L}O}(P)\right)$. By (10.5.4(10.5.4)), the homotopy $\sim$ is an equivalence relation. So, there is a $P[T]$-orientation, $(I, \omega) \in \mathcal{L}O(P[T])$ such that $(I, \omega)_{|T=0} = (I_0, \omega_0)$ and $(I, \omega)_{|T=1} = (I_1, \omega_1)$. By moving lemmas (basic element theory [M2]), we can assume that $(I, \omega) \in \widetilde{\mathcal{L}O}(P[T])$. The projection map $p : X \times \mathbb{A}^1 \longrightarrow X$ induces a pullback map $p^* : \widetilde{CH}^k(X, \mathcal{L}) \xrightarrow{\sim} \widetilde{CH}^k(X \times \mathbb{A}^1, p^*\mathcal{L})$, which is an isomorphism (10.5.2). Consider the commutative diagram

$$\begin{array}{ccccc}
 & & & & \widetilde{CH}^k(X,\mathcal{L}) \\
 & \overset{\Omega}{\nearrow} & & \overset{q\iota\eta}{\nearrow} & \\
\widetilde{\mathcal{L}O}(P) & \xrightarrow{\Phi} & \mathbb{MICM}^k(X,\mathcal{L}) & & \\
\downarrow & & \downarrow{p^*} & & \wr\downarrow{p^*} \\
\widetilde{\mathcal{L}O}(P[T]) & \xrightarrow{\Phi} & \mathbb{MICM}^k\left(X\times\mathbb{A}^1,p^*\mathcal{L}\right) & & \\
 & \underset{\Omega}{\searrow} & & \underset{q\iota\eta}{\searrow} & \\
 & & & & \widetilde{CH}^k(X\times\mathbb{A}^1,p^*\mathcal{L})
\end{array} \tag{10.58}$$

We are required to prove that $\Omega(I_0,\omega_0)=\Omega(I_1,\omega_1)$. Since p^* (the last vertical arrow) is an isomorphism, it is enough to prove that

$$p^*\left(\Omega(I_i,\omega_i)\right)) = \Omega(I,\omega) \qquad \text{for} \quad i=0,1. \tag{10.59}$$

Since T and $T-1$ are interchangeable, it suffices to establish the case $i=0$, which is done subsequently in Lemma 10.5.7. The proof is complete. ∎

The equation (10.59) is established, as follows. In fact, this is an analogue of the corresponding result for Chow groups [Fu, Thm. 3.3].

Lemma 10.5.7. *Consider the setup of* (10.5.5). *Let* $(I,\omega)\in\widetilde{\mathcal{L}O}(P[T])$ *be such that* $(I_0,\omega_0)\in\widetilde{\mathcal{L}O}(P)$. *Let* $p: X\times\mathbb{A}^1\longrightarrow X$ *denote the projection map. Then* $p^*\left(\Omega(I_0,\omega_0)\right)) = \Omega(I,\omega)$.

Proof. The homotopy invariance (10.5.2) of Chow–Witt groups is used.

Let $f_0: P\twoheadrightarrow I_0\cap J_0$ be a lift of $\omega_0: P\twoheadrightarrow\frac{I_0}{I_0^2}$ such that $height(J_0)\geq n$, and $I_0+J_0=A$. By [M4, Lem. 2.2], there is a lift $h(T): P[T]\longrightarrow I$ of ω such that $h(0)=f_0$. Then $I=(h(P[T]),s)$ for some $s\in I^2$. Let $\mathfrak{U}=\{\wp\in\operatorname{Spec}(A[T]): height(\wp)\leq n-1, sT\notin\wp\}$. Now, $(h,sT)\in P[T]^*\oplus A[T]$ is basic on V. By basic element theory [M3], we have that $f:=h+sTg$ is basic on $\mathfrak{U}$ for some $g\in P[T]^*$. It follows that (1) $f(0)=f_0$ and (2) f is a lift of ω and (3) $height(f(P[T]))\geq n$. So, we can write $f(P[T])=I\cap J$, with $height(J)\geq n$ and $I+J=A[T]$. We fix such a lift f of ω. Also, let $\omega_J: P[T]\twoheadrightarrow\frac{J}{J^2}$ and $\omega_{IJ}: P[T]\twoheadrightarrow$

$\frac{IJ}{(IJ)^2}$ be induced by f. Then $(J, \omega_J), (IJ, \omega_{IJ}) \in \widetilde{\mathcal{LO}}(P[T])$. Likewise, denote $(J_0, \omega_{J_0}), (I_0J_0, \omega_{I_0J_0}) \in \widetilde{\mathcal{LO}}(P)$. The Koszul complex of f anf f_0 leads to the symmetric spaces, which may be considered in the $C\mathbb{M}^k(-)$-categories or in the derived categories:

$$\begin{array}{ccccccccccccc}
0 & \longrightarrow & \Lambda^0 P[T]^* & \longrightarrow \cdots \longrightarrow & \Lambda^{k-1}P[T]^* & \longrightarrow & \Lambda^k P[T]^* & \longrightarrow & Ext^k\left(\frac{A[T]}{(I\cap J)}, \mathcal{L}[T]\right) & \longrightarrow & 0 \\
 & & \uparrow \varphi & & \uparrow \varphi & & \uparrow \varphi & & \uparrow \mathfrak{f}_I \oplus \mathfrak{f}_J & & \\
0 & \longrightarrow & \Lambda^k P[T] & \longrightarrow \cdots \longrightarrow & \Lambda^1 P[T] & \xrightarrow[f]{} & \Lambda^0 P[T] & \longrightarrow & \frac{A[T]}{(I\cap J)} & \longrightarrow & 0
\end{array}$$

$$\begin{array}{ccccccccccccc}
0 & \longrightarrow & \Lambda^k P & \longrightarrow \cdots \longrightarrow & \Lambda^1 P & \xrightarrow{f_0} & \Lambda^0 P & \longrightarrow & \frac{A}{(I(0)\cap J(0))} & \longrightarrow & 0 \\
 & & \downarrow \varphi(0) & & \downarrow \varphi(0) & & \downarrow \varphi(0) & & \downarrow \mathfrak{f}_{I_0} \oplus \mathfrak{f}_{J_0} & & \\
0 & \longrightarrow & \Lambda^0 P^* & \longrightarrow \cdots \longrightarrow & \Lambda^{k-1} P^* & \longrightarrow & \Lambda^k P^* & \longrightarrow & Ext^k\left(\frac{A}{(I(0)\cap J(0))}, \mathcal{L}\right) & \longrightarrow & 0
\end{array}$$

Denote $X[T] := X \times \mathbb{A}^1 = \operatorname{Spec}(A[T])$. Denote the symmetric spaces in $\mathscr{D}^{k+1}(\mathscr{P}(X[T])$, as follows:

$$\begin{cases} \mathbf{C}(T\varphi) := (Cone(T), \varphi \wedge T) \cong Cone(T\varphi) & \text{cone of } T\varphi : \Lambda^\bullet P[T] \longrightarrow \Lambda^\bullet P[T] \\ \mathbf{C}(T\varphi(0)[T]) := (Cone(T), \varphi(0)[T] \wedge T) & \text{cone of } T\varphi(0)[T] : \Lambda^\bullet P[T] \longrightarrow \Lambda^\bullet P[T] \end{cases}$$

We claim that these symmetric spaces $\mathbf{C}(T\varphi)$, $\mathbf{C}(T\varphi(0)[T])$ are isometric. Write $f(T) = f_0 + Tf_1 + \cdots + T^m f_m$, with $f_i \in Hom(P, A)$. Consider the map

$$\Delta = \begin{pmatrix} 1_{A[T]} & \frac{f - f_0[T]}{T} \\ 0 & 1_{P[T]} \end{pmatrix} : A[T] \oplus P[T] \xrightarrow{\sim} A[T] \oplus P[T].$$

Then $$\begin{pmatrix} -T & f \end{pmatrix} \begin{pmatrix} 1_{A[T]} & \frac{f - f_0[T]}{T} \\ 0 & 1_{P[T]} \end{pmatrix} = \begin{pmatrix} -T & f_0[T] \end{pmatrix}.$$

So, the diagram
$$\begin{array}{ccccccc}
A[T] \oplus P[T] & \xrightarrow{(-T, f_0[T],)} & A[T] & \longrightarrow & \frac{A[T]}{IJ} & \longrightarrow & 0 \\
\downarrow \Delta & & \downarrow 1 & & \downarrow 1 & & \\
A[T] \oplus P[T] & \xrightarrow[(-T, f)]{} & A[T] & \longrightarrow & \frac{A[T]}{IJ} & \longrightarrow & 0
\end{array}$$
commutes.

We write $Q = A[T] \oplus P[T]$ and $Q_r = \Lambda^r Q$. Then the cone $\mathbf{C}(T\varphi)$ is the Koszul complex of the map $(-T, f) : Q \longrightarrow A[T]$, and

$\mathbf{C}(T\varphi(0)[T])$ is the Koszul complex of the map $(-T, f(0)[T]) : Q \longrightarrow A[T]$.

$$\text{Write} \qquad \begin{cases} M = \frac{A[T]}{(IJ,T)} = \frac{B}{(I_0 J_0, T)} \\ M^{\vee} = Ext^{n+1}(M, \mathcal{L}[T]) \end{cases}$$

Consider the following composition of maps of Koszul complexes and duality:

$$\begin{array}{ccccccccccccc}
0 & \longrightarrow & \Lambda^{n+1}Q & \xrightarrow{\partial_0} \cdots \longrightarrow & \Lambda^2 Q & \longrightarrow & Q & \xrightarrow{(-T,f_0[T])} & A[T] & \longrightarrow & M & \longrightarrow & 0 \\
 & & \| & & \Big\downarrow{\scriptstyle \Lambda^2\Delta} & & \Big\downarrow{\scriptstyle \Delta} & & \| & & \| & & \\
0 & \longrightarrow & \Lambda^{n+1}Q & \xrightarrow[\partial_1]{} \cdots \longrightarrow & \Lambda^2 Q & \longrightarrow & Q & \xrightarrow[(-T,f)]{} & A[T] & \longrightarrow & M & \longrightarrow & 0 \\
 & & \Big\downarrow{\scriptstyle can} & & \Big\downarrow{\scriptstyle can} & & \Big\downarrow{\scriptstyle can} & & \Big\downarrow{\scriptstyle can=1} & & \Big\downarrow{\scriptstyle \mathfrak{f}_I \oplus \mathfrak{f}_J} & & \\
0 & \longrightarrow & A[T]^* & \longrightarrow \cdots \longrightarrow & \Lambda^{n-1}Q^* & \longrightarrow & \Lambda^n Q^* & \xrightarrow{\partial_1^*} & \Lambda^{n+1}Q^* & \longrightarrow & M^{\vee} & \longrightarrow & 0 \\
 & & \| & & \Big\downarrow & & \Big\downarrow{\scriptstyle \Lambda^n\Delta^*} & & \| & & \| & & \\
0 & \longrightarrow & A[T]^* & \longrightarrow \cdots \longrightarrow & \Lambda^{n-1}Q^* & \longrightarrow & \Lambda^n Q^* & \xrightarrow[\partial_0^*]{} & \Lambda^{n+1}Q^* & \longrightarrow & M^{\vee} & \longrightarrow & 0
\end{array}$$

Since $\det(\Delta) = 1$, it follows from (10.5.8) that the composition is the canonical map $can : \Lambda^\bullet Q \xrightarrow{\sim} \Lambda^\bullet Q^*$. Therefore, $\Lambda^\bullet \Delta : \mathbf{C}(T\varphi(0)[T]) \xrightarrow{\sim} \mathbf{C}(T\varphi)$ is an isometry in $\mathscr{D}^{k+1}(\mathscr{P}(X))$.

It follows from this that, under the multiplication map

$$T : \widetilde{CH}^k(X[T], p^*\mathcal{L}) \longrightarrow \widetilde{CH}^{k+1}(X[T], p^*\mathcal{L}) \qquad \text{we have} \quad T\cdot\Omega(I, \omega_I) = T\cdot(p^*\Omega(I_0, \omega_{I_0})).$$

From definition of the retraction maps (inverse of p^*), it follows that

$$\Omega(I, \omega_I) = p^*\Omega(I_0, \omega_{I_0}).$$

For further details, see [M5]. The proof is complete. ∎

Lemma 10.5.8. *Suppose that A is a commutative Noetherian ring and Q is a projective A-module with $rank(Q) = n$ and $\mathcal{L} = \Lambda^n Q$. For any A-module M, we denote $M^* := Hom(M, \mathcal{L})$. Let $0 \le r \le n$ and $can : \Lambda^r Q \longrightarrow \Lambda^{n-r}Q^*$ be the isomorphism induced by the perfect duality $\Lambda^r Q \otimes \Lambda^{n-r}Q \longrightarrow \mathcal{L}$. Let $\Delta : Q \xrightarrow{\sim} Q$ be an isomorphism,* *with* $\det(\Delta) = 1$. *Then the diagram*

$$\begin{array}{ccccc} \Lambda^r Q & \xrightarrow{\Lambda^r \Delta} & \Lambda^r Q & \xrightarrow{can} & \Lambda^{n-r} Q^* \\ & \searrow_{can} & & & \downarrow{\scriptstyle \Lambda^{n-r}\Delta^*} \\ & & & & \Lambda^{n-r} Q^* \end{array} \tag{10.60}$$

commutes.

Proof. First, we prove this for $r = 1$. Let $q \in Q$. Let $can\Delta(q) \in \Lambda^{n-1}Q^*$ be given by $can\Delta(q)(q_2 \wedge \cdots \wedge q_n) = \Delta(q) \wedge q_2 \wedge \cdots \wedge q_n$. Then $\lambda_q := \Lambda^{n-r}\Delta^* can\Delta(q)$ is given by the commutative diagram

$$\begin{array}{ccc} \Lambda^{n-1} Q & \xrightarrow{\Lambda^{n-1} Q} & \Lambda^{n-1} Q \\ & \searrow_{\lambda_q} & \downarrow{\scriptstyle can\Delta(q)} \\ & & \Lambda^n Q \end{array}$$

So,

$$\lambda_q(q_2 \wedge \cdots \wedge q_n) = \Delta(q) \wedge \Lambda^{n-1}\Delta(q_2 \wedge \cdots \wedge q_n) = \Lambda^n(\Delta)(q \wedge q_2 \wedge \cdots \wedge q_n).$$

Since $\det \Delta = 1$, it follows that, the upper-right composition (10.60) sends $q \mapsto (q_2 \wedge \cdots \wedge q_n \mapsto q \wedge q_2 \wedge \cdots \wedge q_n)$, which is the map $can : Q \longrightarrow \Lambda^{n-1}Q^*$.

For the general case, for $q_1, \ldots, q_r \in Q$, let $ev_{q_1 q_2 \ldots q_r} \in \Lambda^{n-r}Q^*$ denote the map $q_{r+1} \wedge \cdots \wedge q_n \mapsto q_1 \wedge \cdots \wedge q_{r+1} \wedge \cdots \wedge q_n$. So, $can\Lambda^r\Delta(q) = ev_{\Delta(q_1)\Delta(q_2)\ldots\Delta(q_r)}$. So, $\Lambda^{n-r}\Delta can \Lambda^r \Delta(q)$ is given by the commutative diagram

$$\begin{array}{ccc} \Lambda^{n-r} Q & \xrightarrow{\Lambda^{n-r} \Delta} & \Lambda^{n-r} Q \\ & \searrow & \downarrow{\scriptstyle ev_{\Delta(q_1)\Delta(q_2)\ldots\Delta(q_r)}} \\ & & \Lambda^n Q \end{array}$$

which sends

$$\begin{aligned} q_{r+1} \wedge \cdots \wedge q_n &\mapsto \Delta(q_1) \wedge \cdots \wedge \Delta(q_r) \wedge \Delta(q_{r+1}) \wedge \cdots \wedge \Delta(q_n) \\ &= \Lambda^n \Delta(q_1 \wedge \cdots \wedge q_r \wedge q_{r+1} \wedge \cdots \wedge q_n). \end{aligned}$$

This is precisely the map $can : \Lambda^r Q \longrightarrow \Lambda^{n-r} Q^*$. The proof is complete. ∎

Definition 10.5.9. Under the notation and hypotheses of (10.5.6), define the Barge–Morel obstruction of P, by

$$\varepsilon_{BM}(P) := \Theta_P\left(\varepsilon(P)\right) \in \widetilde{CH}^k\left(X, \Lambda^k P\right).$$

The question remains whether the Barge–Morel obstruction detects splitting, at least when $2k \geq d+3$ and A is smooth over an infinite perfect field. More generally, one can ask whether the map Θ_P is injective.

Chapter 11

Epilogue: Triangulated Categories

We saw that K-theory was developed in the framework of complicial exact categories (Chapter 9). Then Hermitian K-theory was developed in the framework of dg categories (Chapter 10). There have also been attempts to implement all of it in the paradigm of Triangulated categories. As was evident in Chapters 9 and 10, much of the theory is invariant of the associated triangulated categories. In this chapter, we give some of the definitions we used above. While part of the literature is classical [TT, Kb1], the theory of triangulated Witt groups is of our main interest [Bp, BW]. Throughout this chapter, we assume that 2 is invertible.

11.1. Basic Definitions

For completeness, we include a definition of triangulated category, from [Kb1].

Definition 11.1.1. Suppose that $\mathscr{K}$ is an additive category. By a shift (or translation) functor, we mean an equivalence of categories $T : \mathscr{K} \xrightarrow{\sim} \mathscr{K}$. Given such a shift functor, by a triangle in $\mathscr{K}$, we mean a triple (u, v, w) of composable arrows as follows:

$$(u, v, w) := \left(A \xrightarrow{u} B \xrightarrow{v} C \xrightarrow{w} TA \right) \qquad \text{in } \mathscr{K}. \qquad (11.1)$$

Such a triangle (11.1) is also represented by the diagram

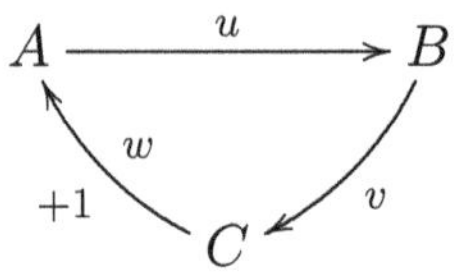

in $\mathscr{K}$. (11.2)

A map $(u, v, w) \longrightarrow (f, g, h)$ of triangles is defined as a commutative diagram of arrows in $\mathscr{K}$:

$$\begin{array}{ccccccc} A & \xrightarrow{u} & B & \xrightarrow{v} & C & \xrightarrow{w} & TA \\ \downarrow{\scriptstyle \varphi_A} & & \downarrow{\scriptstyle \varphi_B} & & \downarrow{\scriptstyle \varphi_C} & & \downarrow{\scriptstyle T\varphi_A} \\ X & \xrightarrow[f]{} & Y & \xrightarrow[g]{} & Z & \xrightarrow[h]{} & TX \end{array} \tag{11.3}$$

A **Triangulated category** is a triple $\mathscr{K} := (\mathscr{K}, T, \Delta\mathscr{K})$, where (1) $\mathscr{K}$ is an additive category (2) $T : \mathscr{K} \longrightarrow \mathscr{K}$ is a shift (or translation) functor, and (3) $\Delta\mathscr{K}$ is a class of triangles, as in (11.1), to be called **distinguished/exact triangles**. This class $\Delta\mathscr{K}$ of exact triangles (11.1) must satisfy the following axioms:

1. (TR0) Consider a map of two triangles, as in (11.3). Assume that $\varphi_A, \varphi_B, \varphi_C$ are isomorphisms. Then (u, v, w) is an exact triangle if and only if (f, g, h) is an exact triangle.
2. (TR1) $\forall A \in Obj(\mathscr{K})$, the triangle $0 \longrightarrow A \xrightarrow{1_A} A \longrightarrow T0$ is an exact triangle. Note that $T0 = 0$.
3. (TR2) (Rotation) If (u, u, w) is exact, then so is $(v, w, -Tu)$.
4. (TR3) Consider two exact triangles (u, v, w), (f, g, h), and the maps φ_A, φ_B, as follows:

$$\begin{array}{ccccccc} A & \xrightarrow{u} & B & \xrightarrow{v} & C & \xrightarrow{w} & TA \\ \downarrow{\scriptstyle \varphi_A} & & \downarrow{\scriptstyle \varphi_B} & & \vdots{\scriptstyle \varphi_C} & & \downarrow{\scriptstyle T\varphi_A} \\ X & \xrightarrow[f]{} & Y & \xrightarrow[g]{} & Z & \xrightarrow[h]{} & TX \end{array}$$

   such that the first rectangle commutes. Then there is a map φ_C such that the other two rectangles commute.

5. (TR4) (**Octahedron Axiom:**) Consider two composable maps $A \xrightarrow{u} B \xrightarrow{v} C$. Then there is a commutative diagram:

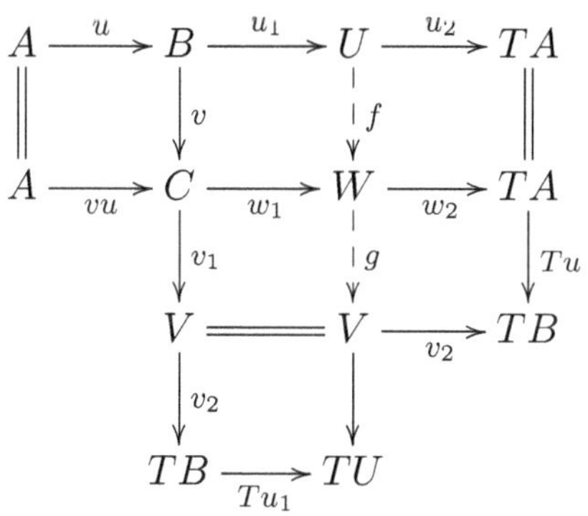

(11.4)

such that the first two horizontal rows and the 2nd, 3rd columns are exact.

There are other definitions of triangulated categories available in the literature which may differ a little. The most comprehensive exposition on this may be [Na1]. Some authors (including [Na1, Bp]) assume that the shift functor T is an automorphism, which we avoid at this time. To align our definition [Kb1] with the others, we include the following.

Exercise 11.1.2. Let $\mathscr{K}$ be a triangulated category, and T be the shift functor. Then

1. Every map $u \in Mor_{\mathscr{K}}(A, B)$ fits in an exact triangle (u, v, w), as in (11.1). Further, $\forall A \in Obj(\mathscr{K})$, the triple $(1_A, 0, 0)$ is an exact triangle.
2. Given an exact triangle (u, v, w), we have $vu = 0$ and $wv = 0$. This is proved by comparing (11.1) with $(1_A, 0, 0)$ using $TR3$.
3. (**Weak kernel and cokernel:**) Let (u, v, w) be an exact triangle. Consider the following two commutative diagrams:

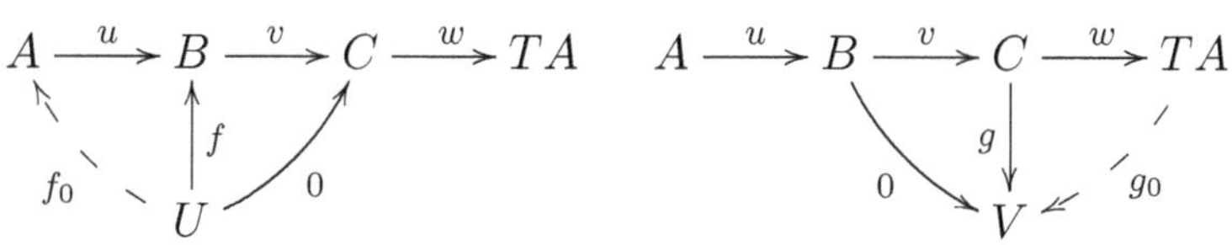

In the first diagram, let f be given with $vf = 0$. Then there is an arrow f_0 with $f = uf_0$. In this case, we say that u is a **weak**

kernel of v. In the second diagram, let g be given with $gv = 0$. Then there is an arrow g_0 with $g = g_0 w$. In this case, we say that w is a **weak cokernel** of v.

4. (**Four Lemma**:) Consider a map of two exact triangles, as in (11.3). If two of the three vertical maps $\varphi_A, \varphi_B, \varphi_C$ are isomorphisms, then so is the third one. In particular, given a morphism u, up to isomorphism, it fits into a unique exact triangle (u, v, w). For this reason C may be referred to as the **Cone** of u.
5. Consider a triangle $(u, v, w) := \left(A \xrightarrow{u} B \xrightarrow{v} C \xrightarrow{w} TA \right)$.
Assume that the triangle $B \xrightarrow{v} C \xrightarrow{w} TA \xrightarrow{-Tu} TB$ is exact. Then (u, v, w) is exact.
Proof. Embed u in an exact triangle (u, v_c, w_c). Consider the diagram

$$\begin{array}{ccccccccc} A & \xrightarrow{u} & B & \xrightarrow{v} & C & \xrightarrow{w} & TA & \xrightarrow{-Tu} & TB \\ \downarrow{\scriptstyle 1_A} & & \downarrow{\scriptstyle 1_B} & & \downarrow{\scriptstyle \varphi_C} & & \downarrow{\scriptstyle 1_{TA}} & & \downarrow{\scriptstyle 1_{TB}} \\ A & \xrightarrow[u]{} & B & \xrightarrow[v_c]{} & Cone(u) & \xrightarrow[w_c]{} & TA & \xrightarrow[-Tu]{} & TB \end{array}$$

Since $(v_c, w_c, -Tu)$ is an exact triangle, by the rotation axiom, the map φ_C is obtained by $TR3$. Now, φ_C is an isomorphism, by (4). Therefore, (u, v, w) is isomorphic to (u, v_c, w_c). So, (u, v, w) is exact, by $TR0$. ■
6. The direct sum of two distinguished triangles is distinguished. **Hint:** See [Na1].

The following lemma upends the rotation axiom to the left.

Lemma 11.1.3. *Let $\mathscr{K}$ be a triangulated category, and T be the shift functor. Let* $(u, v, w) := \left(A \xrightarrow{u} B \xrightarrow{v} C \xrightarrow{w} TA \right)$ *be an exact triangle. Let $\tilde{C} \in Obj(\mathscr{K})$ and $\varphi : T\tilde{C} \xrightarrow{\sim} C$ be an isomorphism. So, $w\varphi = T(\varphi_{-1})$ for some $\varphi_{-1} \in Mor(\tilde{C}, A)$. Then*

$$\tilde{C} \xrightarrow{-\varphi_{-1}} A \xrightarrow{u} B \xrightarrow{\varphi^{-1}v} T\tilde{C} \quad \textit{is exact.}$$

In particular, if T is an automorphism, and if (u, v, w) is an exact triangle, then both $(-T^{-1}w, u, v)$ and $(v, w, -Tu)$ are exact.

Proof. By (11.1.2(5)), it is enough to prove that $A \xrightarrow{u} B \xrightarrow{\varphi^{-1}v} T\tilde{C} \xrightarrow{w\varphi} TA$ is exact. This is valid, since this triangle is isomorphic to (u, v, w). ■

Now we incorporate duality structures in triangulated categories.

Definition 11.1.4. A triangulated category with duality is a triangulated category $\mathscr{K}$, together with (1) an additive functor $\# : \mathscr{K}^{op} \longrightarrow \mathscr{K}$ and (2) two natural isomorphisms

$$\begin{cases} \lambda : \# \xrightarrow{\sim} T\#T \\ \varpi : 1_{\mathscr{K}} \xrightarrow{\sim} \#\# \end{cases} \quad \text{meaning, } \forall X \ \exists \text{ natural isomorphisms}$$

$$\begin{cases} \lambda_X : X^{\#} \xrightarrow{\sim} T\left((TX)^{\#}\right) \\ \varpi : X \xrightarrow{\sim} X^{\#\#} \end{cases}$$

such that

1. For any $A \xrightarrow{u} B \xrightarrow{v} C \xrightarrow{w} TA$ exact triangle, the triangle

$$C^{\#} \xrightarrow{v^{\#}} B^{\#} \xrightarrow{u^{\#}} A^{\#} \xrightarrow{T(w^{\#})o\lambda_A} TC^{\#} \quad \text{is exact.}$$

2. The following diagram commutes:

$$\begin{array}{ccc} T & \longrightarrow & \#\#T \\ {\scriptstyle T\varpi}\downarrow & & \downarrow{\scriptstyle \lambda_{\#T}} \\ T\#\# & \xleftarrow[T\#\lambda]{} & T\#T\#T \end{array} \quad \text{In particular, } \forall X \quad \begin{array}{ccc} TX & \xrightarrow{\varpi_{TX}} & (TX)^{\#\#} \\ {\scriptstyle T(\varpi_X)}\downarrow & & \downarrow{\scriptstyle \lambda_{(TX)^{\#}}} \\ T\left(X^{\#\#}\right) & \xleftarrow[T(\lambda_X^{\#})]{} & T(T((TX)^{\#})^{\#}) \end{array} \quad \text{commutes.}$$

3. $\forall X$, the diagram

$$\begin{array}{ccc} X^{\#} & \xrightarrow{\varpi_{X^{\#}}} & X^{\#\#\#} \\ & \searrow & \downarrow{\scriptstyle \varpi_X^{\#}} \\ & & X^{\#} \end{array} \quad \text{commutes.}$$

We use variations of notation $\mathscr{K} := (\mathscr{K}, \#, \varpi, \lambda) := (\mathscr{K}, T, \Delta\mathscr{K}, \#, \varpi, \lambda)$ to denote triangulated categories with duality.

We define morphisms of triangulated categories, with duality.

Definition 11.1.5. A functor (or morphism) of triangulated categories is a pair $F := (F, \rho) : (\mathscr{K}_1, T_1, \Delta\mathscr{K}_1) \longrightarrow (\mathscr{K}_2, T_2, \Delta\mathscr{K}_2)$ such that

1. $F : \mathscr{K}_1 \longrightarrow \mathscr{K}_2$ is an additive functor,
2. $\rho : FT_1 \xrightarrow{\sim} T_2 F$ is a natural equivalence. In other words, there are natural isomorphisms $\rho_X : FT_1(X) \xrightarrow{\sim} T_2F(X)$ for all $X \in Obj(\mathscr{K}_1)$. We also say that the diagram

$$\begin{array}{ccc} \mathscr{K}_1 & \xrightarrow{F} & \mathscr{K}_2 \\ {\scriptstyle T_1}\downarrow & & \downarrow{\scriptstyle T_2} \\ \mathscr{K}_1 & \xrightarrow[F]{} & \mathscr{K}_2 \end{array} \quad \text{commutes, via the equivalence } \rho.$$

3. F sends exact triangles in $\Delta\mathscr{K}_1$ to exact triangles in $\Delta\mathscr{K}_2$.

Such a functor is also called a **triangle functor** or an exact functor.

A morphism of two triangulated categories with dualities is a triple

$$(F, \rho, \Phi) : (\mathscr{K}_1, \#_1, \varpi_1, \lambda_1) \longrightarrow (\mathscr{K}_2, \#_2, \varpi_2, \lambda_2)$$

where

1. $(F, \rho) : \mathscr{K}_1 \longrightarrow \mathscr{K}_2$ is a triangle functor.
2. $\Phi : F\#_1 \xrightarrow{\sim} \#_2 F$ is a natural equivalence. In other words, there are natural isomorphisms $\Phi_X : F(X^{\#_1}) \xrightarrow{\sim} (FX)^{\#_2}$ for all $X \in Obj(\mathscr{K}_1)$. We also say that the diagram

$$\begin{array}{ccc} \mathscr{K}_1^{op} & \xrightarrow{F} & \mathscr{K}_2^{op} \\ {\scriptstyle \#_1}\downarrow & & \downarrow{\scriptstyle \#_2} \\ \mathscr{K}_1 & \xrightarrow[F]{} & \mathscr{K}_2 \end{array} \quad \text{commutes, via the equivalence } \Phi.$$

3. Φ establishes compatibility with respect to the double dual identifications ϖ_1, ϖ_2:

$$\begin{array}{ccc} F(X) & \xrightarrow{F((\varpi_1)_X)} & F(X^{\#_1\#_1}) \\ {\scriptstyle (\varpi_2)_{FX}}\downarrow & & \downarrow{\scriptstyle \Phi_{X^{\#_1}}} \\ F(X)^{\#_2\#_2} & \xrightarrow[(\Phi_X)^{\#_2}]{} & (F(X^{\#_1}))^{\#_2} \end{array} \quad \text{commutes } \forall X.$$

4. Further, the diagram

$$\begin{array}{ccccc} F\#_1 & \xrightarrow{F(\lambda_1)} & FT\#_1T & \xrightarrow{\rho_{\#_1T}} & TF\#_1T \\ \Phi\downarrow & & & & \downarrow T(\Phi_T) \\ \#_2F & \xrightarrow[\lambda_2 F]{} & T\#_2TF & \xrightarrow[T\#_2(\rho)]{} & T\#_2FT \end{array} \qquad \text{commutes.}$$

We defined (9.3.12) the Grothendieck group $K_0(\mathscr{K})$ of a triangulated category $\mathscr{K}$. Now we define the Grothendieck–Witt group and Witt group of a triangulated category with duality.

Definition 11.1.6. To start with, let $(\mathscr{A}, \#, \varpi)$ be an additive category with duality. By a symmetric morphism, we mean an arrow $u : P \longrightarrow P^\#$ such that $u = u^\# \varpi_P$. In this case, we also write $u = u^\#$. A **symmetric space** is a pair (P, u), where $u : P \xrightarrow{\sim} P^\#$ is a symmetric map and an isomorphism. For symmetric spaces, (P, u), (Q, v), we define the orthogonal sum as $(P, u) \perp (Q, v) := (P \oplus Q, u \oplus v)$.

An isometry $f : (P, u) \xrightarrow{\sim} (Q, v)$ between two symmetric spaces is an isomorphism $f : P \xrightarrow{\sim} Q$, such that $u = f^\# v f$. Note that isometry is an equivalence relation. Isometry classes of (P, u) are denoted by $[(P, u)]$.

Now, let $\mathscr{K} = (\mathscr{K}, T, \Delta\mathscr{K}; \#, \varpi, \lambda)$ be a triangulated category with duality (see Diagram 11.5). Note that $\Delta\mathscr{K}$ inherits a structure of an additive category with duality. So, we can talk about symmetric objects both in $\mathscr{K}$ and in $\Delta\mathscr{K}$.

The Grothendieck–Witt group is defined as the abelian group $GW^0(\mathscr{K})$, generated by the set of isometry classes $[(P, u)]$ of the symmetric spaces, subject to the following relations:

1. For symmetric spaces (P, u), (Q, v), we have $[(P, u) \perp (Q, v)] = [(P, u)] + [(Q, v)]$.
2. Consider a symmetric space $\varphi_\bullet : (u, v, w) \xrightarrow{\sim} (u, v, w)^\#$ in $\Delta\mathscr{K}$, which is given by the following commutative diagram:

$$\begin{array}{ccccccc} P_{-1} & \xrightarrow{u} & P_0 & \xrightarrow{v} & P_1 & \xrightarrow{w} & TP_{-1} \\ \varphi_{-1}\downarrow & & \downarrow\varphi_0 & & \downarrow\varphi_{-1} & & \downarrow T\varphi_{-1} \\ P_1^\# & \xrightarrow[v^\#]{} & P_0^\# & \xrightarrow[u^\#]{} & P_{-1}^\# & \xrightarrow[T(w^\#)\lambda]{} & TP_1^\# \end{array} \quad \ni \quad \begin{cases} \varphi_{-1} = \varphi_1\varpi \\ \varphi_0 = \varphi_0\varpi \\ \varphi_1 = \varphi_{-1}\varpi \end{cases} \qquad (11.5)$$

Given such a symmetric space $\Delta\mathscr{K}$, we have

$$[(P_0, \varphi_0)] = \left[\left(P_{-1} \oplus P_1, \begin{pmatrix} 0 & \varphi_1 \\ \varphi_{-1} & 0 \end{pmatrix}\right)\right].$$

This completes the definition of the **Grothendieck–Witt group** $GW(\mathscr{K})$.

Further, define the Witt group $W(\mathscr{K})$ as the abelian group, generated by the set of isometry classes $[(P, u)]$ of the symmetric spaces, subject to the relation (1) and given a symmetric space in $\Delta\mathscr{K}$, as in (11.5), set

$$[(P_0, \varphi_0)] = 0.$$

This completes the definition of the **Witt group** $W(\mathscr{K})$.

11.2. Triangulated Witt Groups

The study of Witt groups of triangulated categories, crystallized to a complete theory, thanks to [Bp, Bp4, BW] and others.

Remark 11.2.1. Throughout (as always), we assume that 2 is invertible in such triangulated categories $\mathscr{K}$. This means that $\forall A, B \in \mathscr{K}$, the map $2 : Mor_{\mathscr{K}}(A, B) \xrightarrow{\sim} Mor_{\mathscr{K}}(A, B)$ is an isomorphism. We write $1/2 \in \mathscr{K}$ to mean the same. In this section, and subsequently, for a triangulated category $\mathscr{K} := (\mathscr{K}, T, \Delta\mathscr{K})$, the shift functor T is assumed to be an automorphism. In other words, T has an inverse $T^{-1} : \mathscr{K} \longrightarrow \mathscr{K}$.

The definition (11.1.4) of duality $\# : \mathscr{K} \longrightarrow \mathscr{K}$ was modified in [Bp] to deal with issues related to skew dualities and their relationships. The definition is also simplified by replacing the natural isomorphism λ by the identity $T\# = \#T^{-1}$. So, we redefine a triangulated category with δ-duality, as follows.

Definition 11.2.2. Let $\mathscr{K} := (\mathscr{K}, T, \Delta\mathscr{K})$ be a triangulated category. Let $\delta = \pm 1$. We assume that T has an inverse $T^{-1} : \mathscr{K} \longrightarrow \mathscr{K}$ as an additive functor.

An additive functor $\# : \mathscr{K}^{op} \longrightarrow \mathscr{K}$, with

$$\begin{cases} T\# = \#T^{-1}, & \text{and} \\ \varpi : 1_{\mathscr{K}} \xrightarrow{\sim} \#\# & \text{a natural isomorphism} \end{cases}$$

is called a **δ-duality** on $\mathscr{K}$ if

1. For any exact triangle, $A \xrightarrow{u} B \xrightarrow{v} C \xrightarrow{w} TA$, the triangle

$$C\# \xrightarrow{v^\#} B\# \xrightarrow{u^\#} A\# \xrightarrow{\delta T(w^\#)} TC\# \quad \text{is exact.} \tag{11.6}$$

2. For all objects X, we have $T(\varpi_X) = \varpi_{TX}$.
3. For all objects X, the diagram

$$\begin{array}{ccc} X^\# & \xrightarrow{\varpi_{X^\#}} & X^{\#\#\#} \\ & {\searrow}^{1} & \downarrow \varpi_X^\# \\ & & X^\# \end{array} \quad \text{commutes.}$$

From now on, we will be working under such a setup, with such **triangulated categories** $\mathscr{K} := (\mathscr{K}, T, \Delta\mathscr{K}, \#, \varpi) := (\mathscr{K}, \#, \varpi)$ **with δ-duality**. Note that these signs $\delta = \pm 1$ are ingrained in the structure of $\mathscr{K}$ and cannot be switched or replaced. If $\delta = 1$, this is simply referred to as a duality, and if $\delta = -1$, this is referred to as a **skew duality**.

For $i = 1, 2$, let $\mathscr{K}_i := (\mathscr{K}_i, \#_i, \varpi_i)$, be triangulated categories with δ-duality. Functors $F : \mathscr{K}_1 \longrightarrow \mathscr{K}_2$, as defined in (11.1.5), are called **duality preserving triangle functors**.

Exercise 11.2.3. Let $\delta = \pm 1$, and $\mathscr{K} := (\mathscr{K}, T, \Delta\mathscr{K}, \#, \varpi)$ be a triangulated category with δ-duality. Let $n \in \mathbb{Z}$. Then

1. $(\mathscr{K}, T, \Delta\mathscr{K}, T^n o\#, \varpi)$ is a triangulated category with $(-1)^n$ δ-duality.
2. Further, $(\mathscr{K}, T, \Delta\mathscr{K}, \#, -\varpi)$ is a triangulated category with δ-duality.

Definition 11.2.4. Let $\mathscr{K}$ be a triangulated category, with δ-duality. A map $\psi : X \longrightarrow X^\#$ is defined to be symmetric if $\psi = \psi^\# \varpi_X$. A symmetric space is a pair (X, ψ), where $\psi : X \xrightarrow{\sim} X^\#$ is a symmetric isomorphism.

The following is a key theorem in [Bp] for subsequent use.

Theorem 11.2.5. *Let $\delta = \pm 1$, and $\mathscr{K} := (\mathscr{K}, T, \Delta\mathscr{K}, \#, \varpi)$ be a triangulated category with δ-duality. As usual, assume that $1/2 \in \mathscr{K}$. Let $u : X \longrightarrow X^{\#}$ be a symmetric map (i.e. $u = \varpi_X u^{\#}$). Let*

$$X \xrightarrow{u} X^{\#} \xrightarrow{v} C \xrightarrow{w} TX \quad \text{be an exact triangle.} \tag{11.7}$$

Then there is an isomorphism $\psi : C \xrightarrow{\sim} T(C^{\#})$ such that the following diagram

$$\begin{array}{ccccccc} X & \xrightarrow{u} & X^{\#} & \xrightarrow{v} & C & \xrightarrow{w} & TX \\ \downarrow{\scriptstyle \delta\varpi_X} & & \| & & \wr\downarrow{\scriptstyle \psi} & & \downarrow{\scriptstyle \delta T(\varpi_X)} \\ X^{\#\#} & \xrightarrow[\delta u^{\#}]{} & X^{\#} & \xrightarrow[-T(w^{\#})]{} & T(C^{\#}) & \xrightarrow[T(v^{\#})]{} & T(X^{\#\#}) \end{array} \quad \begin{cases} \text{commutes, and} \\ T(\psi^{\#})\varpi_C = -\delta\psi \end{cases} \tag{11.8}$$

The identity $T(\psi^{\#})\varpi_C = -\delta\psi$ is described by the commutative diagram

$$\begin{array}{ccccc} C^{\#\#} & = & T\left(T(C^{\#})^{\#}\right) & \xrightarrow{T(\psi^{\#})} & T\left(C^{\#}\right) \\ \uparrow{\scriptstyle \varpi_C} & & & \nearrow{\scriptstyle -\delta\psi} & \\ C & & & & \end{array}$$

This means that (C, ψ) is a symmetric space in $(\mathscr{K}, T\#, -\delta\varpi)$. Further, other choices of the cone (C, v, w) of u lead to isometric symmetric spaces (C, ψ) in $(\mathscr{K}, T\#, -\delta\varpi)$.

Proof. See [Bp, Thm. 2.6]. Note that the second triangle is obtained from δ-dual (11.6) of the first exact triangle, by necessary sign changes. A map ψ can be obtained by $TR3$, which would necessitate rotations and sign changes. To ensure symmetry, ψ is replaced by the average of ψ and its dual. ■

Given a triangulated category $\mathscr{K}$ with δ-duality, we define n-shifted triangulated category $T^n\mathscr{K}$ for integers n.

Definition 11.2.6. Let $\delta = \pm 1$, and $\mathscr{K} := (\mathscr{K}, T, \Delta\mathscr{K}, \#, \varpi)$ be a triangulated category with δ-duality. Then define the shifted triangulated categories of $\mathscr{K}$ by

$$\begin{cases} T\mathscr{K} := (\mathscr{K}, T, \Delta\mathscr{K}, T\#, (-\delta)\varpi) & \text{with } (-\delta)\text{-duality} \\ T^{-1}\mathscr{K} := (\mathscr{K}, T, \Delta\mathscr{K}, T^{-1}\#, (+\delta)\varpi) & \text{with } (-\delta)\text{-duality.} \end{cases} \tag{11.9}$$

More generally, the process can be iterated, and for $n \in \mathbb{Z}$, the n-shifted triangulated category $T^n\mathscr{K}$, with δ_n-duality by

$$T^n\mathscr{K} = \begin{cases} T(T^{n-1}\mathscr{K}) & \delta_n = -\delta_{n-1} \quad \text{if } n \geq 1 \\ T^{-1}(T^{n+1}\mathscr{K}) & \delta_n = -\delta_{n+1} \quad \text{if } n \leq -1. \end{cases}$$

It is also easy to see that

$$T^n\mathscr{K} = T(T^{n-1}\mathscr{K}) \qquad \forall n \in \mathbb{Z}.$$

In particular,

$$\begin{cases} T^2\mathscr{K} := (\mathscr{K}, T, \Delta\mathscr{K}, T^2\#, -\varpi) & \text{with } (+\delta)\text{-duality} \\ T^{-2}\mathscr{K} := (\mathscr{K}, T, \Delta\mathscr{K}, T^{-2}\#, -\varpi) & \text{with } (+\delta)\text{-duality} \\ \text{If } \delta = 1, \quad T^n\mathscr{K} := \left(\mathscr{K}, T, \Delta\mathscr{K}, T^n\#, (-1)^{\frac{n(n+1)}{2}}\varpi\right) & \text{with } (-1)^n\text{-duality.} \end{cases} \tag{11.10}$$

Definition 11.2.7. Let $\delta = \pm 1$, and $\mathscr{K} := (\mathscr{K}, T, \Delta\mathscr{K}, \#, \varpi)$ be a triangulated category with δ-duality. Let $u : X \longrightarrow X^{\#}$ be a symmetric morphism. Define,

$$Cone(u) = (C, \psi) \qquad \text{as given by Theorem 11.2.5.}$$

For the next definition (11.2.11), note that a symmetric map in $T^{-1}\mathscr{K}$ is a map

$$u : X \longrightarrow T^{-1}X^{\#} \qquad \ni \qquad \begin{array}{ccc} X & \xrightarrow{u} & T^{-1}X^{\#} \\ {\scriptstyle \delta\varpi_X}\downarrow & \nearrow_{u^{\#}} & \\ X^{\#\#} & & \end{array} \qquad \text{commutes.}$$

We define neutral spaces.

Definition 11.2.8. Let $\delta = \pm 1$, and $\mathscr{K} := (\mathscr{K}, T, \Delta\mathscr{K}, \#, \varpi)$ be a triangulated category with δ-duality. A symmetric space (P, ψ) in $\mathscr{K}$ is defined as **neutral** (or **metabolic**) if there is a symmetric space

$$u : X \longrightarrow T^{-1}X^{\#} \quad \text{in} \quad T^{-1}\mathscr{K} \quad \ni \quad Cone(u) = (P, \psi).$$

The following equivalent conditions characterize neutral spaces [Bp, Lem. 3.1].

Exercise 11.2.9. Let $\delta = \pm 1$, and $\mathscr{K} := (\mathscr{K}, T, \Delta\mathscr{K}, \#, \varpi)$ be a triangulated category with δ-duality. Let (P, φ) be a symmetric space. Then the following conditions are equivalent:

1. (P, φ) is neutral.
2. There is an exact triangle

$$T^{-1}L^{\#} \xrightarrow{\ w\ } L \xrightarrow{\ u\ } P \xrightarrow{\ u^{\#}\varphi\ } L^{\#} \quad \ni \quad \begin{array}{ccc} T^{-1}L^{\#} & \xrightarrow{\ w\ } & L \\ & \searrow^{T^{-1}w^{\#}} & \wr\downarrow \varpi \\ & & L^{\#\#} \end{array} \quad \text{commutes}$$

   We also write $w = T^{-1}w^{\#}$. In this case, we say that (L, u, w) is a **Lagrangian** of (P, φ).
3. There is an isomorphism $h : L \xrightarrow{\sim} M$ and an isomorphism of exact triangles, as follows:

$$\begin{array}{ccccccc} T^{-1}M^{\#} & \xrightarrow{\ w\ } & L & \xrightarrow{\ u\ } & P & \xrightarrow{\ v\ } & M^{\#} \\ {\scriptstyle T^{-1}(h^{\#})}\downarrow\wr & & \wr\downarrow h & & \downarrow\varphi & & \wr\downarrow h^{\#} \\ T^{-1}L^{\#} & \xrightarrow[T^{-1}(w^{\#})]{} & M & \xrightarrow[v^{\#}]{} & P^{\#} & \xrightarrow{\ u^{\#}\ } & L^{\#} \end{array}$$

Exercise 11.2.10. Let $\delta = \pm 1$, and $\mathscr{K} := (\mathscr{K}, T, \Delta\mathscr{K}, \#, \varpi)$ be a triangulated category with δ-duality. Let (P, φ) be a symmetric space. Then $(P, \varphi) \perp (P, -\varphi)$ is neutral.

Proof. We prove that the diagonal map $\Delta : P \longrightarrow P \oplus P$ is a Lagrangian of $(P, \varphi) \perp (P, -\varphi)$. So, we prove that the triangle

$$T^{-1}P^{\#} \xrightarrow{\ 0\ } P \xrightarrow{\ \Delta\ } P \oplus P \xrightarrow{\ \Delta^{\#}\varphi\ } P^{\#} \qquad \text{is exact.}$$

First, $0 \longrightarrow P \xrightarrow{1} P \longrightarrow 0$ and $T^{-1}P \longrightarrow 0 \longrightarrow P \xrightarrow{-1} P$ are exact. Consider the following three isomorphic triangles:

$$\begin{array}{ccccccc}
T^{-1}P & \xrightarrow{0} & P & \xrightarrow{(1\,0)} & P \oplus P & \xrightarrow{\begin{pmatrix} 0 \\ -1 \end{pmatrix}} & P \\
\| & & \| & & \downarrow \wr & & \| \\
T^{-1}P & \xrightarrow{0} & P & \xrightarrow{(1\,1)} & P \oplus P & \xrightarrow{\begin{pmatrix} 0 \\ -1 \end{pmatrix}} & P \\
\downarrow & & \| & & \| & & \wr \downarrow \varphi \\
T^{-1}P\# & \xrightarrow{0} & P & \xrightarrow[(1\,1)]{} & P \oplus P & \xrightarrow[\begin{pmatrix} \varphi \\ -\varphi \end{pmatrix}]{} & P\#
\end{array}.$$

The first line being exact, so is the third one. ■

Now we define the Witt group of a triangulated category $\mathscr{K}$ with δ-duality.

Definition 11.2.11. Let $\delta = \pm 1$, and $\mathscr{K} := (\mathscr{K}, T, \Delta\mathscr{K}, \#, \varpi)$ be a triangulated category with δ-duality.

1. The **Witt monoid** $MW(\mathscr{K})$ is defined to be the set of all isometry classes of symmetric spaces $[(P, \psi)]$ in $\mathscr{K}$. Note that $MW(\mathscr{K})$ has a structure of an abelian monoid, under the orthogonal sum. The subset $NW(\mathscr{K}) \subseteq MW(\mathscr{K})$ of all classes $[Cone(u)]$ of neutral spaces is a submonoid.
2. Define the Witt groups of $\mathscr{K}$ by

$$W(\mathscr{K}) := \frac{MW(\mathscr{K})}{NW(\mathscr{K})} \quad \text{as the quotient monoid.}$$

In fact, these are groups by (11.2.10). This definition is also consistent with the definition 11.5, by (11.2.9).

3. For $n \in \mathbb{Z}$, define the n-shifted Witt groups

$$W^n(\mathscr{K}) := W(T^n\mathscr{K}). \tag{11.11}$$

These are also referred to as **Balmer–Witt groups** of $\mathscr{K}$.

Remark 11.2.12. Most importantly, these definitions of shifted Witt groups (11.11) apply to derived categories $\mathbf{D}^b(\mathscr{E}, *, \eta)$ of exact categories $(\mathscr{E}, *, \eta)$ with duality, as in (10.3.15). Among them are the categories $\left(C\mathbb{M}^k(X), \mathcal{E}xt^k(-, \mathcal{L}), \eta\right)$, as in Section 10.4.5.

These groups have four periodicities.

Exercise 11.2.13. Let $\delta = \pm 1$, and $\mathscr{K} := (\mathscr{K}, T, \Delta\mathscr{K}, \#, \varpi)$ be a triangulated category with δ-duality. For $n \in \mathbb{Z}$, the functor $T^2 : T^n\mathscr{K} \longrightarrow T^{n+4}\mathscr{K}$ is a duality preserving isomorphism. Consequently,

$$W^n(\mathscr{K}) \xrightarrow{\sim} W^{n+4}(\mathscr{K}) \qquad \text{is an isomorphism.}$$

For, $n = 0, 1, 2, 3$, the four groups are described as follows:

$$\begin{cases} W^0(\mathscr{K}) = W(\mathscr{K}, T, \Delta\mathscr{K}, \#, \varpi) & \text{we have } \delta \text{ duality} \\ W^1(\mathscr{K}) = W(\mathscr{K}, T, \Delta\mathscr{K}, T\#, -\delta\varpi) & \text{we have } -\delta \text{ duality} \\ W^2(\mathscr{K}) = W(\mathscr{K}, T, \Delta\mathscr{K}, T^2\#, -\varpi) & \text{we have } +\delta \text{ duality} \\ W^3(\mathscr{K}) = W(\mathscr{K}, T, \Delta\mathscr{K}, T^3\#, \delta\varpi) & \text{we have } -\delta \text{ duality.} \end{cases}$$

11.2.1. *Localization*

We define the quotient of triangulated categories and develop exact sequences of shifted Witt groups. First, we have the following definition.

Definition 11.2.14. Let $\mathscr{K} := (\mathscr{K}, T, \Delta\mathscr{K})$ be a triangulated category and $\iota : \mathscr{B} \subseteq \mathscr{K}$ be a triangulated full subcategory. We say that $\mathscr{B}$ is *épaisse* (thick) in $\mathscr{K}$ if every direct summand of an object in $\mathscr{B}$ is in $\mathscr{B}$. See [Na2, Criterion 1.3]).

Given a triangle functor $F : \mathscr{K}_1 \longrightarrow \mathscr{K}_2$, the kernel of F is defined to be the full subcategory,

$$\ker(F) := \{X \in \mathscr{K}_1 : F(X) \xrightarrow{\sim} 0\}.$$

Definition 11.2.15. Let $\mathscr{K} := (\mathscr{K}, T, \Delta\mathscr{K})$ be a triangulated category and $\iota : \mathscr{B} \subseteq \mathscr{K}$ be a triangulated full subcategory. Then there is a triangulated category, to be denoted by $\frac{\mathscr{K}}{\mathscr{B}}$, together with a triangle functor $\mathfrak{q} : \mathscr{K} \longrightarrow \frac{\mathscr{K}}{\mathscr{B}}$, such that $\mathscr{B} \subseteq \ker(\mathfrak{q})$, and $\mathfrak{q}$ is universal among such functors, as follows:

$$\begin{array}{ccccc} \mathscr{B} & \xrightarrow{\iota} & \mathscr{K} & \xrightarrow{\mathfrak{q}} & \frac{\mathscr{K}}{\mathscr{B}} \\ & & & \searrow^{F} & \downarrow \mathfrak{f} \\ & & & & \mathscr{D} \end{array} \tag{11.12}$$

This means that given any other triangle functor F such that $\mathscr{B} \subseteq \ker(F)$, there is a unique functor $\mathfrak{f}$ such that $\mathfrak{f}\mathfrak{q} = F$. The category $\frac{\mathscr{K}}{\mathscr{B}}$ is called the quotient category, also referred to as the **Verdier quotient**. See [Na1, Ch. 2] for details of this construction.

Let $\mathfrak{S}$ be the class of all morphisms u in $\mathscr{K}$ such that $Cone(u)$ is in $\mathscr{B}$. One checks that $\mathfrak{S}$ satisfies the calculus of fractions (A.5.2). Then the quotient category is defined as the category $\mathscr{T} := \frac{\mathscr{K}}{\mathscr{B}} := \mathfrak{S}^{-1}\mathscr{K}$, together with the canonical functor $\mathfrak{q} : \mathscr{K} \longrightarrow \mathscr{T}$.

1. It follows that $\ker(\mathfrak{q})$ is *épaisse* (thick). In fact, $\ker(\mathfrak{q})$ is the smallest thick subcategory containing $\mathscr{B}$. Further, $\ker(\mathfrak{q})$ is a full subcategory of objects in $\mathscr{K}$, which are the direct summands of objects in $\mathscr{B}$.
2. Under this setup, as above, $\mathscr{T}$ has a structure of a triangulated category. The shift functor T in $\mathscr{K}$ induces the shift functor in $\mathscr{T}$, which we continue to denote by T. A triangle $A \longrightarrow B \longrightarrow C \longrightarrow TA$ in $\mathscr{T}$ is defined as exact if it is isomorphic to the image of an exact triangle in $\Delta\mathscr{K}$.
3. We enlarge $\mathfrak{S}$ and define $\tilde{\mathfrak{S}}$ to be the set of all morphisms f such that $\mathfrak{q}(f)$ is an isomorphism. Then $\frac{\mathscr{K}}{\mathscr{B}} \cong \mathfrak{S}^{-1}\mathscr{K} \cong \tilde{\mathfrak{S}}^{-1}\mathscr{K}$. Note that $\tilde{\mathfrak{S}}$ is saturated, meaning for two composable morphisms $X \xrightarrow{u} Y \xrightarrow{v} Z$, if two of u, v, vu are in $\tilde{\mathfrak{S}}$, then so is the third one. As a default, we would replace $\mathfrak{S}$ by $\tilde{\mathfrak{S}}$ and assume that $\mathfrak{S}$ is the set of all morphisms f such that $\mathfrak{q}(f)$ is an isomorphism.
4. A sequence $\mathfrak{T}_0 \xrightarrow{i} \mathfrak{T}_1 \xrightarrow{q} \mathfrak{T}_2$ of triangle functors is said to be an **exact sequence** if (1) i is a full subcategory inclusion, (2) $\mathfrak{T}_0 = \ker(\mathfrak{q})$ (up to equivalence), (3) the induced functor $\mathfrak{f} : \frac{\mathfrak{T}_2}{\mathfrak{T}_0} \xrightarrow{\sim} \mathfrak{T}_2$ is an equivalence of categories.

   If condition (3) is replaced by requiring that $\mathfrak{f}$ is cofinal, we say that the sequence is **exact up to factors (direct sum)**.

Now assume that

$$\mathscr{B} \xrightarrow{\iota} \mathscr{K} \xrightarrow{\mathfrak{q}} \mathfrak{T} \quad \text{is exact. So,} \quad \mathfrak{T} := \frac{\mathscr{K}}{\mathscr{B}}, \ \mathscr{B} = \ker(\mathfrak{q}),$$

$$\mathfrak{S} = \{f : \mathfrak{q}(f) \text{ is an isomorphism}\}.$$

Further, assume that $\mathscr{K} := (\mathscr{K}, T, \Delta\mathscr{K}, \#, \varpi)$ is a triangulated category δ-duality, such that $\#(\mathfrak{S}) \subseteq \mathfrak{S}$. Then by the universal property (11.12), the quotient $\mathfrak{T} = \frac{\mathscr{K}}{\mathscr{B}}$, and $\mathscr{B}$, inherit a δ-duality structure, as in the commutative diagram

$$\begin{array}{ccccc} \mathscr{B}^{op} & \xrightarrow{\iota} & \mathscr{K}^{op} & \xrightarrow{\mathfrak{q}} & \mathfrak{T}^{op} \\ \# \downarrow & & \downarrow \# & & \downarrow \# \\ \mathscr{B} & \xrightarrow[\iota]{} & \mathscr{K} & \xrightarrow[\mathfrak{q}]{} & \mathfrak{T} \end{array} \tag{11.13}$$

Lemma 11.2.16. *Let $\delta = \pm 1$. Let $F : \mathscr{K}_1 \longrightarrow \mathscr{K}_2$ be a functor of triangulated category with δ-duality, where $\mathscr{K}_i := (\mathscr{K}_i, T_i, \Delta\mathscr{K}_i, \#_i, \varpi_i)$ are triangulated categories with δ-duality. Let $F : \mathscr{K}_1 \longrightarrow \mathscr{K}_2$ be a duality-preserving triangle functor. Then for all $n \in \mathbb{Z}$, F induces natural homomorphisms,*

$$W^n(F) : W^n(\mathscr{K}_1) \longrightarrow W^n(\mathscr{K}_2).$$

Before we state the localization theorem, we state the Enriched octahedron axiom [Bp, TR4+].

Definition 11.2.17 (Tr4+). (Enriched Octahedron Axiom.) Suppose that $\mathscr{K} := (\mathscr{K}, T, \Delta\mathscr{K})$ is a triangulated category. Then given two composable arrows u, v, consider the diagram of $TR4$ (11.4). Then f and g can be chosen, such that the triangles

$$B \xrightarrow{w_1 v} W \xrightarrow{\begin{pmatrix} -w_2 \\ g \end{pmatrix}} T(A) \oplus V \xrightarrow{\begin{pmatrix} Tu & v_2 \end{pmatrix}} TB$$

$$B \xrightarrow[\begin{pmatrix} v \\ u_1 \end{pmatrix}]{} C \oplus U \xrightarrow[\begin{pmatrix} w_1 & -f \end{pmatrix}]{} W \xrightarrow[T(u)w_2]{} TB$$

are exact.

(11.14)

We use this in the context of derived categories of exact categories, in which case this axiom is satisfied.

The following is the localization theorem of Balmer–Witt groups [Bp, Thm. 6.2], [BW].

Theorem 11.2.18. *Let* $\delta = \pm 1$, *and* $\mathscr{K} := (\mathscr{K}, T, \Delta\mathscr{K}, \#, \varpi)$ *be a triangulated category with* δ*-duality. Assume that* $1/2 \in \mathscr{K}$, *and* $\mathscr{K}$ *satisfies the* $TR4+$ *axiom. Let* $\iota : \mathscr{B} \hookrightarrow \mathscr{K}$ *be a full subcategory, and* $\mathfrak{T} := \frac{\mathscr{K}}{\mathscr{B}}$ *be the quotient category such that*

$$\mathscr{B} \xrightarrow{\ \iota\ } \mathscr{K} \xrightarrow{\ \mathfrak{q}\ } \mathfrak{T} \quad \text{is an exact sequence.}$$

As commented above, both $\mathscr{B}$ *and* $\mathfrak{T}$ *inherit the duality structure, as in* (11.12)*. Then there is an exact sequence of shifted Witt groups as follows:*

$$\cdots \longrightarrow W^n(\mathscr{B}) \xrightarrow{W^n(\iota)} W^n(\mathscr{K}) \xrightarrow{W^n(\mathfrak{q})} W^n(\mathfrak{T}) \xrightarrow{\ \partial\ } W^{n+1}(\mathscr{B}) \longrightarrow \cdots$$

where $n \in \mathbb{Z}$.

Proof. For a complete proof, refer to [Bp, Thm. 6.2]. The maps $W^n(\iota)$, $W^n(\mathfrak{q})$ are clear (11.2.18). We outline the definition of the connecting homomorphisms ∂. We can assume that $n = 0$. Suppose that $u : X \longrightarrow X^\#$ is a symmetric morphism in $\mathscr{K}$ such that $u \in \mathfrak{S}$. This means that $u = u^\# \varpi_X$, and u induces an isomorphism in $\mathfrak{T}$. Let $(P, \varphi) := Cone(u)$. It follows (11.2.7), that (P, φ) is a symmetric space in $T^1\mathscr{B}$. It is clear that (X, u) represents a symmetric space in $\mathscr{K}$. We define

$$\partial([(X, u)]) = [(P, \varphi)] \in W^1(\mathscr{B}).$$

This extends to a well-defined homomorphism $\partial : W^0(\mathscr{K}) \longrightarrow W^1(\mathscr{B})$. ■

Remark 11.2.19. Considering $W^n(\mathscr{K}) \cong W^{n+4}(\mathscr{K})$, under the setup of (11.2.18), there is a 12-term exact sequence [Bp, Cor. 6.6]

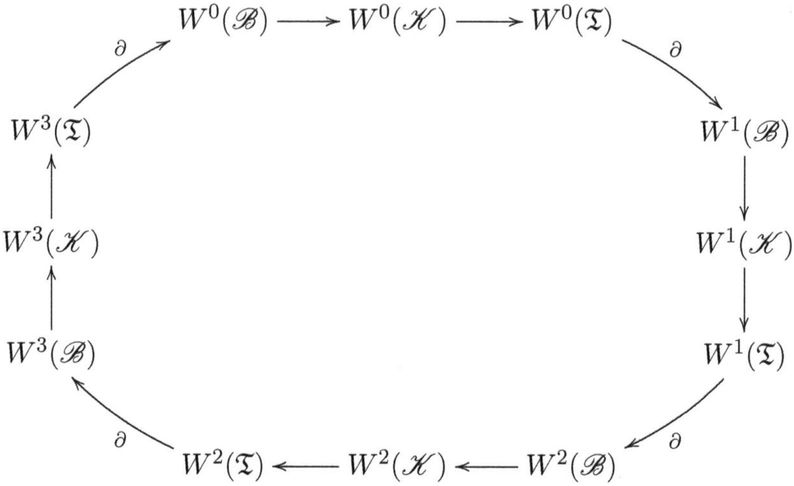

11.2.2. *Derived categories of exact categories and agreement*

This section is from [Bp4, Na2]. The definition of derived categories of abelian categories is available in the standard sources (e.g. [Wc1, §10.4]). The definition of derived categories of an exact category needs a little more care than that of an abelian category. Among the original sources for this is [Na2].

Definition 11.2.20. Let $\mathscr{E}$ be an exact category. As usual, $Ch(\mathscr{E})$ (resp. $Ch^b(\mathscr{E})$) denotes the category of the chain complexes (bounded). The functor $T : Ch(\mathscr{E}) \xrightarrow{\sim} Ch(\mathscr{E})$, given by $TP_\bullet = P_\bullet[-1]$, where $P[-1]_n = P_{n-1}\ \forall n$, is a natural shift functor. For morphisms, $f_\bullet : P_\bullet \longrightarrow Q_\bullet$, in $Ch(\mathscr{E})$, define $Cone(f_\bullet)$, as in (9.2.20). By a standard triangle, we mean the following:

$$P_\bullet \xrightarrow{f_\bullet} Q_\bullet \longrightarrow Cone(f_\bullet) \longrightarrow TP_\bullet \tag{11.15}$$

For maps $f_\bullet, g_\bullet : P_\bullet \longrightarrow Q_\bullet$, in $Ch(\mathscr{E})$, a homotopy from $f_\bullet$ to $g_\bullet$ is defined as a map $\tau_\bullet : P_\bullet \longrightarrow Q_\bullet$ of degree one such that $\tau_\bullet d^P + d^Q \tau_\bullet = f_\bullet - g_\bullet$. This is sometimes displayed as follows:

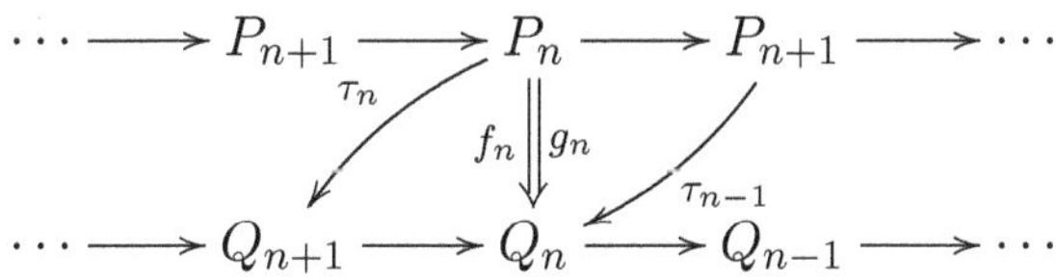

One checks that the homotopy is an equivalence relation on $Hom\,(P_\bullet, Q_\bullet)$. The homotopy category $\mathcal{K}(\mathscr{E})$ is defined as the category whose objects are the same as that of $Ch(\mathscr{E})$, and $Mor_{\mathcal{K}(\mathscr{E})}(P_\bullet, Q_\bullet) = \frac{Hom(P_\bullet, Q_\bullet)}{\sim}$, the set of equivalence classes of maps. Then $\mathcal{K}(\mathscr{E})$ has a natural structure of an additive category. We also define the homotopy category $\mathcal{K}^b(\mathscr{E})$ of bounded chain complexes as the full subcategory whose objects are the same as that of $Ch^b(\mathscr{E})$.

1. Both $\mathcal{K}(\mathscr{E})$, $\mathcal{K}^b(\mathscr{E})$ are triangulated categories, with shift functor $TP_\bullet = P_\bullet[-1]$. An exact triangle in $\mathcal{K}(\mathscr{E})$ is one that is isomorphic to a standard triangle (11.15).
2. A complex

$$\cdots \longrightarrow P_{n+1} \longrightarrow P_n \xrightarrow{d_n} P_{n-1} \longrightarrow \cdots \tag{11.16}$$

   is defined to be acyclic, if $\forall n,\ d_n$ decomposes as

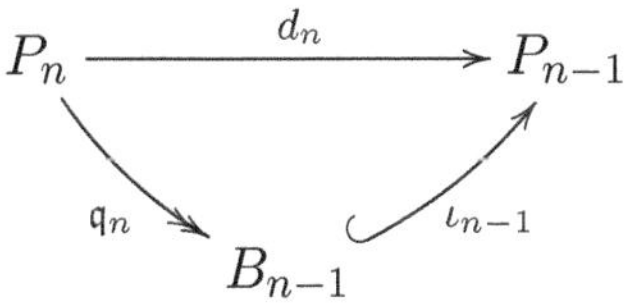

   where $\begin{cases} q_n \text{ is admissible epi} \\ \iota_{n-1} \text{ is admissible mono.} \end{cases}$

   In other words, if $P_\bullet$ is obtained by concatenating conflations in $\mathscr{E}$.
3. Let $\mathcal{K}_{ac}(\mathscr{E}) \subseteq \mathcal{K}(\mathscr{E})$, $\mathcal{K}^b_{ac}(\mathscr{E}) \subseteq \mathcal{K}^b(\mathscr{E})$, respectively, denote the full subcategory of acyclic complexes.

The following is a slightly more relaxed condition than idempotent completeness.

Definition 11.2.21. First, idempotent complete exact categories are also called **saturated** exact categories. Let $\mathscr{E}$ be an exact category. As always, a morphism $p : E \longrightarrow F$ is called a split epimorphism if $p\sigma = 1_F$ for some $\sigma : F \longrightarrow E$. We say that $\mathscr{E}$ is **semisaturated** if every split epimorphism is an admissible epimorphism.

1. If $\mathscr{E}$ is idempotent complete, then it is semisaturated.
2. If $\mathscr{E}$ is semisaturated, then there is a fully faithful embedding $\mathscr{E} \hookrightarrow \mathscr{A}$, where $\mathscr{A}$ is an abelian category such that
   (a) The embedding $\mathscr{E} \hookrightarrow \mathscr{A}$ is extension closed.
   (b) The embedding $\mathscr{E} \hookrightarrow \mathscr{A}$ preserves and detects exactness.
   (c) If $p : E \longrightarrow F$ is a morphism in $\mathscr{E}$, and if it is an epimorphism in $\mathscr{A}$, then it is admissible epi in $\mathscr{E}$.

Definition 11.2.22. Under the setup of (11.2.20), $\mathcal{K}_{ac}(\mathscr{E})$ and $\mathcal{K}^b_{ac}(\mathscr{E})$ are triangulated subcategories. Further,

1. If $\mathscr{E}$ is idempotent complete, then $\mathcal{K}_{ac}(\mathscr{E})$ is *épaisse* in $\mathcal{K}(\mathscr{E})$. The derived category of $\mathscr{E}$ is defined as the Verdier quotient as follows:

$$\mathbf{D}(\mathscr{E}) := \frac{\mathcal{K}(\mathscr{E})}{\mathcal{K}_{ac}(\mathscr{E})}.$$

   If $\mathscr{E}$ is not idempotent complete, define

$$\mathbf{D}(\mathscr{E}) := \frac{\mathcal{K}(\mathscr{E})}{(\mathcal{K}_{ac}(\mathscr{E}))^{ep}} \quad \text{where} \quad (\mathcal{K}_{ac}(\mathscr{E}))^{ep} \quad \text{is the } \textit{épaisse} \text{ completion.}$$

2. If $\mathscr{E}$ is semisaturated, then $\mathcal{K}^b_{ac}(\mathscr{E})$ is *épaisse* in $\mathcal{K}^b(\mathscr{E})$. Consequently, the Verdier quotient exists, and we define the bounded derived category as a Verdier quotient as follows:

$$\mathbf{D}^b(\mathscr{E}) := \frac{\mathcal{K}^b(\mathscr{E})}{\mathcal{K}^b_{ac}(\mathscr{E})}.$$

   This is obtained by inverting quasi-isomorphisms in $\mathcal{K}^b(\mathscr{E})$.

These definitions of derived categories cover the situations we discussed in (9.2.19). Note that, for a Noetherian scheme X, all the subcategories of $\mathcal{K}(QCoh(X))$ we discussed in Section 9.2.3 are idempotent complete, which makes things much more straightforward.

Proof. See [Na2]. ■

We will mainly be interested in the bounded derived category $\mathbf{D}^b(\mathscr{E})$, and will assume that $\mathscr{E}$ is a semisaturated exact category.

Definition 11.2.23. Let $(\mathscr{E}, *, \eta)$ be an exact category with duality. Assume that $\mathscr{E}$ is semisaturated. Then the bounded derived category $\mathbf{D}^b(\mathscr{E})$ inherits a natural duality (with $\delta = +1$). The duality is defined as in (10.10). We denote this by $\left(\mathbf{D}^b(\mathscr{E}), \#, \varpi\right)$.

We state the agreement theorem [Bp4, Thm. 4.3] as follows.

Theorem 11.2.24. *Let $(\mathscr{E}, *, \eta)$ be an exact category with duality. We assume that $1/2 \in \mathscr{E}$, and $\mathscr{E}$ is semisaturated. Then the degree-zero map $(\mathscr{E}, *, \eta) \longrightarrow \left(\mathbf{D}^b(\mathscr{E}), \#, \varpi\right)$ induces a natural isomorphism*

$$W(\mathscr{E}, *, \eta) \xrightarrow{\sim} W^0\left(\mathbf{D}^b(\mathscr{E}), \#, \varpi\right)$$

where the group $W(\mathscr{E}, *, \eta)$ was defined in (10.1.6).

Proof. See [Bp4, Thm. 4.3]. ■

Notation 11.2.25. Use the set of (11.2.24). In the light of (11.2.24), for integers $n \in \mathbb{Z}$, we denote

$$W^n(\mathscr{E}, *, \eta) := W^n\left(\mathbf{D}^b(\mathscr{E}), \#, \varpi\right)$$

Note that $\begin{cases} W(\mathscr{E}) := W^0(\mathscr{E}, *, \eta) \text{ is as defined in (10.1.6).} \\ W^2(\mathscr{E}, *, \eta) = W^0(\mathscr{E}, *, -\eta) \text{ corresponds to the skew duality.} \end{cases}$

11.3. Derived Witt Groups of Schemes

Let X be a Noetherian scheme, with $1/2 \in \mathcal{O}_X$, and $\mathcal{L}$ be an invertible sheaf. Then $P \mapsto Hom(P, \mathcal{L})$ induces a duality on the exact category $\mathscr{P}(X)$. The following is an immediate consequence of the above.

Theorem 11.3.1. *Let X be a regular scheme, and $\mathcal{L}$ be an invertible sheaf. Assume that $1/2 \in \mathcal{O}_X$. Let $\mathbf{D}^b(X) := \mathbf{D}^b(X, \mathcal{L}) := (\mathbf{D}^b(\mathscr{P}(X), \#, \varpi)$ denote the bounded derived category with duality* *induced by* $P \mapsto P^* := Hom(P, \mathcal{L})$. *All the subsequent dualities, in a variety of categories considered, are induced by this duality* $P \mapsto P^* := Hom(P, \mathcal{L})$ *on* $\mathscr{P}(X)$.

Let $U \subseteq X$ *be an open subset, and* $Z := X - U$. *Let* $\mathbf{D}^b_Z(X) \subseteq \mathbf{D}^b(X)$ *be the full subcategory of objects* $P_\bullet$ *such that* $(P_\bullet)_{|U}$ *is acyclic. Then the sequence*

$$\mathbf{D}^b_Z(X, \mathcal{L}) \longrightarrow \mathbf{D}^b(X, \mathcal{L}) \longrightarrow \mathbf{D}^b\left(U, \mathcal{L}_{|U}\right) \quad \text{is exact.}$$

We use the notation

$$W^n(X) := W^n(X, \mathcal{L}) := W^n\left(\mathbf{D}^b(X, \mathcal{L})\right),$$

$$W^n(X, Z) := W^n(X, \mathcal{L}, Z) := W^n\left(\mathbf{D}^b_Z(X, \mathcal{L})\right).$$

Consequently, there is a long exact sequence of shifted Witt groups:

$$\cdots \longrightarrow W^n(X, \mathcal{L}, Z) \longrightarrow W^n(X, \mathcal{L}) \longrightarrow$$

$$W^n\left(U, \mathcal{L}_{|U}\right) \longrightarrow W^{n+1}(X, \mathcal{L}, Z) \longrightarrow \cdots$$

Proof. Follows from (11.2.18). See [Bp4, Rem. 4.9]. ∎

The following is an important perspective from [Bp4, Thm. 5.6].

Remark 11.3.2. Let $(A, \mathfrak{m})$ be a local Noetherian commutative ring, with $1/2 \in A$, and $X := \mathrm{Spec}(A)$. We use the usual duality induced by $P \mapsto Hom(P, A)$. Then

$$W^1(X) = W^2(X) = W^3(X) = 0$$

The following is the Dévissage theorem for the Witt theory [BM].

Proposition 11.3.3. *Let* $(A, \mathfrak{m})$ *be a regular local ring containing a field* κ, *with* $1/2 \in \kappa$ *and* $\dim A = d$. *Write* $X = \mathrm{Spec}(A)$. *Let* $\kappa(\mathfrak{m}) = \frac{A}{\mathfrak{m}}$ *be the residue field. Then*

$$W(\kappa(\mathfrak{m})) \xrightarrow{\sim} W\left(C\mathbb{M}^d(X)\right) \cong W^0\left(\mathbf{D}^b(C\mathbb{M}^d(X))\right) \cong W^d\left(\mathscr{D}^d(X)\right).$$

Proof. For the first isomorphism, see [BW, Prop. 6.2]. The second isomorphism follows from (11.2.24). The third isomorphism follows from (10.4.33). ∎

The following is from [BW, §7].

Theorem 11.3.4. *Let X be a regular scheme, and $\mathcal{L}$ be an invertible sheaf. Assume that $1/2 \in \mathcal{O}_X$. All the subsequent dualities we consider are induced by the duality $P \mapsto P^* := Hom(P, \mathcal{L})$ on $\mathscr{P}(X)$. For integers $k \geq 0$, from* (9.5.6) *and with the same notation, the sequence*

$$\mathscr{D}^{k+1}(\mathscr{P}(X)) \longrightarrow \mathscr{D}^{k}(\mathscr{P}(X)) \longrightarrow \coprod_{x\in X^{(k)}} \mathscr{D}^{k}(\mathscr{P}(X_x))$$

is an exact sequence

of triangulated categories with dualities. Further,

1. *There is a long exact sequence of shifted Witt groups:*

$$\cdots \longrightarrow W^n\left(\mathscr{D}^{k+1}(\mathscr{P}(X))\right) \longrightarrow W^n\left(\mathscr{D}^{k}(\mathscr{P}(X))\right) \longrightarrow \bigoplus_{x\in X^{(k)}} W^n\left(\mathscr{D}^{k}(\mathscr{P}(X_x))\right)$$

$$\longrightarrow W^{n+1}\left(\mathscr{D}^{k+1}(\mathscr{P}(X))\right) \longrightarrow \cdots \qquad n \in \mathbb{Z}$$

2. *For $x \in X^{(k)}$, and $n \in \mathbb{Z}$, we have*

$$W^{n+k}\left(\mathscr{D}^{k}(\mathscr{P}(X_x))\right) \cong W^n\left(C\mathbb{M}^k(X_x)\right)$$

where the right-hand side is interpreted as in (11.2.24). Since X is regular, $C\mathbb{M}^k(X_x)$ is the exact category of finite length modules, with duality $M \mapsto Ext^k(M, \mathcal{L}_x)$.

Proof. (1) is immediate from (11.2.18) and (2) follows from the derived category version of (10.4.33) and from using the notation in (11.2.24). ■

Following the same methods to develop the Gersten complex (10.44) for $\mathbb{GW}$-groups, we obtain a Gersten complex, as follows [BW], of Witt groups. Further, similar to the Gersten GW spectral sequence (10.4.40), we obtain a spectral sequence of Witt groups. Under the hypotheses and notation, as in (11.3.4), we can define the spectral sequence [BW, Thm. 7.2]:

$$E(X)_1^{kq} := EW(X)_1^{kq} = \begin{cases} \bigoplus_{x\in X^{(k)}} W^{k+q}\left(\mathscr{D}^k\mathscr{P}(X_x)\right) & 0 \leq k \leq d \\ 0 & \text{otherwise} \end{cases}$$

Compare this with other such spectral sequences (6.79, 9.36, 10.46).

The line $q = 0$ of this spectral sequence gives the Gersten–Witt complex, as follows.

Corollary 11.3.5. *Let X be as in (11.3.4), with $d = \dim X$. Let K be the function field of X. Then there is a Gersten complex of the Witt groups, of exact categories, as follows:*

$$\begin{array}{c} 0 \\ \downarrow \\ W(X, \mathcal{L})) \\ \downarrow \\ W(K) \\ \downarrow \\ \bigoplus_{x \in X^{(1)}} W\left(C\mathbb{M}^1(X_x)\right) \\ \downarrow \\ \cdots \\ \downarrow \\ \bigoplus_{x \in X^{(k-1)}} W\left(C\mathbb{M}^{k-1}(X_x)\right) \\ \downarrow \\ \bigoplus_{x \in X^{(k)}} W\left(C\mathbb{M}^{k}(X_x)\right) \\ \downarrow \\ \bigoplus_{x \in X^{(k+1)}} W\left(C\mathbb{M}^{k+1}(X_x)\right) \\ \downarrow \\ \cdots \\ \downarrow \\ \bigoplus_{x \in X^{(d)}} W\left(C\mathbb{M}^{d}(X_x)\right) \\ \downarrow \\ 0 \end{array} \tag{11.17}$$

Compare this with other Gersten complexes (6.4.5, 9.35, 10.45).

Proof. Follows from (11.3.4), following the standard construction, as in (10.44). In fact, repeating the process, in (10.44), we obtain the differential

$$\begin{array}{ccc} \bigoplus_{x\in X^{(k)}} W^k\left(\mathscr{D}^k(X_x)\right) & \longrightarrow & W^{k+1}\left(\mathscr{D}^{k+1}(X)\right) \\ & \searrow & \downarrow \\ & & \bigoplus_{x\in X^{(k+1)}} W^{k+1}\left(\mathscr{D}^{k+1}(X_x)\right). \end{array}$$

The desired complex is obtained, by (11.3.4(2)), by identifying

$$W^k\left(\mathscr{D}^k(X_x)\right) \cong W^0(\mathbf{D}^b(C\mathbb{M}^k(X))) \cong W(C\mathbb{M}^k(X)).$$

The proof is complete. ■

Remark 11.3.6. The complex (11.17) is called the Gersten–Witt complex. Further, for $x \in X^{(k)}$, one can identify

$W(C\mathbb{M}^k(X_x) \cong W(\kappa(x))$ where $\kappa(x)$ denotes the residue field.

Refer to [BW, Prop. 6.2], for a proof.

The following is the Witt-analogue of the exactness theorem of Gersten K-theory complex (6.4.9). This was proved in [BPW].

Theorem 11.3.7. *Let* $(A, \mathfrak{m})$ *be a regular local ring, containing a field* κ*, with* $1/2 \in \kappa$*. Write* $X = \operatorname{Spec}(A)$*. Then the Gersten–Witt complex* (11.17) *is exact.*

Proof. See [BPW]. ■

Exercise 11.3.8. Let X be a Noetherian scheme with an ample family of line bundles. Let $Z \subseteq X$, and $U = X - Z$. Then $\mathbf{D}^b_Z(X) \longrightarrow \mathbf{D}^b(X) \longrightarrow \mathbf{D}^b(U)$ is exact up to factor. **Hint:** [Sm1, A.4.7].

11.4. Revisit Chow–Witt Groups

We defined Chow Witt groups (10.47, 10.48, 10.50) as certain cohomology groups of the Gersten GW-complexes. In this section, we give the original definition of Chow Witt groups, due to Barge and Morel [BM]. Again, a good reference is [Fj]. We define the fundamental ideal of the Witt groups (rings).

Definition 11.4.1. Let $(A, \mathfrak{m})$ be a regular local ring, containing a field F, with $1/2 \in F$ and $\dim A = d$. Write $X = \mathrm{Spec}\,(A)$. Let $\kappa(\mathfrak{m}) = \frac{A}{\mathfrak{m}}$ denote the residue field. By the Dévissage theorem (11.3.3),

$$\chi : W(\kappa(\mathfrak{m})) \xrightarrow{\sim} W\left(C\mathbb{M}^d(X)\right)$$

in an isomorphism. The fundamental ideal $I(\kappa(\mathfrak{m})) \subseteq W(\kappa(\mathfrak{m}))$ is defined as the ideal of $[(V,\eta)]$, where (V,η) is an even-dimensional symmetric space [Lt]. For integers $n \geq 0$, define $I^n(X) := I^n(C\mathbb{M}^d(X)) := \chi(I^n(\kappa(\mathfrak{m}))$. For $n \leq 0$, define $I^n(X) := I^n(C\mathbb{M}^d(X)) = W\left(C\mathbb{M}^d(X)\right)$.

We fix some notation, as follows.

Definition 11.4.2. We denote the base field by F, and assume $1/2 \in F$. We work with a Noetherian regular separated scheme, to be denoted by X over F, with $\dim X = d$. We fix a locally free sheaf $\mathcal{L}$. All dualities considered are induced by the duality $P \mapsto P^* := Hom(P, \mathcal{L})$ on $\mathscr{P}(X)$.

For $x \in X$, denote the residue field, at x, by $\kappa(x)$.

With these notation, we introduce the original definition of the Chow Witt groups $\widetilde{C\mathfrak{H}}^k(X, \mathcal{L})$, as follows. Here we temporarily use the notation $\widetilde{C\mathfrak{H}}^k$ to distinguish it from other notation used in (10.5.1). Consider the diagram

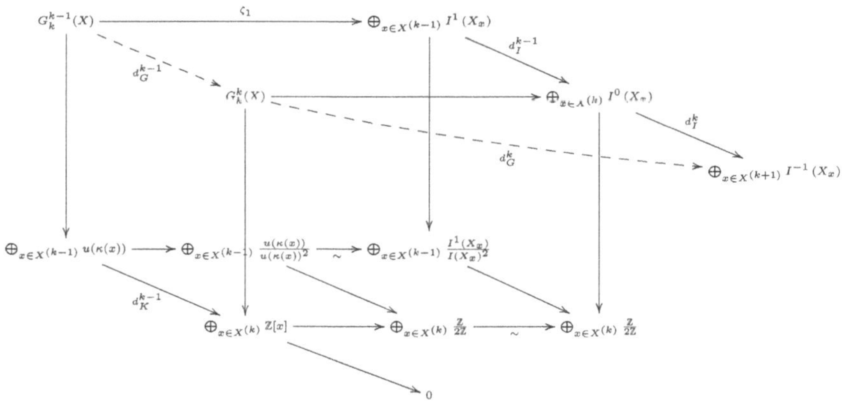

(11.18)

Here, $\begin{cases} \forall\ x \in X^{(k-1)} & I^1(X_x) \subseteq W\left(C\mathbb{M}^{k-1}(X_x), \mathcal{L}_x\right) \quad \text{is the fundamental ideal.} \\ \forall\ x \in X^{(k)} & I^0(X_x) = W\left(C\mathbb{M}^{k}(X_x), \mathcal{L}_x\right) \\ \forall\ x \in X^{(k+1)} & I^{-1}(X_x) = W\left(C\mathbb{M}^{k+1}(X_x), \mathcal{L}_x\right) \end{cases}$

1. In the lower left diagonal, $u(\kappa(x)) = K_1(\kappa(x))$ denotes the group of units of the residue field $\kappa(x)$, and $\mathbb{Z}[x] = K_0(\kappa(x))$ is the K_0-group of $\kappa(x)$.
2. At the lower right, we identify $\frac{\mathbb{Z}}{2\mathbb{Z}} = \frac{W(C\mathbb{M}^k(X_x))}{I^1(C\mathbb{M}^k(X_x))}$ for all $x \in X^{(k)}$ [Lt, Cor. 1.6]. It is also not very difficult to see that

$$\frac{u(\kappa(x))}{u(\kappa(x))^2} \xrightarrow{\sim} \frac{I^1(X_x)}{I^2(X_x)} \qquad \text{is an isomorphism [Lt, Cor. 2.3].}$$

3. In the lower left corner, d_K^{k-1} is the usual map induced by order maps (6.59), in the definition of Chow groups. In good cases (6.4.14), this is the tail end, the degree $k-1, k$ portion, of the Gersten complex for K-theory (6.50, 6.51). In the original definition [BM] this is the tail end of Milnor K-theory complex, which is similar to Gersten complex, while not always the same.

The upper right complex is derived from the Gersten Witt complex (11.17), which is only the degree $k-1, k, k+1$ portion of the longer complex. The upper left sequence is obtained by patching the lower left and upper right sequence. This is also the degree $k-1, k, k+1$ portion of a longer sequence. For the purpose of defining Chow Witt groups $\widetilde{C\mathfrak{H}}^k(X, \mathcal{L})$, we focus on this degree $k-1, k, k+1$ portion of a more elaborate construction [BM, Mf]. Thus, things are much elementary in our situation, as is evident above (2).

4. The upper left diagonal of the diagram (11.18) is obtained by Cartesian product. So, for fixed $0 \leq k \leq d$, we have constructed a complex

$$G_k^{k-1}(X) \xrightarrow{d_G^{k-1}} G_k^k(X) \xrightarrow{d_G^k} G_k^{k+1}(X). \tag{11.19}$$

The Chow Witt groups in [BM] were defined as the cohomology of these complexes (11.19), as follows:

$$\widetilde{C\mathfrak{H}}^k(X, \mathcal{L}) := \frac{\ker\left(d_G^k\right)}{Image\left(d_G^{k-1}\right)}. \tag{11.20}$$

5. While we focus on this definition (11.20), the complex (11.19) is the degree $k-1, k, k+1$ portion of a complex of length d, as follows:

$$0 \longrightarrow G_k^0(X) \longrightarrow \cdots \longrightarrow G_k^{k-1}(X) \xrightarrow{d_G^{k-1}} G_k^k(X) \xrightarrow{d_G^k} G_k^{k+1}(X)$$

$$\cdots \longrightarrow G_k^d(X) \longrightarrow 0 \tag{11.21}$$

obtained, as described in [BM, Mf], by patching the full Milnor K-theory complexes, and the full complexes of fundamental ideals. We denote the complex by $G_k^\bullet(X)$. We caution readers that Milnor K-theory complex (groups) may differ from the Gersten complexes (6.4.14) (of Quillen K-groups).

We compare this complex (11.21) with the Gersten GW-complexes (10.45). For $0 \le n \le d$, write

$$\mathcal{G}W_k^n(X) = \bigoplus_{x \in X^{(n)}} GW_{k-n}^{[k-n]}(dgC\mathbb{M}^n(X_x)).$$

So, Gersten GW-complexes (10.45) have a notation $\mathcal{G}W_k^\bullet$, which can be written as

$$0 \longrightarrow \mathcal{G}W_k^0(X) \longrightarrow \cdots \longrightarrow \mathcal{G}W_k^{k-1}(X) \xrightarrow{d_{GW}^{k-1}} \mathcal{G}W_k^k(X) \xrightarrow{d_{GW}^{k}} \mathcal{G}W_k^{k+1}(X)$$

$$\cdots \longrightarrow \mathcal{G}W_k^d(X) \longrightarrow 0 \tag{11.22}$$

6. We look at the degree $k-1, k, k+1$ portion of these complexes. From (10.48), this portion of the complex $\mathcal{G}W_k^\bullet(X)$ can be rewritten as

$$\bigoplus\nolimits_{x \in X^{(k-1)}} GW_1^{[1]}\left(dgC\mathbb{M}^{k-1}(X_x)\right) \longrightarrow \bigoplus\nolimits_{x \in X^{(k)}} GW\left(C\mathbb{M}^k(X_x)\right)$$

$$\longrightarrow \bigoplus\nolimits_{x \in X^{(k+1)}} W\left(C\mathbb{M}^{k+1}(X_x)\right). \tag{11.23}$$

In fact, $G_k^{k+1}(X) = \mathcal{G}W_k^{k+1}(X)$. There are natural maps as follows:

$$\begin{array}{ccccc}
\bigoplus_{x \in X^{(k-1)}} I^1(X_x) & \xrightarrow{d_I^{k-1}} & \bigoplus_{x \in X^{(k)}} I^0(X_x) & \xrightarrow{d_I^{k}} & \bigoplus_{x \in X^{(k+1)}} I^{-1}(X_x) \\
\uparrow \vdots & & \uparrow & & \uparrow 1 \\
\mathcal{G}W_k^{k-1}(X) & \longrightarrow & \mathcal{G}W_k^k(X) & \longrightarrow & \mathcal{G}W_k^{k+1}(X) \\
\vdots \downarrow & & \downarrow & & \\
\bigoplus_{x \in X^{(k-1)}} K_1(X_x) & \xrightarrow[d_K^{k-1}]{} & \bigoplus_{x \in X^{(k)}} K_0(X_x) & \xrightarrow[d_K^{k}]{} & 0.
\end{array}$$

The first two vertical maps follow from the description of $\mathcal{G}W_k^{k-1}(X)$, in terms of generators and relations. Readers are referred to [FS, p. 321], [BL] for details. Also, see the comments following the exact sequence (10.3). Other vertical maps are obvious. By properties of the cartesian product, we obtain the following maps of complexes:

$$\begin{array}{ccccc} \mathcal{G}W_k^{k-1}(X) & \longrightarrow & \mathcal{G}W_k^{k}(X) & \longrightarrow & \mathcal{G}W_k^{k+1}(X) \\ \downarrow & & \downarrow & & \downarrow \\ G_k^{k-1}(X) & \xrightarrow[d_G^{k-1}]{} & G_k^{k}(X) & \xrightarrow[d_G^{k}]{} & G_k^{k+1}(X) \end{array} \tag{11.24}$$

Theorem 11.4.3. *Consider the notation and the setup of* (11.4.2). *Fix an integer* $0 \leq k \leq d$. *Then the map of the complexes* (11.24) *induces an isomorphism*

$$\widetilde{CH}^k(X, \mathcal{L}) \xrightarrow{\sim} \widetilde{C\mathfrak{H}}^k(X, \mathcal{L}),$$

where the Chow-Witt group $\widetilde{CH}^k(X, \mathcal{L})$ *was defined in* (10.5.1).

Proof. See [FS, Thm. 33].

Exercise 11.4.4. Consider the notation and the setup of (11.4.2). Give a proof of (10.5.6, 10.5.7) using the description (11.20), $\widetilde{CH}^k(X, \mathcal{L}) \cong \widetilde{C\mathfrak{H}}^k(X, \mathcal{L})$. In other words, establish the natural map $\pi_0(\mathcal{L}O(P)) \longrightarrow \widetilde{C\mathfrak{H}}^k(X, \mathcal{L})$ using the cartesian product diagram (11.18).

Appendix A

Category Theory and Exact Categories

The importance of Category Theory as a learning tool may sometimes be understated. Category Theory is a way to learn how analogy works in mathematics, from one area to the others. This is a good way to understand different areas of mathematics just by learning one single area. Among the sources used for this chapter are [HS, Sv, Wc1, Sw3]. First, recall some of the basic definitions from [HS].

A.1. Main Definitions

Definition A.1.1. A **Category** $\mathscr{C}$ consists of the following:

1. A class of objects, denoted by $Obj(\mathscr{C})$.
2. For any two objects A, B in $Obj(\mathscr{C})$, a set of morphisms, from A to B, to be denoted by $Mor(A, B)$, more precisely by $Mor_{\mathscr{C}}(A, B)$. (*Depending on the context, several other notation are used for* $Mor(A, B)$. *One of them is* $Hom(A, B)$. *Morphisms are also called "arrows," "maps" and other things appropriate to the specific context.*)
3. To each triple A, B, C of objects, **a law of composition**

$$Mor(A, B) \times Mor(B, C) \longrightarrow Mor(A, C) \qquad \text{sending} \quad (f, g) \mapsto gf.$$

The above data satisfy the following Axioms:

A 1 $Mor(A, B)$ are mutually disjoint.
A 2 For composable morphisms f, g, h, the associativity holds: $h(gf) = (hg)f$.
A 3 For each object A, there is a morphism $1_A : A \longrightarrow A$, to be called the **identity map**, such that $1_A f = f\ \forall f \in Mor(X, A)$ and $g1_A = g\ \forall g \in Mor(A, X)$.

Although $Obj(\mathscr{C})$ is not a set, in general, we write $X \in Obj(\mathscr{C})$ to mean that X is an object in $\mathscr{C}$.

Definition A.1.2. Given a category $\mathscr{C}$, the following are some obvious definitions:

1. A morphism $f : X \longrightarrow Y$ in $\mathscr{C}$ is said to be an **isomorphism** if there is a morphism $g : Y \longrightarrow X$ such that $fg = 1_Y$ and $gf = 1_X$. In this case, g is called the inverse of f.
2. A morphism $f : X \longrightarrow Y$ in $\mathscr{C}$ is called a **monomorphism** if for all pairs of morphisms $\forall\ g_1, g_2 \in Mor(A, X)$, $fg_1 = fg_2 \Longrightarrow g_1 = g_2$. Monomorphisms are also called **monic** or **injective maps.** A morphism $f : X \longrightarrow Y$ in $\mathscr{C}$ is called an **epimorphism** or a **surjective map** $\forall\ g_1, g_2 \in Mor(Y, B)$, $g_1 f = g_2 f \Longrightarrow g_1 = g_2$. An epimorphism is also called an **epi**.
3. An object L in $\mathscr{C}$ is called an **initial object** if $Mor(L, X)$ is a singleton for all X in $Obj(\mathscr{C})$. An object R in $\mathscr{C}$ is called a **final object** (or **terminal object**) if $Mor(X, R)$ is a singleton for all X in $Obj(\mathscr{C})$. An object $\mathbf{0}$ in $\mathscr{C}$ is called a **zero object** if it is both an initial and a final object.
4. For a category $\mathscr{C}$, define the opposite category $\mathscr{C}^{opp}$, where $Obj(\mathscr{E}) = Obj(\mathscr{C}^{opp})$, and for two objects in $\mathscr{C}$, $Mor_{\mathscr{C}^{opp}}(A, B) = Mor_{\mathscr{C}}(B, A)$. We also use the notation $\mathscr{C}^o$ or $\mathscr{C}^{op}$ to denote the category $\mathscr{C}^{opp}$.

Definition A.1.3. A **covariant functor** $F : \mathscr{C} \longrightarrow \mathscr{D}$ from a category $\mathscr{C}$ to another category $\mathscr{D}$ is defined by the following data:

1. To each object $A \in \mathscr{C}$, it associates an object $FA \in \mathscr{D}$.
2. For objects A, B, there is a set theoretic map

$$F : Mor_{\mathscr{E}}(A, B) \longrightarrow Mor_{\mathscr{D}}(FA, FB).$$

3. Further, $F(fg) = F(f)F(g)$ for any two composable morphisms f, g.

A **contravariant functor** $F : \mathscr{C} \longrightarrow \mathscr{D}$ is a covariant functor $\mathscr{C}^{opp} \longrightarrow \mathscr{D}$. By a functor, we usually mean a covariant functor.

Definition A.1.4. Let $F, G : \mathscr{C} \longrightarrow \mathscr{D}$ be two covariant functors. A **natural transformation** $\tau : F \longrightarrow G$ associates to each object X in $\mathscr{C}$ a morphism $\tau_X \in Mor_{\mathscr{D}}(F(X), G(X))$ such that $\forall\ f \in Mor_{\mathscr{C}}(X, Y)$ the diagram

$$\begin{array}{ccc} F(X) & \xrightarrow{\tau_X} & G(X) \\ {\scriptstyle F(f)}\downarrow & & \downarrow{\scriptstyle G(f)} \\ F(Y) & \xrightarrow[\tau_Y]{} & G(Y) \end{array} \qquad \text{commutes.}$$

If τ_X is an isomorphism for all objects X, then τ is called a **natural equivalence**.Natural transformations and natural equivalences between two contravariant functors are defined likewise.

Example A.1.5. (*Refer to* (A.1.7) *for the definition of the category* **Sets** *of sets.*) Let $\mathscr{C}$ be a category and $A \in Obj(\mathscr{C})$. Then

$$\begin{cases} X \mapsto Mor_{\mathscr{C}}(A, X) & \text{is a covariant functor} \quad \mathscr{C} \longrightarrow \textbf{Sets}. \\ X \mapsto Mor_{\mathscr{C}}(X, A) & \text{is a contravariant functor} \quad \mathscr{C} \longrightarrow \textbf{Sets}. \end{cases}$$

They are called **representable** functors. These are among the most common and important examples of functors.

The following are some special functors:

Definition A.1.6. Let $F : \mathscr{C} \longrightarrow \mathscr{D}$ be a functor.

1. F is said to be **full** if

$$\forall\ X, Y \in Obj(\mathscr{C}) \quad \text{the map}$$
$$F : Mor_{\mathscr{C}}(X, Y) \longrightarrow Mor_{\mathscr{D}}(FX, FY) \quad \text{is surjective.}$$

2. F is called **faithful** if

$$\forall\ X, Y \in Obj(\mathscr{C}) \quad \text{the map}$$
$$F : Mor_{\mathscr{C}}(X, Y) \longrightarrow Mor_{\mathscr{D}}(FX, FY) \quad \text{is injective.}$$

3. F is called **fully faithful**, if

$$\forall\ X, Y \in Obj(\mathscr{C}) \quad \text{the map}$$
$$F : Mor_{\mathscr{C}}(X, Y) \longrightarrow Mor_{\mathscr{D}}(FX, FY) \quad \text{is bijective.}$$

4. Given a category $\mathscr{C}$ by a **subcategory**, we mean a functor $\iota : \mathscr{C}_0 \longrightarrow \mathscr{C}$ such that (1) ι is one-to-one on $Obj(\mathscr{C}_0)$ and (2) $\forall\ X, Y \in Obj(\mathscr{C}_0)$, the map $Mor_{\mathscr{C}_0}(X, Y) \longrightarrow Mor_{\mathscr{C}}(X, Y)$ is injective. So, subcategories are, in particular, faithful functors. However, not all subcategories are full subcategory.
5. Given another functor $G : \mathscr{D} \longrightarrow \mathscr{E}$ the composition GF is defined naturally.

Refer to (A.1.7) for examples.

A.1.1. *Classical and standard examples*

We recall some of the classical and Standard examples.

Examples A.1.7. We recall a few examples and establish notation.

1. The category of sets will be denoted by **Sets**. The class $Obj(\textbf{Sets})$ of objects of **Sets** is the class of all sets. For two sets X, Y, a morphism $f : X \longrightarrow Y$ is a set theoretic map.
2. (*Refer to Section B.1 for the definition of topological spaces.*) The category of Topological spaces is denoted by **Top**. The class $Obj(\textbf{Top})$ of objects of **Top** is the class of all topological spaces. For two topological spaces X, Y the set of morphisms $Mor(X, Y)$ is the set of all continuous maps $f : X \longrightarrow Y$. Note that the empty set ϕ is the initial object and singleton sets $\{x\}$ are the terminal objects in **Top**. In Topology, there are two more important categories:
   (a) The category of Pointed Topological spaces is denoted by $\textbf{Top}_\bullet$. The class $Obj(\textbf{Top}_\bullet)$ of objects of $\textbf{Top}_\bullet$ is the class of all pairs (X, x), where X is a topological space and $x \in X$ is a point, to be called the **base point**. For two objects $(X, x), (Y, y)$ the set of morphisms $Mor((X, x), (Y, y))$ is the set of all continuous maps $f : X \longrightarrow Y$, with $f(x) = y$. Note that $\textbf{Top}_\bullet$ has **zero objects**, namely $(\{x\}, x)$, where $\{x\}$ is any singleton set.

(b) The category of Topological Pairs is denoted by $\mathbf{Top^2}$. An object in $\mathbf{Top^2}$ is a pair (X, A) of topological spaces such that $A \subseteq X$ is a topological subspace. For two objects $(X, A), (Y, B)$ in $\mathbf{Top^2}$, the set of morphisms $Mor((X, A), (Y, B))$ is defined as the set of all continuous maps $f : X \longrightarrow Y$, with $f(A) \subseteq B$.

The following is a commutative diagram of natural functors:

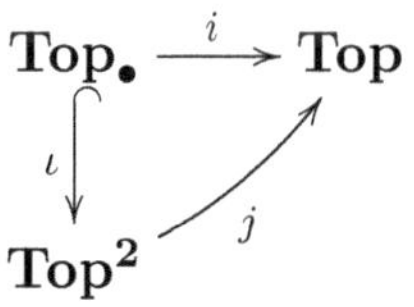

where
$$\begin{cases} \iota(X, x) = (X, x) \\ i(X, x) = X \\ j(X, A) = X \end{cases}$$

where the vertical functor ι is a full subcategory, while i, j may be called the "forgetful functors". Note that i, j are not full, though they are faithful.

3. The category of Groups will be denoted by $\mathbf{Gr}$. The class $Obj(\mathbf{Gr})$ of objects of $\mathbf{Gr}$ is the class of all groups. For two groups X, Y the set of morphisms $Mor(X, Y)$ is the set of all group homomorphisms $f : X \longrightarrow Y$.
4. The category of Abelian Groups is denoted by $\mathbf{Ab}$. The class $Obj(\mathbf{Ab})$ of objects of $\mathbf{Ab}$ is the class of all abelian groups. For two abelian groups G, H the set of morphisms $Mor_{\mathbf{Ab}}(G, H)$ is the set of all group homomorphisms $f : G \longrightarrow H$. In this case, we also write $Hom(G, H) := Mor_{\mathbf{Ab}}(G, H)$. (*One distinguishing feature regarding* $\mathbf{Ab}$ *is that* $Hom(G, H)$ *also has a natural structure of an abelian group. Subsequently, in Section A.2, we will define additive and abelian categories, and* $\mathbf{Ab}$ *is an example of both.*) Note that $\mathbf{Ab} \hookrightarrow \mathbf{Gr}$ is a full subcategory.
5. The category of Rings with unity and $0 \neq 1$ will be denoted by $\mathbf{Ring}$. The class $Obj(\mathbf{Ring})$ of objects of $\mathbf{Ring}$ is the class of all Rings with unity and $0 \neq 1$. For two Rings A, B the set of morphisms $Mor(A, B)$ is the set of all ring homomorphisms $f : A \longrightarrow B$, sending $1 \mapsto 1$.
6. The category of commutative Rings with unity and $0 \neq 1$ will be denoted by $\mathbf{cRing}$. The class $Obj(\mathbf{cRing})$ of objects of $\mathbf{cRing}$ is the class of all commutative Rings, with unity and $0 \neq 1$. For two commutative Rings A, B the set of morphisms $Mor(A, B)$ is

the set of all ring homomorphisms $f : A \longrightarrow B$, sending $1 \mapsto 1$. Note that **cRing** $\hookrightarrow$ **Ring** is a full subcategory.

7. The category of Schemes will be denoted by **Sch**. The class $Obj(\mathbf{Sch})$ of objects of **Sch** is the class of all Schemes. For two Schemes $(X, \mathcal{O}_X), (Y, \mathcal{O}_Y)$ the set of morphisms $Mor((X, \mathcal{O}_X), (Y, \mathcal{O}_Y))$ is the set of all morphisms $f : (X, \mathcal{O}_X) \longrightarrow (Y, \mathcal{O}_Y)$ of schemes (see [Hr, Sec II.2]).
8. Let A be a commutative ring.

   (a) Then $QCoh(A)$ or $Mod(A)$ will denote the category of all A-modules. The class $Obj(Mod(A))$ of objects of $Mod(A)$ is the class of all A-modules. Given two A-modules M, N, a morphism $f : M \longrightarrow N$ is an A-linear homomorphism. In particular, if $A = k$ is a field, the $Mod(k)$ is the category of all vector spaces.
   (b) $Coh(A)$ or $fMod(A)$ will denote the category of all finitely generated A-modules. Objects of $fMod(A)$ are the finitely generated A-modules. Given two finitely generated A-modules M, N, a morphism $f : M \longrightarrow N$ is an A-linear homomorphism. In particular, if $A = k$ is a field, then $fMod(k)$ is the category of all vector spaces, of finite dimension.
   (c) $\mathscr{P}(A)$ will denote the category of all finitely generated projective A-modules. Objects of $\mathscr{P}(A)$ are the finitely generated Projective A-modules. Given two finitely generated A-modules P, Q, a morphism $f : P \longrightarrow Q$ is an A-linear homomorphism.

   So, $\mathscr{P}(A) \hookrightarrow Coh(A) \hookrightarrow QCoh(A)$ are full subcategories.
9. Let $(X, \mathcal{O}_X)$ be a Scheme (see [Hr, Sec II.5]) .

   (a) Then $QCoh(X)$ will denote the category of all quasi-coherent $\mathcal{O}_X$-modules. The class $Obj(Coh(X))$ of objects of $QCoh(X)$ is the class of Quasi-Coherent $\mathcal{O}_X$-modules. Given two Quasi-Coherent $\mathcal{O}_X$-modules $\mathcal{F}, \mathcal{G}$, a morphism $f : \mathcal{F} \longrightarrow \mathcal{G}$ is a morphism of $\mathcal{O}_X$-modules.
   (b) $Coh(X)$ will denote the category of all coherent $\mathcal{O}_X$-modules. The class $Obj(Coh(X))$ of objects of $Coh(X)$ is the class of all Coherent $\mathcal{O}_X$-modules. Given two Coherent $\mathcal{O}_X$-modules $\mathcal{F}, \mathcal{G}$, a morphism $f : \mathcal{F} \longrightarrow \mathcal{G}$ is a morphism of $\mathcal{O}_X$-modules.

(c) $\mathscr{P}(X)$ will denote the category of all locally free $\mathcal{O}_X$-modules. Objects of $\mathscr{P}(X)$ are the locally free $\mathcal{O}_X$-modules. Given two locally free $\mathcal{O}_X$-modules $\mathcal{F}, \mathcal{G}$, a morphism $f : \mathcal{F} \longrightarrow \mathcal{G}$ is a morphism of $\mathcal{O}_X$-modules.

So, $\mathscr{P}(X) \hookrightarrow Coh(X) \hookrightarrow QCoh(X)$ are full subcategories.

10. Let $(S, \leq)$ be a partially ordered set. Then $(S, \leq)$ is a category. Here, $Obj(S, \leq) = S$, and $\forall\ x, y \in S$, define $Mor(x, y) = \{\leq\}$, a singleton set, if $x \leq y$, else define $Mor(x, y) = \phi$, the empty set.

The above examples of categories lead to a number of inclusion functors. Readers are well advised to think about them. The following is a basic example of duality and natural transformation.

Example A.1.8. Let A be a Noetherian commutative ring. Given a finitely generated projective A-module P, let $P^* = Hom_A(P, A)$. Then

$$P \mapsto P^* \quad \text{defines a contravariant functor} \quad \mathscr{P}(A) \longrightarrow \mathscr{P}(A).$$

Composition of functors are functors. So, $P \longrightarrow P^{**}$ is a covariant functor $\mathscr{P}(A) \longrightarrow \mathscr{P}(A)$. For $P \in Obj(\mathscr{P}(A))$ and for $p \in P$, define the evaluation map $ev_p : P^* \longrightarrow A$ by $ev_p(f) = f(p)$. Then $ev_p \in P^{**}$. So, we have a homomorphism

$$ev : P \longrightarrow P^{**}. \quad \text{In fact, it is an isomorphism.}$$

Let $1_{\mathscr{P}(A)}$ denote the identity functor. Also, let $D(P) = P^{**}$ be the double dual functor. Then $ev : 1_{\mathscr{P}(A)} \longrightarrow D$ is a natural equivalence.

A.1.2. *Pullback, Pushforward, kernel, and cokernel*

We define pullback and pushforward.

Definition A.1.9. Let $B \xrightarrow{g} X \xleftarrow{f} A$ be two morphisms in a category $\mathscr{C}$. Consider the diagram

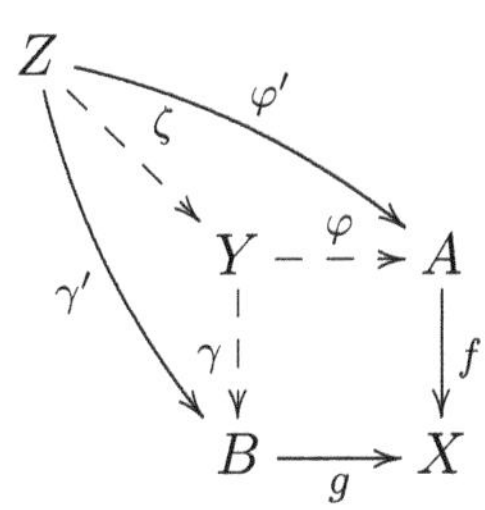

We say that (Y, φ, γ), in short Y, is the **pullback** of f, g if

1. The rectangle commutes, i.e. $f\varphi = g\gamma$, and
2. Given morphisms φ', γ' such that $f\varphi' = g\gamma'$, there is a unique morphism ζ such that the outer triangles commute.

A standard notation for pullback is $A \times_X B := (Y, \varphi, \gamma)$. Likewise, the **pushout** or **pushforward** of morphisms $f : X \longrightarrow A$, $g : X \longrightarrow B$ in the category $\mathscr{C}$ is defined by the commutative diagram

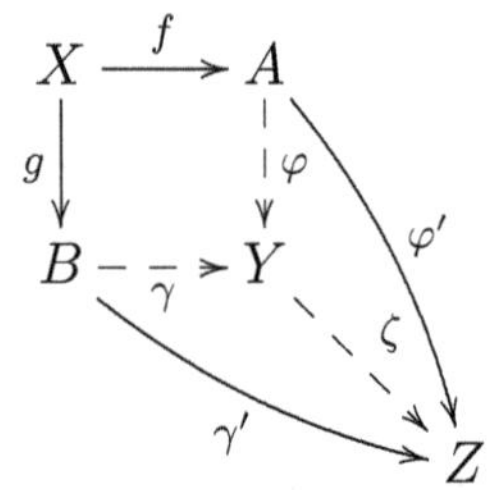

That means we say that (Y, φ, γ), in short Y, is the pushout of (f, g) if

1. The rectangle commutes, i.e. $\varphi f = \gamma g$, and
2. Given morphisms φ', γ' such that $\varphi' f = \gamma' g$, there is a unique morphism ζ such that the outer triangles commute.

A standard notation for pushforward is $A \coprod_X B := (Y, \varphi, \gamma)$.

Remarks:

1. The definitions of pullback and pushout make sense in any category. However, their existence is not always guaranteed. But, if they exist, they are unique up to isomorphism.
2. It is easy to see that pullback and pushout exist in the following categories: (1) **Gr**, (2) **Ab**, (3) $QCoh(A)$, where A is a commutative ring.

Very similar and related is the concept of direct product and direct coproduct (sum).

Definition A.1.10. Let $\mathscr{C}$ be a category and $A, B \in Obj(\mathscr{C})$. Consider the diagram

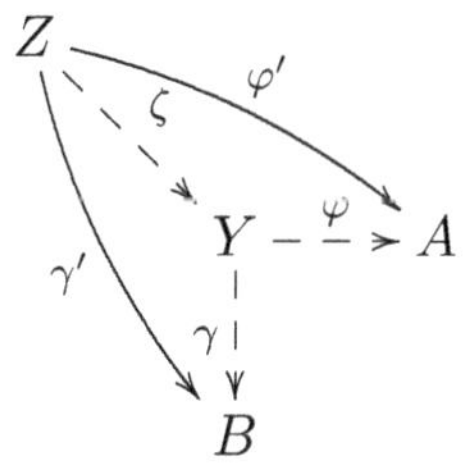

We say that (Y, φ, γ), in short Y, is a **direct product** of A, B if for given morphisms φ', γ' such that $f\varphi' = g\gamma'$, there is a unique morphism ζ such that the outer triangles commute.

The standard notation for the direct product of A, B is $A \times B := Y$. Again, for any $A, B \in Obj(\mathscr{C})$, the definition of $A \times B$ makes sense, while the existence is not guaranteed. If $A \times B$ exists, then it is unique up to isomorphism. If $\mathscr{C}$ has a terminal object 0_T, then the direct product is, indeed, the pullback of the diagram

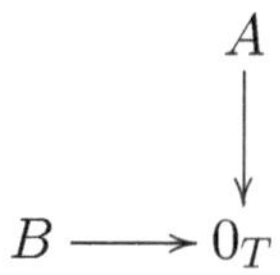

Likewise, the **coproduct** of $A, B \in Obj(\mathscr{C})$ in the category $\mathscr{C}$ is defined by the commutative diagram

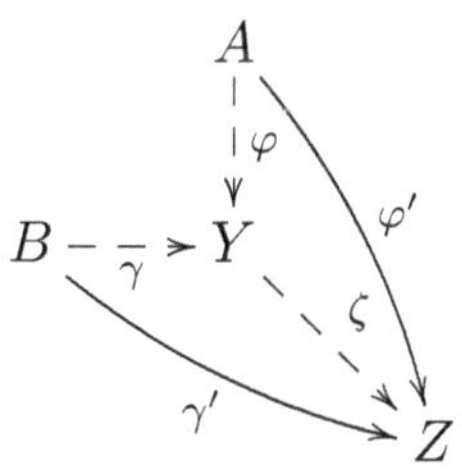

That means we say that (Y, φ, γ), in short Y, is the direct coproduct of (A, B) if for given morphisms φ', γ' such that $\varphi' f = \gamma' g$, there is a unique morphism ζ such that the outer triangles commute.

The standard notation for the direct coproduct of A, B is $A \coprod B := Y$. The coproduct $A \coprod B$ is defined up to isomorphisms, while existence is not guaranteed. Depending on the context (category), the direct coproduct is also called the **direct sum**, or **disjoint union**. If $\mathscr{C}$ has an initial object 0_I, then the direct coproduct is, indeed, the pushout of the diagram

$$\begin{array}{ccc} 0_I & \longrightarrow & A \\ \downarrow & & \\ B & & \end{array}$$

Next we define kernels and cokernels.

Definition A.1.11. Suppose that $\mathscr{C}$ is a category with a zero object **0**. Given $X, Y \in Obj(\mathscr{C})$, there is a unique "zero element" $0_{XY} \in Mor(X, Y)$ given by the commutative diagram:

$$\begin{array}{ccc} X & \longrightarrow & 0 \\ & \searrow^{0_{XY}} & \downarrow \\ & & Y \end{array} \quad \ni \quad \begin{cases} \forall\ f \in Mor(W, X)\ 0_{XY} f = 0_{WY}, \text{ and} \\ \forall\ g \in Mor(Y, Z) \quad g 0_{XY} = 0_{XZ}. \\ \text{We denote} \quad 0 := 0_{XY} \end{cases}$$

Suppose that $\sigma : X \longrightarrow Y$ is a morphism.

1. The **Kernel** of $\sigma : X \longrightarrow Y$ is defined to be a morphism $\iota : K \longrightarrow X$ such that
   (a) $\sigma\iota = 0$ and
   (b) Further,

   $$\forall\ \tau \in Mor(L, X) \text{ with } \sigma\tau = 0 \quad \exists \text{ a unique } \tau_0 \in Mor(L, K) \ \ni\ \tau = \iota\tau_0.$$

   The same is demonstrated by the following commutative diagram:

   $$\begin{array}{ccccc} L & & & & \\ {\scriptstyle \tau_0}\downarrow{\scriptstyle \exists!} & \searrow^{\tau} & & \searrow^{0} & \\ K & \xrightarrow{\ \iota\ } & X & \xrightarrow{\ \sigma\ } & Y \\ & & \underset{0}{\longrightarrow} & & \end{array} \quad \begin{cases} \text{We denote } \ker(\sigma) := (K, \iota) \\ \text{It is defined up to isomorphisms} \\ \text{Also, note that } \ker(\sigma) = X \times_Y 0. \end{cases}$$

2. The **cokernel** of $\sigma : X \longrightarrow Y$ is defined as a morphism $\zeta : Y \longrightarrow C$ such that

(a) $\zeta\sigma = 0$ and
(b) Further,

$$\forall\ \psi \in Mor(Y, Z) \text{ with } \psi\sigma = 0 \quad \exists \text{ a unique}$$
$$\eta \in Mor(C, Z) \ni \ \eta\zeta = \psi.$$

The same is demonstrated by the following commutative diagram:

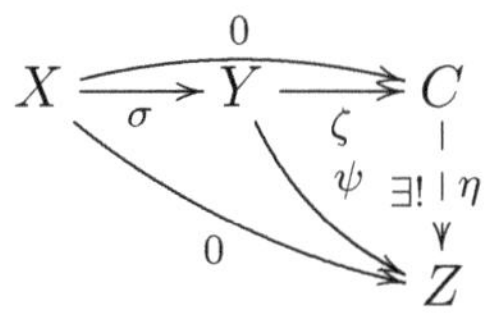

$\begin{cases} \text{We denote } co\ker(\sigma) := (C, \zeta) \\ \text{It is defined up to isomorphisms} \\ \text{Also, note that } co\ker(\sigma) = Y \coprod_X 0. \end{cases}$

A.1.3. *Equivalence and adjoint functors*

We give two basic definitions. First, we define equivalence of categories.

Definition A.1.12. A functor $F : \mathscr{C} \longrightarrow \mathscr{D}$ is called an **equivalence of categories** if there is a functor $G : \mathscr{D} \longrightarrow \mathscr{C}$ such that FG is naturally equivalent to $1_{\mathscr{D}}$ (*the identity functor*) and GF is naturally equivalent to $1_{\mathscr{C}}$ (as in A.1.4). If $\mathscr{C}$ and $\mathscr{D}$ are small, F is an **equivalence of categories** if and only if the following two conditions hold:

1. (**Essentially Surjective**): $\forall\ Z \in Obj(\mathscr{D})$ there is an isomorphism $FX \xrightarrow{\sim} Z$ for some object $X \in Obj(\mathscr{C})$.
2. (**Fully Faithful**): For all objects $X, Y \in Obj(\mathscr{C})$, the map

$$Mor_{\mathscr{C}}(X, Y) \xrightarrow{\sim} Mor_{\mathscr{D}}(FX, FY) \quad \text{is a bijection.}$$

We will recall the following example.

Example A.1.13. Suppose that A is a commutative ring. Then the following contravariant functor

$$QCoh(A) \xrightarrow{\sim} QCoh(\text{Spec}\,(A)) \qquad \text{is an equivalence of categories.}$$

Proof. See [Hr, Sec. II.5]. ■

We define adjoint of a functor next.

Definition A.1.14. Let $F : \mathscr{C} \longrightarrow \mathscr{D}$ and $G : \mathscr{D} \longrightarrow \mathscr{C}$ be two covariant functors. We say that F is a **left adjoint** of G and/or that G is a **right adjoint** of F if $\forall\ X \in Obj(\mathscr{C})$, $\forall\ A \in Obj(\mathscr{D})$, there is a bijection

$$\eta := \eta_{XA} : Mor_{\mathscr{D}}(FX, A) \xrightarrow{\sim} Mor_{\mathscr{C}}(X, GA), \quad \text{which is natural,}$$

in the sense that $\forall\ f \in Mor_{\mathscr{C}}(X, Y)$, $\forall\ g \in Mor_{\mathscr{D}}(A, B)$, the diagrams:

$$\begin{array}{ccc} Mor_{\mathscr{D}}(FX, A) & \xrightarrow[\sim]{\eta_{XA}} & Mor_{\mathscr{C}}(X, GA) \\ {\scriptstyle g_*}\downarrow & & \downarrow{\scriptstyle G(g)_*} \\ Mor_{\mathscr{D}}(FX, B) & \xrightarrow[\eta_{XB}]{\sim} & Mor_{\mathscr{C}}(X, GB) \end{array} \qquad \begin{array}{ccc} Mor_{\mathscr{D}}(FX, A) & \xrightarrow[\sim]{\eta_{XA}} & Mor_{\mathscr{C}}(X, GA) \\ {\scriptstyle F(f)^*}\uparrow & & \uparrow{\scriptstyle f^*} \\ Mor_{\mathscr{D}}(FY, A) & \xrightarrow[\eta_{YA}]{\sim} & Mor_{\mathscr{C}}(Y, GA) \end{array}$$

commute. In other words, the following two functors:

$$\mathscr{C}^{opp} \times \mathscr{D} \longrightarrow \textbf{Sets} \quad \begin{cases} (X, A) \mapsto Mor_{\mathscr{D}}(FX, A) \\ (X, A) \mapsto Mor_{\mathscr{C}}(X, GA) \end{cases}$$

are naturally equivalent.

The Left and Right Adjoint of contravariant functors are defined likewise.

Lemma A.1.15. *Suppose $F : \mathscr{C} \longrightarrow \mathscr{D}$ is a covariant functor.*

1. *Then F has a left adjoint if and only if, given $Y \in Obj(\mathscr{D})$, there is an object $L(Y) \in Obj(\mathscr{C})$, and a universal arrow $\chi_Y : Y \longrightarrow FL(Y)$ in $\mathscr{D}$. The universal property of χ_Y is explained as follows:*

$$\begin{cases} \forall\ X \in Obj(\mathscr{C}) \\ \forall \varphi \in Mor_{\mathscr{D}}(Y, FX) \\ \exists \text{ unique } \Phi \in Mor_{\mathscr{D}}(LY, X) \end{cases} \ni \quad \begin{array}{ccc} Y & \xrightarrow{\chi_Y} & FL(Y) \\ & {\scriptstyle\varphi}\searrow & \downarrow{\scriptstyle F(\Phi)} \\ & & FX \end{array} \quad \textit{commutes.}$$

2. *Similarly, F has a right adjoint if and only if, given $Y \in Obj(\mathscr{D})$, there is an object $R(Y) \in Obj(\mathscr{C})$, and a universal arrow $\chi_Y : FR(Y) \longrightarrow Y$ in $\mathscr{D}$. The universal property of χ_Y is explained as follows:*

$$\begin{cases} \forall\ X \in Obj(\mathscr{C}) \\ \forall \varphi \in Mor_{\mathscr{D}}(FX, Y) \\ \exists \text{ unique } \Phi \in Mor_{\mathscr{D}}(X, RY) \end{cases} \ni \quad \begin{array}{ccc} FX & & \\ {\scriptstyle F(\Phi)}\downarrow & \searrow{\scriptstyle\varphi} & \\ FRY & \xrightarrow[\chi_Y]{} & Y \end{array} \quad \textit{commutes.}$$

(These statements are more transparent when $F : \mathscr{C} \hookrightarrow \mathscr{D}$ is a subcategory.)

Proof. We will only outline the proof of (1). Suppose that $L : \mathscr{D} \longrightarrow \mathscr{C}$ is a left adjoint of F. Then for $X \in Obj(\mathscr{C})$, $Y \in Obj(\mathscr{D})$, there are natural bijections

$$\begin{cases} \eta_{XY} : Mor_{\mathscr{C}}(LY, X) \xrightarrow{\sim} Mor_{\mathscr{D}}(Y, FX). & \text{Taking } X = LY \\ \eta_{LYY} : Mor_{\mathscr{C}}(LY, LY) \xrightarrow{\sim} Mor_{\mathscr{D}}(Y, FLY). \end{cases}$$

Define $\chi_Y := \eta_{LYY}(1_{LY})$. Conversely, suppose that the universal property is satisfied. Then $Y \mapsto L(Y)$ defines a functor $\mathscr{D} \longrightarrow \mathscr{C}$. So, for $g \in Mor_{\mathscr{D}}(Y, Y_1)$, we define $L(g)$, as follow. There is a unique $\Phi \in Mor_{\mathscr{D}}(LY, LY_1)$ such that the diagram

$$\begin{array}{ccc} Y & \xrightarrow{\chi} & FLY \\ {\scriptstyle g}\downarrow & & \downarrow{\scriptstyle F(\Phi)} \\ Y_1 & \xrightarrow[\chi]{} & FLY_1 \end{array} \quad \text{commutes. Define} \quad L(g) := \Phi$$

Now, the required bijections are defined by the universal property. ■

A.2. Additive and Abelian Categories

Next we define additive and abelian categories.

Definition A.2.1. A category $\mathscr{C}$ is called an **additive category** if (1) $\mathscr{C}$ has a zero object $\mathbf{0}$, (2) for $X, Y \in Obj(\mathscr{C})$, the direct product $X \times Y$ exists, (3) for $X, Y \in Obj(\mathscr{C})$, the set $Mor(X, Y)$ has the structure of an abelian group such that the composition maps are bilinear. That means that, for objects X, Y, Z in $\mathscr{C}$ and $f \in Mor(X, Y)$, $g \in Mor(Y, Z)$, the composition maps

$$\begin{cases} g_* : Mor(X, Y) \longrightarrow Mor(X, Z) & h \mapsto gh \\ f^* : Mor(Y, Z) \longrightarrow Mor(X, Z) & h \mapsto hf \end{cases}$$

are group homomorphisms.

In additive categories, the finite coproduct $\coprod$ exists and is denoted by $\oplus$. In fact, $X \oplus Y = X \times Y$ (Ex. A.6.1). For $i = 1, 2$, let $f_i \in$

$Mor_{\mathscr{C}}(M_i, N)$ and $g_i \in Mor_{\mathscr{C}}(M, N_i)$, we use matrix notation

$$\begin{cases} (f_1\ f_2) : M_1 \oplus M_2 \longrightarrow N \\ \begin{pmatrix} g_1 \\ g_2 \end{pmatrix} : M \longrightarrow N_1 \oplus N_2 \end{cases}$$

for the maps given by universal properties of coproduct and product.

Definition A.2.2. Suppose that $\mathscr{C}, \mathscr{D}$ are two additive categories. A functor $F : \mathscr{C} \longrightarrow \mathscr{D}$ is called a **functor of additive categories** if for all objects X, Y of $\mathscr{C}$, the map

$$F : Mor_{\mathscr{C}}(X, Y) \longrightarrow Mor_{\mathscr{D}}(FX, FY) \quad \text{is a group homomorphism.}$$

Such a functor is also called an **additive functor**.

The following is a reinterpretation of the definition of kernel and cokernel of an arrow.

Lemma A.2.3. *Suppose that $\mathscr{C}$ is an additive category and $\sigma : X \longrightarrow Y$ is a morphism. Then a morphism $\iota : K \longrightarrow X$ defines the kernel of σ if and only if $\forall\ A \in Obj(\mathscr{C})$ the sequence of groups*

$$0 \longrightarrow Mor(A, K) \xrightarrow{\iota_*} Mor(A, X) \xrightarrow{\sigma_*} Mor(A, Y) \quad \textit{is exact.}$$

Likewise, a morphism $\zeta : Y \longrightarrow C$ defines the cokernel of σ if and only if $\forall\ Z \in Obj(\mathscr{C})$ the sequence of groups

$$0 \longrightarrow Mor(C, Z) \xrightarrow{\zeta^*} Mor(Y, Z) \xrightarrow{\sigma^*} Mor(X, Z) \quad \textit{is exact.}$$

The following from [HS] makes a lot of intuitive sense.

Exercise A.2.4. Let $\mathscr{A}$ be an additive category. For $i = 1, 2$, let $f_i \in Mor_{\mathscr{A}}(M_i, M)$ and $g_i \in Mor_{\mathscr{A}}(P, M_i)$. Then the diagram

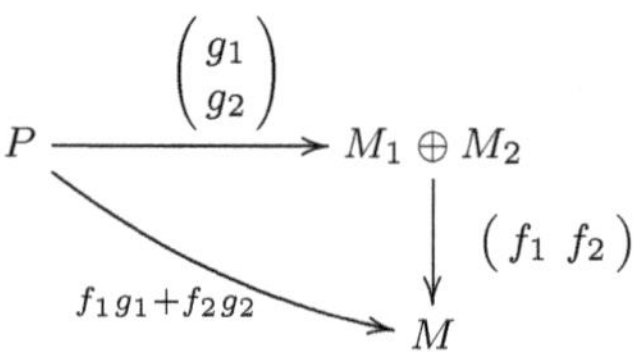

commutes, i.e. $(f_1\ f_2)\begin{pmatrix} g_1 \\ g_2 \end{pmatrix} = f_1g_1 + f_2g_2$

Proof. See [HS, Prop. II.9.3]. ∎

A.2.1. *Abelian categories*

Definition A.2.5. Suppose that $\mathscr{C}$ is an additive category. Further, assume that every morphism $\sigma : X \longrightarrow Y$ has a kernel and a cokernel. Define the image $Im(\sigma)$ and coImage $coIm(\sigma)$ as follows:

$$\begin{cases} Im(\sigma) := \ker\left(coKernel(\sigma)\right) \\ coIm(\sigma) := coKernel\left(\ker(\sigma)\right) \end{cases}$$

Then there is a natural map, $\quad coIm(\sigma) \longrightarrow Im(\sigma)$.

Now we define abelian category [HS, p. 78].

Definition A.2.6. Suppose that $\mathscr{A}$ is an additive category. Then $\mathscr{A}$ is called an **Abelian Category** if

1. Every morphism $\sigma : X \longrightarrow Y$ has a kernel and a cokernel.
2. Every monomorphism is the kernel of its cokernel.
   Every epimorphism is the cokernel of its kernel.
3. Every morphism is expressible as a composition of an epimorphism and a monomorphism.

We quote the following proposition from [HS, Prop 9.6, p. 78].

Proposition A.2.7. *Suppose that $\mathscr{A}$ is an abelian category and $\sigma \in Mor_{\mathscr{A}}(X, Y)$. Then we have the diagram of arrows:*

$$\begin{array}{ccccc} K & \overset{\mu}{\hookrightarrow} & X & \overset{\eta}{\twoheadrightarrow} & I \\ & & & \searrow^{\sigma} & \downarrow \nu \\ & & & & Y \\ & & & & \downarrow \epsilon \\ & & & & C \end{array} \qquad where \qquad \begin{cases} \mu = \ker(\sigma) \\ \epsilon = co\ker(\sigma) \\ \eta = co\ker(\mu) =: coIm(\sigma) \\ \nu = \ker(\epsilon) =: Im(\sigma) \\ \sigma = \nu\eta \\ By\ definition,\ I := image(\sigma) \end{cases}$$

Further, this decomposition $\sigma = \nu\eta$, as a composition of an epimorphism and a monomorphism, is unique up to isomorphism.

Proof. See [HS, Prop. 9.6, p. 78].

The following characterization of abelian categories is given as a definition of the same by some authors [Sv, p. 409].

Lemma A.2.8. *Suppose that $\mathscr{A}$ is an additive category. Then $\mathscr{A}$ is an **Abelian Category** if and only if the following two conditions hold:*

1. *Every morphism $\sigma : X \longrightarrow Y$ has a kernel and a cokernel.*
2. *And, $\forall\ \sigma \in Mor_{\mathscr{A}}(X,Y)$, the canonical arrow $coIm(\sigma) \longrightarrow Im(\sigma)$ is an isomorphism.*

Proof. Immediate from (A.2.7). ∎

Remark A.2.9. For an arrow $\sigma : X \longrightarrow Y$ in an abelian category $\mathscr{A}$, $coIm(\sigma) \xrightarrow{\sim} Im(\sigma)$ is an isomorphism. Because of this, we will identify $coIm(\sigma)$ and $Im(\sigma)$ without further clarifications.

Lemma A.2.10. *Suppose that $\mathscr{A}$ is an abelian category and $\sigma : X \longrightarrow Y$ is a morphism. Let $\nu : I \hookrightarrow Y$ be the image of σ and $\mu : X \twoheadrightarrow I$ be the corresponding epi morphism* (A.2.7). *Then there is a factorization $\sigma = \mu\nu$, where μ is epi and ν is mono. This factorization is unique, up to isomorphism in the following sense:*

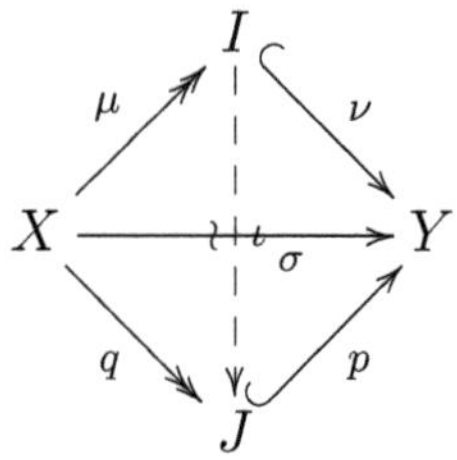

*Given any other factorization $\sigma = pq$, where q is epi and p is mono, there is an isomorphism $\iota : I \xrightarrow{\sim} J$ such that the diagram commutes. We say that the **image of** σ is $Im(\sigma) = I$, or $Im(\sigma) = \nu$, or $Im(\sigma) = (\nu, I)$.*

Proof. Suppose that $t : \ker(\sigma) \hookrightarrow X$ is the kernel of σ. Then $pqt = \sigma t = 0$. Since p is mono, it follows $qt = 0$. Since $\mu : X \longrightarrow I$ is the cokernel of t, there is an arrow $\iota : I \longrightarrow J$ such that $\iota\mu = q$. Also, $(p\iota)\mu = pq = \sigma = \nu\mu$. Since μ is mono, $p\iota = \nu$. This establishes that the diagram commutes. Now, ι is epi because so is q, and ι is mono because so is ν. By Lemma A.3.5, ι is an isomorphism. ∎

We introduce the following definitions for future reference.

Definition A.2.11. Let $\mathscr{A}$ be an abelian category. Let M be an object in $\mathscr{A}$. By a subobject of M in $\mathscr{A}$, we mean an injective arrow $\iota : N \hookrightarrow M$. Often, we just write $N \subseteq M$ and ignore mentioning ι. Now suppose that $M_1 \subseteq M$ and $M_2 \subseteq M$ are two subobjects. We define intersection and sum

$$M_1 \cap M_2 = \ker\left(M \longrightarrow \frac{M}{M_1} \oplus \frac{M}{M_2} \right), M_1 + M_2 = Image\left(M_1 \oplus M_2 \longrightarrow M \right).$$

There are various other equivalent but natural ways to define the same.

Example A.2.12. Here are the main examples and comments:

1. The category **Ab** of abelian groups is an abelian category.
2. Suppose that A is a commutative ring. Then $Mod(A) = QCoh(A)$, $fMod(A) = Coh(A)$ are abelian categories. (*Perhaps the definition of the abelian category was modeled after these two examples, along with* **Ab**.) More generally, let X be a scheme. Then $QCoh(X)$ and $Coh(X)$ are abelian categories.
3. We remark, for a commutative ring A, that the category $\mathscr{P}(A)$, of finitely generated projective A-modules, is not necessarily an abelian category. This category $\mathscr{P}(A)$ is an Exact category, which we define subsequently.

A.3. Frequently Used Lemmas

In this section, we list a few frequently used lemmas in K-Theory.

A.3.1. *Pullback and Pushforward Lemmas*

The following on pullback and pushforward are widely used in K-Theory. Most of the material in the section can be traced back to [HS].

Lemma A.3.1. *Suppose that $\mathscr{C}$ is a category with a zero object. Let $X \xrightarrow{v} Z \xleftarrow{h} Y$ be two arrows in $\mathscr{C}$. Let*

$$\begin{array}{ccc} P & \overset{p_1}{\dashrightarrow} & X \\ {\scriptstyle p_2}\downarrow & & \downarrow{\scriptstyle v} \\ Y & \underset{h}{\longrightarrow} & Z \end{array} \quad \text{be the pullback.} \quad \begin{cases} \text{Assume that } p_2 \text{ has a kernel,} \\ \quad \text{and } (K,\eta) = \ker(p_2) \\ \text{where } \eta \in Mor(K,X). \\ \text{Then we have } \ker(v) = p_1\eta. \end{cases}$$

Conversely, suppose that v has a kernel, and $\iota : K \longrightarrow X$ is the kernel of v. Then there is an induced arrow $\eta : K \longrightarrow P$, given by *the following commutative diagram:*

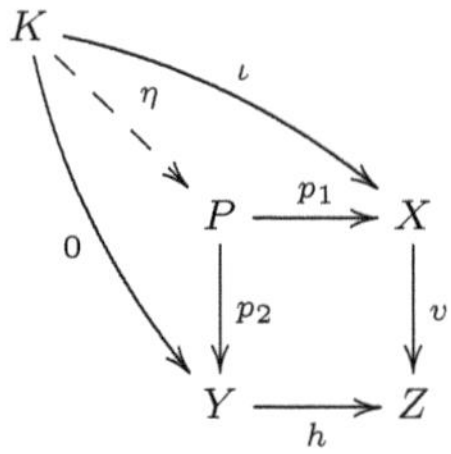

$\left\{\begin{array}{l}\textit{Then } (K,\eta)=\ker(p_2).\ \textit{This is} \\ \textit{better expressed by the diagram}\end{array}\right.$

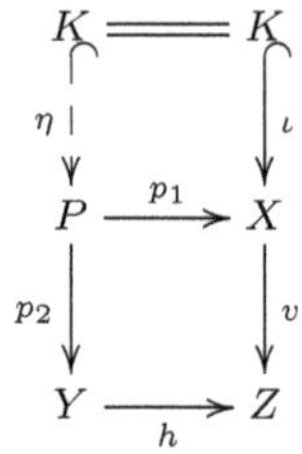

One can write down the same, in terms of the kernels of the horizontal maps p_1 and h.

Proof. Use the definition of pullback and the kernel. See [HS, Thm II.6.2]. ■

Reversing the arrows, we have the following.

Lemma A.3.2. *Suppose that $\mathscr{C}$ is a category with zero. Let $X \xleftarrow{h} Z \xrightarrow{v} Y$ be two arrows in $\mathscr{C}$. Let*

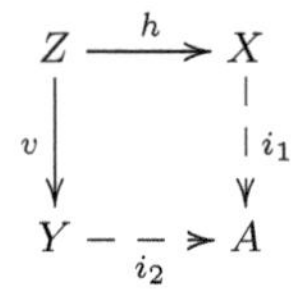

be the pushforward. $\left\{\begin{array}{l}\textit{Assume that } i_2 \textit{ has a cokernel, and} \\ \quad (j,C)=\operatorname{coker}(i_2), \\ \textit{where } j\in Mor(A,C) \\ \textit{Then we have } ji_1=\operatorname{coker}(h).\end{array}\right.$

Conversely, suppose that h has a cokernel and $\zeta : X \longrightarrow C$ is the cockerel of h. Then the following commutative diagram defines a map $\beta : A \longrightarrow C$:

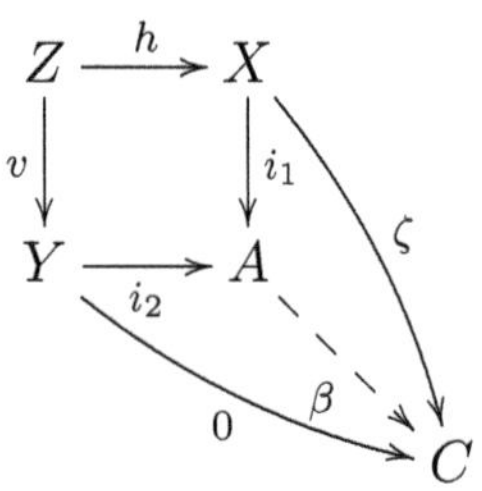

Then (C,β) is the cokernel of i_2. The same is better expressed by the diagram

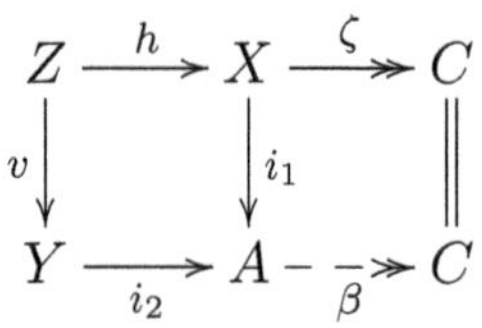

Proof. Reverse the arrow in the proof of (A.3.1). ■

Lemma A.3.3. *Let $\mathscr{A}$ be an abelian category. Consider the following rectangle (not necessarily commutative) of arrows, and the related sequence:*

$$\begin{array}{ccc} P & \xrightarrow{g_1} & M_1 \\ {\scriptstyle g_2}\downarrow & & \downarrow{\scriptstyle f_1} \\ M_2 & \xrightarrow[f_2]{} & M \end{array}, \qquad P \xrightarrow{\begin{pmatrix} g_1 \\ -g_2 \end{pmatrix}} M_1 \oplus M_2 \xrightarrow{(f_1\ f_2)} M \tag{A.1}$$

Then

1. *The rectangle commutes if and only if the composition* $(f_1\ f_2)\begin{pmatrix} g_1 \\ -g_2 \end{pmatrix} = 0.$
2. *The rectangle is a pullback diagram if and only if* $\left(P, \begin{pmatrix} g_1 \\ -g_2 \end{pmatrix}\right) = \ker \begin{pmatrix} g_1 \\ -g_2 \end{pmatrix}.$
3. *The rectangle is a pushforward diagram if and only if* $\left(\begin{pmatrix} g_1 \\ -g_2 \end{pmatrix}, M\right) = co\ker \begin{pmatrix} g_1 \\ -g_2 \end{pmatrix}.$

In particular, pullback and pushforward exist in abelian categories.

Proof. First, (1) follows from Proposition A.2.4. Then (2) follows from the definitions of pullback and kernel. Likewise, (3) follows from definitions of pushforward and cokernel. The proof is complete. ■

Corollary A.3.4. *Let $\mathscr{A}$ be an abelian category. Consider the following commutative rectangle of arrows:*

$$\begin{array}{ccc} P & \xrightarrow{g_1} & M_1 \\ {\scriptstyle g_2}\downarrow & & \downarrow{\scriptstyle f_1} \\ M_2 & \xrightarrow[f_2]{} & M \end{array}$$

If the rectangle is a pullback diagram and f_1 is epi, then the diagram is a pushforward diagram and g_2 is epi. Likewise, if the rectangle is a *pushforward diagram and g_1 is monic, then the diagram is a pullback diagram and f_2 is monic.*

Proof. Consider the sequence as in (A.1). Suppose that the rectangle is a pullback and f_1 is epi. Note that $\begin{pmatrix} f_1 & f_2 \end{pmatrix}\begin{pmatrix} 1 \\ 0 \end{pmatrix} = f_1$ is epi. So, $\begin{pmatrix} f_1 & f_2 \end{pmatrix}$ is epi. Further, by (A.3.3), $\left(P, \begin{pmatrix} g_1 \\ -g_2 \end{pmatrix}\right) = \ker\begin{pmatrix} g_1 \\ -g_2 \end{pmatrix}$. So $\left(\begin{pmatrix} g_1 \\ -g_2 \end{pmatrix}, M\right) = co\ker\begin{pmatrix} g_1 \\ -g_2 \end{pmatrix}$. Again, by (A.3.3), the rectangle is a pushforward. By (A.3.2), $co\ker(g_1) = co\ker(f_1) = 0$. So, g_1 is epi. This completes the proof of the first assertion. The latter assertion is proved similarly (or by reversing arrows). The proof is complete. ■

Lemma A.3.5. *Suppose that $\mathscr{A}$ is an abelian category and $f : M \longrightarrow N$ is an arrow. Then*

(1) *f is mono if and only if it is the kernel of its cokernel.*
(2) *f is epi if and only if it is cokernel of its kernel.*
(3) *f is an isomorphism if and only if f is epi and mono.*

Proof. By definition of cokernel, we have the pushforward diagram

$$\begin{array}{ccc} M & \xrightarrow{f} & N \\ \downarrow & & \downarrow p \\ 0 & \longrightarrow & co\ker(f) \end{array}$$

If f is monic, by (A.3.4), this is a pullback diagram. So, $f = \ker(p)$. Conversely, if $f = \ker(p)$, then f is monic. This completes the proof of (1). We prove (2) likewise.

It is clear that, if f is an isomorphism, it is both mono and epi. Now assume that f is both mono and epi. To prove the converse, denote the terminal arrow by $j_N : N \longrightarrow 0$. Since f is epi, $co\ker(f) = p = 0$. Again, since f is mono, $f = \ker(p)$. Also, $1_N = \ker(p)$. So, by the uniqueness property of kernel, we have an isomorphism, g, as in the following commutative diagram:

$$\begin{array}{ccccc} M & \xrightarrow{f} & N & \xrightarrow{p} & 0 \\ {\scriptstyle g}\downarrow\wr & & \| & & \| \\ N & \xrightarrow[1_N]{} & N & \xrightarrow{p} & 0 \end{array}$$

So, g is the inverse of f. The proof is complete. ∎

A.3.2. *The Snake Lemma*

A frequently used lemma in K-Theory is a classical result in Homological Algebra that is known as the Snake Lemma. Before we state and prove the Snake Lemma, we prove the following elementary lemma.

Lemma A.3.6. *Suppose that $\mathscr{A}$ is an abelian category. Consider the following diagram of arrows:*

$$K \xrightarrow{f} M \xrightarrow{g} C \qquad \ni \qquad fg = 0.$$

Let $t : \ker(g) \hookrightarrow M$ be the kernel of g and $s : Im(f) \hookrightarrow M$ be the monomorphism, corresponding to $Im(f)$. Then there is a unique arrow $\iota : Im(f) \longrightarrow ker(g)$ such that the diagram

$$\begin{array}{ccc} Im(f) & \overset{s}{\hookrightarrow} & M \\ {\scriptstyle \iota}\downarrow & \nearrow{\scriptstyle t} & \\ \ker(g) & & \end{array} \qquad \textit{commutes.}$$

In fact, ι is a mono.

Proof. Since, s is mono, ι is also mono if it exists. Consider the diagram

$$\begin{array}{ccccccc} & & & \overset{f}{\frown} & & & \\ K & \xrightarrow[u]{} & Im(f) & \underset{s}{\hookrightarrow} & M & \xrightarrow{g} & C \\ & & & & \| & & \| \\ K & \xrightarrow{v} & \ker(g) & \overset{t}{\hookrightarrow} & M & \xrightarrow{g} & C \\ & & & \underset{f}{\smile} & & & \end{array}$$

Since $gsu = gf = 0$ and u is epi, $gs = 0$. Now, from definitions of kernel s, there is a morphism $\iota : Im(f) \hookrightarrow \ker(g)$, as required. ■

The most elementary version of the Snake Lemma is stated for Modules over a ring. The following is one of the most general versions.

Lemma A.3.7. *Suppose that $\mathscr{A}$ is an abelian category. Consider the commutative diagram*

$$\begin{array}{ccccccc} & & K & \xrightarrow{i} & M & \xrightarrow{j} & L & \longrightarrow & 0 \\ & & \downarrow{\scriptstyle f} & & \downarrow{\scriptstyle g} & & \downarrow{\scriptstyle h} & & \\ 0 & \longrightarrow & A & \xrightarrow[\iota]{} & B & \xrightarrow[\zeta]{} & C & & \end{array} \qquad in \quad \mathscr{A},$$

where the rows are exact in $\mathscr{A}$. Then there is a natural exact sequence

$$\ker(f) \xrightarrow{i_k} \ker(g) \xrightarrow{j_k} \ker(h) \xrightarrow{\delta} co\ker(f) \xrightarrow{\iota_c}$$
$$co\ker(g) \xrightarrow{\zeta_c} co\ker(h), \qquad \text{(A.2)}$$

where the maps $i_k, j_k, \iota_c, \zeta_c$ are induced by the respective morphisms i, j, ι, ζ. The morphism δ is defined in the proof.

Proof. This proof is classical and many versions are available in the net. We extend the above diagram (A.2) and label the morphisms as follows:

$$\begin{array}{ccccccccc} & & \ker(f) & \xrightarrow{i_k} & \ker(g) & \xrightarrow{j_k} & \ker(h) & & \\ & & \downarrow{\scriptstyle f_k} & & \downarrow{\scriptstyle g_k} & & \downarrow{\scriptstyle h_k} & & \\ & & K & \xrightarrow{i} & M & \xrightarrow{j} & L & \longrightarrow & 0 \\ & & \downarrow{\scriptstyle f} & & \downarrow{\scriptstyle g} & & \downarrow{\scriptstyle h} & & \\ 0 & \longrightarrow & A & \xrightarrow[\iota]{} & B & \xrightarrow[\zeta]{} & C & & \\ & & \downarrow{\scriptstyle f_c} & & \downarrow{\scriptstyle g_c} & & \downarrow{\scriptstyle h_c} & & \\ & & co\ker(f) & \xrightarrow[\iota_c]{} & co\ker(g) & \xrightarrow[\zeta_c]{} & co\ker(h) & & \end{array}$$

Construction of the connecting arrow δ: Let $(P, u, v) := Pullback(j, h_k)$ and $(Q, \varphi, \eta) := Pushforward(\iota_c, f_c)$, which we display in the following diagram:

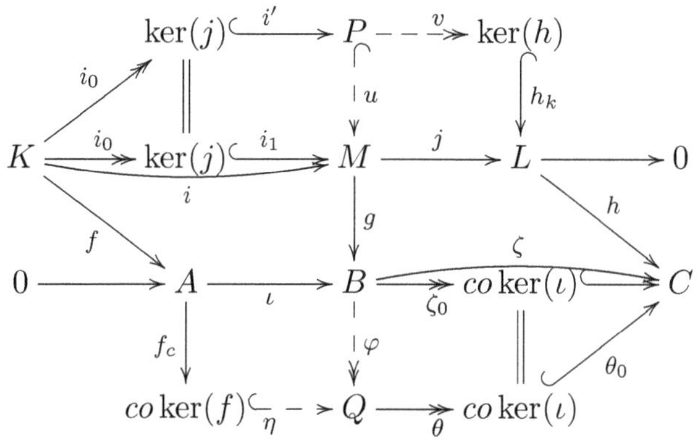

By exactness properties of abelian categories or by (A.3.1), (A.3.2), η, u are injective and v, φ are surjective. Also, by (A.3.1), (A.3.2), we have $\ker(j) \cong \ker(v) =: i'$ and $coker(\eta) = co\ker(\iota)$. We look at the vertical composition $\varphi g u$. Since i_0 is surjective and θ_0 is injective, we obtain two (dual) identities:

$$\begin{cases} (\varphi g u) i' i_0 = \varphi g i = \varphi \iota f = \eta f_c f = \eta 0 = 0 & \Longrightarrow \varphi g u i' = 0 \\ \theta_0 \theta(\varphi g u) = t\zeta_0 g u = h j u = h h_k v = 0 v = 0 \Longrightarrow \theta(\varphi g u) = 0. \end{cases}$$

Since $\eta = \ker(\theta)$, there is an arrow $\beta : P \longrightarrow co\ker(f)$ such that

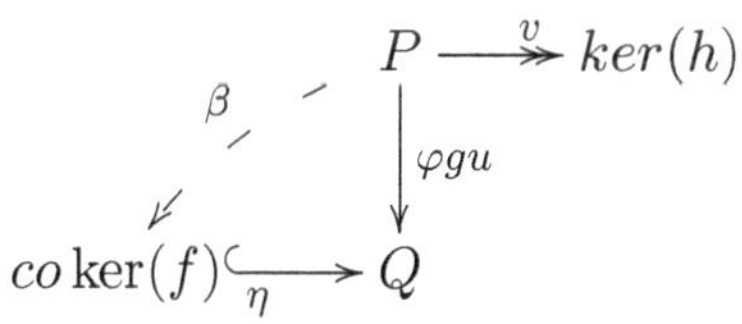

commutes.

Now, $\eta\beta i' = (\varphi g u) i' = 0$. Since η is injective, $\beta i' = 0$. So, β factors through $\ker(h)$, and we obtain the arrow δ, as in the commutative diagram

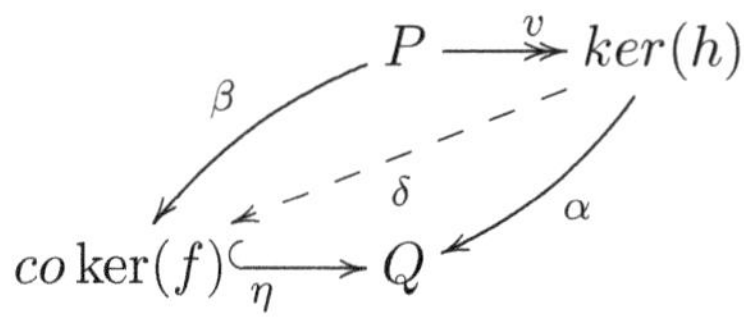

with $\alpha := \eta\delta$.

Thus, we obtain the connecting arrow δ.

Now, we proceed to prove exactness.

1. **Exactness at** $\ker(g)$**:** By (A.3.6), there is a monomorphism $\omega : Im(i_k) \hookrightarrow \ker(j_k)$ such that $s\omega = t$, where $s : ker(j_k) \hookrightarrow \ker(g)$ is the kernel of j_k, and $t : Im(i_k) \hookrightarrow \ker(g)$ is the monomorphism corresponding to the image of i_k:

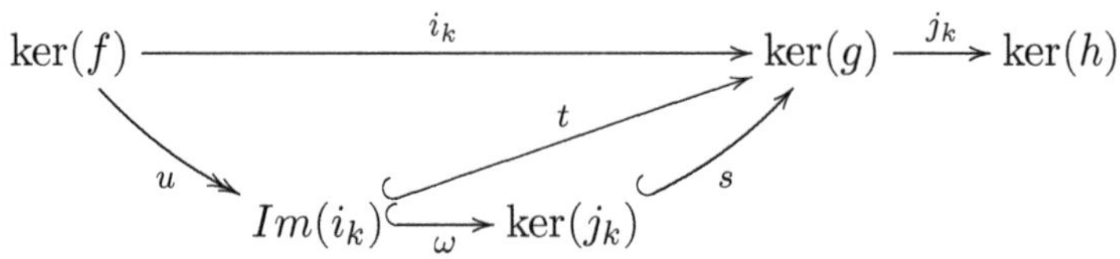

We prove that ω is an isomorphism. To define the morphism in the opposite direction, consider the following commutative diagram and label the morphisms:

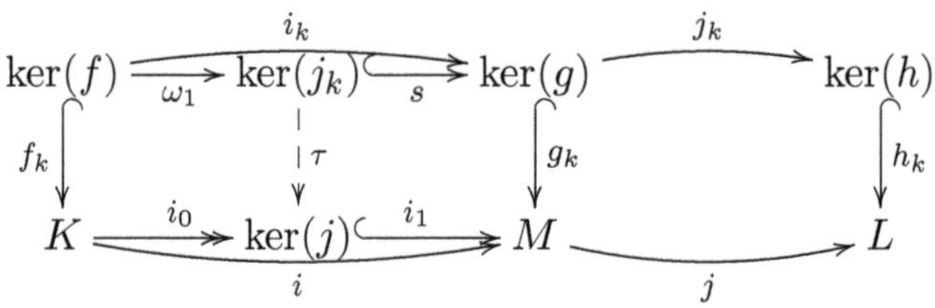

In this diagram the morphism τ is defined by the property of kernel $\ker(j)$ because $jg_ks = j_kg_ks = 0$. Let $(U, \lambda, \mu) := K \times_{\ker(j)} \ker(j_k)$. Consider the diagram

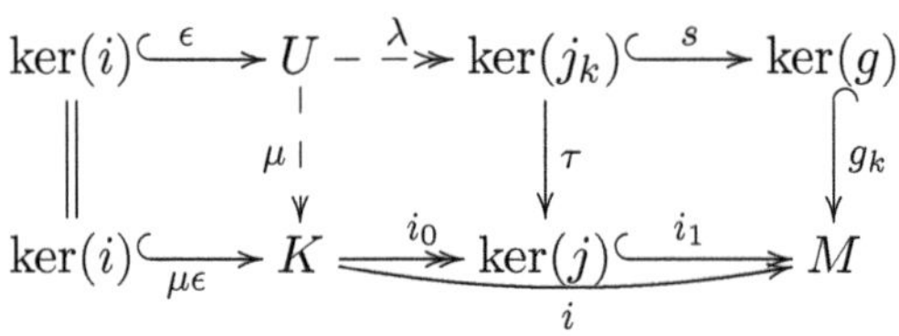

Note λ is epi. We have

$$i_0\mu = \tau\lambda \Longrightarrow i\mu = i_1i_0\mu = i_1\tau\lambda = g_ks\lambda \Longrightarrow$$
$$gi\mu = 0 \Longrightarrow \iota f\mu = 0 \Longrightarrow f\mu = 0,$$

because ι is mono. By property of kernel f_k, we have an arrow ψ as follows:

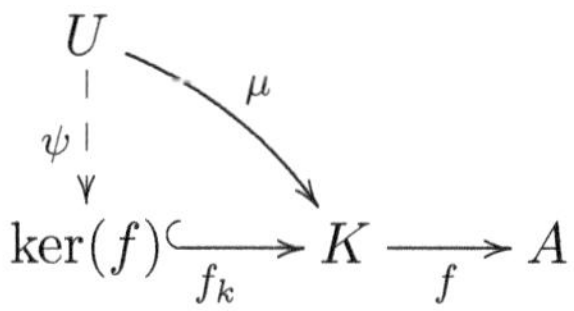

So, $g_k i_k \psi = i f_k \psi = i\mu = i_1 i_0 \mu = i_1 \tau \lambda = g_k s \lambda \Longrightarrow i_k \psi = s\lambda$. Diagramatically,

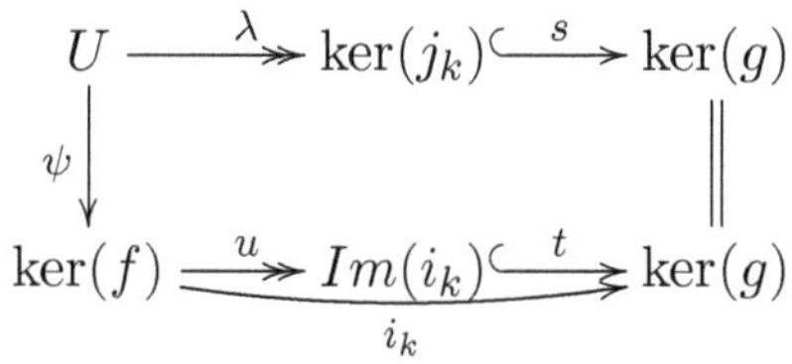

We have $i_k \psi \epsilon = s\lambda\epsilon = 0$. Since, $\epsilon = \ker(\lambda)$, there is an arrow Ψ, as in the commutative diagram

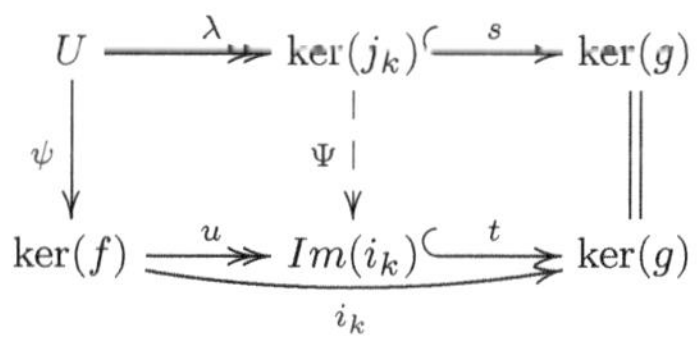

$$\begin{cases} \text{In particular } t\Psi = s. \\ \text{So, } s\omega\Psi = s, \text{ and } t\Psi\omega = s\omega = t. \\ \text{Hence, } \omega\Psi = 1, \ \Psi\omega = 1. \end{cases}$$

This completes the proof of exactness at $\ker(g)$.

2. **Exactness at** $\ker(h)$**:** We have the following commutative diagram:

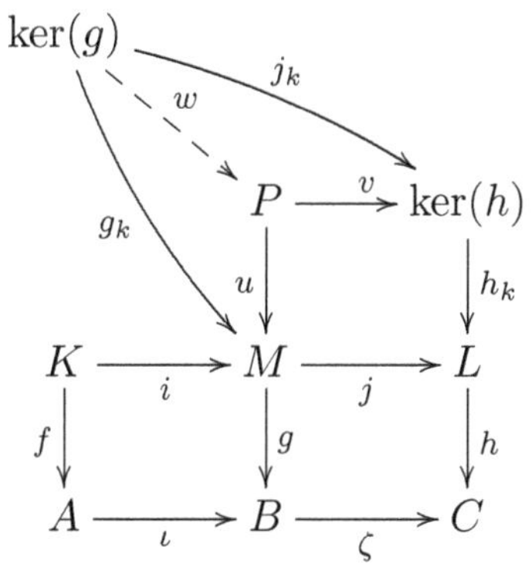

First, prove $\delta j_k = 0$. Since η is monic, we prove $\eta\delta j_k = 0$, or $\eta\delta v w = 0$. However, $\eta\delta v w = \eta\beta w = \varphi g u w = \varphi g g_k = 0$.

By (A.3.6), there is an arrow $\Psi : Im(j_k) \hookrightarrow \ker(\delta)$ such that $t\Psi = s$, where s, t are the monomorphisms, as in the commutative diagram

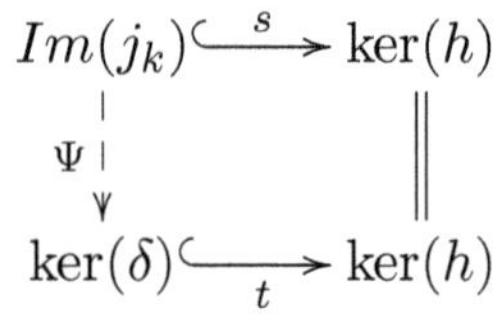

Now we proceed to define the inverse of Ψ. Consider the diagram

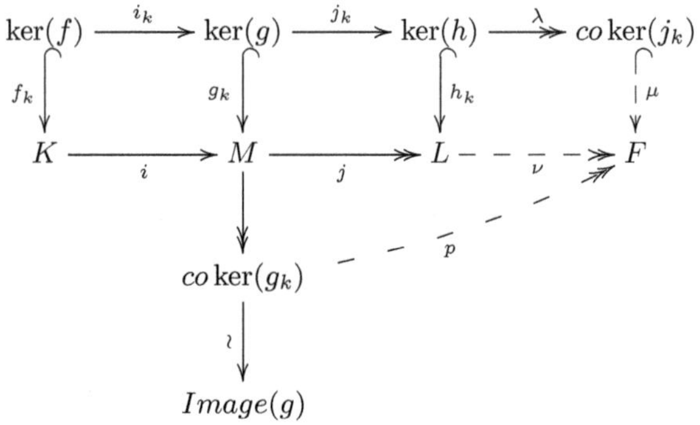

where $F = L \coprod_{\ker(h)} co\ker(j_k)$

It follows that μ is mono and ν is epi. It follows that $\nu j g_k = 0$. So, νj factors through $co\ker(g_k)$. We further extend the above diagram as follows:

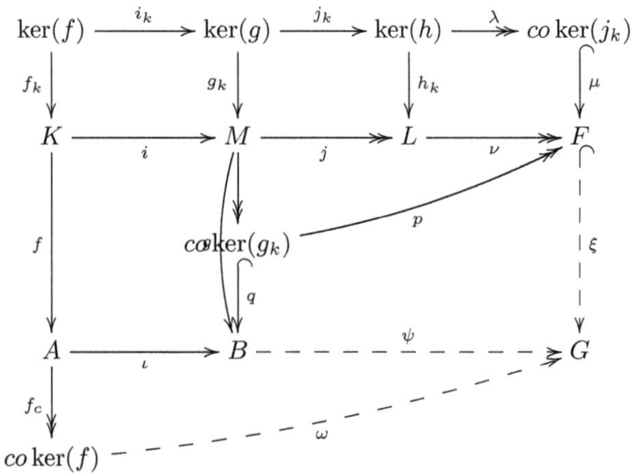

where $(G, \xi, \psi) := B \times_{co\ker(g_k)} F$. Since q is mono, so is ξ. By the similar argument, as above, $\psi\iota$ factors through $co\ker(f)$. Now, we have the pushforward diagram, which defines an arrow τ.

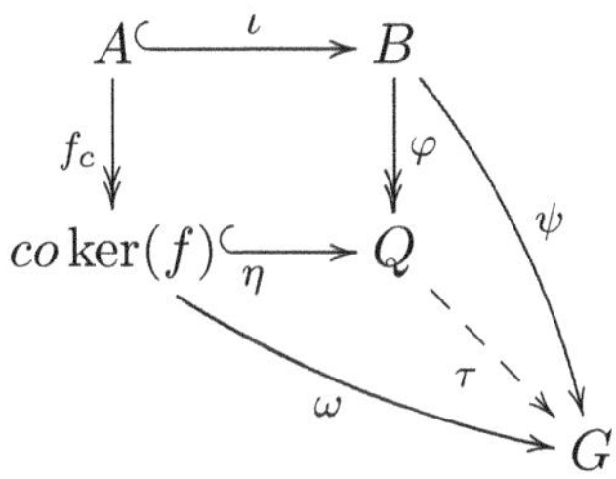

Now,

$$\omega\delta v = \tau\eta\delta v = \tau\eta\beta = \tau\varphi g u = \psi g u = \xi\nu j u = \xi\nu h_k v = \xi\mu\lambda v.$$

Since v is epi, $\omega\delta = \xi\mu\lambda$. In summary,

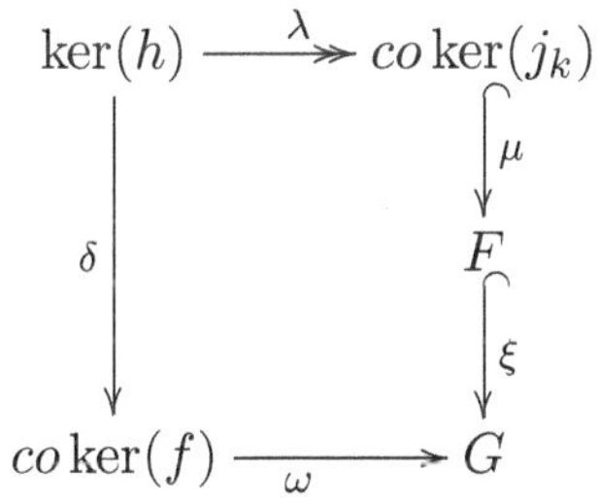

With $t = \ker(\delta)$, we decompose as follows:

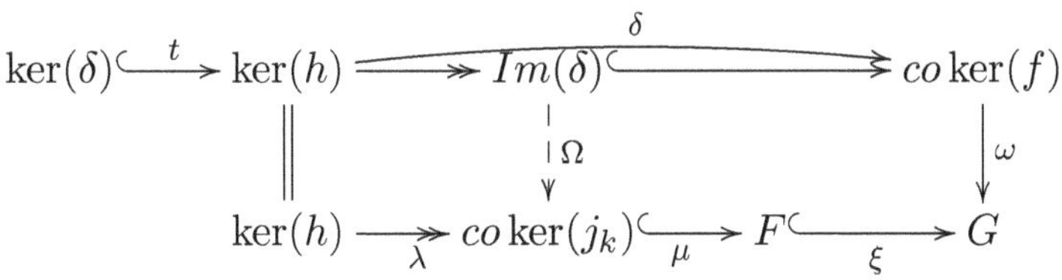

We have $\xi\mu\lambda t = \omega\delta t = 0$. So, $\lambda t = 0$. Since $Im(\delta) = co\ker(t)$, there is a unique morphism Ω, making the first rectangle commutative.

Again, since $\lambda t = 0$, by property of kernel s, we have the following commutative diagram for some arrow Θ:

$$\begin{array}{ccccccccc}
\ker(\delta) & \xrightarrow{t} & \ker(h) & \xrightarrow[e]{} & Im(\delta) & & \xrightarrow[m]{} & & co\ker(f) \\
\Theta \downarrow & & \| & & \downarrow \Omega & & & & \downarrow \omega \\
Im(j_k) & \xrightarrow[s]{} & \ker(h) & \xrightarrow[\lambda]{} & co\ker(j_k) & \xrightarrow[\mu]{} & F & \xrightarrow[\xi]{} & G
\end{array}$$

(with $\delta : \ker(h) \to co\ker(f)$ the composite me)

So, we have $s\Theta = t$. Since $t\Psi = s$, and s, t are mono, we have $\Theta\Psi = Id$ and $\Psi\Theta = Id$. This establishes the exactness at $\ker(h)$.

3. **Exactness at** $co\ker(f)$: This follows from the dual argument of the proof of exactness at $\ker(h)$.
4. **Exactness at** $co\ker(g)$: This follows from the dual argument of the proof of exactness at $\ker(g)$.

The proof is complete. ■

The following is a lighter version of the Snake Lemma.

Corollary A.3.8. *Suppose that $\mathscr{A}$ is an abelian category. Consider the commutative diagram*

$$\begin{array}{ccccc}
K & \xrightarrow{i} & M & \xrightarrow{j} & L \\
f \downarrow & & \downarrow g & & \downarrow h \\
A & \xrightarrow[\iota]{} & B & \xrightarrow[\zeta]{} & C
\end{array} \qquad \text{in} \qquad \mathscr{A},$$

where the first row is exact at M and the second row is exact at B. Then

$$\begin{cases}
\ker(f) \xrightarrow{i_k} \ker(g) \xrightarrow{j_k} \ker(h) & \textit{is exact if } \iota \textit{ is mono} \\
co\ker(f) \xrightarrow[\iota_c]{} co\ker(g) \xrightarrow[\zeta_c]{} co\ker(h) & \textit{is exact if } j \textit{ is epi}
\end{cases}$$

Proof. Follows from the proof of (A.3.7). ■

A.4. Exact Categories

We define Exact Categories as follows [Q, p. 99].

Definition A.4.1. Consider an additive category $\mathscr{E}$, together with a family of sequences of morphisms

$$X \overset{i}{\hookrightarrow} Y \overset{p}{\twoheadrightarrow} Z \qquad \text{to be called } \textbf{conflations}. \tag{A.3}$$

$$\text{We also say} \quad 0 \longrightarrow X \overset{i}{\longrightarrow} Y \overset{p}{\longrightarrow} Z \longrightarrow 0 \quad \text{is } \textbf{exact}. \tag{A.4}$$

The left part $X \hookrightarrow Y$ of such a conflation is called an **inflation** and is denoted as such. Likewise, the right part $Y \twoheadrightarrow Z$ is called a **deflation** and is denoted as such.

Such an additive category $\mathscr{E}$, together with such a family of conflations (A.3), is called an **Exact Category** if the following conditions are satisfied:

1. In a conflation (A.3), the inflation i is the kernel of p, and p is a cokernel of i.
2. Conflations are closed under isomorphisms.
3. Inflations and deflations are closed under composition.
4. Any diagram $X \longrightarrow Z \overset{p}{\twoheadleftarrow} Y$ of morphisms, with p deflation, can be completed to a pullback diagram

$$\begin{array}{ccc} W & \dashrightarrow & Y \\ {\scriptstyle q}\,\downarrow & & \downarrow\,{\scriptstyle p} \\ X & \longrightarrow & Z \end{array} \qquad \begin{cases} \text{with } q \text{ deflation.} \\ \text{We say that deflations are closed under base change.} \end{cases}$$

5. Dually, any diagram $Z \longleftarrow X \overset{i}{\hookrightarrow} Y$ of morphisms, with i inflation, can be completed to a pushforward diagram

$$\begin{array}{ccc} X & \overset{i}{\hookrightarrow} & Y \\ \downarrow & & \downarrow \\ Z & \underset{j}{\hookrightarrow} & W \end{array} \qquad \begin{cases} \text{with } j \text{ inflation.} \\ \text{We say that inflations are closed} \\ \quad \text{under cobase change.} \end{cases}$$

6. For all objects $X, Y \in Obj(\mathscr{C})$, the following

$$X \longrightarrow X \oplus Y \longrightarrow Y \quad \text{is a conflation.}$$

Further, an inflation is also called an **admissible monomorphism**, and a deflation is also called an **admissible epimorphism**.

Definition A.4.2. Suppose that $\mathscr{E}, \mathscr{B}$ are exact categories and that $F : \mathscr{B} \longrightarrow \mathscr{E}$ is a functor of additive categories. We say that F is an **exact functor** if it sends conflations to conflations.

Example A.4.3. Here is a list of examples.

1. To start with, any abelian category is an exact category. Where a sequence

$$0 \longrightarrow K \xrightarrow{\iota} M \xrightarrow{p} C \longrightarrow 0 \tag{A.5}$$

is declared a conflation (or exact) if $\ker(p) = \iota$ and $co\ker(\iota) = p$. This follows from (A.3.4).

   (a) So, for a ring R, the category $Mod(R)$ of all (left) R-modules is an abelian category and is hence an exact category. Conflations are defined as above (A.5).
   (b) For a commutative Noetherian ring A, $QCoh(A)$ and $Coh(A)$ are abelian categories. More generally, for a Noetherian scheme X, $QCoh(X)$, $Coh(X)$ are abelian categories. Hence they are exact categories.

2. For a Noetherian scheme X, the category $\mathscr{P}(X)$ of locally free sheaves forms an exact category. If fact, $\mathscr{P}(X)$ is a fully exact subcategory of $Coh(X)$. In particular, for a commutative Noetherian ring A, the category $\mathscr{P}(A)$ of finitely generated projective A-modules forms an exact category.
3. Suppose that $\mathscr{E}$ is an exact category. Let $Ch(\mathscr{E})$ be the category of chain complexes. The objects of $Ch(\mathscr{E})$ are chain complexes

$$(A_\bullet, d_\bullet) : \cdots \longrightarrow A_{i+1} \longrightarrow A_i \xrightarrow{d_i} A_{i-1} \longrightarrow \cdots$$
$$\ni \quad dd = 0.$$

Morphisms in $Ch(\mathscr{E})$ are chain complex maps, given by commutative diagrams as follows:

$$f^\bullet : (A_\bullet, d_\bullet) \longrightarrow (B_\bullet, \partial_\bullet) : \quad \begin{array}{ccccccccc} \cdots & \longrightarrow & A_{i+1} & \longrightarrow & A_i & \xrightarrow{d_i} & A_{i-1} & \longrightarrow & \cdots \\ & & \downarrow & & \downarrow{\scriptstyle f_i} & & \downarrow & & \\ \cdots & \longrightarrow & B_{i+1} & \longrightarrow & B_i & \xrightarrow[\partial_i]{} & B_{i-1} & \longrightarrow & \cdots \end{array}$$

$$\text{A sequence } (A_\bullet, d_\bullet) \longrightarrow (B_\bullet, d_\bullet) \longrightarrow (C_\bullet, d_\bullet) \text{ is declared}$$

$$\text{exact if } \forall\ i \in \mathbb{Z} \quad A_i \longrightarrow B_i \longrightarrow C_i \quad \text{is exact.}$$

Then $Ch(\mathscr{E})$ is an exact category. Its full subcategory $Ch^b(\mathscr{E})$ of bounded chain complexes is also an exact subcategory. A complex $(A_\bullet, d_\bullet) \in Ch^b(\mathscr{E})$, if $A_i = 0\ \forall\ i \gg 0,\ i \ll 0$.

Definition A.4.4. Suppose that $\mathscr{E}$ is an exact category and $\mathscr{B} \subseteq \mathscr{E}$ is a full subcategory. We say that $\mathscr{B}$ is **closed under extensions** if $\mathscr{B}$ has a zero object and if $\mathscr{B}$ is closed under extensions in $\mathscr{E}$, which means that for any conflation (A.3) in $\mathscr{E}$, if X, Z are isomorphic to an object in $\mathscr{B}$, then so is Y.

1. This gives a structure of an exact category to $\mathscr{B}$, by declaring a sequence in $\mathscr{B}$ is a conflation in $\mathscr{B}$ if it is isomorphic to an exact sequence in $\mathscr{E}$ (A.3).
2. Inclusion $\mathscr{B} \subseteq \mathscr{E}$ preserves and detects conflations.
3. In this case, we also say that $\mathscr{B} \subseteq \mathscr{E}$ is an **extension-closed** exact subcategory [Q, p. 108], or a **fully exact subcategory**. Further, the inclusion functor $\iota : \mathscr{B} \hookrightarrow \mathscr{E}$ is an exact functor.

Lemma A.4.5. *Every small exact category $\mathscr{E}$ can be embedded into an abelian category $\mathscr{C}$ as a fully exact, extension-closed subcategory.*

Proof. See [Q, p. 100] ∎

Remark A.4.6. While we skipped the proof, Lemma A.4.5 can be a very useful tool. One reason for this is that the Snake Lemma (A.3.7) becomes available. Most applications of K-Theory concern exact subcategories of the abelian category $QCoh(X)$, where X is a scheme (*or some similar variation*). However, some may feel the impulse to derive facts about Exact categories from the axioms of Exact categories (A.4.1).

A.5. Localization and Quotient Categories

This section is of some special interest to us. Inverting multiplicative subsets S of a commutative ring A is routinely covered in commutative algebra classes. However, this process extends, almost routinely, to inversion of a class of morphisms in a category, as follows.

A.5.1. *Calculus of fractions*

Among the original references to this section are [GZ], [Hr2, p. 28]. Other helpful references are [Sm1, Wc1].

Definition A.5.1. Let $\mathscr{C}$ be a category and S be a class of morphisms in $\mathscr{C}$. (*It is best to think of S as a subcategory, while we intend to invert all the morphisms in S.*) The **localization of $\mathscr{C}$ with respect to** S is a functor $\iota : \mathscr{C} \longrightarrow S^{-1}\mathscr{C}$, where $S^{-1}\mathscr{C}$ is a category with the following universal property. Given any functor $F : \mathscr{C} \longrightarrow \mathscr{D}$ such that $\forall\ f \in S,\ F(f)$ is an isomorphism in $\mathscr{D}$,

$$\exists \text{ a unique functor } \varphi : S^{-1}\mathscr{C} \dashrightarrow \mathscr{D} \quad \ni \quad \begin{array}{ccc} \mathscr{C} & \xrightarrow{\iota} & S^{-1}\mathscr{C} \\ & {\scriptstyle F}\searrow & \Big\downarrow{\scriptstyle \varphi} \\ & & \mathscr{D} \end{array} \quad \text{commutes.}$$

As always, the existence of such a localization $S^{-1}\mathscr{C}$ is questionable, and if it exists, it is unique up to a unique equivalence.

We work toward giving a sufficient condition for existence of such localized categories.

Definition A.5.2. Let $\mathscr{C}$ be a category and S be a class of morphisms in $\mathscr{C}$. We say that S satisfies the **calculus of right fractions** if the following three conditions hold:

1. The class S is closed under composition. The identity 1_X is in S for all $X \in Obj(\mathscr{C})$.
2. (Ore conditions) The following first diagram can be completed to the second commutative diagram:

$$\begin{array}{ccc} & & X \\ & & \Big\downarrow{\scriptstyle f} \\ Y & \xrightarrow[s]{} & Z \end{array} \quad \text{with } s \in S, \qquad \begin{array}{ccc} W & \overset{t}{\dashrightarrow} & X \\ {\scriptstyle g}\Big\downarrow & & \Big\downarrow{\scriptstyle f} \\ Y & \xrightarrow[s]{} & Z \end{array} \quad \text{with } t \in S. \tag{A.6}$$

3. Let $f, g : X \longrightarrow Y$ be two morphisms in $\mathscr{C}$ and $s : Y \longrightarrow Z$ in S such that $sf = sg$. Then there is a morphism $t : W \longrightarrow X$ in S such that $ft = gt$.

If the dual of these three conditions hold, then S is said to satisfy the **calculus of left fractions**. If S satisfies both the conditions, then we say that S satisfies the **calculus of fractions**.

The following existence theorem suffices our purposes.

Lemma A.5.3. *Suppose that $\mathscr{C}$ is a category with a set of isomorphism classes of objects. Let S be a class of morphisms in $\mathscr{C}$. Assume that S satisfies the calculus of right fractions or of left fractions. Then the localization category $S^{-1}\mathscr{C}$ exists and is constructed as in the proof.*

Proof. We assume that S satisfies the calculus of right fractions. The objects of $S^{-1}\mathscr{C}$ are the same as those of $\mathscr{C}$. To define morphisms, for two objects X, Y, in view of (A.6), let

$$Pair(X,Y) = \\ \{(s,f) \in Mor_{\mathscr{C}}(U,X) \times Mor_{\mathscr{C}}(U,Y) : U \in Obj(\mathscr{C}), s \in S\}.$$

Under the hypothesis, we can treat $Pair(X,Y)$ as a set, by restricting U to a set of representatives of isomorphism classes of objects. Refer to the elements $(s,f) \in Pair(X,Y)$, by the diagrams $X \xleftarrow{s} U \xrightarrow{f} Y$. Two such diagrams, $X \xleftarrow{s} U \xrightarrow{f} Y$, $X \xleftarrow{t} V \xrightarrow{g} Y$ in $Pair(X,Y)$, are said to be equivalent if

$$\exists \quad \begin{cases} \bar{t} : W \longrightarrow U \\ \bar{s} : W \longrightarrow V \end{cases} \text{either } \bar{t} \in S \text{ or } \bar{s} \in S \quad \ni$$

$$\begin{array}{ccccc} & & U & & \\ & \swarrow^{s} & \uparrow{\scriptstyle \bar{t}} & \searrow^{f} & \\ X & & W & & Y \\ & \nwarrow_{t} & \downarrow{\scriptstyle \bar{s}} & \nearrow_{g} & \\ & & V & & \end{array} \quad \text{commutes.}$$

One checks that this is an equivalence relation on $Pair(X,Y)$. Let fs^{-1} denote the equivalence class of the diagram $X \xleftarrow{s} U \xrightarrow{f} Y$

and let

$$Mor_{S^{-1}\mathscr{C}}(X,Y) := \{fs^{-1} : (s,f) \in Pair(X,Y)\}.$$

Let $fs^{-1} \in Mor_{S^{-1}\mathscr{C}}(X,Y)$, $gt^{-1} \in Mor_{S^{-1}\mathscr{C}}(Y,Z)$ be given by the diagrams $X \xleftarrow{s} U \xrightarrow{f} Y$, $Y \xleftarrow{t} V \xrightarrow{g} Z$. To define compositions, by the conditions of the calculus of right fractions, we complete the following commutative diagram:

$$\begin{array}{ccccc} & & W & \xrightarrow{\varphi} & V & \xrightarrow{g} & Z \\ & & {\scriptstyle\tau}\downarrow & & \downarrow{\scriptstyle t} & & \\ X & \xleftarrow[s]{} & U & \xrightarrow[f]{} & Y & & \end{array}$$

with $\tau \in S$. Define $(gt^{-1})(fs^{-1}) := (g\varphi)(s\tau)^{-1}$.

Again, one checks that this composition is well defined. Further, it is clear that the functor $\iota : \mathscr{C} \longrightarrow S^{-1}\mathscr{C}$ satisfies the universal property. The proof is complete. ■

Remark A.5.4. Recall that for a commutative ring A, we can invert any subset $S \subseteq A$. Likewise, barring set theoretic issues, $S^{-1}\mathscr{C}$ exists for any class of morphisms S in a category $\mathscr{C}$, whether S satisfies the calculus of (either or both) fractions or not. The condition in (A.5.3) that $\mathscr{C}$ has a set of isomorphism classes is required to avoid any set theoretic issues. There are a variety of ways to relax this condition (see [Wc2, p. 191]).

A.5.2. *Quotient of abelian categories*

Among the original and comprehensive references to this subsection is the book of Swan [Sw3]. The quotient of abelian (*or additive*) categories by a subcategory is defined similarly (A.5.1) by remaining within the universe of abelian categories, as follows.

Definition A.5.5. (see [Sw3, p. 2].) Suppose that $\mathscr{A}$ is an abelian category and $\mathscr{B} \subseteq \mathscr{A}$ is an abelian subcategory. The quotient of $\mathscr{A}$ by $\mathscr{B}$ is an abelian category $\frac{\mathscr{A}}{\mathscr{B}}$, together with an exact functor $\iota : \mathscr{A} \longrightarrow \frac{\mathscr{A}}{\mathscr{B}}$ such that $\iota(X) = 0$ for all $X \in Obj(\mathscr{B})$, with the following universal property: Given any exact functor $F : \mathscr{A} \longrightarrow \mathscr{D}$, where $\mathscr{D}$ is an abelian category such that $F(X) = 0$, for all objects in $\mathscr{B}$,

$$\exists \text{ a unique additive functor } \varphi : \frac{\mathscr{A}}{\mathscr{B}} \longrightarrow \mathscr{D} \quad \ni$$

$$\begin{array}{ccc} \mathscr{C} & \xrightarrow{\iota} & \frac{\mathscr{A}}{\mathscr{B}} \\ & {\scriptstyle F}\searrow & \downarrow{\scriptstyle \varphi} \\ & & \mathscr{D} \end{array} \quad \text{commutes.}$$

If the quotient $\frac{\mathscr{A}}{\mathscr{B}}$ exists, then it is unique up to a unique equivalence. However, existence of such a quotient as an abelian category is not guaranteed. We proceed to give a sufficient condition for existence of the quotient.

Definition A.5.6. (See [Sw3, p. 1].) Let $\mathscr{A}$ be an abelian category. A full subcategory $\mathscr{B} \subseteq \mathscr{A}$ is said to be a **Serre Subcategory** if $\mathscr{B}$ is nonempty and if for any conflation

$$K \hookrightarrow M \twoheadrightarrow C \quad \text{in } \mathscr{A}$$

$$K, C \in Obj(\mathscr{B}) \iff M \in Obj(\mathscr{B}) \text{ (up to isomorphisms).}$$

That means that if and only if $\mathscr{B}$ is closed under subobjects, quotient objects, and extensions. A Serre subcategory is an abelian subcategory.

Lemma A.5.7. *Let $\mathscr{A}$ be an abelian category (small) and $\mathscr{B} \subseteq \mathscr{A}$ be a Serre subcategory. Then the quotient category $\frac{\mathscr{A}}{\mathscr{B}}$ exists.*

Proof. A morphism f in $\mathscr{A}$ is defined as a **mod-$\mathscr{B}$ isomorphism** if $\ker(f), co\ker(f) \in \mathscr{B}$. Let S be the set of all mod-$\mathscr{B}$ isomorphisms. Then S satisfies the calculus of fractions:

1. Let $f : X \longrightarrow Y$, $g : Y \longrightarrow Z$ be in S. We show that $gf \in S$.

$$\ker(f) \hookrightarrow \ker(gf) \twoheadrightarrow \ker(g) \cap Im(f) \quad \text{is a conflation.} \tag{A.7}$$

   Since $\ker(g) \in Obj(\mathscr{B})$, $\ker(g) \cap Im(f) \in Obj(\mathscr{B})$. Also, since $\ker(f) \in Obj(\mathscr{B})$, $\ker(gf) \in Obj(\mathscr{B})$. By reversing the arrows, $co\ker(gf) \in Obj(\mathscr{B})$. So, S is closed under composition.

2. Now, we check the Ore condition of the calculus of right fractions (A.6). Let $M_2 \xrightarrow[f]{} N \xleftarrow[s]{} M_1$ be arrows in $\mathscr{A}$ with $s \in S$. Consider the pullback diagram

$$\begin{array}{ccc} M_1 \times_N M_2 & \overset{g}{\dashrightarrow} & M_1 \\ {\scriptstyle t}\,\big\downarrow & & \big\downarrow{\scriptstyle s} \\ M_2 & \xrightarrow[f]{} & N \end{array}$$

By Lemma A.3.1, $\ker(t) = \ker(s) \in Obj(\mathscr{B})$. One checks that the map $co\ker(t) \hookrightarrow co\ker(s)$ is an injective map. In fact, $Im(t) = f^{-1}Im(s)$, So, $co\ker(t) = \frac{M_2}{Im(t)} = \frac{M_2}{f^{-1}Im(s)}$, and $co\ker(s) = \frac{N}{Im(s)}$. (*It is easier to check the same module theoretically!*) So, $t \in S$.

By reversing the arrows, it follows that the corresponding condition of left fractions is satisfied.

3. To check the third condition, let $f, g : M \longrightarrow N$ be two maps in $\mathscr{A}$ and $s : N \longrightarrow W$ be a map in S, such that $sf = sg$. With $F = f - g$, we have $sF = 0$. It follows that there is an arrow $F_0 : M \longrightarrow \ker(s)$ such that $F = F_0\iota$, as in the diagram

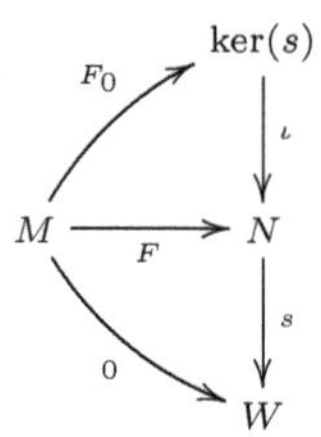

This leads to

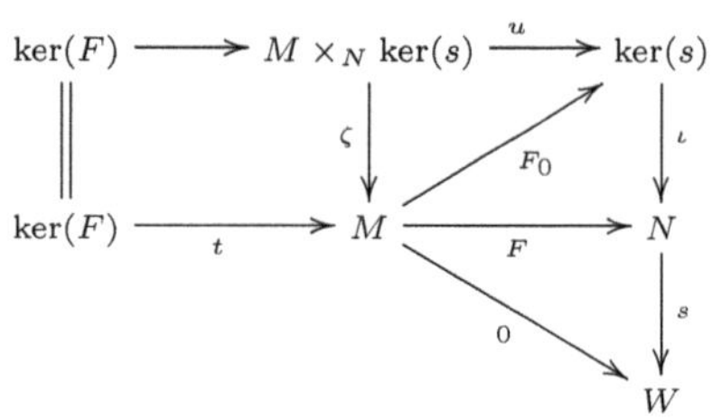

Claim that t is in S. Clearly, $\ker(t) = 0 \in \mathscr{B}$. Now, $co\ker(t) = Im(F) \subseteq \ker(s) \in \mathscr{B}$. Hence, $co\ker(t)$ is in $\mathscr{B}$. So, $t \in S$. The dual of this property is checked by reversing the arrows.

This establishes that S satisfies the calculus of fractions. Hence, $S^{-1}\mathscr{A}$ exists. One checks, that $S^{-1}\mathscr{A} =: \frac{\mathscr{A}}{\mathscr{B}}$ is an abelian category and satisfies the required universal property (Exercise A.6.8). The proof is complete. ■

A.5.3. *Quotient of exact categories*

Until Section 9.4, we only considered the quotient of abelian categories by Serre subcategories (A.5.7). The question of the quotient of exact categories was considered in ([Sm2]). First, we give a definition ([Sm2]).

Definition A.5.8. Suppose that $\mathscr{E}$ is an exact category and $\mathscr{P} \subseteq \mathscr{E}$ is an extension-closed full subcategory. An exact category $\mathscr{Q}$, together with an exact functor $\iota : \mathscr{E} \longrightarrow \mathscr{Q}$, is called a quotient of $\mathscr{E}$ by $\mathscr{P}$ if ι satisfies the following conditions:

1. $\iota(P) \simeq 0$ for all $P \in Obj(\mathscr{P})$.
2. The functor ι is universal in the following sense: suppose that $\mathscr{D}$ is an exact category, and $\zeta : \mathscr{E} \longrightarrow \mathscr{D}$ is an exact functor, such that $\zeta(P) \simeq 0 \ \ \forall\, P \in Obj(\mathscr{P})$. Then

$$\exists \text{ Unique exact functor } \varphi : \mathscr{Q} \longrightarrow \mathscr{D} \qquad \begin{array}{ccc} \mathscr{E} & \xrightarrow{\ \iota\ } & \mathscr{Q} \\ & \searrow^{\zeta} & \downarrow{\scriptstyle \varphi} \\ & & \mathscr{D} \end{array} \qquad \text{commutes.}$$

As before, the existence of such a quotient $\mathscr{Q}$ is not guaranteed. However, when it exists, it is unique up to isomorphism (*functorial equivalence*). When it exists, we denote $\mathscr{Q} := \frac{\mathscr{E}}{\mathscr{P}}$.

A sufficient condition for the existence of such a quotient $\frac{\mathscr{E}}{\mathscr{P}}$ of exact categories is given in [Sm2], which is too technical to discuss here. The sufficient conditions were designed to apply for a situation, as follows.

Example A.5.9. Suppose that $\mathscr{E}$ is a small exact category and that $\mathscr{P} \subseteq \mathscr{E}$ is an extension-closed full subcategory. A morphism $f : M \longrightarrow N$ in $\mathscr{E}$ is called a **weak isomorphism** if $f = f_n f_{n-1} \cdots f_1$ is a finite composition of morphisms, $f_i : M_{i-1} \longrightarrow M_i$ in $\mathscr{E}$ for $i = 1, \ldots, n$, with $M_0 = M$ and $M_n = N$, such that

1. either f_i is a deflation, with $\ker(f_i) \in Obj(\mathscr{P})$,
2. or f_i is an inflation, with $co\ker(f_i) \in Obj(\mathscr{P})$.

Let Σ denote the class of all weak isomorphisms. Then

1. If Σ satisfies the calculus of fractions (left or right), define

$$\frac{\mathscr{E}}{\mathscr{P}} := \Sigma^{-1}\mathscr{E} \qquad \text{obtained by the calculus of fractions.}$$

   It is not asserted, at this stage, that $\frac{\mathscr{E}}{\mathscr{P}}$ has a structure of an exact category. See [Sm2, Lem. 1.13] for details.
2. Under certain further conditions, $\frac{\mathscr{E}}{\mathscr{P}}$ inherits a structure of an exact category from $\mathscr{E}$. A sequence in $\frac{\mathscr{E}}{\mathscr{P}}$ is defined as exact if it is isomorphic to the image of an exact sequence in $\mathscr{E}$. In these cases, $\frac{\mathscr{E}}{\mathscr{P}}$ is the quotient category, in the sense of (A.5.8). See [Sm2, Prop. 1.16] for details.

We will revisit this construction much later, in Definition 9.4.6.

A.6. Exercises

Exercise A.6.1. Suppose that $\mathscr{C}$ is an additive category. Then $\forall\ X, Y \in Obj(\mathscr{C})$, the direct sum exists, and $X \times Y = X \oplus Y$. (See [HS, Prop. II.9.1].)

Exercise A.6.2. Suppose that $\mathscr{C}$ is an additive category and $X, Y \in Obj(\mathscr{C})$. Then for $\forall\ f, g \in Mor(X, Y)$, the sum $f + g$ is given by the following commutative diagram:

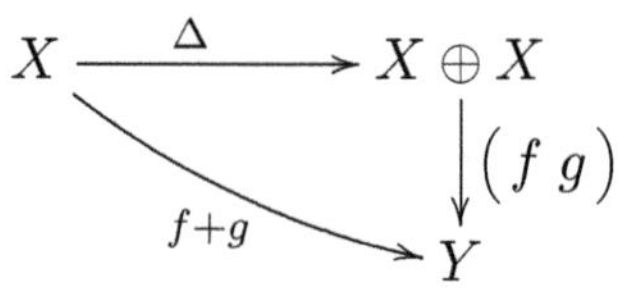

Hint: Use (A.2.4).

Exercise A.6.3. Suppose that $\mathscr{A}$ is an abelian category, and M is an object. Let $M_1, M_2 \subseteq M$ be two subobjects. (*Refer to Def. A.2.11.*) Then

1. The following

$$\begin{array}{ccc} M_1 \cap M_2 & \hookrightarrow & M_1 \\ \downarrow & & \downarrow \\ M_2 & \hookrightarrow & M \end{array}$$

is a pullback diagram.

2. There is an exact sequence

$$0 \longrightarrow M_1 \cap M_2 \longrightarrow M_1 \oplus M_2 \longrightarrow M_1 + M_2 \longrightarrow 0.$$

Exercise A.6.4. Suppose that $\mathscr{A}$ is an abelian category.

1. Let $f : M \longrightarrow N$ be an arrow. Then f is epi if and only if the cokernel of f is $N \longrightarrow 0$. And, f is mono if and only if the kernel of f is $0 \hookrightarrow M$.
2. Suppose that for composable arrows f, g, if fg is epi, then so is f. Similarly, if fg is mono, so is g.
3. For $i = 1, 2$, let $f_i \in Mor_{\mathscr{A}}(M_i, M)$, and $g_i \in Mor_{\mathscr{A}}(N, N_i)$. If f_1 is epi, then the induced map $\begin{pmatrix} f_1 & f_2 \end{pmatrix} : M_1 \oplus M_2 \longrightarrow M$ is epi. Dually, if g_1 is mono, then so is $\begin{pmatrix} g_1 \\ g_2 \end{pmatrix} : N \longrightarrow N_1 \oplus N_2$.

Exercise A.6.5. [Sw3, p 1.] Let $T : \mathscr{A} \longrightarrow \mathscr{M}$ be an exact functor of the abelian categories. Let $\mathscr{B} \subseteq \mathscr{A}$ be the full subcategory of objects $M \in \mathscr{A}$, such that $T(M)$ is isomorphic to zero. Then $\mathscr{B}$ is a Serre subcategory of $\mathscr{A}$.

Exercise A.6.6. [Sw3, p. 2.] Let $\mathscr{A}$ be an abelian category, and $\mathscr{B} \subseteq \mathscr{A}$ be a nonempty full subcategory. Then $\mathscr{B}$ is a Serre subcategory of $\mathscr{A}$ if and only if

$$\forall \text{ exact sequences } \quad M_1 \xrightarrow{f} M \xrightarrow{g} M_2$$
$$\text{in} \quad \mathscr{A}, \quad M_1, M_2 \in \mathscr{B} \Longrightarrow M \in \mathscr{B}.$$

Exercise A.6.7. Let $\mathscr{E}$ be an exact category. (*Some of the following is obvious by* (A.4.5), *sometimes in conjunction with the Snake Lemma* (A.3.7). *It would be good to obtain a proof from the axioms of Exact categories only.*)

1. Consider the map of two conflations:

$$\begin{array}{ccccc} K_1 & \hookrightarrow & M_1 & \twoheadrightarrow & C_1 \\ {\scriptstyle f_K}\downarrow & & \downarrow{\scriptstyle f_M} & & \downarrow{\scriptstyle f_C} \\ K_2 & \hookrightarrow & M_2 & \twoheadrightarrow & C_2 \end{array}$$

   If two of the vertical maps are isomorphisms, then so is the third one.
2. The sequence $0 \longrightarrow M \xrightarrow{p} C$ is a conflation if and only if p is an isomorphism. Similarly, $K \xrightarrow{i} M \xrightarrow{p} 0$ is a conflation if and only if i is an isomorphism.

Exercise A.6.8. Let $\mathscr{A}$ be an abelian category and $\mathscr{B} \subseteq \mathscr{A}$ be a Serre subcategory. Refer to (A.5.7). *A morphism $f : M \longrightarrow N$ in $\mathscr{A}$ is called a mod-$\mathscr{B}$ isomorphism, if* $\ker(f), co\ker(f) \in Obj(\mathscr{B})$. Let S be the set (class) of all such mod-$\mathscr{B}$ isomorphisms (*the same set as in the proof of* (A.5.7)). Let $\mathfrak{q} : \mathscr{A} \longrightarrow \frac{\mathscr{A}}{\mathscr{B}}$ be the quotient functor. Then

1. Define $\frac{\mathscr{A}}{\mathscr{B}}$ as an additive category by defining the addition of morphisms. Then prove that $\frac{\mathscr{A}}{\mathscr{B}}$ is an abelian category.
2. For $t \in Mor_{\mathscr{A}}(M, N)$, $\mathfrak{q}(t)$ is an isomorphism if and only if t is a mod-$\mathscr{B}$ isomorphism.
3. A morphism $f \in Mor_{\frac{\mathscr{A}}{\mathscr{B}}}(M, N)$ is an isomorphism if and only if $f = st^{-1}$ (equivalently, if $f = s^{-1}t$) for some mod-$\mathscr{B}$-isomorphisms s, t.
4. Prove that the functor $\mathfrak{q}$ satisfies the universal property of the quotient abelian categories (A.5.5).

Exercise A.6.9. [Sw3, p. 115.] Let R be a Noetherian commutative ring and $S \subseteq R$ be a multiplicatively closed subset. Let $\mathcal{T}or(S) \subseteq Coh(R)$ be the full subcategory of $M \in Coh(R)$ such that $M_S = 0$. Then $\mathcal{T}or(S) \subseteq Coh(R)$ is a Serre subcategory and the functor $\frac{Coh(R)}{\mathcal{T}or(S)} \longrightarrow Coh(R_S)$ is an equivalence of categories.

Exercise A.6.10. [Q, p. 119.] Suppose that X is a Noetherian scheme, and $Z \subseteq X$ is a closed subscheme. Let U be the open complement of Z. Let $Coh_Z(X) \subseteq Coh(X)$ be the full subcategory of objects $\mathscr{F} \in Coh(X)$, with support in Z. Then $Coh_Z(X)$ is a

Serre subcategory of $Coh(X)$ and $\frac{Coh(X)}{Coh_Z(X)} \longrightarrow Coh(U)$ is an equivalence of categories. (Hint: *Identify the mod-$Coh_Z(X)$ isomorphisms in $Coh(X)$. Use [Hr, Ex. II.5.15].*)

Exercise A.6.11. [Q, p. 123.] Let A be a Noetherian commutative ring and $X = \mathrm{Spec}\,(A)$. For integers $k \geq 0$, let $X^{(k)} = \{\wp \in X : height(\wp) = k\}$. Further, let $Coh^k(A) \subseteq Coh(A)$ be the full subcategory of A-modules M with $co\dim(Supp(M)) \geq k$. Then $Coh^{k+1}(A) \hookrightarrow Coh^k(A)$ is a Serre subcategory, and $\frac{Coh^k(A)}{Coh^{k+1}(A)} \cong \coprod_{\wp \in X^{(k)}} Coh^k(A_\wp)$.

Hint: The functor $Coh^k(A) \longrightarrow \coprod_{x \in X^{(k)}} Coh^k\,(A_\wp)$ vanishes on $Coh^{k+1}(A)$. Prove that the induced functor, from the quotient category, is essentially surjective and fully faithful.

Appendix B

Homotopy Theory

Much of the material in this chapter is available in standard textbooks on Topology [W, S]. However, some of the not-so-standard material was taken from [H]. We shall begin this chapter with the definition of topological spaces.

B.1. Elements of Topological Spaces

The basic definitions included in this section are omnipresent in the literature and on the internet. Topology is the study of continuous functions.

Definition B.1.1. Suppose that X is a nonempty set. A topology on X is a set $\mathcal{U}$ of subsets of X, to be called **open sets** of X, such that

1. The empty set ϕ and X are open. That is $\phi, X \in \mathcal{U}$.
2. If $\{U_i : i \in I\} \subseteq \mathcal{U}$, then $\bigcup_{i\in I} U_i \in \mathcal{U}$. So, the union (finite or infinite) of open sets is open.
3. If $U_1, U_2, \ldots, U_n \in \mathscr{U}$, then $\cap_{i=1}^{n} U_i \in \mathcal{U}$. So, the finite intersection of open sets is open.

A set X, with a topology $\mathcal{U}$ on X, is called a **topological space**. We also say that $\mathcal{U}$ is a topology on X. Often, we just say that X is a topological space, provided that the set $\mathcal{U}$ of open sets is understood. If it is clear from context, a topological space is referred to as a "space".

Now, assume that $(X,\mathcal{U})$ is a topological space. Here are a few more definitions:

1. A subset $C \subseteq X$ is called a **closed set** if its complement $C^c := X \setminus C$ is open. Therefore,
   (a) The intersection (finite or infinite) of closed sets is closed.
   (b) The finite union of closed sets is closed.

   So, a topological space can, equivalently, be defined by specifying the closed sets instead of the open sets.
2. Let $x \in X$ be a point. A subset $U \subseteq X$ is called a **neighborhood** of x if there is an open subset V such that $x \in V \subseteq U$. If U itself is open, we say that U is an **open neighborhood** of x.
3. A set of open sets $\mathcal{B} \subseteq \mathcal{U}$ is called a **basis** (or Base) of X if, given any open neighborhood $x \in U$ of a point x, there is an open set $V \in \mathscr{B}$ such that $x \in V \subseteq U$.
   So, given any open set U, we have $U = \cup_{i \in I} V_i$, where $V_i \in \mathcal{B}$ for all $i \in I$.
4. Given a subset $A \subseteq X$, then

$$\overline{A} := \bigcap \{C : A \subseteq C, \text{ and } C \text{ is closed}\}$$

   is called the **closure** of A. So, $\overline{A}$ is the smallest closed set containing A.
5. A subset $D \subseteq X$ is called **dense** in X if $D \cap U \neq \phi$ for all nonempty open sets U. It follows that D is a dense subset of X if and only if the closure $\overline{D} = X$.

We intend to define the category of Topological spaces **Top**, whose objects are the topological spaces X. The morphisms in the category **Top** are the continuous functions, as defined in what follows.

Definition B.1.2. Suppose that $(X,\mathcal{U})$ and $(Y,\mathcal{V})$ are two topological spaces. A set theoretic map $f : X \longrightarrow Y$ is called a **continuous function** if $f^{-1}(V) \in \mathcal{U}$ for all $V \in \mathcal{V}$.

1. If follows that the collection of all topological spaces, together with continuous functions, form a category (A.1.7), to be denoted by **Top**. Usually, we write X for a topological space, while $\mathcal{U}$ is understood. So, for $X, Y \in Obj(\mathbf{Top})$, $Mor_{\mathbf{Top}}(X,Y)$ is defined as the set of all continuous functions $X \longrightarrow Y$.

2. A continuous function $f :\longrightarrow Y$ is called a **homeomorphism** if $f : X \longrightarrow Y$ is bijective and $f^{-1} : Y \longrightarrow X$ is continuous. So, the homeomorphisms f are the "isomorphisms" in the category **Top**.

A continuous function may sometimes be referred to as a "map", when the context is obvious.

Examples B.1.3. Here are some examples of topological spaces.

1. **(Indiscrete Topology).** Let X be a set and $\mathcal{U} = \{\phi, X\}$. Then $(X,\mathcal{U})$ is a topological space. This is called the **trivial** or the **Indiscrete topology** on X. In this case, given any topological space $(Y, \mathcal{V})$, any set theoretic function $f : Y \longrightarrow X$ is continuous.
2. **(Discrete Topology).** Let X be a set and $\mathcal{P}(X)$ be the power set of X (i.e. $\mathcal{P}(X)$ *is the set of all subsets of* X). Then $(X, \mathcal{P}(X))$ is a topological space, to be called the **discrete topology** on X. We have the following comments:
   (a) In the discrete topology on X, all subsets of X are open (and also closed).
   (b) $\mathcal{B} = \{\{x\} : x \in X\}$ is the set of all singleton subsets of X. Then $\mathcal{B}$ is a base for this topology.
   (c) Let $(Y, \mathcal{V})$ be any other topological space and $f : X \longrightarrow Y$ be any set theoretic function. Then $f : (X, \mathcal{P}(X)) \longrightarrow (Y, \mathcal{V})$ is continuous.
3. **(Induced Topology).** Let $(X,\mathcal{U})$ be a topological space and $Y \subseteq X$ be a subset. Let

   $$\mathcal{V} = \{Y \cap U : U \in \mathcal{U}\}. \quad \text{Then } (Y, \mathcal{V}) \text{ is a topological space.}$$

4. **(Usual Topology on $\mathbb{R}^n$).** For our purpose, the most important topology is the usual topology on $\mathbb{R}^n$. For any $x_0 \in \mathbb{R}^n$ and $r > 0$, define the open ball of radius r as

   $$B(x_0, r) = \{x \in \mathbb{R}^n :\| x - x_0 \|< r\}.$$

   A subset $U \subseteq \mathbb{R}^n$ is defined as open, in the usual topology, if U is a union of open balls (*possibly an infinite union*). Let $\mathcal{U}$ be the collection of all such open sets. Then $(\mathbb{R}^n,\mathcal{U})$ is a topological space.

(a) Suppose that $X \subseteq \mathbb{R}^n$ is a nonempty subset. Then the usual topology on $\mathbb{R}^n$ induces a topology on X (as in (3)). We continue to call it the **usual topology** on X. The concept of continuity, taught in undergraduate classes, coincides with the continuity defined by Definition B.1.2.
(b) The usual topology on $\mathbb{C}^n$ is defined in the same way.
(c) (**Exercise**). Prove that the set $\mathbb{Q}$ of rational numbers is a dense subset of $\mathbb{R}$.

5. **Examples from Algebra**:

(a) (**Affine Schemes**). Suppose that A is a commutative ring. Let $\mathrm{Spec}\,(A)$ be the set of all prime ideals of A. For any ideal $I \subset A$, let $V(I) = \{\wp \in \mathrm{Spec}\,(A) : I \subseteq \wp\}$. The **Zariski Topology** on $\mathrm{Spec}\,(A)$ is defined as the one where closed sets are $V(I)$, where I runs through all the ideals of A.

i. So, open sets in Zariski Topology on $\mathrm{Spec}\,(A)$ are $\mathrm{Spec}\,(A) \setminus V(I)$, where I is an ideal.
ii. For $f \in A$, define $D(f) = \{\wp \in \mathrm{Spec}\,(A) : f \notin \wp\}$. Then $\mathcal{B} = \{D(f) : f \in A\}$ forms a basis for this topology.

(b) (**Projective Schemes**). Suppose that $S = S_0 \oplus S_1 \oplus S_2 \oplus \cdots$ is a graded commutative ring. Let $S_+ = S_1 \oplus S_2 \oplus \cdots$ be called the irrelevant ideal. Let

$$\begin{aligned}\mathrm{Proj}\,(S) = \{&\wp \in \mathrm{Spec}\,(S) : \\ &\wp \text{ is a homogeneous prime ideal}, S_+ \not\subseteq \wp\}.\end{aligned}$$

For any homogeneous ideal $I \subset S$, let $V(I) = \{\wp \in \mathrm{Proj}\,(S) : I \subseteq \wp\}$. The **Zariski Topology** on $\mathrm{Proj}\,(S)$ is defined as the one where closed sets are $V(I)$, where I runs through all the homogeneous ideals of A.

i. So, open sets in Zariski Topology on $\mathrm{Proj}\,(S)$ are $\mathrm{Proj}\,(S) \setminus V(I)$, where I is a homogeneous ideal.
ii. For a homogeneous element $f \in S_n$, define $D(f) = \{\wp \in \mathrm{Proj}\,(S) : f \notin \wp\}$. Then $\mathcal{B} = \{D(f) : f \in S_n,\ n \geq 0\}$ forms a basis for this topology.

(As a "scheme" [Hr], $\mathrm{Spec}\,(A)$ and $\mathrm{Proj}\,(S)$ have more structures than what we gave above.)

The following is about the existence of product and coproduct in the category **Top**.

Theorem B.1.4. Let X_1, X_2 be topological spaces.

1. The cartesian product $X_1 \times X_2$ is given the smallest topology where the projection maps $X_1 \times X_2 \longrightarrow X_i$ are continuous for $i = 1, 2$. With this topology, $X_1 \times X_2$ represents the product of X_1, X_2 in **Top**. A basis for this topology is formed by $U_1 \times U_2$, where U_i runs through the set of all open sets in X_i for $i = 1, 2$.
2. Let $X_1 \coprod X_2$ denote the disjoint union. A subset $U \subseteq X_1 \coprod X_2$ is defined to be open if $X_i \cap U$ is open in X_i for $i = 1, 2$. With this topology, $X_1 \coprod X_2$ represents the coproduct of X_1, X_2 in **Top**.

The Hausdorff property is a fundamental concept in topology, as follows.

Definition B.1.5. Suppose that X is a topological space. We say that X is a **Hausdorff space** if, for any two distinct points $x, y \in X$, $x \neq y$, there are two open neighborhoods $x \in U_x$ and $y \in U_y$ such that $U_x \cap U_y = \phi$. Hausdorff spaces are also called T_2-spaces.

Clearly, $\mathbb{R}^n$ is a Hausdorff space. Indeed, in a wide range of mainstream mathematics, all topological spaces considered are Hausdorff. However, for commutative Noetherian rings A, any two nonempty Zariski open subsets of Spec (A) intersect as long as Spec (A) is connected. What follows is a list of fundamental properties of Hausdorff spaces.

We define compactness property, which is another fundamental concept in topology.

Definition B.1.6. Suppose that X is a topological space. An open cover of X is a collection of open sets $\{U_i : i \in I\}$ such that $X = \bigcup_{i \in I} U_i$.

1. X is said to be **compact** if every open cover $\{U_i : i \in I\}$ of X has a finite subcover. This means that there are $i_1, \ldots, i_n \in I$ such that $X = \bigcup_{j=1}^{n} U_{i_j}$.
2. A subset $Y \subseteq X$ is called the compact subset of X if, with its induced topology, Y is a compact topological space.

The following two are well-known results.

Lemma B.1.7. *Suppose that X is a Hausdorff space and $K \subseteq X$ is a compact subset. Then K is closed.* (This fails if X is not Hausdorff.)

Proof. We prove that $Y := X \setminus K$ is open. Fix $y \in Y$. Then $\forall\, x \in K$, there are open sets U_x, V_x such that $x \in U_x$, $y \in V_x$ and $U_x \cap V_x = \phi$. Since $\{U_x \cap K : x \in K\}$ is an open cover of K, and K is compact $K \subseteq U_{x_1} \cup \cdots \cup U_{x_n}$ for finitely many $x_1, \ldots x_n \in K$. Write $V = V_{x_1} \cap \cdots \cap V_{x_n}$. Then V is open, $y \in V$ and $K \cap V = \phi$. So, $y \in V \subseteq Y$. Therefore, Y is open. ∎

We define locally compact topologies.

Definition B.1.8. A topological space X is called **locally compact** if X is Hausdorff and if every point $x \in X$ has a compact neighborhood.

Given two topological spaces, X, Y, the set $Mor_{\mathbf{Top}}(X, Y)$ of all continuous functions $X \longrightarrow Y$ is called a **function space**. We will reserve the notation Y^X for the set of all set-theoretic maps $X \longrightarrow Y$. We topologize $Mor_{\mathbf{Top}}(X, Y)$, as follows.

Definition B.1.9. Let X, Y be two topological spaces. For subsets $A \subseteq X$, $B \subseteq Y$, let

$\mathbf{C}(A, B) = \{f \in Mor_{\mathbf{Top}}(X, Y) : f(A) \subseteq B\}$.

$\mathcal{B} =$

$$\{\mathbf{C}(K, U) : K \text{ is a compact subset of } X, \text{ and } U \text{ is an open subset of } Y\}.$$

The **Compact-Open Topology** on $Mor_{\mathbf{Top}}(X, Y)$ is the topology generated by $\mathcal{B}$. Unless stated otherwise, $Mor_{\mathbf{Top}}(X, Y)$, together with compact-open topology, will be denoted by the same notation $Mor_{\mathbf{Top}}(X, Y)$. That means that compact-open topology on $Mor_{\mathbf{Top}}(X, Y)$ is the smallest topology on $Mor_{\mathbf{Top}}(X, Y)$ containing $\mathcal{B}$. One can see an open set in this topology is a union of finite intersection of members of $\mathcal{B}$. We are mainly interested in the case when X is compact.

We remark that, for $x_0 \in X$, evaluation map $ev_{x_0} : Mor_{\mathbf{Top}}(X, Y) \longrightarrow Y$ is continuous.

The following is among the standard theorems on the function spaces.

Theorem B.1.10. *Let X, Y, Z be topological spaces. We have a bijection* (set theoretic)

$$Z^{X\times Y} \xrightarrow{\sim} \left(Z^Y\right)^X \quad sending \quad F \mapsto \tilde{F} \quad where \quad \tilde{F}(x)(y) = F(x,y). \tag{B.1}$$

1. *Fix a continuous function $F : X \times Y \longrightarrow Z$. Then $\forall\ x \in X$, the function $\tilde{F}(x) \in Mor_{\mathbf{Top}}(Y, Z)$, meaning $\tilde{F}(x)$ is continuous. So, this defines a function*

$$\tilde{F} : X \longrightarrow Mor_{\mathbf{Top}}(Y, Z) \quad In\ fact,\ \tilde{F}\ is\ continuous.$$

2. *Given a (set theoretic) function $G : X \longrightarrow Mor_{\mathbf{Top}}(Y, Z) \subseteq Z^Y$, the bijection* (B.1) *defines the function $\tilde{G} : X \times Y \longrightarrow Z$. If X is locally compact Hausdorff, and G is continuous, then $\tilde{G}$ is also continuous.*

Proof. The proof is available in standard textbooks. See [Ha, Prop. A.14]. ■

The following is helpful to put things in context.

Theorem B.1.11. *Suppose that X is a topological space. We consider two functors*

$$\begin{cases} L : \mathbf{Top} \longrightarrow \mathbf{Top}\ L(Y) = X \times Y \\ R : \mathbf{Top} \longrightarrow \mathbf{Top}\ R(Y) = Mor_{\mathbf{Top}}(X, Y). \end{cases}$$

If X is locally compact Hausdorff, then R is right adjoint to L.

Proof. Follows from (B.1.10). ■

The following lemma is about the Hausdorff property of the compact-open topology.

Lemma B.1.12. *If X, Y are Hausdorff, then the compact-open topology on $Mor_{\mathbf{Top}}(X, Y)$ is also Hausdorff.*

Proof. Let $f, g \in Mor_{\mathbf{Top}}(X, Y)$ and $f \neq g$. So, there is $x_0 \in X$ such that $y_1 := f(x_0) \neq g(x_0) =: y_2$. Since Y is Hausdorff, there are open neighborhoods $y_1 \in V_1$ and $y_2 \in V_2$ such that $V_1 \cap V_2 = \phi$. It follows that $\mathbf{C}(\{x_0\}, V_1) \cap \mathbf{C}(\{x_0\}, V_2) = \phi$ and $f \in \mathbf{C}(\{x_0\}, V_1)$, $g \in \mathbf{C}(\{x_0\}, V_2)$. The proof is complete. ■

B.1.1. *Compactly generated topologies*

Compactly generated topologies play a subtle role in K-Theory because of the subtle behavior of the product in the category of CW complexes (see Theorem C.2.2). Standard textbooks [W, p. 18] define compactly generated topologies for Hausdorff spaces only. For better functorial consistencies, we take the following definition from [H, Def. 2.4.21].

Definition B.1.13. Recall that Hausdorff spaces are also called T_2-spaces. The full subcategory of **Top**, whose objects are Compact T_2-spaces, will be denoted by **C**.

1. A topological space X is defined to be a **weak T_2-space** if, $\forall f \in Mor_{\mathbf{Top}}(K, X)$ with $K \in \mathbf{C}$, the image $f(K)$ is closed.
2. Let X be a topological space.
   (a) A subset $A \subseteq X$ is defined as **compactly closed** if $\forall f \in Mor_{\mathbf{Top}}(K, X)$ with $K \in \mathbf{C}$, $f^{-1}A$ is closed.
   (b) We say that X is a **Kelley space** if every compactly closed set in X is closed. The full subcategory of **Top**, whose objects are Kelley spaces, will be denoted by **K**.
   (c) Define a new topological space kX, whose underlying set is X, and closed sets are all the compactly closed sets in X. One checks that $kX \in \mathbf{K}$. Further, $kX \longrightarrow X$ is continuous.
3. We say that X is **compactly generated** if $X \in \mathbf{K}$ and is a weak T_2-space. The full subcategory of **Top**, whose objects are compactly generated spaces, will be denoted by **T**.

So, we have subcategories $\mathbf{C} \hookrightarrow \mathbf{T} \hookrightarrow \mathbf{K} \hookrightarrow \mathbf{Top}$. Note that objects in **T** need not be Hausdorff.

Lemma B.1.14. *Let X be locally compact. Then X is compactly generated.*

Lemma B.1.15. *Suppose that X is a T_2-space. Then $A \subseteq X$ is compactly closed if and only if $A \cap K$ is closed for all compact subspaces $K \subseteq X$.*

Lemma B.1.16. *Let $X \in \mathbf{K}$. Then $f : X \longrightarrow Y$ is continuous if and only if the composition*

$$\begin{array}{ccc} K & \xrightarrow{\iota} & X \\ & \searrow & \downarrow f \\ & & Y \end{array} \quad \text{is continuous } \forall K \in \mathbf{C},\ \forall\, \iota \in Mor_{\mathbf{Top}}(K, X).$$

Proof. One way is obvious. Suppose that the compositions $f\iota$ are continuous. Let $B \subseteq Y$ be closed. We need to prove $f^{-1}B$ is closed. It suffices to prove that $\iota^{-1}(f^{-1}B)$ is closed. But $\iota^{-1}(f^{-1}B) = (f\iota)^{-1}B$ is closed. ∎

Lemma B.1.17. *The association $X \mapsto kX$ defines a well-defined functor $k : \mathbf{Top} \longrightarrow \mathbf{K}$. We also have $kX = X$ for all $X \in \mathbf{K}$.*

For future references, we include the following, on the associated compactly generated spaces of product spaces.

Lemma B.1.18. *Recall (A.1.7) the category $\mathbf{Top}_\bullet$ of pointed topological spaces. Let $(X, \star), (Y, \star) \in \mathbf{Top}_\bullet$.*

1. *Let $p_1 : k(X \times X) \longrightarrow X$ and $p_2 : k(X \times X) \longrightarrow Y$ be the projection maps. Then p_1, p_2 are continuous.*
2. *Let $\iota_1 : X \longrightarrow k(X \times Y)$ be the map $x \mapsto (x, \star)$, and $\iota_2 : Y \longrightarrow k(X \times Y)$ be the map $y \mapsto (\star, y)$. Then both ι_1, ι_2 are continuous.*
3. *The diagonal map $\Delta : X \longrightarrow k(X \times X)$ is also continuous.*
4. *For continuous maps $\alpha, \beta : X \longrightarrow Y$, the product map $\alpha \times \beta : k(X \times Y) \longrightarrow k(Y \times Y)$ is continuous.*

Proof. Note that p_1 is a composition of two continuous functions

$$\begin{array}{ccc} k(X \times Y) & \longrightarrow & X \times X \\ & \searrow^{p_1} & \downarrow \\ & & X \end{array}$$

So, p_1, and p_2 are continuous. We prove that ι_1 is continuous. Let $\iota : K \longrightarrow X$ be continuous, with $K \in \mathbf{C}$, and consider the following commutative diagram:

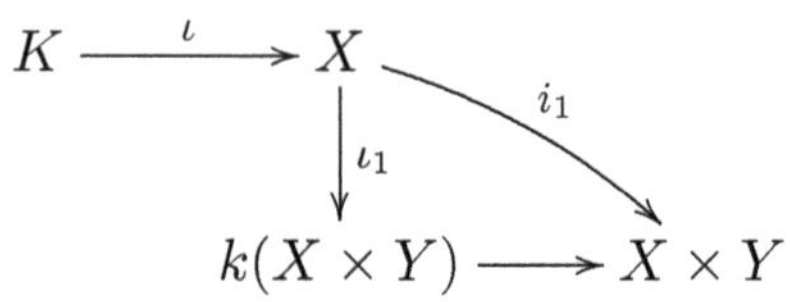

Let $C \subseteq k(X \times X)$ be closed. Since $i_1\iota$ is continuous, by definition, $(i_1\iota)^{-1}(C)$ is closed. But set theoretically $(i_1\iota)^{-1}(C) = (\iota_1\iota)^{-1}(C)$. So, ι_1, ι_2 are continuous. By the same reasoning, as above, Δ is continuous. The proof of (4) follows from (B.1.17). The proof is complete. ■

Lemma B.1.19. *Suppose that* $X, Y \in$ **Top**. *Then* $k(X \times Y) \cong k(kX \times kY)$.

Proof. We have the commutative diagram

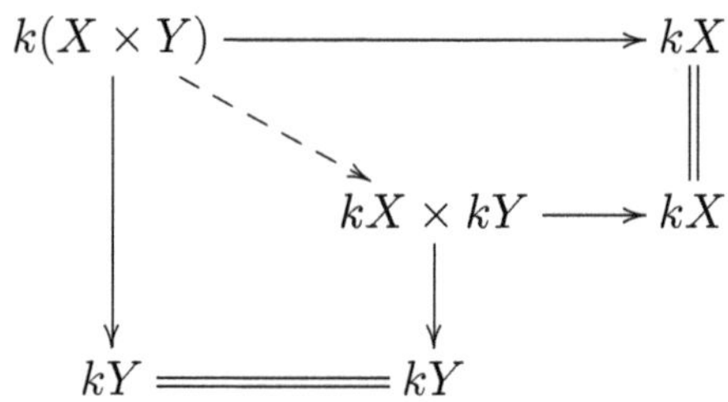

The broken arrow factors through a continuous map $k(X \times Y) \longrightarrow k(kX \times kY)$, by (B.1.17). In the opposite direction, as well, we have the following commutative diagram of maps:

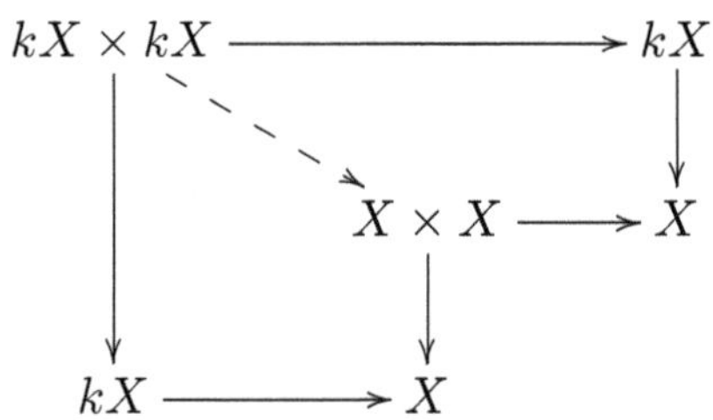

The broken arrow factors through the continuous map $k(kX \times kX) \longrightarrow k(X \times X)$. The proof is complete. ■

We finish this section with the following important notation [W, p. 20].

Notation B.1.20. Let $X, Y \in \mathbf{Top}$ and $Mor_{\mathbf{Top}}(X,Y)$ be the function space (B.1.9). Then $\mathbf{F}(X,Y)$ denotes the compactly generated topological space $k\left(Mor_{\mathbf{Top}}(X,Y)\right)$.

B.2. Homotopy

Among the standard references on Homotopy are [W, S]. The set of all real numbers is denoted by $\mathbb{R}$. Denote the unit sphere and the unit disc/ball, as follows:

$$\mathbb{S}^n = \left\{x \in \mathbb{R}^{n+1} : \parallel x \parallel = 1\right\}, \quad \mathbb{D}^n = \left\{x \in \mathbb{R}^n : \parallel x \parallel \leq 1\right\}.$$

For $n = 0$, we let $\mathbb{D}^0 = \{0\}$ and $\mathbb{S}^{-1} = \phi$. Let $I = [0,1]$ denote the unit interval. So, we have inclusions $\mathbb{S}^{n-1} \subseteq \mathbb{D}^n \subseteq \mathbb{R}^n$. Note that $\mathbb{D}^n \cong I^n$ are homeomorphic.

Definition B.2.1. We give a variety of definitions and notation related to homotopy. Let X, Y be topological spaces and $f, g : X \longrightarrow Y$ be two continuous maps. We say that f is homotopic to g if

$$\exists \text{ a continuous map } H : X \times I \longrightarrow Y \quad \ni \quad \forall\, x \in X \begin{cases} H(x,0) = f(x) \\ H(x,1) = g(x). \end{cases} \tag{B.2}$$

Further, given a subspace $A \subseteq X$, we say that H is a homotopy **relative to** A, if H is constant on $A \times I$, that is, if $H(x,t) = f(x) = g(x)\ \forall\ t \in I,\ x \in I$. Homotopy is an equivalence relation, and so is homotopy relative to A.

1. (**Notation**): Given a homotopy H, as in (B.2), for $t \in I$, $H_t : X \longrightarrow Y$ denotes the map $x \mapsto H(x,t)$.
2. Define the **Homotopy Category** $\mathscr{H}$**omoTop** of topological spaces as follows:
   (a) The objects of $\mathscr{H}$**omoTop** are the topological spaces.
   (b) For topological spaces X, Y and continuous maps $f : X \longrightarrow Y$, let $[f]$ denote the homotopy equivalence class of f. The set of morphisms in $\mathscr{H}$**omoTop** is given by

$$Mor_{\mathscr{H}\mathbf{omoTop}}(X,Y) := Mor(X,Y) :=$$
$$\{[f] : f : X \longrightarrow Y \text{ is a continuous map}\}.$$

3. Now we consider Homotopy of maps in the category $\mathbf{Top}_\bullet$ Let $(X, x_0), (Y, y_0) \in Obj(\mathbf{Top}_\bullet)$ be pointed topological spaces. Assume that $f, g : X \longrightarrow Y$ preserve base points. (*In other words, $f, g \in Mor_{\mathbf{Top}_\bullet}((X, x_0), (Y, y_0))$.*) Then we say that f is homotopic to g, if f and g are homotopic, relative to x_0. That means that, if

$$\exists \text{ continuous map } H : X \times I \longrightarrow Y \quad \ni \begin{cases} H(x,0) = f(x) & x \in X \\ H(x,1) = g(x) & x \in X \\ H(x_0,t) = y_0 & t \in I \end{cases} \tag{B.3}$$

We say that H is a base point preserving homotopy. This homotopy of base point preserving maps is also an equivalence relation. Define the **Homotopy Category $\mathscr{H}$omoTop$_\bullet$** of pointed topological spaces in the same way as **$\mathscr{H}$omoTop**. So, (1) Objects of **$\mathscr{H}$omoTop$_\bullet$** are pointed topological spaces, and (2) a morphism $[f] \in Mor_{\mathscr{H}\mathbf{omoTop}_\bullet}((X, x_0), (Y, y_0))$ is a homotopy equivalence class of maps f in $Mor_{\mathbf{Top}_\bullet}((X, x_0), (Y, y_0))$.

4. Recall the category $\mathbf{Top^2}$ of topological pairs (see Example A.1.7(2b))). We consider Homotopy in the category $\mathbf{Top^2}$ of topological pairs (X, X_0). A map $f : (X, X_0) \longrightarrow (Y, Y_0)$ in $\mathbf{Top^2}$ is a continuous map $f : X \longrightarrow Y$ such that $f(X_0) \subseteq Y_0$.
Given two maps $f, g : (X, X_0) \longrightarrow (Y, Y_0)$ of topological pairs, f is defined to be homotopic to g if

$$\exists \text{ continuous map } H : X \times I \longrightarrow Y \quad \ni \begin{cases} H(x,0) = f(x) & x \in X \\ H(x,1) = g(x) & x \in X \\ H(x,t) \in Y_0 & x \in X_0, t \in I \end{cases} \tag{B.4}$$

In particular, homotopy $H : (X \times I, X_0 \times I) \longrightarrow (Y, Y_0)$ is a map in $\mathbf{Top^2}$. In this case also, the homotopy of topological pairs is an equivalence relation. Define the **Homotopy Category** of topological pairs, to be denoted by **$\mathscr{H}$omoTop^2**, in the same way as **$\mathscr{H}$omoTop**. As in the case of parent categories, we have a commutative diagram of functors as follows:

$$\begin{array}{ccc} \mathscr{H}\mathbf{omoTop}_\bullet & \xrightarrow{\bar{i}} & \mathscr{H}\mathbf{omoTop} \\ \bar{\iota}\downarrow & \nearrow_{\bar{j}} & \\ \mathscr{H}\mathbf{omoTop^2} & & \end{array}$$

Here the vertical functor is a full subcategory.

5. Fix a base point $* \in \mathbb{S}^n$, say $* = (1, 0, \ldots, 0) \in \mathbb{S}^n$. Given a pointed space (X, x_0), $\pi_n(X, x_0)$ denotes the set of all equivalence classes of maps $(\mathbb{S}^n, *) \longrightarrow (X, x_0)$. That means that

$$\begin{cases} \pi_n(X, x_0) := Mor_{\mathscr{H}\mathbf{omoTop}_\bullet}((\mathbb{S}^n, *), (X, x_0)) & \text{is the representable functor,} \\ \text{from} \quad \mathscr{H}omoTop_\bullet \longrightarrow \mathbf{Set} & \text{corresponding to the object} \quad (\mathbb{S}^n, *). \end{cases} \tag{B.5}$$

Subsequently, we establish a group structure on $\pi_n(X, x_0)$ for $n \geq 1$. These sets $\pi_n(X, x_0)$ are referred to as the nth-**homotopy groups** (at x_0) for $n \geq 1$ or homotopy sets for $n = 0$. The study of these groups/sets $\pi_n(X, x_0)$ constitutes the **Classical Homotopy Theory**.

Remark. *The interpretation* (B.5) *of* π_n *as a representable functor provides a clue that one can replace* $(\mathbb{S}^n, *)$ *by any sequence* $(\mathbb{T}_n, *)$ *of pointed spaces, and develop a theory considering the sets* $\tau_n(X, x_0) := Mor_{\mathscr{H}\mathbf{omoTop}_\bullet}((\mathbb{T}^n, *), (X, x_0))$, *parallel to the Classical Homotopy Theory. For example, one can consider hyperbolic spaces* $\sum_{i=1}^{n} x_i y_i + z^2 = 1$, $\sum_{i=1}^{n+1} x_i y_i = 1$.

(a) So, $\pi_0(X, x_0)$ is the set of all path-connected components of X.
(b) Note that $\pi_1(X, x_0)$ is called the **Fundamental Group** of X.
(c) **Important Fact:** For fixed n, the association $(X, x_0) \mapsto \pi_n(X, x_0)$ is, in fact, a covariant functor from $\mathbf{Top}_\bullet \longrightarrow \mathbf{Set}$, and we have the following commutative diagram of functors:

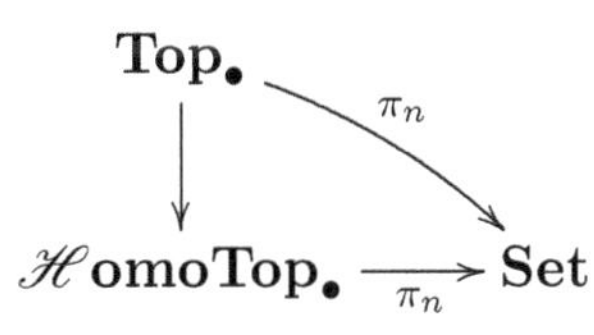

More explicitly, given a map $\varphi : (X, x_0) \longrightarrow (Y, y_0)$ in $\mathbf{Top}_\bullet$, the map

$$\pi_n(\varphi) : \pi_n(X, x_0) \longrightarrow \pi_n(Y, y_0) \quad \text{is defined by} \quad \pi_n(\varphi)([f]) = [\varphi f]$$

where $[f]$ denotes the homotopy equivalence class of $f : (\mathbb{S}^n, *) \longrightarrow (X, x_0)$.

Lemma B.2.2. *Let $\partial I^n = \{(t_1, \ldots, t_n) \in I^n : \exists\, i \ni t_i = 0, 1\} \subseteq I^n$ denote the boundary of I^n. Then the set $\pi_n(X, x_0)$ is in bijection with the set of homotopy equivalence classes of maps $(I^n, \partial I^n) \longrightarrow (X, x_0)$ of pairs of topological spaces. In other words, there is a bijection*

$$\pi_n(X, x_0) \xrightarrow{\sim} Mor_{\mathcal{H}\mathbf{omoTop}^2}\left((I^n, \partial I^n), (X, x_0)\right). \tag{B.6}$$

Proof. First, consider the quotient map $\lambda : I^n \twoheadrightarrow \frac{I^n}{\partial I^n}$. Let $* = [\partial I^n] \in \frac{I^n}{\partial I^n}$ be the base point. Then

1. There is a homeomorphism $\theta : (\mathbb{S}^n, *) \xrightarrow{\sim} \left(\frac{I^n}{\partial I^n}, *\right)$ (*see (B.6.8) for quotient spaces*).
2. There is a bijection

$$\begin{cases} Mor_{\mathbf{Top_\bullet}}\left(\left(\frac{I^n}{\partial I^n}, *\right), (X, x_0)\right) \xrightarrow{\sim} Mor_{\mathbf{Top^2}}\left((I^n, \partial I^n), (X, x_0)\right) \\ \text{sending} \quad f \mapsto f\lambda \end{cases}$$

3. A continuous map $f : (\mathbb{S}^n, *) \longrightarrow (X, x_0)$ in $\mathbf{Top_\bullet}$ gives rise to a map $f\theta^{-1}\lambda : (I^n, \partial I^n) \longrightarrow (X, x_0)$.

This defines a bijection

$$Hom_{\mathbf{Top_\bullet}}\left((\mathbb{S}^n, *), (X, x_0)\right) \longrightarrow Hom_{\mathbf{Top^2}}\left((I^n, \partial I^n), (X, x_0)\right).$$

So, we have established something stronger. One further checks that this bijection respects homotopy. ■

Definition B.2.3. Using this description (B.2.2) of $\pi_n(X, x_0)$, it follows that, for $n \geq 1$, the set $\pi_n(X, x_0)$ has a group structure, which is abelian if $n \geq 2$. For $f, g : (I^n, \partial I^n) \longrightarrow (X, x_0)$, the product (or sum) is given by the class of $(fg) : (I^n, \partial I^n) \longrightarrow (X, x_0)$, defined as follows:

$$(fg)(t_1, \ldots, t_n) = \begin{cases} f(2t_1, t_2, \ldots, t_n) & 0 \leq t_1 \leq \frac{1}{2} \\ g(2t_1 - 1, t_2, \ldots, t_n) & \frac{1}{2} \leq t_1 \leq 1 \end{cases} \tag{B.7}$$

One checks that this induces a well-defined product on $\pi_n(X, x_0)$ for $n \geq 1$. Further, the product (B.7) can be defined using any coordinate instead of the 1st without changing the end result. Obviously, the constant map x_0 represents the identity of this product. For $n \geq 2$, the groups $\pi_n(X, x_0)$ are abelian. All these need checking or a proof, which is skipped [Ha, S, W]. When $n \geq 2$, the availability of a second coordinate is instrumental.

Lemma B.2.4. *Suppose that* $\varphi, \psi : (X, x_0) \longrightarrow (Y, y_0)$ *are homotopic, in the sense of* (B.3). *Then*

$$\pi_n(\varphi) = \pi_n(\psi) : \pi_n(X, x_0) \longrightarrow \pi_n(Y, y_0) \quad \forall\ n \geq 0.$$

Proof. This is a reinterpretation of the fact (B.5) that

$$\pi_n(X, x_0) := Mor_{\mathscr{H}\mathbf{omoTop}_\bullet}((\mathbb{S}^n, *), (X, x_0)) \quad \text{is the representable functor.}$$

Lemma B.2.5. *Let* $X \in \mathbf{Top}$ *and* $\gamma : [0, 1] \longrightarrow X$ *be a path, with* $\gamma(0) = x_0$, $\gamma(1) = x_1$. *Then* $\forall\ n \geq 0$, γ *induces an isomorphism,* $\pi_n(X, x_0) \xrightarrow{\sim} \pi_n(X, x_1)$.

Proof. First, we define the map $\pi_n(X, x_0) \xrightarrow{\sim} \pi_n(X, x_1)$. Let $f : (I^n, \partial I^n) \longrightarrow (X, x_0)$ be a map in $\mathbf{Top^2}$. Define

$$\mu : I^n \longrightarrow [0, 1] \quad \text{by} \quad \mu(t) = \max\{|\ t_i - .5\ |\}, \quad \text{where} \quad t = (t_1, \ldots, t_n) \in I^n.$$

Note that $\mu(t) \leq \frac{1}{2}\ \forall\ t \in I^n$. Now define

$$f * \gamma : (I^n, \partial I^n) \longrightarrow (X, x_1) \quad \text{by} \quad f * \gamma(t) = \begin{cases} f(2t - .5) & \text{if } \mu(t) \leq \frac{1}{4} \\ \gamma(4\mu(t) - 1) & \text{if } \frac{1}{4} \leq \mu(t) \leq \frac{1}{2} \end{cases}$$

Note that $f * \gamma : (I^n, \partial I^n) \longrightarrow (X, x_1)$ is a map in $\mathbf{Top^2}$.

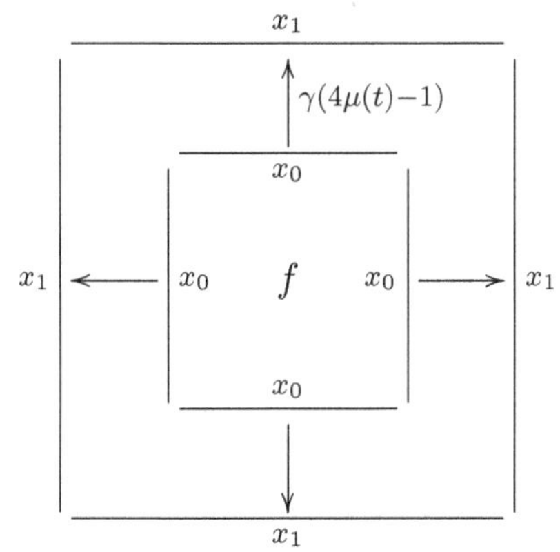

One checks that, if f, g are homotopic in $\mathbf{Top^2}$, then $f * \gamma$ is homotopic to $g * \gamma$. This establishes that γ induces a map

$$\Psi_\gamma : \pi_n(X, x_0) \longrightarrow \pi_n(X, x_1).$$

In fact, Ψ_γ is a homomorphism for $n \geq 1$.

Let $\gamma'(t) = \gamma(1-t)$ be the path x_1 to x_0, opposite γ. Since the concatenation $\gamma\gamma'$ is homotopic to the constant map, it follows that Ψ_γ and $\Psi_{\gamma'}$ are inverses of each other. ∎

Lemma B.2.6. *Suppose that $H : X \times I \longrightarrow Y$ is a homotopy, from $H_0 = f$ to $H_1 = g$. Then for $x_0 \in X$,*

$\alpha(t) = H(x_0, t) : I \longrightarrow Y$ is the trajectory of x_0 $\quad y_0 := f(x_0) \mapsto y_1 = g(x_0)$.

Then the conjugation by α defines an isomorphism

$$h_\alpha : \pi_n(Y, f(x_0)) \longrightarrow \pi_n(Y, g(x_0)).$$

It follows that the diagram

$$\begin{array}{ccc} \pi_n(X, x_0) & \xrightarrow{\pi_n(f)} & \pi_n(Y, y_0) \\ & \searrow^{\pi_n(g)} & \wr \downarrow h_\alpha \\ & & \pi_n(Y, y_1) \end{array} \qquad \textit{commutes.}$$

Before we close this section, we provide a basic definition.

Definition B.2.7. A space $X \in \mathbf{Top}$ is defined to be **contractible** if 1_X is homotopic to a constant map $X \longrightarrow \{x_0\}$.

B.2.1. *Relative homotopy groups*

We briefly need to talk about the homotopy groups of topological pairs. These are among the machinery needed to prove Theorem B (2.2.3). We mainly follow [Ha, W, S] for this section.

Definition B.2.8. In analogy to the category $\mathbf{Top^2}$ of topological pairs (see A.1.7(2b)), we define $\mathbf{Top^n}$, while we do not have any interest beyond $\mathbf{Top^3}$. By a topological n-tuple, we mean a sequence $\mathbf{X} := (X_1, X_2, \ldots, X_n)$ of topological spaces, where $X_{i+1} \subseteq X_i$ are subspaces, for $i = 1, \ldots, n-1$. Given two topological n-tuples, $\mathbf{X} := (X_1, X_2, \ldots, X_n)$ and $\mathbf{Y} := (Y_1, Y_2, \ldots, Y_n)$, a map $f : \mathbf{X} \longrightarrow \mathbf{Y}$ is a map $f : X_1 \longrightarrow Y_1$ in **Top**, with $f(X_i) \subseteq Y_i$, for $i = 1, \ldots, n$. Topological n-tuples, together with such maps, form a category, to be denoted by $\mathbf{Top^n}$.

1. Given $\mathbf{X} := (X_1, X_2, \ldots, X_n)$ in $\mathbf{Top^n}$, for convenience let $\mathbf{X} \times I := (X_1 \times I, X_2 \times I, \ldots, X_n \times I)$, which is an object in $\mathbf{Top^n}$.
2. For two maps $f, g : \mathbf{X} \longrightarrow \mathbf{Y}$ in $\mathbf{Top^n}$, a homotopy from f to g is a map
$$H : \mathbf{X} \times I \longrightarrow \mathbf{Y} \quad \text{is a map in } \mathbf{Top^n} \ \ni \qquad \begin{cases} H(-,0) = f \\ H(-,1) = g. \end{cases}$$
As before (B.2.1), homotopy establishes an equivalence relation on $Mor_{\mathbf{Top^n}}(\mathbf{X}, \mathbf{Y})$.
3. Let $\mathscr{H}\mathbf{omTop^n}$ denote the category whose objects are the same as that of $\mathbf{Top^n}$, and $Mor_{\mathscr{H}\mathbf{omTop^n}}(\mathbf{X}, \mathbf{Y})$ denote the equivalence classes of maps in $Mor_{\mathbf{Top^n}}(\mathbf{X}, \mathbf{Y})$.
4. For a topological pair (X, A) and $x_0 \in A$, we proceed to define the relative homotopy groups (sets) $\pi_n(X, A, x_0)$ for integers $n \geq 1$. Identify I^{n-1} with $I^{n-1} \times 0 \subseteq I^n$. Let $J^{n-1} := I^{n-1} \times 1 \cup \partial I^{n-1} \times I$, as shown in the following diagram:

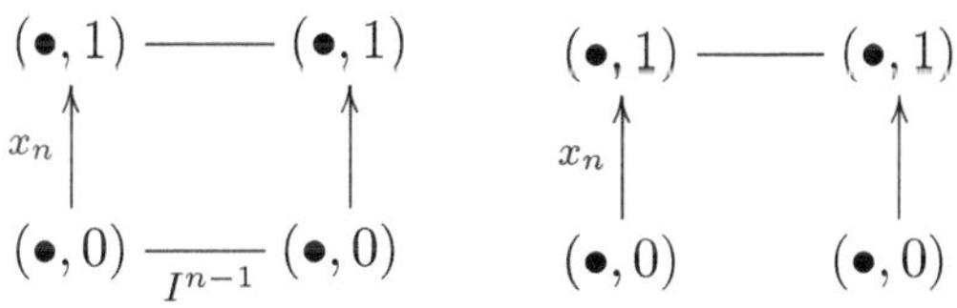

Here the first rectangle represents I^n, and the second one represents J^{n-1}, where • represents the boundary ∂I^{n-1} of I^{n-1}.
5. Now, for $n \geq 1$, define
$$\begin{aligned} \pi_n(X, A, x_0) &= Mor_{\mathscr{H}\mathbf{omoTop^3}}\left((I^n, \partial I^n, J^{n-1}), (X, A, x_0)\right) \\ &= Mor_{\mathscr{H}\mathbf{omoTop^3}}\left((\mathbb{D}^n, \mathbb{S}^{n-1}, *), (X, A, x_0)\right) \end{aligned} \tag{B.8}$$
where the second equality follows, as in (B.6). For completeness, define
$$\pi_0(X, A, x_0) := \frac{\pi_0(X, x_0)}{Image\left(\pi_0(A, x_0)\right)}.$$

So, for $x, z \in X$, the class $[x]$ is neutral in $\pi_0(X, A, x_0)$, if and only if x is in the path component, in X, of some point in A. If $[x] = [z]$ and is nonneutral in $\pi_0(X, A, x_0)$, then $[x] = [z]$ in $\pi_0(X, x_0)$.

(a) In particular, $\pi_1(X, A, x_0) = Mor_{\mathscr{H}\mathbf{omTop^3}}((I, \{0\}, \{1\}), (X, A, x_0))$ is the set of homotopy classes of paths in X starting from a point in A to x_0.
(b) Also, we have $\pi_n(X, \{x_0\}, x_0) = \pi_n(X, x_0)$ for all $n \geq 0$.
(c) For $n \geq 2$, define addition on the set $\pi_n(X, A, x_0)$ using the same formula above (B.7). While the formula (B.7) uses the first coordinate, one can use any of the first $n-1$ coordinates. It follows that $\pi_n(X, A, x_0)$ is a group for $n \geq 2$ and they are abelian groups for $n \geq 3$. Let $C_{x_0} : \mathbb{D}^n \longrightarrow X$, defined by $C_{x_0}(w) = x_0 \ \forall w \in \mathbb{D}^n$, denote the constant map. Then the class $[C_{x_0}] \in \pi_n(X, A, x_0)$ is called the neutral element in $\pi_n(X, A, x_0)$, which is the identity element for $n \geq 2$. While, $\pi_1(X, A, x_0)$, $\pi_0(X, A, x_0)$ are sets only, they are treated as pointed sets, $[C_{x_0}]$ being the distinguished points.

6. The category $\mathbf{Top}^2_\bullet$ of pointed topological pairs can be defined as the full subcategory of $\mathbf{Top}^3$ consisting of the objects (X, A, x_0), with $x_0 \in A \subseteq X$. Likewise, one can define the homotopy category $\mathscr{H}\mathbf{omoTop}^2_\bullet$ of pointed topological pairs as a full subcategory of $\mathscr{H}\mathbf{omoTop}^3$.

For $n \geq 1$, any map $\varphi : (X, A, x_0) \longrightarrow (Y, B, y_0)$ in $\mathbf{Top}^2_\bullet$ induces a map $\varphi_* : \pi_n(X, A, x_0) \longrightarrow \pi_n(Y, B, y_0)$. Such maps factor through maps in $\mathscr{H}\mathbf{omoTop}^2_\bullet$.

It follows that for $n \geq 1$, the associations

$$\begin{cases} (X, A, x_0) \mapsto \pi_n(X, A, x_0) \\ \varphi \mapsto \varphi_* \end{cases}$$

define covariant functors, as in the following commutative diagram:

$$\begin{array}{ccc} \mathbf{Top}^2_\bullet & \xrightarrow{\pi_n} & \mathscr{C}_n \\ \downarrow & \nearrow_{\pi_n} & \\ \mathscr{H}\mathbf{omoTop}^2_\bullet & & \end{array}$$

where $\mathscr{C}_1 = \mathbf{Sets}$ is the category of sets, $\mathscr{C}_2 = \mathbf{Gr}$ is the category of groups, and $\mathscr{C}_n = \mathbf{Ab}$ is the category of abelian groups for $n \geq 3$.

7. Let $n \geq 2$, $(X, A, B) \in \mathbf{Top}^3$ be an object, and $* \in B$. Given a map $f : (I^n, \partial I^n, J^{n-1}) \longrightarrow (X, A, *)$, the restriction of f to I^{n-1} is a map $f_{|I^{n-1}} : (I^{n-1}, \partial I^{n-1}) \longrightarrow (A, *)$. This association $f \mapsto f_{|I^{n-1}}$ defines a map
$$\delta : \pi_n(X, A, *) \longrightarrow \pi_{n-1}(A, *).$$
Now define the connecting map ∂ by the composition (see [S, p. 378]):
$$\begin{array}{ccc} \pi_n(X, A, *) & \xrightarrow{\delta} & \pi_{n-1}(A, *) \\ & \searrow^{\partial} & \downarrow \\ & & \pi_{n-1}(A, B, *) \end{array} \tag{B.9}$$
where the vertical map is induced by the inclusion $(A, *) \hookrightarrow (A, B)$. For $n \geq 3$, the connecting map ∂ is a group homomorphism. Further, when $B = \{*\}$, $\partial = \delta$ is a group homomorphism if $n \geq 2$.

These relative homotopy groups (or sets) fit into a long exact sequence, as follows.

Proposition B.2.9. *Let (X, A, B) be an object in $\mathbf{Top}^3$ and $* \in B$. Then there is a long exact sequence, as follows:*
$$\begin{array}{l} \cdots \longrightarrow \pi_n(A, B, *) \xrightarrow{\iota_*} \pi_n(X, B, *) \xrightarrow{\zeta_*} \pi_n(X, A, *) \xrightarrow{\partial} \pi_{n-1}(A, B, *) \\ \\ \cdots \longrightarrow \pi_1(A, B, *) \longrightarrow \pi_1(X, B, *) \longrightarrow \pi_1(X, A, *) \end{array} \tag{B.10}$$
ending at $n = 1$. Here ι and ζ are the inclusion maps, and ∂ is the connecting map (B.9)*. With $B = \{*\}$, the same* (B.10) *specializes and extends to the following long exact sequence:*
$$\begin{array}{l} \cdots \longrightarrow \pi_n(A, *) \xrightarrow{\iota_*} \pi_n(X, *) \xrightarrow{\zeta_*} \pi_n(X, A, *) \xrightarrow{\partial} \pi_{n-1}(A, *) \\ \\ \cdots \longrightarrow \pi_0(A, *) \longrightarrow \pi_0(X, *) \end{array} \tag{B.11}$$

ending at $n = 0$. (The exactness at the tail end of (B.10, B.11) means that the inverse image of the distinguished point is the image of the preceding map.)

Proof. For a complete proof, see [S, §7.2 Theorem 15]. The proof of exactness of (B.10) uses that of (B.11). The proof of exactness of (B.11) is also available in [Ha, Theorem 4.3]. It is fairly easy to see that the compositions of the successive maps are trivial. The rest of the proof involves elementary arguments, including the following lemma (B.2.10). ■

The following is the frequently used **compression criterion** for neutrality in the relative homotopy groups (sets).

Lemma B.2.10. *Let* (X, A) *be a topological pair and* $x \in A$. *Let* $f : (\mathbb{D}^n, \partial\mathbb{D}^n, \star) \longrightarrow (X, A, x)$ *be a map in* $\mathbf{Top}^3$, *where* $\star$ *denotes the base point (the north pole) in* $\partial\mathbb{D}^n = \mathbb{S}^{n-1}$. *Then the following conditions are equivalent:*

1. *There is a map* $g : \mathbb{D}^n \longrightarrow A$ *such that* f *and* g *are homotopic, relative to* $\partial\mathbb{D}^n$.
2. *There is a map* $g : \mathbb{D}^n \longrightarrow A$, *with* $g(\star) = x$, *such that* f *and* g *are homotopic, as maps* $(\mathbb{D}^n, \partial\mathbb{D}^n, \star) \longrightarrow (X, A, x)$ *in* $\mathbf{Top}^3$.
3. f *is homotopic to the constant map* $C_x : \mathbb{D}^n \longrightarrow X$, *as maps* $(\mathbb{D}^n, \partial\mathbb{D}^n, \star) \longrightarrow (X, A, x)$ *in* $\mathbf{Top}^3$.
4. *The class* $[f] \in \pi_n(X, A, x)$ *is the neutral element.*

Proof. We will prove $(1) \Longrightarrow (2) \Longrightarrow (3) \Longrightarrow (4) \Longrightarrow (1)$. For $n = 0$, since we require $\star \in \partial\mathbb{D}^0$, we interpret $\mathbb{D}^0 = \partial\mathbb{D}^0 = \{\star\}$. So, the equivalence of these statements is obvious for $n = 0$. So, we assume $n \geq 1$.

($(1) \Longrightarrow (2)$): Such a relative homotopy H is, indeed, $H : (\mathbb{D}^n, \partial\mathbb{D}^n, \star) \times I \longrightarrow (X, A, x)$, a homotopy in $\mathbf{Top}^3$, and $H_1(\star) = f(\star) = x$. Therefore, this implication is obvious.

($(2) \Longrightarrow (3)$): Assume that (2) holds. Replacing f by g, we can assume that $f(\mathbb{D}^n) \subseteq A$. Since $\mathbb{D}^n$ is contractible, relative to $\star$, there is a homotopy, $H : \mathbb{D}^n \times I \longrightarrow \mathbb{D}^n$, such that $\forall\ z \in \mathbb{D}^n$, $t \in I$, we have $H(z, 0) = z$, $H(z, 1) = \star$, $H(\star, t) = \star$. Then $fH(\mathbb{D}^n \times I) \subseteq A$, and $\forall\ z \in \mathbb{D}^n$, $t \in I$, we have, $fH(\star, t) = f(\star) = x$, $fH(z, 0) = f(z)$, and $fH(z, 1) = f(\star) = x$. So, $fH : (\mathbb{D}^n, \partial\mathbb{D}^n, \star) \times I \longrightarrow (X, A, x)$ is a homotopy in $\mathbf{Top}^3$ that takes f to the constant map. So, (3) holds.

((3) $\Longrightarrow$ (4)): This follows from the definition of the neutral element.

((4) $\Longrightarrow$ (1)): Assume that there is a homotopy $H : (\mathbb{D}^n, \partial\mathbb{D}^n, \star) \times I \longrightarrow (X, A, x)$ such that $H_0 = C_x$ is the constant map x, $H_1 = f$. By Lemma B.2.11, there is a continuous map

$$F : \mathbb{D}^n \times I \longrightarrow \mathbb{D}^n \times I \quad \ni \begin{cases} F(z,0) \in (\mathbb{D}^n \times 0) \cup (\partial\mathbb{D}^n \times I) & \forall\ z \in \mathbb{D}^n \\ F(z,t) = (z,1) & \forall\ (z,t) \in \partial\mathbb{D}^n \times I \\ F(z,1) = (z,1) & \forall\ z \in \mathbb{D}^n \end{cases} \tag{B.12}$$

Let $\widetilde{H} = HF : \mathbb{D}^n \times I \longrightarrow X$. Then

$$\begin{cases} \widetilde{H}(z,0) \in H((\mathbb{D}^n \times 0) \cup (\partial\mathbb{D}^n \times I)) \subseteq A & \forall\ z \in \mathbb{D}^n \\ \widetilde{H}(z,t) = H(z,1) = f(z) & \forall\ z \in \partial\mathbb{D}^n, t \in I \\ \widetilde{H}(z,1) = H(z,1) = f(z) & \forall\ z \in \mathbb{D}^n \end{cases}$$

So, $\widetilde{H}$ is a homotopy, from $\widetilde{H}_0$ to f, relative to $\partial\mathbb{D}^n$. So, (1) holds.

The proof is complete. ■

Lemma B.2.11. *There is a continuous map* $F : \mathbb{D}^n \times I \longrightarrow: \mathbb{D}^n \times I$ *with properties* (B.12).

Proof. Define two functions

$$\begin{cases} \mu : \mathbb{R} \times I \longrightarrow \mathbb{R} \text{ by } \mu(w,t) = \frac{2w-t+1}{t+1}\ \forall\ (w,t) \in \mathbb{R} \times I \\ \tau : \mathbb{R} \longrightarrow \mathbb{R} \text{ by } \quad \tau(w) = 2w-1 \quad \forall\ w \in \mathbb{R}. \end{cases}$$

Consider the horizontal slice $\mathbb{D}^n \times t \subseteq \mathbb{D}^n \times I$. For fixed $t \in I$, we have

$$\begin{cases} \mu(t,t) = 1, & \mu\left(t + \frac{1-t}{2}, t\right) = \mu\left(\frac{t+1}{2}, t\right) = \frac{2}{t+1} \\ \tau\left(t + \frac{1-t}{2}\right) = \tau\left(\frac{t+1}{2}\right) = t, & \tau(1) = 1. \end{cases}$$

For $z \in \mathbb{R}^n$, denote the length of z by $\ell(z)$. Now define $F : \mathbb{D}^n \times I \longrightarrow \mathbb{D}^n \times I$ by

$$F(z,t) = \begin{cases} (z,t) & \text{if } \ell(z) \le t \\ (\mu(\ell(z),t)z, t) & t \le \ell(z) \le \frac{t+1}{2} \\ \left(\frac{z}{\ell(z)}, \tau(\ell(z))\right) & \frac{t+1}{2} \le \ell(z) \le 1. \end{cases}$$

The following diagram is helpful:

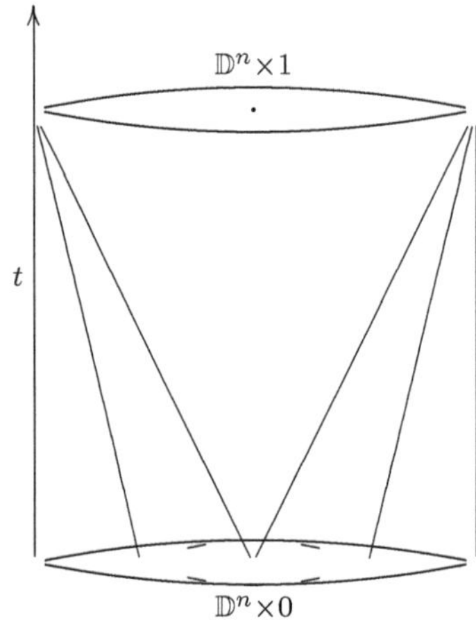

$$\begin{cases} \cdot & \text{is the origin} \\ \text{dotted circle} & \text{has radius } \frac{1}{2} \\ \ell(z) = t & \text{is the inner line} \\ \ell(z) = \frac{t+1}{2} & \text{is the 2nd line} \end{cases}$$

Note that the function is well-defined and continuous.

The proof is complete. ■

B.2.2. *Excision: Dold–Thom Theorem*

The proof of Theorem B (2.2.3), in [Q], uses a theorem (B.2.14) of Dold and Thom [DT] on excision. We proceed to prove the same. Before we state (B.2.14), we prove the following lemma (B.2.12).

Lemma B.2.12. *Suppose that $\iota : (X, U) \hookrightarrow (Y, V)$ is an inclusion to topological pairs. Fix an integer $n \geq 1$. For points $w \in \mathbb{R}^{n+1}$, label the coordinates as $w = (w_0, w_1, \ldots, w_n)$. Let $\partial_+\mathbb{D}^n = \{w \in \partial\mathbb{D}^n, w_n \geq 0\}$, and $\partial_-\mathbb{D}^n = \{w \in \partial\mathbb{D}^n, w_n \leq 0\}$ denote the upper and lower hemispheres, and $\partial\mathbb{D}^n_0 = \partial_+\mathbb{D}^n \cap \partial_-\mathbb{D}^n \cong \mathbb{S}^{n-2}$ be the equator. Also, let $\star \in \partial\mathbb{D}^n_0$ denote the base point. Then the following conditions are equivalent:*

1. *The maps*

$$\begin{cases} \iota_* : \pi_{n-1}(X, U, x) \longrightarrow \pi_{n-1}(Y, V, x) & \textit{has trivial kernel}, \\ \iota_* : \pi_n(X, U, x) \longrightarrow \pi_n(Y, V, x) & \textit{is surjective}, \end{cases} \qquad \forall\, x \in U.$$

2. *Let* f_0 *and* h *be maps in* $\mathbf{Top}^2$*, as follows:*

$$\begin{cases} f_0 : (\mathbb{D}^n, \partial_-\mathbb{D}^n) \times 0 \to (Y,V) \\ h : (\partial_+\mathbb{D}^n, \partial\mathbb{D}^n_0) \times I \to (Y,V) \end{cases} \ni \begin{cases} h_{|\partial_+\mathbb{D}^n \times 0} = (f_0)_{|\partial_+\mathbb{D}^n \times 0} \\ h_1 : (\partial_+\mathbb{D}^n, \partial\mathbb{D}^n_0) \times 1 \to (X,U) \end{cases} \tag{B.13}$$

In other words, we are given a map

$$\widetilde{h} : (\mathbb{D}^n \times 0 \cup \partial_+\mathbb{D}^n \times I, \partial\mathbb{D}^n \times 0 \cup \partial\mathbb{D}^n_0 \times I) \longrightarrow (Y,V),$$

$$\ni \quad \widetilde{h}(\partial_+\mathbb{D}^n \times 1, \partial\mathbb{D}^n_0 \times 1) \subseteq (X,U).$$

Such a map extends to a map

$$H : (\mathbb{D}^n, \partial_-\mathbb{D}^n) \times I \longrightarrow (Y,V) \quad \ni \quad H_1 : (\mathbb{D}^n, \partial_-\mathbb{D}^n) \longrightarrow (X,U).$$

3. *The same as (2), with the additional hypothesis that* h *is independent of* t*. So,* $\forall\ (x,t) \in \partial_+\mathbb{D}^n \times I,\ h(w,t) = f_0(w)$*. (So, only* f_0 *is given, with* $f_0((\mathbb{D}^n, \partial_-\mathbb{D}^n, \partial_+\mathbb{D}^n, \partial\mathbb{D}^n_0)) \subseteq (Y,V,X,U)$*.)*

Proof. ((3) $\Longrightarrow$ (1)): Assume that (3) holds. Fix $x_0 \in U$. Suppose that $[f] \in \pi_{n-1}(X,U,x_0)$ and $\iota_*([f]) = 0 \in \pi_{n-1}(Y,V,x_0)$ (*zero means trivial for* $n = 1,2$). We treat f as a map $f : (\partial_+\mathbb{D}^n, \partial\mathbb{D}^n_0, \star) \longrightarrow (X,U,x)$.

So, there is a homotopy $\Phi : \partial_+\mathbb{D}^n, \times I \longrightarrow Y$, relative to $\partial\mathbb{D}^n_0$ such that $\Phi_1 = c_{x_0}$ is the constant map, and $\Phi_0 = f$. For $w = (w_1, \ldots, w_n) \in \mathbb{D}^n$, let $\widetilde{w}_n = \sqrt{1 - \sum_{i=1}^{n-1} w_i^2}$. Define the projection p, the scaling ℓ, and the extension f_0, as follows:

$$\begin{cases} p : \mathbb{D}^n \to \partial_+\mathbb{D}^n \text{ by } \quad p(w) = (w_1, \ldots w_{n-1}, \widetilde{w}_n) \\ \ell : \mathbb{D}^n \longrightarrow [0,1] \text{ by } \ \ell(w) = \begin{cases} \frac{w_n - \widetilde{w}}{2\widetilde{w}} \text{ if } \widetilde{w}_n \neq 0 \quad i.e. \quad w \notin \partial\mathbb{D}^n_0 \\ 1/2 \quad \text{if } \widetilde{w}_n = 0; \quad i.e. \quad w \in \partial\mathbb{D}^n_0 \end{cases} \\ f_0 : \mathbb{D}^n \to Y \quad \text{by} \quad f_0(w) = \Phi\left(p(w), \ell(w)\right). \end{cases} \tag{B.14}$$

Note that $(f_0)_{|\partial_-\mathbb{D}^n} = c_{x_0}$. Let $h := f \times 1_I : \partial_+\mathbb{D}^n \times I \longrightarrow X$. By (3), the map $\widetilde{h}$ as defined above extends to a homotopy

$$H : (\mathbb{D}^n, \partial_-\mathbb{D}^n) \times I \longrightarrow (Y,V) \quad \ni \quad H_1 : (\mathbb{D}^n, \partial_-\mathbb{D}^n) \longrightarrow (X,U).$$

So, H_1 is an extension of f. Now, vertically downward, from upper to lower hemispheres, write a homotopy $\mathcal{H}$ from f to $(H_1)_{|\partial_-\mathbb{D}^n}$, as follows:

$$\mathcal{H} : \partial_+\mathbb{D}^n \times I \longrightarrow X \quad \text{by}$$
$$\mathcal{H}((w_1, \ldots, w_{n-1}, w_n), t) = H_1(w_1, \ldots, w_{n-1}, (1 - 2t)w_n).$$

So, $\mathcal{H}_0 = f$, $\mathcal{H}_1(\partial_+\mathbb{D}^n) = H_1(\partial_-\mathbb{D}^n) \subseteq U$. For points on the equator $\partial\mathbb{D}^n_0$, $w_n = 0$, and hence $\mathcal{H}$ is a homotopy relative to $\partial\mathbb{D}^n_0$. Therefore, by compression criterion (B.2.10(1)), $[f] = 0 \in \pi_{n-1}(X, U, x_0)$. So, ι_* has a trivial kernel.

To check surjectivity, let $[f_0] \in \pi_n(Y, V, x_0)$. We treat f_0 as a function $f_0 : (\mathbb{D}^n, \partial\mathbb{D}^n, \partial_+\mathbb{D}^n) \times 0 \longrightarrow (Y, V, x_0)$. Let $h : \partial_+\mathbb{D}^n \times I \longrightarrow Y$ be the constant map $h(w, t) = x$. By (3), there is a homotopy

$$H : (\mathbb{D}^n, \partial\mathbb{D}^n, \partial_+\mathbb{D}^n) \times I \longrightarrow (Y, V, x)$$
$$\ni \begin{cases} H_0 = f_0 \\ H_{\mathbb{D}^n_+ \times I} = h \\ H_1 : (\mathbb{D}^n, \partial\mathbb{D}^n, \partial_+\mathbb{D}^n) \longrightarrow (X, U, x). \end{cases}$$

So, $\iota_*([H_1]) = [f_0]$. So, ι_* is surjective. So, (1) is established.

((1) $\implies$ (2)): Suppose that f_0 and h are given as in (B.13). Since h_0 extends to f_0 on $\mathbb{D}^n$, as given, we have $[h_0] = 0 \in \pi_{n-1}(Y, V, h_0(\star))$. Fix a base point $\star \in \partial\mathbb{D}^n_0$. Under the isomorphism $\pi_{n-1}(Y, V, h_1(\star)) \xrightarrow{\sim} \pi_{n-1}(Y, V, h_0(\star))$, the image of $[h_1]$ is $[h_0] = 0$. Therefore, $[h_1] = 0 \in \pi_{n-1}(Y, V, h_1(\star))$.

By the injectivity property in (1), $[h_1] = 0 \in \pi_{n-1}(X, U, h_1(\star))$. By compression (B.2.10(1)), there is a homotopy $\Phi : (\partial_+\mathbb{D}^n, \partial\mathbb{D}^n_0, \star) \times I \longrightarrow (X, U, x)$, relative to $\partial\mathbb{D}^n_0$, such that $\Phi_0 = h_1$ and $\Phi_1(\partial_+\mathbb{D}^n) \subseteq U$. By the same argument, as in (B.14), h_1 extends to a map $f_1 : (\mathbb{D}^n, \partial_-\mathbb{D}^n) \to (X, U)$.

Consider f_1 as a function $f_1 : \mathbb{D}^n \times 1 \longrightarrow (X, U)$. Let $\mathcal{X} := \mathbb{D}^n \times \{0, 1\} \cup \partial\mathbb{D}^n_+ \times I$. (*So, $\mathcal{X}$ is obtained from an empty can, by removing half of its side.*) Let $\widetilde{h} \vee f_1 : \mathcal{X} \longrightarrow (Y, V)$ be a function defined by $\widetilde{h}$ and f_1. It remains to be shown that $\widetilde{h} \vee f_1$ extends to a map $H : \mathbb{D}^n \times I \longrightarrow (Y, V)$, while other conditions are already embedded in $\widetilde{h} \vee f_1$. There is a homeomorphism (B.2.13)

$$\lambda : \mathbb{D}^n \times I \xrightarrow{\sim} \mathbb{D}^n \times I \quad \ni \quad \lambda(\mathbb{D}^n \times 0) = \mathcal{X}. \tag{B.15}$$

Fix a base point $\star \in \partial\mathbb{D}^n$ such that $\lambda(\star) \in \partial\mathbb{D}^n_0 \times 1$. So, $x_0 := (\widetilde{h} \vee f_1)\lambda_0(\star) \in U$. Consider $(\widetilde{h} \vee f_1)\lambda_0 : (\mathbb{D}^n, \partial\mathbb{D}^n, \star) \longrightarrow (Y, V, x_0)$.

$$\begin{array}{ccc} \mathbb{D}^n & \xrightarrow{\lambda_0} & \mathcal{X} \\ & \searrow^{(\widetilde{h}\vee f_1)\lambda_0} & \downarrow{\scriptstyle \widetilde{h}\vee f_1} \\ & & Y \end{array}$$

Recall that $\pi_n(Y, V, x_0) = Hom_{\mathscr{H}\mathbf{omoTop}}((\mathbb{D}^n, \partial\mathbb{D}^n, \star), (Y, V, x_0))$.

So, $(\widetilde{h} \vee f_1)\lambda_0$ represents an element in $\pi_n(Y, V, x_0)$. Since $\pi_n(X, U, x_0) \longrightarrow \pi_n(Y, V, x_0)$ is surjective, there is a homotopy

$$\mathcal{H} : (\mathbb{D}^n, \partial\mathbb{D}^n, \star) \times I \longrightarrow (Y, V, x_0) \quad \ni \quad \begin{cases} \mathcal{H}_0 = (\widetilde{h} \vee f_1)\lambda_0 \\ \mathcal{H}_1 : (\mathbb{D}^n, \partial\mathbb{D}^n, \star) \longrightarrow (X, U, x_0). \end{cases}$$

Now $H := \mathcal{H}\lambda^{-1} : \mathbb{D}^n \times I \longrightarrow (Y, V)$ is an extension of $\widetilde{h} \vee f_1$. So, (2) is established.

It is obvious that (2) $\Longrightarrow$ (3). So, the lemma is established. ■

Lemma B.2.13. *For $n \geq 1$, there is a homeomorphism $\lambda : \mathbb{D}^n \times I \xrightarrow{\sim} \mathbb{D}^n \times I$, satisfying the above condition* (B.15).

Proof. Let $\mathbb{E}^n \subseteq Int(\partial_-\mathbb{D}^n \times I)$ be any disc (i.e $\cong \mathbb{D}^n$), within the interior of $\partial_-\mathbb{D}^n \times I$. To be specific, let $\mathbb{E}^n := \partial\mathbb{D}^n_{w_n \leq -1/2} \times [1/4, 3/4]$, as shown in the diagram

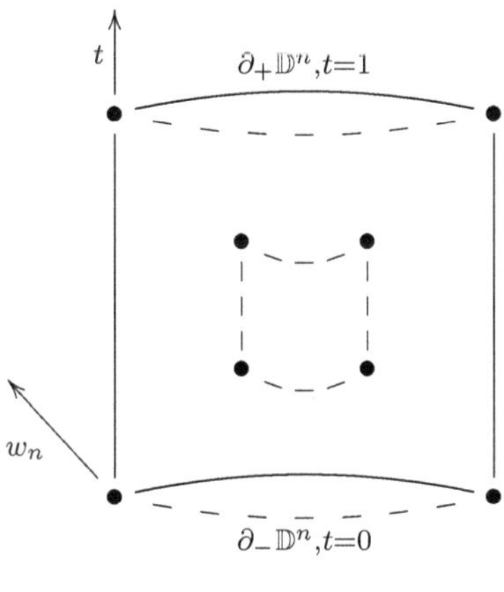

where the • represents $\mathbb{D}^n_{|w_n=0}$ and the dotted rectangle represents $\mathbb{E}^n$.

Recall that $\mathcal{X} := \mathbb{D}^n \times \{0,1\} \cup \partial\mathbb{D}^n_+ \times I$. We will prove that there is a homeomorphism

$$\lambda : \mathbb{D}^n \times I \xrightarrow{\sim} \mathbb{D}^n \times I \quad \ni \quad \begin{cases} \lambda(\mathbb{D}^n \times 0) = \mathcal{X}, \\ \lambda(\mathbb{D}^n \times 1) = \mathbb{E}^n. \end{cases}$$

Let $\mathbb{H}^n_L = \{w \in \mathbb{D}^n : w_n \leq 0\}$ be the lower half of the disc. For $w = (w_1, \ldots, w_n) \in \mathbb{D}^n$, denote $\widetilde{w}_n := \sqrt{1 - \sum_{i=1}^{n-1} w_i^2}$ and $pr(w) = (w_1, \ldots, w_{n-1})$. Compress the disc $\mathbb{D}^n$ to the half disc $\mathbb{H}^n_L$, as follows:

$$\beta_1 : \mathbb{D}^n \longrightarrow \mathbb{H}^n_L \quad \text{by} \quad \beta_1(w) = (w_1, \ldots, w_n) = \left(w_1, \ldots, w_{n-1}, \frac{w_n - \widetilde{w}_n}{2}\right).$$

Since β_1 is continuous and bijective, β_1 is a homeomorphism (using compactness). Further, $(\beta_1)_{|\partial_-\mathbb{D}^n} = 1_{|\partial_-\mathbb{D}^n}$ and $\beta_1(\partial_+\mathbb{D}^n) = \mathbb{D}^n_{|w_n=0} \subsetneq \partial\mathbb{H}^n_L$ is the disc $\mathbb{D}^{n-1}$. Let $\gamma_1 := \beta_1 \times 1_I$. So, $\gamma_1 : \mathbb{D}^n \times I \xrightarrow{\sim} \mathbb{H}^n_L \times I$ is a homeomorphism, such that $\gamma_1(\partial_+\mathbb{D}^n \times I) = \partial\mathbb{D}^n_{|w_n=0} \times I$, and the restriction of γ_1 to $\partial_-\mathbb{D}^n \times I$ is identity. We have $\gamma_1(\mathcal{X}) = \mathbb{H}^n_L \times \{0,1\} \cup \mathbb{D}^n_{|w_n=0} \times I =: \mathcal{X}_1$. Define $\gamma_2 : \mathbb{H}^n_L \times I \longrightarrow \mathbb{R}^{n+1}$ by

$$\gamma_2(w,t) = \begin{cases} \left(pr(w), w_n \cos\left(\frac{\pi}{2} - 2\pi t\right), t + w_n \sin\left(\frac{\pi}{2} - 2\pi t\right)\right) & \text{if } 0 \leq t \leq 1/4 \\ (w,t) & \text{if } 1/4 \leq t \leq 3/4 \\ \left(pr(w), w_n \cos\left(-\frac{3\pi}{2} + 2\pi t\right), t - w_n \sin\left(-\frac{3\pi}{2} + 2\pi t\right)\right) & \text{if } 3/4 \leq t \leq 1 \end{cases}$$

Write $\Omega = \gamma_2(\mathbb{H}^n_L \times I))$, and $\mathcal{X}_2 := \gamma_2(\mathcal{X}_1)$. We list the following observations:

1. $\gamma_2 : \mathbb{H}^n_L \times \times I \xrightarrow{\sim} \Omega$ is a homeomorphism. In fact, Ω has the shape of a (solid) boat. The top side of the boat is shown in the following diagram:

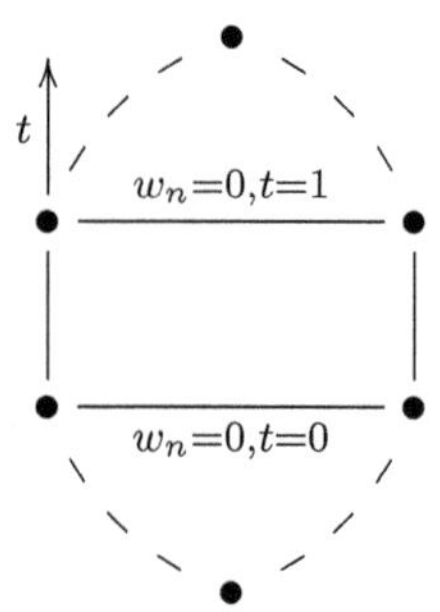

So, this (top) section of the boat Ω in the $w_n = 0$-plane is obtained by attaching two halves of the disc, to the two sides of $(\mathbb{D}^n_{|w_n=0} \times I)$. The maximum width on t-direction is 3, ranging from $t = -1$ to $t = 2$.

Further, note that $\mathcal{X}_2 := \gamma_2(\mathcal{X}_1)$ is precisely this top part of the boat.

2. γ_2 is identity when $w_n = 0$, meaning on $\mathbb{D}^n_{|w_n=0} \times I$. Also, γ_2 is identity on $\mathbb{H}^n_L \times [1/4, 3/4]$. So, γ_2 is identity on $\mathbb{E}^n$. So, $\mathbb{E}^n$ remains in the interior of the bottom of the boat Ω.

Now it is transparent enough that there is a homeomorphism

$$\gamma_3 : \Omega \xrightarrow{\sim} \mathbb{D}^n \times I \quad \ni \quad \gamma_3(\mathcal{X}_2) = \mathbb{D}^n \times 0, \ \gamma_3(\mathbb{E}^n) = \mathbb{D}^n \times 1.$$

Now, $\lambda = (\gamma_3\gamma_2\gamma_1)^{-1}$ satisfies the conditions of the lemma. The proof is complete. ■

The following result on relative homotopy is classical and is due to Dold and Thom [DT]. This may be referred to as an Excision theorem. It is used in the proof of Theorem B (2.2.3), in [Q]. Our proof here improvises on the proof given in [Ha, Proposition 4K.1] and other resources available on the internet.

Proposition B.2.14. *Let $f : X \longrightarrow Y$ be a map in* **Top**. *Let U_1, U_2 be subsets of X, and V_1, V_2 be subsets of Y, with $f(U_i) \subseteq V_i$ for $i = 1, 2$. Further assume that X is the union of the interiors of U_1, U_2, and Y is the union of the interiors of V_1, V_2, For a fixed integer n, assume that, $\forall\ x_0 \in U_1 \cap U_2$, and $i = 1, 2$, we also have that*

$$\begin{cases} f_* : \pi_j\left(U_i, U_1 \cap U_2, x_0\right) \longrightarrow \pi_j\left(V_i, V_1 \cap V_2, f(x_0)\right) \ \textit{are bijections} \ \forall\ j \leq n-1 \\ f_* : \pi_n\left(U_i, U_1 \cap U_2, x_0\right) \longrightarrow \pi_n\left(V_i, V_1 \cap V_2, f(x_0)\right) \ \textit{is a surjection.} \end{cases}$$

Then for $i = 1, 2$,

$$\begin{cases} f_* : \pi_i\left(X, U_j, x_0\right) \longrightarrow \pi_j\left(Y, V_i, f(x_0)\right) \ \textit{are bijections} \ \forall\ j \leq n-1; \ \forall\ x_0 \in U_i \\ f_* : \pi_n\left(X, U_i, x_0\right) \longrightarrow \pi_n\left(Y, V_i, f(x_0)\right) \ \textit{is surjection} \ \forall\ x_0 \in U_i. \end{cases}$$

Proof. We reduce the proof to the case when $X \hookrightarrow Y$ is a subspace. Let $C(f) := \frac{(X \times I) \coprod Y}{\{(x,1) \sim f(x)\ \forall\ x \in X\}}$ be the mapping cylinder of f. We have

the commutative diagram

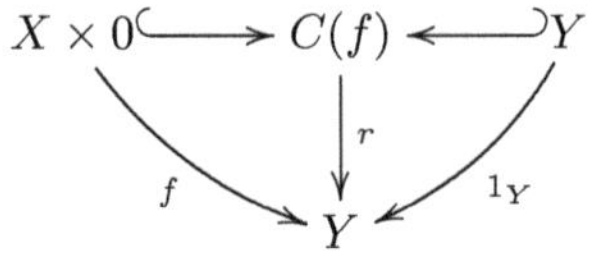

$$\begin{cases} V_1' = C(f_{|U_1}) \cup f^{-1}V_1 \times (1/2, 1) \subseteq C(f) \\ V_2' = C(f_{|U_2}) \cup f^{-1}V_2 \times (1/2, 1) \subseteq C(f). \end{cases}$$

One checks that the interior of V_1', V_2' covers $C(f)$. Since r is a deformation retract, it is enough to prove the proposition for the inclusion $X \hookrightarrow C(f)$. So, we assume that $f : X \hookrightarrow Y$ is a subspace map. So, we have the commutative diagram of inclusions

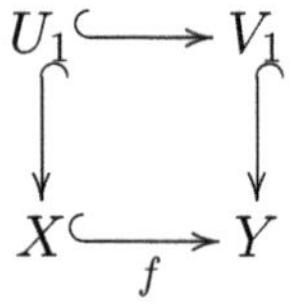

We prove the case $i = 1$ only. For $n = 0$, we prove that $f_* : \pi_0(X, U_1, x_0) \longrightarrow \pi_0(Y, V_1, f(x_0))$ is surjective. Let $y \in Y$. We assume that $[y] \in \pi_0(Y, V_1, f(x_0))$ is nontrivial. In particular, $y \in V_2$. By hypothesis, there is $x \in U_2$ such that $[x] = [y] \in \pi_0(V_2, V_1 \cap V_2, f(x_0))$, and is nontrivial. So, there is a path in V_2 and hence in Y from x to y. This establishes the proposition for $n = 0$.

Now, we prove the proposition for $n \geq 1$. We assume that condition (2) of Proposition B.2.12 holds for the inclusions $(U_i, U_1 \cap U_2) \hookrightarrow (V_i, V_1 \cap V_2)$, $i = 1, 2$. We prove that the condition (3) of Proposition B.2.12 holds for the inclusions $(X, U_1) \hookrightarrow (Y, V_1)$.

For this part of the argument, we replace $(\mathbb{D}^n, \partial_-\mathbb{D}^n, \partial_+\mathbb{D}^n)$ by $(I^n, I^{n-1} \times 1, J^{n-1})$, where J^{n-1} is the closure of $\partial I^n \setminus I^{n-1} \times 1$. (*Note a minor modification of the notation in (B.2.8)(4).*) For notational analogy, let $(I^n, \partial_- I^n, \partial_+ I^n) := (I^n, I^{n-1} \times 1, J^{n-1})$, and $\partial(I_0^n) := \partial_+ I^n \cap \partial_- I^n$. Consider the maps (compare (B.13))

$$\begin{cases} f_0 : (I^n, \partial_- I^n) \times 0 \longrightarrow (Y, V_1) \\ h := (f_0)_{|\partial_+ I^n \times 0} \times 1_I \end{cases} \ni f_0(I^n, \partial_- I^n, \partial_+ I^n, \partial(I_0^n)) \subseteq (Y, V_1, X, U_1). \tag{B.16}$$

Let $\widetilde{h} : I^n \times 0 \cup \partial_+ I^n \times I \longrightarrow Y$ be the map defined by f_0 and h. We are required to extend $\widetilde{h}$ to a homotopy $H : (I^n, \partial_- I^n) \times I \longrightarrow (Y, V_1)$, such that $H_1(I^n, \partial_- I^n) \subseteq (X, U_1)$, as required in condition (3) of (B.2.12).

For an integer $N \geq 2$ and for integers $0 \leq k_1, k_2, \ldots, k_n \leq N-1$, let

$$I^n_{k_1k_2\ldots k_n} = \left\{ (x_1, \ldots, x_n) \in I^n : \frac{k_i}{N} \leq x_i \leq \frac{k_i+1}{N} \right\}.$$

Since Y is a union of interiors of V_1 and V_2 when $N \gg 0$, $\forall (k_1, \ldots, k_n)$, either $f_0(I^n_{k_1k_2\ldots k_n} \times 0) \subseteq V_1$ or $f_0(I^n_{k_1k_2\ldots k_n}) \subseteq V_2$ (*consult the Lebesgue Lemma*). We fix such an integer $N \gg 0$.

Let Λ be the (finite) set of all subcubes (i.e. faces) $K \subseteq I^n_{k_1k_1\ldots k_n}$ of all dimensions $0 \leq \dim K \leq n$ for all k_i, as above. For $0 \leq d \leq n$, denote $\Lambda_d = \{K \in \Lambda : \dim K = d\}$. Inductively, $\forall\ K \in \Lambda$, we define a map $H_K : K \times I \longrightarrow Y$ such that

$$\begin{cases} (H_K)_{|K\times 0} = (f_0)_{|K\times 0} & \text{if } \forall\ K \in \Lambda \\ (H_K)_{|K\times I} = h_{|K\times I} & \text{if } K \subseteq \partial_+ I^n \\ (H_K)_{|\Delta\times I} = H_\Delta & \forall \text{ faces } \Delta \subseteq K \\ H_K(K\times I, K\times 1) \subseteq (V_1, U_1) \text{ if } \widetilde{h}(K\times 0) = f_0(K\times 0) \subseteq V_1 \\ H_K(K\times I, K\times 1) \subseteq (V_2, U_2) \text{ if } \widetilde{h}(K\times 0) = f_0(K\times 0) \subseteq V_2. \end{cases} \tag{B.17}$$

By default, for $K \in \Lambda$, with $K \subseteq \partial_+ I^n = J^{n-1}$, we define $H_K := h_{|K\times I}$, which satisfies (B.17), by the properties of f_0 and h. Note that I^n and $\partial_+ I^n = J^{n-1}$ have the same set of vertices. Therefore, if $\dim K = 0$, H_K is already defined.

Let $K \in \Lambda$. Assume that H_Δ has been defined for all faces of $\Delta \subseteq K$, satisfying (B.17). Combining such maps H_Δ on $\Delta \times I$, we have maps

$$\begin{cases} H_+ : \partial_+ K \times I \longrightarrow Y \\ H_- : \partial_- K \times I \longrightarrow Y \text{ where } \partial_- K = \{x \in K : x_n = \max\}. \end{cases}$$

Likewise, we also have a map $H_- : \partial_- K \times I \longrightarrow Y$. By hypothesis, the restriction of H_+, H_- to $-\times 0$ is that of f_0. In fact, $K \subseteq I^n_{k_1,\ldots,k_n}$ is a face for some $k_1, \ldots, k_n$. Therefore, either $f_0(K \times 0) \subseteq V_1$ or $f_0(K \times 0) \subseteq V_2$. We consider various cases:

1. Assume that $H_-(\partial_- K \times 0) \subseteq V_1 \cap V_2$. (*Recall that, on $\partial_- K \times 0$, H_- and f_0 agree.*)
   By (B.17), $H_-(\partial_- K \times I, \partial_- K \times 1) \subseteq U_1 \cap U_2$.

(a) Assume that $f_0(K \times 0) \subseteq V_1$, and $f_0(K \times 0) \not\subseteq V_2$. By (B.17), $H_+(\partial_+ K \times I, \partial_+ K \times 1) \subseteq (V_1, U_1)$. So, $(f_0)_{|K\times 0}(K \times 0, \partial_- K \times 0) \subseteq (V_1, V_1 \cap V_2)$. Also, $H_+(\partial_+ K \times 1, (\partial_+ K \cap \partial_- K) \times 1) \subseteq (U_1, U_1 \cap U_2)$.

Apply (2) of B.2.12 for the pair $(f_0)_{K\times 0}, H_+$, and the inclusion $(U_1, U_1 \cap U_2) \hookrightarrow (V_1, V_1 \cap V_2)$. So, there is a map $H_K : (K, \partial_- K) \times I \longrightarrow (V_1, V_1 \cap V_2)$ such that $H_K((K, \partial_- K) \times 1) \subseteq (U_1, U_1 \cap U_2)$, and it extends the pair $(f_0)_{|K\times 0}, H_+$. (*It appears that* $(H_K)_{|\partial_- K \times I}$ *may differ from the already existing map* H_-*. However,* $(H_K)_{|\partial_- K\times I}$ *and* H_- *agree on the* $\Delta \times I$ *for all faces* $\Delta \subseteq \partial_- K$ *So, we can replace* H_- *by* $(H_K)_{|\partial_- K\times I}$*.*)

(b) The cases $f_0(K \times 0) \subseteq V_2$ and $f_0(K \times 0) \not\subseteq V_1$ are dealt with similarly.

(c) Now, consider the case $f_0(K \times 0) \subseteq V_1 \cap V_2$. As before, by (B.17),

$$\begin{cases} (f_0)_{|K\times 0}(K \times 0, \partial_- K \times 0) \subseteq (V_1 \cap V_2, V_1 \cap V_2) \\ H_+(\partial_+ K \times I, \partial_+ K \times 1) \subseteq (V_1 \cap V_2, U_1 \cap U_2) \\ H_+(\partial_+ K \times 1, (\partial_+ K \cap \partial_- K) \times 1) \subseteq (U_1 \cap U_2, U_1 \cap U_2). \end{cases}$$

Write $\mathcal{K} := K \times 0 \cup \partial_+ K \times I$. We want to define a suitable retraction $\rho : K \times I \longrightarrow \mathcal{K}$. For this, let z_0 be the midpoint of the bottom of $\partial_+ K = J$. Then $z = (z_0, 1)$ falls outside K, directly above the bottom of $\partial_+ K = J$. Now, let $\rho : K \times I \longrightarrow \mathcal{K}$ be the projection from $(z, 2) = (z_0, 1, 2)$, as displayed in the following diagram:

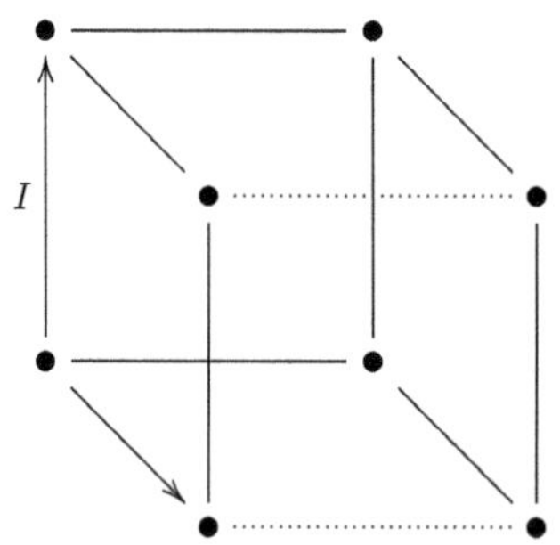

$$\begin{cases} K \times 0 \text{ is the represented bottom of the cube.} \\ \partial_+ K \times 0 = J \times 0 \text{ all three sides, except the front.} \\ \mathcal{K} \text{ consists of the four sides of } K \times I \text{ except front and top.} \\ \rho \text{ is a projection from a suitable point} \\ \text{at the front at the plane } t = 1 \end{cases}$$

from $(z,1)$ to $\mathcal{K}$. Then ρ is a retraction of $\mathcal{K}$, as follows:

$$\begin{cases} \rho : K \times I \longrightarrow \mathcal{K} \\ \ni \ \rho(K \times 1) \subseteq \partial_+ K \times 1 \end{cases} \quad \text{Define } H_K \text{ by} \quad \begin{array}{ccc} K \times I & \xrightarrow{\rho} & \mathcal{K} \\ & {\scriptstyle H_K} \searrow & \downarrow {\scriptstyle \widetilde{H}} \\ & & V_1 \cap V_2 \end{array} \tag{B.18}$$

where $\widetilde{H} : \mathcal{K} \longrightarrow V_1 \cap V_2$ is defined by combining f_0 and H_+. Note that $H_K(K \times 1) \subseteq \widetilde{H}(\partial_+ K \times 1) \subseteq U_1 \cap U_2$, as required. We suggest the same remedy in case $(H_K)_{|\partial_- K \times I}$, defined above, differs from H_-, already defined.

2. Now consider the case, $H_-(\partial_- K \times 0) \nsubseteq V_1 \cap V_2$. So, either $H_-(\partial_- K \times 0) \nsubseteq V_1$ or $H_-(\partial_- K \times 0) \nsubseteq V_2$. We assume that $H_-(\partial_- K \times 0) \nsubseteq V_2$, and hence $f_0(K \times 0) \nsubseteq V_2$. Therefore, $f_0(K \times 0) \subseteq V_1$. So, $H_+(\partial_+ K \times I) \subseteq V_1$. Use the retraction ρ and define $H_K : K \times I \longrightarrow V_1$, as in (B.18), while $\widetilde{H} : \mathcal{K} \longrightarrow V_1$.

Inductively, it follows that f_0, h, given in (B.16), extends to $H : (I^n, \partial_- I^n) \times I \longrightarrow (Y, V_1)$, with $H((I^n, \partial_- I^n) \times 1) \subseteq (X, U_1)$. The proof is complete. ■

B.3. Fibrations

Fibrations are some basic arrows in **Top** that lead to long exact sequences of homotopy groups. They are about the homotopy lifting property, as defined in the following.

Definition B.3.1. Let $\mathbb{J}$ denote the set of inclusions $\mathbb{D}^n \hookrightarrow \mathbb{D}^n \times I$ sending $x \mapsto (x, 0)$, where $n \in \mathbb{N}$. A map $f : X \longrightarrow Y$ is called a **fibration** if f has the **right lifting property**, with respect to every map in $\mathbb{J}$, in the sense of the following commutative diagram:

$$\begin{array}{ccc} \mathbb{D}^n & \longrightarrow & X \\ \big\downarrow & \overset{\varphi}{\nearrow} & \big\downarrow {\scriptstyle f} \\ \mathbb{D}^n \times I & \longrightarrow & Y \end{array} \quad \text{Equivalently,} \quad \begin{array}{ccc} I^n & \longrightarrow & X \\ \big\downarrow & \overset{\varphi}{\nearrow} & \big\downarrow {\scriptstyle f} \\ I^n \times I & \longrightarrow & Y \end{array} \tag{B.19}$$

This means, given the outer commutative diagram of continuous maps, there is a continuous map φ making all the triangles commute. Such fibrations are often referred to as **weak fibrations** or

Serre fibrations. We remark that such a qualification, as a weak fibration or a Serre fibration, is sometimes necessary to distinguish it from other definitions available in the literature [S, W], which we recall in what follows (B.3.3). For us, a "fibration" always means a **weak fibration**. It is also noteworthy that such a right lifting property is also called **Homotopy lifting property**, for obvious reasons.

Example B.3.2. The following are some useful examples of weak fibrations.

1. The constant map $X \longrightarrow \{x_0\}$ is trivially a fibration. The singleton $\{x_0\}$ being the terminal object in **Top**, one says, every topological space X is **fibrant**.
2. Any projection map $p : Y \times B \longrightarrow B$ is a fibration.

Proof. Obvious! ■

For completeness, we recall two other definitions of fibrations, from [S, W].

Definition B.3.3. Let $p : E \longrightarrow B$ be a continuous map.

1. (See [S, p. 66]): We say that p is a **topological fibration** if p has the homotopy lifting property, with respect to every space X, interpreted as in (B.19):

$$\begin{array}{ccc} X & \longrightarrow & E \\ {\scriptstyle i_0}\downarrow & \nearrow & \downarrow{\scriptstyle p} \\ X \times I & \xrightarrow[F]{} & B \end{array},$$

   In other words, p has the **homotopy lifting property**, with respect to any $X \in$ **Top**. Such a topological fibration is also called a **Hurewicz fibration**.
2. (See [W, p. 29]): Under the same notation, we say that p is a **White fibration** if the homotopy lifting property holds for all compactly generated T_2-spaces X. (See Lemma B.1.15, while no notation was used in (B.1.13) for the category of such spaces.)

3. Therefore,

$$f \text{ is a topological fibration} \Longrightarrow f \text{ is a White fibration}$$
$$\Longrightarrow f \text{ is a Weak fibration.}$$

Theorem B.3.4. *The composition of fibrations* (of any kind) *is a fibration* (of the same kind).

Proof. Completed by routine diagram chasing. ■

Theorem B.3.5. *Let $p : X \longrightarrow B$ be a fibration* (of any kind) *and $f : B' \longrightarrow B$ be continuous. Then the base change map $p' : X' \longrightarrow B'$, obtained by pullback as follows, is a fibration* (of the same kind)*:*

$$\begin{array}{ccc} X' & \longrightarrow & X \\ {\scriptstyle p'}\downarrow & & \downarrow \\ B' & \underset{f}{\longrightarrow} & B \end{array} \qquad \text{where} \quad X' := B' \times_B X.$$

In particular, for $A \subseteq B$, the map $p^{-1}A \longrightarrow A$ is a fibration.

Proof. Completed by routine diagram chasing. ■

Before we give Hurewicz criteria for fibrations, we define paracompact spaces.

Definition B.3.6. Suppose that X is a topological space. An open cover $\{V_i : i \in \mathcal{I}\}$ of X is said to be **locally finite** if $\forall x \in X$ there is open neighborhood $x \in U$ such that $V_i \cap U \neq \phi$ only for finitely many $i \in \mathcal{I}$.

An open cover $\{V_i : i \in \mathcal{I}\}$ of X is said to be a refinement of another open cover $\{U_j : i \in \mathcal{J}\}$ of X if, given $i \in \mathcal{I}$, $V_i \subseteq U_j$ for some $j \in \mathcal{J}$.

The space X is said to be **paracompact** if, given any open cover $\{U_j : i \in \mathcal{J}\}$ of X, there is a locally finite refinement $\{V_i : i \in \mathcal{I}\}$ of $\{U_j : i \in \mathcal{J}\}$. Usually, X is assumed to be Hausdorff.

Theorem B.3.7 (Hurewicz). *Let $p : X \longrightarrow B$ be continuous and B be paracompact. Then p is a* (White) *fibration if there is an open cover $B = \bigcup_{i \in I} V_i$ such that $p^{-1}V_i \longrightarrow V_i$ are fibrations $\forall i \in I$.*

The following is a foundational theorem in Homotopy theory.

Lemma B.3.8. *Let $p : X \longrightarrow Y$ be a weak fibration, and $x_0 \in X$, $y_0 = p(x_0)$. Let $F = p^{-1}(p(x_0))$, and let $\iota : F \hookrightarrow X$ denote the inclusion map. Then there is a long exact sequence*

$$\begin{array}{llll}
\cdots \longrightarrow & \pi_{n+1}(F, x_0) \xrightarrow{\iota_*} & \pi_{n+1}(X, x_0) \xrightarrow{p_*} & \pi_{n+1}(Y, y_0) \\
\xrightarrow{\partial_{n+1}} & \pi_n(F, x_0) \xrightarrow{\iota_*} & \pi_n(X, x_0) \xrightarrow{p_*} & \pi_n(Y, y_0) \\
\longrightarrow & \cdots & \cdots & \cdots \\
\longrightarrow & \pi_1(F, x_0) \longrightarrow & \pi_1(X, x_0) \longrightarrow & \pi_1(Y, y_0) \\
\longrightarrow & \pi_0(F, x_0) \longrightarrow & \pi_0(X, x_0) \longrightarrow & \pi_0(Y, y_0).
\end{array} \tag{B.20}$$

Here ∂_n is a group homomorphism, $\forall\ n \geq 2$. For $n = 0$, $\pi_0(-, \star)$ is to be understood as a pointed set, where the path component of $\star$ is the base point. (By the kernel of a map of pointed sets, we mean the inverse image of the base point. The exactness of a sequence of maps of pointed sets means that the image is the same as the kernel.)

Proof. See [S, Theorem 7.2.10] for a detailed proof. The maps ι_* and p_* are the obvious functorial maps. We define the connecting homomorphism $\partial_{n+1} : \pi_{n+1}(Y, y_0) \longrightarrow \pi_n(F, x_0)$. Let $\alpha = [\gamma] \in \pi_{n+1}(Y, y_0)$, represented by a map $\gamma : (I^{n+1}, \partial I^{n+1}) \longrightarrow (Y, y_0)$. Let $\gamma_0 : I^n \times 0 \longrightarrow X$ be the constant map x_0. Then we have the following commutative diagram:

$$\begin{array}{ccc}
I^n \times 0 & \xrightarrow{\gamma_0} & X \\
\downarrow & \overset{g}{\nearrow} & \downarrow p \\
I^n \times I & \xrightarrow[\gamma]{} & Y
\end{array}$$

where g is obtained by the homotopy lifting property of the fibration p. It follows that $pg(\partial I^n) \subseteq \gamma(\partial I^{n+1}) = \{y_0\}$. Therefore, $g(\partial I^n) \subseteq F$.

Define $g_1 : I^n \longrightarrow F$ by $g_1(t) = g(t,1)$. Then $g_1(\partial I^n) = x_0$. Therefore, $g_1 : (I^n, \partial I^n) \longrightarrow (F, x_0)$. Define $\partial_{n+1}(\alpha) = [g_1] \in \pi_n(F, x_0)$. One checks that ∂_{n+1} is well-defined, and one checks the exactness. ■

B.4. Construction of Fibrations

Not surprisingly, we have some interest in the long exact sequence (B.20), while it does not appear to be easy to detect whether a continuous map $p : X \longrightarrow Y$ is a fibration or not. Given a continuous map $p : X \longrightarrow Y$, we replace p by a fibration $\mathcal{X} \longrightarrow Y$, together with a homotopy equivalence $X \xrightarrow{\sim} \mathcal{X}$. Consequently, we can write down a long exact sequence, similar to (B.20). Materials in this section are classical. Among the standard sources are [S, §2.8], [W, §I.7]. We establish some notation.

Notation B.4.1. As usual $I = [0,1]$. For a space B, let $\mathbf{P}(B) := \mathbf{P}(I, B) := \mathbf{F}(I, B)$ be the space of all paths in B. Note that $\mathbf{F}(I, B)$ has compactly generated topology (see B.1.9 and B.1.20). Suppose that $f : X \longrightarrow B$ is a continuous map. Let

$$\mathbf{I}^f = \{(x,u) \in X \times \mathbf{P}(B) : p(x) = u(0)\}. \qquad \text{So,} \quad \mathbf{I}^f \subseteq X \times \mathbf{P}(B).$$

So, $\mathbf{I}^f$ is the set of pairs (x, γ) where $x \in X$ and γ is a path in B starting at $p(x)$.

We list a few immediate classical results from [S, §2.8], [W, §I.7].

Proposition B.4.2. *With the notation as in* (B.4.1)*, the following are White fibrations.*

$$\begin{cases} p : \mathbf{P}(X) \longrightarrow X \times X & p(u) = (u(0), u(1)) \\ p_0 : \mathbf{P}(X) \longrightarrow X & p_0(u) = u(0) \\ p_1 : \mathbf{P}(X) \longrightarrow X & p_1(u) = u(1) \end{cases}$$

Corollary B.4.3. *Let $f : X \longrightarrow B$ be a map in* **Top**. *Then the projection map $\mathbf{I}^f \longrightarrow X$ is a* (white) *fibration.*

Proof. The following

$$\begin{array}{ccc} \mathbf{I}^f & \xrightarrow{f'} & \mathbf{P}(B) \\ {\scriptstyle p'_0}\downarrow & & \downarrow{\scriptstyle p_0} \\ X & \xrightarrow[f]{} & B \end{array} \qquad \text{is a pullback diagram.}$$

Since p_0 is a fibration, so is p'_0 by (B.3.5). The proof is complete. ■

For a map $f : X \longrightarrow B$ in **Top**, we define a few maps in **Top**:

$$\begin{cases} \text{For } y \in B \quad e_y : I \longrightarrow B \text{ the constant map } y & \\ \varepsilon : B \longrightarrow \mathbf{P}(B) & \varepsilon(y) = e_y \\ p_0 : \mathbf{P}(B) \longrightarrow B & p_0(\gamma) = \gamma(0) \\ p_1 : \mathbf{P}(B) \longrightarrow B & p_{(}\gamma) = \gamma(1) \\ f' : \mathbf{I}^f \longrightarrow \mathbf{P}(B) & f'(x,\gamma) = \gamma \\ p'_0 : \mathbf{I}^f \longrightarrow X & p'_0(x,\gamma) = x \\ \lambda : X \longrightarrow \mathbf{I}^f & \lambda(x) = (x, \varepsilon f) \\ p : \mathbf{I}^f \longrightarrow B & p = p_1 f' \end{cases}$$

The following are some helpful commutative diagrams:

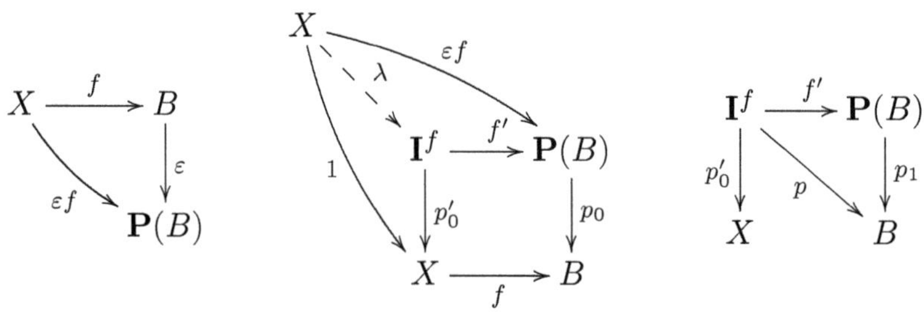

Note that the map $f : X \longrightarrow B$ does not fit in the bottom of the last diagram.

Theorem B.4.4. *With the notation as above, we have*

1. *The map* $p = p_1 f' : \mathbf{I}^f \longrightarrow B$ *is a White fibration. Indeed,*

$$p(x,\gamma) = \gamma(1) \qquad \textit{the other end of } \gamma.$$

2. *Then*

$$\begin{cases} p_0'\lambda = 1_X, \ \lambda p_0' \simeq 1_{\mathbf{I}^f} \\ p\lambda = f, \qquad fp_0' \simeq p \end{cases} \qquad \textit{where} \ \simeq \ \textit{means homotopic.}$$

3. *Therefore, p_0' and λ are homotopy equivalences.*

Proof. Consider the commutative diagram:

$$\begin{array}{ccc} Z \times 0 & \xrightarrow{g} & \mathbf{I}^f \\ \downarrow & & \downarrow p \\ Z \times I & \xrightarrow[H]{} & B \end{array}$$

Write $g(z,0) = (g_1(z), g_2(z)(s)) \in \mathbf{I}^f \subseteq X \times \mathbf{P}(B)$ with $s \in I$.

Since the path $g_2(z)(s)$ starts at $f(g_1(z))$ and because of the commutativity, we have

$$g_2(z)(0) = f(g_1(z)) \qquad g_2(z)(1) = H(z,0)$$

For $(z,t) \in Z \times I$, define a path

$$\gamma(z,t) : I \longrightarrow B \qquad \gamma(z,t)(s) = \begin{cases} g_2(z)(s) & \text{if } t = 0 \\ H\left(z, \frac{(s-t)t}{1-t}\right) & \text{if } 0 < t < s \\ g_2(z)(s/t) & \text{if } t \neq 0, \ s \leq t \end{cases}$$

Then $\gamma(z,t)$ is a path from $g_2(z)(0) = f(g_1(z))$ to $\gamma(z,t)(1) = H(z,t)$. Now, define

$$G : Z \times I \longrightarrow \mathbf{I}^f \quad G(z,t) = (g_1(z), \gamma(z,t)).$$

So, H lifts to a homotopy G. This establishes that p is a White fibration.

The equalities in (2) are obvious. The last equivalence follows from the third and second: $p\lambda = f \Longrightarrow p\lambda p_0' = fp_0' \Longrightarrow p \simeq fp_0'$. It remains to be proven that $\lambda p_0' \simeq 1_X$. Note that $\lambda p_0'(x,\gamma) = (x, e_{\gamma(0)})$. Now,

$$\text{with} \quad F((x,\gamma(s)),t) = (x,\gamma(st)) \quad \begin{cases} F(-,0) = \lambda p_0' \\ F(-,1) = 1 \end{cases}$$

This completes the proof of (2). Now, (3) follows from (2). ■

The following is a fundamental concept in Homotopy theory.

Definition B.4.5. Let $f : X \longrightarrow B$ be a map in **Top**, and $y \in B$ be a base point. Let $p : \mathbf{I}^f \longrightarrow B$, and other notation be as above. The **Homotopy fiber** of f is defined as $\mathscr{F}(f, y) := p^{-1}(y)$.

So, $$\mathscr{F}(f, y) = p^{-1}(y) = \{(x, \gamma) \in X \times \mathbf{P}(B) : \gamma(1) = y,\ \gamma(0) = f(x)\}.$$

So, $(x, \gamma) \in \mathscr{F}(f, y)$ if γ is a path in B from $f(x) \mapsto y$. We write $\mathscr{F}(f) := \mathscr{F}(f, y)$, when the base point y is understood.

Following is a summary of the above.

Theorem B.4.6. *Let $f : X \longrightarrow B$ be a map in* **Top**, *$y_0 \in B$. Let $p : \mathbf{I}^f \longrightarrow B$ and other notation be as above. Consider the diagram*

$$\begin{array}{ccccc} \mathscr{F}(f, y_0) & \xrightarrow{\iota} & \mathbf{I}^f & \xrightarrow{p} & Y \\ & \searrow^{q} & \downarrow{\scriptstyle p_0' \,\wr} & \nearrow_{f} & \\ & & X & & \end{array}$$

$$\begin{cases} \textit{Here, } \iota \textit{ is the inclusion map, and } \quad q := p\iota \\ \textit{The triangle on the right commutes up to homotopy.} \end{cases}$$

Further, p is a White fibration and hence a weak fibration. So, up to homotopy, any continuous map f is equivalent to the Fibration p.

Now, let $f(x_0) = y_0$, and $z_0 := (x_0, \gamma_0) \in \mathscr{F}(f, y_0)$ for some γ (i.e. γ_0 is a loop at y_0, possibly, the constant loop.) *It follows from Lemma B.3.8 that there is a long exact sequence*

$$\begin{array}{llll} \cdots \longrightarrow \pi_n(\mathscr{F}(f), z_0) & \longrightarrow \pi_n(X, x_0) & \longrightarrow \pi_n(B, y_0) & \\ \xrightarrow{\partial_n} \pi_{n-1}(\mathscr{F}(f), z_0) & \longrightarrow \pi_{n-1}(X, x_0) & \longrightarrow \pi_{n-1}(B, y_0) & \\ \longrightarrow \cdots & \quad \cdots & \quad \cdots & \quad (B.21) \\ \longrightarrow \pi_1(\mathscr{F}(f), z_0) & \longrightarrow \pi_1(X, x_0) & \longrightarrow \pi_1(B, y_0) & \\ \longrightarrow \pi_0(\mathscr{F}(f), z_0) & \longrightarrow \pi_0(X, x_0) & \longrightarrow \pi_0(B, y_0). & \end{array}$$

Definition B.4.7. Let $B \in \mathbf{Top}$ and $y \in B$. Let $\star$ denote the singleton set and $c : \star \longrightarrow Y$ be the map, sending $\star \mapsto y$. With other notation as above, write

$$\Omega B := \Omega(B, y) := \mathscr{F}(c) \quad \text{to be called the } \textbf{loop space} \text{ of } B \text{ at } y.$$

The loop space ΩY is to be considered a pointed space, with constant path c_y at y as the base point. It follows from the long exact sequence (B.4.6) that

$$\pi_n(\Omega B, c_y) = \pi_{n+1}(B, y) \qquad \forall\, n \geq 0.$$

Note that $\Omega B = \Omega(B, y)$ is the space of all loops in B at y. Often, the base point y is understood or B is assumed to be path connected, and we write ΩB for what should be $\Omega(B, y)$.

B.5. The Quasi-Fibrations

A substantial part of the foundation of Algebraic K-theory of Quillen resides on the celebrated Theorem B (2.2.3) in [Q]. This leads to the long exact sequences of the K-groups, as a consequence of the long exact sequences of homotopy groups (B.21). The proof of Theorem B (2.2.3) requires some background on quasi-fibrations. Essentially, all the main points in this section date back to the paper of Dold and Thom [DT]. Some readers may decide to postpone reading this section until we reach Theorem B (2.2.3).

First, we recall the definition of the homotopy cartesian product.

Definition B.5.1. Consider a commutative diagram of continuous maps

$$\begin{array}{ccc} E' & \xrightarrow{h'} & E \\ {\scriptstyle g'}\downarrow & & \downarrow{\scriptstyle g} \\ Y & \xrightarrow[h]{} & B \end{array} \tag{B.22}$$

Write

$$\mathcal{E}(g, h) := E \times_B \mathbf{F}(I, B) \times_B Y :=$$

$$\left\{ (z, \gamma, y) \in E \times \mathbf{F}(I, B) \times Y : \begin{array}{c} \gamma(0) = g(z) \\ \gamma(1) = h(y), \end{array} \right\}. \tag{B.23}$$

The diagram (B.22) naturally extends to the following commutative diagram:

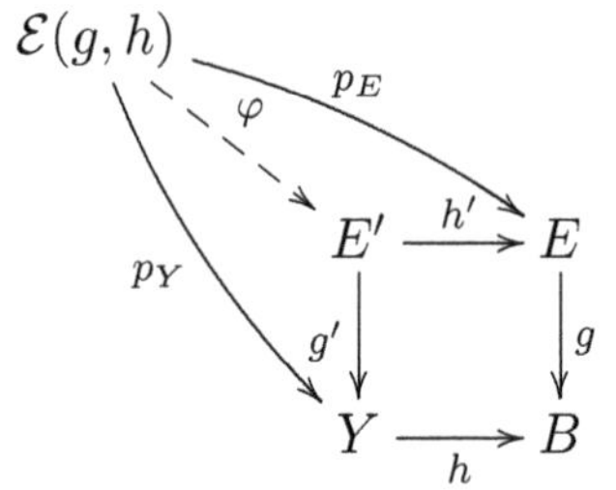

(B.24)

$$\text{where}\begin{cases} p_E : \mathcal{E}(g,h) \longrightarrow E & \text{defined by} \quad p_E(z,\gamma,y) = z \\ p_Y : \mathcal{E}(g,h) \longrightarrow Y & \text{defined by} \quad p_Y(z,\gamma,y) = y \\ \varphi : E' \longrightarrow \mathcal{E}(g,h) & \text{defined by} \quad \varphi(z) = \big(h'(z), c_{hg'(z)}, g'(z)\big) \\ \text{while } c_{hg'(z)} : [0,1] \longrightarrow B \text{ denotes the constant path } hg'(z) \end{cases} \tag{B.25}$$

The rectangle (B.22) is defined as a **Homotopy cartesian** square if φ is a homotopy equivalence.

Remark B.5.2. In general, $\mathcal{E}(g,h)$ does not have the properties of the Pullback of (g,h) in the category $\mathscr{H}$**omoTop**. However, we have the following. Let

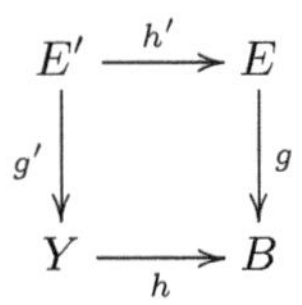

be a diagram in **Top**, which commutes in $\mathscr{H}$**omoTop**.

This means that there is a homotopy

$$H : E' \times [0,1] \longrightarrow B \qquad \ni \qquad \begin{cases} H(-,0) = gh' \\ H(-,1) = hg' \end{cases}$$

Define $\varphi : E' \longrightarrow \mathcal{E}(g,h)$ by

$$\varphi(z) = (g'(z), H(z,-), h'(z)). \quad \text{Note} \quad H(z,0) = gh'(z), \ H(z,1) = hg'(z).$$

Therefore, in the following diagram:

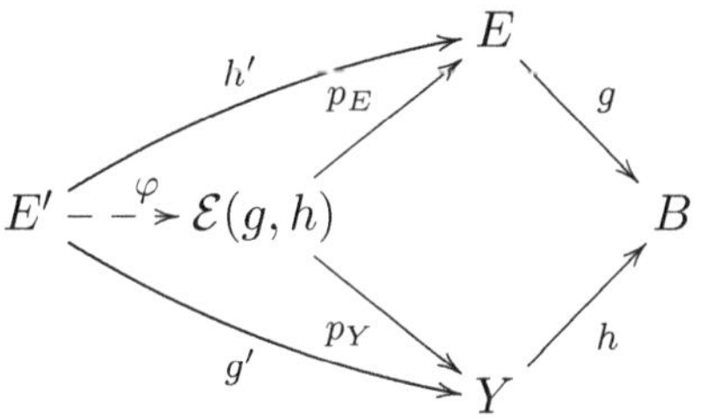

the right-hand rectangle commutes in the category $\mathscr{H}$**omoTop**, and the two triangles on the left (actually) commute. While the existence of φ is exhibited here, its uniqueness in $\mathscr{H}$**omoTop** is not guaranteed. We are mainly interested when $Y = \{b\}$ is a singleton, with $b \in B$.

Definition B.5.3. Let $g : E \longrightarrow B$ be a map in **Top** and $b \in B$. The **Homotopy Fiber** $\mathcal{F}(g, b)$ was defined in (B.4.5). In the cartesian product notation, the homotopy fiber can be described as follows:

$$\mathcal{F}(g, b) = E \times_B B^I \times_B \{b\} =$$
$$\{(z, \gamma, b) \in E \times B^I \times \{b\} : \gamma \in B^I, \gamma(1) = b, \ \gamma(0) = g(z)\}.$$

We will often drop the third coordinate b, and write

$$\mathcal{F}(g, b) = \{(z, \gamma) \in E \times B^I : \gamma \in B^I, \gamma(1) = b, \ \gamma(0) = g(z)\}.$$

Further, a sequence $F \xrightarrow{f} E \xrightarrow{g} B$ of maps in **Top** is called a **Homotopy fibration**, (or *Homotopy fibration sequence*) if the composition $gf(x) = b \in B \ \forall \ x \in F$ is the constant map, and the natural map $\varphi : F \longrightarrow \mathcal{F}(g, b)$, as defined in (B.25), is a homotopy equivalence. In other words, if

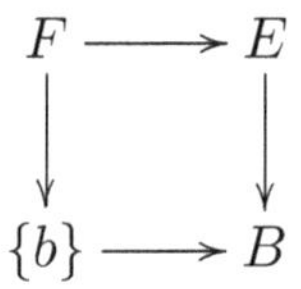

is homotopy cartesian.

Weak equivalence is a concept of fundamental nature, that we define next.

Definition B.5.4. A map $f : X \longrightarrow Y$ in **Top** is defined as a **weak homotopy equivalence** or simply a **weak equivalence** if

$$\forall\, x_0 \in X,\ \forall\, n \geq 0 \quad \pi_n(f) : \pi_n(X, x_0) \longrightarrow \pi_n(Y, f(x_0)) \quad \text{is a bijection.}$$

Clearly, any homotopy equivalence is a weak homotopy equivalence. In fact, if X and Y are CW complexes, the converse is also true (Theorem C.3.8, *which we discuss later*). Indeed, we have rare occasions, in this theory [Q], to consider any space that is not a CW complex.

Corollary B.5.5. *Let* $F \xrightarrow{\iota} E \xrightarrow{g} B$ *be a homotopy fibration, with* $gf(x) = b\ \forall\, x \in F$. *Then* $\forall\, x_0 \in F$, *there is a long exact sequence*

$$\cdots \longrightarrow \pi_n(F, x_0) \longrightarrow \pi_n(E, \iota(x_0)) \longrightarrow \pi_n(B, b) \longrightarrow \pi_{n-1}(F, x_0) \longrightarrow \cdots$$

ending at $n = 0$. *In particular, if* B *is a contractible space, then* $\iota : F \longrightarrow E$ *is a weak homotopy equivalence.*

Proof. The existence of the long exact sequence is an immediate consequence of the same in Corollary B.4.6. When B is contractible, $\pi_n(B, b)$ is trivial, and hence the maps $\pi_n(F, x_0) \longrightarrow \pi_n(E, \iota(x_0))$ are isomorphisms. Therefore, ι is a weak homotopy equivalence. ■

Now we define quasi-fibrations, which are the main topic of discussions in this section.

Definition B.5.6. A map $g : E \longrightarrow B$ in **Top** is defined as a **quasi-fibration** if for all $b \in B$, the induced maps $p^{-1}(b) \longrightarrow \mathcal{F}(g, b)$ are weak equivalences (B.5.4).

Remark B.5.7. The following are two immediate remarks.

1. Let $g : E \longrightarrow B$ be a map in **Top**. Then g is a quasi-fibration if and only if, for each path component C of B, the restriction $g^{-1}C \longrightarrow C$ is a quasi-fibration.
2. Let $g : E \longrightarrow B$ be a fibration. Assume that g is surjective. Then g is a quasi-fibration. In particular, the projection maps $X \times B \longrightarrow B$ are quasi-fibrations.

Proof. The proof of (1) is obvious. The proof of (2) follows by comparing the long exact sequences (B.20) and (B.21), using the Five Lemma arguments. ■

For the convenience of subsequent discussions, we introduce the following notation.

Notation B.5.8. For a topological space X, let $\alpha : I \longrightarrow X$ be a path. The opposite path of α is denoted by $\overline{\alpha}$. That means that $\overline{\alpha}(t) = \alpha(1-t)\ \forall\ t \in I$. Also, for another path $\gamma : I \longrightarrow X$, with $\alpha(1) = \gamma(0)$, the concatenation is denoted by $\gamma\alpha$. So,

$$\gamma\alpha(t) = \begin{cases} \alpha(2t) & \text{if } t \leq 1/2 \\ \gamma(2t-1) & \text{if } 1/2 \leq t \leq 1. \end{cases}$$

The following is our first lemma on quasi-fibration.

Lemma B.5.9. *Suppose that $g : E \longrightarrow B$ is a map in* **Top**. *Let $\alpha : I \longrightarrow B$ be a path such that $\alpha(0) = b_0$ and $\alpha(1) = b_1$. Then α defines a homotopy equivalence $\mathcal{F}(g, b_1) \longrightarrow \mathcal{F}(g, b_0)$. In particular, if g is a quasi-fibration, then there is a zigzag sequence of weak equivalences:*

$$\begin{array}{ccc} p^{-1}(b_1) & \dashrightarrow & p^{-1}(b_0) \\ \wr\downarrow & & \downarrow\wr \\ \mathcal{F}(g, b_1) & \xrightarrow{\sim} & \mathcal{F}(g, b_0) \end{array}$$

Further, if B is path connected, then $p^{-1}(b)$ are weak homotopy invariant.

Proof. Recall that $\mathcal{F}(g, b_0) = \{(z, \gamma) \in E \times B^I : \gamma(0) = g(z), \gamma(1) = b_0\}$. Define the map

$$\omega_\alpha : \mathcal{F}(g, b_0) \longrightarrow \mathcal{F}(g, b_1) \quad \text{by} \quad \omega_\alpha(z, \gamma) = (z, \alpha\gamma).$$

Then $\omega_{\overline{\alpha}} : \mathcal{F}(g, b_0) \longrightarrow \mathcal{F}(g, b_1)$. We have

$$\omega_{\overline{\alpha}}\omega_\alpha : \mathcal{F}(g, b_0) \longrightarrow \mathcal{F}(g, b_0). \quad \text{And} \quad \omega_{\overline{\alpha}}(\omega_\alpha(z, \gamma)) = (z, \overline{\alpha}(\alpha\gamma)).$$

Since $\overline{\alpha}\alpha$ is homotopic to the constant path, it follows that $\omega_{\overline{\alpha}}\omega_\alpha$ is homotopic to the identity map. Likewise, $\omega_\alpha\omega_{\overline{\alpha}}$ is homotopic to the identity map. So, ω_α is a homotopy equivalence. The proof is complete. ■

The following is an equivalent way to characterize quasi-fibrations.

Proposition B.5.10. *Let $f : E \longrightarrow B$ be a map in* **Top**. *Then f is a quasi-fibration if and only if, $\forall\, b \in B$, $\forall\, x \in f^{-1}b$, $\forall\, n \geq 0$, the maps $\pi_n\left(E, f^{-1}b, x\right) \longrightarrow \pi_n(B, b)$ are isomorphisms.*

Proof. Because of Remark B.5.7, we can assume that B is path connected. We use the notation from Section B.4. We have $\mathbf{I}^f = \{(x,u) \in E \times \mathbf{P}(B) : u(0) = f(x)\}$. Fix $b \in B$, write $E_b = f^{-1}(b)$, and $x \in E_b$. Consider the following commutative diagram:

$$\begin{array}{ccccc}
E_b & \overset{\iota}{\hookrightarrow} & E & \overset{f}{\longrightarrow} & B \\
\Big\downarrow{\scriptstyle \lambda_b} & & \Big\downarrow{\scriptstyle \lambda} & & \Big\| \\
\mathscr{F}(f,b) & \underset{\zeta}{\hookrightarrow} & \mathbf{I}^f & \underset{p}{\longrightarrow} & B
\end{array}$$

where p is given by $(x,u) \mapsto u(1)$. The long exact sequences (B.11, B.21) lead to the following commutative diagram (*with labels added*) of homotopy groups/sets:

$$\begin{array}{ccccccccc}
\pi_n(E_b,x)) & \overset{\iota_*}{\longrightarrow} & \pi_n(E,x) & \overset{\beta_*}{\longrightarrow} & \pi_n(E,E_b,x) & \overset{\partial}{\longrightarrow} & \pi_{n-1}(E_b,x) & \longrightarrow & \cdots \\
\Big\downarrow{\scriptstyle (\lambda_b)_*} & & \wr\Big\downarrow{\scriptstyle \lambda_*} & & \Big\downarrow{\scriptstyle f_*} & & \Big\downarrow & & \\
\pi_n(\mathscr{F}(f,b),*) & \underset{\zeta_*}{\longrightarrow} & \pi_n\left(\mathbf{I}^f,*\right) & \underset{p_*}{\longrightarrow} & \pi_n(B,b) & \underset{d}{\longrightarrow} & \pi_{n-1}(\mathscr{F}(f,b),*) & \longrightarrow & \cdots
\end{array}$$

Here $* = (x, c_b)$, where c_b is the constant path b, and β_* denotes the map induced by the inclusion $\beta : (E,x,x) \hookrightarrow (E,E_b,x)$. Since λ is a homotopy equivalence (Theorem B.4.4), λ_* is an isomorphism for all n. The rest of the argument is a repetition of the Five Lemma arguments (A.3.7), which is that $(\lambda_b)_*$ is an isomorphism for all $n \geq 0$ if and only if f_* is. However, since π_0 and $\pi_1(E,E_b,x)$ are not groups, further care is necessary, as follows.

Assume that f is a quasi-fibration. Since B is assumed to be path connected and $\pi_0(E,E_b,x)$ was defined to be the quotient, it follows that $\pi_0(E,E_b,x) \longrightarrow \pi_0(B,b)$ is a bijection of singleton sets. Now consider $n = 1$:

$$\begin{array}{ccccccccc}
\pi_1(E_b,x)) & \overset{\iota_*}{\longrightarrow} & \pi_1(E,x) & \overset{\beta_*}{\longrightarrow} & \pi_1(E,E_b,x) & \overset{\partial}{\longrightarrow} & \pi_0(E_b,x)) & \longrightarrow & \pi_0(E,x) \\
\Big\downarrow{\scriptstyle (\lambda_b)_*} & & \wr\Big\downarrow{\scriptstyle \lambda_*} & & \Big\downarrow{\scriptstyle f_*} & & \wr\Big\downarrow{\scriptstyle (\lambda_b)_*} & & \wr\Big\downarrow{\scriptstyle \lambda_*} \\
\pi_1(\mathscr{F}(f,b),*) & \underset{\zeta_*}{\longrightarrow} & \pi_1\left(\mathbf{I}^f,*\right) & \underset{p_*}{\longrightarrow} & \pi_1(B,b) & \underset{d}{\longrightarrow} & \pi_0(\mathscr{F}(f,b),*) & \longrightarrow & \pi_0\left(\mathbf{I}^f,*\right).
\end{array}$$

We prove that f_* is a bijection, given that four other vertical maps are bijections, while π_0-terms are pointed sets. Let $y := [\alpha] \in \pi_1(B,b)$, where α is a loop at b. So, $d(y)$ is the path component of (x,α). By the usual arguments, it follows that there is $[G] \in \pi_1(E,E_b,x)$ such that $d(y) = df_*([G])$. It follows that $d(y - f_*([G])) = d(f_*([G]) - f_*([G]))$ is neutral. So, $y - f_*([G]) = p_*(z)$ for some $z \in \pi_1(\mathbf{I}^f,*)$. So, there is $[g] \in \pi_1(E,x) = \pi_1(E,x,x)$ such that $\lambda_*([g]) = z$. We have

$$\begin{cases} g : (I,\{0,1\},\{1\}) \longrightarrow (E,x,x) \subseteq (E,E_b,x) \\ G : (I,\{0,1\},\{1\}) \longrightarrow (E,E_b,x). \end{cases}$$

The concatenation $gG : (I,\{0,1\},\{1\}) \longrightarrow (E_0,E_b,x)$ is defined. So, $f_*([gG]) = f_*([g]) + f_*([G]) = p_*(z) + f_*([G]) = y$. So, f_* is surjective.

To see injectivity, for $i = 1,2$, let $w_i : (I,\{0,1\},\{1\}) \longrightarrow (E,E_b,x)$ and $f_*([w_1]) = f_*([w_2])$. We have $w_1(1) = w_2(1) = x$. It follows that $w_1(0), w_2(0)$ are in the same path components of E_b. Modifying by any path, in E_b, from $w_1(0)$ to $w_2(0)$, we can assume that $w_1(0) = w_2(0)$. The concatenation $\omega := w_1\overline{w}_2 : I \longrightarrow E$ is a loop at x. Now $f_*\beta_*([\omega]) = f_*([w_1]) - f_*([w_2]) = 0$. Again, it follows from the usual arguments of the Five Lemma that $[\omega]$ is in the image of ι_*, and hence ω represents the neutral element in $\pi_1(E,E_b,x)$. In other words, ω is homotopic to the constant path x. Therefore, $[w_2] = [\omega w_2] = [w_1] \in \pi_1(E,E_b,x)$. This establishes that f_* is bijective, for $n = 1$. Now, by the Five Lemma of group theory, it follows that f_* is bijective for all $n \geq 0$.

Conversely, assume that f_* is bijective for all $n \geq 0$. We prove that f is a quasi-fibration. We have the commutative diagram of exact sequences as follows:

$$\begin{array}{ccccccccc}
\pi_1(E_b,x)) & \xrightarrow{\iota_*} & \pi_1(E,x) & \xrightarrow{\beta_*} & \pi_1(E,E_b,x) & \xrightarrow{\partial} & \pi_0(E_b,x)) & \twoheadrightarrow & \pi_0(E,x) \\
\downarrow{\scriptstyle(\lambda_b)_*} & & \wr\downarrow{\scriptstyle\lambda_*} & & \wr\downarrow{\scriptstyle f_*} & & \downarrow{\scriptstyle(\lambda_b)_*} & & \wr\downarrow{\scriptstyle\lambda_*} \\
\pi_1(\mathscr{F}(f,b),*) & \xrightarrow[\zeta_*]{} & \pi_1(\mathbf{I}^f,*) & \xrightarrow[p_*]{} & \pi_1(B,b) & \xrightarrow[d]{} & \pi_0(\mathscr{F}(f,b),*) & \twoheadrightarrow & \pi_0(\mathbf{I}^f,*).
\end{array}$$

Since B is assumed to be path connected, $\pi_0(E,E_b,x) \cong \pi_0(B,b_0)$ is a singleton. Therefore, the last two horizontal arrows are surjective. To prove $(\lambda_b)_*$ is surjective, let $y = [(z,\gamma)] \in \pi_0(\mathscr{F}(f,b),*)$ be the path component of $(z,\gamma) \in \mathscr{F}(f,b)$. By the usual argument, there is $x_1 \in E_b$ such that $(\lambda_b)_*([x_1])$ and y have the same image in $\pi_0\left(\mathbf{I}^f, *\right)$. For $n = 0$, $(\lambda_b)_*$ is independent of the choice of the base point x. By changing the base point, we can assume that $x_1 = x$. Hence, y maps to the distinguished point in $\pi_0\left(\mathbf{I}^f, *\right)$. Now it follows that y is in the image of $(\lambda_b)_*$. To prove injectivity, let $(\lambda_b)_*[w_1] = (\lambda_b)_*[w_2]$ for some $w_1, w_2 \in E_b$. Again, we can change the base point, and assume that $(\lambda_b)_*[w_1] = (\lambda_b)_*[w_2] = [*]$. We prove that $[w_1] = [w_2] = [x] \in \pi_0(E_b, x))$. This follows by the routine Five Lemma arguments.

From $n \geq 1$, as well, it follows from the group theoretic Five Lemma (A.3.7) that $(\lambda_b)_*$ are isomorphisms. So, f is a quasi-fibration. The proof is complete. ■

Subsequently, we give three sufficient conditions for a map to be a quasi-fibration, due to Dold and Thom [DT]. The book of Hatcher [Ha, Lemma 4k.3; p. 480] also gives a nice exposition of the same. The following is the first one [DT, 2.2 Satz].

Proposition B.5.11. *Let $g : E \longrightarrow B$ be a map in* **Top**. *Let V_1, V_2 be open subsets of B, and $B = V_1 \cup V_2$. Assume that the restrictions of g*

$$\begin{cases} g^{-1}(V_1) \longrightarrow V_1 \\ g^{-1}(V_2) \longrightarrow V_2 \\ g^{-1}(V_1 \cap V_2) \longrightarrow V_1 \cap V_2 \end{cases} \qquad \textit{are quasi-fibrations.}$$

Then g is a quasi-fibration.

Proof. For $k = 1, 2$, write $U_k = g^{-1}V_k$. Fix $b \in V_1 \cap V_2$. For $k = 1, 2$, consider the maps of the triple $(U_k, U_1 \cap U_2, g^{-1}b) \longrightarrow (V_k, V_1 \cap V_2, b)$. Fix $x \in g^{-1}b$. We first prove that

$$\forall\ n \geq 0 \ \text{ the maps } \begin{cases} \pi_n\left(U_1, U_1 \cap U_2, x\right) \longrightarrow \pi_n\left(V_1, V_1 \cap V_2, b\right) \\ \pi_n\left(U_2, U_1 \cap U_2, x\right) \longrightarrow \pi_n\left(V_2, V_1 \cap V_2, b\right) \end{cases} \text{ are bijective.} \qquad \text{(B.26)}$$

We will only prove the first one, and subsequently label the map as f_3. Consider the corresponding long exact sequences (*with labels added*), which end at $n = 1$:

$$\begin{array}{ccccccc} \pi_n\left(U_1 \cap U_2, g^{-1}b, x\right) & \xrightarrow{\iota_U} & \pi_n\left(U_1, g^{-1}b, x\right) & \xrightarrow{\varsigma_U} & \pi_n\left(U_1, U_1 \cap U_2, x\right) & \xrightarrow{\partial_U} & \pi_{n-1}\left(U_1 \cap U_2, g^{-1}b, x\right) \\ \wr\downarrow f_1 & & \wr\downarrow f_2 & & \downarrow f_3 & & \wr\downarrow \\ \pi_n\left(V_1 \cap V_2, b\right) & \xrightarrow[\iota_V]{} & \pi_n\left(V_1, b\right) & \xrightarrow[\varsigma_V]{} & \pi_n\left(V_1, V_1 \cap V_2, b\right) & \xrightarrow[\partial_V]{} & \pi_{n-1}\left(V_1 \cap V_2, b\right). \end{array} \qquad \text{(B.27)}$$

By hypothesis, f_1, f_2 are bijections $\forall\ n \geq 0$. Note that for $n = 1$, bijections f_1, f_2 induce group structures on their respective domains. So, by the Five Lemma, the third vertical maps f_3 are isomorphisms for all $n \geq 2$. We check the same for $n = 0, 1$. For $n = 0$, by the definition of relative homotopy π_0, we have the following commutative diagram:

$$\begin{array}{ccccc}
\pi_0(U_1 \cap U_2, x) & \longrightarrow & \pi_0(U_1, x) & \twoheadrightarrow & \pi_0(U_1, U_1 \cap U_2, x) \\
\downarrow & & \downarrow & & \| \\
\pi_0\left(U_1 \cap U_2, g^{-1}b, x\right) & \longrightarrow & \pi_0\left(U_1, g^{-1}b, x\right) & \twoheadrightarrow & \pi_0(U_1, U_1 \cap U_2, x) \\
\wr\downarrow f_1 & & \wr\downarrow f_2 & & \downarrow f_3 \\
\pi_0(V_1 \cap V_2, b) & \longrightarrow & \pi_0(V_1, b) & \twoheadrightarrow & \pi_0(V_1, V_1 \cap V_2, b)
\end{array}$$

It follows immediately that f_3 is surjective. To prove injectivity, let $x_1, x_2 \in U_1$ and $f_3([x_1]) = f_3([x_2]) =: y$. So, $y = [g(x_1)] = [g(x_2)] \in \pi_0(V_1, V_1 \cap V_2, b)$.

1. Assume that $y = [b] \in \pi_0(V_1, V_1 \cap V_2, b)$ is neutral. In particular, there is $w_1 \in U_1 \cap U_2$ such that $g(w_1)$ and $g(x_1)$ are in the same path components in V_1. It follows that $[w_1] = [x_1] \in \pi_0\left(U_1, g^{-1}b, x\right)$. We have two cases. If $[w_1] = [x_1]$ is neutral in $\pi_0\left(U_1, g^{-1}b, x\right)$, then $[x_1]$ is neutral in $\pi_0(U_1, U_1 \cap U_2, x)$. In case $[w_1] = [x_1]$ is nonneutral in $\pi_0\left(U_1, g^{-1}b, x\right)$, then x_1 and w_1 are in the same path component in U_1. Hence, $[x_1] = [x] \in \pi_0(U_1, U_1 \cap U_2, x)$. Likewise, $[x_2] = [x] = [x_1] \in \pi_0(U_1, U_1 \cap U_2, x)$.
2. Now assume that $y \neq [b] \in \pi_0(V_1, V_1 \cap V_2, b)$ is nonneutral. In this case, $g(x_1)$ and $g(x_2)$ are in the same path components in V_1. So, $[g(x_1)] = [g(x_2)] \in \pi_0(V_1, b)$, and hence, $[x_1] = [x_2] \in \pi_0\left(U_1, g^{-1}b, x\right)$. Hence, $[x_1] = [x_2] \in \pi_0(U_1, U_1 \cap U_2, x)$.

This completes the proof that f_3 is a bijection for $n = 0$. Now we consider the case $n = 1$ and improvise the diagram (B.27), as follows:

$$\begin{array}{ccccccccc}
\pi_1\left(U_1 \cap U_2, g^{-1}b, x\right) & \xrightarrow{\iota_U} & \pi_1\left(U_1, g^{-1}b, x\right) & \xrightarrow{\zeta_U} & \pi_1(U_1, U_1 \cap U_2, x) & \dashrightarrow & \pi_0\left(U_1 \cap U_2, g^{-1}b, x\right) & \longrightarrow & \pi_0\left(U_1, g^{-1}b, x\right) \\
\wr\downarrow f_1 & & \wr\downarrow f_2 & & \downarrow f_3 & & \wr\downarrow & & \wr\downarrow \\
\pi_1(V_1 \cap V_2, b) & \xrightarrow[\iota_V]{} & \pi_1(V_1, b) & \xrightarrow[\zeta_V]{} & \pi_1(V_1, V_1 \cap V_2, b) & \xrightarrow[\partial_V]{} & \pi_0(V_1 \cap V_2, b) & \longrightarrow & \pi_0(V_1, b).
\end{array}$$

Usually, there is no well-defined map at the place of the broken arrow, though it is defined here by composition. Both the lines are exact, though we need a proof that the first line is exact at $\pi_1(U_1, U_1 \cap U_2, x)$ and at $\pi_0\left(U_1 \cap U_2, g^{-1}b, x\right)$. Clearly, the compositions are constant maps. Let $\alpha : (I, \{0,1\}, 1) \longrightarrow (U_1, U_1 \cap U_2, x)$ be such that $[\alpha]$ is neutral in $\pi_0\left(U_1 \cap U_2, g^{-1}b, x\right)$. There is $x_1 \in g^{-1}b$ such that $\alpha(0)$ and x_1 are in the same path component of $U_1 \cap U_2$. Adjusting, we assume that $\alpha : (I, \{0,1\}, 1) \longrightarrow (U_1, g^{-1}b, x)$. This establishes that the first line is exact at $\pi_1(U_1, U_1 \cap U_2, x)$.

To prove exactness at $\pi_0\left(U_1 \cap U_2, g^{-1}b, x\right)$, let $z \in U_1 \cap U_2$, which represents the neutral element in $\pi_0\left(U_1, g^{-1}b, x\right)$. By definition, there is a path $\alpha : I \longrightarrow U$ such that $\alpha(0) = z$ and $\alpha(1) =: z_1 \in g^{-1}b$. So, $[\alpha] \in \pi_1(U_1, U_1 \cap U_2, x)$ maps to $[z]$. This establishes that the first line is exact at $\pi_0\left(U_1 \cap U_2, g^{-1}b, x\right)$.

Now, to prove f_3 is injective, let $\alpha, \beta : (I, \{0,1\}, 1) \longrightarrow (U_1, U_1 \cap U_2, x)$ be such that $f_3([\alpha]) = f_3([\beta])$.

1. Let $f_3([\alpha]) = f_3([\beta])$ be neutral. In this case, the usual Five Lemma arguments work, and hence both $[\alpha]$, $[\beta]$ are neutral.
2. Now assume that $f_3([\alpha]) = f_3([\beta])$ is nonneutral.
   (a) Assume that $\partial_V f_3([\alpha]) = \partial_V f_3([\beta])$ is neutral. Then there is $z, y \in \pi_1\left(U_1, g^{-1}b, x\right)$ such that $\zeta_U z = [\alpha]$, $\zeta_U y = [\beta]$. Use the group structures on $\pi_1\left(U_1 \cap U_2, g^{-1}b, x\right)$, $\pi_1\left(U_1, g^{-1}b, x\right)$ induced by f_1, f_2. It follows that $\iota_V f_2(z - y)$ is neutral. By the Five Lemma argument $z - y = \iota_U(w)$ for some $w \in \pi_1\left(U_1 \cap U_2, g^{-1}b, x\right)$. So, the image of $z - y$ in $\pi_1(U_1, U_1 \cap U_2, x)$ is homotopic to the constant path.
   $$[\alpha] = \zeta_U(z) = \zeta_U((z-y)+y) = \zeta_U(y) = [\beta].$$
   (b) Assume that $\partial_V f_3([\alpha]) = \partial_V f_3([\beta])$ are nonneutral. So, $\alpha(0)$ and $\beta(0)$ are in the same path components in $U_1 \cap U_2$. By adjusting, we can assume that $\alpha(0) = \beta(0)$. So, $\alpha\overline{\beta}$ is a well-defined loop at x in U_1. It follows that $f_3\zeta_U([\alpha\overline{\beta}])$ is neutral. By the Five Lemma arguments, $\iota_U(z) = [\alpha\overline{\beta}]$ for some z. So, $[\alpha\overline{\beta}]$ is neutral in $\pi_1(U_1, U_1 \cap U_2, x)$. Hence, $[\beta] = [\alpha\overline{\beta}]\beta] = [\alpha]$.

This completes the proof that f_3 is injective.

To prove the surjectivity of f_3, let $\eta : (I, \{0,1\}, 1) \longrightarrow (V_1, V_1 \cap V_2, b)$. By the usual Five Lemma arguments, there is $\alpha : (I, \{0,1\}, 1) \longrightarrow (U_1, U_1 \cap U_2, x)$, with $\partial_V f_3([\alpha]) = \partial_V([\eta])$. So, $\eta(0)$ and $g\alpha(0)$ have the same path component in $V_1 \cap V_2$. So, by adjusting η, we can assume that $g\alpha(0) = \eta(0)$. Write $x_0 := \alpha(0)$ and $b_0 := g(x_0) = \eta(0)$. Now, re-do the above commutative diagram, for x_0, b_0:

$$\begin{array}{ccccccccc}
\pi_1(U_1 \cap U_2, g^{-1}b_0, x_0) & \xrightarrow{\iota_U} & \pi_1(U_1, g^{-1}b_0, x_0) & \xrightarrow{\varsigma_U} & \pi_1(U_1, U_1 \cap U_2, x_0) & \dashrightarrow & \pi_0(U_1 \cap U_2, g^{-1}b_0, x_0) & \longrightarrow & \pi_0(U_1, g^{-1}b_0, x_0) \\
\wr\downarrow f_1 & & \wr\downarrow f_2 & & \downarrow f_3 & & \wr\downarrow & & \downarrow\wr \\
\pi_1(V_1 \cap V_2, b_0) & \xrightarrow[\iota_V]{} & \pi_1(V_1, b_0) & \xrightarrow[\varsigma_V]{} & \pi_1(V_1, V_1 \cap V_2, b_0) & \xrightarrow[\partial_V]{} & \pi_0(V_1 \cap V_2, b_0) & \longrightarrow & \pi_0(V_1, b_0).
\end{array}$$

Now $(g\overline{\alpha})\eta$ is a loop in V_1 at b_0. There is $\alpha_1 : (I, \{0,1\}1) \longrightarrow (U_1, g^{-1}b_0, x_0)$ such that $f_2([\alpha_1]) = [(g\overline{\alpha})\eta]$. Since $\alpha_1(1) = x_0 = \alpha(0)$, the concatenation $\alpha\alpha_1 : (I, \{0,1\}, 1) \longrightarrow (U_1, U_1 \cap U_2, x)$ is defined. It follows that $f_3([\alpha\alpha_1]) = [(g\alpha)(g\alpha_1)] = [(g\alpha)((g\overline{\alpha})\eta)] = [\eta]$. So, f_3 is surjective, and hence bijective for $n = 1$. Therefore, f_3 is bijective for all $n \geq 0$.

So, the claim (B.26) is established. From (B.2.14), it follows that, for $k = 1, 2$, the maps

$$\pi_n(E, U_k, x) \longrightarrow \pi_n(B, V_k, g(x)) \quad \text{are bijections } \forall\, x \in U_k,\ \forall\, n \geq 0. \tag{B.28}$$

Now consider the map $(E, U_1, g^{-1}b) \longrightarrow (B, V_1, b)$ and let $x \in g^{-1}b$. We have the commutative diagram of long exact sequence (*with labels added*), which ends at $n = 1$:

$$\begin{array}{ccccccc}
\pi_n(U_1, g^{-1}b, x) & \xrightarrow{\partial_U} & \pi_n(E, g^{-1}b, x) & \xrightarrow{\iota_U} & \pi_n(E, U_1, x) & \longrightarrow & \pi_{n-1}(U_1, g^{-1}b, x) \\
\wr\downarrow f_2 & & \downarrow f_4 & & \wr\downarrow f_5 & & \wr\downarrow \\
\pi_n(V_1, b) & \xrightarrow[\partial_V]{} & \pi_n(B, b) & \xrightarrow[\iota_V]{} & \pi_n(B, V_1, b) & \longrightarrow & \pi_{n-1}(V_1, b).
\end{array}$$

The first vertical arrow f_2 is bijective, by hypothesis, and the third one f_5, was established (B.28) to be a bijection. So, the second vertical arrow is a bijection, by the Five Lemma, when $n \geq 2$. For $n = 0$, the above diagram takes the following form:

$$\begin{array}{ccccccc}
\pi_1(E, U_1, x) & \overset{\partial_U}{\dashrightarrow} & \pi_0(U_1, g^{-1}b, x) & \xrightarrow{\iota_U} & \pi_0(E, g^{-1}b, x) & \longrightarrow & \pi_0(E, U_1, x) \\
\wr \downarrow f_5 & & \wr \downarrow f_2 & & \downarrow f_4 & & \wr \downarrow \\
\pi_1(B, V_1, b) & \xrightarrow[\partial_V]{} & \pi_0(V_1, b) & \xrightarrow[\iota_V]{} & \pi_0(B, b) & \longrightarrow & \pi_0(B, V_1, b).
\end{array} \tag{B.29}$$

The lower line is exact, while complete information about the top line seems unavailable. The broken arrow is defined by composition. We claim that the top line is also exact. We only need to check the exactness at $\pi_0\left(U_1, g^{-1}b, x\right)$. Let $z \in U_1$ and $\iota_U([z])$ be neutral. This means that there is a path $\gamma : I \longrightarrow E$ such that $\gamma(0) = z$ and $\gamma(1) =: z_1 \in g^{-1}b \subseteq U_1$. So, $[\gamma] \in \pi_1(E, U_1, x)$ maps to $[z]$. So, both the lines are exact.

To prove f_4 is injective, let $f_4([z_1]) = f_4([z_2])$ for $z_1, z_2 \in E$. So, $[g(z_1)] = [g(z_2)] \in \pi_0(B, b)$, meaning that $g(z_1)$ and $g(z_2)$ are connected by a path in B.

1. If $[g(z_1)] = [g(z_2)]$ is nonneutral in $\pi_0(B, V_1, b)$, then $[z_1] = [z_2]$ is nonneutral in $\pi_0(E, U_1, x)$. So, z_1 and z_2 are in the same path component in E. So, $[z_1] = [z_2]$ in $\pi_0\left(E, g^{-1}b, x\right)$.
2. Now assume that $[g(z_1)] = [g(z_2)]$ is neutral in $\pi_0(B, V_1, b)$. Then $[z_1] = [z_2]$ is neutral in $\pi_0(E, U_1, x)$. So, $\iota_U([w_1]) = [z_1]$ and $\iota_U([w_2]) = [z_2]$ for some $w_1, w_2 \in U_1$. In other words, we can assume that $z_1, z_2 \in U_1$.

   (a) If $[g(z_1)] = [g(z_2)]$ is neutral in $\pi_0(B, b)$, then by the routine Five lemma arguments, it follows that $[z_1] = [z_2]$ is neutral in $\pi_0\left(E, g^{-1}b, x\right)$.
   (b) So, assume that $[g(z_1)] = [g(z_2)]$ is nonneutral in $\pi_0(B, b)$. Write $b_1 = g(z_1) \in V_1$, $b_2 = g(z_2) \in V_1$, and change the base points, as follows:

$$\begin{array}{ccccccc}
\pi_1(E, U_1, z_2) & \overset{\partial_U}{\dashrightarrow} & \pi_0(U_1, g^{-1}b_2, z_2) & \xrightarrow{\iota_U} & \pi_0(E, g^{-1}b_2, z_2) & \longrightarrow & \pi_0(E, U_1, z_2) \\
\wr \downarrow f_5 & & \wr \downarrow f_2 & & \downarrow f_4 & & \wr \downarrow \\
\pi_1(B, V_1, b_2) & \xrightarrow[\partial_V]{} & \pi_0(V_1, b_2) & \xrightarrow[\iota_V]{} & \pi_0(B, b_2) & \longrightarrow & \pi_0(B, V_1, b_2).
\end{array}$$

   Note that $[g(z_1)]$ is neutral in $\pi_0(B, b_2)$. Again, by the Five lemma arguments, $[z_1]$ is neutral in $\pi_0\left(E, g^{-1}b_2, z_2\right)$. So, z_1 is connected by a path in E to z_2', with $g(z_2) = g(z_2') = b_2$.

Now, refer back to the diagram (B.29). Since $[b_2]$ is nonneutral in $\pi_0\,(B, b)$, it is nonneutral in $\pi_0\,(V_1, b)$. Since f_2 is an isomorphism, $[z_2'] = [z_2] \in \pi_0\left(U_1, g^{-1}b, x\right)$ are nonneutral. Therefore, z_2 and z_2' are connected by a path in U_1. Hence, z_1 and z_2 are connected by a path in E.

This completes the proof that f_4 is injective for $n = 0$. To prove that f_4 is surjective, let $y \in B$. By the usual Five Lemma arguments, there is $z \in E$ such that $[g(z)] = [y] \in \pi_0\,(B, V_1, b)$. If $[y]$ is neutral in $\pi_0\,(B, V_1, b)$, then $[y]$ is in the image of ι_V and hence in the image of f_4. If $[y]$ is nonneutral in $\pi_0\,(B, V_1, b)$, then $g(z)$ is path connected to y in B. So, $f_4([z]) = [g(z)] = [y] \in \pi_0\,(B, b)$. So, f_4 is surjective. This completes the proof that f_4 is bijective for $n = 0$.

For $n = 1$, we reproduce the above diagram of exact sequences:

$$\begin{array}{ccccccccc}
\pi_2\,(E, U_1, x) & \longrightarrow & \pi_1\,(U_1, g^{-1}b, x) & \xrightarrow{\iota_U} & \pi_1\,(E, g^{-1}b, x) & \xrightarrow{\zeta_U} & \pi_1\,(E, U_1, x) & \overset{\partial_U}{\dashrightarrow} & \pi_0\,(U_1, g^{-1}b, x) \\
\wr\downarrow & & \wr\downarrow f_2 & & \downarrow f_4 & & \wr\downarrow f_5 & & \wr\downarrow \\
\pi_2\,(B, V_1, b) & \longrightarrow & \pi_1\,(V_1, b) & \xrightarrow[\iota_V]{} & \pi_1\,(B, b) & \xrightarrow[\zeta_V]{} & \pi_1\,(B, V_1, b) & \xrightarrow[\partial_V]{} & \pi_0\,(V_1, b).
\end{array}$$

The broken arrow is defined by composition. Both lines are exact, while it needs a proof that the second line is exact at $\pi_1\,(E, U_1, x)$, which follows immediately. Note that $\pi_1\left(U_1, g^{-1}b, x\right)$ also has a group structure. To prove that f_4 is surjective, let $[\alpha] \in \pi_1\,(B, b)$.

By the usual arguments, there is a path $\beta : (I, \{0, 1\}, 1) \longrightarrow (E, g^{-1}b, x)$ such that $\zeta_V([\alpha]) = \zeta_V([g\beta])$. So, $\zeta_V([(\overline{g\beta})\alpha])$ is neutral. For the first line in the diagram (B.30), change the base point to $\beta(0) =: x_0$.

$$\begin{array}{ccccccccc}
\pi_2\,(E, U_1, x_0) & \longrightarrow & \pi_1\,(U_1, g^{-1}b, x_0) & \xrightarrow{\iota_U} & \pi_1\,(E, g^{-1}b, x_0) & \xrightarrow{\zeta_U} & \pi_1\,(E, U_1, x_0) & \overset{\partial_U}{\dashrightarrow} & \pi_0\,(U_1, g^{-1}b, x_0) \\
\wr\downarrow & & \wr\downarrow f_2 & & \downarrow f_4 & & \wr\downarrow f_5 & & \wr\downarrow \\
\pi_2\,(B, V_1, b) & \longrightarrow & \pi_1\,(V_1, b) & \xrightarrow[\iota_V]{} & \pi_1\,(B, b) & \xrightarrow[\zeta_V]{} & \pi_1\,(B, V_1, b) & \xrightarrow[\partial_V]{} & \pi_0\,(V_1, b).
\end{array}$$

So, there is a path $\varphi : (I, \{0, 1\}, 1) \longrightarrow (U_1, g^{-1}b, x_0)$ such that $f_4([\varphi]) = [(\overline{g\beta})\alpha])]$. Now, $\beta(0) = x_0 = \varphi(1)$. So, $\beta\varphi$ is defined, starting from $\varphi(0)$, ending at $\beta(1) = x$. So, $[\beta\varphi] \in \pi_1\left(E, g^{-1}b, x\right)$, and $f_4([\beta\varphi]) = [(g\beta)(\overline{g\beta})\alpha] = [\alpha]$. So, f_4 is surjective.

To prove injectivity, let $[\beta_0], [\beta_1] \in \pi_1\left(E, g^{-1}b, x\right)$ be such that $f_4([\beta_0]) = f_4([\beta_1])$. Write $x_0 := \beta_0(0)$ and $x_1 := \beta_1(0)$. So, $\overline{\beta}_0\beta_1$

is a path from x_1 to x_0. Therefore, $[\overline{\beta}_0\beta_1] \in \pi_1(E, g^{-1}b, x_0)$, and $f_4([\overline{\beta}_0\beta_1]) = 0 \in \pi_1(B, b)$. Consider the above diagram, with base point x_0. It follows from the Five Lemma argument that $[\overline{\beta}_0\beta_1] = 0 \in \pi_1(E, g^{-1}b, x_0)$ is neutral. Now, it follows that $[\beta_0] = [\beta_0(\overline{\beta}_0\beta_1)] = [\beta_1]$ in $\pi_1(E, g^{-1}b, x)$. So, f_4 is injective for $n = 1$, hence bijective.

Now, the proof that f_4 is bijective for all $n \geq 0$ is complete. Therefore, g is a quasi-fibration. The proof is complete. ∎

The proof of Theorem B (2.2.3) also uses the following sufficient condition, due to Dold and Thom [DT, 2.15 Satz], for a map to be a quasi-fibration.

Proposition B.5.12. *Let $g : E \longrightarrow B$ be a map in* **Top**. *Let there be an increasing sequence $B_1 \subseteq B_2 \subseteq B_3 \subseteq \cdots$ of subspaces of B such that*

1. $B = \cup_{k=1}^{\infty} B_k$.
2. *Given any compact subset $C \subseteq B$, $C \subseteq B_k$ for some $k \geq 1$.*
3. *For all $k \geq 1$, the restriction maps $g^{-1}B_k \longrightarrow B_k$ are quasi-fibrations.*

Then g is a quasi-fibration.

Proof. Write $E_k := g^{-1}B_k$. It follows from condition (2) that any compact set D in E is contained in E_k for some $k \geq 1$. Therefore, for $b \in B$ and $x \in g^{-1}b \subseteq E$, we have

$$\forall\, n \geq 0 \quad \pi_n(E, g^{-1}b, x) = \lim_k \pi_n(E_k, g^{-1}b, x), \quad \pi_n(B, b) = \lim_k \pi_n(B_k, b).$$

In these equations, we assumed that $b \in B_1$. By hypothesis, the maps $\pi_n(E, g^{-1}b, x) \xrightarrow{\sim} \pi_n(B_k, b)$ are isomorphisms for all $k \geq 1$ and $n \geq 0$. These isomorphisms induce isomorphisms in the limit. The proof is complete. ∎.

We are about to state (B.5.14) and prove the third of the three sufficient conditions for a map to be a quasi-fibration. This is also due to Dold and Thom [DT, 2.10 Hilfssatz]. For the benefit of the statement, we introduce the following definition.

Definition B.5.13. Let $(X, A) \in \mathbf{Top}^2$. Suppose that there is a homotopy $H : X \times [0,1] \longrightarrow X$, with $H_0 = 1_X$, $H_1(X) \subseteq A$ and $H_t(A) \subseteq A$ for all $t \in [0,1]$. In this case, we say that A is a **weak deformation retract** of X (provided by H). It is easy to prove that, if A is a weak deformation retract of X, the inclusion map $A \hookrightarrow X$

is a homotopy equivalence (see Exercise B.6.14). See (C.3.1) for the definition of deformation retractions.

Now we state the theorem of Dold and Thom.

Proposition B.5.14. *Let $p : E \longrightarrow B$ be a map in* **Top**. *Let $B' \subseteq B$ be a subspace and $E' := p^{-1}B'$. Assume that there are two homotopies:*

$$\begin{cases} D : E \times I \longrightarrow E \text{ which provides a weak deformation retraction of } E' \subseteq E \\ d : B \times I \longrightarrow B \text{ which provides a weak deformation retraction of } B' \subseteq B \end{cases} \tag{B.30}$$

Further assume that $pD_1 = d_1 p$, and the maps $D_1 : p^{-1}b \longrightarrow p^{-1}(d_1 b)$ are weak equivalences $\forall\, b \in B$. Now, if the restriction map $p_{|E'} : E' \longrightarrow B'$ is a quasi-fibration, then p is also a quasi-fibration.

Proof. For $b \in B$ and $x \in p^{-1}b$, and $n \geq 0$, consider the following commutative diagram:

$$\begin{array}{ccc} \pi_n\left(E, p^{-1}b, x\right) & \longrightarrow & \pi_n\left(E', p^{-1}(d_1 b), D_1(x)\right) \\ \big\downarrow & & \big\downarrow \\ \pi_n(B, b) & \longrightarrow & \pi_n\left(B', d_1(b)\right) \end{array} \tag{B.31}$$

Since $p_{|E'} : E' \longrightarrow B'$ is a quasi-fibration, the right vertical map is a bijection. The inclusions $E' \hookrightarrow E$ and $B' \hookrightarrow B$ are homotopy equivalences (see Exercise B.6.14). Therefore, the bottom arrow is also a bijection. To complete the proof, it is enough to prove that the top horizontal arrow is a bijection.

Corresponding to the map $D_1 : (E, p^{-1}b) \longrightarrow (E', p^{-1}(d_1(b))$ in $\mathbf{Top}^2$, we have the map of the long exact sequences as follows:

$$\begin{array}{ccccccc} \pi_n\left(p^{-1}b, x\right) & \longrightarrow & \pi_n(E, x) & \longrightarrow & \pi_n\left(E, p^{-1}b, x\right) & \longrightarrow & \pi_{n-1}\left(p^{-1}b, x\right) \\ \wr\big\downarrow & & \big\downarrow\wr & & \big\downarrow & & \big\downarrow \\ \pi_n\left(p^{-1}(d_1(b), D_1(x)\right) & \longrightarrow & \pi_n\left(E', D_1(x)\right) & \longrightarrow & \pi_n\left(E', p^{-1}(d_1(b), D_1(x)\right) & \longrightarrow & \pi_{n-1}\left(p^{-1}(d_1(b), D_1(x)\right). \end{array}$$

The first vertical map is a bijection, by hypothesis. The second vertical map is also a bijection, because $E' \hookrightarrow E$ is a homotopy equivalence. Therefore, by Five Lemma arguments, the third vertical map is a bijection. The proof is complete. ■

B.6. Exercises

Exercise B.6.1. Prove the statement in (a) of (B.1.3). That is, suppose that $f : \mathbb{R}^n \longrightarrow \mathbb{R}$ is a function (set theoretic). Prove that f is continuous, in the sense it is defined in the undergraduate calculus courses, if and only if f is continuous with the usual topology on $\mathbb{R}^n$ and $\mathbb{R}$.

Also prove that the set of rational numbers $\mathbb{Q}$ is dense in $\mathbb{R}$.

Exercise B.6.2. Suppose that $K \subseteq \mathbb{R}^n$ is a subset. Then K is compact if and only if K is closed and bounded.

Exercise B.6.3. Let A be an integral domain and $X = \mathrm{Spec}\,(A)$. Let $U \subseteq X$ be any nonempty open subset. Prove that U is dense in X. Likewise, let $S = \bigoplus_{n \geq 0} S_n$ be a graded commutative ring and $Y = \mathrm{Proj}\,(S)$. Assume that S is an integral domain. Let $V \subseteq Y$ be any nonempty open subset. Prove that V is dense in Y.

Exercise B.6.4. Let X be a Hausdorff space.

1. Then for any point $x \in X$, prove that $\{x\}$ is a closed set.
2. Let $f, g : Z \longrightarrow X$ be two continuous maps. Prove that the set $\{z \in Z : f(z) = g(z)\}$ is a closed subset of Z.

Exercise B.6.5. Let $(X, x_0) \in \mathbf{Top}_\bullet$ and refer to (B.2.3). Let $n \geq 2$ be an integer. Then

1. Prove that the product on $\pi_n\,(X, x_0)$, as defined in (B.7), is well-defined.
2. Prove that the product structure on $\pi_n\,(X, x_0)$ is independent of the choice of coordinate used to define it in (B.7).
3. Prove that $\pi_n\,(X, x_0)$ is an abelian group.

Exercise B.6.6. Let X be a locally compact T_2-space. Then X is compactly generated.

Exercise B.6.7. Let X, Y be T_2-spaces. If X is locally compact and Y is compactly generated, then $X \times Y$ is compactly generated.

Exercise B.6.8. Let X be a topological space and $f : X \twoheadrightarrow Y$ be a surjective map (set theoretic). A subset $V \subseteq Y$ is defined as open if $f^{-1}V$ is open in X. Prove that this defines a topology on Y.

1. Prove that this is the largest topology on Y that makes the function f continuous. Further, a set theoretic map $g : Y \longrightarrow Z$ is continuous if and only if gf is continuous.
2. Now, suppose that there is an equivalence relation $\sim$ on X, and that Y is the set of equivalence classes. Then the surjective map $X \twoheadrightarrow Y$ defines a topology on Y. Sometimes, Y is denoted by $\frac{X}{\sim}$.
3. Let $A \subseteq X$. For $x, y \in X$, define $x \sim y$ if $x, y \in A$ or if $x = y$. This is an equivalence relation on X. So, the **quotient space** X/A is defined as a topological space. We say that X/A is obtained by collapsing A to a point.

Exercise B.6.9. Refer to (B.2.9) and prove that the sequence (B.10) of homotopy groups is exact.

Exercise B.6.10. Refer to (B.3.4) and complete the proof that a composition of two fibrations is a fibration.

Exercise B.6.11. Refer to (B.3.5) and complete the proof that fibrations are closed under pullback.

Exercise B.6.12. Refer to (B.3.7) and complete the proof of the local criterion for a map to be a fibration when the base space is paracompact.

Exercise B.6.13. Refer to (B.3.8) and complete the proof that the sequence (B.20) of homotopy groups is exact.

Exercise B.6.14. [Ha, p. 18.] Let $(X, A) \in \mathbf{Top}^2$ and A be a weak deformation retraction of X (see B.5.13). Then the inclusion $\iota : A \hookrightarrow X$ is a homotopy equivalence.

Appendix C

CW Complexes

The category of CW complexes is an important subcategory of **Top**. Almost never in Quillen's work [Q], or more generally in K-theory, do we consider a space that is not a CW complex (*barring spaces in scheme theory*). Consequently, K-theory literature is heavily dependent on the topology of CW complexes.

C.1. Elements of CW Complexes

We start with establishing some notation and a basic definition.

Definition C.1.1. As before (Section B.2), for integers $n \geq 0$, $\mathbb{D}^n \subseteq \mathbb{R}^n$ will denote the unit disc and $\partial\mathbb{D}^n := \mathbb{S}^{n-1} \subseteq \mathbb{D}^n$ will denote the unit sphere, which is the boundary of $\mathbb{D}^n$. The interior of $\mathbb{D}^n$, the open disc, is denoted by $\mathbb{U}^n = \{x \in \mathbb{D}^n : \| x \| < 1\}$.

Suppose that X is a topological space. An n-**cell** of X is a subspace $\mathfrak{e} \subseteq X$, which is homeomorphic of $\mathbb{U}^n$. A cell in X is an n-cell for some $n \geq 0$. An n-cell $\mathfrak{e}$ is sometimes denoted by $\mathfrak{e}^n$, with the superscript indicating the dimension. We also write $\dim(\mathfrak{e}) = n$ for an n-cell $\mathfrak{e}$.

We introduce the mechanism of attaching cells.

Definition C.1.2. Let (X, A) be a topological pair and $f : A \longrightarrow Y$ be a continuous map. Define the topological space $X \coprod_f Y$ by the following pushforward diagram:

$$\begin{array}{ccc} A & \xrightarrow{\quad f \quad} & Y \\ \downarrow & & \downarrow{\scriptstyle \iota_Y} \\ X & \xrightarrow[\iota_X]{} & X \coprod_f Y \end{array} \qquad \text{in} \qquad \textbf{Top}.$$

So, $X \coprod_f Y$ is constructed, set theoretically, from the disjoint union $X \coprod Y$ by identifying $a \sim f(a)\ \forall \in A$. So, $\iota_Y : Y \hookrightarrow X \coprod_f Y$ is one-to-one. The topology of $X \coprod_f Y$ is the quotient topology (B.6.8). In fact, it is the largest topology on $X \coprod_f Y$, so that the maps ι_Y and $\iota_X : X \longrightarrow X \coprod_f Y$ are continuous. A set theoretic map $g : X \coprod_f Y \longrightarrow Z$ is continuous if and only if $g\iota_X$ and $g\iota_Y$ are continuous. A subset $W \subseteq X \coprod_f Y$ is open if and only if $\iota_X^{-1} W$ and $\iota_Y^{-1} W$ are open. We say that $X \coprod_f Y$ is obtained from Y by **attaching** X **along** A **via** f. Note that ι_X is one-to-one on $X - A$. If A is closed, then $X - A \xrightarrow{\sim} \iota_X(X - A)$ is a homeomorphism and $\iota_X(X - A)$ is open.

Of our particular interest is the case when $(X, A) = (\mathbb{D}^n, \mathbb{S}^{n-1})$. **Attaching an** n**-cell to** Y means that (1) there is a continuous map $\varphi : \mathbb{S}^{n-1} \longrightarrow Y$, and (2) a space is obtained by pushforward

$$\begin{array}{ccc} \mathbb{S}^{n-1} & \xrightarrow{\quad \varphi \quad} & Y \\ \downarrow & & \downarrow \\ \mathbb{D}^n & \xrightarrow[\Phi]{} & \mathbb{D}^n \coprod_f Y \end{array} \qquad \text{in} \quad \textbf{Top}. \quad \text{Note that}$$

$$\begin{cases} \mathfrak{e} := \Phi(\mathbb{U}^n) \text{ is open} \\ \Phi_{|\mathbb{U}^n} : \mathbb{U}^n \xrightarrow{\sim} \mathfrak{e} \ \text{ is a homeomorphism} \\ \text{With } Z = \mathbb{D}^n \coprod_f Y, \text{ we say that} \\ Z \text{ is obtained by attaching the cell } \mathfrak{e}. \end{cases}$$

Further, if $\varphi(\mathbb{S}^{n-1})$ is closed in Y, then $\Phi(\mathbb{D}^n)$ is closed. Recall that this is always the case if Y is a T_2-space.

A CW complex is defined as follows.

Definition C.1.3. A **CW complex** is a topological space X, together with a sequence of subspaces $X^0 \subseteq X^1 \subseteq \cdots \subseteq X^n \cdots \subseteq X$, such that the following conditions hold:

1. $X = \cup_{i=0}^{\infty} X^n$. Let X^0 be given the discrete topology.
2. Inductively, X^n is obtained from X^{n-1} by attaching n-cells $\mathfrak{e}_\alpha^n$ via continuous maps $\varphi_\alpha : \mathbb{S}^{n-1} \longrightarrow X^{n-1}$. This means that there is an indexing set $\mathcal{I}^n$ and a family of continuous maps $\{\varphi_\alpha : \mathbb{S}^{n-1} \longrightarrow X^{n-1} : \alpha \in \mathcal{I}^n\}$ such that

$$\begin{array}{ccc} \coprod_{\alpha\in\mathcal{I}^n} \mathbb{S}^{n-1} & \xrightarrow{\coprod \varphi_\alpha} & X^{n-1} \\ \downarrow & & \downarrow \\ \coprod_{\alpha\in\mathcal{I}^n} \mathbb{D}^n & \xrightarrow[\coprod \Phi_\alpha]{} & X^n \end{array} \quad \text{is a pushforward in } \mathbf{Top}. \tag{C.1}$$

   By definition (C.1.2), the topology on X^n is determined by this diagram. We write $\mathfrak{e}_\alpha^n = \Phi_\alpha(\mathbb{U}^n)$.
3. X is given the weak topology, meaning that a subset $A \subseteq X$ is defined as open (resp. closed) if and only if $A \cap X^n$ is open (resp. closed) in X^n for all $n \geq 0$. (This condition is redundant if $X = X^n$ for some n.)

Remark C.1.4. The construction (C.1.3) of the CW complexes X is no more than a purely set theoretic process. One could start with a discrete set X^0 and, inductively, provide the patching maps $\{\varphi_\alpha : \mathbb{S}^{n-1} \longrightarrow X^{n-1} : \alpha \in \mathcal{I}^n\}$ for $n \geq 0$ to obtain X^n. The final product $X = \bigcup X^n$ of this process is defined as a CW complex. If we want a topology on X (*which we do*) it is gotten only at the expense of the usual topology on $\mathbb{D}^n$. Conversely, to endow a CW structure on a given topological space X (*which is called triangulation*), we identify a discrete subspace $X^0 \subseteq X$, develop an inductive process as above, with $X = \bigcup X^n$, and check that the CW topology on X agrees with the topology on X. The basic ingredient of this process is the sequence of families of maps $\{\varphi_\alpha : \mathbb{S}^{n-1} \longrightarrow X^{n-1} : \alpha \in \mathcal{I}^n\}$ for $n \geq 0$ along with the inductive process. The following are further comments and definitions.

1. The subspace X^n is called the n-**skeleton** of X, which is closed in X. It follows that $X^{n-1} \subseteq X^n$ is closed in X^n, as is evident from the pushforward diagram (C.1).
2. For $\alpha \in \mathcal{I}^n$, the maps $\Phi_\alpha : \mathbb{D}^n \longrightarrow X^n \subseteq X$ are obtained inductively, as in the diagram (C.1). We use the same notation Φ_α for both the maps $\Phi_\alpha : \mathbb{D}^n \longrightarrow X^n$ and $\Phi_\alpha : \mathbb{D}^n \longrightarrow X$.

They are called **Characteristic maps**. For $\alpha \in \mathcal{I}^n$, with $\mathfrak{e}_\alpha^n := \mathfrak{e}_\alpha := \Phi_\alpha(\mathbb{U}^n)$, it follows that $(\Phi_\alpha)_{|\mathbb{U}^n} : \mathbb{U}^n \xrightarrow{\sim} \mathfrak{e}_\alpha$ is a homeomorphism and $\Phi_\alpha(\mathbb{S}^{n-1}) \subseteq X^{n-1}$.

3. We have that

$$X = \coprod_{\alpha \in \mathcal{I}^m, m \geq 0} \mathfrak{e}_\alpha^m, \qquad X^n = \coprod_{\alpha \in \mathcal{I}^m, m \leq n} \mathfrak{e}_\alpha^m$$

$$\text{are disjoint union, as sets.} \tag{C.2}$$

We say that $\mathscr{E} := \{\mathfrak{e}_\alpha \subseteq X : \alpha \in \mathcal{I}^n, n \geq 0\}$ is a **cell decomposition** of X. A cell decomposition $\mathscr{E}$ is not a complete description of a cell structure in the absence of the attaching maps φ_α or the characteristic maps Φ_α.

4. It follows that a subset $A \subseteq X$ is open (resp. closed) if and only if $\Phi_\alpha^{-1}(A)$ is open (resp. closed) in $\mathbb{D}^n$ $\forall\, \alpha \in \bigcup_{n=0}^\infty \mathcal{I}^n$. This follows from the fact that a subset $A \subseteq X$ is closed in X if and only if $A \cap X^n$ is closed in X^n.
5. **Notational convention:** It is customary to just say, "*X is a CW complex*", while the cell structure (C.1.3) remains at the background. We will do the same. All the notation above, X^n for the skeletons, $\mathscr{E} := \{\mathfrak{e}_\alpha \subseteq X : \alpha \in \mathcal{I}\}$ for the cell structure, φ_α for the patching maps, and Φ_α for the characteristic maps, are used, without further clarifications.

A topological space X can have more than one CW structures. An easy example of this is the real spheres $\mathbb{S}^n$. The definition (C.1.3) of a CW complex X is consistent with the one in [H, p. 528]. However, alternate and equivalent definitions of the same are available that start with a cell decomposition $\mathscr{E} := \{\mathfrak{e}_\alpha \subseteq X : \alpha \in \mathcal{I}^n, n \geq 0\}$ of X, and the characteristic maps Φ_α.

The following are further definitions.

Definition C.1.5. Let X be a CW complex (notation as in C.1.4).

1. A closed subspace $Y \subseteq X$ is called a **subcomplex** of X if Y is a union of a subfamily of cells in X. So, $Y = \bigcup_{\alpha \in \mathcal{J}} \mathfrak{e}_\alpha$ for some $\mathcal{J} \subseteq \mathcal{I}$. Such a subcomplex $Y \subseteq X$ is called a **finite subcomplex** if Y is the union of finitely many cells in X.

2. Let Z be another CW complex and $f : X \longrightarrow Z$ be a continuous map. We say that f is a **cellular map** if $\forall\, n \geq 0,\ f(X^n) \subseteq Z^n$, where X^n and Z^n are the n-skeletons, respectively, of X and Z.

CW complexes satisfy the so-called first separation axiom, as follows.

Lemma C.1.6. *Suppose that X is a CW complex and $x_0 \in X$. Then $\{x_0\}$ is closed.*

Proof. Use the notation as in C.1.4. Suppose that $x_0 \in X^n \setminus X^{n-1}$. Then $\{x_0\} \cap X^{n-1} = \phi$, and $x_0 \in \mathfrak{e}_\alpha^n$ for some unique $\alpha \in \mathcal{I}^n$. Now, it follows from pushout topology on X^n that $\{x_0\}$ is closed in X^n and hence in X. ■

The following lemma on CW complexes will be of our interest in the future.

Lemma C.1.7. *Let $X = \bigcup_{n=0}^{\infty} X^n$ be a CW complex* (notation as in C.1.4). *Let Y be a compact subspace of X. Then $Y \subseteq X^n$ for some $n \geq 0$.*

Proof. The proof is a consequence of the separation axiom (C.1.6). To prove this by contrapositive, assume that $Y \not\subseteq X^n$, $\forall\, n \geq 0$. For $n \geq 0$, pick a point $x_n \in Y \setminus X^n$. For $p \geq 0$, let $S_p = \{x_p, x_{p+1}, \ldots\}$. For $n \geq 0$, the set $X^n \cap S_p \subseteq \{x_0, \ldots, x_{n-1}\}$ is finite. Therefore, $S_p \cap X^n$ is a closed subset of X^n (C.1.6). Hence, S_p is closed in X. So, $S_p \subseteq Y$ is a decreasing sequence of closed subsets of X.

Fix integer $q \geq 0$. Then $x_q \in X^p$ for some $p \geq q+1$. Therefore, $x_q \neq x_n\ \forall\, n \geq p$. So, $x_q \notin S_p$. Hence, $\cap_{p=0}^{\infty} S_p = \phi$ is empty. This is a contradiction that Y is compact. The proof is complete. ■

The following is stronger than (C.1.7), which will also be of some future interest.

Lemma C.1.8. *Suppose that X is a CW complex* (notation as in C.1.4). *Let $Y \subseteq X$ be a compact subset. Then Y is contained in a finite subcomplex of X.*

Proof. First, we show that Y can only meet finitely many cells in X. Assume the contrary, that Y meets infinitely many distinct cells $\mathfrak{e}_{\alpha_1}, \mathfrak{e}_{\alpha_2}, \ldots$. Choose $x_i \in \mathfrak{e}_{\alpha_i} \cap Y$ and let $Z = \{x_1, x_2 \ldots, \}$. We claim that Z is closed. Inductively, assume that $Z \cap X^{n-1}$ is closed.

It suffices to show that $\forall\ \alpha \in \mathcal{I}^n$, $\Phi_\alpha^{-1}(X^n \cap Z)$ is closed in $\mathbb{D}^n$. Now,

$$\Phi_\alpha^{-1}(X^n \cap Z) = \Phi_\alpha^{-1}(X^{n-1} \cap Z) \cup \Phi_\alpha^{-1}(\mathfrak{e}_\alpha^n \cap Z).$$

By induction, the first one is a closed set in $\mathbb{S}^{n-1}$ and the second one is a singleton or null. This establishes that Z is closed in X. The same argument shows that any subset of Z is closed. Hence, Z is a discrete compact set. Since Z is a closed subset of the compact set Y, Z is also compact. Therefore, Z is a finite set. This is a contradiction.

Therefore, Y meets only finitely many cells of X. Hence, Y is contained in finitely many cells. We show that the finite union $E = \cup_{i=1}^k \mathfrak{e}_i^{n_i}$ of finitely many cells is contained in a finite sub complex B of $\dim B \leq N(E)$. We use induction on $N(E) = \max\{n_1, \ldots, n_k\}$.

Clearly, the zero cells $\mathfrak{e}_\alpha^0 \subseteq X^0$ are closed in X^0, and hence closed in X. So, it is established for $N(E) = 0$. For the inductive step, assume that it is proved for $N(E) \leq n - 1$. Note that a finite union of finite subcomplexes is a finite subcomplex. So, we will prove that $\mathfrak{e}_\alpha^n$ is contained in a finite subcomplex B, with $\dim B \leq n$. Consider the characteristic map $\Phi_\alpha : \mathbb{D}^n \longrightarrow X$. Then $\Phi_\alpha(\mathbb{D}^n) = \varphi_\alpha(\mathbb{S}^{n-1}) \cup \mathfrak{e}_\alpha^n$. Now, $\varphi_\alpha(\mathbb{S}^{n-1}) \subseteq X^{n-1}$. Now, $\varphi_\alpha(\mathbb{S}^{n-1})$ being compact, $\varphi_\alpha(\mathbb{S}^{n-1}) \subseteq \cup_{i=1}^m \mathfrak{e}_{\alpha_i}^{n_i}$, with $n_i \leq n - 1$ (*by the first paragraph*). By induction, $\varphi_\alpha(\mathbb{S}^{n-1}) \subseteq A$, where $A \subseteq X^{n-1}$ is a finite subcomplex, with $\dim A \leq n - 1$.

Now, $\Phi_\alpha(\mathbb{D}^n) \subseteq A \cup \mathfrak{e}_\alpha^n = A \cup \Phi_\alpha(\mathbb{D}^n)$. Now, $X^{n-1} \cap (A \cup \mathfrak{e}_\alpha^n) = A$ is closed. Then $\Phi_\alpha^{-1}(A \cup \mathfrak{e}_\alpha^n) = \mathbb{D}^n$. For $\beta \in \cup_{m \leq n} \mathcal{I}^m$, $\alpha \neq \beta$, we have $\Phi_\beta^{-1}(A \cup \mathfrak{e}_\alpha^n) = \varphi_\beta^{-1}(A)$, which is closed. Therefore, $A \cup \mathfrak{e}_\alpha$ is closed in X^n, and hence in X. So, $A \cup \mathfrak{e}_\alpha$ is a subcomplex.

The proof is complete. ■

Before we prove that CW complexes are Hausdorff, we give the following useful construction of open sets.

Construction C.1.9. Suppose that X is a CW complex (notation as in C.1.4).

1. Let $U \subseteq X^{n_0}$ be an open subset of X^{n_0}. Then there are open sets $N(U) \subset X$ such that $N(U) \cap X^{n_0} = U$. The following proof will provide further explicit information regarding such open sets.
2. Let $U, V \subseteq X^{n_0}$ be open subsets in X^{n_0}, with $U \cap V = \phi$. Then there are open sets $N(U), N(V) \subset X$ such that $N(U) \cap X^{n_0} = U$, $N(V) \cap X^{n_0} = V$ and $N(U) \cap N(V) = \phi$.

Proof. Suppose that an open set $N^n(U) \subseteq X^n$ in X^n has been defined.

Fix any function $\mathcal{I} \longrightarrow [0,1)$, denoted by $\alpha \mapsto r_\alpha$. Define

$$\begin{cases} U_\alpha = \Phi_\alpha\left(\{rx : \varphi_\alpha(x) \in N^n(U), x \in \mathbb{S}^n, r_\alpha < r \leq 1\}\right) \\ N^{n+1}(U) = N^n(U) \cup \left(\bigcup_{\alpha \in \mathcal{I}^{n+1}} U_\alpha\right) \\ N(U) = \bigcup_{n \geq n_0} N^{n+1}(U) \quad \text{with} \quad N^{n_0}(U) = U. \end{cases} \tag{C.3}$$

So, $N^{n+1}(U) \cap X^n = N^n(U)$ is open. For $\alpha \in \mathcal{I}^{n+1}$, in the radial coordinate, we have $\Phi_\alpha^{-1}\left(N^{n+1}(U)\right) = \varphi_\alpha^{-1} N(U) \times (r_\alpha, 1]$, which is open. So, $N^{n+1}(U)$ is open in X^{n+1}. Thus, $N(U)$ is open in X.

Now, let $U, V \subseteq X^{n_0}$, such that $U \cap V = \phi$. It follows from the construction that $N(U) \cap N(V) = \phi$. The proof is complete. ∎

Proposition C.1.10. *Suppose that X is a CW complex* (notation as in C.1.4). *Then X is Hausdorff.*

Proof. Let $x \neq y \in X$. Then, $x \in e_\alpha^m$ and $y \in e_\beta^n$. We assume that $m \leq n$. If $n = m$, we can find open sets U, V in X^n, with $x \in U$, $y \in V$ and $U \cap V = \phi$ (*whether $\alpha = \beta$ or not*). Let $N(U)$, $N(V)$ be the open sets, as in (C.1.9). Then $x \in N(U)$, $y \in N(V)$ and $N(U) \cap N(V) = \phi$ as required.

Now suppose that $m \leq n - 1$. Let $x \in W$ be open in X^m. Let $U := N^{n-1}(W)$ be the open set in X^{n-1}, given by (C.1.9). Then $x \in W \subseteq U$.

We have $\Phi_\beta(y_0) = y$ for unique $y_0 \in \mathbb{U}^n$. Let $\| y_0 \| < r_\beta < 1$. As in (C.3), define

$$\begin{cases} U' := U_\beta = \Phi_\beta\left(\{rx : \varphi_\beta(x) \in U, x \in \mathbb{S}^{n-1}, r_\beta < r \leq 1\}\right) \\ \text{and} \quad V := V_\beta = \Phi_\beta\left(\{z \in \mathbb{D}^n : \| z \| < r_\beta\}\right). \end{cases}$$

So, $x \in U'$, $y \in V$ are open in X^n, $U_\beta \cap V_\beta = \phi$. Now, $N(U'), N(V)$ are open in X, separating x, y. The proof is complete. ∎

Having established that CW complexes possess the basic desirable property, the following is another useful property for our purposes.

Proposition C.1.11. *Let X be a CW complex* (notation as in C.1.4). *Then X is compactly generated.*

Proof. (See B.1.13 and B.1.15.) We have X is Hausdorff. Let $A \subseteq X$ be a subset such that $A \cap C$ is closed for all compact subsets C of X.

For $\alpha \in \mathcal{I}^n$, $\Phi(\mathbb{D}^n)$ is compact. By hypothesis, $A' := A \cap \Phi_\alpha(\mathbb{D}^n)$ is closed. Therefore, $\Phi_\alpha^{-1}(A) = \Phi_\alpha^{-1}(A')$ is closed. So, A is closed in X. The proof is complete. ■

Remark C.1.12. It is not necessary that CW complexes are locally compact.

C.2. Product of CW Complexes

Let X, Y be two CW complexes. Then it is desirable that the topological space $X \times Y$ have a natural CW structure. The answer to this question is in the affirmative, with a caveat. We set up some of the notation on the CW structures of X, Y (C.1.4).

Notation C.2.1. Let X, Y be two CW complexes. For X, as usual, X^n denotes the n-skeleton, the characteristic maps are denoted by $\Phi_\alpha : \mathbb{D}^n \longrightarrow X$, the attaching maps by $\varphi_\alpha : \mathbb{S}^{n-1} \longrightarrow X^{n-1}$, and the indexing set by $\mathcal{I} = \bigcup_{n\geq 0} \mathcal{I}^n$. Likewise, for Y, Y^n denotes the n-skeleton, the characteristic maps are denoted by $\Psi_\beta : \mathbb{D}^n \longrightarrow X$, the attaching maps by $\psi_\beta : \mathbb{S}^{n-1} \longrightarrow X^{n-1}$, and the indexing set by $\mathcal{J} = \bigcup_{n\geq 0} \mathcal{J}^n$. Also, denote the cell structures as follows:

$$\mathscr{E} = \{\mathfrak{e}_\alpha^n : \alpha \in \mathcal{I}\}, \qquad \mathscr{F} = \left\{\varepsilon_\beta^n : \beta \in \mathcal{J}\right\}.$$

1. It may be more intuitive to identify $\mathbb{D}^n \cong I^n$ and $\mathbb{S}^{n-1} \cong \partial I^n$. So, a characteristic map will (sometimes) be considered as $\Phi_\alpha : I^n \longrightarrow X$ or $\Psi_\beta : I^n \longrightarrow Y$.
2. For $\alpha \in \mathcal{I}^m, \beta \in \mathcal{J}^n$, there are the natural maps $\Phi_\alpha \times \Psi_\beta : I^{m+n} \longrightarrow X \times Y$. Note that the restrictions $(\Phi_\alpha \times \Psi_\beta)_{|(0,1)^{m+n}} \xrightarrow{\sim} \mathfrak{e}_\alpha^m \times \varepsilon_\beta^n$ are homeomorphisms.
3. Let $(\mathcal{I} \star \mathcal{J})^n = \cup_{i+j=n} \mathcal{I}^i \times \mathcal{J}^j$, $\mathcal{I} \star \mathcal{J} = \cup_{n\geq 0} (\mathcal{I} \star \mathcal{J})^n$, and

$$\mathscr{E} \star \mathscr{F} = \left\{\mathfrak{e}_\alpha^i \times \varepsilon_\beta^j : (\alpha, \beta) \in \mathcal{I} \star \mathcal{J}\right\}.$$

The following is on CW structures of the product of CW complexes, which are also useful in Quillen K-Theory.

Theorem C.2.2. *Let X, Y be CW complexes, and all the other notation be as in* (C.2.1). *As before, $k(X \times Y)$ denotes the compactly* *generated space, corresponding to* $X \times Y$*. Then* $k(X \times Y)$ *has a CW structure, given by characteristic maps* $\Phi_\alpha \times \Psi_\beta$*, where* (α, β) *runs through the set* $\mathcal{I} \star \mathcal{J}$. (Consequently (B.6.7, C.1.11), if X or Y is locally compact, then $X \times Y = k(X \times Y)$ owns the CW structure.)

Proof. Write $Z := k(X \times Y)$, and let

$$Z^n := \bigcup \left\{ \mathfrak{e}_\alpha^i \times \varepsilon_\beta^j : (\alpha, \beta) \in (\mathcal{I} \star \mathcal{J})^m, m \le n \right\}.$$

Denote

$$(\Phi \star \Psi)_n = \coprod_{(\alpha,\beta)\in(\mathcal{I}\star\mathcal{J})^n} \Phi_\alpha \times \Psi_\beta : \coprod_{(\alpha,\beta)\in(\mathcal{I}\star\mathcal{J})^n} I^n \longrightarrow Z^n.$$

Note that the restriction of $(\Phi \star \Psi)_n$ to ∂I^n maps into Z^{n-1}. One needs to prove that

$$\begin{array}{ccc} \coprod_{(\alpha,\beta)\in(\mathcal{I}\star\mathcal{J})_n} \partial I^n & \longrightarrow & Z^{n-1} \\ \downarrow & & \downarrow \\ \coprod_{(\alpha,\beta)\in(\mathcal{I}\star\mathcal{J})_n} I^n & \xrightarrow[(\Phi\star\Psi)_n]{} & Z^n \end{array} \qquad \text{is a pushout.}$$

Set theoretically, this is obvious. The rest of the questions are topological. The CW topology on $X \times Y$ is built from bottom up, while the topology on $Z = k(X \times Y)$ is obtained from the product topology on $X \times Y$. We need to check that these two topologies are same.

1. Suppose that $A \subseteq X \times Y$ is closed in $Z = k(X \times Y)$. For each $(\alpha, \beta) \in (\mathcal{I} \star \mathcal{J})^n$ $n \ge 0$, the map $\Phi_\alpha \times \Psi_\beta : I^n \longrightarrow X \times Y$ is continuous, with product topology on $X \times Y$. The intersection $A \cap (\Phi_\alpha \times \Psi_\beta(I^n))$ is closed in the product topology. So, $(\Phi_\alpha \times \Psi_\beta)^{-1}(A)$ is closed. So, A is closed in the CW topology.
2. Suppose that $A \subseteq X \times Y$ is closed in the CW topology. Let $C \subseteq X \times Y$ be a compact set. Then $C \subseteq C_1 \times C_2$, where $C_1 \subseteq X$, $C_2 \subseteq Y$ are projections of C. Then C_1 and C_2 are compact. So, both C_1 and C_2 are, respectively, contained in finite subcomplexes $X_1 = \cup_{i=1}^{r_1} e_{\alpha_i}^{n_i} \subseteq X$, $Y_1 = \cup_{j=1}^{r_2} \varepsilon_{\beta_j}^{m_j} \subseteq Y$. It follows that $C \subseteq X_1 \times Y_1 = \bigcup_{i=1}^{r_1} \bigcup_{j=1}^{r_2} e_{\alpha_i}^{n_i} \times \varepsilon_{\beta_j}^{m_j}$. To prove that $A \cap C$ is closed in

$X \times Y$, we can replace X by X_1 and Y by Y_1 and assume X, Y are finite. It follows that

$$X_1 \times Y_1 = \bigcup_{i=1,j=2}^{r_1,r_2} e_{\alpha_i}^{n_i} \times \varepsilon_{\beta_j}^{m_j} = \bigcup_{i=1,j=2}^{r_1,r_2} \Phi_{\alpha_i} \times \Psi_{\beta_j}(I^{n_i+m_j})$$

Since A is closed in CW topology, $(\Phi_{\alpha_i} \times \Psi_{\beta_j})^{-1}(A)$ is closed in $I^{n_i+m_j}$. Therefore,

$$A \cap (\Phi_{\alpha_i} \times \Psi_{\beta_j})(I^{n_i+m_j}) = (\Phi_{\alpha_i} \times \Psi_{\beta_j})\left((\Phi_{\alpha_i} \times \Psi_{\beta_j})^{-1}(A)\right)$$

is closed in $X_1 \times Y_1$, since it is T_2.

$$A \cap (X_1 \times Y_1) = \bigcup_{i=1,j=2}^{r_1,r_2} (\Phi_{\alpha_i} \times \Psi_{\beta_j})(I^{n_i+m_j}) \cap (A \cap (X_1 \times Y_1))$$

which is closed in $X_1 \times Y_1 \subseteq X \times Y$. Hence, $A \cap C$ is closed in $X \times Y$.

So, it is established that two topologies coincide. The proof is complete. ■

C.3. Frequently Used Results

In this section, we discuss the Cellular Approximation Theorem (C.3.7), the Whitehead Theorem (C.3.8) on weak equivalences, and a few more standard results on CW complexes. At this point, we recall the definitions regarding retraction of a space [S].

Definition C.3.1. Let X be a topological space. A subspace $\iota : A \hookrightarrow X$ is called a **retract** of X if there is a continuous map $\rho : X \longrightarrow A$ such that the restriction $\rho_{|A} = 1_A$. In this case, ρ is called a **retraction**. Such a retract $\iota : A \hookrightarrow X$ is called a **deformation retract** if there is a homotopy $H : X \times I \longrightarrow X$ such that $H_0 = 1_X$ and $H_1 = \rho$. Therefore, a deformation retract $\iota : A \hookrightarrow X$ is a homotopy equivalence. A deformation retract $\rho : X \longrightarrow A$ (*with a retraction* ρ) is called a **strong deformation retract** if the homotopy H can be chosen to be a homotopy relative to A. So, for a strong deformation retract $H_0 = 1_X$, $H_1 = \rho$ and $H(x,t) = x \; \forall \; (x,t) \in A \times I$.

For completeness, we define and discuss the homotopy extension property.

Definition C.3.2. Let (X, A) be a pair in $\mathbf{Top}^2$. The pair (X, A) is said to have the **homotopy extension property** (HEP) if for any continuous map $f : X \longrightarrow Y$, and a homotopy $H : A \times I \longrightarrow Y$ with $H_0 = f_{|A}$ there is a homotopy $\mathcal{H} : X \times I \longrightarrow Y$, such that $\mathcal{H}_0 = f$ and the restriction $\mathcal{H}_{|A\times I} = H$.

If A is assumed to be closed, (X, A) has the homotopy extension property if and only if any continuous map $h : (X \times 0) \cup A \times I \longrightarrow Y$ extends to a continuous map $\mathcal{H} : X \times I \longrightarrow Y$.

The following is an equivalent condition for the homotopy extension property.

Lemma C.3.3. *Let X be a T_2 space and $A \subseteq X$ be a subspace. Write $W := (X \times 0) \cup (A \times I)$ and let $\iota : W \hookrightarrow X \times I$ be the inclusion. Then (X, A) has the homotopy extension property if and only if ι has a retraction $r : X \times I \longrightarrow W$.*

Proof. Assume that (X, A) has the homotopy extension property. Take $Y = W$, $f : X \longrightarrow W$ to be the map $f(x) = (x, 0)\ \forall\, x \in X$. Let $H : A \times I \longrightarrow W$ be the inclusion map. Then there is a continuous map $\mathcal{H} : X \times I \longrightarrow W$, with $\mathcal{H}_0 = f$ and $\mathcal{H}_{|A\times I} = H$. So, $\mathcal{H}$ is a retraction.

Conversely, let $r : X \times I \longrightarrow W$ be a retraction. We consider the map r as a map $r : X \times I \longrightarrow X \times I$. We have $W = \{z \in X \times I : z = r(z)\}$. By (B.6.4), W is closed. Consider the map $\iota_1 : X \longrightarrow X \times I$, given by $x \mapsto (x, 1)$. So, $\iota_1^{-1} W = A$ is closed in X.

Let $f : X \longrightarrow Y$ be a map and $H : A \times I \longrightarrow Y$ be a homotopy, with $H_0 = f_{|A}$. Define $F : W \longrightarrow Y$ by $F(x, 0) = f(x)\ \forall\ x \in X$, and $F_{|A\times I} = H$. Since A is closed, F is continuous. Consider the composition

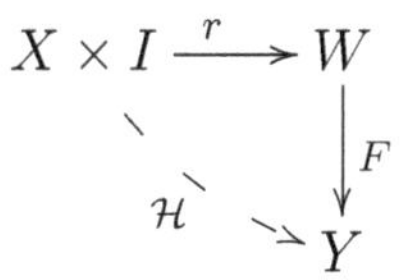

Then $\mathcal{H}$ has the required property.

The proof is complete. ■

Remark C.3.4. Let X be a T_2 space and $A \subseteq X$ be a subspace. The proof of (C.3.3) shows that if (X, A) has the homotopy extension property, then A is closed.

The following is on the homotopy extension property of CW pairs.

Lemma C.3.5. *Let X be a CW complex and $A \subseteq X$ be a subcomplex. Then (X, A) has the homotopy extension property.*

Proof. First, recall that CW complexes are Hausdorff (C.1.10), and hence we can use (C.3.3). Since X^0 has discrete topology, it follows that (X^0, A^0) has the homotopy extension property. Accordingly, there is a retraction $\rho_0 : X^0 \times I \longrightarrow X^0 \times 0 \cup A^0 \times I$ of the inclusion $X^0 \times 0 \cup A^0 \times I \subseteq X^0 \times I$, defined in the obvious way. We will first prove that, for all $k \geq 1$, the pair $(X^k, X^{k-1} \cup A^k)$ has the homotopy extension property. With $W_k = (X^k \times 0) \cup (X^{k-1} \cup A^k) \times I$, equivalently by (C.3.3), it is enough to construct a retraction $\rho_k : X^k \times I \longrightarrow W_k$, of the inclusion $W_k \hookrightarrow X^k \times I$. Fix an integer $k \geq 1$.

Let $C_k := (\mathbb{D}^k \times 0) \cup (\mathbb{S}^{k-1} \times I)$ be the open-top can. Let $C_k \hookrightarrow \mathbb{D}^k \times I$ be the inclusion. For $w \in \mathbb{D}^k \times I$, let $r_k(w) \in C_k$ be the unique point in C_k, on the line through $(0, 2) \in \mathbb{D}^k \times \mathbb{R}$ and w. Now, $r_k : \mathbb{D}^k \times I \longrightarrow C_k$ defines a retraction of the inclusion $C_k \subseteq \mathbb{D}^k \times I$.

Use the notation (C.1.4) for the CW structure of X. So, $\mathcal{I}^k$ corresponds to the k-cells of X, and let $\mathcal{J}^k \subseteq \mathcal{I}^k$ be the same for A. For $i \in \mathcal{I}^k$, let $(\Phi_i, \varphi_i) \longrightarrow (X^k, X^{k-1})$ denote the characteristic map. Consider the following commutative diagram:

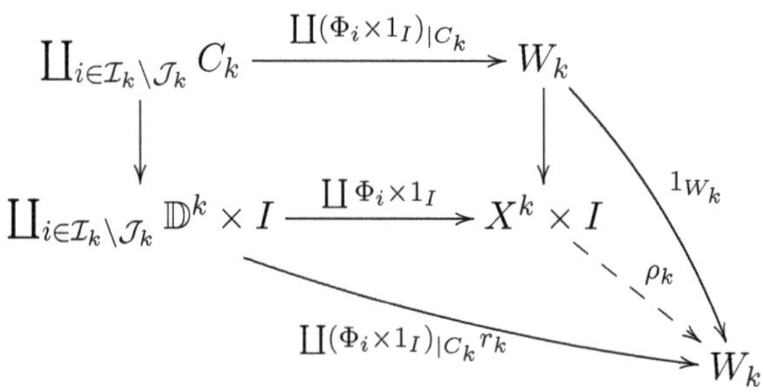

The rectangle is, indeed, a pushforward diagram, and ρ_k is defined by properties of pushforward diagrams. Clearly, ρ_k is a retraction of the inclusion $W_k \subseteq X^k \times I$. Therefore, $(X^k, X^{k-1} \cup A^k)$ has the Homotopy extension property.

Now we will prove that (X, A) has the homotopy extension property. Let $f : X \longrightarrow Y$ be a continuous map and $H : A \times I \longrightarrow Y$ be a homotopy, with $H_0 = f_{|A}$. Since (X^0, A^0) has the homotopy extension property, there is a homotopy $\mathcal{H}^0 : X^0 \times I \longrightarrow Y$ with $\mathcal{H}^0_{|A^0 \times I} = H_{|A^0 \times I}$ and $\mathcal{H}^0_{|t=0} = f_{|X^0}$.

Fix $k \geq 1$. By induction, we construct a sequence of homotopies $\mathcal{H}^k : X^k \times I \longrightarrow Y$ such that $\mathcal{H}^k_{|X^{k-1} \times I} = \mathcal{H}^{k-1}$, $\mathcal{H}^k_{|A^k \times I} = H_{|A^k \times I}$, and $\mathcal{H}^k_{|t=0} = f_{|X^k}$. Assume that $\mathcal{H}^{k-1} : X^{k-1} \times I \longrightarrow Y$ has been constructed, as proposed. In particular, $\mathcal{H}^{k-1}_{|A^{k-1} \times I} = H_{|A^{k-1} \times I}$ and $\mathcal{H}^{k-1}_{|t=0} = f_{|X^{k-1}}$. Define

$$H^k : (X^{k-1} \cup A^k) \times I \longrightarrow Y \text{ by, } H^k(x,t)$$
$$= \begin{cases} H(x,t) & \text{if } (x,t) \in A^k \times I \\ \mathcal{H}^{k-1}(x,t) & \text{if } (x,t) \in X^{k-1} \times I. \end{cases}$$

Since $A^k \times I$, $X^{k-1} \times I$ are closed subsets of $X \times I$, and since H^k is well defined, H^k is continuous. We also have $H^k_{|t=0} = f_{|(X^{k-1} \cup A^k)}$. Since $(X^k, X^{k-1} \cup A^k)$ has the homotopy extension property, there is a homotopy $\mathcal{H}^k : X^k \times I \longrightarrow Y$ such that $\mathcal{H}^k_{|(X^{k-1} \cup A^k) \times I} = H^k$, and $\mathcal{H}^k_{|t=0} = f_{|X^k}$. This completes the induction step. Now, define

$$\mathcal{H} : X \times I \longrightarrow Y \quad \text{by } \mathcal{H}_{|(X^k \times I)} = \mathcal{H}^k.$$

Since $\mathcal{H}^k$ is an extension of $\mathcal{H}^{k-1}$, $\mathcal{H}$ is well-defined. Note that $X \times I$ has a CW structure (C.2.2). The $k+1$ skeleton $(X \times I)^{k+1} = X^{k+1} \times 0 \cup X^{k+1} \times 1 \cup X^k \times I$ is the union of three closed sets. So, the restriction of $\mathcal{H}$ to its skeletons is continuous. Therefore, $\mathcal{H}$ is continuous. The proof is complete. ■

The following lemma (C.3.6) is an important step forward, for the proof of the celebrated Cellular Approximation theorem (C.3.7). While the statement of the lemma may appear fairly intuitive to some readers, the proof is somewhat technical.

Lemma C.3.6. *Let Z be a topological space, obtained by attaching an n-cell to a subspace $W \subseteq Z$. So, we have the characteristic map $(\Phi, \varphi) : (\mathbb{D}^n, \partial\mathbb{D}^n) \longrightarrow (Z, W)$, and*

$$\text{the diagram}\quad \begin{array}{ccc} \partial\mathbb{D}^n & \xrightarrow{\ \varphi\ } & W \\ \big\downarrow & & \big\downarrow \\ \mathbb{D}^n & \xrightarrow[\ \Phi\ ]{} & Z \end{array} \quad \text{is a pushforward diagram.}$$

It follows, with $\mathfrak{e} = \Phi(\mathbb{U}^n)$, that the restriction $\Phi_{|\mathbb{U}^n} : \mathbb{U}^n \xrightarrow{\sim} \mathfrak{e}$ is a homeomorphism. Let $f : (I^m, \partial I^m) \longrightarrow (Z, W)$ be a continuous map, with $m < n$. Then f is homotopic, relative to ∂I^m, to a map $F : I^m \longrightarrow W$. Equivalently, the same is valid for continuous maps $f : (\mathbb{D}^m, \partial\mathbb{D}^m) \longrightarrow (Z, W)$, with $m < n$.

Proof. The proof is given by induction, on n. For the initial step, let $n = 1$, and $m = 0$. So, $\Phi : (I, \{0, 1\}) \longrightarrow (Z, W)$ and $f : (I^0, \phi) \longrightarrow (Z, W)$, where $I^0 = \{\star\}$ is a singleton. If $f(\star) \in W$, there is nothing to prove. If $f(\star) \notin W$, then $f(\star) = \Phi(t_0)$, where $0 < t_0 < 1$. Define homotopy $H : I^0 \times I \longrightarrow Z$ by $H(\star, s) = \Phi(st_0)$. Then $H(\star, 0) = \Phi(0) \in W$ and $H(\star, 1) = \Phi(t_0) = f(\star)$.

Fix $n \geq 2$. We construct a homotopy $H : I^m \times I \longrightarrow Z$, relative to ∂I^m, such that $H_0 = f$ and $\Phi(0) \notin H_1(I^m)$. Before that, we prove that such a map H_1 is homotopic to a map $F : I^m \longrightarrow W$, as desired, which establishes the Lemma. To see this, write $Z_0 = Z \setminus \{\Phi(0)\}$, $f_1 := H_1$, and consider the inclusion $\iota : W \hookrightarrow Z_0$. So, $f_1 : (I^m, \partial I^m) \longrightarrow (Z_0, W)$. The map

$$T : \mathbb{D}^n \setminus \{0\} \times I \longrightarrow \mathbb{D}^n \setminus \{0\}, \text{ by } \quad T(x, t) = \frac{(1-t) \parallel x \parallel + t}{\parallel x \parallel} x$$

is a strong deformation retraction $r = T_1$, of the inclusion $\partial\mathbb{D}^n \hookrightarrow \mathbb{D}^n$. This induces a strong deformation retraction $\rho : Z_0 \longrightarrow W$ of the inclusion $W \hookrightarrow Z_0$, as we demonstrate. Now consider the pushforward diagram

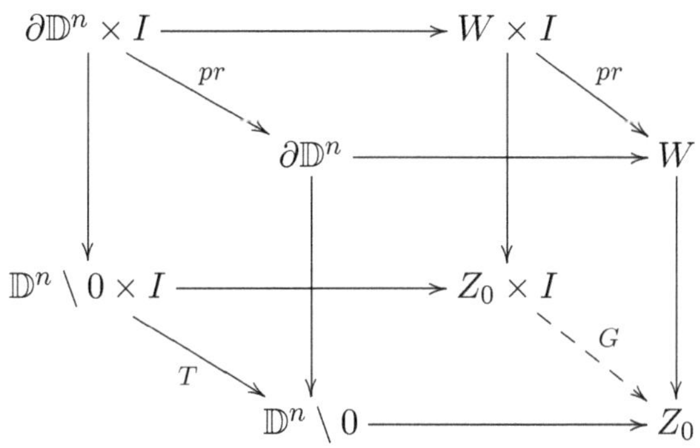

Here, G is induced by T. We have $G_0 = 1_{Z_0}$, $G_{|W\times I} = 1_{|W\times I}$, and $\rho := G_1 : Z_0 \longrightarrow W$. So, ρ is a strong deformation retraction of $\iota : W \hookrightarrow Z_0$. Define $\mathcal{H}$ by the composition

$$\begin{array}{ccc} I^m \times I & \xrightarrow{f_1 \times 1_I} & Z_0 \times I \\ & \searrow^{\mathcal{H}} & \downarrow G \\ & & Z_0 \end{array} \qquad \text{Then} \quad \begin{cases} \mathcal{H}_0 = f_1 \\ \mathcal{H}_1 = \rho f_1 \\ \mathcal{H}_{|\partial I^m \times I} = 1_{\partial I^m \times I} \end{cases} \qquad \text{Note that } \rho f_1(I^m) \subseteq W.$$

Therefore, it is enough to construct H, as proposed above, for all $m < n$. We do this by induction on n. Assume that $H := H^{n-1}$ has already been constructed for $n-1$, and hence the lemma is established for $n-1$. For any choice of the base point $\star \in \mathbb{S}^{n-1}$, we can apply the lemma for the following pushforward diagram:

$$\begin{array}{ccc} \partial\mathbb{D}^{n-1} & \longrightarrow & \star \\ \big\downarrow & & \big\downarrow \\ \mathbb{D}^{n-1} & \longrightarrow & \mathbb{S}^{n-1} \end{array} \qquad \text{collapsing } \partial\mathbb{D}^{n-1} \text{ to } \star\,.$$

It follows from the validity of the lemma, for $n-1$, that $\forall\ k < n-1$:

1. For any choice of base point $\star \in \mathbb{S}^{n-1}$, any map $(I^k, \partial I^k) \longrightarrow (\mathbb{S}^{n-1}, \star)$ is homotopic to the constant map $\star$, relative to ∂I^k.

2. Let $a < b$ and $\gamma : \partial I^{k+1} \longrightarrow \mathbb{S}^{n-1} \times (a, b)$ be any continuous map. Claim that γ extends to a map $I^{k+1} \longrightarrow \mathbb{S}^{n-1} \times (a, b)$. Instead, we prove that any map $\gamma : \mathbb{S}^k \longrightarrow \mathbb{S}^{n-1} \times (a, b)$ extends to a map $\mathbb{D}^{k+1} \longrightarrow \mathbb{S}^{n-1} \times (a, b)$. Since the projection map $p : \mathbb{S}^{n-1} \times (a, b) \longrightarrow \mathbb{S}^{n-1}$ is a homotopy equivalence, it follows from (1) that $\gamma : \mathbb{S}^k \longrightarrow \mathbb{S}^{n-1} \times (a, b)$ is homotopic to the constant map, relative to $\star$. Let $\Gamma : \mathbb{S}^k \times I \longrightarrow \mathbb{S}^{n-1} \times (a, b)$ be a homotopy, from γ to $\star$. Collapsing $\mathbb{S}^k \times 1 \subseteq \mathbb{S}^k \times I$, there is map $\widetilde{\Gamma} : \frac{\mathbb{S}^k \times I}{\mathbb{S}^k \times 1} \longrightarrow \mathbb{S}^{n-1} \times (a, b)$, induced by Γ. The map $\mathbb{S}^k \times I \longrightarrow \mathbb{D}^{k+1}$ sending $(x, t) \mapsto (1 - t)x$ induces a homeomorphism $\alpha : \frac{\mathbb{S}^k \times I}{\mathbb{S}^k \times 1} \xrightarrow{\sim} \mathbb{D}^{k+1}$. One checks that $\widetilde{\Gamma}\alpha^{-1} : \mathbb{D}^{k+1} \longrightarrow \mathbb{S}^{n-1} \times (a, b)$ is an extension of γ.

For $\epsilon \geq 0$, denote $\mathbb{U}^n_\epsilon := \{x \in \mathbb{R}^n : \| x \| < \epsilon\}$, $\mathbb{V}^n_\epsilon := \{x \in \mathbb{D}^n : \| x \| > \epsilon\}$. Let $U = \Phi\left(\mathbb{U}^n_{2/3}\right) \subseteq \mathfrak{e} \subseteq Z$, $V = W \cup_{\partial\mathbb{D}^n} \Phi(\mathbb{V}^n_{1/3} \cap \mathbb{D}^n) \subseteq Z$. Clearly, $Z = U \cup V$ and there are homeomorphisms $\mathbb{S}^{n-1} \times (1/3, 2/3) \cong \{x \in \mathbb{D}^n : 1/3 < \| x \| < 2/3\} \cong U \cap V$, while the second homeomorphism is given by the restriction of Φ.

So, $f^{-1}U$, $f^{-1}V$ is an open cover of the compact metric space I^n. Fix an integer $N \geq 2$. For integers $0 \leq k_1, k_2, \ldots, k_m \leq N - 1$, let

$$I^m_{k_1 k_2 \ldots k_m} = \left\{(x_1, \ldots, x_m) \in I^m : \frac{k_i}{N} \leq x_i \leq \frac{k_i + 1}{N}\right\}.$$

When $N \gg 0$, $\forall (k_1, \ldots, k_m)$, either $f(I^m_{k_1 k_2 \ldots k_m}) \subseteq U$ or $f(I^m_{k_1 k_2 \ldots k_m}) \subseteq V$ (*consult the Lebesgue Lemma*). We fix such an integer $N \gg 0$. Inductively, define a filtration of I^m by closed subsets:

$$\partial I^m \subseteq \mathscr{I}_{-1} \subseteq \mathscr{I}_0 \subseteq \mathscr{I}_1 \subseteq \cdots \subseteq \mathscr{I}_{m-1} \subseteq \mathscr{I}_m = I^m \quad \text{as follows.}$$

To do this, let Λ be the (finite) set of all subcubes (i.e. faces) $\lambda \subseteq I^m_{k_1 k_1 \ldots k_m}$ of all dimensions $0 \leq \dim \lambda \leq m$, for all k_i as above. For $0 \leq d \leq m$, denote $\Lambda_d = \{\lambda \in \Lambda : \dim \lambda = d\}$.

$$\text{Inductively, let} \quad \begin{cases} \mathscr{I}_{-1} = \bigcup\{\lambda \in \Lambda : f(\lambda) \subseteq V\}. \\ \quad \text{So,} \quad \partial I^m \subseteq \mathscr{I}_{-1}. \\ \mathscr{I}_d = \mathscr{I}_{d-1} \bigcup (\bigcup\{\lambda \in \Lambda_d : f(\lambda) \nsubseteq V\}) \\ \quad \text{So,} \quad \mathscr{I}_m = I^m. \end{cases}$$

This provides a CW structure on $(I^m, \mathscr{I}_{-1})$. For $-1 \leq d \leq m$, we define

$$F_d : \mathscr{I}_d \longrightarrow Z \quad \supset \quad F_{-1} = f_{|\mathscr{I}_{-1}},$$
$$(F_d)_{|\mathscr{I}_{d-1}} = F_{d-1}, \ F_d(\lambda) \subseteq U \cap V, \ \forall \, \lambda \in \Lambda_d, \text{ with } f(\lambda) \nsubseteq V.$$

Note that $\mathscr{I}_0 \setminus \mathscr{I}_{-1} = \{v_1, \ldots, v_r\}$ consists of finitely many vertices $v_i \in I^n_{k_1 \ldots k_m}$, with $f(x_i) \notin V$. Hence, $\| f(x_i) \| \leq 1/3$ and, in particular, $f(x_i) \in U$. Fix a point $y \in U \cap V$, and let $\gamma_i : I \longrightarrow \Phi(\mathbb{U}^n_{2/3}) \subseteq \mathfrak{e}$ be a path such that $\gamma_i(0) = f(x_i)$ and $\gamma(1) = y \in U \cap V$. Define $F_0 : \mathscr{I}_0 \longrightarrow V$ by $(F_0)_{|\mathscr{I}_{-1}} = f_{|\mathscr{I}_{-1}}$ and $F_0(x_i) = y$ for $i = 1, \ldots, r$. So, $F_0(\mathscr{I}_0) \subseteq V$. Note that $\mathscr{I}_d$ are closed subsets of I^n. Note that $\mathscr{I}_0 \setminus \mathscr{I}_{-1}$ is a discrete closed subset of $\mathscr{I}_0$. So, F_0 is continuous.

Now suppose that $F_{d-1} : \mathscr{I}_{d-1} \longrightarrow Z$ has already been defined. Note that $\mathscr{I}_d = \mathscr{I}_{d-1} \bigcup (\bigcup_{i=1}^r \lambda_i)$, where, for $i = 1, \ldots, r$, $\lambda_i \in \Lambda_d$, and $f(\lambda_i) \nsubseteq V$. Note, λ_i is a subcube of $I^m_{k_1 k_2 \ldots k_m}$ for some $0 \leq k_1, \ldots, k_m \leq N - 1$. Since $f(\lambda_i) \nsubseteq V$, we have $f(I^m_{k_1 k_2 \ldots k_m}) \subseteq U$. So, $f(\lambda_i) \subseteq U$ and hence $f(\partial \lambda_i) \subseteq U$, for all $i = 1, \ldots, r$.

On the other hand, let β be a face of $\partial \lambda_i$. So, $\beta \in \Lambda_{d-1}$. If $f(\beta) \nsubseteq V$, then by inductive hypothesis, $F_{d-1}(\beta) \subseteq U \cap V$. If $f(\beta) \subseteq V$, then $\beta \subseteq \mathscr{I}_{-1}$. So, $F_{d-1}(\beta) = f(\beta) \subseteq f(\lambda_i) \subseteq U$. So, in either case, $F_{d-1}(\beta) \subseteq U \cap V$. Therefore, $F_{d-1}(\partial \lambda_i) \subseteq U \cap V \cong \mathbb{S}^{n-1} \times (1/3, 2/3)$ for all $i = 1, \ldots, r$. From the consequence (2) of the inductive hypothesis, it follows that the restriction map $(F_{d-1})_{|\partial \lambda_i}$ extends as follows:

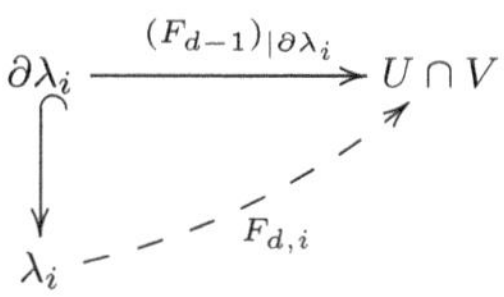

Define F_d by

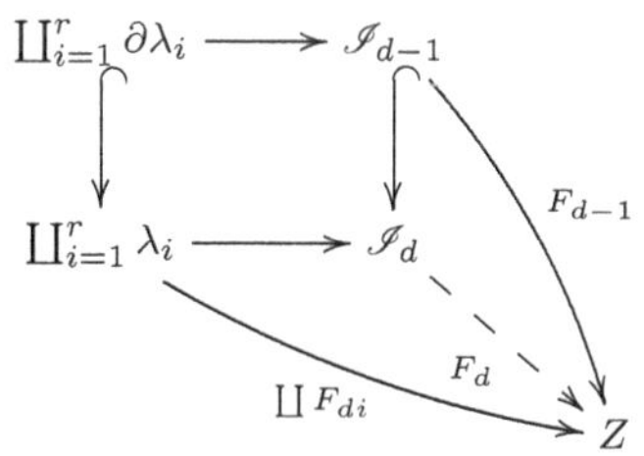

where the rectangle in the second diagram is a pushforward diagram, and F_d is defined by properties of such diagrams. It is clear that F_d has all the properties, as proposed. This completes the inductive step of the construction of F_d.

Let $F := F_m : I^m \longrightarrow Z$. It also follows, inductively, that $F(I^m) \subseteq V$, and hence $\Phi(0) \notin F(I^m)$. It remains to be shown that F is homotopic to f, relative to ∂I^m. By construction, $F_{|\mathscr{I}_{-1}} = f_{|\mathscr{I}_{-1}}$.

Also, by induction, we see that $f(I^m \setminus \mathscr{I}_{-1}) \subseteq U$ and $F(I^m \setminus \mathscr{I}_{-1}) \subseteq U$, while the restriction $\Phi : \mathbb{U}^n_{2/3} \xrightarrow{\sim} U \subseteq \mathfrak{e}$ is a homeomorphism. Use the convexity of $\mathbb{U}^n_{2/3}$, and define a homotopy

$$H : I^m \times I \longrightarrow Z \quad \text{by} \quad H(x,t)$$
$$= \begin{cases} f(x) & \text{if } x \in \mathscr{I}_{-1} \\ \Phi[(1-t)\Phi^{-1}(f(x)) + t\Phi^{-1}(F(x))] & x \notin \mathscr{I}_{-1} \end{cases}$$

The proof is complete. ∎

The following is the celebrated **Cellular Approximation Theorem.**

Theorem C.3.7. *Let X, Y be two CW complexes and $f : X \longrightarrow Y$ be a continuous map. Then f is homotopic to a cellular map $g : X \longrightarrow Y$* (C.1.5). *For a subcomplex $A \subseteq X$, if the restriction $f_{|A} : A \longrightarrow Y$ is already a cellular map, then the homotopy can be taken to be relative to A.*

Proof. We use the notation, as in (C.2.1), for the cell structures of X and Y. So, $\mathcal{I} = \bigcup \mathcal{I}^n$ denotes the indexing set for the cells in X, and $\mathcal{J} = \bigcup \mathcal{J}^n$ denotes the indexing set for the cells in Y. In addition, $\mathcal{I}(A) = \bigcup \mathcal{I}^n(A) \subseteq \mathcal{I}$ denotes the indexing set for the cell structure of A. Write $Z^{-1} = A$ and for $n \geq 0$, write $Z^n = A \cup X^n$. In fact, the following:

$$\begin{array}{ccc} \coprod_{i\in\mathcal{I}^n\setminus\mathcal{I}^n(A)} \partial\mathbb{D}^n & \longrightarrow & Z^{n-1} \\ \big\downarrow & & \big\downarrow \\ \coprod_{i\in\mathcal{I}^n\setminus\mathcal{I}^n(A)} \mathbb{D}^n & \xrightarrow[\coprod \Phi_i]{} & Z^n \end{array}$$

is a pushforward diagram $\forall\, n$.

We take $f_{-1} = f$. For $n \geq 0$, we construct maps $f_n : X \longrightarrow Y$, $H^n : X \times I \longrightarrow Y$, such that

1. $H^n_0 = f_{n-1}$, $H^n_1 = f_n$. Further, H^n is a homotopy, relative to Z^{n-1}. So, $\forall\, t \in I$, $(H^n_t)_{|Z^{n-1}} = (f_{n-1})_{|Z^{n-1}} = (f_n)_{|Z^{n-1}}$.
2. $f_n(\mathfrak{e}_i) \subseteq Y^n$, for all $i \in J_n(X) \setminus J_n(A)$.

We use induction and assume that f_{n-1}, H^{n-1} have already been constructed, as proposed.

Fix $i \in \mathcal{I}^n \setminus \mathcal{I}^n(A)$. We construct a map $f_{n,i} : \mathbb{D}^n \longrightarrow Y^n$ such that $f_{n,i}$ is homotopic to the restriction $f_{n-1}\Phi_i$, relative to $\partial\mathbb{D}^n$. If $f_{n-1}\Phi_i(\mathbb{D}^n) \subseteq Y^n$, we take $f_{n,i} = f_{n-1}\Phi_i$. So, assume $f_{n-1}\Phi_i(\mathbb{D}^n) \not\subseteq Y^n$ (or equivalently $f_{n-1}(\mathfrak{e}_i) \not\subseteq Y^n$). By (C.1.8) $f_{n-1}\Phi_i(\mathbb{D}^n) \subseteq \varepsilon_1^{n_1} \cup \cdots \cup \varepsilon^{n_r} = Y'$, where $\varepsilon_i^{n_i}$ is an n_i-cell in Y. Assume that $n_1 \leq n_2 \leq \cdots \leq n_r$. By our assumption, $n_r \geq n+1$. With $W_r = Y^{n_r} \setminus \varepsilon_r^{n_r}$, we have the pushforward diagram

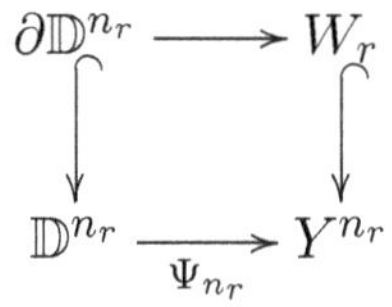

By (C.3.6), there is a map $g_r : \mathbb{D}^n \longrightarrow W_r$ and a homotopy $G^r : \mathbb{D}^n \times I \longrightarrow Y^{n_r}$, relative to $\partial\mathbb{D}^n$, with $G_0^r = f_{n-1}\Phi_i$ and $G_1^r = g_r$. By repeating this process, we obtain a map $f_{ni} : \mathbb{D}^n \longrightarrow Y^n$ and a homotopy $\Gamma^i : \mathbb{D}^n \times I \longrightarrow Y^{n_r}$, relative to $\partial\mathbb{D}^n$, with $\Gamma_0^i = f_{n-1}\Phi_i$ and $\Gamma_1^i = f_{n,i}$. Consider the pushforward diagram

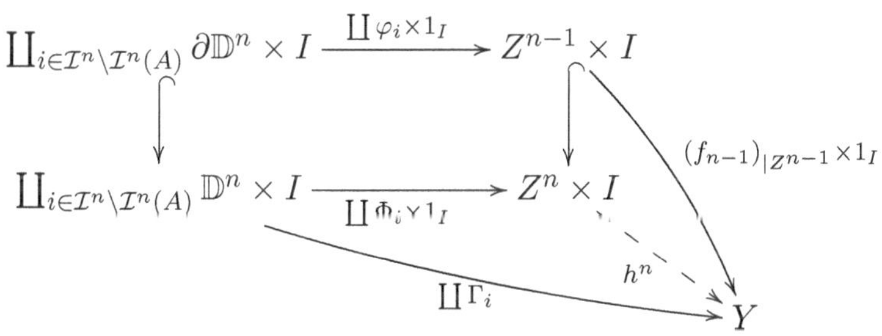

where h^n is defined by the properties of pushforward diagrams. It follows that $h_0^n = (f_{n-1})_{|Z^n}$, $h_1^n(X^n) \subseteq Y^n$, and h^n is a homotopy, relative to Z^{n-1}. By the homotopy extension property (C.3.5) of (X, Z^n), h^n extends to a homotopy $H^n : X \times I \longrightarrow Y$, with $H_0^n = f_{n-1}$. With $f_n := H_1^n$, for $i \in \mathcal{I}^n(X) \setminus \mathcal{I}^n(A)$, we have $f_n(\mathfrak{e}_i) = h_1^n(\mathfrak{e}_i) = f_{n,i}(\mathfrak{e}_i) \subseteq Y^n$. So, the inductive step of the construction of f_n and H^n is complete.

Recall that $H_0^1 = f$. To complete the proof, we consider H^n a homotopy on $\left[\frac{1}{2^{n-1}}, \frac{1}{2^n}\right]$. Define homotopy

$$H : X \times I \longrightarrow Y \quad \text{by}$$

$$H(x,t) = \begin{cases} H^n(x,t) & \text{if } t \in \left[\frac{1}{2^{n-1}}, \frac{1}{2^n}\right]. \\ H^N(x,1) = f_N(x) & \text{if } t = 1,\ x \in X^N \end{cases}$$

Note that H is well-defined and continuous. Note that $g := H_1$ is a cellular map.

The proof is complete. ■

The following is the celebrated theorem of J. H. C. Whitehead, establishing that, for CW complexes, weak equivalences (B.5.4) coincide with homotopy equivalences. We also refer to this theorem as the **Weak equivalence theorem** for CW complexes.

Theorem C.3.8. *Let X, Y be two connected CW complexes. Then a continuous map $f : X \longrightarrow Y$ is a weak equivalence if and only if f is a homotopy equivalence.*

Proof. If f is a homotopy equivalence, it is obviously a weak equivalence. Assume that f is a weak equivalence, and we prove that f is a homotopy equivalence. By Cellular Approximation (C.3.7), we can assume that f is a cellular map. First assume that $f : X \hookrightarrow Y$ is a subcomplex. We are required to prove that $f : X \hookrightarrow Y$ is a deformation retract. In fact, we prove that $f : X \hookrightarrow Y$ is a strong deformation retract. So, we exhibit a map $\rho : Y \longrightarrow X$, such that $\rho_{|X} = 1_X$ and there is a homotopy $\mathcal{H} : Y \times I \longrightarrow Y$, relative to X, from 1_Y to ρ. By hypothesis, it immediately follows (B.11) that $\pi_n(Y, X, x) = 0$ for all $x \in X$, $n \geq 0$. For $n \geq 0$, we define maps $\rho^n : Y \longrightarrow Y$, $H^n : Y \times I \longrightarrow Y$ such that (1) the restriction $\rho^n_{|X} = 1_X$, (2) $H^n_0 = \rho^{n-1}$, (3) $\rho^n := H^n_1$, (4) $\rho^n(Y^n) \subseteq X$, and (5) H^n is a homotopy, relative to $X \cup Y^{n-1}$.

We write $\rho^{-1} = 1_Y : Y \longrightarrow Y$, and do the above construction by induction on $n \geq 0$. Assume that $\rho^{n-1} : Y \longrightarrow Y$, $H^{n-1} : Y \times I \longrightarrow Y$ have been defined, as proposed.

Let $\mathcal{I}^n$ be the indexing set for the n-cells of Y, and $\mathcal{I}^n(X) \subseteq \mathcal{I}^n$ be the same for X. For $i \in \mathcal{I}^n$, let $(\Phi_i, \varphi_i) : (\mathbb{D}^n, \partial\mathbb{D}^n) \longrightarrow (Y^n, Y^{n-1})$ be the characteristic map. As usual, $\mathfrak{e}_i \subseteq Y^n$ denotes the n-cell, and $\Phi_i : \mathbb{U}^n \xrightarrow{\sim} \mathfrak{e}_i \subseteq Y^n$ is a homeomorphism. If $i \in \mathcal{I}^n(X)$, then $\mathfrak{e}_i \subseteq X^n$ is an n-cell of X, and if $i \notin \mathcal{I}^n(X)$, then $\mathfrak{e}_i \cap X = \phi$ is empty.

1. For $i \in \mathcal{I}^n(X)$, write $\omega_i = \Phi_i = \rho^{n-1}\Phi_i$. Also, let $F_i : \mathbb{D}^n \times I \longrightarrow X \subseteq Y$ be the constant homotopy $(F_i)_t = \Phi_i \ \forall\, t \in I$.
2. Now, let $i \notin \mathcal{I}^n(X)$. As before, let $\star$ denote the base point (*the north pole*) of $\partial\mathbb{D}^n$, and $\Phi_i(\star) = x$. Consider $\omega_i := \rho^{n-1}\Phi_i : (\mathbb{D}^n, \partial\mathbb{D}^n) \longrightarrow (Y, X)$. (*For* $n = 0$, $\omega_i = \Phi_i$.) By the homotopy

exact sequence for (Y, X), it follows that $\pi_n(Y, X, x)$ is a singleton. So, ω_i represents the neutral element. By the compression criterion (B.2.10), there is a homotopy $F_i : \mathbb{D}^n \times I \longrightarrow Y$, relative to $\partial\mathbb{D}^n$, such that $(F_i)_0 = \omega_i$, and $(F_i)_1 : \mathbb{D}^n \longrightarrow X$.

$$\begin{aligned}\text{So, } &\forall\ (z,t) \in \partial\mathbb{D}^n \times I,\ F_i(z,t) = F_i(z,0) = \omega_i(z)\\ &= \rho^{n-1}(\varphi_i(z)) = H^{n-1}(\rho^{n-1}\varphi_i(z), t).\end{aligned}$$

Consider the following pushforward diagram:

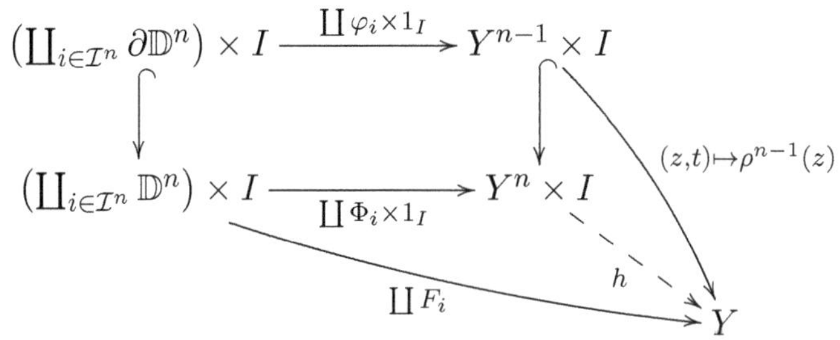

We have

$$\begin{cases} \forall\ z \in \mathbb{D}^n,\ h_0(\Phi_i(z)) = F_i(z,0) = \omega_i(z) = \rho^{n-1}(\varphi_i(z)); \\ \quad \forall z \in Y^{n-1}\ h_0(z) = \rho^{n-1}(z) \in X \\ \forall\ z \in \mathbb{D}^n,\ h_1(\Phi_i(z)) = F_i(z,1) \in X; \\ \quad \forall z \in Y^{n-1}\ h_1(z) = \rho^{n-1}(z) \in X \\ \text{So, } h_0 = \rho^{n-1}_{|Y^n}, \quad \text{and} \quad h_1(Y^n) \subseteq X \\ \forall\ (z,t) \in X^n \times I \text{ ,if } z \in X^{n-1}\ h(z,t) = \rho^{n-1}(z) = z; \\ \text{if } z \notin X^{n-1},\ \exists i \in \mathcal{I}^n(X), z \in \mathfrak{e}_i,\ h(z,t) = F_i(\Phi_i^{-1}(z),t) = z. \\ \forall\ (y,t) \in Y^{n-1} \times I, \quad h(z,t) = \rho^{n-1}(z) \\ \text{So, } \quad h \text{ is a homotopy, relative to } \quad X^n \cup Y^{n-1}, \\ \quad \text{from } \rho^{n-1}_{|Y^n} \text{ to } h_1 : Y^n \longrightarrow X. \end{cases}$$

$$\text{Define} \quad H : (Y^n \cup X) \times I \longrightarrow Y \ \text{ by } H(z,t) = \begin{cases} h(z,t) & \text{if } z \in Y^n \\ z & \text{if } z \in X \end{cases}$$

Since h is a homotopy, relative to $X^n \cup Y^{n-1}$, H is well-defined and continuous. So, H is a homotopy, relative to $X \cup Y^{n-1}$, from the restriction $\rho^{n-1}_{|(Y^n \cup X)}$ to the map $H_1 : Y^n \cup X \longrightarrow X$. Define

$$\widetilde{H} : Y \times 0 \cup [(Y^n \cup X) \times I] \longrightarrow Y \text{ by}$$

$$\widetilde{H}(y,t) = \begin{cases} H(y,t) \ \forall (y,t) \in (Y^n \cup X) \times I \\ \rho^{n-1}(y) \ \forall \ y \in Y, \ t = 0. \end{cases}$$

By Lemma C.3.5, $\widetilde{H}$ extends to a homotopy $H^n : Y \times I \longrightarrow Y$. Let $\rho^n := H^n_{t=1}$. So, H^n is a homotopy, relative to $X \cup Y^{n-1}$, from ρ^{n-1} to ρ^n. So, the induction step for the construction of ρ^n and H^n, as proposed, is complete.

To complete the proof, consider H^n as a homotopy on the interval $\left[1 - \frac{1}{2^n}, 1 - \frac{1}{2^{n+1}}\right]$.

Note that $\forall \ y \in Y^n, \forall \ n \geq n+1$

$$\rho^n(y) = H^n\left(y, 1 - \frac{1}{2^{n+1}}\right) = H^{n+1}\left(y, 1 - \frac{1}{2^{n+2}}\right) = \cdots$$

Define, $\mathcal{H} : Y \times I \longrightarrow Y$ by

$$\mathcal{H}(y,t) = \begin{cases} H^n(y,t) \text{ if } t \in \left[1 - \frac{1}{2^n}, 1 - \frac{1}{2^{n+1}}\right] \\ \rho^n(y) \quad \text{ if } y \in Y^n, t = 1 \end{cases}$$

So, $\mathcal{H}$ is a well-defined homotopy, from $\rho^0 = 1_Y$ to $\rho := \mathcal{H}_1$, relative to X. Also, note that the restriction $\rho_{|X} = 1_X$. Also, for $y \in Y^n$, we have

$$\rho(y) = \rho^n(y) = H^N(y,t) \qquad \forall N \geq n+1, t \in \left[1 - \frac{1}{2^N}, 1 - \frac{1}{2^{N+1}}\right].$$

This completes the proof, when $f : X \hookrightarrow Y$ is a subcomplex.

Now, assume that $f : X \longrightarrow Y$ is cellular and not necessarily a subcomplex. Let $C(f) := \frac{(X \times I) \coprod Y}{\{(x,1) \sim f(x) \ \forall \ x \in X\}}$ be the mapping cylinder of f, where $\coprod$ denotes a disjoint union. We have the commutative diagram

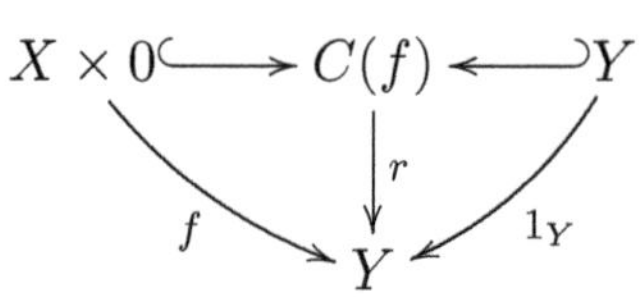

The vertical map is a deformation retract, given by

$$H : C(f) \times I \longrightarrow C(f)$$
$$\begin{cases} F([(x,t)],s) = [(x,\max\{t,s\})] & \forall\, x \in X, \quad s,t \in I \\ F([y],s) = [y] & \forall y \in Y, \quad s \in I \end{cases}$$

So, it is enough to prove that $X \cong X \times 0 \hookrightarrow C(f)$ is a homotopy equivalence. However, $C(f)$ has a CW structure, inherited from $(X \times I) \coprod Y$. Further, $X \hookrightarrow C(f)$ is a subcomplex. Therefore, it follows from the above that $X \hookrightarrow C(f)$ is a deformation retract. The proof is complete. ■

The following property of CW complexes is another tool used in [Q].

Theorem C.3.9. *Let X be a CW complex, X^n be the n-skeleton, and $x \in X^n$. Then*

$$\begin{cases} \text{the map} \quad \pi_k(X^n, x) \xrightarrow{\sim} \pi_k(X, x) \\ \qquad \text{is an isomorphism} \quad \forall\, k \leq n-1 \\ \text{the map} \quad \pi_n(X^n, x) \twoheadrightarrow \pi_n(X, x) \\ \qquad \text{is surjective.} \end{cases}$$

Proof. Since any point in X^n is path connected to a point in X^0, we assume that $x \in X^0$. From the homotopy exact sequence (B.11) for (X, X^n), it is enough to prove that $\pi_k(X, X^n, x) = 0$ (singleton), $\forall\, k \leq n$. Fix $k \leq n$. Let $\star$ denote the base point (*the north pole*) of $\mathbb{S}^k$. Let $[\omega] \in \pi_k(X, X^n, x)$, where $\omega : (\mathbb{D}^k, \mathbb{S}^{k-1}, \star) \longrightarrow (X, X^n, x)$ is a map. We construct a map $\Omega : \mathbb{D}^n \longrightarrow X^n$, which is homotopic to ω, relative to $\mathbb{S}^{k-1}$. By the Compression criterion (B.2.10), this will complete the proof that $[\omega] = 0$. This is done by repeated application of Cellular Approximation (C.3.7).

First, by (C.3.7), the restriction $\omega_{|\mathbb{S}^{k-1}}$ is homotopic to a cellular map $\alpha : (\mathbb{S}^{k-1}, \star) \longrightarrow (X^{k-1}, x)$. So, there is a homotopy $H : (\mathbb{S}^{k-1}, \star) \times I \longrightarrow (X^n, x)$ such that $H(\star, t) = x \;\forall\, t \in I$, $H_0 = \omega_{|\mathbb{S}^{k-1}}$, $H_1 = \alpha$.

We use some standard argument again. Let $C := (\mathbb{D}^k \times 0) \cup (\mathbb{S}^{k-1} \times I)$ be the open-top can, and $\iota : C \hookrightarrow \mathbb{D}^k \times I$ be the inclusion.

The projection $\rho : \mathbb{D}^k \times I \longrightarrow C$, from $(0,2)$, defines a retraction of C. Define

$$\widetilde{H} : C \longrightarrow X \quad \text{by} \quad \widetilde{H}(z,t) = \begin{cases} H(z,t) & \text{for } (z,t) \in \mathbb{S}^{k-1} \times I \\ \omega(z) & \text{for } t = 0. \end{cases}$$

Let $\mathcal{H} := \widetilde{H}\rho : \mathbb{D}^k \times I \longrightarrow X$. Therefore, $\mathcal{H}(z,0) = \omega(z)\ \forall\ z \in \mathbb{D}^k$, and $\mathcal{H}(\mathbb{S}^{k-1} \times I) = H(\mathbb{S}^{k-1} \times I) \subseteq X^n$. Let $\omega_1 := \mathcal{H}_1 : \mathbb{D}^k \longrightarrow X$. Then the restriction $(\omega_1)_{|S^{k-1}} = \alpha$ is a cellular map. So, $[\omega] = [\omega_1] \in \pi_k(X, X^n, x)$. By replacing ω by ω_1, we assume that the restriction map $\omega_{|\mathbb{S}^{k-1}}$ is cellular.

By another application of (C.3.7), there is a cellular map $\Omega : \mathbb{D}^k \longrightarrow X$ that is homotopic to ω, relative to $\mathbb{S}^{k-1}$. In particular, $\Omega(\mathbb{D}^k) \subseteq X^k \subseteq X^n$. The proof is complete. ■

C.3.1. *A triangle of fibrations*

The homotopy fibrations form a triangle, in the following sense.

Lemma C.3.10. *Let (X, x_0), (Y, y_0) be two pointed CW complexes, and $f : X \longrightarrow Y$ be a pointed map. Let $F \xrightarrow{\iota} X \xrightarrow{f} Y$ be a homotopy fibration. Then there is a homotopy fibration $\Omega Y \longrightarrow F \xrightarrow{\iota} X$ of pointed topological spaces.*

Proof. By Cellular approximation (C.3.7), we can assume that f is a cellular map. Since $F \simeq \mathscr{F}(f)$ are homotopically equivalent, we can assume that $F = \mathscr{F}(f, y_0)$. Let ω_0 be the constant loop y_0. Then ω_0 is the chosen base point of ΩY. Note that there is an injective map $\tilde{f} : \Omega Y \longrightarrow \mathscr{F}(f, y_0)$, sending $\gamma \mapsto (x_0, \gamma)$. Consider the diagram

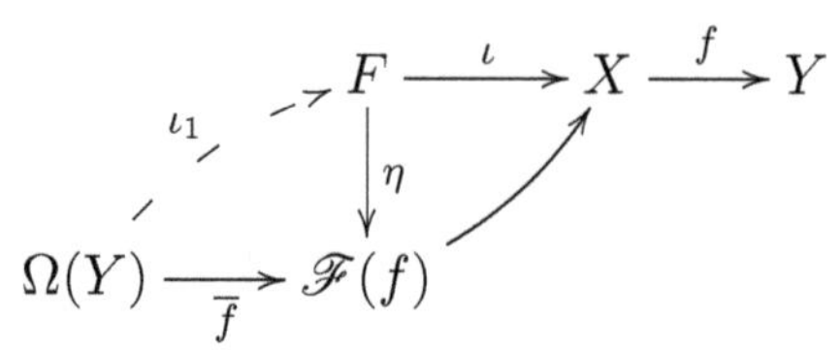

where ι_1 is obtained by combining $\overline{f}$ with the homotopy inverse of η. Thus, we have a sequence $\Omega Y \xrightarrow{\iota_1} F \xrightarrow{\iota} X$ of pointed spaces.

Correspondingly, there is a map $\zeta : (\Omega Y, \omega_0) \longrightarrow (\mathscr{F}(\iota), \star)$, where $\star$ is the image of ω_0. Consider the commutative diagram

$$\begin{array}{ccccccc}
\pi_{n+1}(F, z_0) & \longrightarrow & \pi_{n+1}(X, x_0) & \longrightarrow & \pi_{n+1}(Y, y_0) & \longrightarrow & \pi_n(F, z_0) \\
\| & & \| & & \wr \big\downarrow \partial & & \| \\
\pi_{n+1}(F, z_0) & \longrightarrow & \pi_{n+1}(X, x_0) & \longrightarrow & \pi_n(\Omega Y, \omega_0) & \longrightarrow & \pi_n(F, z_0) \\
\| & & \| & & \big\downarrow \zeta_* & & \| \\
\pi_{n+1}(F, z_0) & \longrightarrow & \pi_{n+1}(X, x_0) & \longrightarrow & \pi_n(\mathscr{F}(\iota), \star) & \longrightarrow & \pi_n(F, z_0)
\end{array}$$

The first and the third line are exact, by (B.4.6). The second horizontal line is exact because the first line is. So, by the Five Lemma argument, ζ_* is an isomorphism for all n. One checks that both ΩY and $\mathscr{F}(\iota, z_0)$ are CW complexes. Now, by Whitehead's Theorem (C.3.8), ζ is a homotopy equivalence. The proof is complete. ■

C.4. Exercises

The following problems are classical.

Exercise C.4.1. Let $(X, x_0), (Y, y_0) \in \mathbf{Top}_\bullet$ be CW complexes, with $x_0 \in X$ and $y_0 \in Y$ being zero cells. Then the Wedge sum $X \vee Y := \frac{X \coprod Y}{x_0 = y_0}$ inherits a CW structure.

Exercise C.4.2. Let $(X, x_0), (Y, y_0) \in \mathbf{Top}_\bullet$ be CW complexes, with $x_0 \in X$ and $y_0 \in Y$ being zero cells. Then the **smash product** $X \wedge Y := \frac{X \times Y}{X \vee Y}$ inherits a CW structure.

Exercise C.4.3. Give an explicit cell structure on the torus $\mathbb{S}^1 \times \mathbb{S}^1$.

Exercise C.4.4. Give an explicit CW pair structure on $(\mathbb{S}^N, p)$, where $\mathbb{S}^N$ is the real sphere and p is the north pole. Prove that $\pi_n\left(\mathbb{S}^N, \star\right) = 0$ for all $n < N$.

Exercise C.4.5. Let (X, A) be a CW pair. Give a CW structure on the quotient space X/A.

Exercise C.4.6. Let X be a CW Complex. Give a CW structure on the **suspension** $\mathbb{S}X := \frac{X \times I}{X \times 0, X \times 1}$, meaning that $\mathbb{S}X$ is obtained from $X \times I$ by collapsing $X \times 0$ to a point (south pole), and $X \times 1$ to a point (north pole). Likewise, give a CW structure on the **cone** $\mathbf{C}X := \frac{X \times I}{X \times 1}$.

Exercise C.4.7. Suppose that X is a CW complex. Show that X is path connected if and only if the 1-skeleton X^1 is path connected.

Exercise C.4.8. Let X be a CW complex (connected), and X^n be the n-skeleton. Would it be true or false to say

$$\pi_k(X^n, \star) = 0 \qquad \forall k \geq n+1?$$

Hint: See some examples!

Exercise C.4.9. Suppose that X is a CW complex. Show that X is locally connected if and only if each point has a neighborhood that meets only finitely many cells.

Exercise C.4.10. [H, Lem. 4.6, p. 346.] Suppose that (X, A) is a CW pair and (Y, B) is in $\mathbf{Top}^2$, with $B \neq \phi$. Assume that $\forall\, n \geq 0$, if $X - A$ has a n-cell, then $\pi_n(Y, B, b) = 0$ for all $b \in B$. Let $f : (X, A) \longrightarrow (Y, B)$ be a map in $\mathbf{Top}^2$. Then there is a map $g : X \longrightarrow B$ such that f is homotopic to g, relative to A. (**Hint**: *Repeat the first part of the proof of (C.3.8). Consequently, that part of the proof of (C.3.8) can be substituted, by an application of this statement, with* $f = 1_Y$.)

Problem C.4.11. Let $f : X \longrightarrow Y$ be a continuous map and $C(f) := \frac{(X \times I) \coprod Y}{\{(x,1) \sim f(x)\ \forall\ x \in X\}}$ be the mapping cylinder. Assume X, Y are CW complexes and f is a cellular map. Show that $(C(f), X)$ has a structure of a CW pair.

Problem C.4.12. Let $f : X \longrightarrow B$ be a map of CW complexes, and $b \in B$. Then the homotopy fibre $\mathscr{F}(f, b)$ has a CW structure. In particular, for CW complexes B, the loop space $\Omega(B, b)$ is a CW complex. (**Hint:** See [Mi3].)

References

[ABS] Atiyah, M. F.; Bott, R.; Shapiro, A. Clifford modules. *Topology* 3 (1964), no. suppl. 1, 3–38.

[AB] Auslander, M.; Bridger, M., Stable module theory. Memoirs of the American Mathematical Society, No. 94. American Mathematical Society, Providence, R.I. 1969.

[Bp] Balmer, P. Triangular Witt groups. I. The 12-term localization exact sequence. *K-Theory* 19 (2000), no. 4, 311–363.

[BS] Balmer, P.; Schlichting, M. Idempotent completion of triangulated categories. *J. Algebra* 236 (2001), no. 2, 819–834.

[Bp4] Balmer, P. Triangular Witt groups. II. From usual to derived. *Math. Z.* 236 (2001), no. 2, 351–382.

[Bp2] Balmer, P. Supports and filtrations in algebraic geometry and modular representation theory. *Am. J. Math.* 129 (2007), no. 5, 1227–1250.

[Bp3] Balmer, P. Niveau spectral sequences on singular schemes and failure of generalized Gersten conjecture. *Proc. Amer. Math. Soc.* 137 (2009), no. 1, 99–106.

[BW] Balmer, P.; Walter, C. A Gersten-Witt spectral sequence for regular schemes. *Ann. Sci. École Norm. Sup.* (4) 35 (2002), no. 1, 127–152.

[BPW] Balmer, P.; Gille, S.; Panin, I.; Walter, C. The Gersten conjecture for Witt groups in the equicharacteristic case. *Doc. Math.* 7 (2002), 203–217.

[B] Bass, H. *Algebraic K-Theory.* W. A. Benjamin, Inc., New York and Amsterdam, 1968.

[BR] Bass, H.; Roy, A. Lectures on topics in algebraic K-theory. Notes by Amit Roy. Tata Institute of Fundamental Research Lectures on Mathematics, No. 41, Tata Institute of Fundamental Research, Bombay, 1967.

[BM] Barge, J.; Morel, F. Groupe de Chow des cycles orientés et classe d'Euler des fibrés vectoriels. (French) [The Chow group of oriented cycles and the Euler class of vector bundles] *C. R. Acad. Sci. Paris S'er. I Math.* 330 (2000), no. 4, 287–290.

[BL] Barge, J.; Lannes, J. *Suites de Sturm, indice de Maslov et périodicité de Bott.* (French) [*Sturm Sequences, Maslov Index and Bott Periodicity*] Progress in Mathematics, 267. Birkhäuser Verlag, Basel, 2008. viii+199 pp.

[BK] Bhatwadekar, S. M.; Keshari, M. K. A question of Nori: Projective generation of ideals. *K-Theory* 28 (2003), no. 4, 329–351.

[BS2] Bhatwadekar, S. M.; Sridharan, R. Projective generation of curves in polynomial extensions of an affine domain and a question of Nori. *Invent. Math.* 133 (1998), no. 1, 161–192.

[BS1] Bhatwadekar, S. M.; Sridharan, R. On Euler classes and stably free projective modules. *Algebra, Arithmetic and Geometry, Part I, II*, Mumbai, 2000, 139–158, Tata Inst. Fund. Res. Stud. Math., 16, Tata Institute of Fundamental Research, Bombay, 2002.

[BH] Bruns, W.; Herzog, J. Cohen-Macaulay rings. *Cambridge Studies in Advanced Mathematics, 39.* Cambridge University Press, Cambridge, 1993.

[CHSW] Cortiñas, G.; Haesemeyer, C.; Schlichting, M.; Weibel, C. Cyclic homology, cdh-cohomology and negative K-theory. *Ann. Math.* (2) 167 (2008), no. 2, 549–573.

[DHM] Dutta, S. P.; Hochster, M.; McLaughlin, J. E. Modules of finite projective dimension with negative intersection multiplicities. *Invent. Math.* 79 (1985), no. 2, 253–291.

[DT] Dold, A.; Thom, R. Quasifaserungen und unendliche symmetrische Produkte. (German) *Ann. Math.* (2) 67 1958 239–281.

[E] Eisenbud, D. *Commutative Algebra. With a View toward Algebraic Geometry.* Graduate Texts in Mathematics, 150. Springer-Verlag, New York, 1995.

[Fj] Fasel, J. *Groupes de Chow-Witt.* (French) [*Chow-Witt Groups*] Mém. Soc. Math. Fr. (N.S.) No. 113 (2008), viii+197 pp.

[FS] Fasel, J.; Srinivas, V. Chow-Witt groups and Grothendieck–Witt groups of regular schemes. *Adv. Math.* 221 (2009), no. 1, 302–329.

[F] Fossum, Robert M. Vector bundles over spheres are algebraic. *Invent. Math.* 8 (1969), 222–225.

[Fu] Fulton, W. Intersection theory. *Ergebnisse der Mathematik und ihrer Grenzgebiete (3)* [*Results in Mathematics and Related Areas (3)*], 2. Springer-Verlag, Berlin, 1984.

[FL] Fulton, W.; Lang, S. Riemann-Roch algebra. *Grundlehren der Mathematischen Wissenschaften* [*Fundamental Principles of Mathematical Sciences*], 277. Springer-Verlag, New York, 1985.

[GZ] Gabriel, P.; Zisman, M. *Calculus of Fractions and Homotopy Theory.* Ergebnisse der Mathematik und ihrer Grenzgebiete, Band 35, Springer-Verlag New York, Inc., New York, 1967.

[GG] Gillet, H.; Grayson, D. R. The loop space of the Q-construction. *Illinois J. Math.* 31 (1987), no. 4, 574–597.

[G2] Gersten, S. M. Higher K-theory of rings. *Algebraic K-Theory, I: Higher K-Theories (Proc. Conf. Seattle Res. Center, Battelle Memorial Inst., 1972)*, pp. 3–42. Lecture Notes in Math., Vol. 341, Springer, Berlin, 1973.

[Gd1] Grayson, D. Higher algebraic K-theory. II (after Daniel Quillen). *Algebraic K-theory (Proc. Conf., Northwestern Univ., Evanston, IL., 1976)*, pp. 217–240. Lecture Notes in Math., Vol. 551, Springer, Berlin, 1976.

[Gd3] Grayson, D. R. Projections, cycles, and algebraic K-theory. *Math. Ann.* 234 (1978), no. 1, 69–72.

[Gd2] Grayson, D. R. Localization for flat modules in algebraic K-theory. *J. Algebra* 61 (1979), no. 2, 463–496.

[Gd4] Grayson, D. R. Exact sequences in algebraic K-theory. *Illinois J. Math.* 31 (1987), no. 4, 598–617.

[Ga] Grothendieck, A. Éléments de géométrie algébrique. III. Étude cohomologique des faisceaux cohérents. I. *Inst. Hautes Études Sci. Publ. Math.* 11 (1961).

[GJ] Goerss, P. G.; Jardine, J. F. *Simplicial Homotopy Theory.* Reprint of the 1999. Modern Birkhäuser Classics. Birkhäuser Verlag, Basel, 2009.

[Ha] Hatcher, A. *Algebraic Topology.* Cambridge University Press, Cambridge, 2002. xii+544 pp.

[HS] Hilton, P. J.; Stammbach, U. *A Course in Homological Algebra.* 2nd edn. Graduate Texts in Mathematics, 4. Springer-Verlag, New York, 1997. xii+364 pp.

[H] Hovey, M. *Model Categories.* Mathematical Surveys and Monographs, 63. American Mathematical Society, Providence, RI, 1999. xii+209 pp.

[H2] Hovey, M. Spectra and symmetric spectra in general model categories. *J. Pure Appl. Algebra* 165 (2001), no. 1, 63–127.

[HSS] Hovey, M.; Shipley, B.; Smith, J. Symmetric spectra. *J. Am. Math. Soc.* 13 (2000), no. 1, 149–208.

[Hr2] Hartshorne, R. *Residues and Duality.* Lecture Notes in Mathematics, no. 20 Springer-Verlag, Berlin and New York, 1966.

[Hr] Hartshorne, R. *Algebraic geometry.* Graduate Texts in Mathematics, no. 52. Springer-Verlag, New York and Heidelberg, 1977. xvi+496 pp.

[Hu] Husemoller, D. *Fibre Bundles.* 3rd edn. Graduate Texts in Mathematics, 20. Springer-Verlag, New York, 1994. xx+353 pp.

[Kb1] Keller, B. Derived categories and their uses. *Handbook of Algebra*, Vol. 1, 671-701, Handb. Algebr., 1, Elsevier/North-Holland, Amsterdam, 1996.

[Km] Kervaire, Michel A. Multiplicateurs de Schur et K-théorie. (French) 1970. *Essays on Topology and Related Topics* (*Mémoires dédiés à Georges de Rham*), pp. 212—225 Springer, New York.

[K] Kunz, E. *Kähler Differentials.* Advanced Lectures in Mathematics. *Friedr.* Vieweg & Sohn, Braunschweig, 1986.

[Lt] Lam, T. Y. *Introduction to Quadratic Forms over Fields.* Graduate Studies in Mathematics, 67. American Mathematical Society, Providence, RI, 2005.

[L] Lindel, H. On the Bass-Quillen conjecture concerning projective modules over polynomial rings. *Invent. Math.* 65 (1981/82), no. 2, 319–323.

[Mc] MacLane, S. *Categories for the Working Mathematician.* Graduate Texts in Mathematics, Vol. 5. Springer-Verlag, New York and Berlin, 1971. ix+262 pp.

[M] Mandal, S. Witt, GW, K-theory of quasi-projective schemes. *J. Pure Appl. Algebra* 221 (2017), no. 2, 286–303.

[M2] Mandal, S. *Projective Modules and Complete Intersections.* Lecture Notes in Mathematics, 1672. Springer-Verlag, Berlin, 1997. viii+114 p.

[M4] Mandal, S. Homotopy of sections of projective modules. With an appendix by Madhav V. Nori. *J. Algebraic Geom.* 1 (1992), no. 4, 639–646.

[M3] Mandal, S. Derived Witt group formalism. *J. Pure Appl. Algebra* 219 (2015), no. 8, 3518–3534.

[M5] Mandal, S. Convergence of two obstructions for projective modules, arXiv:2001.09561.

[MM] Mandal, S.; Mishra, B. The monoid structure on homotopy obstructions. *J. Algebra* 540 (2019), 168–205.

[MV] Mandal, S.; Varma, P. L. N. On a question of Nori: The local case. *Comm. Algebra* 25 (1997), no. 2, 451–457.

[Mh] Matsumura, H. *Commutative Algebra.* 2nd edn. Mathematics Lecture Note Series, 56. Benjamin/Cummings Publishing Co., Inc., Reading, MA., 1980.

[Mi2] Milnor, J. The geometric realization of a semi-simplicial complex. *Ann. of Math. (2)* 65 (1957), 357–362.

[Mi3] Milnor, J. On spaces having the homotopy type of a CW-complex. *Trans. Am. Math. Soc.* 90 (1959), 272–280.

[Mi] Milnor, J. *Introduction to Algebraic K-Theory.* Annals of Mathematics Studies, No. 72. Princeton University Press, Princeton, NJ; University of Tokyo Press, Tokyo, 1971. xiii+184 pp.

[Mp] May, J. P. *The Geometry of Iterated Loop Spaces.* Lectures Notes in Mathematics, Vol. 271. Springer-Verlag, Berlin and New York, 1972.

[Mf] Morel, F. $\mathbb{A}^1$-*Algebraic Topology over a Field.* Lecture Notes in Mathematics, 2052. Springer, Heidelberg, 2012. x+259 pp.

[N] Nenashev, A. K_1 by generators and relations. *J. Pure Appl. Algebra* 131 (1998), no. 2, 195–212.

[Na2] Neeman, A. The derived category of an exact category. *J. Algebra* 135 (1990), no. 2, 388–394.

[Na1] Neeman, A. *Triangulated Categories.* Annals of Mathematics Studies, 148. Princeton University Press, Princeton, NJ, 2001. viii+449 p.

[GM] Gupta, S. K.; Murthy, M. P. *Suslin's Work on Linear Groups over Polynomial Rings and Serre Problem.* ISI Lecture Notes, 8. Macmillan Co. of India, Ltd., New Delhi, 1980.

[P] de Seguins Pazzis, C. The geometric realization of a simplicial Hausdorff space is Hausdorff. *Topology Appl.* 160 (2013), no. 13, 1621–1632.

[Pd] Popescu, D. General Néron desingularization. *Nagoya Math. J.* 100 (1985), 97–126.

[Q] Quillen, D. Higher algebraic K-theory. I. *Algebraic K-theory, I: Higher K-theories* (*Proc. Conf., Battelle Memorial Inst., Seattle, WA, 1972*), pp. 85–147. Lecture Notes in Math., Vol. 341, Springer, Berlin, 1973.

[R] Rosenberg, J. *Algebraic K-Theory and Its Applications.* Graduate Texts in Mathematics, 147. Springer-Verlag, New York, 1994.

[Sm1] Schlichting, M. *Higher Algebraic K-Theory.* Topics in Algebraic and Topological K-Theory, 167–241, Lecture Notes in Math., 2008, Springer, Berlin, 2011.

[Sm2] Schlichting, M. Delooping the K-theory of exact categories. *Topology* 43 (2004), no. 5, 1089–1103.

[Sm3] Schlichting, M. Hermitian K-theory, derived equivalences and Karoubi's fundamental theorem. *J. Pure Appl. Algebra* 221 (2017), no. 7, 1729–1844.

[Sm4] Schlichting, M. Hermitian K-theory of exact categories. *J. K-Theory* 5 (2010), no. 1, 105–165.

[Sm5] Schlichting, M. Negative K-theory of derived categories. *Math. Z.* 253 (2006), no. 1, 97–134.

[Sm6] Schlichting, M. The Mayer-Vietoris principle for Grothendieck–Witt groups of schemes. *Invent. Math.* 179 (2010), no. 2, 349–433.

[Sg] Segal, Graeme Configuration-spaces and iterated loop-spaces. *Invent. Math.* 21 (1973), 213–221.

[S] Spanier, E. H. *Algebraic topology.* Corrected reprint. Springer-Verlag, New York and Berlin, 1981. xvi+528 pp.

[Sv] Srinivas, V. *Algebraic K-Theory.* 2nd edn. Progress in Mathematics, 90. Birkhäuser Boston, Inc., Boston, MA, 1996. xviii+341 p.

[St] Steenrod, N. *The Topology of Fibre Bundles.* Reprint of the 1957 edition. Princeton Landmarks in Mathematics. Princeton Paperbacks. Princeton University Press, Princeton, NJ, 1999. viii+229 p.

[Sw3] Swan, R. G. *Algebraic K-Theory.* Lecture Notes in Mathematics, No. 76. Springer-Verlag, Berlin and New York, 1968.

[Sw4] Swan, R. G. Topological examples of projective modules. *Trans. Amer. Math. Soc.* 230 (1977), 201–234.

[Sw1] Swan, R. G. K-theory of quadric hypersurfaces. *Ann. of Math.* (2) 122 (1985), no. 1, 113–153.

[Sw2] Swan, R. G. *Vector Bundles, Projective Modules and the K-Theory of Spheres. Algebraic Topology and Algebraic K-Theory* (Princeton, NJ, 1983), 432–522, Ann. of Math. Stud., 113, Princeton Univ. Press, Princeton, NJ, 1987.

[T] Tate, J. Homology of Noetherian rings and local rings. *Illinois J. Math.* 1 (1957), 14–27.

[TT] Thomason, R. W.; Trobaugh, T. Higher algebraic K-theory of schemes and of derived categories. *The Grothendieck Festschrift, Vol. III*, 247–435, Progr. Math., 88, Birkhäuser Boston, Boston, MA, 1990.

[V] Vorst, T. The general linear group of polynomial rings over regular rings. *Comm. Algebra* 9 (1981), no. 5, 499-509.

[Wh] Waldhausen, F. *Algebraic K-Theory of Spaces. Algebraic and geometric topology* (New Brunswick, NJ, 1983), 318–419, Lecture Notes in Mathematics, 1126, Springer, Berlin, 1985.

[Wc1] Weibel, C. A. *An Introduction to Homological Algebra.* Cambridge Studies in Advanced Mathematics, 38. Cambridge University Press, Cambridge, 1994. xiv+450 p.

[Wc2] Weibel, C. A. *The* K-Book. *An Introduction to Algebraic* K-*Theory*. Graduate Studies in Mathematics, 145. American Mathematical Society, Providence, RI, 2013. xii+618 p.

[W] Whitehead, G. W. *Elements of Homotopy Theory*. Graduate Texts in Mathematics, 61. Springer-Verlag, New York and Berlin, 1978. xxi+744 p.

Index

Milton Keynes UK
Ingram Content Group UK Ltd.
UKHW021411100823
426656UK00009B/100

9 789811 269387